# Mathematics Reference Book
## for
## Scientists and Engineers

by J.H. Heinbockel
Emeritus Professor of Mathematics
Old Dominion University

Copyright © 2009 by John H. Heinbockel.

All rights reserved. No part of this publication may be reproduced, stored in a retrieval system, or transmitted, in any form or by any means, electronic, mechanical, photocopying, recording, or otherwise, without the written prior permission of the author.

The cover picture is courtesy of the NASA web site.
http://www.nasa.gov/multimedia/imagegallery/image feature 1072.html
The web site has the following information concerning the cover picture:
*"A Place in the Universe*
*This montage of planetary images was taken by spacecraft managed by NASA's*
*Jet Propulsion Laboratory. Included are (from top to bottom) images of Mercury,*
*Venus , Earth (and moon), Mars, Jupiter, Saturn, Uranus and Neptune.*
*The spacecraft responsible for these images are as follows:*
*1. The Mercury image was taken by Mariner 10,*
*2. The Venus image by Magellan,*
*3. The Earth image by Galileo,*
*4. The Mars image by Viking, and*
*5. The Jupiter, Saturn, Uranus and Neptune were images taken by Voyager."*

The photos of Aristotle, Archimedes, Plato, Newton, Leibnitz and Gauss courtesy of Wikimedia Commons, the free media repository.

| | | |
|---|---|---|
| Library of Congress Control Number: | | 2008911469 |
| ISBN: | Hardcover | 978-1-4363-9181-8 |
| | Softcover | 978-1-4363-9180-1 |

This book was printed in the United States of America.

**To order additional copies of this book, contact:**
Xlibris Corporation
1-888-795-4274
www.Xlibris.com
Orders@Xlibris.com
56523

To
Jack, Anna, Will,
Matthew, Jake and Emilie
and all future generations.
Remember that mathematics is to be found everywhere.
You just have to look for it.

## List of Tables

| Table | Title | Page |
|---|---|---|
| 2.1 | Special Values for the Trigonometric Functions | 73 |
| 4.1 | Substitutions for Evaluating Integrals | 199 |
| 4.2 | Partial Fraction Expansions for $\frac{P(x)}{Q(x)} = \frac{P_\ell(x)}{P_m(x)}$, $\ell < m$ | 200 |
| 7.1 | Bessel Function Properties | 336 |
| 7.2 | Orthogonal Functions | 368 |
| 7.3 | Laplace Transform Properties | 378 |
| 7.4 | Short Table of Laplace Transforms | 379 |
| 7.5 | Fourier Exponential Transform Properties | 381 |
| 7.6 | Fourier Exponential Transforms | 382 |
| 7.7 | Fourier Sine Transforms | 383 |
| 7.8 | Fourier Cosine Transforms | 384 |
| 8.1 | United States Corn, Soybean and Wheat Production | 386 |
| 8.2 | Systolic Blood Pressure (mmHg) measurements taken from 200 Random Individuals | 386 |
| 8.3 | Frequency Table | 387 |
| 8.4 | Area Under Standard Normal Curve | 420 |
| 8.5(a)(b) | Critical Values for the Chi-Square Distribution with $\nu$ Degrees of Freedom | 421 |
| 8.6 | Critical Values for $t_{\alpha,n}$ for the Student's t Distribution with $n$ Degrees of Freedom | 423 |
| 8.7(a) | Critical Values of the F-Distribution for $\alpha = 0.1$ | 424 |
| 8.7(b) | Critical Values of the F-Distribution for $\alpha = 0.05$ | 424 |
| 8.7(c) | Critical Values of the F-Distribution for $\alpha = 0.025$ | 424 |
| 8.7(d) | Critical Values of the F-Distribution for $\alpha = 0.01$ | 424 |
| 8.7(e) | Critical Values of the F-Distribution for $\alpha = 0.005$ | 424 |
| 9.1 | The Chemical Elements by Atomic Number | 469 |
| 9.2 | Electron Shells and Subshells | 472 |

# Preface

The main purpose of this book is to provide under one cover selected elementary mathematics topics appended to more advanced mathematics topics found at the college/university level to, (i) serve as a reference book for scientists and engineers who have a need for mathematics fundamentals in their work, (ii) provide students with some insight into the kinds of mathematics they will be expected to know if they undertake a mathematics curriculum at the college/university level, (iii) introduce students to higher level mathematics which can be further investigated at the graduate level and (iv) summarize previous mathematics courses.

There are many courses at the college/university level which require a prerequisite mathematics course. Many times a student technically meets the prerequisite requirement but for some reason or another may need a quick review of fundamentals from one or more mathematics background courses. This text is designed to aid students who need a quick review of a subject matter by providing under one cover basic fundamentals from numerous mathematics subjects.

The material presented in these pages has been selected from both elementary and advanced mathematics sources. It summarizes basic facts, definitions, formulas and rules that are most frequently used by students and researchers who have a need for mathematics. There are three appendices. The appendix A contains units of measurement from the Système International d'Unitès (designated SI in all Languages). The appendix B contains tables of integrals for both indefinite and definite integrals, with over 850 integrals presented. The appendix C contains miscellaneous topics that students entering the fields of science or engineering should be aware of. There is also an extensive index to aid in finding information about a specific topic.

In short, this is a reference book for scientists, engineers and undergraduate college/university level students containing basic mathematics fundamentals from selected mathematical areas. These fundamentals can be expanded upon by obtaining more detailed information on the particular subject of interest.

I would like to thank John E. Nealy for reading parts of the manuscript and for his suggestions in improving the presentation of the material.

J.H. Heinbockel<br>
December 2008

# Mathematics Reference Book for Scientists and Engineers

**Chapter 1    Preliminaries** ............................................. 1

Introduction, History of Mathematics, Growth of Mathematics, Selected Contributors to Mathematical Development, Mathematical Specialty Areas, More Information, Mathematical Notation, Number Representation, Sufficient and Necessary Conditions, Indirect Proofs, Mathematical Induction, Arithmetic, Rule of Signs, The Rule of Nine, Arithmetic Operations, Calculators, Number Theory, The Greek Alphabet, Elementary Business Mathematics, Solutions of Linear and Quadratic Equations, Cubic Equation, Polynomial Equation, Special Products and Factors, Group, Algebraic Field.

**Chapter 2    Selected Fundamental Concepts** ................. 49

Permutations and Combinations, Elementary Probability, The Binomial Formula, Multinomial Expansion, Theory of Proportions, Sum of Arithmetic Progression, Sum of Geometric Progression, Equations Containing Radicals, Matrix Algebra, Special Matrices and Properties, Linear Dependence and Independence, Properties of Matrix Multiplication, Vector Norms, System of Linear Equations, Determinant of Order n, Pythagorean Theorem, Trigonometry, Trigonometric Functions, Cofunctions, Trigonometric Functions Defined for Other Angles, Sign Variation of the Trigonometric Functions, Graphs of the Trigonometric Functions, Trigonometric Functions of Sums and Differences, Double-angle Formulas, Half-angle Formulas, Product, Sum and Difference Formula, Simple Harmonic Motion, Inverse Functions, Inverse Trigonometric Functions, Principal Value Properties, Hyperbolic Functions, Hyperbolic Identities, Properties of Hyperbolic Functions, Inverse Hyperbolic Functions, Complex Numbers, Summary of Properties of Trigonometric and Hyperbolic Functions

**Chapter 3    Geometry** ..................................................... 97

Classification of Triangles, Similar Triangles, Congruent Triangles, Golden Ratio, Medians and Perpendicular Bisectors, Law of sines for Triangle, Law of Cosines for Triangle, Law of Tangents for Triangle, Area of Triangle, Miscellaneous Properties of a Triangle, Two-dimensional Rectangular coordinates, Translation and Rotation of Axes, Vectors, Vector Components, Direction Cosines, Properties of Vectors, Curve Sketching, Special Graph Paper, Straight Lines, Use of Determinants, Polar Coordinates, Curve Sketching in Polar Coordinates, Geometric Shapes, Polyhedron, Conic Sections, Determinants and Conic Sections, Conic Sections in Polar Coordinates, Functions and Graphs, Plane Curves , Plane Curves in Polar and Parametric Form, Spirals, Cycloids, Epicycloids, Hypocycloid, The Ovals of Cassini, Solid Analytic Geometry, Geometry and Graphics

Table of Contents

# Chapter 4    Calculus ...... 153

Limits, Properties of Limits, Continuity, $\epsilon - \delta$ Definition of Continuity, Intermediate Value Theorem, Derivatives, Basic differentiation Rules, Differentials, Properties of Differentials, Higher Derivatives, Parametric Functions, Leibnitz Rule for Differentiating Products, Partial Derivatives, Implicit Differentiation, Mean Value Theorem for Derivatives, Rolle's Theorem, Cauchy's Generalized Mean Value Theorem, Indeterminate Forms $\frac{0}{0}$, $\frac{\infty}{\infty}$, $0 \cdot \infty$, $\infty - \infty$, $0^0$, $\infty^0$, $1^\infty$, Maximum and Minimum Values, Taylor Series for Functions of One Variable, Examples of Series Expansions, Taylor Series for Functions of Two Variables, Summary of Differentiation Rules, The Indefinite Integral, Rules for Integration, Differentiation and Integration, The Exponential and Logarithmic Functions, The Definite Integral, Improper Integrals, Bliss's Theorem, Arc Length Formula, Area Formulas, Average Value of a Function, Volumes, Surface Area, Double Integrals, Double Integrals in Polar Coordinates, Triple Integrals, Evaluation of Integrals $\int f(x)\,dx$, Substitutions, Reduction Formula, Differentiation and Integration of Arrays, Inequalities Involving Integrals, General Series, Weierstrass Approximation Theorems, Infinite Series of Functions, Numerical Integration

# Chapter 5    Vector Calculus ...... 209

Introduction, Vector Addition and Subtraction, Unit Vectors, Scalar or Dot Product (inner product), Direction Cosines Associated With Vectors, The Cross Product or Outer Product, Properties of the Cross Product, Scalar and Vector Fields, The Gradient, Integration of Vectors, Line Integrals of Scalar and Vector Functions, Representation of Line Integrals, Surface and Volume Integrals, Surface Placed in a Scalar Field, Surface Placed in a Vector Field, Surface Area from Parametric Form, Volume Integrals, Divergence, Gauss Divergence Theorem, Physical Interpretation of Divergence, Green's Theorem in the Plane, Solution of Differential Equations by Line Integrals, Change of Variable in Integration, The Curl of a Vector Field, Physical Interpretation of Curl, Stokes' Theorem, Related Integral Theorems, Region of Integration, Green's First and Second Identities, Additional Operators, Properties of the Del Operator, Curvilinear Coordinates, Left and Right-handed Coordinate Systems, Gradient, Divergence, Curl and Laplacian

# Chapter 6    Ordinary Differential Equations ...... 269

Linear Differential Operators, Solutions to Differential Equations, Existence and Uniqueness of Solutions, Solutions Containing a Complex Number, Differential Equations Easily Solved, Exact Differential Equations, Linear First Order Differential Equations, Dependent Variable Absent, Independent Variable Absent, Parametric Solutions to Differential Equations, Linear $n$th Order Differential Equations , Solutions to $n$th Order Linear Homogeneous Differential Equations, Wronskian Determinant, Characteristic Equation, The Phase Plane, Boundary Value Problems, Solution of Nonhomogeneous Linear Differential Equations, Method of Undetermined Coefficients, Variation of Parameters, Differential Equations with Variable Coefficients, Cauchy or Euler Equations, Second Order Exact Differential Equations, Adjoint Operators, Forms Associated with Second Order Differential Equations, Abel's Formula, Self Adjoint Form, Nonhomogeneous Equations, Series Representation, Solution of Differential Equations by Series Methods, Numerical Solution to First Order Differential Equations

### Table of Contents

## Chapter 7    Special Functions ................................. 331

Integer Function, Heaviside Step Function, Impulse Function and Dirac Delta Function, Inverse Trigonometric Functions, The Gamma Function, The Beta Function, Bessel functions, Modified Bessel's Equation, The hypergeometric function, Recursion Formula, Generalized hypergeometric function, The Riemann Differential Equation, Generating Functions, Orthogonal Functions, Sturm-Liouville systems, Orthogonality, Fourier Trigonometric Series, Sturm-Liouville Theorem, Fourier Integral Theorem, The Fresnel Integrals, Integral Equations, The Error Function, The Sine, Cosine and Exponential Integrals, Elliptic Integral of the First Kind, Elliptic Integral of the Second Kind, Elliptic Integral of the Third Kind, Riemann Zeta Function, The Laplace Transform, The Fourier Transforms

## Chapter 8    Probability and Statistics ...................... 385

Introduction, The Representation of Data, Tabular Representation of Data, Arithmetic Mean or Sample Mean, Median, Mode and Percentiles, The Geometric and Harmonic Mean, The Root Mean Square (RMS), Mean Deviation and Sample Variance, Probability, Probability Fundamentals, Probability of an Event, Conditional Probability, Discrete and Continuous Probability Distributions, Scaling, The Normal Distribution, Standardization, The Binomial Distribution, The Multinomial Distribution, The Poisson Distribution, The Hypergeometric Distribution, The Exponential Distribution, The Gamma Distribution, Chi-Square $\chi^2$ Distribution, Student's t-Distribution, The F-Distribution, The Uniform Distribution, Confidence Intervals, Least Squares Curve Fitting, Linear Regression, Monte Carlo Methods, Linear Interpolation

## Chapter 9    Selected Applied Mathematics Topics ......... 429

Motion of a Particle (Dynamics), Kepler's Laws, Moment of a Force, Center of Mass for System of Particles, Center of Mass for Plane Areas, Centroids and Volumes, Second Moments or Moments of Inertia, Angular Velocity, Angular Momentum, Moments and Newton's Second Law, Impulse-Momentum Laws, Euler Angles, Space Curves, Curvature, and Torsion, Velocity and Acceleration, Relative Motion, Mechanical Vibrations, Phenomenon of Beats, Vibrations of a Spring Mass System, Mechanical Resonance, Torsional Vibrations, Solid Angles, Laplace's Equation, The Periodic Table of Elements, Modeling of Chemical Kinetics, Thermodynamics, Electrical Circuits, Four-terminal networks, Prisms

**Bibliography** ................................................................. 487

**Appendix A Units of Measurement** ............................... 489

**Appendix B Table of Integrals** ..................................... 491

**Appendix C Miscellaneous Topics** ................................ 545

**Index** .............................................................................. 546

# Chapter 1
# Preliminaries

## Introduction

This is a mathematical reference book for anyone who needs to look up fundamental concepts from the subject areas of arithmetic, geometry, algebra, calculus, differential equations, vector analysis, linear algebra, probability and statistics, series, transform methods and special functions. These are fundamental areas of applied mathematics used by engineers and scientists in their undergraduate college/university course work. Selected fundamentals from physics and chemistry are also presented as well as a chapter on selected applications of mathematics.

A short history of mathematics is presented highlighting those mathematicians who have helped to bring mathematical knowledge to the point where it is today and gives the reader a glance as to how mathematics has developed and expanded over the last 2000 years. This brief historical background is then followed by selected fundamental concepts from the mathematics courses available to college/university level students. There is an extensive index to aid in determining where specific concepts can be found. In brief, this mathematical reference book contains chapters presenting a review of selected material from different subject areas of applied mathematics available to undergraduate students in a science or engineering curriculum at a college or university. It contains many illustrative examples to aid in understanding basic mathematical concepts.

This reference book also gives selected derivations of concepts and theorems and at times even provides examples as to how a concept is applied. There are numerous figures and tables presenting concepts and ideas showing how mathematics is applied in science and engineering. There are three Appendices. The Appendix A contains units of measurement from the Système International d'Unités designated SI in all languages together with selected conversion factors associated with units used in science and engineering. The Appendix B contains a table of integrals and the Appendix C provides some advice on miscellaneous topics.

## History of Mathematics

The early history of mathematics consists of recordings on clay tablets, stones and monuments that date back thousands of years before 2000 BCE.[1] Knowledge of mathematics before

---

[1] CE stands for "Common Era." A new term that has replaced AD for "Anno Domini" in Latin or "the year of the Lord" in English. The terms CE and AD have the same numerical value and only the notation used differs. The word "common" is based upon the Gregorian Calendar.

BCE stands for "Before the common era" and is new terminology which replaces BC or "Before Christ". The notations BC and BCE have the same numerical values associated with their usage.

# 2

3500 BCE consists of isolated facts surrounded by suppositions, theories and inferences. One can say that counting is the earliest mathematics recorded. Archaeological records indicate that counting is found in the caves and dwellings of prehistoric man. Numbers were created possibly to keep a written record of possessions or animals in a collection or herd. Many cultures developed some type of counting board or abacus type of board. Archaeological recordings from Europe, Asia and Africa indicate applications of mathematics are to be found in the areas of geometry, use of number systems and the recording of the movement of planets and position of the Sun and stars. Geometry was developed to settle land disputes and used for the construction of dwellings and monuments. The figure 1-1 illustrates some of the numerical symbols used by the Sumerians of 3400 BCE to record numbers.

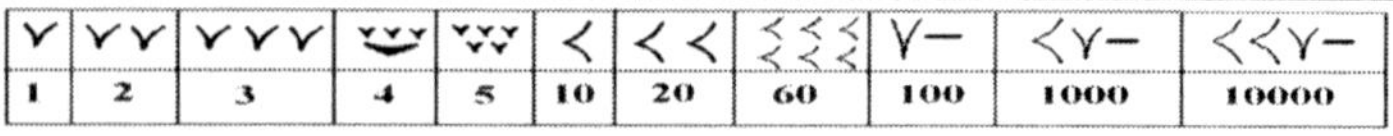

**Figure 1-1.** Sumerian counting system of 3400 BCE

Archaeological evidence of very early mathematics is limited. Detailed information concerning the early development of mathematics really starts with the Babylonian culture. Knowledge of Babylonian mathematics comes from the discovery of over 400 clay tablets dating back to around 2000 BCE. These clay tablets show that the early Babylonians used a base-60 or sexagesimal number system. This system is still in use today with our usage of 60 seconds in a minute, 60 minutes in an hour and 360 degrees in a circle. To represent a number in the sexagesimal number system the Babylonians used a notation similar to the following. The squares of numbers were represented

$$8^2 = 1, 4 \text{ meaning } 64 = 1 \times 60 + 4$$
$$9^2 = 1, 21 \text{ meaning } 81 = 1 \times (60) + 21$$
$$10^2 = 1, 40 \text{ meaning } 100 = 1 \times (60) + 40$$
$$\vdots$$
$$17^2 = 4, 49 \text{ meaning } 289 = 4 \times (60) + 49$$

Tablets containing the squares of the numbers from 1 to 59 are among the 400 clay tablets discovered. The Babylonians used squares of numbers for multiplication. They knew that the product $ab$ of the two numbers $a$ and $b$ could be calculated by subtracting certain square numbers. They usually used one of the identities

$$ab = \left[(a+b)^2 - a^2 - b^2\right]/2 \quad \text{or} \quad ab = \left[(a+b)^2 - (a-b)^2\right]/4$$

Division $a/b$ was multiplication by a reciprocal and treated as $a \times 1/b$. This is inferred from reciprocal tables found among the 400 clay tablets. These tablets showed that the Babylonian mathematics could handle fractions and multiplication. Babylonian mathematics was the norm from about 2000 BCE to around 500 BCE. During this time the Babylonians calculated numbers like $\sqrt{2}$ and $\pi$ to several decimal places. They had knowledge of geometry and had knowledge of the Pythagorean theorem. All this knowledge was for practical applications and there are no records of any proofs associated with the mathematics they used. There are also implications that the Babylonians used some form of trigonometry.

The oldest known Egyptian mathematics document in the form of a text is a papyrus which is known as the Moscow mathematical papyrus or Golenischev papyrus which dates back to around 1850 BCE. The Moscow mathematical papyrus can be found in the Pushkin State Museum of Fine Arts in Moscow. The papyrus is 18 feet long and varies in width from 1.5 to 3 inches. It consists of twenty-five mathematical problems with solutions. A sample problem is given in the figure 1-2.

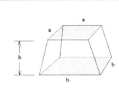

The 14th problem in the Moscow papyrus is a problem requiring the calculation of the volume of a frustum of a pyramid. The height $h$ is given as 6, the top is a square with sides of length 2 and the bottom is a square with sides of length 4. The correct answer of 56 is given. Using modern methods one can show the answer in general is given by
$$V = \frac{h}{3}\left(a^2 + ab + b^2\right)$$

**Figure 1-2** The 14th problem in the Moscow papyrus.

The next oldest Egyptian mathematics document is the Rhind papyrus which was purchased in the year 1858 by a Scotsman by the name of Alexander Henry Rhind. The papyrus was obtained from an Egyptian excavation during that period. The papyrus is approximately 18 feet long and 13 inches wide. Rhind died in 1864 and the papyrus, together with an Egyptian Mathematical Leather Roll, wound up in the British museum. These documents can be found in the Rhind collection of the British Museum. The Rhind papyrus was written by the Egyptian priest Ahmes (A'h-mosé) and dates back to around 1650-1700 BCE. The Rhind papyrus contains a collection of over 80 mathematical problems, with solutions (and a couple of errors in solutions). The solutions require knowledge of only arithmetic and geometry. The papyrus is said to be an Egyptian handbook of practical mathematics. The papyrus contains many examples on the use of fractions and uses the verbal rule $(\frac{8}{9}d)^2$ for the area of a circle, where $d$ is the diameter of the circle. Many other papyri containing mathematics have been discovered, for example, the Reisner papyrus, the Berlin papyrus and the Kahn papyrus. These other papyri are not as old as the Moscow and Rhind papyri. There is a Cairo papyrus, written around 300 BCE which shows that the early Egyptians were capable of creating structures which required extensive use of geometry and shows that they were

# 4

able to solve linear equations $ax = b$, quadratic equations $ax^2 + bx + c = 0$ and simple systems of equations. However, most of the Egyptian mathematics were verbal statements of problems which were also solved verbally without the use of symbols.

The mathematics of the early Babylonians, Egyptians and Greeks was mainly number representations, counting, geometry for construction and the recording of the position of stars. These civilizations were marvelous at creating geometric structures using only a straight edge and compass. They knew about sequences and series, square numbers, triangular numbers and were greatly influenced by patterns which occurred in nature. It is doubtful that they knew the theory behind their mathematics. Theoretical considerations about mathematics came much later.

The use of counting boards, abacus, tally sheet or tally stick are found in many different cultures and have been around for thousands of years. It is inferred that these instruments led to our current positional notation for representing numbers. This is a realistic assumption because an examination of the tally sheet in figure 1-3 can be translated to the markings representing the number 1629 in a decimal system.

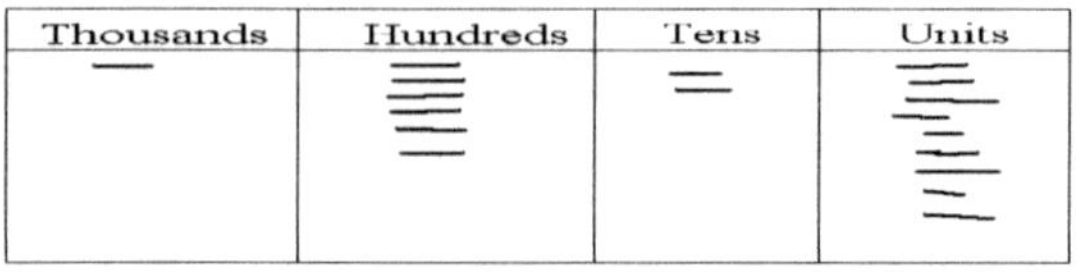

| Thousands | Hundreds | Tens | Units |
|---|---|---|---|
| — | ≡≡≡≡ | ≡ | ≡≡≡ |

**Figure 1-3.** Tally sheet markings

Make note of the fact that if a column in a tally sheet was blank, then this would suggest the need for a symbol to represent zero or an empty column. The zero was introduced in India. The Indian term *sunya* for zero is translated as meaning empty or blank. The above simple example illustrates that counting boards or tally sheets have played an important part in the development of early mathematics. The adoption of a positional notation for representing numbers and the use of zero took many hundreds of years before it was fully accepted. It wasn't until the sixteenth century that arithmetic and numeration reached a form that is even close to what is used in our school systems today.

## Growth of Mathematics

The following three pages show a time line of major contributors to mathematical discoveries with approximations as to the time frame where mathematicians made discoveries. It is a chronological development of some of the more easily recognizable areas of mathematics with apologies to the numerous mathematicians that I have omitted from the time line.

## The Development of Mathematics

| Year | Comments |
|---|---|
| 3500 BCE | Numbers and counting in most cultures |
| ↓ | Babylonian priests study celestial mechanics |
| ↓ | |
| ↓ | |
| ↓ | |
| 3000 BCE | |
| ↓ | Stonehenge built |
| ↓ | |
| ↓ | |
| ↓ | Pyramids of Egypt built |
| 2500 BCE | |
| ↓ | Calendar invented by the Egyptians. |
| ↓ | In Egypt, Babylon, Greece, India and China there developed |
| ↓ | various number systems and counting with written records in some form. |
| ↓ | Geometric patterns and surveying found in most cultures. |
| 2000 BCE | Babylonian mathematics, addition, multiplication, subtraction, |
|  | division, simple linear equations. Early trigonometry. |
| ↓ | Babylonians have approximation for $\pi$ and $\sqrt{2}$. The Moscow papyrus written. |
| ↓ | Concept of infinity developed in India. |
| ↓ | The Rhind papyrus written by Ahmes. |
| ↓ | |
| 1500 BCE | Development of arithmetic along with geometry and trigonometry |
| ↓ | Different cultures exchange/borrow mathematical ideas. |
|  | Chinese astronomers record lunar eclipse |
| ↓ | Geometry and trigonometry used for tracking the sun and moon. |
| ↓ | |
|  | Beginning algebra. The word "algebra" comes from the Arabic word |
|  | "Al-jabra" meaning restoration. Algebra develops with time and is |
| ↓ | driven by construction of buildings, land disputes, wars and astronomy. |
| 1000 BCE | Egyptians study fractions, addition of fractions and series. |
| ↓ | |
|  | The Greeks develop geometry and introduce deductive reasoning and |
| ↓ | use symbols in representing mathematical concepts. |
| ↓ | Hindu algebra introduces base 10 and positional notation . |
|  | Geometry and algebra continue to develop at a more advanced level. |
| ↓ | Pythagoras of Samos introduces the word "mathematics". |
|  | Solar eclipse predicted |

**6** *The Development of Mathematics*

| Year | Comments |
|---|---|
| 500 BCE | Greek philosophers Hippocrates of Chios, Theodorus of Cyrene. |
| ↓ | Zeno of Eleas and his paradoxes, Socrates, Plato, Aristotle. |
| |     Aristotle    Archimedes    Plato |
| ↓ | Democritus of Abdera introduces atomic theory. Appollonius studies conic sections. Cairo papyrus. Archimedes of Syracuse. |
| ↓ | The Egyptians solve linear equations, quadratic equations and simple systems of equations. |
| ↓ | Notion of zero introduced in India. Euclid and geometry. Eratosthenes of Cyrene studies prime numbers. Trigonometry used for astronomy |
| 100 BCE | Greek mathematics dominates for the next 500 years. |
| ↓ | Heron of Alexandria. |
| 100 CE | Diophantus uses symbols for unknowns. |
| ↓ | Negative numbers begin to be accepted. |
| ↓ | Number theory and geometry develop |
| ↓ | |
| 500 | Trigonometric functions studied and used in navigation |
| ↓ | Arabic algebra. Written documents passed to future generations. |
| ↓ | Use of textbooks |
| ↓ | Greek geometry, deductive reasoning and use of symbols in mathematics. |
| ↓ | |
| 1000 | |
| ↓ | Fibbonacci numbers. |
| ↓ | Series expansions and infinite series used for trigonometric functions. |
| ↓ | Spherical geometry begins to be developed. |
| ↓ | Leonardo da Vinci, Nicolas Copernicus, Albert Dürer, |
| | Multiplication and division methods are more fully developed |
| | Roger Bacon and the scientific method. |
| 1500 | Tartaglia, Cardano, Bombelli use complex numbers. |
| | Mercator uses projections for maps. Francis Vieta, Tycho Brahe |

*The Development of Mathematics*

| Year | Comments |
|---|---|
| 1600 | John Napier logarithms, Johannes Kepler, Rene Descartes, Pierre de Fermat, John Wallis, Blaise Pascal studies probability. Isaac Barrow, Christopher Wren, Robert Hooke, James Gregory, Isaac Newton, Gottfried Leibniz, Leibniz and Newton share in the development of the differential and integral calculus. |

Newton     Leibniz

James Bernoulli, Johann Bernoulli, Edmund Halley, Abraham De Moivre
Colin Maclaurin

| Year | Comments |
|---|---|
| 1700 | Brook Taylor develops Taylor Series DeMoivre relates trigonometric functions and complex numbers. Differential equations are investigated. Bernoulli family. Pierre Simon de Laplace, Simeon Poisson Leonhard Euler, Joseph Lagrange, Carl Friedrich Gauss |
| 1800 | Joseph Fourier, Augustin Louis Cauchy, Niels Henrik Abel, Carl Gustav Jacob Jacobi, William Rowan Hamilton, Karl Weierstrass, Ludwig Boltzmann, Jules Henri Poincaré, Alfred North Whitehead |
| 1900 | Bertrand Russell, Albert Einstein, Niels Henrik Bohr, Erwin Schrödinger Louis de Broglie, Wolfgang Pauli, Paul Adrien Maurice Dirac, John von Neumann, Werner Heisenberg |
| 2000 | |
| ↓ | |
| ↓ | |
| ↓ | New advances in computer usage and solving of mathematical problems |
| ↓ | |
| 2500 | Mathematics applied for study of the human genome and epigenetic processes |
| ↓ | |
| ↓ | |
| ↓ | New mathematical discoveries |
| ↓ | |

# 8

Note that the time line presented goes into the future. It is not difficult to project that new mathematical areas will evolve and that the current mathematics will change to meet new conditions. History tells us that new ideas pop up continually. Just look at how mathematics has grown and expanded over the last two millennia.

As an example of what to look for in the future, focus on the new particle accelerators being constructed. These accelerators, when finished, will produce new physics and associated mathematics. New mathematics is always created when new frontiers of research are investigated.

The previous chronological listing of mathematical development highlights only selected events and discoveries. For a more detailed chronological development of mathematics the reader is referred to the references. Also note that there are numerous individuals from various nations of the world that have contributed to the development of mathematics and whose names do not appear in the previous chronological listing.

In the chronological development of mathematics presented on the previous pages there may be some names that are unfamiliar to the reader. Therefore, the following listing of each of the names occurring in the time line is presented along with a very brief description of the individuals contribution to the development of mathematics.

## Selected Contributors to Mathematical Development

### Pythagoras of Samos (569 BCE - 475 BCE)
Studied mathematics, music and astronomy. Proved the theorem which has his name. Note the Pythagorean theorem is believed to have been known to the Babylonians hundreds of years earlier. The Chinese had a geometric proof of the Pythagorean theorem hundreds of years before Pythagoras.

### Hippocrates of Chios (470 BCE - 410 BCE)
Greek mathematician who studied geometry. Not to be confused with Hippocrates of Kos, the father of medicine.

### Theodorus of Cyrene (465 BCE - 396 BCE)
Greek philosopher and teacher who was tutor of Plato.

### Zeno of Eleas (490 BCE - 430 BCE)
Formed paradoxes which made mathematicians think. For example, Zeno's first paradox as recorded by Aristotle: "The first is the one on the non-existence of motion, on the ground that what moves must always attain the middle point sooner than the end point."

### Socrates (470 BCE - 399 BCE)
Greek philosopher who developed ethics, logic and epistemology. Socrates was one of Plato's teachers.

### Plato (428 BCE - 347 BCE)
Teacher and founder of the Academy at Athens. The Academy developed ideas for philosophy, politics, mathematics and the sciences. Plato was a teacher of Aristotle.

**Aristotle (384 BCE - 322 BCE)**
  Greek philosopher who developed deductive logic and studied mathematics and the sciences.

**Democritus of Abdera (460 BCE - 370 BCE)**
  Studied geometry and developed atomic theory.

**Appolonius (262 BCE - 190 BCE)**
  Greek geometer who studied conic sections.

**Euclid (325 BCE - 265 BCE)**
  Greek mathematician and geometer who wrote *The Elements*. Euclid's book *The Elements* consists of 13 books on plane geometry, algebra and number theory. It has been translated from the Greek into many other languages.

**Archimedes of Syracuse (287 BCE - 212 BCE)**
  One of the great mathematicians of history who developed many concepts which if derived today would require knowledge of calculus. The Archimedes "method" used a form of calculus involving an infinite process. How knowledge of Archimedes and some of his work survived the centuries can be found in the Netz-Noel reference.

**Eratosthenes of Cyrene (276 BCE - 194 BCE)**
  Greek mathematician who studied prime numbers.

**Heron of Alexandria (10 CE -75 CE)**
  Studied geometry and developed formula for determining the area of a triangle knowing its sides.

**Diophantus of Alexandria (200 CE - 284 CE)**
  Greek mathematician who is sometimes referred to as the 'father of algebra'.

**Leonardo Fibbonacci (1170 CE - 1250 CE)**
  Hand written books introduced Europeans to ancient mathematical techniques and decimal system.

**Leonardo da Vinci (1452 CE - 1519 CE)**
  Italian artist and scholar who studied geometry, mathematics, mechanical machines, optics and anatomy.

**Nicolaus Copernicus (1473 CE -1543 CE)**
  Polish astronomer and mathematician.

**Albrecht Dürer (1471 CE-1528 CE)**
  German artist and engraver who helped develop descriptive geometry.

**Roger Bacon (1214 CE-1292 CE)**
  English mathematician who applied geometry to optics. He was an advocate of the scientific method.

**Nicolo Fontana Tartaglia (1500 CE-1576 CE)**
  Italian doctor who studied mathematics.

**Girolamo Cardano (1526 CE -1572 CE)**
  Italian scholar who helped develop complex numbers and algebra.

**Rafael Bombelli (1526 CE-1572 CE)**
  Italian scholar who helped develop complex numbers and algebra.

# 10

**Francis Vièta (1540 CE-1603 CE)**
Mathematician and astronomer who wrote book *In artem analyticam isagage* which introduced algebraic notation.

**Tycho Brahe (1546 CE-1601 CE)**
Danish astronomer who aided Kepler by providing astronomical data.

**Gerardus Mercator (1512 CE- 1594 CE)**
Flemish geographer who studied projections and developed maps.

**John Napier (1550 CE-1617 CE)**
Scottish mathematician who developed logarithms and studied spherical triangles.

**Johannes Kepler (1571 CE- 1630 CE)**
German mathematician who developed three laws of planetary motion using only algebra and geometry.

**René Descartes (1596 CE-1650 CE)**
Wrote *La géométrie* which applied algebra to geometry. Today Cartesian geometry is a reminder of his early work.

**Pierre de Fermat (1601 CE-1665 CE)**
French lawyer who studied number theory and helped develop foundations of calculus. Famous for 'Fermat's last theorem'.

**John Wallis (1616 CE-1703 CE)**
English mathematician who is known for developing interpolation and integration techniques.

**Blaise Pascal (1623 CE-1662 CE)**
French mathematician who developed foundations of probability theory.

**Isaac Barrow (1630 CE-1677 CE)**
Helped lay foundations for the calculus while Lucasian professor of mathematics at Oxford. He was a teacher of Isaac Newton.

**Christopher Wren (1632 CE-1723 CE)**
English architect and mathematician.

**Robert Hooke (1635 CE-1703 CE)**
English mathematician who studied optics, mechanics and astronomy.

**James Gregory (1638 CE-1675 CE)**
Scottish mathematician who worked with infinite series.

**Isaac Newton (1643 CE-1727 CE)**

English mathematician who is accredited with developing calculus and its applications. Made major contributions to physics, mechanics, astronomy and optics as well as mathematics. He was also a philosopher and alchemist. He is famous for his treatise *Philosophi Naturalis Principia Mathematica*, which was published in 1687, and which greatly influenced the development of science over the next 300 years.

**Gottfried Wilhelm von Leibniz (1646 CE-1716 CE)**

German mathematician who, independent of Newton, developed calculus. His calculus notation is still used today. Leibniz made major contributions to the scientific areas of philosophy, physics, biology, medicine, geology, probability, linguistics, and information science. He is famous not only for developing calculus but of having hundreds of letters written on many wide ranging subjects, such as politics, law, ethics, theology, history, and philology.

**Bernoulli Family**

Mathematical family of scientists from Switzerland.

Jacob (1654 CE-1705 CE) Mathematician

Johann (1667 CE -1748 CE) Mathematician who studied optics

Johann II (1710 CE-1790 CE) Studied heat and light

Johann III (1744 CE-1807 CE) Studied probability

Jacob II (1759 CE -1789 CE) Mathematical physics

Daniel (1700 CE-1782 CE) Bernoulli principle of fluid flow

Nicolaus I (1687 CE-1759 CE) Mathematician

Nicolaus II (1695 CE-1726 CE) Studied calculus

**Edmund Halley (1656 CE-1742 CE)**

English astronomer and supporter of Isaac Newton.

**Abraham De Moivre (1667 CE-1746 CE)**

French mathematician who worked mainly in analytic geometry and probability.

**Colin Maclaurin (1698 CE-1746 CE)**

Scottish mathematician studied series.

**Brook Taylor (1685 CE-1731 CE)**

English mathematician who worked with series and finite differences.

**Pierre Simon de Laplace (1749 CE-1827)**

French mathematician who made major contributions to many areas of mathematics.

**Leonhard Euler (1707 CE-1783 CE)**

Swiss mathematician who made major contributions to analytic geometry, number theory and calculus.

**Johann Carl Friedrich Gauss (1777-1855 CE)**

German mathematician ranked with Archimedes and Newton as one of the world's great mathematicians of history. He contributed to numerous scientific and mathematical fields of study, such as differential geometry, number theory, complex analysis, statistics, astronomy and optics. He worked with Wilhelm Webber on developing electricity and magnetism. Some textbooks refer to Gauss as "the prince of mathematicians" or as "the greatest mathematician since antiquity".

He produced a large number of original contributions to mathematics and greatly influenced the development of many different areas of science and mathematics.

**Siméon Denis Poisson (1781 CE-1840 CE)**

French mathematician who was student of Laplace and Lagrange who made contributions to ordinary and partial differential equations.

**Augustin Louis Cauchy (1789 CE-1857 CE)**

French mathematician who helped develop complex variable theory and made major contributions in other areas of mathematics.

**Niels Henrik Abel (1802 CE-1829 CE)**

Danish mathematician who studied series and special functions.

**Carl Gustov Jacob Jacobi (1804 CE-1851 CE)**

German mathematician famous for developing theory of elliptic functions.

**Sir William Rowan Hamilton (1805 CE-1865 CE)**

Irish mathematician who developed quaternions which can be described as a noncommutative algebra with many applications.

**Karl Theodor Wilhelm Weierstrass (1815 CE- 1897 CE)**

German mathematician who helped develop complex variable theory.

**Ludwig Boltzmann (1844 CE-1906 CE)**

Austrian mathematician/physicist who developed statistical mechanics and worked in area of thermodynamics. The Boltzmann equation is well known in fluids and particle physics.

**Jules Henri Poincaré (1854 CE- 1912 CE)**

French mathematician who developed algebraic topology.

**Alfred North Whitehead (1861 CE-1947 CE)**

English mathematician best known for *Principia Mathematica* a three volume treatise on symbolic and mathematical logic.

**Bertrand Arthur William Russell (1872 CE-1970 CE)**

English mathematician best known for his work with Whitehead in developing *Principia Mathematica*.

**Albert Einstein (1874 CE- 1955 CE)**
German mathematician and physicist who is famous for developing the special and general theory of relativity. Einstein won the Noble prize in physics for developing the theory of the photo-electric effect and the quantum nature of light.

**Niels Henrik Bohr (1885 CE- 1962 CE)**
Danish mathematician/physicist who helped develop quantum mechanics.

**Erwin Rudolf Josef Alexander Schrödinger (1887 CE-1961 CE)**
Austrian mathematician/physicist who helped develop quantum mechanics.

**Louis Victor Pierre Raymond duc de Broglie (1892 CE-1987 CE)**
French mathematician/physicist who developed duality of wave and particle motion.

**Wolfgang Ernst Pauli (1900 CE-1958 CE)**
Austrian mathematician/physicist best known for the Pauli exclusion principle. Helped develop quantum mechanics.

**Paul Adrien Maurice Dirac (1902 CE- 1984 CE)**
English mathematician/physicist who helped develop quantum mechanics. Won Noble prize in physics for predicting the positron.

**John Von Neumann (1903 CE-1957 CE)**
Hungarian mathematician/physicist who worked in area of quantum mechanics, game theory and computer science.

**Werner Karl Heisenberg (1901 CE-1976 CE)**
Heisenberg uncertainty principle. Helped develop quantum mechanics.

## Mathematical Specialty Areas

Over the last two thousand years new mathematical areas have been created and much specialization and research has been done to develop and refine these new areas of study. The new mathematical areas have spread from country to country and mathematics is now pretty much standardized throughout the world. Mathematics is truly a universal language. The figure 1-4 is an attempt to illustrate how mathematics has grown and expanded over the years. Mathematics is still growing and the figure 1-4 is just a small sampling of the specialty areas which exist today. There is some type of mathematics to be found in just about everything. Mathematics will continue to expand and grow because in order to find solutions to new and challenging problems requires innovative applications of mathematics. I believe it was Heisenberg who is quoted as having said that "Progress is made by solving problems." This statement can be applied to any scientific discipline.

It was the discovery of calculus and its many applications that opened the doors to developing new ideas and methods in mathematics. In particular, mathematical physics developed where calculus provided the means for the representation of the basic laws of physics in a mathematical form.

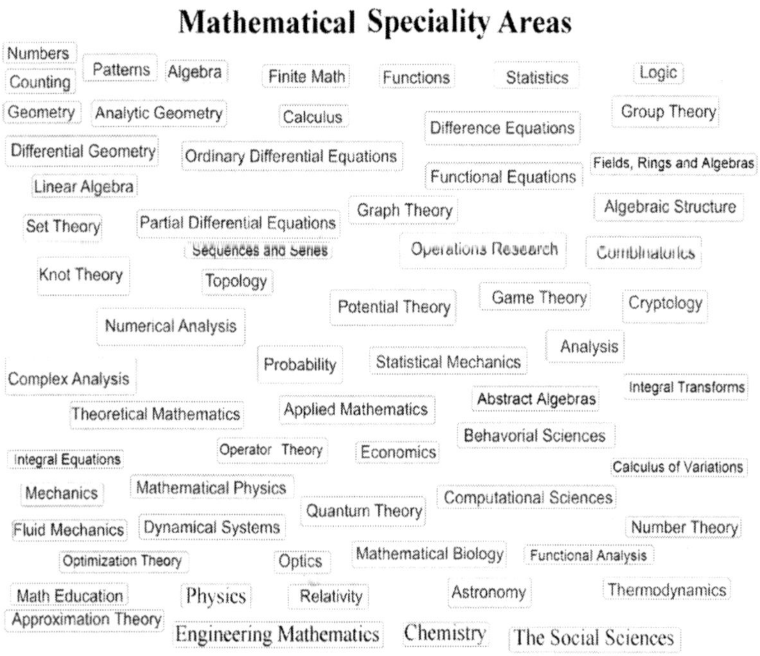

## Mathematical Speciality Areas

Numbers, Counting, Patterns, Algebra, Finite Math, Functions, Statistics, Logic, Geometry, Analytic Geometry, Calculus, Difference Equations, Group Theory, Differential Geometry, Ordinary Differential Equations, Functional Equations, Fields, Rings and Algebras, Linear Algebra, Graph Theory, Algebraic Structure, Set Theory, Partial Differential Equations, Sequences and Series, Operations Research, Combinatorics, Knot Theory, Topology, Potential Theory, Game Theory, Cryptology, Numerical Analysis, Analysis, Complex Analysis, Probability, Statistical Mechanics, Integral Transforms, Theoretical Mathematics, Applied Mathematics, Abstract Algebras, Behavioral Sciences, Integral Equations, Operator Theory, Economics, Calculus of Variations, Mechanics, Mathematical Physics, Computational Sciences, Fluid Mechanics, Dynamical Systems, Quantum Theory, Number Theory, Optimization Theory, Optics, Mathematical Biology, Functional Analysis, Math Education, Physics, Relativity, Astronomy, Thermodynamics, Approximation Theory, Engineering Mathematics, Chemistry, The Social Sciences

**Figure 1-4.** Mathematics developed over the last two thousand years without any ordering of the categories.

Other areas of science found applications for mathematics. The advanced mathematics in the years 1900-2000 consisted of an introduction to the study areas of pure and applied mathematics, the use of statistics, computing and computational methods, the actuarial sciences and operations research. During this time frame there developed applications of statistics to the study areas of economics, psychological measurements, and biology to create the mathematical areas of econometrics, psychometrics and biometrics. Mathematical methods were also applied to develop the social sciences, physical chemistry and the engineering sciences. All these study areas employ many forms of mathematics, statistics or computing and in turn these areas have broadened the development and expansion of mathematics and have introduced new mathematical ideas.

## More Information

Much additional information on the history of mathematics and the development of mathematics is now available on the world wide web. The following is a small sampling of such web sites which give additional time lines and mathematical references.

- http://en.wikipedia.org/wiki/Timeline_of_mathematics
- http://aleph0.clarku.edu/~djoyce/mathhist/chronology.html
- http://www.ams.org/msc/
- http://www-groups.dcs.st-and.ac.uk/ history/BiogIndex.html

Additional references can be found in the Bibliography at the end of this book.

## Mathematical Notation

Various shorthand notations and abbreviations have developed to represent ideas and concepts in mathematics, the sciences and engineering. Any new symbolism that is introduced is usually defined to have a precise meaning. If the symbolism proves to be popular it is accepted over time and becomes known as a "standardized" representation of the idea or concept. There are many notations that are not standardized. For example, the error function of $x$ is denoted by the notation erf(x), however different textbooks use different definitions for the error function. Some texts define it as

$$\operatorname{erf}(x) = \frac{2}{\sqrt{\pi}} \int_0^x e^{-\xi^2}\, d\xi \qquad (1.1)$$

while other textbooks use the definition

$$\operatorname{erf}(x) = \int_0^x e^{-\xi^2}\, d\xi \qquad (1.2)$$

As another example the operator $\nabla$ called nabla or the del operator is used to represent the gradient in the study of vectors, but it is used as a backward difference operator in the study area of numerical analysis. In general, to avoid difficulty in symbolism and use of mathematical notation, carefully define the symbolism being employed.

In applied mathematics it is important that equations are dimensionally consistent. That is, each term in an equation must have the same dimensions. To emphasize the dimensions of a quantity many texts use the notation[2] $[A]$ to denote the dimensions of the quantity $A$. For example, $[y]$ = meters, is read, "The dimension of $y$ is meters."

Mathematical notation can at times lead to difficulty in understanding a concept. Whenever this happens, look for alternative presentations of the subject matter. The following is a tabular listing of mathematical abbreviations, notations and functions. This list is given with the proviso that the reader should note that some notations have multiple uses and/or definitions. Many concepts in mathematics can be represented in a variety of ways.

---

[2] Notation introduced by J.B.J. Fourier, theorie analytique de la chaleur, Paris 1822.

| Mathematical Notation and Abbreviations ||
|---|---|
| Notation | Meaning |
| °, ′, ″ | degree, minute, second measures of angle in sexagesimal system |
| + | Plus sign, $1 + 2 = 3$ |
| − | Minus sign, $3 - 2 = 1$ |
| / or ÷ or − | Divide sign, $3/2 = 1.5$ or $3 \div 2 = 1.5$ or $\frac{3}{2} = 1.5$ |
| · or ∗ or × | Multiply sign, $3 \cdot 2 = 6$ or $3 \ast 2 = 6$ or $3 \times 2 = 6$ |
| ± | Plus or minus sign |
| ∓ | Minus or plus sign |
| = | Equal sign, $a = b$ |
| ... | three dots meaning 'and so forth' or 'and so forth up to' |
| △ | Triangle, $\triangle ABC$ is triangle with vertices $A, B$ and $C$. |
| ≅ | Is congruent to, $\triangle ABC \cong \triangle DEF$ |
| $A \sim B$ | Asymptotic to, $A$ is asymptotic to $B$ or the ratio $A/B$ approaches 1. |
| < | Less than, $A < B$ |
| ≤ | Less than or equal to, $A \leq B$ |
| > | Greater than, $A > B$ |
| ≥ | Greater than or equal to, $A \geq B$ |
| ≠ | Not equal to, $A \neq B$ |
| $\{A, B, C, D\}$ or $\{S\}$ | The set of objects $A, B, C, D$ or the set $S$ |
| ∈ | Belongs to or is a member of, $a \in \{A\}$ |
| ∉ | Does not belong to, not in, $b \notin \{A\}$ |
| ∅ | The null or empty set |
| ∩ | Intersection of two sets, $A \cap B$, elements in both $A$ and $B$ |
| ∪ | Union of two sets, $A \cup B$, elements in $A$ or in $B$ or both. |
| ⊂ | Inclusion, $A \subset B$, here $A$ is contained in $B$ as proper subset |
| \ | $A \setminus B$ denotes the set of elements in $A$ that are not in $B$ |
| ⊆ | $A \subseteq B$ denotes $A$ is a subset of $B$ with all elements in $A$ also in $B$ |
| ⊇ | $A \supseteq B$, $A$ is a superset of $B$ with all elements in $B$ also in $A$ |
| ⊃ | $A \supset B$, here $A$ contains $B$ as proper subset |
| ≈ | Approximately, $\pi \approx 3.14159$ |
| $\overline{AB}$ or $AB$ | Length of line segment from $A$ to $B$ |
| ⊥ | Perpendicular, $\ell_1 \perp \ell_2$, line $\ell_1$ is perpendicular to line $\ell_2$ |
| √ | Radical sign, $\sqrt{a}$=square root of $a$, $\sqrt[n]{a}$ is $n$th root of $a$ |
| $\|a\|$ | Absolute value of $a$, $\|a\| = \begin{cases} +a, & \text{if } a > 0 \\ -a, & \text{if } a < 0 \end{cases}$ |

| Mathematical Notation and Abbreviations | |
|---|---|
| Notation | Meaning |
| $\Sigma$ | Summation sign, $\sum_{n=1}^{m} a_n = a_1 + a_2 + a_3 + \cdots + a_m$ |
| $\Pi$ | Product sign, $\prod_{n=1}^{m} a_n = a_1 a_2 \cdots a_m$ |
| $\operatorname{sgn}(x)$ | Signum function, sign of $x$, $\operatorname{sgn}(x) = \frac{x}{|x|}$ or $\operatorname{sgn}(x) = \frac{2}{\pi} \int_0^\infty \frac{\sin(xt)}{t} dt$ |
| $A = \begin{pmatrix} a_{11} & a_{12} \\ a_{21} & a_{22} \\ a_{31} & a_{32} \end{pmatrix}$ | $A$ is called a 3 by 2 matrix of elements. (3 rows, 2 columns) |
| $A^T$ | Transpose of matrix $A$, where rows and columns are interchanged |
| $A \sim B$ | The matrix $A$ is equivalent to matrix $B$ |
| $I$ | Identity matrix, $I = \begin{pmatrix} 1 & 0 & 0 \\ 0 & 1 & 0 \\ 0 & 0 & 1 \end{pmatrix}$ is the 3 by 3 identity matrix |
| $A^{-1}$ | Inverse of matrix $A$, $AA^{-1} = A^{-1}A = I$ |
| $\det A = \begin{vmatrix} a_{11} & a_{12} \\ a_{21} & a_{22} \end{vmatrix}$ | Determinant of matrix $A = \begin{pmatrix} a_{11} & a_{12} \\ a_{21} & a_{22} \end{pmatrix}$, $\det A = a_{11}a_{22} - a_{12}a_{21}$ |
| $y = a_0 + \frac{1}{a_1+} \frac{1}{a_2+} \cdots$ | Short hand notation for continued fraction $y = a_0 + \cfrac{1}{a_1 + \cfrac{1}{a_2 + \cfrac{1}{a_3 + \cdots}}}$ |
| $\infty$ | Infinity, concept of something increasing without bound. |
| $\forall$ | for all |
| $\exists$ | there exists |
| $\ni$ | such that |
| $\therefore$ | therefore or thus |
| $\because$ | because or due to |
| $\operatorname{Re}\{\ \}$ | Real part of |
| $\operatorname{Im}\{\ \}$ | Imaginary part of |
| $|$ | such that, $\{x \mid a < x < b\}$, the set of $x$ such that $x$ lies between $a$ and $b$. |
| $n!$ | $n$-factorial or factorial $n$, $n! = n(n-1)\cdots 3 \cdot 2 \cdot 1$ |
| $\binom{n}{m}$ | Binomial coefficients, $\binom{n}{m} = \frac{n!}{m!(n-m)!}$ |
| $a^n$ | $\underbrace{a \cdot a \cdot a \cdots a}_{n-times}$ is $n$th power of $a$ |
| $a^{-n}$ | Reciprocal of $a^n$, $a^{-n} = \frac{1}{a^n}$ |
| $z = x + iy$ | Complex variable where $i^2 = -1$ |
| $\bar{z} = x - iy$ | Complex conjugate of complex variable $z$ |
| $\iff$ | if and only if |

| Mathematical Notation and Abbreviations | |
|---|---|
| Notation | Meaning |
| $\Longrightarrow$ | $A$ implies $B$, $\quad A \Longrightarrow B$ |
| $\log_b N$ | logarithm of $N$ to base $b$, $x = \log_b N \iff b^x = N$ |
| $\log N$ or $\ln N$ | Natural logarithm of $N$, (to base e) |
| $\log_{10} N$ | Common logarithm (to base 10) |
| rad | radian measure of angle, $180° = \pi$ radians |
| $(a, b)$ or $[a, b]$ | Open or closed interval from a to b |
| $\{a_n\}$ | sequence of terms with general term $u_n$ |
| $\lim_{n \to \infty} a_n$ | limit of a sequence as $n$ increases without bound |
| $\limsup_{n \to \infty} a_n$ | The greatest limit point of a sequence $\{a_n\}$ |
| $\liminf_{n \to \infty} a_n$ | The smallest limit point of sequence $\{a_n\}$ |
| $f(x), g(x)$ | functions of a single variable $x$ |
| $C^n(X)$ | The set of all functions possessing $n$ continuous derivatives on $X$ |
| $\max_{a \le x \le b} f(x)$ | maximum of $f(x)$ on interval $[a, b]$ |
| $\min_{a \le x \le b} f(x)$ | minimum of $f(x)$ on interval $[a, b]$ |
| $\sup_{a \le x \le b} f(x)$ | supremum or the least upper bound of $f(x)$ on interval $[a, b]$ |
| $\inf_{a \le x \le b} f(x)$ | infimum or the greatest lower bound of $f(x)$ on interval $[a, b]$ |
| $\lim_{x \to x_0} f(x) = f(x_0)$ | The function $f(x)$ has limit $f(x_0)$ as $x$ approaches $x_0$ |
| $\lim_{x \to x_0^+} f(x) = A$ | Right-hand limit of $f(x)$ as $x$ approaches $x_0$ with $x > x_0$ |
| $\lim_{x \to x_0^-} f(x) = B$ | Left-hand limit of $f(x)$ as $x$ approaches $x_0$ with $x < x_0$ |
| $\lim_{x \to \infty} f(x) = C$ | As $x$ increases without bound, $f(x)$ approaches $C$ |
| $\max(a_1, a_2, \ldots, a_n)$ | The largest number from the set $a_1, a_2, \ldots, a_n$ |
| $\min(a_1, a_2, \ldots, a_n)$ | The smallest number from the set $a_1, a_2, \ldots, a_n$ |
| $\frac{dy}{dx}, y', f'(x), \frac{df(x)}{dx}$ | The first derivative of $y = f(x)$ |
| $\dot{y}, \frac{dy}{dt}, \dot{x}, \frac{dx}{dt}$ | Derivatives with respect to time $t$ |
| $\frac{d^n y}{dx^n}, y^{(n)}, f^{(n)}(x), \frac{d^n f(x)}{dx^n}$ | The $n$th derivative of $y$ with respect to $x$, where $n = 2, 3, \ldots$ |
| $dy, df(x)$ | Differential of function $y = f(x)$ |
| $\delta y, \delta f(x)$ | The first variation of function $y = f(x)$ |
| $\frac{\partial z}{\partial x}, \frac{\partial f}{\partial x}$ | The first partial derivative of $z = f(x, y)$ with respect to $x$ |
| $\frac{\partial z}{\partial y}, \frac{\partial f}{\partial y}$ | The first partial derivative of $z = f(x, y)$ with respect to $y$ |
| $\frac{\partial^2 z}{\partial x^2}, \frac{\partial^2 z}{\partial y^2}, \frac{\partial^2 z}{\partial x \partial y}$ | Second partial derivatives of $z = f(x, y)$ |

| Mathematical Notation and Abbreviations ||
|---|---|
| Notation | Meaning |
| $J = \frac{\partial(x,y)}{\partial(u,v)}, \quad J\left(\frac{x,y}{u,v}\right)$ | Jacobian determinant $\begin{vmatrix} \frac{\partial x}{\partial u} & \frac{\partial x}{\partial v} \\ \frac{\partial y}{\partial u} & \frac{\partial y}{\partial v} \end{vmatrix}$ |
| $\int f(x)\,dx$ | Indefinite integral of $f(x)$ |
| $\int_a^b f(x)\,dx$ | Definite integral of $f(x)$ with limits $a$ to $b$ |
| $\int_a^\infty f(x)\,dx$ | Improper integral, $\int_a^\infty f(x)\,dx = \lim_{T\to\infty} \int_a^T f(x)\,dx$ |
| $F(x)\big|_a^b$ | $F(x)\big|_a^b = F(b) - F(a)$ |
| $\iint_R f(x,y)\,dxdy$ | Double integral of $f(x,y)$ over region $R$ |
| $\iiint_V f(x,y,z)\,dxdydz$ | Triple integral of $f(x,y,z)$ over region $V$ |
| $\iint \cdots \int_{V_n} f(x_1, x_2, \ldots, x_n)\,dx_1 \cdots dx_n$ | n-tuple integral of $f(x_1, \ldots, x_n)$ over region $V_n$ |
| $(f, g)$ | Weighted inner product of two functions $f(x)$ and $g(x)$, $(f,g) = \int_a^b w(x) f(x) g(x)\,dx$, $w(x) > 0$ is weight function |
| $\|f\|$ | The norm of function $f(x)$, $\|f\| = \sqrt{(f,f)}$ |
| $\hat{e}_1, \hat{e}_2, \hat{e}_3$ or $\hat{i}, \hat{j}, \hat{k}$ | Unit vectors in Cartesian coordinates |
| $\vec{V} = V_1\hat{e}_1 + V_2\hat{e}_2 + V_3\hat{e}_3$ | Vector in three dimensions |
| $|\vec{V}|$ | $|\vec{V}| = \sqrt{V_1^2 + V_2^2 + V_3^2}$ length of vector $\vec{V}$ |
| $\vec{A} \times \vec{B}$ | Cross product of vectors $\vec{A}$ and $\vec{B}$ $\vec{A} \times \vec{B} = \begin{vmatrix} \hat{e}_1 & \hat{e}_2 & \hat{e}_3 \\ A_1 & A_2 & A_3 \\ B_1 & B_2 & B_3 \end{vmatrix}$ |
| $k\vec{V}$ | scalar $k$ times vector $\vec{V}$ |
| $\vec{A} \cdot \vec{B}$ | dot product of vectors $\vec{A}, \vec{B}$, $\vec{A} \cdot \vec{B} = A_1B_1 + A_2B_2 + A_3B_3$ |
| $\vec{A} \cdot (\vec{B} \times \vec{C})$ | triple scalar product $\vec{A} \cdot (\vec{B} \times \vec{C}) = \vec{B} \cdot (\vec{C} \times \vec{A}) = \vec{C} \cdot (\vec{A} \times \vec{B})$ |
| $\frac{d\vec{A}}{dt}$ | derivative of vector $\vec{A}$, $\frac{d\vec{A}}{dt} = \frac{dA_1}{dt}\hat{e}_1 + \frac{dA_2}{dt}\hat{e}_2 + \frac{dA_3}{dt}\hat{e}_3$ |
| $\vec{r}$ | position vector to point $(x,y,z)$, $\vec{r} = x\hat{e}_1 + y\hat{e}_2 + z\hat{e}_3$ |
| $\nabla$ | nabla operator $\nabla = \frac{\partial}{\partial x}\hat{e}_1 + \frac{\partial}{\partial y}\hat{e}_2 + \frac{\partial}{\partial z}\hat{e}_3$ |
| $\nabla^2$ | Laplacian operator $\nabla^2 \phi = \frac{\partial^2 \phi}{\partial x^2}\hat{e}_1 + \frac{\partial^2 \phi}{\partial y^2}\hat{e}_2 + \frac{\partial^2 \phi}{\partial z^2}\hat{e}_3$ |
| | In n-dimensions $\nabla^2 = \sum_{i=1}^n \frac{\partial^2}{\partial x_i^2}$ |

| Mathematical Notation and Abbreviations | |
|---|---|
| Notation | Meaning |
| $\operatorname{grad} \phi$, $\nabla \phi$ | gradient of $\phi$, $\quad \nabla \phi = \dfrac{\partial \phi}{\partial x}\hat{e}_1 + \dfrac{\partial \phi}{\partial y}\hat{e}_2 + \dfrac{\partial \phi}{\partial z}\hat{e}_3$ |
| $\operatorname{div} \vec{V}$, $\nabla \cdot \vec{V}$ | divergence of $\vec{V}$, $\quad \nabla \cdot \vec{V} = \dfrac{\partial V_1}{\partial x} + \dfrac{\partial V_2}{\partial y} + \dfrac{\partial V_3}{\partial z}$ |
| $\operatorname{curl} \vec{V}$, $\nabla \times \vec{V}$ | curl of vector $\vec{V}$, $\quad \nabla \times \vec{V} = \begin{vmatrix} \hat{e}_1 & \hat{e}_2 & \hat{e}_3 \\ \frac{\partial}{\partial x} & \frac{\partial}{\partial y} & \frac{\partial}{\partial z} \\ V_1 & V_2 & V_3 \end{vmatrix}$ |
| $\delta_{ij}$ | The Kronecker delta $\delta_{ij} = \begin{cases} 1, & \text{if } i = j \\ 0, & \text{if } i \neq j \end{cases}$ |
| $\sin x$ | The sine function of $x$ |
| $\cos x$ | The cosine function of $x$ |
| $\tan x$ | The tangent function of $x$ |
| $\cot x$ | The cotangent function of $x$ |
| $\sec x$ | The secant function of $x$ |
| $\csc x$ | The cosecant function of $x$ |
| $\arcsin x$, $\sin^{-1} x$ | The arc sine of $x$, angle whose sine is $x$ |
| $\arccos x$, $\cos^{-1} x$ | The arc cosine of $x$, angle whose cosine is $x$ |
| $\arctan x$, $\tan^{-1} x$ | The arc tangent of $x$, angle whose tangent is $x$ |
| $\operatorname{arccot} x$, $\cot^{-1} x$ | The arc cotangent of $x$, angle whose cotangent is $x$ |
| $\operatorname{arcsec} x$, $\sec^{-1} x$ | The arc secant of $x$, angle whose secant is $x$ |
| $\operatorname{arccsc} x$, $\csc^{-1} x$ | The arc cosecant of $x$, angle whose cosecant is $x$ |
| $\sinh x$ | The hyperbolic sine of $x$ |
| $\cosh x$ | The hyperbolic cosine of $x$ |
| $\tanh x$ | The hyperbolic tangent of $x$ |
| $\coth x$ | The hyperbolic cotangent of $x$ |
| $\operatorname{sech} x$ | The hyperbolic secant of $x$ |
| $\operatorname{csch} x$ | The hyperbolic cosecant of $x$ |
| $\sinh^{-1} x$ | The inverse hyperbolic sine of $x$ |
| $\cosh^{-1} x$ | The inverse hyperbolic cosine of $x$ |
| $\tanh^{-1} x$ | The inverse hyperbolic tangent of $x$ |
| $\coth^{-1} x$ | The inverse hyperbolic cotangent of $x$ |
| $\operatorname{sech}^{-1} x$ | The inverse hyperbolic secant of $x$ |
| $\operatorname{csch}^{-1} x$ | The inverse hyperbolic cosecant of $x$ |
| $e^x$, $\operatorname{Exp}(x)$ | The exponential function, $\quad e = 2.71828\ldots$ |
| $\Gamma(x)$ | The Gamma function |

| Mathematical Notation and Abbreviations ||
|:---:|:---:|
| Notation | Meaning |
| $B(p,q)$ | The Beta function |
| $J_n(x)$ | The Bessel function of the first kind of order $n$ |
| $Y_n(x)$ | The Bessel function of the second kind of order $n$ |
| $j_n(x), n_n(x)$ | Spherical Bessel functions of first and second kind |
| $P_n(x)$ | The Legendre polynomial of degree $n$ |
| $P_n^m(x)$ | The associated Legendre functions |
| $T_n(x)$ | The Chebyshev polynomial of degree $n$ |
| $L_n(x)$ | The Laguerre polynomial of degree $n$ |
| $L_n^m(x)$ | The associated Laguerre polynomials |
| $H_n(x)$ | The Hermite polynomial of degree $n$ |
| $Y_\ell^m(\theta,\phi)$ | Spherical harmonic function of degree $\ell$ and order $m$ |
| $F(k,\phi)$ | Legendre elliptic integral of first kind |
| $E(k,\phi)$ | Legendre elliptic integral of second kind |
| $\pi(k,n,\phi)$ | Legendre elliptic integral of third kind |
| $\mathrm{erf}(x)$ | The error function of $x$ |
| $\mathrm{erfc}(x)$ | The complimentary error function of $x$ |
| $\mathrm{sn}(u,k)$ | The Jacobian elliptic function sn |
| $\mathrm{cn}(u,k)$ | The Jacobian elliptic function cn |
| $\mathrm{dn}(u,k)$ | The Jacobian elliptic function dn |
| $Si(x)$ | The sine integral, $Si(x) = \int_0^x \frac{\sin t}{t} dt$ |
| $Ci(x)$ | The cosine integral, $Ci(x) = \int_x^\infty \frac{\cos t}{t} dt$ |
| $\ell i(x)$ | The logarithmic integral, $\ell i(x) = \int_0^x \frac{dt}{\ln t}$ |
| $Ei(x)$ | The exponential integral, $Ei(x) = \int_x^\infty \frac{e^{-t}}{t} dt$ |
| $\phi(N)$ | The Euler $\phi$ function |
| $S(x)$ | The Fresnel sine integral, $S(x) = \sqrt{\frac{2}{\pi}} \int_0^x \sin t^2 \, dt$ |
| $C(x)$ | The Fresnel cosine integral $C(x) = \sqrt{\frac{2}{\pi}} \int_0^x \cos t^2 \, dt$ |
| $\zeta(x)$ | Riemann zeta function $\zeta(x) = \frac{1}{1^x} + \frac{1}{2^x} + \frac{1}{3^x} + \cdots$ |
| $\wp(x)$ | The Weierstrassian elliptic function |

## Mathematical Notation and Abbreviations

| Notation | Meaning |
|---|---|
| $\mathcal{F}_e\{f(x); x \to \omega\}$ | The Fourier exponential transform of $f(x)$ |
| $\mathcal{F}_s\{f(x); x \to \omega\}$ | The Fourier sine transform of $f(x)$ |
| $\mathcal{F}_c\{f(x); x \to \omega\}$ | The Fourier cosine transform of $f(x)$ |
| $\mathcal{L}\{f(t); t \to s\}$ | The Laplace transform of $f(t)$ |
| GCD | Greatest Common Divisor |
| | Let $N_1$ have set of prime factors $S_1$ and $N_2$ have a set of prime factors $S_2$, then determine $S_1 \cap S_2$ and multiply all numbers common to both $S_1$ and $S_2$. The largest number that divides evenly into each of a given set of numbers. Also called greatest common factor (GCF), highest common factor (HCF). |
| LCM | Lowest Common Multiple |
| | Let $N_1$ have set of prime factors $S_1$ and $N_2$ have a set of prime factors $S_2$, then form Venn diagram for $S = S_1 \cup S_2$ and multiply all prime factors in $S$. The smallest quantity that is divisible by two or more given quantities without a remainder. Also called least common multiple. |
| LCD | Lowest Common Denominator |
| | Is calculated by finding lowest common multiple of the denominators |
| HCF | Highest Common Factor. See GCD |
| RMS | Root Mean Square $\text{RMS} = \sqrt{\frac{x_1^2 + x_2^2 + \cdots + x_n^2}{n}}$ |
| AM | Arithmetic mean, $\text{AM} = \dfrac{x_1 + x_2 + x_3 + \cdots + x_n}{n}$ |
| GM | Geometric mean, $\text{GM} = \sqrt[n]{x_1 x_2 \cdots x_n}$ Each $x_i \geq 0$ |
| HM | Harmonic mean, $\dfrac{1}{\text{HM}} = \dfrac{1}{n}\left(\dfrac{1}{x_1} + \dfrac{1}{x_2} + \cdots + \dfrac{1}{x_n}\right)$, Each $x_i > 0$ |
| IVP | Initial Value Problem |
| BVP | Boundary Value Problem |
| $W(y_1, y_2)$ | Wronskian determinant of $y_1(x)$ and $y_2(x)$, $W = \begin{vmatrix} y_1(x) & y_2(x) \\ y_1'(x) & y_2'(x) \end{vmatrix}$ |
| $\dfrac{dy}{dx} = f(x,y)$ | First order ordinary differential equation |
| $\dfrac{\partial u}{\partial x} = f(x,y)$ | First order partial differential equation |

| Mathematical Notation and Abbreviations ||
|---|---|
| Notation | Meaning |
| $\Box^2$ | D'Alembertian operator, $\Box^2 = -\dfrac{\partial^2}{\partial x_1^2} + \sum_{i=2}^{n} \dfrac{\partial^2}{\partial x_i^2}$ |
| $\dfrac{\partial^2 u}{\partial x^2} = k^2 \dfrac{\partial u}{\partial t}$ | One dimensional heat equation |
| $\dfrac{\partial^2 u}{\partial t^2} - c^2 \dfrac{\partial^2 u}{\partial x^2} = 0$ | One dimensional wave equation |
| $\dfrac{\partial^2 u}{\partial x^2} + \dfrac{\partial^2 u}{\partial y^2} = 0$ | Laplace equation describing potential functions |
| $\dfrac{\partial^2 u}{\partial x^2} + \dfrac{\partial^2 u}{\partial y^2} = f(x,y)$ | Poisson's equation |
| $y = \dfrac{1}{\sqrt{2\pi}} e^{-t^2/2}$ | Standard form for normal probability density (Bell shaped curve) |
| $\Phi(x) = \dfrac{1}{\sqrt{2\pi}} \int_{-\infty}^{x} e^{-t^2/2}\, dt$ | Standard normal distribution function |
| $(a)_n$ | The factorial function $(a)_n = a(a+1)(a+2)\cdots(a+n-1)$ |
| $\arg z$ | Argument of $z$ |
| $\operatorname{am} z$ | The amplitude of the complex number $z$ |
| $\mathfrak{B}_n$ | $n$th Bernoulli number |
| $\mathfrak{E}_n$ | $n$th Euler number |
| $E[g(x)]$ | Expected value of $g(x)$ |
| $S_n^{(m)}$ | Stirling number of first kind |
| $\mathfrak{S}_n^{(m)}$ | Stirling number of second kind |
| $ce_n(x,q)$ | Mathieu function first kind – cosine-elliptic |
| $se_n(x,q)$ | Mathieu function first kind– sine-elliptic |
| $Ce_n(x,q)$ | Modified Mathieu function-cosine |
| $Se_n(x,q)$ | Modified Mathieu function-sine |
| $j_n(z)$ | Spherical Bessel function of first kind |
| $k_\nu(z)$ | The Bateman function |
| $K_\nu(z)$ | Modified Bessel function |
| $\ker_\nu(x), \operatorname{kei}_\nu(x)$ | The Kelvin functions |
| $H_\nu(z)$ | Struve function |
| $L_\nu(z)$ | Modified Struve function |
| $M_{x,\nu}(z)$ | The Whittaker function |
| $\lambda_1, \lambda_2, \ldots$ | Eigenvalues |

| Mathematical Notation and Abbreviations ||
|---|---|
| Notation | Meaning |
| $P(X \leq x)$ | The probability that $X$ is less than or equal to $x$ |
| $\mathbf{E}_\nu(z)$ | The Webber function |
| $\mathrm{E}_\nu^{(m)}(z)$ | The Webber parabolic cylinder function |
| $y[x_i, x_{i+1}]$ | First divided difference |
| $y[x_i, x_{i+1}, x_{i+2}]$ | Second divided difference |
| $Z[f(n)] = F(z)$ | The Z-transform of $f(n)$ is $F(z)$ |
| $H(t)$ | The Heaviside unit step function $H(t) = \begin{cases} 1, & t > 0 \\ 0, & t < 0 \end{cases}$ |
| $n^{[k]}$ | Factorial polynomial $n^{[k]} = n(n-1)(n-2)\cdots(n-k+1)$ |
| $D = \frac{d}{dx}$ | Differential operator |
| $I = \int (\ )\, dx$ | Integral operator |
| $RK-4$, $RK4$ | Runge-Kutta fourth order method |
| $MC$ | Monte Carlo |
| $MPI$ | Message Passing Interface |
| $B_{i,n}(t)$ | B-splines |
| $\mathbb{N}$ | The set of natural numbers $N_0 = \{0, 1, 2, \ldots\}$, $N^* = \{1, 2, \ldots\}$ |
| $\mathbb{C}$ | The set of complex numbers |
| $\mathbb{Z}$ | The set of integer numbers |
| $\mathbb{Q}$ | The set of rational numbers |
| $\mathbb{P}$ | The set of prime numbers $\{2, 3, 5, 7, 11, \ldots\}$ |
| $\mathbb{R}$ | The set of real numbers |
| $\mathbb{R}^n$ | The set of ordered $n$-tuples of real numbers |
| $i$ or $j$ | Imaginary unit $i^2 = -1$ or $j^2 = -1$ |
| $gd(z)$ | Gudermannian $\operatorname{gd} u = \sin^{-1}\tanh u = \cos^{-1}\operatorname{sech} u = \tan^{-1}\sinh u$ $\operatorname{gd} u = \cot^{-1}\operatorname{csch} u = \sec^{-1}\cosh u = \csc^{-1}\coth u$ |
| $O(\ )$ | Landau symbol "big Oh" |
| $o(\ )$ | Landau symbol "little Oh" |
| Q.E.D. | Quod erat demonstrandum—which was to be proved |
| Q.E.F. | Quod erat faciendum—which was to be constructed |

## Number Representation

A base b number representation of a number $N$ in terms of various powers of b is a series having the form

$$N_b = \cdots + \alpha_n b^n + \alpha_{n-1} b^{n-1} + \cdots + \alpha_3 b^3 + \alpha_2 b^2 + \alpha_1 b^1 + \alpha_0 b^0 + \beta_1 b^{-1} + \beta_2 b^{-2} + \beta_3 b^{-3} + \cdots$$

where the quantities $\alpha_0, \alpha_1, \ldots, \alpha_n$ and $\beta_1, \beta_2, \beta_3, \ldots$ are coefficients representing a digit belonging to the set $S = \{0, 1, 2, 3, \ldots, b-2, b-1\}$. Whenever the base b is 10, the number system is called a decimal system and the set $S$ has the representation $S = \{0, 1, 2, 3, 4, 5, 6, 7, 8, 9\}$ and in this case the subscript is not used as the base is understood to be 10. Some examples of numbers represented in different base numbering systems are given in the following table.

| Examples of Number Respresentations | | | |
|---|---|---|---|
| Base b | Base b representation | Base 10 number | Set $S$ |
| 10 | $453.781 = 4(10)^2 + 5(10)^1 + 3(10)^0 + 7(10)^{-1} + 8(10)^{-2} + 1(10)^{-3}$ | 453781/1000 | $\{0,1,2,3,4,5,6,7,8,9\}$ |
| 2 | $101.011_2 = 1(2)^2 + 0(2)^1 + 1(2)^0 + 0(2)^{-1} + 1(2)^{-2} + 1(2)^{-3}$ | 43/8 | $\{0,1\}$ |
| 3 | $201.121_3 = 2(3)^2 + 0(3)^1 + 1(3)^0 + 1(3)^{-1} + 2(3)^{-2} + 1(3)^{-3}$ | 529/27 | $\{0,1,2\}$ |
| 5 | $432.123_5 = 4(5)^2 + 3(5)^1 + 2(5)^0 + 1(5)^{-1} + 2(5)^{-2} + 3(5)^{-3}$ | 14663/125 | $\{0,1,2,3,4\}$ |
| 7 | $5066.606_7 = 5(7)^3 + 0(7)^2 + 6(7)^1 + 6(7)^0 + 6(7)^{-1} + 0(7)^{-2} + 6(7)^{-3}$ | 605051/343 | $\{0,1,2,3,4,5,6\}$ |

In a hexadecimal number system the base is $b = 16$ and the digits are from the set $S = \{0, 1, 2, 3, 4, 5, 6, 7, 8, 9, A, B, C, D, E, F\}$. Whenever the base $b$ is larger than 10 it is customary to use the letters $A - Z$ to represent the digits. Here $A$ is the digit representing 10, $B$ for 11, $C$ for 12, etc. An example of a hexadecimal number is

$$4A3C.E_{16} = 4(16)^3 + A(16)^2 + 3(16)^1 + C(16)^0 + E(16)^{-1} = 605051/343$$

Recall the Babylonians had a sexagesimal number system or base 60 number system. This required 60 symbols to represent the digits in this system.

To convert a number $N$ from a decimal base to base b it is necessary to calculate the coefficients $\alpha_n, \alpha_{n-1}, \ldots, \alpha_1, \alpha_0, \beta_1, \beta_2, \ldots$ in the base b number system, where $N$ has the representation $N = \alpha_n b^n + \alpha_{n-1} b^{n-1} + \cdots + \alpha_2 b^2 + \alpha_1 b^1 + \alpha_0 b^0 + \beta_1 b^{-1} + \beta_2 b^{-2} + \cdots$. The integer part of $N$ is determined by the $\alpha$-coefficients and is denoted using the notation $I[N]$ for "integer part of $N$". The integer part of $N$ can be expressed in the factored form

$$I[N] = \alpha_0 + b(\alpha_1 + b(\alpha_2 + b(\alpha_3 + b(\alpha_4 + \cdots))))$$

so that if the integer part of $N$ is divided by b, then $\alpha_0$ is the remainder and the quotient is $Q_1 = \alpha_1 + b(\alpha_2 + b(\alpha_3 + b(\alpha_4 + \cdots))))$. If $Q_1$ is divided by b, then the remainder is $\alpha_1$ and the new quotient is $Q_2 = \alpha_2 + b(\alpha_3 + b(\alpha_4 + \cdots))$. Continuing this process and saving the remainders $\alpha_0, \alpha_1, \ldots, \alpha_n$ the coefficients for the integer part of $N$ can be determined. The fractional part of $N$ is determined by the $\beta$-coefficients and is denoted using the notation $F[N]$ to denote "the fractional part of $N$". The fractional part of $N$ can be expressed in the form

$$F[N] = \beta_1 b^{-1} + \beta_2 b^{-2} + \beta_3 b^{-3} + \cdots$$

Observe that if $F[N]$ is multiplied by b, there results

$$bF[N] = \beta_1 + \beta_2 b^{-1} + \beta_3 b^{-2} + \cdots$$

so that $\beta_1$ is the integer part of $bF[N]$ and the term $\beta_2 b^{-1} + \beta_3 b^{-2} + \cdots$ represents the fractional part of $bF[N]$. Hence if one continues to multiply the resulting fractional parts by b, then one can calculate the coefficients $\beta_1, \beta_2, \beta_3, \ldots$ associated with the fractional part of the base b representation of $N$.

**Example 1-1.** (**Number conversion.**)
Convert the number $N = 123.640625$ to a base 2 representation.
**Solution:** Start with the integer part of $N$ and write $I[N] = 123 = \sum_{i=0}^{n} \alpha_i (2)^i$. Now divide 123 by 2 and save the remainder $R$. Continue to divide the resulting quotients and save the remainders as the remainders give us the coefficients $\alpha_0, \alpha_1, \ldots, \alpha_n$ in the base 2 representation. One can construct the following table to find the coefficients.

| N | I[N/2] =Q | R |
|---|---|---|
| 123 | I[123/2]=61 | $1 = \alpha_0$ |
| 61 | I[61/2]=30 | $1 = \alpha_1$ |
| 30 | I[30/2]=15 | $0 = \alpha_2$ |
| 15 | I[15/2]=7 | $1 = \alpha_3$ |
| 7 | I[7/2]=3 | $1 = \alpha_4$ |
| 3 | I[3/2]=1 | $1 = \alpha_5$ |
| 1 | I[1/2]=0 | $1 = \alpha_6$ |

| N | 2N | F[2N] | I[2N] |
|---|---|---|---|
| 0.640625 | 1.28125 | 0.28125 | $1 = \beta_1$ |
| 0.28125 | 0.5625 | 0.5625 | $0 = \beta_2$ |
| 0.5625 | 1.125 | 0.125 | $1 = \beta_3$ |
| 0.125 | 0.25 | 0.25 | $0 = \beta_4$ |
| 0.25 | 0.50 | 0.50 | $0 = \beta_5$ |
| 0.50 | 1.00 | 0.00 | $1 = \beta_6$ |

The integer part of $N$ can now be represented $I[N] = 123 = 1111011_2$ The fractional part of $N$ is written in the form $F[N] = 0.640625 = \sum_{i=1}^{\infty} \beta_i (2)^{-i}$. Now continue to multiply the fractional part by 2 and save the integer part each time. These integer parts represent the coefficients $\beta_1, \beta_2, \ldots$. One can construct the above table for determining these coefficients.

One finds the fractional part of $N$ can be represented $F[N] = 0.640625 = 0.101001_2$ and consequently the original number $N$ has the base 2 representation $N = 123.640625 = 1111011.101001_2$.

∎

## Sufficient and Necessary Conditions

"If $A$ holds, then $B$ is valid", is a statement where the hypothesized condition $A$ is said to be a sufficient condition for the conclusion $B$. That is, the sufficient condition guarantees that the conclusion $B$ is satisfied or is valid. As an example, one can say, $N$ being a positive and divisible by two is a sufficient condition for $N$ to be an even integer.

Conversely, if the validity of the conclusion requires the condition $A$ be satisfied and is such that the condition is derivable from the conclusion, then the condition $A$ is said to be a necessary condition. For example, the condition that $N$ is an integer greater than two and an odd number, is a necessary condition for $N$ to be a prime number greater than two. When mathematicians make a statement like, "$P$ is a necessary and sufficient condition for $Q$", then one should be able to show that (i) $P$ is necessary for $Q$ and (ii) $Q$ is necessary for $P$, which is the same as saying $P$ is sufficient for $Q$ to be valid. Mathematicians sometimes use the phrase "if and only if" instead of the phrase "necessary and sufficient".

## Indirect Proofs

Indirect proofs, or the Latin reductio ad absurdum, are often used in proving mathematical statements. If a supposition $P$ leads to a wrong or self-contradictory conclusion $Q$, then $P$ must be false. In terms of logic principles, one can say, if $P$ implies $Q$ and $Q$ is false, then $P$ is false.

## Mathematical Induction

In order to prove some mathematical statement or proposition $P(n)$, which depends upon a positive integer $n$, one must first prove that the proposition $P(n)$ is true for $n = 1$. If one can show that from the assumption $P(n)$ is valid for $n = k$, it must follow that $P(n)$ is also valid for $n = k + 1$, then the principle of mathematical induction states that $P(n)$ is valid for all positive integers $n$.

## Arithmetic

Basic arithmetic consists of learning the operations of addition, subtraction, multiplication, division and root finding.

*The Basic Laws of Arithmetic and Algebra*

1. $a = a$ This is known as the reflexive law
2. If $a = b$, then $b = a$, this is known as the symmetric law
3. If $a = b$ and $b = c$, then $a = c$, this is known as the transitive law
4. If $a = \alpha$ and $b = \beta$, then $a + b = \alpha + \beta$ or equals added to equals the results are equal.
5. If $a = \alpha$ and $b = \beta$, $ab = \alpha\beta$ or equals multiplied by equals the results are equal Real numbers also obey the following properties under addition and multiplication.
6. The sum $a + b$ of two real numbers is also a real number. This is known as the closure property under addition.
7. $(a + b) + c = a + (b + c)$ This is know as the associative law under addition.

8. $a + 0 = 0 + a = a$, Here zero is the identity element under addition.
9. $a + (-a) = (-a) + a = 0$, Here $-a$ is called the additive inverse of $a$ or negative of $a$
10. $a + b = b + a$, This is known as the commutative law for addition.

Multiplication has the properties

11. The product $ab$ of two real numbers is also a real number. This is known as the closure property under multiplication.
12. $(ab)c = a(bc)$, This is the associative law for multiplication of real numbers
13. There exists a real number one $(1 \neq 0)$ called the identity element under multiplication with the property that $a \cdot 1 = 1 \cdot a = a$ for each real number $a$
14. Corresponding to each nonzero real number $a$, there exists a real number $\frac{1}{a} = a^{-1}$ called the multiplicative inverse of $a$. An alternative name for the multiplicative inverse of $a$ is 'reciprocal of $a$'.
15. The commutative law for multiplication is written $\alpha\beta = \beta\alpha$ for all real numbers $\alpha$ and $\beta$. This shows that the order of multiplication is not important when dealing with real numbers.
16. The left distributive law requires that $a(b + c) = ab + ac$
17. The right distributive law requires that $(b + c)a = ba + ca$

Arithmetic involves operations using numbers. The numbers being operated on are positive and negative integers, zero, rational numbers $p/q$, and irrational numbers. A number is called rational if and only if it can be expressed as the ratio $p/q$ of two integers $p$ and $q$, where $q \neq 0$. One can say that $p/q$ is the unique number $x$ satisfying the equation $qx = p$, if such a number exists. If $q \neq 0$, then the number $1/q$ is called the reciprocal of $q$.

Real positive numbers can be represented as points on a line as illustrated in the figure 1-5. The integer numbers are marked off from some origin. If a real number $N$ is multiplied by $-1$, then it is rotated $180°$ or $\pi$ radians in a counterclockwise direction about the origin. This produces the negative numbers along the left-hand side of the line. Note that if $(-N)$ is multiplied by $(-1)$ it is rotated $180°$ counterclockwise about the origin back to the point $N$. This shows $(-N)(-1) = N$ and gives the special case $(-1)(-1) = +1$. The origin is called the point zero, separating the positive and negative numbers.

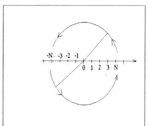

**Figure 1-5.** The number line.

## Rule of Signs

If $a, b$ are real positive numbers, then

$(+a)(+b) = (+ab),$   $(-a)(-b) = (+ab),$   $(+a)(-b) = (-ab),$   $(-a)(+b) = (-ab)$

The number zero separating the positive and negative integers satisfies the properties

$$(+a)(0) = (0)(+a) = 0. \qquad (-a)(0) = (0)(-a) = 0. \qquad (0)(0) = (0)$$

## The Rule of Nine

The rule of nine is very old dating back to early Greek mathematics.

*If the sum of all the digits of a number is a multiple of 3 or 9, then the number itself is also a multiple of 3 or 9.*

$9547839219987 = 3(3182613073329)$

$9547839219987 = 9(1060871024443)$

## Arithmetic Operations $+, -, \times, \div$

There is no one "correct way" to perform each of the operations of addition, subtraction, multiplication and division. Throughout history many different techniques have been developed to accomplish the tasks of basic arithmetic. The following discussions are a small sampling representing some of the techniques developed for performing arithmetic operations.

### Addition $(a + b)$

Consider the addition of the column of numbers on the right. In the "standard" method of addition one aligns the decimal points, places the numbers in rows, draws a line under the bottom number and then starts by adding the numbers in the right-hand column, like six plus four plus three, etc and then moves on to the addition of the next column being sure to include any "carry overs". This procedure is just an algorithm for the addition of the above numbers. Many other techniques for addition can be devised.

```
8176
 314
1243
6547
 765
9854
5114
4138
```

The Trachtenberg system, see references, for adding the above column of numbers uses the rule of 11 and considers only sums adding to 11. If a sum is above 11, one makes a tick mark (') and carries the remainder to the addition of the remaining numbers in the column. One can start with any column and begin by adding the numbers in that column. Let us select the second column from the right as an example. Adding the numbers in the second column, starting at the top gives 7 plus 1 plus 4 equal to 12, which is higher than 11 by 1. Make a tick mark at the 4 and continue to count starting with the carry over of 1 to get 1 plus 4 plus 6 equal 11 with a carry over of zero. Make a tick mark at the 6 and continue counting with 5 plus 1 plus 3 equal to 9. Do this counting method for each column and then record the tick marks and carry overs to obtain

|              |    |    | 8  | 1  | 7  | 6  |
|--------------|----|----|----|----|----|----|
|              |    |    |    | 3  | 1  | 4  |
|              |    |    | 1  | 2  | 4' | 3' |
|              |    |    | 6' | 5' | 4  | 7  |
|              |    |    |    | 7  | 6' | 5' |
|              |    |    | 9' | 8' | 5  | 4  |
|              |    |    | 5  | 1  | 1  | 4' |
|              |    |    | 4' | 1  | 3  | 8  |
| Carry over:  |    |    | 0  | 6  | 9  | 8  |
| Tick marks:  |    |    | 3  | 2  | 2  | 3  |

The numbers in the last two rows are added using the following rules. The last two numbers on the right-hand side are added. Write down the right-hand digit of the sum and use the left-hand digit as a carry over to the next sum. All additions after the first use the following "L-shaped" addition pattern $\begin{smallmatrix}a\\b\ \ c\end{smallmatrix} = a+b+c$. After each sum write down the right-hand digit of the sum and treat the left-hand digit of the sum as a carry over to the next L-shaped sum. Using the above column of numbers as an example one finds the following sums from which to obtain an answer to our addition pattern.

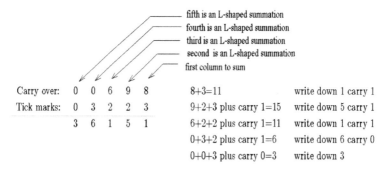

| Carry over: | 0 | 0 | 6 | 9 | 8 | 8+3=11              | write down 1 carry 1 |
|-------------|---|---|---|---|---|---------------------|----------------------|
| Tick marks: | 0 | 3 | 2 | 2 | 3 | 9+2+3 plus carry 1=15 | write down 5 carry 1 |
|             | 3 | 6 | 1 | 5 | 1 | 6+2+2 plus carry 1=11 | write down 1 carry 1 |
|             |   |   |   |   |   | 0+3+2 plus carry 1=6 | write down 6 carry 0 |
|             |   |   |   |   |   | 0+0+3 plus carry 0=3 | write down 3         |

**Subtraction** $(a - b)$

Subtraction is the inverse operation of addition. Usually the larger number is placed in the first row and the smaller number, with the digits aligned properly, are placed in the second row along with a minus sign and then a line is drawn and under the line the answer is written. Various algorithms can be devised to perform the operation of subtraction. Two such algorithms illustrating the calculation of the difference (12345 − 3496) are as follows.

**Change the top row of numbers**

$$\begin{array}{r} 1^0 \ 2^1 \ 3^2 \ 4^3 \ 5 \\ - \ \ \ 0 \ \ 3 \ \ 4 \ \ 9 \ \ 6 \\ \hline 0 \ \ 8 \ \ 8 \ \ 4 \ \ 9 \end{array}$$

The subtraction begins with the right-hand column. Observe that one cannot subtract 6 from 5 and so cross out the 4 in the top left column and change it to a 3 and carry 10 over to the right-hand column making it a 15. Now 6 from 15 leaves 9. One then moves to the left one column and repeats this process. One cannot subtract 9 from 3 and so a 10 is borrowed from the top left column by crossing out the 3 and making it a 2. Now 9 from 13 gives 4.

Continuing this algorithm of borrowing from the top left column one obtains the difference between the numbers 12345 and 3496 as illustrated.

Instead of decreasing the numbers in the top row one can select to increase the numbers in the bottom row. The subtraction begins just like the previous method. One starts with the right-hand column and tries to subtract 6 from 5 and determines that a 1 must be borrowed from the column on the left. This gives 6 from 15 or 9.

**Change the bottom row of numbers**

$$\begin{array}{r} 1 \ \ 2 \ \ 3 \ \ 4 \ \ 5 \\ - \ \ 0_1 \ 3_4 \ 4_5 \ 9_{10} \ 6 \\ \hline 0 \ \ 8 \ \ 8 \ \ 4 \ \ 9 \end{array}$$

If a one is borrowed from the left column then the number in the bottom row is increase by 1. So one crosses out the 9 and puts a 10. Moving one column to the left one finds 10 from 4 is not possible, so one must borrow from the left-column and in doing so changes the 4 in column 3 to a 5 in the bottom row. Now 10 from 14 gives 4. Continuing to the next column of numbers one tries to subtract 5 from 3 and finds that one must borrow from the left column and consequently the bottom row 3 in column 4 is increased to 4 because of the borrowing. This produces 5 from 13 or 8 in the third column. The next column gives 4 from 12 or 8 and one obtains the same answer as in the previous example. Note that when a borrow occurs, then the digit in the bottom row is increased by 1.

**Multiplication** ($a \times b$, $a \cdot b$, $a * b$)

Multiplication as done by the early Greeks was a process of many additions. Again, the steps one has learned to multiply numbers is just some algorithm that educators find convenient and easy to teach. There are many algorithms for multiplying numbers. Here are three examples illustrating how to calculate the product $32 \times 47$.

*Multiplication 1* One can apply the distributive laws of algebra and write

$$32 \times 47 = (30 + 2) \times (40 + 7) = 1200 + 80 + 210 + 14 = 1504$$

*Multiplication 2*

Multiply 32 by 47. On begins by placing the numbers one under the other and then multiply 7 by 32 and write down the results. Then multiply 4 by 32 and write down the result shifted because of the digit positions of units, tens, hundreds, thousands,... The results are then added as illustrated.

|   |   |   | 3 | 2 |
|---|---|---|---|---|
|   |   |   | 4 | 7 |
|   |   | 2 | 2 | 4 |
|   | 1 | 2 | 8 |   |
|   | 1 | 5 | 0 | 4 |

*Multiplication 3*

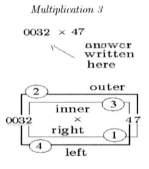

The Trachtenberg speed system for the product of two digit numbers combines four operations so that one can write down the answer immediately. The four operations are numbered in the left-hand figure. The first multiplication involves a product of the right-most digits of the numbers being multiplied. The right most digit of the first product is written down and the left digit is used as a carry over. Next is an inner and outer product being added plus the carry over from the first multiplication. Here inner and outer refer to the position of the digits being multiplied and is illustrated in the figure. Again the right most digit of the result is written down and the left digit is a carry over.

The last operation is a left-product of the digits plus the carry over from the previous operation with the result written down. The left, right, inner, outer product of digits is illustrated in the pattern on the left. An example illustrating these operations follows.

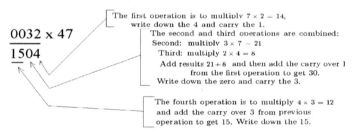

**Division** $(a \div b,\ a/b,\ \frac{a}{b})$

Division is the inverse operation of multiplication. The so called long division method is based upon the division theorem which states that when $a$, the dividend, is divided by $b$, the divisor, there must exist unique numbers $q$, the quotient and $r$ the remainder such that

$$a = bq + r, \quad 0 \le r \le b \quad \text{or} \quad \frac{a}{b} = q + \frac{r}{b}$$

There are many division algorithms from which the following are selected.

**Division 1**

Most texts teach a trial and error method for division requiring the determination of how many times one number goes into another number without going over. For example, to divide 543 by 4 one would ask the question, "How many times does 4 go into 5?", then write this down over the first number and then multiply the $4 \times 1$ and write it under the 5 and then subtract to get 1. Then bring down the 4 and repeat the process. Ask the question, "How many times does 4 go into 14?", write this down over the 4 in the dividend. Then multiply 4 by 3 and put the result under the 14 and subtract, then bring down the 3 and repeat the process. Determine the number of times 4 goes into 23 and write the result above the 3 in the dividend. Multiply 5 by 4 and put the result under the 23 and then subtract leaving a remainder

```
        1   3   5
    4 | 5   4   3
        4
        ‾‾‾
        1   4
        1   2
        ‾‾‾‾‾‾‾
            2   3
            2   0
            ‾‾‾‾‾‾‾
                3
```

$\Longrightarrow \quad \frac{543}{4} = 135 + \frac{3}{4}$

**Division 2** (The double-division method)

Start as in the previous method and write $1_r =$ to the left of the divisor and under that write in consecutive rows $2_r =$, $4_r =$ and $8_r =$ for two times, four times and eight times the divisor of 4 to get the numbers 4,8,16,32. This gives a divisor column that will be used over and over. Note that the divisors have been doubled each time to form a divisor column between the dividend and the multiplication factors 1,2,4 and 8.

The double-division method is as follows. Write under the 5 in the dividend the largest number from the divisor column which goes into 5. One finds 4 and writes this down under the 5 in the dividend and then fill to the right with zero's. Next write the multiplier (1,2,4 or 8) used, followed by the same number of zero's used in the fill of the previous step. Write the scaled multiplication factor to the extreme right-hand side of the 400. Next, subtract the 400 from the 543 to get 143 and now repeat the above process using the number 143 as the starting number. That is, determine the largest divisor from the divisor column which goes into 14 and write this down and fill in the appropriate number of zero's to the right. Write the multiplier (1,2,4 or 8) followed by the same number of zeros used in the fill process on the right-hand side and then subtract 80 from 143 to get 63. Now repeat the above steps again using 63 as the starting number. Determine the largest divisor into 6, which is 4.

# 34

Write down the 4 under the six and fill with zero's. Next write the multiplier (1,2,4 or 8) followed by the same number of zero's to the extreme right-hand side. This produces a 10 to the right-hand side of 40. Now subtract to get 23 and start the process all over again. By repeating the above steps find 16 is the largest divisor into 23 and write the multiplier to the right-hand side and then subtract 16 from 23 to get 7. The largest divisor into 7 is 4 with multiplication factor of 1 to the right-hand side. Subtract 4 from 7 to get 3 which is less than all the divisors and so stop and add the column of numbers on the right-hand side

```
                                      ┌─── Column to be added
                                      ↓
   1_z  =   4  | 5    4   3
   2_z  =   8  |-4    0   0      100
   4_z  =  16  | 1    4   3
   8_z  =  32  |-8    0         20
                 6    3
                -4    0         10                543        3
                ─────                         →   ───  = 135 + ─
                 2    3                            4           4
                -1    6          4
                ─────
                 7
                -4             +1
                ───            ───
                 3             135
```

Divisor Column

**Square Root** ( $\sqrt{N}$ ) There are numerous algorithms for the calculation of the square root of a number. The following two algorithms might be familiar to some.

**Square root algorithm 1**

The following is an example showing the steps used to calculate the square root of a number. For illustrative purposes find $\sqrt{543}$. The general procedure is to group the digits of the number into pairs starting from the right-hand side of the number. This gives $\sqrt{(5)(43)}$ Now find an integer which when squared is less than or equal to the first digit(s). Trial gives $2^2 = 4 < 5$ and write 2 above the first digit(s) to obtain

$$\frac{2}{2\sqrt{(5)(43).(00)(00)}}$$

Now square the 2 and put it under the 5 and subtract and then bring down the next grouping of two digits. Double the number 2 on top of the 5 and place (4x) to the left of the 143.

```
                   2     x
           2  | (5)    (43) .  (00)    (00)
                4
         (4x)   1       43
```

The next objective is to find the digit $x$ such that $(4x)$ times $x$ is close to 143. Note that $(42)(2) = 84$, $(43)(3) = 129$ and $(44)(4) = 176$, so 3 is the digit required. Write 3 above the next grouping along side of the 2. Multiply 3 times 43 and place the result under the 143, then subtract the numbers and bring down the next group of two numbers to obtain

$$\begin{array}{r|llll} & 2 & 3 & y & \\ 2 & \overline{(5)} & (43)\phantom{.} & (00) & (00) \\ & 4 & & & \\ (43) & 1 & 43 & & \\ & 1 & 29 & & \\ (46y) & & 14 & 00 & \end{array}$$

Now double the 23 to get 46 and write $(46y)$ to the left of the 1400. Find a digit $y$ such that $(46y)$ times $y$ does not exceed 1400. Note that $(462)(2) = 934$, $(463)(3) = 1389$ and $(464)(4) = 1856$ and so 3 is the desired digit. This gives 23.3 as an approximation to $\sqrt{543}$.

**Square root algorithm 2**

The introduction of computers and hand held calculators has enabled the Newton's method to be employed to calculate a square root. Newton's method to calculate $\sqrt{N}$ is an iterative method having the form

$$x_{n+1} = x_n - \frac{x_n^2 - N}{2x_n}, \qquad \text{for } n = 0, 1, 2, \ldots \tag{1.3}$$

One guesses at a value $x_0$ for $\sqrt{N}$ and substitutes $x_0$ into the right-hand side of equation (1.3). This gives a value $x_1$, which is a much improved value for the square root of $N$. Substituting the value $x_1$ into the right-hand side of equation (1.3) produces a value $x_2$, a still more improved value for square root of $N$. This iterative process continues until convergence is achieved where the iterative values are all the same.

As an example, take a bad guess for the $\sqrt{543}$, say $x_0 = 16$. Substitute $x_0 = 16$ into the right-hand side of equation (1.3) and show $x_1 = 24.97$. Now substitute $x_1$ into the right-hand side of equation (1.3) and show $x_2 = 23.358$. Continuing this iterative process shows $x_3 = 23.3024$ and $x_4 = 23.3024$ rounded to 4 digits after the decimal point.

## Calculators

Hand held calculators can perform the operations of $+, -, \times, \div$ and $\sqrt{\phantom{x}}$, and usually come with either a Polish notation, sometimes referred to as a prefix notation (PN), or a reverse Polish notation (RPN). The notation results from the symbolic logic invented by the Polish mathematician Jan Lukasiewicz. In the Polish notation, the order of the operators and operands is what determines the results.

The reverse Polish notation does calculations much faster and uses a stack for storing information. The data is placed in a stack with one piece of data on top of another piece of data. Stack size can vary from small to large depending upon the computer. The RPN calculators do calculations faster then PN calculators. However, they are more expensive and require an "enter" key. RPN calculators do not enter the operations of $+, -, \times, \div$ onto the stack but they perform an operation on the stack entries. Many main frame computers use the stack design concept in their construction.

An example illustrating the difference between PN and RPN calculators.

You can type into a PN calculator "7+3+((3+4)*5)-17" and then hit = to get 28. To solve the same problem using a RPN calculator one would enter

$$7\ 3\ 3\ 4\ +\ 5\ *\ +\ +\ 17\ -$$

to get the same answer. This can be better understood by examining what is happening in the stack.

| Input | Operation | Stack Contents | Comments |
|---|---|---|---|
| 7 | enter | 7 | |
| 3 | enter | 7,3 | |
| 3 | enter | 7,3,3 | |
| 4 | | 7,3,3,4 | |
| + | Add | 7,3,7 | bottom two entries added |
| 5 | | 7,3,7,5 | |
| * | Multiply | 7,3,35 | bottom two entries multiplied |
| + | Add | 7,38 | bottom two entries added |
| + | Add | 45 | bottom two entries added |
| 17 | | 45,17 | |
| - | Subtract | 28 | bottom two entries subtracted |

**Number Theory**

Number theory involves developing properties of the integers. The natural numbers in the decimal system are represented $\mathbb{N}= \{0, 1, 2, 3, 4, \ldots\}$ which is based upon representing numbers as linear combinations of powers of 10.

$10^{-3}$ 1 thousandth    $10^4$ 1 ten thousand    $10^{21}$ Sextillion

$10^{-2}$ 1 hundreth    $10^5$ 1 hundred thousand    $10^{24}$ Septillion

$10^{-1}$ 1 tenth    $10^6$ 1 million    $10^{27}$ Octillion

$10^0$ 1 unit    $10^9$ 1 billion    $10^{30}$ Nonillion    $10^{100}$ =google

$10^0$ 1 ten    $10^{12}$ 1 trillion    $10^{33}$ Decillion    $10^{10^{100}}$ =googolplex

$10^2$ 1 hundred    $10^{15}$ 1 quadrillion    $10^{36}$ Undecillion

$10^3$ 1 thousand    $10^{18}$ 1 quintillion    $10^{39}$ Duodecillion

Note the European billion is $10^{12}$ and $10^9$ is called a thousand million.

Many theoretical properties of the integers can be obtained by just playing around with the numbers and then using mathematical induction or some other technique to generalize your results. For example, the reader does not need to know any special mathematics to derive the following results.

$$1 + 2 + 3 + 4 + \cdots + (n-1) + n = \frac{n(n+1)}{2}$$

$$1^2 + 2^2 + 3^2 + 4^2 + \cdots + (n-1)^2 + n^2 = \frac{n(n+1)(2n+1)}{6}$$

$$1^3 + 2^3 + 3^3 + 4^3 + \cdots + (n-1)^3 + n^3 = \left(\frac{n(n+1)}{2}\right)^2$$

$$1^4 + 2^4 + 3^4 + 4^4 + \cdots + (n-1)^4 + n^4 = \frac{n(n+1)(2n+1)(3n^2+3n-1)}{30}$$

$$1 + 3 + 5 + 7 + \cdots + (2n-3) + (2n-1) = n^2$$

### Prime Numbers

The set of prime numbers $\mathbb{P}$ consists of numbers $N$ which have two and only two divisors. The two divisors are the number 1 and itself or $\{1, N\}$. Nonprime numbers have more than two divisors and can be represented as a product of prime factors. The number 1 is not considered a prime number. The primes less than 100 are as follows.

$$2, 3, 5, 7, 11, 13, 17, 19, 23, 29, 31, 37, 41, 43, 47, 53, 59, 61, 67, 71, 73, 79, 83, 89, 97, \ldots$$

### Fundamental Theorem of Arithmetic

Every number $N$ can be expressed as a product of positive prime numbers $P_1, P_2, \ldots, P_k$ having the form

$$N = P_1^{n_1} P_2^{n_2} \cdots P_k^{n_k}$$

where $n_1, n_2, \ldots, n_k$ are integers representing the multiplicities of the prime factors of $N$.

**Rational Numbers** A rational number is the ratio of two integers $\frac{m}{n}$, $n \neq 0$. Noninteger rational numbers like the ratio $\frac{\pi}{e}$ are called fractions. If $\mathbb{Z}$ denotes the set of integers, then the set of rational numbers can be denoted $\mathbb{Q} = \{\frac{m}{n}\}$ where $m, n \in \mathbb{Z}$ and $n \neq 0$. Here the $\mathbb{Q}$ represents quotient.

### Irrational Numbers

The following are some preliminary results needed to show the existence of irrational numbers.

An integer is called even if it can be expressed in the form $2n$ where $n$ is some integer. An integer is called odd if it can be expressed in the form $2n+1$ where $n$ is some integer.

**Properties of even and odd integers**

> If $p$ is an even number, then $p^2$ is also an even number.
> If $q$ is an odd number, then $q^2$ is also an odd number.
> If $n$ is an integer and $n^2$ is even, then $n$ is even.

Euclid used an indirect proof to show that the number representing the square root of 2 cannot be represented as a rational number. His proof starts by letting $p$ and $q$ be integers having no common factors except ±1. The assumption $p/q$ represents $\sqrt{2}$ gives the equation $p/q = \sqrt{2}$. By squaring both sides one gets $p^2/q^2 = 2$ or $p^2 = 2q^2$. Here $q$ and $q^2$ are integers and $p^2$ is an even integer. This implies $p$ is also an even integer. The previous assumption that $p$ and $q$ have no common factors implies $q$ must be odd, for if $q$ were even, then $p$ and $q$ would have a common factor of 2. Because $p$ is even, one can write $p = 2n$ where $n$ is an integer. This implies $p^2 = (2n)^2 = 2q^2$, which implies $q^2 = 2n^2$ which says $q^2$ is even and so $q$ is even, which is a contradiction.

The above proof, by Euclid, implies the following. If $p$ is not the perfect $n$th power of some integer or rational number, then the equation $x^n = p$ is such that $x$ cannot be a rational number. Such numbers were called irrational numbers. Some examples of irrational numbers are $\sqrt{2}, \sqrt{3}, \sqrt{5}, \pi, e, \sqrt[3]{5}$.

**Congruences**

One can develop the theory of congruences to study more advanced properties of numbers.

**Definition of Congruence** *Two integers $a$ and $b$ are said to be congruent modulo $m$ if there exists an integer $k$ such that $a = b + km$. The integers $a, b$ and $k$ can be positive, negative or zero. If $a$ and $b$ are congruent modulo $m$, then this is written*

$$a \equiv b \bmod m \quad \Longrightarrow \quad a = b + km \quad \text{for some integer } k$$

Congruences have the following properties.

1. If $a \equiv c \bmod m$ and $b \equiv c \bmod m$, then $a \equiv b \bmod m$
2. If $a \equiv b \bmod m$ and $\alpha \equiv \beta \bmod m$, then $a \pm \alpha \equiv b \pm \beta \bmod m$
3. If $a \equiv b \bmod m$ and $j$ is an integer, then $ja \equiv jb \bmod m$
4. If $a \equiv b \bmod m$ and $\alpha \equiv \beta \bmod m$, then $a\alpha \equiv b\beta \bmod m$
5. If $a \equiv b \bmod m$ and $n$ is an integer, then $a^n \equiv b^n \bmod m$
6. If $a \equiv b \bmod m$ and $f(x)$ is any polynomial in $x$ having integer coefficients which can be negative, zero or positive, then $f(a) \equiv f(b) \bmod m$
7. If $a \equiv b \bmod m$ and $d$ is a divisor or $m$, then $a \equiv b \bmod d$
8. If $a \equiv b \bmod m_1$ and $a \equiv b \bmod m_2$, and $m$ is a least common multiple of $m_1$ and $m_2$, then $a \equiv b \bmod m$
9. If $a \equiv b \bmod m$, and if $P(x)$ is a polynomial in $x$ with integral coefficients, then one can show $P(a) \equiv P(b) \bmod m$
10. If $ax \equiv b \bmod m$, then $a$ must not have $m$ as a factor. Any integer $x_1$ which satisfies $ax_1 \equiv b \bmod m$ is said to be a solution of $ax \equiv b \bmod m$. Hence any number $x_2 \equiv x_1 \bmod m$ is also a solution. One way to obtain solutions is by trial where one assigns values to $x$ from the set $\{1, 2, 3, \ldots, m-1\}$.

11. The congruence $ax \equiv b \bmod m$ has a solution for $x$ if and only if the greatest common divisor of $a$ and $m$ divides $b$. If $d = (a, m)$, then there is a unique solution mod $m_1 = \frac{m}{d}$ and hence $d$ solutions mod $m$.
12. The integer $a'$ is called the reciprocal of $a$ mod $m$ if $aa' \equiv 1 \bmod m$. For example, if $3x \equiv 1 \bmod 5$, then $x = 2$ is a solution or 2 is the reciprocal of 3 mod 5.

## The Greek Alphabet

There are many occasions in mathematics where it is convenient to use letters from the Greek alphabet to denote variables, subscripts or superscripts.

| A | $\alpha$ | Alpha | H | $\eta$ | Eta | N | $\nu$ | Nu | T | $\tau$ | Tau |
|---|---|---|---|---|---|---|---|---|---|---|---|
| B | $\beta$ | Beta | $\Theta$ | $\theta$ | Theta | $\Xi$ | $\xi$ | Xi | $\Upsilon$ | $\upsilon$ | Upsilon |
| $\Gamma$ | $\gamma$ | Gamma | I | $\iota$ | Iota | O | $o$ | Omicron | $\Phi$ | $\phi$ | Phi |
| $\Delta$ | $\delta$ | Delta | K | $\kappa$ | Kappa | $\Pi$ | $\pi$ | Pi | X | $\chi$ | Chi |
| E | $\epsilon$ | Epsilon | $\Lambda$ | $\lambda$ | Lambda | P | $\rho$ | Rho | $\Psi$ | $\psi$ | Psi |
| Z | $\zeta$ | Zeta | M | $\mu$ | Mu | $\Sigma$ | $\sigma$ | Sigma | $\Omega$ | $\omega$ | Omega |

**Figure 1-6.** The Greek alphabet

## Algebra

The following presentations follow the standard procedure that letters from the beginning of an alphabet, such as a,b,c,..., or $\alpha, \beta, \gamma,...$ are used to denote real constants and letters at the end of an alphabet, such as, u,v,w,x,y,z, are used to denote real variables or unknown quantities. Whenever there are many constants or variables required in a formula, then subscripts are assigned to the symbols. For example, $x_1, x_2, x_3, \ldots, x_n$ might denote $n$ unknowns and $a_0, a_1, a_2, \ldots, a_n$ might denote $n + 1$ constant values, where $n$ is some positive integer.

### The Basic Laws of Algebra

1. $a = a$ This is known as the reflexive law
2. If $a = b$, then $b = a$, this is known as the symmetric law
3. If $a = b$ and $b = c$, then $a = c$, this is known as the transitive law
4. If $a = \alpha$ and $b = \beta$, then $a + b = \alpha + \beta$ or equals added to equals the results are equal.
5. If $a = \alpha$ and $b = \beta$, $ab = \alpha\beta$ or equals multiplied by equals the results are equal.

Real numbers also obey the following properties under addition and multiplication.

6. The sum $a + b$ of two real numbers is also a real number. This is known as the closure property under addition.
7. $(a + b) + c = a + (b + c)$ This is know as the associative law under addition.

8. $a + 0 = 0 + a = a$, Here zero is the identity element under addition.
9. $a + (-a) = (-a) + a = 0$, Here $-a$ is called the additive inverse of $a$ or negative of $a$
10. $a + b = b + a$, This is known as the commutative law for addition.

Similarly, for multiplication

11. The product $ab$ of two real numbers is also a real number. This is known as the closure property under multiplication.
12. $(ab)c = a(bc)$, This is the associative law for multiplication of real numbers
13. There exists a real number one $(1 \neq 0)$ called the identity element under multiplication with the property that $a \cdot 1 = 1 \cdot a = a$ for each real number $a$
14. Corresponding to each nonzero real number $a$, there exists a real number $\frac{1}{a} = a^{-1}$ called the multiplicative inverse of $a$. An alternative name for the multiplicative inverse of $a$ is 'reciprocal of $a$'.
15. The commutative law for multiplication is written $\alpha\beta = \beta\alpha$ for all real numbers $\alpha$ and $\beta$. This shows that the order of multiplication is not important when dealing with real numbers.
16. The left distributive law requires that $a(b + c) = ab + ac$
17. The right distributive law requires that $(b + c)a = ba + ca$

Any mathematical system satisfying the above properties 1 through 15 is called a field.

### Fractions

In algebra, a fraction is a symbol $\frac{a}{b}$, sometimes expressed $a/b$, where $a$ and $b$ represent real numbers with $a$ called the numerator and $b$ called the denominator of the fraction. Fractions obey the following laws.

1. If $a, b, c$ are real nonzero constants, then $\frac{a}{b} = \frac{ac}{bc}$
2. Two fractions $a/b$ and $c/d$ are equal if and only if $ad = bc$
3. Two fractions $a/b$ and $c/d$ can be multiplied to give the product $\frac{a}{b} \cdot \frac{c}{d} = \frac{ac}{bd}$
4. Two fractions $a/b$ and $c/d$ can be divided to produce the quotient $\frac{a}{b} \div \frac{c}{d} = \frac{\frac{a}{b}}{\frac{c}{d}} = \frac{ad}{bc}$
5. Two fractions $a/b$ and $c/d$ can be added after first getting a common denominator. Thus,

$$\frac{a}{b} + \frac{c}{d} = \frac{ad}{bd} + \frac{bc}{bd} = \frac{ad + bc}{bd}$$

6. Two fractions $a/b$ and $c/d$ can be subtracted after first getting a common denominator. This is written

$$\frac{a}{b} - \frac{c}{d} = \frac{ad}{bd} - \frac{bc}{bd} = \frac{ad - bc}{bd}$$

### Inequalities

Using the symbolic notations $<$, $=$ and $>$, where $<$ is for less than, $=$ is for equal and $>$ is for greater than, the properties of order can be expressed.

**Order properties of real numbers**

1. If $a < b$, then $a + c < b + c$ for all constants $c$
2. If $a < b$, and $c > 0$, then $ac < bc$ for all constants $c$
3. The transitive law states that if $a < b$ and $b < c$, then $a < c$
4. The law of trichotomy state that if $a, b$ are real numbers, then one of the following must hold true. $a = b$, $\quad a < b$, $\quad b < a$
5. If $a < b$ and $a = A$ and $b = B$, then $A < B$

The following relationships are implied relations. Use the symbol $\Rightarrow$ to mean 'implies'.

| | | |
|---|---|---|
| $a < b$ | $\Rightarrow$ | $0 < b - a$ |
| $0 < a$ and $0 < b$ | $\Rightarrow$ | $0 < a + b$ |
| $0 < a$ and $b < 0$ | $\Rightarrow$ | $0 < ab$ |
| $0 < a$ and $b > 0$ | $\Rightarrow$ | $ab < 0$ |
| $a < b$ and $c > 0$ | $\Rightarrow$ | $ac > bc$ |
| $a > 0$ and $b > 0$ with $a < b$ | $\Rightarrow$ | $\sqrt{a} < \sqrt{b}$ |

**Absolute Value**

The absolute value of a real number $x$ is written $|x|$ and is defined as follows.

If $x = 0$, then $|x| = 0$.

If $x \neq 0$, then $|x| = \begin{cases} x, & \text{if } x > 0 \\ -x, & \text{if } x < 0 \end{cases}$

If $|x| < a$, then this implies that $-a < x < a$

If $|x - a| < b$ then this implies that $-b < x - a < b$ which implies $a - b < x < a + b$

**Exponents**

For $n$ a positive integer the product of $x$ with itself $n$ times is written $\underbrace{x \cdot x \cdot x \cdots x}_{n \text{ products}} = x^n$

If $n$ is a positive integer and $x \neq 0$, one can write $x^{-n} = \frac{1}{x^n}$ and for the special case $x \neq 0$ define $x^0 = 1$. The laws for exponents are as follows.

| Law of Exponents | Example |
|---|---|
| $x^m x^n = x^{m+n}$ | $x^6 x^{-1} = x^5$ |
| $(xy)^m = x^m y^m$ | $(xy)^3 = x^3 y^3$ |
| $(x^m)^n = x^{mn}$ | $(x^3)^2 = x^6$ |
| $\left(\dfrac{x}{y}\right)^m = \dfrac{x^m}{y^m}, \quad y \neq 0$ | $\left(\dfrac{x}{y}\right)^3 = \dfrac{x^3}{y^3}, \quad y \neq 0$ |
| $\dfrac{x^m}{x^n} = x^{m-n}, \quad x \neq 0$ | $\dfrac{x}{x^3} = x^{-2} = \dfrac{1}{x^2}, \quad x \neq 0$ |

Also by definition, if $x^r = y$, where $r$ is a positive integer greater than 1, then $x$ is called the $r$th root of $y$ and is written in one of the following forms.

$$x = y^{1/r} \quad \text{or} \quad x = \sqrt[r]{y}$$

For $m$ and $n$ positive integers

$$x^{m/n} = (x^{1/n})^m = (\sqrt[n]{x})^m \quad \text{or} \quad x^{m/n} = (x^m)^{1/n} = \sqrt[n]{x^m}$$

Note that since the square root function is multiple valued, one defines the principal square root of the number $a^2$ as $\sqrt{a^2} = |a|$.

**Logarithms**

A logarithm is an exponent. Thus, if $x = b^y$, then $y$ is called the logarithm of $x$ to the base $b$ and one can employ the equivalent statements

$$x = b^y \quad \Leftrightarrow \quad y = \log_b x$$

Historically, logarithms have been used to multiply, divide and raise numbers to a power as well as extracting roots of a number. The common logarithm with base 10 is written $\log_{10}$. A logarithm is composed of two parts called a characteristic and a mantissa. The mantissa is that part of the logarithm to the right of the decimal point while the characteristic is that part of the logarithm to the left of the decimal point. The advantage of using common logarithms is that knowing the digits of a number one can find the mantissa without regard to where the decimal point occurs in the number. Note that in tables of logarithms only the mantissa of a number is given. For example, the numbers 1,230,000,000.0, 1,230.0, 12.30, 1.23 all have the same mantissa. An examination of a table of logarithms one can find the mantissa of the above numbers is given by 0.089905. The characteristic of the logarithm is determined by the position of the decimal point in the number. For numbers greater than one the characteristic is one less than the number of digits to the left of the decimal point. Thus, one would write

$$\log_{10} 1,230,000,000.0 = 9.089905, \quad \log_{10} 1,230.0 = 3.089905, \quad \log_{10} 12.30 = 1.089905$$

For numbers less than one the characteristic must be negative. In using a logarithmic table, where only mantissas are given, is customary to write the logarithm of a number as follows. The characteristic to the left of the decimal point is obtained by subtracting the number of zeros immediately to the right of the decimal point from the number 9 (or 19 or 29, etc.). It is then customary to subtract 10 (or 20 or 30, etc.) from the right of the number. Some examples are,

$$\log_{10} 0.00123 = 7.089905 - 10, \quad \log_{10} 0.00000123 = 4.089905 - 10, \quad \log_{10} 0.00000000000123 = 8.089905 - 20$$

The above examples apply when using tables. Hand held calculators and computers provide the logarithms directly without having to remember the rules for using tables of logarithms. For example, using a hand held calculator one finds $\log_{10} 0.0000123 = -5.910095$, with the result being the same as the logarithm $4.089905 - 10$ given above.

The following are the basic laws of logarithms

| Law of Logarithms | Example |
|---|---|
| $\log_b MN = \log_b M + \log_b N$ | $\log_{10} 3 \cdot 2 = \log_{10} 3 + \log_{10} 2$ |
| $\log_b \frac{M}{N} = \log_b M - \log_b N$ | $\log_{10} \frac{9}{5} = \log_{10} 9 - \log_{10} 5$ |
| $\log_b N^p = p \log_b N$ | $\log_{10} 5^2 = 2 \log_{10} 5$ |
| $\log_b \sqrt[m]{N} = \frac{1}{m} \log_b N$ | $\log_{10} \sqrt{5} = \frac{1}{2} \log_{10} 5$ |

For $b$ a positive constant, the function $y = b^x$ is called the exponential function and the function $y = \log_b x$ is called the logarithm function. The natural logarithm uses the base $e = 2.7182818\ldots$ and is usually written in one of the forms

$$y = \log_e x = \log x \quad \text{or} \quad y = \log_e x = \ln x$$

That is, whenever the base of the logarithm is not specified it is understood to represent the base $e = 2.7182818\ldots$. The system of common logarithms uses the base 10 and is written $y = \log_{10} x$, while the natural logarithm uses the base $e$ and is written $y = \log x$ or $y = \ln x$. The exponential function $y = e^{\alpha x}$, $\alpha$ constant, and the logarithmic function $y = \log x = \ln x$ are illustrated in the figures 1-7 (a) and (b).

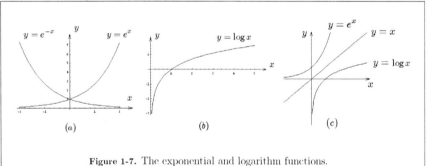

Figure 1-7. The exponential and logarithm functions.

The functions $e^x$ and $\log x$ or $\ln x$ are inverse functions of one another so that

$$e^{\log x} = x \quad \text{and} \quad \log e^x = x \quad \text{or} \quad e^{\ln x} = x \quad \text{and} \quad \ln e^x = x$$

Note that inverse functions are symmetric about the line $y = x$.

If $\log_a N = x$, then $N = a^x$ so that by taking the logarithm to the base $b$ of this equation gives

$$\log_b N = \log_b a^x = x \log_b a \qquad (1.4)$$

But, $x = \log_a N$ and so equation (1.4) gives us the relationship between logarithms of a number $N$ with respect to different bases $a$ and $b$. Substituting for $x$ in equation (1.4) gives

$$\log_b N = (\log_a N)(\log_b a) \qquad (1.5)$$

## Elementary Business Mathematics

When money is invested in a savings certificate, one obtains a certain amount of money over a specified period of time. The amount received is called interest. The sum of money invested is called the principal. Whenever the interest is added to the principal at the end of equal intervals of time, called the interest period, then the interest is said to be compounded or converted to principal. Interest periods can be weekly, bi-weekly, monthly, quarterly or yearly depending upon the type of investment. The interest rate $r$ is usually quoted on a yearly basis, so that if interest is compounded more often, then the interest rate $r$ must be divided by the number of interest periods per year.

Let $P$ denote the principal and $r$ the rate of interest per each compound period, then the amount of money received at the end of the first compound period is $P + rP = P(1+r)$. The amount received at the end of the second compound period is $P(1+r) + rP(1+r) = P(1+r)^2$. At the end of the third period the amount received is $P(1+r)^3$ and in general one can show that the amount received after $n$ compound periods is

$$A = P(1+r)^n \qquad (1.6)$$

The borrowing of $D$ dollars at $I\%$ interest/year with the promise to repay the debt by making periodic monthly payments of $R$ dollars over a period of $Y$ years, is called an amortization. The monthly interest rate is $i = \frac{I\%}{100*12}$ and there are $N = 12*Y$ periodic monthly payments. Amortization is the periodic payment of an amount which reduces a debt, where debt equals principal plus interest. The periodic payment is broken into two parts. One part reduces the principal and the other part pays the interest. If $D$ is the debt to be repaid subject to compound interest charges at a rate of $i$ per conversion period, then what must the periodic payment $R$ be in order to pay off the debt after $N$ payments? One method of solving this problem is to use difference equations. Let $R$ denote the amount of the periodic payment to be made at the end of each conversion period and let $P_n$ denote the principal remaining after the $n$th payment and before the $(n+1)$st payment is made. At this time the debt $D$ increases by an amount equal to the interest due on the principal $P_n$. After the $(n+1)$st payment of $R$, the new debt is $P_{n+1}$. Here

$$P_0 = D, \qquad P_{n+1} = P_n + iP_n - R, \qquad \text{for} \quad n = 0, 1, 2, \ldots, N$$

The difference equation $P_{n+1} = (1+i)P_n - R$ subject to the initial condition $P_0 = D$ has the solution

$$P_n = (1+i)^n D - R\left[\frac{(1+i)^n - 1}{i}\right] \qquad (1.7)$$

To determine the periodic payment $R$, to amortize a debt $D$ in exactly $n = N$ payments, set $P_N = 0$ in equation (1.7) and then solve for $R$. Some algebra produces the periodic payments given by

$$R = \frac{Di}{1 - (1+i)^{-N}} \qquad (1.8)$$

## Solution of Linear and Quadratic Equations

### linear equations

If $ax + b = c$, with $a, b, c$ constants, then solving for $x$ produces the solution $x = \frac{c-b}{a}$, provided $a \neq 0$.

### quadratic equations

To solve the quadratic equation $ax^2 + bx + c = 0$, for $a \neq 0$, one can divide by $a$ and write the given equation in the form $x^2 + \frac{b}{a}x + \frac{c}{a} = 0$. At this point it is customary to complete the square by taking one-half of the $x$-coefficient, squaring it, and then adding it to both sides of the equation. This produces the result

$$x^2 + \frac{b}{a}x + \left(\frac{b}{2a}\right)^2 = -\frac{c}{a} + \left(\frac{b}{2a}\right)^2 \qquad (1.9)$$

The left-hand side of equation (1.9) is now a perfect square so that one can write this equation in the form

$$\left(x + \frac{b}{2a}\right)^2 = \frac{b^2 - 4ac}{4a^2} \qquad (1.10)$$

Take the square root of both sides of equation (1.10) and rearrange terms to obtain the solutions

$$x = -\frac{b}{2a} \pm \frac{\sqrt{b^2 - 4ac}}{2a} \qquad (1.11)$$

If $x_1, x_2$ are the roots of the quadratic equation $x^2 + \alpha x + \beta = 0$, then the equation can be represented in the alternative form $(x - x_1)(x - x_2) = 0$, where $x_1 + x_2 = -\alpha$ and $x_1 x_2 = \beta$. The roots $x_1$ and $x_2$ are given by

$$x_1 = -\frac{\alpha}{2} + \frac{1}{2}\sqrt{\alpha^2 - 4\beta} \quad \text{and} \quad x_2 = -\frac{\alpha}{2} - \frac{1}{2}\sqrt{\alpha^2 - 4\beta}$$

The quantity $\Delta = \alpha^2 - 4\beta$ is called the discriminant of the quadratic equation and for $\Delta > 0$ there are two distinct roots, for $\Delta = 0$ there is one real double root and for $\Delta < 0$ there are two conjugate complex roots.

## Cubic Equation

A cubic equation is usually placed in the basic form $x^3 + ax^2 + bx + c = 0$, where $a, b, c$ are real quantities and then one can make the substitution $x = y - a/3$ to reduce the cubic equation to the form

$$y^3 + \alpha y + \beta = 0, \quad \text{where} \quad \alpha = \frac{1}{3}(3b - a^2) \text{ and } \beta = \frac{2}{27}a^3 - \frac{1}{3}ab + c$$

If $x_1, x_2, x_3$ are the roots of the cubic equation, then the cubic equation can be represented in the form

$$(x - x_1)(x - x_2)(x - x_3) = 0$$

where

$$x_1 + x_2 + x_3 = -a, \quad x_1x_2 + x_2x_3 + x_1x_3 = b, \quad x_1x_2x_3 = -c$$

The quantity $D = \left(\frac{\alpha}{3}\right)^3 + \left(\frac{\beta}{2}\right)^2$ is called the discriminant of the cubic equation and is used together with the quantities

$$u = \sqrt[3]{-\frac{\beta}{2} + \sqrt{D}}, \quad v = \sqrt[3]{-\frac{\beta}{2} - \sqrt{D}}$$

to give the roots

$$x_1 = u + v, \quad x_2 = -\frac{u+v}{2} + \frac{u-v}{2}\sqrt{-3}, \quad x_3 = -\frac{u+v}{2} - \frac{u-v}{2}\sqrt{-3}$$

If the discriminant is positive there is one real and two conjugate imaginary roots. If the discriminant equals zero there will be three real roots, two of which are the same. If the discriminant is negative, then there will be three real distinct roots.

## Polynomial Equation

Any equation of the form

$$P_n(x) = a_0x^n + a_1x^{n-1} + a_2x^{n-2} + \cdots + a_{n-2}x^2 + a_{n-1}x + a_n = 0 \tag{1.12}$$

where $a_0$ is a nonzero constant, is known as a polynomial equation of degree $n$.

**Theorem** *A polynomial equation of degree $n$, $P_n(x) = 0$, has $n$-roots $\{x_1, x_2, \ldots, x_n\}$ and consequently if these roots are known, then the polynomial equation can be written in the factored form*

$$P_n(x) = a_0(x - x_1)(x - x_2) \cdots (x - x_n) = 0$$

The coefficients $a_0, a_1, \ldots, a_n$ of the polynomial equation are related to the roots $x_1, x_2, \ldots, x_n$ of the polynomial equation by the relations

$$-\frac{a_1}{a_0} = \sum_{i=1}^{n} x_i, \quad \sum_{\text{all distinct products}} x_i x_j = \frac{a_2}{a_0}, \quad \sum_{\text{all distinct products}} x_i x_j x_k = -\frac{a_3}{a_0}, \ldots,$$

$$\ldots, \sum_{\text{all distinct products}} \underbrace{x_i x_j \cdots x_k}_{m \text{ products}} = (-1)^m \frac{a_m}{a_0}, \ldots, x_1 x_2 \cdots x_n = (-1)^n \frac{a_n}{a_0} \quad (1.13)$$

For example, the quartic equation $x^4 + ax^3 + bx^2 + cx + d = 0$ with roots $x_1, x_2, x_3, x_4$ can be expressed in the factored form

$$(x - x_1)(x - x_2)(x - x_3)(x - x_4) = 0$$

where

$$x_1 + x_2 + x_3 + x_4 = -a$$

$$x_1 x_2 + x_2 x_3 + x_3 x_4 + x_1 x_4 + x_1 x_3 + x_2 x_4 = b$$

$$x_1 x_2 x_3 + x_2 x_3 x_4 + x_1 x_2 x_4 + x_1 x_3 x_4 = -c$$

$$x_1 x_2 x_3 x_4 = d$$

## Special Products and Factors

The following are some special relations with factorization that occur quite frequently in the simplification of equations. These results can be generalized to higher order polynomials.

$$(x+a)(x+b) = x^2 + (a+b)x + ab$$

$$(x+a)(x+b)(x+c) = x^3 + (a+b+c)x^2 + (ab+ac+bc)x + abc$$

$$(x+a)(x+b)(x+c)(x+d) = x^4 + (a+b+c+d)x^3 + (ab+ac+bc+ad+bd+cd)x^2$$
$$+ (abc+abd+acd+bcd)x + abcd$$

Some additional factored forms are

$$(x+a)(x-a) = x^2 - a^2, \qquad (x+a)^2 = (x+a)(x+a) = x^2 + 2ax + a^2$$

$$(x-a)^2 = (x-a)(x-a) = x^2 - 2ax + a^2, \qquad x^2 + \frac{b}{a}x + \frac{c}{a} = \left(x + \frac{b}{2a}\right)^2 + \frac{4ac-b^2}{4a^2}$$

$$(\alpha x + a)(\beta x + b) = \alpha\beta x^2 + (a\beta + b\alpha)x + ab, \qquad (x-a)^3 = x^3 - 3ax^2 + 3a^2 x - a^3$$

$$(x+a)^3 = x^3 + 3ax^2 + 3a^2 x + a^3, \qquad x^3 - a^3 = (x-a)(x^2 + ax + a^2)$$

$$x^3 + a^3 = (x+a)(x^2 - ax + a^2), \qquad x^4 - y^4 = (x-y)(x+y)(x^2+y^2)$$

$$x^5 - y^5 = (x-y)(x^4 + x^3 y + x^2 y^2 + x y^3 + y^4)$$

$$x^5 + y^5 = (x+y)(x^4 - x^3 y + x^2 y^2 - x y^3 + y^4)$$

## Group

A group is defined as a nonempty set $G$ containing elements $a, b, c, \ldots$, an equals relation and some operation $\circ$ such that the following four postulates are satisfied.

(i) If $a, b \in G$, then $a \circ b \in G$ (Closure property)

(ii) The associative law holds under the operation so that $(a \circ b) \circ c = a \circ (b \circ c)$

(iii) There exists an identity element $e$ such that $a \circ e = e \circ a = a$ for all $a \in G$

(iv) Every element $a \in G$ has an inverse element $a^{-1}$ such that $a \circ a^{-1} = e = a^{-1} \circ a$

If also the operation $\circ$ satisfies $a \circ b = b \circ a$, then the group is called commutative or Abelian. If the group contains a finite number of elements, then it is called a finite group.

## Algebraic Field

A field $F = \{a, b, c, \ldots\}$ contains elements $a, b, c, \ldots$, an equals relation and two well-defined operations $\oplus$ and $\odot$ such that the following eleven postulates are satisfied.

| | |
|---|---|
| The elements $a, b, c, \ldots$ constitute an Abelian group under the operation of $\oplus$ with the identity element being denoted by $z$ and the inverse of $a$ being denoted by $a^*$. | The elements $a, b, c, \ldots$, with $z$ omitted, constitute an Abelian group under the operation $\odot$ with the identity element denoted by $e$ and the inverse of an element $a$ being denoted by $a^{-1}$. |
| Closure: $\qquad a \oplus b \in F$ | Closure: $\qquad a \odot b \in F$ |
| Associativity: $(a \oplus b) \oplus c = a \oplus (b \oplus c)$ | Associativity: $(a \odot b) \odot c = a \odot (b \odot c)$ |
| $\oplus$ identity: There exists a $z \in F$ such that $b \oplus z = z \oplus b = b$ | $\odot$ identity: There exists a element $e \in F$ such that $b \odot e = e \odot b = b$ |
| $\oplus$ inverse: For each element $a \in F$ there exists an element $a^*$ such that $a \oplus a^* = a^* + a = z$ | $\odot$ inverse: For each $b \in F$, where $b \neq z$, there exists and element $b^{-1}$ such that $b \odot b^{-1} = b^{-1} \odot b = e$ |
| Commutativity: $\quad a \oplus b = b \oplus a$ | Commutativity: $\quad a \odot b = b \odot a$ |

The distributive law connects the two operations of $\oplus$ and $\odot$

$$a \odot (b \oplus c) = a \odot b \oplus a \odot c \quad \text{and} \quad (a \oplus b) \odot c = a \odot c \oplus b \odot c$$

for every $a, b, c, \ldots \in F$.

A field is usually denoted using the notation $(F, \oplus, \odot)$. Two examples of fields are $(\mathbb{R}, +, \cdot)$ and $(\mathbb{C}, +, \cdot)$.

# Chapter 2
# Selected Fundamental Concepts

## Permutations and Combinations

The order of arrangement of a set of things is called a permutation of the set. For example, the numbers $\{1,2,3\}$, taken as a set can be arranged in the following six permutations.

$$123 \quad 132 \quad 312 \quad 321 \quad 231 \quad 213$$

In the ordering 123, if one interchanges two consecutive terms, this is called a transposition. The six permutations of the numbers 123 can be divided into those with an even number of transpositions 123, 312, 231 and those with an odd number of transpositions 132, 321, 213.

In dealing with permutations and combinations the following fundamental principle is often used.

> *If 'something' can be done in $m_1$-different ways, then after it has been done in any one of these ways, there is a second 'something' that can be done in $n_2$-different ways, then the product $m_1 n_2$ represents the ways that the two 'somethings' can be done in the order stated.*

## Example 2-1.

Given 5 numbers $x_1, x_2, x_3, x_4, x_5$, and suppose someone asks the question, "How many distinct products of two numbers can be formed from this collection?"

**Solution:** There are 5 choices for the first number $x_i$ which leaves 4 choices for the second number $x_j$. This gives $4 \cdot 5 = 20$ ways to form the product $x_i x_j$. Observe that the product $x_i x_j$ is the same as the product $x_j x_i$ and so for distinct products, do not count these products twice. Hence, there are $\frac{5 \cdot 4}{2} = 10$ distinct products of two numbers. These 10 distinct products can be represented by the set of numbers

$$S = \{x_1 x_2, \; x_1 x_3, \; x_1 x_4, \; x_1 x_5, \; x_2 x_3, \; x_2 x_4, \; x_2 x_5, \; x_3 x_4, \; x_3 x_5, \; x_4 x_5\} \tag{2.1}$$

The notation $\sum_{\substack{\text{all distinct} \\ \text{products}}} x_i x_j$ is used to denote the summation of the terms contained in the set S.

∎

Given a set of different objects or different things, then these things can be arranged in some order. For example, different ways of arranging books on a shelf. Each arrangement of a set of things is called a permutation of the set. The number of permutations of n things take r at a time is denoted using the notation $_nP_r$.

(1) There are $n$ choices for the first thing in the arrangement.

(2) There are $n-1$ choices for the second thing in the arrangement.
(3) There are $n-2$ choices for the third thing in the arrangement.
$\vdots$
(r) There are $n-(r-1)$ choices for the $r$th thing in the arrangement.

Using the above fundamental principle, the number of permutations of $n$ things taken $r$ at a time is given by the product

$$_nP_r = n(n-1)(n-2)\cdots(n-(r-1)) \tag{2.2}$$

Note the special case when $r = n$ gives

$$_nP_n = n(n-1)(n-2)\cdots 3\cdot 2\cdot 1 = n! \tag{2.3}$$

where $n!$ is called $n$ factorial or factorial $n$ and represents the product of the first $n$ integers. Note that $0! = 1$ by definition.

Suppose in the set of $n$ objects or things there are repeats. For example, given the set of 6 objects $\{A, A, A, B, B, C\}$. What is the number of permutations of these six letters. Consider the permutation $ABACAB$ and note that if the $A$'s were different, say $A_1, A_2, A_3$, then there would be $3! = 3\cdot 2\cdot 1$ permutations involving the $A$'s similar to the above, namely

$$A_1BA_2CA_3B \qquad A_2BA_1CA_3B \qquad A_3BA_1CA_2B$$
$$A_1BA_3CA_2B \qquad A_2BA_3CA_1B \qquad A_3BA_2CA_1B$$

If the subscripts are dropped, there results $3!$ repeats of the original permutation. For the six letters $\{A_1, A_2, A_3, B_1, B_2, C\}$ there are $6!$ permutations. There are $3!$ repeats for the $A$'s if they are alike and $2!$ repeats for the $B$'s if they are alike. This reduces the number of permutations to $\frac{6!}{3!2!} = \frac{6\cdot 5\cdot 4\cdot 3\cdot 2\cdot 1}{(3\cdot 2\cdot 1)(2\cdot 1)} = 60$ in order that the repeats are not counted as a new arrangement when all the $A$'s and $B$'s are the same.

In general, if there are $n$-things with $m_1$-repeats of one kind, $m_2$-repeats of a second kind,...,$m_r$-repeats of the $r$th kind, then the number of permutations of these things taken all at a time is given by

$$P = \frac{n!}{m_1!m_2!\cdots m_r!} \tag{2.4}$$

When a collection of $n$-objects or things are examined without regard to the order in which they are arranged, one says a combination of $n$-things is being examined. The process of selecting a subset of $m$ objects taken from a collection of $n$ objects, $m \leq n$, without regard to the ordering is called a combination of $n$ objects taken $m$ at a time and is denoted using one of the notations $_nC_m$ or $C(n,m)$. Note that for each combination of $m$ things selected there are $m!$ permutations which are repeats as the order is not considered when dealing with combinations. Therefore, the number of permutations of $n$ things taken $m$ at a time can be written

$$_nP_m = m!\,_nC_m = m!\,C(n,m) \tag{2.5}$$

and consequently

$$_nC_m = C(n,m) = \frac{_nP_m}{m!} = \frac{n(n-1)(n-2)\cdots(n-(m-1))}{m!} \quad (2.6)$$

Multiplying both the numerator and denominator of the right-hand side of equation (2.6) by the term $(n-m)!$ gives

$$_nC_m = C(n,m) = \frac{n!}{m!(n-m)!} = \binom{n}{m} \quad (2.7)$$

It is an easy exercise to show that $_nC_m = {_nC_{n-m}}$. Here $\binom{n}{m}$ is the notation for the binomial coefficients in the expansion

$$(a+b)^n = a^n + \binom{n}{1}a^{n-1}b + \binom{n}{2}a^{n-2}b^2 + \binom{n}{3}a^{n-3}b^3 + \cdots + \binom{n}{m}a^{n-m}b^m + \cdots + \binom{n}{n}b^n \quad (2.8)$$

In the special case where $a = b = 1$, the equation (2.8) reduces to

$$(1+1)^n = 2^n = 1 + \binom{n}{1} + \binom{n}{2} + \cdots + \binom{n}{n}$$

Therefore, the total number of combinations of $n$ things taken $1, 2, 3, \ldots, n-1, n$ at a time is given by

$$\binom{n}{1} + \binom{n}{2} + \cdots + \binom{n}{n} = 2^n - 1 \quad (2.9)$$

## Elementary Probability

If one assumes that there are h-ways that an event can happen and f-ways an event can fail and these ways are all equally likely, then the probability that the event will happen is given by

$$p = \frac{h}{h+f} \quad (2.10)$$

The probability that the event will fail is given by

$$q = \frac{f}{h+f} = 1 - p \quad (2.11)$$

Note that $p + q = 1$. An alternative way of saying this is to let E denote an event which can happen h-ways out of a total of $n$ equally likely ways. The probability of the event happening is sometimes written as

$$p = P(E) = \frac{h}{n} \quad (2.12)$$

and the probability of the event not happening is written

$$q = P(\text{not } E) = \frac{n-h}{n} = 1 - P(E) \quad (2.13)$$

Note that the notations $\overline{E}, \tilde{E}, \tilde{\phantom{E}}E$ are sometimes employed to denote the event "not E".

The above terms 'equally likely' can at times be confusing and so statistician usually use a limiting process to define probability. An empirical definition of probability of an event is taken as the relative frequency of occurrence of the event and is based upon taking a large number of observations. The probability is then taken as the limit of the relative frequency as the number of observations increases without bound.

## The Binomial Formula

Examine the binomial expansion $(a+b)^n$ for the values $n = 0, 1, 2, 3, \ldots$

| Binomial Formula | Pascal's Triangle |
|---|---|
| $(a+b)^0 = 1$ | 1 |
| $(a+b)^1 = a+b$ | 1   1 |
| $(a+b)^2 = a^2 + 2ab + b^2$ | 1   2   1 |
| $(a+b)^3 = a^3 + 3a^2b + 3ab^2 + b^3$ | 1   3   3   1 |
| $(a+b)^4 = a^4 + 4a^3b + 6a^2b^2 + 4ab^3 + b^4$ | 1   4   6   4   1 |
| $(a+b)^5 = a^5 + 5a^4b + 10a^3b^2 + 10a^2b^3 + 5ab^4 + b^5$ | 1   5   10   10   5   1 |

The above binomial expansions can be verified by direct multiplication. The coefficients of the binomial expansions, when written down as illustrated above, give rise to the Pascal triangle where the odd and even rows are staggered with ones at each end such that adding two consecutive numbers of a row gives the number in the next row. Using the factorial notation

$$n! = n(n-1)(n-2)\cdots 3 \cdot 2 \cdot 1, \qquad 0! = 1 \quad \text{by definition} \tag{2.14}$$

the binomial expansion can be expressed by the more general form

$$(a+b)^n = \sum_{m=0}^{n} \binom{n}{m} a^{n-m} b^m \tag{2.15}$$

where $\binom{n}{m} = \dfrac{n!}{m!(n-m)!}$ are called the binomial coefficients.

## Multinomial Expansion

Expansions of the form

$$(x_1 + x_2 + \cdots + x_r)^n = \sum \frac{n!}{n_1! n_2! \cdots n_r!} x_1^{n_1} x_2^{n_2} \cdots x_r^{n_r} \tag{2.16}$$

are called multinomial expansions. In the expansion given by equation (2.16) the summation is taken over all nonnegative integers $n_1, n_2, \ldots, n_r$ with sum $n_1 + n_2 + \cdots + n_r = n$

Some examples are

$(x_1 + x_2 + x_3)^2 = x_1^2 + 2x_1x_2 + x_2^2 + 2x_1x_3 + 2x_2x_3 + x_3^2$

$(x_1 + x_2 + x_3)^3 = x_1^3 + 3x_1^2 x_2 + 3x_1 x_2^2 + x_2^3 + 3x_1^2 x_3 + 6x_1 x_2 x_3 + 3x_2^2 x_3 + 3x_1 x_3^2 + 3x_2 x_3^2 + x_3^3$

$(x_1 + x_2 + x_3 + x_4)^2 = x_1^2 + 2x_1 x_2 + x_2^2 + 2x_1 x_3 + 2x_2 x_3 + x_3^2 + 2x_1 x_4 + 2x_2 x_4 + 2x_3 x_4 + x_4^2$

$(x_1 + x_2 + x_3 + x_4)^3 = x_1^3 + 3x_1^2 x_2 + 3x_1 x_2^2 + x_2^3 + 3x_1^2 x_3 + 6x_1 x_2 x_3 + 3x_2^2 x_3 + 3x_1 x_3^2 + 3x_2 x_3^2$
$\qquad + x_3^3 + 3x_1^2 x_4 + 6x_1 x_2 x_4 + 3x_2^2 x_4 + 6x_1 x_3 x_4 + 6x_2 x_3 x_4 + 3x_3^2 x_4$
$\qquad + 3x_1 x_4^2 + 3x_2 x_4^2 + 3x_3 x_4^2 + x_4^3$

## Theory of Proportions

If $\dfrac{a}{\alpha} = \dfrac{b}{\beta} = \dfrac{c}{\gamma} = k$, then it is possible to select any nonzero weighting factors $\lambda, \mu, \nu$ and show that

$$k = \frac{\lambda a + \mu b + \nu c}{\lambda \alpha + \mu \beta + \nu \gamma} \tag{2.17}$$

## Sum of Arithmetic Progression

An arithmetic progression has the form

$$S = a + (a+d) + (a+2d) + (a+3d) + \cdots + [a+(n-1)d] \tag{2.18}$$

where $a$ is known as the first term of the progression, $d$ is known as the difference between successive terms within the progression and $n$ is the number of terms in the series. If one reverses the order in the way the arithmetic progression is written one finds

$$S = [a+(n-1)d] + [a+(n-2)d] + \cdots + [(a+2d)] + [(a+d)] + a \tag{2.19}$$

Now adding the series given by equations (2.18) and (2.19) one obtains the result

$$2S = n[a + a + (n-1)d] \quad \text{or} \quad 2S = n(a + \ell) \tag{2.20}$$

Here $\ell$ denotes the last term and $n$ denotes the number of terms. This result shows that the sum of an arithmetic progression is $n/2$ times the sum of the first plus last term with the sum $S$ given by either of the relations

$$S = \frac{n}{2}[a + a + (n-1)d] \quad \text{or} \quad S = \frac{n}{2}(a+\ell) \quad \text{where } \ell = a + (n-1)d \tag{2.21}$$

## Sum of Geometric Progression

A geometric progression has the form

$$S = a + ar + ar^2 + ar^3 + \cdots + ar^{n-1} \tag{2.22}$$

where $a$ is the first term of the progression, $r$ is the common ratio of successive terms and $n$ is the number of terms in the series. If one multiplies both sides of equation (2.22) by the common ratio $r$ one obtains

$$rS = ar + ar^2 + ar^3 + \cdots + ar^{n-1} + ar^n \tag{2.23}$$

Now subtract equation (2.22) from the equation (2.23) to obtain

$$rS - S = ar^n - a \quad \text{which gives} \quad S = a\left(\frac{r^n - 1}{r - 1}\right) \tag{2.24}$$

If the common ratio $r$ is such that $|r| < 1$, then the above result is written in the form $S = \dfrac{a(1-r^n)}{1-r}$ and in the limit as $n \to \infty$ one obtains the result

$$S = a + ar + ar^2 + ar^3 + \cdots = \frac{a}{r-1}, \qquad |r| < 1 \tag{2.25}$$

since $r^n \to 0$ with increasing $n$ if $|r| < 1$.

**Equations Containing Radicals**

Equations containing radicals having the form

$$\sqrt{\alpha x + \beta} + \gamma = \sqrt{\delta x} \tag{2.26}$$

where $\alpha, \beta, \gamma, \delta$ are constants, The radicals can be removed by squaring both sides of the equation (2.26). (Note that sometimes the squaring is done more than once.) Because square roots are multiple valued functions the squaring process can sometimes lead to equations which are not equivalent to the original equation. Consequently, the validity of any answer obtained must be tested by substituting into the original equation to see if it satisfies the equation. Squaring both sides of the equation (2.26) gives

$$\alpha x + \beta + 2\gamma\sqrt{\alpha x + \beta} + \gamma^2 = \delta x \tag{2.27}$$

which simplifies to

$$2\gamma\sqrt{\alpha x + \beta} = (\delta - \alpha)x - \beta - \gamma^2 \tag{2.28}$$

If one now squares both sides of the equation (2.28) one obtains

$$4\gamma^2(\alpha x + \beta) = \left[(\delta - \alpha)x - \beta - \gamma^2\right]^2 \tag{2.29}$$

which simplifies to a quadratic equation in the variable $x$ which can be easily solved.

**Matrix Algebra**

A rectangular array of numbers having $m$-rows and $n$-columns of the form

$$\begin{pmatrix} a_{11} & a_{12} & \cdots & a_{1j} & \cdots & a_{1n} \\ a_{21} & a_{22} & \cdots & a_{2j} & \cdots & a_{2n} \\ \vdots & \vdots & \ddots & \vdots & \ddots & \vdots \\ a_{i1} & a_{i2} & \cdots & a_{ij} & \cdots & a_{in} \\ \vdots & \vdots & \ddots & \vdots & \ddots & \vdots \\ a_{m1} & a_{m2} & \cdots & a_{mj} & \cdots & a_{mn} \end{pmatrix} \tag{2.30}$$

is called a $(m \times n)$ matrix (read "m times n matrix") where the quantities $a_{11}, a_{12}, \ldots, a_{mn}$ are called the elements of the matrix. A matrix is usually denoted by a capital letter and sometimes expressed for brevity as $A = (a_{ij})$ or $B = (b_{ij})$ by writing in parenthesis the general element in the $i$th row and $j$th column of the matrix. In the double subscript notation $a_{ij}$, the first subscript $i$ denotes the row number and the second subscript $j$ denotes the column number. The special $(1 \times n)$ matrix $(a_{11}, a_{12}, a_{13}, \ldots, a_{1n})$ is called a row vector and the special $(n \times 1)$ matrix $\begin{pmatrix} b_{11} \\ b_{21} \\ b_{31} \\ \vdots \\ b_{n1} \end{pmatrix}$ is called a column vector. To conserve space, column vectors are written $\text{col}(b_{11}, b_{21}, b_{31}, \ldots, b_{n1})$.

Two $m \times n$ matrices $A = (a_{ij})$ and $B = (b_{ij})$ are equal if $a_{ij} = b_{ij}$ for all values of $i = 1, \ldots, m$ and $j = 1, \ldots, n$. Equality is expressed $A = B$. Addition of matrices is defined as the sum of corresponding elements. Thus if $C = A + B$, this means $c_{ij} = a_{ij} + b_{ij}$ for $i = 1, \ldots, m$ and $j = 1, \ldots, n$. The difference of two matrices of the same dimensions is obtained by subtracting corresponding elements. For example, if $D = A - B$, then $d_{ij} = a_{ij} - b_{ij}$ for $i = 1, \ldots, m$ and $j = 1, \ldots, n$. If a matrix $A$ is multiplied by a scalar $m$, then $mA = (ma_{ij})$. That is, each element $a_{ij}$ of $A$ gets multiplied by the scalar $m$.

Matrices $A, B, C$, all having the same dimension $(m \times n)$, satisfy the following properties under addition and scalar multiplication.

| | |
|---|---|
| commutative law | $A + B = B + A$ |
| associative law | $A + (B + C) = (A + B) + C$ |
| scalar multiplication | $m(A + B) = mA + mB = (A + B)m$ |
| | $(m + n)A = mA + nA$ |
| | $m(nA) = (mn)A$ |
| There exists a matrix $D$ such that | $A + D = B$ |

### Dot product of vectors

The dot product of a row vector and a column vector with the same number of elements is a single number given by

$$(a_{11}, a_{12}, a_{13}, \ldots, a_{1n}) \begin{pmatrix} b_{11} \\ b_{21} \\ b_{31} \\ \vdots \\ b_{n1} \end{pmatrix} = a_{11}b_{11} + a_{12}b_{21} + a_{13}b_{31} + \cdots + a_{1n}b_{n1} = \sum_{j=1}^{n} a_{1j} b_{j1} \qquad (2.31)$$

Note that corresponding elements are multiplied and then added. The length of the row vector and the length of the column vector must be the same in order for the dot product to exist.

A matrix $A$ of dimension $(m \times \ell)$ can be viewed as consisting of $m$-row vectors each having a length $\ell$. Similarly, a matrix $B$ of dimension $(\ell \times n)$ can be viewed as consisting of $n$-column vectors of length $\ell$. A matrix $A$ is said to be conformable to a matrix $B$ if the length $\ell$ of the row vectors of $A$ is the same as the length $\ell$ of the column vectors of $B$. A matrix $A$ of dimension $(m \times \ell)$ can be multiplied by a matrix $B$ of dimension $\ell \times n$ to obtain a matrix $C$ of dimension $(m \times n)$. Matrix multiplication $AB = C$ is defined as follows.

$$\begin{pmatrix} a_{11} & a_{12} & a_{13} & \cdots & a_{1\ell} \\ a_{21} & a_{22} & a_{23} & \cdots & a_{2\ell} \\ \vdots & \vdots & \vdots & \ddots & \vdots \\ a_{i1} & a_{i2} & a_{i3} & \cdots & a_{i\ell} \\ \vdots & \vdots & \vdots & \ddots & \vdots \\ a_{m1} & a_{m2} & a_{m3} & \cdots & a_{m\ell} \end{pmatrix} \begin{pmatrix} b_{11} & b_{12} & \cdots & b_{1j} & \cdots & b_{1n} \\ b_{21} & b_{22} & \cdots & b_{2j} & \cdots & b_{2n} \\ b_{31} & b_{32} & \cdots & b_{3j} & \cdots & b_{3n} \\ \vdots & \vdots & \ddots & \vdots & \ddots & \vdots \\ b_{\ell 1} & b_{\ell 2} & \cdots & b_{\ell j} & \cdots & b_{\ell n} \end{pmatrix} = \begin{pmatrix} c_{11} & c_{12} & \cdots & c_{1j} & \cdots & c_{1n} \\ c_{21} & c_{22} & \cdots & c_{2j} & \cdots & c_{2n} \\ \vdots & \vdots & \ddots & \vdots & \ddots & \vdots \\ c_{i1} & c_{i2} & \cdots & c_{ij} & \cdots & c_{in} \\ \vdots & \vdots & \ddots & \vdots & \ddots & \vdots \\ c_{m1} & c_{m2} & \cdots & c_{mj} & \cdots & c_{mn} \end{pmatrix}$$

$$(m \times \ell) \qquad\qquad (\ell \times n) \qquad\qquad\qquad (m \times n)$$

The element $c_{ij}$ is obtained by taking the dot product of the $i$th row vector from $A$ and dotting it with the $j$th column vector from $B$ to obtain the element $c_{ij}$ in the $i$th row, $j$th column of the matrix $C$. This dot product can be represented

$$c_{ij} = a_{i1}b_{1j} + a_{i2}b_{2j} + a_{i3}b_{3j} + \cdots + a_{i\ell}b_{\ell j} = \sum_{k=1}^{\ell} a_{ik}b_{kj}$$

Perform all possible dot products for $i = 1, 2, \ldots, m$ and $j = 1, 2, \ldots, n$ to calculate the elements of the matrix $C$.

**Example 2-2.**

If $A = \begin{pmatrix} 1 & 2 & 3 \\ 4 & 5 & 6 \\ 7 & 8 & 9 \end{pmatrix}$ and $B = \begin{pmatrix} 1 & 4 & 7 \\ 2 & 5 & 8 \\ 3 & 6 & 9 \end{pmatrix}$ are two $3 \times 3$ square matrices, then the matrix product is given by

$$AB = \begin{pmatrix} 1 & 2 & 3 \\ 4 & 5 & 6 \\ 7 & 8 & 9 \end{pmatrix} \begin{pmatrix} 1 & 4 & 7 \\ 2 & 5 & 8 \\ 3 & 6 & 9 \end{pmatrix} = \begin{pmatrix} c_{11} & c_{12} & c_{13} \\ c_{21} & c_{22} & c_{23} \\ c_{31} & c_{32} & c_{33} \end{pmatrix} = C = \begin{pmatrix} 14 & 32 & 50 \\ 32 & 77 & 122 \\ 50 & 122 & 194 \end{pmatrix}$$

where

$c_{11} =$ (row 1 of $A$) $\cdot$ (column 1 of $B$) $= (1)(1) + (2)(2) + (3)(3) = 14$

$c_{12} =$ (row 1 of $A$) $\cdot$ (column 2 of $B$) $= (1)(4) + (2)(5) + (3)(6) = 32$

$c_{13} =$ (row 1 of $A$) $\cdot$ (column 3 of $B$) $= (1)(7) + (2)(8) + (3)(9) = 50$

$c_{21} =$ (row 2 of $A$) $\cdot$ (column 1 of $B$) $= (4)(1) + (5)(2) + (6)(3) = 32$

$c_{22} =$ (row 2 of $A$) $\cdot$ (column 2 of $B$) $= (4)(4) + (5)(5) + (6)(6) = 77$

$c_{23} =$ (row 2 of $A$) $\cdot$ (column 3 of $B$) $= (4)(7) + (5)(8) + (6)(9) = 122$

$c_{31} =$ (row 3 of $A$) $\cdot$ (column 1 of $B$) $= (7)(1) + (8)(2) + (9)(3) = 50$

$c_{32} =$ (row 3 of $A$) $\cdot$ (column 2 of $B$) $= (7)(4) + (8)(5) + (9)(6) = 122$

$c_{33} =$ (row 3 of $A$) $\cdot$ (column 3 of $B$) $= (7)(7) + (8)(8) + (9)(9) = 194$

∎

## Special Matrices and Properties

1. The zero matrix $\underline{0}$ has all zeros for its elements. Trace of $A = \operatorname{Tr} A = \sum_{i=1}^{n} a_{ii}$ is the sum of the diagonal elements of a matrix.

2. The identity matrix $I$ is a square matrix having 1's down the main diagonal and zeros everywhere else and can be expressed $I = (\delta_{ij})$ where $\delta_{ij} = \begin{cases} 0, & i \neq j \\ 1, & i = j \end{cases}$ is the Kronecker delta symbol. A $3 \times 3$ identity matrix can be written $I = \begin{pmatrix} 1 & 0 & 0 \\ 0 & 1 & 0 \\ 0 & 0 & 1 \end{pmatrix}$

   If one of the following operations is performed upon a $n \times n$ identity matrix I,

   (i) Interchange any two rows

   (ii) Multiply any row by a constant

   (iii) Multiply any row by a scalar $\alpha$ and add the result to some other row,

   then the resulting matrix is denoted by $E$ and is called an elementary matrix. If a matrix $B$ can be obtained from a matrix $A$ by performing a sequence of elementary row operations, $B = E_m E_{m-1} \cdots E_2 E_1 A$, then $A$ and $B$ are called equivalent matrices and one writes $A \sim B$.

3. A $n \times n$ square matrix $U = (u_{ij})$ where $u_{ij} = 0$ for $i > j$, is called an upper triangular matrix having the form

$$U = \begin{pmatrix} u_{11} & u_{12} & u_{13} & \cdots & u_{1n} \\ 0 & u_{22} & u_{23} & \cdots & u_{2n} \\ 0 & 0 & u_{33} & \cdots & u_{3n} \\ \vdots & \vdots & \vdots & \ddots & \vdots \\ 0 & 0 & 9 & \cdots & u_{nn} \end{pmatrix}$$

   with all zeros below the main diagonal.

4. The $n \times n$ square matrix $L = (\ell_{ij})$ where $\ell_{ij} = 0$ for $i < j$ is called a lower triangular matrix. This matrix has the form

$$L = \begin{pmatrix} \ell_{11} & 0 & 0 & \cdots & 0 \\ \ell_{21} & \ell_{22} & 0 & \cdots & 0 \\ \ell_{31} & \ell_{32} & \ell_{33} & \cdots & 0 \\ \vdots & \vdots & \vdots & \ddots & \vdots \\ \ell_{n1} & \ell_{n2} & \ell_{n3} & \cdots & \ell_{nn} \end{pmatrix}$$

   with all zeros above the main diagonal.

5. The $n \times n$ square matrix $A = (a_{ij})$ with the property that

$$a_{ij} = \begin{cases} 0, & \text{for} \quad i+s \leq j, \quad 1 < s < n \\ 0, & \text{for} \quad j+t \leq i, \quad 1 < t < n \end{cases}$$

   is called a banded matrix with band width $w = s + t - 1$. For example, the $5 \times 5$ tridiagonal matrix

$$A = \begin{pmatrix} a_{11} & a_{12} & 0 & 0 & 0 \\ a_{21} & a_{22} & a_{23} & 0 & 0 \\ 0 & a_{32} & a_{33} & a_{34} & 0 \\ 0 & 0 & a_{43} & a_{44} & a_{45} \\ 0 & 0 & 0 & a_{54} & a_{55} \end{pmatrix}$$

   is a banded matrix with band width $w = 3$.

Another example of a banded matrix is a diagonal matrix with band width $w = 1$. An example of a $5 \times 5$ diagonal matrix is

$$D = \begin{pmatrix} a_{11} & 0 & 0 & 0 & 0 \\ 0 & a_{22} & 0 & 0 & 0 \\ 0 & 0 & a_{33} & 0 & 0 \\ 0 & 0 & 0 & a_{44} & 0 \\ 0 & 0 & 0 & 0 & a_{55} \end{pmatrix}$$

6. If $A$ is a $m \times n$ matrix, then the matrix constructed by interchanging the rows and columns of $A$ is called the transpose matrix and it is written as $A^T$. Note that $A^T$ is a $n \times m$ matrix and if $A = (a_{ij})$, then $A^T = (a_{ji})$ for $i = 1, \ldots, m$ and $j = 1, \ldots, n$. This can be written in expanded form as

$$A = \begin{pmatrix} a_{11} & a_{12} & \cdots & a_{1m} \\ a_{21} & a_{22} & \cdots & a_{2m} \\ \vdots & \vdots & \ddots & \vdots \\ a_{n1} & a_{n2} & \cdots & a_{nm} \end{pmatrix} \quad \text{with} \quad A^T = \begin{pmatrix} a_{11} & a_{21} & a_{31} & \cdots & a_{n1} \\ a_{12} & a_{22} & a_{32} & \cdots & a_{n2} \\ \vdots & \vdots & \vdots & \ddots & \vdots \\ a_{1m} & a_{2m} & a_{3m} & \cdots & a_{nm} \end{pmatrix}$$

If $A^T = A$, then $A$ is called a symmetric matrix. If $A^T = -A$, then $A$ is called a skew-symmetric matrix. As an exercise show that $\frac{1}{2}(A+A^T)$ is a symmetric matrix and $\frac{1}{2}(A-A^T)$ is a skew-symmetric matrix. Note also that $(A^T)^T = A$, $(mA)^T = mA^T$, $m$ a scalar, the transpose of a sum is $(A + B)^T = A^T + B^T$, and the transpose of a product of matrices is the product of the transpose matrices in reverse order $(AB)^T = B^T A^T$.

7. If $A$ is a $n \times n$ square matrix and there exists a $n \times n$ matrix $B$ such that $AB = BA = I$, then $B$ is called the inverse matrix of $A$ and is written $B = A^{-1}$ (read "B equals A inverse"). The inverse matrix satisfies $AA^{-1} = A^{-1}A = I$.

If $A$ and $B$ are $n \times n$ square matrices with inverses $A^{-1}$ and $B^{-1}$, then $(AB)^{-1} = B^{-1}A^{-1}$. Note that the inverse of the product of two matrices is the product of the inverses in reverse order. Also $A^{-n} = (A^{-1})^n$.

If $A^2 = I$, then $A$ is called involutory. That is, $A$ is its own inverse matrix.

If $AA^T = A^T A = I$, then $A$ is called an orthogonal matrix. In this special case $A^T = A^{-1}$

If there are elementary matrices $E_1, E_2, \ldots, E_k$ such that $E_k E_{k-1} \cdots E_3 E_2 E_1 A = I$, then $A^{-1} = E_k E_{k-1} \cdots E_3 E_2 E_1$ can be represented as a product of elementary matrices.

8. Given $A$ and $B$ are $n \times n$ square matrices.
    (a) If $AB = BA$, then $A$ and $B$ are called commutative matrices.
    (b) If $AB \neq BA$, then $A$ and $B$ are called noncommutative matrices.
    (c) If $AB = -BA$, then $A$ and $B$ are called anti-commutative matrices.

9. Given that $A$ is a $n \times n$ square matrix.
    (a) If $A^{k+1} = A$ for some least positive integer $k$, then $A$ is called a periodic matrix with period $k$.
    (b) If $k = 1$, so that $A^2 = A$, then $A$ is called idempotent.
    (c) If there exists a positive integer $n$ such that $A^n = \underbrace{AA \cdots A}_{n \text{ times}} = \underline{0}$, then $A$ is called nilpotent of index $n$.

10. The determinant of a $n \times n$ matrix $A$ is a single number denoted by $|A|$ or $\det A$. Consider all possible products that can be obtained by selecting one and only one element from each row and column of $A$. These products are then multiplied by either $+1$ or $-1$ depending upon the order of the elements selected. The determinant of $A$ is the sum of these products and has the form

$$\det A = |A| = \sum_{\overline{m}} (-1)^m a_{1m_1} a_{2m_2} \cdots a_{nm_n}$$

where $\overline{m} = (m_1, m_2, \ldots, m_n)$ denotes a permutation of the integers $(1, 2, \ldots, n)$ and $\sum$ is a sum over all possible permutations. Here $m$ is even if $\overline{m}$ is an even permutation and $m$ is odd if $\overline{m}$ is an odd permutation.

11. The minor $m_{ij}$ of an element $a_{ij}$ of an $n \times n$ square matrix $A$ is defined as the value of the $(n-1) \times (n-1)$ subdeterminant formed by deleting all the elements in row $i$ and column $j$ of $A$. The cofactor $c_{ij}$ associated with each element $a_{ij}$ of $A$ is defined $c_{ij} = (-1)^{i+j} m_{ij}$. The cofactor matrix associated with $A$ is written $\text{cof} A = (c_{ij})$. The adjoint of $A$ is defined as $\text{Adj} A = (\text{cof} A)^T$. The matrix A, the cofactor of $A$ and the adjoint of $A$ satisfy the properties

$$A(\text{Adj} A) = A(\text{cof} A)^T = (\det A)I$$

and are related to the inverse matrix by

$$A^{-1} = \frac{1}{\det A} (\text{cof} A)^T, \qquad \det A \neq 0$$

If the determinant of $A$ is different from zero, then $A$ is called a nonsingular matrix and in this case $A^{-1}$ will exist. However, if the determinant of $A$ is zero, then $A$ is called a singular matrix and is such that the inverse matrix will not exist.

12. The vector equation $\vec{y} = A\vec{x}$ where $\vec{y}$ and $\vec{x}$ are $n \times 1$ column vectors and $A$ is a $n \times n$ square matrix can be viewed as a matrix operation of multiplication which transforms the vector $\vec{x}$ into the vector $\vec{y}$. In terms of an input/output system, the vector $\vec{x}$ is the input vector and the vector $\vec{y} = A\vec{x}$ is the output vector.

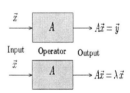

In the special case there exists a nonzero input vector $\vec{x}$ which gets transformed to an output vector $\vec{y} = A\vec{x}$ which is proportional to the input vector $\vec{x}$, such that $\vec{y} = A\vec{x} = \lambda \vec{x}$ for some scalar $\lambda$, then the vector $\vec{x}$ is called an eigenvector and the scalar $\lambda$ is called an eigenvalue. The equation $A\vec{x} = \lambda \vec{x}$ is equivalent to the equation

$$A\vec{x} - \lambda \vec{x} = (A - \lambda I)\vec{x} = \vec{0} \tag{2.32}$$

and this equation has a nonzero solution if and only if

$$\det(A - \lambda I) = |A - \lambda I| = 0 \tag{2.33}$$

The equation (2.33) when expanded is a polynomial equation in $\lambda$ of degree $n$ having the form

$$C(\lambda) = |A - \lambda I| = (-\lambda)^n + c_{n-1}(-\lambda)^{n-1} + \cdots + c_1(-\lambda) + c_0 = 0 \tag{2.34}$$

called the characteristic equation associated with the square matrix $A$. The roots of the characteristic equation (2.34) give the eigenvalues and the equation (2.32) gives the nonzero eigenvectors.

**Linear Dependence and Independence**

If $c_1, c_2, \ldots, c_m$ are constants, not all zero, and $\vec{a}_i$, $i = 1, \ldots, m$ is a set of vectors, then the vector $\vec{b} = c_1\vec{a}_1 + c_2\vec{a}_2 + \cdots + c_m\vec{a}_m$ is a linear combination of the set of vectors $\{\vec{a}_1, \ldots, \vec{a}_m\}$. The set of nonzero row vectors $\vec{a}_i$, $i = 1, \ldots, m$ of a matrix are linearly dependent if there exists constants $c_1, c_2, \ldots, c_m$, such that $c_1\vec{a}_1 + c_2\vec{a}_2 + \cdots + c_m\vec{a}_m = \vec{0}$, where not all the constants $c_i$, $i = 1, \ldots, m$ are zero at the same time. Then at least one of the vectors from the set can be represented as a linear combination of the other vectors in the set. The set of vectors $\vec{a}_i$, $i = 1, \ldots, m$ is said to be linearly independent if the only way that the linear combinations $c_1\vec{a}_1 + c_2\vec{a}_2 + \cdots + c_m\vec{a}_m = \vec{0}$ vanishes is for $c_1 = c_2 = \cdots = c_m = 0$. Note that whenever the number of vectors $m$ is greater than the dimension $n$ of the vectors, then the set of vectors will always be linearly dependent. A basis of an $n$-dimensional vector space $\mathbb{R}^n$ is a set of linearly independent vectors such that every vector in the vector space can be constructed as a linear combination of the base vectors.

**Properties of Matrix Multiplication**

Conformable matrices $A, B, C$ have the following properties.

$$\begin{array}{ll} \text{Left distributive law} & A(B+C) = AB + AC \\ \text{Right distributive law} & (A+B)C = AC + BC \\ \text{Associative law} & A(BC) = (AB)C \end{array}$$

In general conformable matrices $A$ and $B$ are noncommutative so that $AB \neq BA$. Make note of the following (i) The product $AB = \underline{0}$ does not necessarily imply that $A = \underline{0}$ or $B = \underline{0}$ (ii) The matrix equation $AB = AC$ does not necessarily imply that $B = C$.

**Example 2-3.**

Given the matrices $A = \begin{pmatrix} 1 & 2 \\ 3 & 6 \end{pmatrix}$, $B = \begin{pmatrix} -6 & -4 \\ 3 & 2 \end{pmatrix}$ and $C = \begin{pmatrix} -2 & -6 \\ 1 & 3 \end{pmatrix}$. Verify that $AB = \begin{pmatrix} 1 & 2 \\ 3 & 6 \end{pmatrix}\begin{pmatrix} -6 & -4 \\ 3 & 2 \end{pmatrix} = \begin{pmatrix} 0 & 0 \\ 0 & 0 \end{pmatrix} = \underline{0}$ and $AC = \begin{pmatrix} 1 & 2 \\ 3 & 6 \end{pmatrix}\begin{pmatrix} -2 & -6 \\ 1 & 3 \end{pmatrix} = \begin{pmatrix} 0 & 0 \\ 0 & 0 \end{pmatrix} = \underline{0}$ Also calculate $BA = \begin{pmatrix} -6 & -4 \\ 3 & 2 \end{pmatrix}\begin{pmatrix} 1 & 2 \\ 3 & 6 \end{pmatrix} = \begin{pmatrix} -18 & -30 \\ 9 & 18 \end{pmatrix}$ to show $BA \neq AB$. ∎

**Example 2-4.** Given the matrix $A = \begin{pmatrix} 1 & 0 & -1 \\ 1 & 2 & 1 \\ 2 & 2 & 3 \end{pmatrix}$ find the eigenvalues and eigenvectors.

**Solution:** If $A\vec{x} = \lambda \vec{x}$, then

$$(A - \lambda I)\vec{x} = \left[ \begin{pmatrix} 1 & 0 & -1 \\ 1 & 2 & 1 \\ 2 & 2 & 3 \end{pmatrix} - \lambda \begin{pmatrix} 1 & 0 & 0 \\ 0 & 1 & 0 \\ 0 & 0 & 1 \end{pmatrix} \right] \begin{bmatrix} x_1 \\ x_2 \\ x_3 \end{bmatrix} = \begin{bmatrix} 0 \\ 0 \\ 0 \end{bmatrix} \qquad (2.35)$$

In order for this system of equations to have a nonzero solution it is necessary that

$$\det(A - \lambda I) = \vec{0} \quad \text{or} \quad \begin{vmatrix} 1-\lambda & 0 & -1 \\ 1 & 2-\lambda & 1 \\ 2 & 2 & 3-\lambda \end{vmatrix} = 0$$

This produces the characteristic equation

$$|A - \lambda I| = -\lambda^3 + 6\lambda^2 - 11\lambda + 6 = -(\lambda-1)(\lambda-2)(\lambda-3) = 0$$

with characteristic roots $\lambda = 1, 2, 3$. Substituting the characteristic roots into the equation (2.35) gives the following equations

$\lambda = 1$

$\begin{pmatrix} 0 & 0 & -1 \\ 1 & 1 & 1 \\ 2 & 2 & 2 \end{pmatrix} \begin{pmatrix} x_1 \\ x_2 \\ x_3 \end{pmatrix} = \begin{pmatrix} 0 \\ 0 \\ 0 \end{pmatrix}$

$\implies x_3 = 0, \; x_1 = -x_2$

$(-1, 1, 0)$ is an eigenvector

$\lambda = 2$

$\begin{pmatrix} -1 & 0 & -1 \\ 1 & 0 & 1 \\ 2 & 2 & 1 \end{pmatrix} \begin{pmatrix} x_1 \\ x_2 \\ x_3 \end{pmatrix} = \begin{pmatrix} 0 \\ 0 \\ 0 \end{pmatrix}$

$\implies x_1 = -x_3, \; x_1 = -2x_2$

$(-2, 1, 2)$ is an eigenvector

$\lambda = 3$

$\begin{pmatrix} -2 & 0 & -1 \\ 1 & -1 & 1 \\ 2 & 2 & 0 \end{pmatrix} \begin{pmatrix} x_1 \\ x_2 \\ x_3 \end{pmatrix} = \begin{pmatrix} 0 \\ 0 \\ 0 \end{pmatrix}$

$\implies x_1 = -x_2, \; x_3 = -2x_1$

$(-1, 1, 2)$ is an eigenvector

∎

### Vector Norms

A vector or matrix norm $\| \cdot \|$ is a mapping from $R^n$ to $R$ which is a measure of distance associated with vectors $\vec{x} = (x_1, x_2, \ldots, x_n)^T$. Vector norms have the following properties
(i) $\| \vec{x} \| \geq 0$. (ii) $\| \alpha \vec{x} \| = |\alpha| \| \vec{x} \|$ where $\alpha$ is a scalar. (iii) $\| \vec{x} + \vec{y} \| \leq \| \vec{x} \| + \| \vec{y} \|$ for all $\vec{x}, \vec{y}$ in $R^n$
(iv) $\| \vec{x} \| = 0$ if and only if $\vec{x} = 0$. Three norms used quite often are the Euclidean norm $\| \cdot \|_2$, the $\ell_\infty$ norm and the $\ell_p$ norm defined by

$$\| \vec{x} \|_2 = \left[ \sum_{i=1}^{n} x_i^2 \right]^{1/2}, \qquad \| \vec{x} \|_\infty = \max_{1 \leq i \leq n} |x_i|, \qquad \| \vec{x} \|_p = \left[ \sum_{i=1}^{n} |x_i|^p \right]^{1/p}$$

### System of Linear Equations

Let us examine three methods to solve the linear system of equations

$$\begin{matrix} \alpha x + \beta y = \gamma \\ ax + by = c \end{matrix} \quad \text{or} \quad \begin{pmatrix} \alpha & \beta \\ a & b \end{pmatrix} \begin{pmatrix} x \\ y \end{pmatrix} = \begin{pmatrix} \gamma \\ c \end{pmatrix} \qquad (2.36)$$

where $a, b, c, \alpha, \beta, \gamma$ are given constants and $x, y$ are unknowns to be determined. The three methods considered are

(i) A graphical method, (ii) The elimination method and (iii) Cramer's rule.

**Graphical method**

Plot the straight lines $\alpha x + \beta y = \gamma$ and $ax + by = c$ on the same graph paper. If the lines intersect, then there is one unique solution to the linear system. The unique solution is found where the two lines intersect. This represents a common value for $(x, y)$ satisfying both equations. If the plotted lines are parallel, then no intersection can occur, so no solution will exist. If the lines lie on top of one another, then there is an infinite number of solutions possible. The following example illustrates these three cases.

**Example 2-5.**

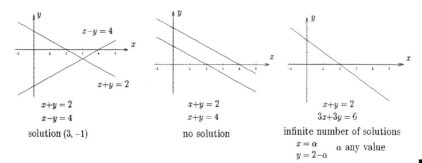

$x+y=2$
$x-y=4$
solution $(3, -1)$

$x+y=2$
$x+y=4$
no solution

$x+y=2$
$3x+3y=6$
infinite number of solutions
$x = \alpha$
$y = 2-\alpha$    $\alpha$ any value

**The elimination method**

Consider the system of equations

$$\alpha x + \beta y = \gamma \quad (2.37)$$
$$ax + by = c \quad (2.38)$$

Multiply the equation (2.37) by $a/\alpha$, $\alpha \neq 0$ to obtain the equivalent system

$$ax + (a/\alpha)\beta y = (a/\alpha)\gamma \quad (2.39)$$
$$ax + by = c \quad (2.40)$$

Subtract equation (2.39) from equation (2.40) to eliminate the variable $x$. There results the new system

$$ax + (a/\alpha)\beta y = (a/\alpha)\gamma \quad (2.41)$$
$$(b - (a/\alpha)\beta)y = c - (a/\alpha)\gamma \quad (2.42)$$

Solve the equation (2.42) for $y$ and then substitute $y$ into equation (2.41) and solve for $x$. This process is called back-substitution. One finds after simplification the solutions

$$x = \frac{\gamma b - \beta c}{ab - a\beta} \quad \text{and} \quad y = \frac{ac - a\gamma}{ab - a\beta} \tag{2.43}$$

**Cramer's rule**

Cramer's rule to solve the system of equations

$$\begin{aligned} ax + \beta y &= \gamma \\ ax + by &= c \end{aligned} \quad \text{or} \quad \begin{pmatrix} a & \beta \\ a & b \end{pmatrix} \begin{pmatrix} x \\ y \end{pmatrix} = \begin{pmatrix} \gamma \\ c \end{pmatrix}$$

is the representation of the solutions $x, y$ as the ratio of certain determinants. Recall that a mnemonic to evaluate the $2 \times 2$ determinant $\begin{vmatrix} a & b \\ c & d \end{vmatrix}$ one constructs the diagonals illustrated.

$$\begin{vmatrix} a & b \\ c & d \end{vmatrix} = ad - bc$$

then multiply the elements along the diagonals and assign a positive sign to one product and a negative sign to the other product.

Using Cramer's rule one sets up the ratio of determinants $x = \dfrac{\begin{vmatrix} \gamma & \beta \\ c & b \end{vmatrix}}{\begin{vmatrix} a & \beta \\ a & b \end{vmatrix}}$, $y = \dfrac{\begin{vmatrix} a & \gamma \\ a & c \end{vmatrix}}{\begin{vmatrix} a & \beta \\ a & b \end{vmatrix}}$

where the denominator determinant always contains the coefficients of the linear system of equations occurring on the left-hand side of the system of equations. The numerator determinant for $x$ is created from the denominator determinant by removing the first column of elements and replacing them by the elements which occur on the right-hand side of the system of equations. The numerator determinant for $y$ is created from the denominator determinant by removing the second column of elements and replacing them by the elements occurring on the right-hand side of the system of equations. Evaluating the determinants one obtains the same solutions as given by equations (2.43).

The Cramer's rule for the solution of the $3 \times 3$ system of equations

$$\begin{aligned} a_1 x + b_1 y + c_1 z &= d_1 \\ a_2 x + b_2 y + c_2 z &= d_2 \\ a_3 x + b_3 y + c_3 z &= d_3 \end{aligned} \quad \text{or} \quad \begin{pmatrix} a_1 & b_1 & c_1 \\ a_2 & b_2 & c_2 \\ a_3 & b_3 & c_3 \end{pmatrix} \begin{pmatrix} x \\ y \\ z \end{pmatrix} = \begin{pmatrix} d_1 \\ d_2 \\ d_3 \end{pmatrix} \tag{2.44}$$

also requires one to set up the ratio of certain determinants to obtain the solution. These determinants are construct in a manner similar to the previous $2 \times 2$ system of equations.

The solutions to the above $3 \times 3$ system of equations is by Cramer's rule

$$x = \frac{\begin{vmatrix} d_1 & b_1 & c_1 \\ d_2 & b_2 & c_2 \\ d_3 & b_3 & c_3 \end{vmatrix}}{\begin{vmatrix} a_1 & b_1 & c_1 \\ a_2 & b_2 & c_2 \\ a_3 & b_3 & c_3 \end{vmatrix}}, \quad y = \frac{\begin{vmatrix} a_1 & d_1 & c_1 \\ a_2 & d_2 & c_2 \\ a_3 & d_3 & c_3 \end{vmatrix}}{\begin{vmatrix} a_1 & b_1 & c_1 \\ a_2 & b_2 & c_2 \\ a_3 & b_3 & c_3 \end{vmatrix}}, \quad z = \frac{\begin{vmatrix} a_1 & b_1 & d_1 \\ a_2 & b_2 & d_2 \\ a_3 & b_3 & d_3 \end{vmatrix}}{\begin{vmatrix} a_1 & b_1 & c_1 \\ a_2 & b_2 & c_2 \\ a_3 & b_3 & c_3 \end{vmatrix}} \tag{2.45}$$

Note the denominator determinants are all the same and the rows and columns are obtained from the elements representing the coefficients of the given system of equations. In solving for $x$, note the determinant of the numerator is such that the first column of the denominator determinant is removed and replaced by the elements on the right-hand side of the system of equations. A similar column replacement is done when solving for $y$ and $z$. A mnemonic device to evaluate the $3 \times 3$ determinant $D = \begin{vmatrix} a & b & c \\ d & e & f \\ g & h & i \end{vmatrix}$ is to append the first two columns to the end of the determinant and then draw diagonal lines through the elements as illustrated below. One then multiplies the elements along the diagonals and assigns appropriate plus and minus signs to the products. A summation of the products gives the value of the determinant.

$$\begin{vmatrix} a & b & c \\ d & e & f \\ g & h & i \end{vmatrix} \begin{matrix} a & b \\ d & e \\ g & h \end{matrix} = \begin{array}{l} aei + bfg + cdh \\ - gec - hfa - idb \end{array}$$

Using the above mnemonic device the determinants in equation (2.45) can be evaluated to obtain the solution to the $3 \times 3$ system of equations.

Note the above mnemonic devices for the evaluation of $2 \times 2$ and $3 \times 3$ determinants do not work for higher ordered determinants.

**Determinant of Order n**

A determinant $|A|$ of order n is a square array of numbers or $n \times n$ array of numbers having the form

$$|A| = \begin{vmatrix} a_{11} & a_{12} & a_{13} & \cdots & a_{1n} \\ a_{21} & a_{22} & a_{23} & \cdots & a_{2n} \\ a_{31} & a_{32} & a_{33} & \cdots & a_{3n} \\ \vdots & \vdots & \vdots & \ddots & \vdots \\ a_{n1} & a_{n2} & a_{n3} & \cdots & a_{nn} \end{vmatrix} \tag{2.46}$$

There are $n$ rows and $n$ columns in a $n \times n$ determinant. Here the notation $a_{ij}$ denotes the element in the $i$th row and $j$th column. That is, the first subscript denotes the row number and the second subscript denotes the column number of the element. The minor of an element $a_{ij}$ is another element $m_{ij}$ obtained from the determinant resulting after crossing out the elements of $|A|$ in the $i$th row and $j$th column. Multiplying the minor of the element $a_{ij}$ by $(-1)^{i+j}$ gives the cofactor element $c_{ij} = (-1)^{i+j} m_{ij}$.

**Determinant $|A|$**

To evaluate the determinant $|A|$, select any row (or column) of $|A|$ and then multiply each element in the row (or column) by its own cofactor and then perform a summation of these products. This is called *expansion by cofactors*.

For example, by selecting the first row of $|A|$, the determinant is obtained from the expansion

$$|A| = a_{11}C_{11} + a_{12}C_{12} + a_{13}C_{13} + \cdots + a_{1n}C_{1n} = \sum_{k=1}^{n} a_{1k}C_{1k} \qquad (2.47)$$

In the special case of a $3 \times 3$ determinant the expansion by cofactors along the first row gives

$$|A| = \begin{vmatrix} a_{11} & a_{12} & a_{13} \\ a_{21} & a_{22} & a_{23} \\ a_{31} & a_{32} & a_{33} \end{vmatrix} = a_{11}(-1)^{1+1}\begin{vmatrix} a_{22} & a_{23} \\ a_{32} & a_{33} \end{vmatrix} + a_{12}(-1)^{1+2}\begin{vmatrix} a_{21} & a_{23} \\ a_{31} & a_{33} \end{vmatrix} + a_{13}(-1)^{1+3}\begin{vmatrix} a_{21} & a_{22} \\ a_{31} & a_{32} \end{vmatrix} \qquad (2.48)$$

For determinants of order $n > 4$ the method of cofactor expansion for the evaluation of a determinant is not recommended because it is too time consuming. Evaluation of cofactors requires the evaluation of approximately $n!$ arithmetic operations. Use instead elementary row operations to reduce the matrix of the determinant to a diagonal form. The determinant is then the product of the diagonal elements.

## Pythagorean Theorem

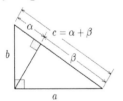

**Theorem** Let $a$ and $b$ denote the legs of a right triangle and let $c$ denote the hypotenuse, then $a^2 + b^2 = c^2$.

**Proof:** For the right triangles illustrated, show using similar triangles $\dfrac{b}{c} = \dfrac{\alpha}{b}$ and $\dfrac{a}{c} = \dfrac{\beta}{a}$ or $b^2 = \alpha c$ and $a^2 = \beta c$ and consequently by addition $b^2 + a^2 = c(\alpha + \beta) = c \cdot c = c^2$

Certain special right triangles having integer values for their sides have been found in Egyptian and Babylonian documents. However, there is no written record of any proofs of the Pythagorean theorem from these cultures. Pythagoras is thought to be the first European to have proved this result, giving us the theorem bearing his name. However, the Chinese knew a geometric proof of this theorem many hundreds of years before Pythagoras.

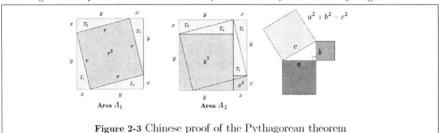

Figure 2-3 Chinese proof of the Pythagorean theorem

The Chinese proof is illustrated in the figure 2-3 where a square of side $r$ is constructed inside a larger square with sides of length $x+y$, and there are four congruent right triangles, with legs $x$ and $y$, labeled $T_1, T_2, T_3, T_4$, also within the larger square. Now move the triangles $T_1$ and $T_4$ to the positions indicated in the middle sketch in figure 2-3 and note that now $x^2 + y^2 = r^2$. Equating area $A_1$ to area $A_2$, $A_2 = (x+y)^2 = x^2 + y^2 + 4\left(\frac{1}{2}xy\right) = r^2 + 4\left(\frac{1}{2}xy\right) = A_1$ which implies $x^2 + y^2 = r^2$. This was the Chinese proof of the Pythagorean theorem. It was proved by Pythagoras using the ratio of areas of similar triangles. It is estimated that there are over 400 different proofs of the Pythagorean theorem. The right-hand sketch in figure 2-3 is a geometric representation of the Pythagorean theorem.

There are special right triangles having integer values for the sides. The integer values are known as Pythagorean triples. The Euclid formula for Pythagorean triples associated with a right triangle with sides $x$ and $y$ and hypotenuse $r$ is written

$$x = m^2 - n^2, \quad y = 2mn, \quad r = m^2 + n^2 \quad \text{which satisfies} \quad x^2 + y^2 = r^2 \qquad (2.49)$$

where $m$ and $n$ are integers such that $x$ is positive. For example, with $m = 2$ and $n = 1$ there results the Pythagorean triple $(x, y, r) = (3, 4, 5)$. Other selected Pythagorean triples are $(5, 12, 13)$, $(7, 24, 25)$, $(9, 40, 41)$, $(8, 15, 17)$, and $(11, 60, 61)$.

An application of the Pythagorean theorem is used to calculate the distance between two points in the plane. With reference to the accompanying figure, if $(x_1, y_1)$ and $(x_2, y_2)$ are the coordinates of two points in the plane, then by connecting these two points with a straight line and drawing horizontal

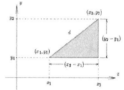

lines $y = y_1$ and $y = y_2$ together with the vertical lines $x = x_1$ and $x = x_2$ one can construct the right triangle illustrated having sides with distances $(x_2 - x_1)$ and $(y_2 - y_1)$ and hypotenuse $d$. The Pythagorean theorem requires that $d^2 = (x_2 - x_1)^2 + (y_2 - y_1)^2$, from which there results the formula for distance between two points in the plane

$$d = \sqrt{(x_2 - x_1)^2 + (y_2 - y_1)^2}$$

## Trigonometry

The word trigonometry comes from the Greek language and means 'measure of triangles'. Plane trigonometry deals with triangles in a plane and spherical trigonometry deals with spherical triangles on a sphere.

In plane trigonometry if a line $\overline{0A}$ is drawn through the origin of a coordinate system and then rotated counterclockwise about the origin to the position $\overline{0B}$, then the positive angle $\angle A0B = \theta$ is said to have been generated. If $\overline{0B}'$ is the reflection of $\overline{0B}$ about the $x$-axis, then when the line $\overline{0A}$ is rotated in a clockwise direction to the position $\overline{OB}'$, a negative angle $-\theta$ is said to be generated. The situation is illustrated in the accompanying figure.

By definition if two angles add to 90°, then they are called complementary angles and if two angles add to 180°, then the angles are called supplementary. Angles are sometimes measured in degrees (°), minutes ('), and seconds (") where there are 360° in a circle, 60 minutes in one degree and 60 seconds in one minute. Another unit of measurement for the angle is the radian. One radian is the angle subtended by an arc on a circle which has the same length as the radius of the circle as illustrated in the above figure. The circumference of a circle is given by the formula $C = 2\pi r$, where $r$ is the radius of the circle. In the special case $r = 1$,

$$(a) \quad 2\pi \text{ radians} = 360° \qquad (c) \quad 1 \text{ radian} = \frac{180}{\pi}°$$
$$(b) \quad \pi \text{ radians} = 180° \qquad (d) \quad \frac{\pi}{180} \text{ radians} = 1° \qquad (2.50)$$

Thus, to convert 30° to radians, just multiply both sides of equation (2.50) (d) by 30 to get the result that $30° = \pi/6$ radians.

An angle is classified as acute if it is less than 90° and called obtuse if it lies between 90° and 180°. An angle $\theta$ is called a right angle when it has the value of 90° or $\frac{\pi}{2}$ radians.

In scientific computing one always uses the radian measure in the calculations. Also note that most hand-held calculators have a switch for converting from one angular measure to another and so owners of such calculators must learn to set them appropriately before doing any calculations.

## Trigonometric Functions

The first recorded trigonometric tables comes from the Greeks around 150 BCE. However, it is inferred that trigonometry was used in sailing, astronomy and construction going back thousands of years before this time, but only limited records are available.

The ratio of sides of a right triangle are used to define the six trigonometric functions associated with one of the acute angles of the right triangle. These definitions can then be extended to apply to positive and negative angles associated with a point moving on a unit circle.

The six trigonometric functions associated with a right triangle are

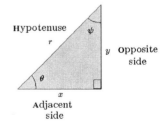

    sine       tangent      secant

    cosine    cotangent   cosecant

which are abbreviated respectively as

    sin, tan, sec, cos, cot , and csc

and are defined

$$\sin\theta = \frac{y}{r} = \frac{\text{Opposite side}}{\text{Hypotenuse}} \quad \cos\theta = \frac{x}{r} = \frac{\text{Adjacent side}}{\text{Hypotenuse}} \quad \tan\theta = \frac{y}{x} = \frac{\text{Opposite side}}{\text{Adjacent side}}$$
$$\csc\theta = \frac{r}{y} = \frac{\text{hypotenuse}}{\text{opposite side}} \quad \sec\theta = \frac{r}{x} = \frac{\text{hypotenuse}}{\text{adjacent side}} \quad \cot\theta = \frac{x}{y} = \frac{\text{adjacent side}}{\text{opposite side}} \quad (2.51)$$

A mnemonic devise to remember the sine, cosine and tangent definitions is the expression "**S**ome **O**ld **H**orse **C**ame **A** **H**opping **T**owards **O**ur **A**lley."

Observe that the following functions are reciprocals

$$\sin\theta \text{ and } \csc\theta, \quad \cos\theta \text{ and } \sec\theta, \quad \tan\theta \text{ and } \cot\theta \qquad (2.52)$$

and satisfy the relations $\quad \csc\theta = \dfrac{1}{\sin\theta}, \quad \sec\theta = \dfrac{1}{\cos\theta}, \quad \cot\theta = \dfrac{1}{\tan\theta} \qquad (2.53)$

In addition to the above 6 trigonometric functions, the functions versine, coversine and haversine are sometimes used when dealing with trigonometric relations. These functions do not have standardize abbreviations and can be found under different names in different books. These functions are defined

$$\text{versed sine } \theta = \text{versine}\,\theta = \text{vers}\,\theta = \text{ver}\,\theta = 1 - \cos\theta$$

$$\text{versed cosine } \theta = \text{coversed sine}\,\theta = \text{coversine}\,\theta = \text{covers}\,\theta = \text{cov}\,\theta = 1 - \sin\theta \qquad (2.54)$$

$$\text{haversine } \theta = \text{hav}\,\theta = \frac{1}{2}\text{ver}\,\theta = \frac{1}{2}(1 - \cos\theta)$$

The figure 2-4 gives a graphical representation of the above trigonometric functions in terms of distances associated with the unit circle.

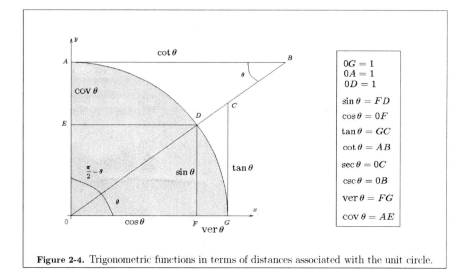

**Figure 2-4.** Trigonometric functions in terms of distances associated with the unit circle.

Consider the special right triangles illustrated

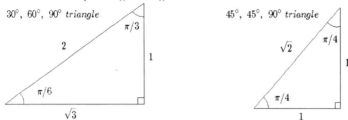

Note that in the 45 degree triangle, the sides opposite the 45 degree angles are equal and so any convenient length can be used to represent the equal sides. If the number 1 is used, then the hypotenuse has the value $\sqrt{2}$. Similarly, in the 30-60-90 degree right triangle, the side opposite the 30 degree angle is always one-half the hypotenuse and so by selecting the value of 2 for the hypotenuse one obtains the sides $2, 1, \sqrt{3}$ as illustrated. By using the trigonometric definitions given by the equations (2.51) together with the special 30-60-90, and 45 degree right triangles, the following table of values result.

| Angle $\theta$ in degrees | Angle $\theta$ in radians | $\sin\theta$ | $\cos\theta$ | $\tan\theta$ | $\cot\theta$ | $\sec\theta$ | $\csc\theta$ |
|---|---|---|---|---|---|---|---|
| 0° | 0 | 0 | 1 | 0 | undefined | 1 | undefined |
| 30° | $\pi/6$ | 1/2 | $\sqrt{3}/2$ | $\sqrt{3}/3$ | $\sqrt{3}$ | $2\sqrt{3}/3$ | 2 |
| 45° | $\pi/4$ | $\sqrt{2}/2$ | $\sqrt{2}/2$ | 1 | 1 | $\sqrt{2}$ | $\sqrt{2}$ |
| 60° | $\pi/3$ | $\sqrt{3}/2$ | 1/2 | $\sqrt{3}$ | $\sqrt{3}/3$ | 2 | $2\sqrt{3}/3$ |
| 90° | $\pi/2$ | 1 | 0 | undefined | 0 | undefined | 1 |

The values for $\theta = 0$ and $\theta = 90$ degrees are special cases which can be examined separately by imagining the triangle changing as the point $P$ moves around a circle having radius $r$. The 45 degree triangle being a special case when $x = 1$, $y = 1$ and $r = \sqrt{2}$. The 30-60-90 degree triangle is a special case when there is a circle with $x = 1$, $r = 2$ and $y = \sqrt{3}$. The trigonometric functions associated with the limiting values of $\theta = 0$ and $\theta = 90$ degrees can be obtained from these special circles by examining the values of $x$ and $y$ associated with the point $P$ for the limiting conditions $\theta = 0$ and $\theta = \pi/2$.

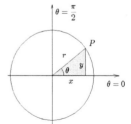

## Cofunctions

In the definitions of the trigonometric functions certain combinations of these functions are known as cofunctions. For example, the sine and cosine functions are known as cofunctions. Thus the cofunction of the sine function is the cosine function and the cofunction of the cosine function is the sine function. Similarly, the functions tangent and cotangent are cofunctions and the functions secant and cosecant are cofunctions.

**Theorem** *Each trigonometric function of an acute angle $\theta$ is equal to the cofunction of the complementary angle $\psi = \frac{\pi}{2} - \theta$.*

The above theorem follows directly from the definitions of the trigonometric functions giving the results

$$\sin\theta = \cos(\frac{\pi}{2} - \theta) \qquad \tan\theta = \cot(\frac{\pi}{2} - \theta) \qquad \sec\theta = \csc(\frac{\pi}{2} - \theta)$$
$$\cos\theta = \sin(\frac{\pi}{2} - \theta) \qquad \cot\theta = \tan(\frac{\pi}{2} - \theta) \qquad \csc\theta = \sec(\frac{\pi}{2} - \theta) \qquad (2.65)$$

The above results are known as the cofunction formulas.

## Trigonometric Functions Defined for Other Angles

Consider a circle drawn with respect to some $xy$ coordinate system and let $P$ denote a point on the circle which moves around the circle in a counterclockwise direction. The figure 2-5 illustrates the point $P$ lying in various quadrants which are denoted by the Roman numerals I,II,III,IV. Figure 2-5(a) denotes the point $P$ lying in the first quadrant with figures 2-5(b)(c) and (d) illustrating the point $P$ lying in the second, third and fourth quadrant respectively.

Let $\theta$ denote the angle swept out as $P$ moves counterclockwise about the circle and define the six trigonometric functions of $\theta$ as follows

$$\sin\theta = \frac{y}{r}, \qquad \cos\theta = \frac{x}{r}, \qquad \tan\theta = \frac{y}{x}, \qquad \cot\theta = \frac{x}{y}, \qquad \sec\theta = \frac{r}{x}, \qquad \csc\theta = \frac{r}{y} \qquad (2.66)$$

Here $y$ denotes the ordinate of the point $P$, $x$ denotes the abscissa of the point $P$ and $r$ denotes the polar distance of the point $P$ from the origin. These distances are illustrated in the figures 2-5 (a)(b)(c) and (d). Note that these definitions imply that

$$\tan\theta = \frac{\sin\theta}{\cos\theta}, \qquad \cot\theta = \frac{\cos\theta}{\sin\theta}, \qquad \sec\theta = \frac{1}{\cos\theta}, \qquad \csc\theta = \frac{1}{\sin\theta}, \qquad \cot\theta = \frac{1}{\tan\theta} \qquad (2.67)$$

Write the Pythagorean theorem, $x^2 + y^2 = r^2$ for the right triangle in each quadrant of figure 2-5. The Pythagorean theorem can be written in any of the alternative forms

$$\left(\frac{x}{r}\right)^2 + \left(\frac{y}{r}\right)^2 = 1, \qquad 1 + \left(\frac{y}{x}\right)^2 = \left(\frac{r}{x}\right)^2, \qquad \left(\frac{x}{y}\right)^2 + 1 = \left(\frac{r}{y}\right)^2$$

which by the above trigonometric definitions become

$$\cos^2\theta + \sin^2\theta = 1, \qquad 1 + \tan^2\theta = \sec^2\theta, \qquad \cot^2\theta + 1 = \csc^2\theta \qquad (2.68)$$

These are fundamental relations between the trigonometric functions and are known as trigonometric identities. The above identities are sometimes called the Pythagorean identities.

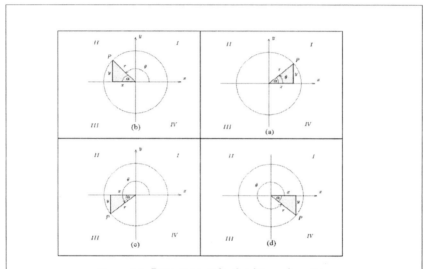

**Figure 2-5.** Point $P$ on circle of radius $r$ where $P$ is in quadrant I, quadrant II, quadrant III and quadrant IV.

### Sign Variation of the Trigonometric Functions

In the figure 2-5 make note of the following.
(i) The six trigonometric functions of $\theta$, as the point $P$ moves about the circle, are sometimes referred to as circular functions.
(ii) The ray $\overline{0P}$ defines not only the angle $\theta$ but also the angles $\theta \pm 2m\pi$ where $m$ is a positive integer or zero.
(iii) As $P$ moves around the circle, the radial distance $r = \overline{0P}$ always remains constant, but the $x$ and $y$ values change sign as $P$ moves through the different quadrants. An analysis of these sign changes produces the table of signs given in the figure 2-6.
(iv) The six trigonometric functions take on special values whenever $x = 0$ or $y = 0$. One of these special values will occur whenever $\theta$ is equal to some multiple of $\frac{\pi}{2}$.

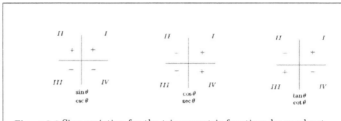

**Figure 2-6** Sign variation for the trigonometric functions by quadrant.

(v) The angle $\alpha$ in the figures 2-5 is the smallest nonnegative angle between the line $\overline{OP}$ and the $x$-axis. Limiting values for $\alpha$ are $\frac{\pi}{2}$ and 0 radians. The angle $\alpha$ is called a reference angle. If the reference angle is different from 0 or $\frac{\pi}{2}$, then it can be viewed as an acute positive angle in the first quadrant. Note that there is then a definite relationship between the six trigonometric functions of $\theta$ and the six trigonometric functions of the reference angle $\alpha$. The six trigonometric functions of the reference angle $\alpha$ are all positive and so one need only add the appropriate sign change to obtain the six trigonometric functions of $\theta$. The appropriate sign changes are given in the figure 2-6 and are used to construct the relations given in the table 2.1.

(vi) The figures 2-5 and 2-6 can be combined into one figure so that by using the definition of the trigonometric functions, the correct sign of a trigonometric function can be determined corresponding to $\theta$ in any quadrant. For example, in quadrant II the functions $\sin\theta$ and $\csc\theta$ are positive. In quadrant III the functions $\tan\theta$ and $\cot\theta$ are positive and in quadrant IV, the functions $\cos\theta$ and $\sec\theta$ are positive.

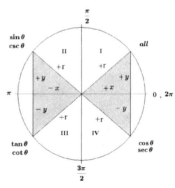

(v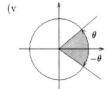

If the angle $\theta$ is a positive acute angle, then $-\theta$ lies in the fourth quadrant. The reference angle is $\alpha = \theta$ is positive and gives the results

$$\begin{aligned} \sin(-\theta) &= -\sin\theta & \cos(-\theta) &= \cos\theta & \tan(-\theta) &= \tan\theta \\ \csc(-\theta) &= -\csc(\theta) & \sec(-\theta) &= \sec\theta & \cot(-\theta) &= -\cot\theta \end{aligned} \qquad (2.69)$$

(viii) Recall that any function $f(\theta)$ satisfying $f(-\theta) = f(\theta)$ is called an even function of $\theta$ and functions satisfying $f(-\theta) = -f(\theta)$ are called odd functions of $\theta$. The above arguments show that the functions sine, tangent, cotangent and cosecant are odd functions of $\theta$ and the functions cosine and secant are even functions of $\theta$. These results are sometimes referred to as even-odd identities.

Special Values

| Angle $\theta$ degrees | Angle $\theta$ radians | $\sin\theta$ | $\cos\theta$ | $\tan\theta$ | $\cot\theta$ | $\sec\theta$ | $\csc\theta$ |
|---|---|---|---|---|---|---|---|
| 0 | 0 | 0 | 1 | 0 | undefined | 1 | undefined |
| 90 | $\pi/2$ | 1 | 0 | undefined | 0 | undefined | 1 |
| 180 | $\pi$ | 0 | $-1$ | 0 | undefined | $-1$ | undefined |
| 270 | $3\pi/2$ | $-1$ | 0 | undefined | 0 | undefined | $-1$ |
| 360 | $2\pi$ | 0 | 1 | 0 | undefined | 1 | undefined |

Quadrant II

| Angle $\theta$ degrees | Angle $\theta$ radians | Reference angle $\alpha$ | $\sin\theta$ | $\cos\theta$ | $\tan\theta$ | $\cot\theta$ | $\sec\theta$ | $\csc\theta$ |
|---|---|---|---|---|---|---|---|---|
| 120 | $2\pi/3$ | $\pi/3$ | $\sqrt{3}/2$ | $-1/2$ | $-\sqrt{3}$ | $-\sqrt{3}/3$ | $-2$ | $2\sqrt{3}/3$ |
| 135 | $3\pi/4$ | $\pi/4$ | $\sqrt{2}/2$ | $-\sqrt{2}/2$ | $-1$ | $-1$ | $-\sqrt{2}$ | $\sqrt{2}$ |
| 150 | $5\pi/6$ | $\pi/6$ | $1/2$ | $-\sqrt{3}/2$ | $-\sqrt{3}/3$ | $-\sqrt{3}$ | $-2\sqrt{3}/3$ | 2 |

Quadrant III

| Angle $\theta$ degrees | Angle $\theta$ radians | Reference angle $\alpha$ | $\sin\theta$ | $\cos\theta$ | $\tan\theta$ | $\cot\theta$ | $\sec\theta$ | $\csc\theta$ |
|---|---|---|---|---|---|---|---|---|
| 210 | $7\pi/6$ | $\pi/6$ | $-1/2$ | $-\sqrt{3}/2$ | $\sqrt{3}/3$ | $\sqrt{3}$ | $-2\sqrt{3}/3$ | $-2$ |
| 225 | $5\pi/4$ | $\pi/4$ | $-\sqrt{2}/2$ | $-\sqrt{2}/2$ | 1 | 1 | $-\sqrt{2}$ | $-\sqrt{2}$ |
| 240 | $4\pi/3$ | $\pi/3$ | $-\sqrt{3}/2$ | $-1/2$ | $\sqrt{3}$ | $\sqrt{3}/3$ | $-2$ | $-2\sqrt{3}/3$ |

Quadrant IV

| Angle $\theta$ degrees | Angle $\theta$ radians | Reference angle $\alpha$ | $\sin\theta$ | $\cos\theta$ | $\tan\theta$ | $\cot\theta$ | $\sec\theta$ | $\csc\theta$ |
|---|---|---|---|---|---|---|---|---|
| 300 | $5\pi/3$ | $\pi/3$ | $-\sqrt{3}/2$ | $1/2$ | $-\sqrt{3}$ | $-\sqrt{3}/3$ | 2 | $-2\sqrt{3}/3$ |
| 315 | $7\pi/4$ | $\pi/4$ | $-\sqrt{2}/2$ | $\sqrt{2}/2$ | $-1$ | $-1$ | $\sqrt{2}$ | $-\sqrt{2}$ |
| 330 | $11\pi/6$ | $\pi/6$ | $-1/2$ | $\sqrt{3}/2$ | $-\sqrt{3}/3$ | $-\sqrt{3}$ | $2\sqrt{3}/3$ | $-2$ |

Table 2.1 Special values for the Trigonometric Functions

## Graphs of the Trigonometric Functions

Consider the special case where the radius $r$ in figure 2-5 is unity. The advantage of setting $r = 1$ in the above definitions for the trigonometric functions is that in this special case the ordinate value $y$ represents $\sin\theta$ and the abscissa value $x$ represents $\cos\theta$. By plotting $y$ vs $\theta$ one obtains the figure 2-7(a). Similarly, by rotating the unit circle it is possible to plot a graph of $\cos\theta$ vs $\theta$ by observing the $x$-value. This is equivalent to plotting the height of the point $P$ as a function of $\theta$ as the point $P$ moves in a counterclockwise direction around the circumference of the unit circle. These graphs are illustrated in the figure 2-7(a) and 2-7(b). Plot the ratio $\tan\theta = \dfrac{\sin\theta}{\cos\theta}$ vs $\theta$ as illustrated in the figure 2-7(c) for a graphical illustration of the tangent function.

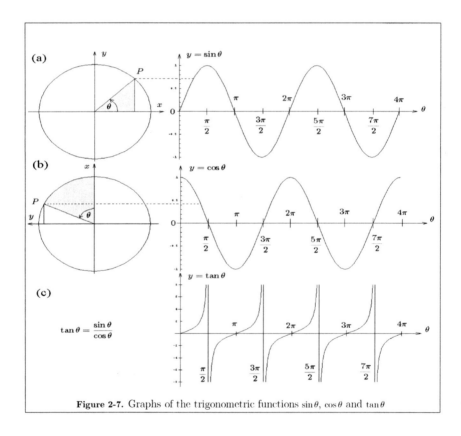

**Figure 2-7.** Graphs of the trigonometric functions $\sin\theta$, $\cos\theta$ and $\tan\theta$

Recall that a continuous function $y = f(x)$, which is well defined everywhere, is said to be a periodic function of period $p$, if $p$ is the smallest number such that $f(x+p) = f(x)$ is satisfied for all values of $x$. In figure 2-7 make note of the following.

(i) The trigonometric functions $\sin\theta$, $\cos\theta$, and consequently the reciprocal functions $\csc\theta$ and $\sec\theta$ are periodic functions of $\theta$ with period $2\pi$ and satisfy

$$\sin(\theta + 2\pi) = \sin\theta, \quad \cos(\theta + 2\pi) = \cos\theta, \quad \csc(\theta + 2\pi) = \csc\theta, \quad \sec(\theta + 2\pi) = \sec\theta$$

(ii) Similarly, the tangent and cotangent functions are periodic functions of $\theta$ with period $\pi$ and satisfy

$$\tan(\theta + \pi) = \tan\theta, \quad \cot(\theta + \pi) = \cot\theta$$

(iii) The function $\tan\theta$ becomes undefined at $\frac{\pi}{2}, \frac{3\pi}{2}, \frac{5\pi}{2}, \ldots$. This is because $\tan\theta = \frac{\sin\theta}{\cos\theta}$ and the cosine function takes on the value of zero at these values.

(iv) The sine and cosine functions oscillate between $+1$ and $-1$ as well as being periodic.

(v) Let $f$ denote any trigonometric function of $\theta$, then note that $f(\theta + 2n\pi) = f(\theta)$, where $n$ is any integer. This is because the angle $\theta \pm 2n\pi$ is coterminal with the angle $\theta$.

(vi) Curves of the form $y = A\sin(\beta x + \gamma)$ or $y = A\cos(\beta x + \gamma)$ are said to have an amplitude $A$ and to be periodic of period $2\pi/\beta$. The graphs of these functions can be obtained by translation of the graphs $y = A\sin\beta x$ and $y = A\cos\beta x$ by a distance of $\gamma/\beta$ in the negative $x$-direction. Another way to view these curves is to set $\theta = \beta x + \gamma$ and then ask the questions, "Where is $\theta$ equal to zero?", and "Where is $\theta$ equal to $2\pi$?". One finds $\theta = 0$ when $x = -\gamma/\beta$ and $\theta = 2\pi$ when $x = (2\pi - \gamma)/\beta$. Subtracting these values gives the period $2\pi/\beta$ and the zero value for $\theta$ gives the translational distance $-\gamma/\beta$. These curves oscillate between $+A$ and $-A$ and are scaled versions of the basic sine and cosine curves.
Graphs for the reciprocal relations

$$\csc\theta = \frac{1}{\sin\theta}, \quad \sec\theta = \frac{1}{\cos\theta}, \quad \text{and} \quad \cot\theta = \frac{1}{\tan\theta}$$

are illustrated in the figure 2-8. To better comprehend how the graphs in figure 2-8 were constructed, sketch on the $y = \csc\theta$ vs $\theta$ graph the function $y = \sin\theta$ and on the $y = \sec\theta$ vs $\theta$ graph, sketch the function $y = \cos\theta$. It is then easier to see that the cosecant curve becomes unbounded as the sine function approaches zero. Similarly, the secant function becomes unbounded when the cosine function approaches zero.

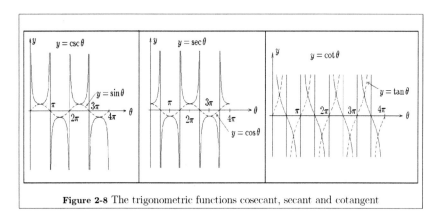

**Figure 2-8** The trigonometric functions cosecant, secant and cotangent

## Trigonometric Functions of Sums and Differences

Consider a unit circle with the points $P_0, P_1, P_2$ lying on the circumference of the circle. Let the rays $\overline{OP_0}, \overline{OP_1}, \overline{OP_2}$ make positive angles of $0, \alpha$ and $\alpha + \beta$ radians respectively with respect to the $x$-axis, as is illustrated in the figure 2-9(a).

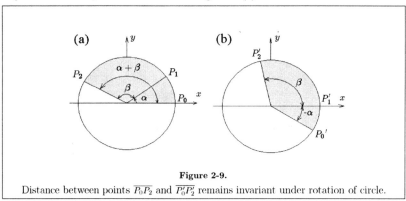

**Figure 2-9.**
Distance between points $\overline{P_0P_2}$ and $\overline{P_0'P_2'}$ remains invariant under rotation of circle.

The point $P_0, P_1, P_2$ are on the unit circle so that the coordinates of these points can be written as

$$
\begin{aligned}
P_0: &\quad (x_0, y_0) = (1, 0), \\
P_1: &\quad (x_1, y_1) = (\cos \alpha, \sin \alpha), \\
P_2: &\quad (x_2, y_2) = (\cos(\alpha + \beta), \sin(\alpha + \beta))
\end{aligned}
\qquad (2.65)
$$

If the points in figure 2-9(a) are rotated clockwise through an angle $\alpha$, the points $P_0, P_1, P_2$ move to the primed positions illustrated in the figure 2-9(b). The coordinates of the primed points are given by

$$P'_0 : \quad (x'_0, y'_0) = (\cos\alpha, -\sin\alpha),$$
$$P'_1 : \quad (x'_1, y'_1) = (1, 0), \qquad (2.66)$$
$$P'_2 : \quad (x'_2, y'_2) = (\cos\beta, \sin\beta).$$

Now the distance $\overline{P_2P_0}$ remains invariant under a rotation of axis and so this distance must be the same as the distance $\overline{P'_2P'_0}$. This requires that

$$\overline{P_2P_0} = \overline{P'_2P'_0}$$
$$\sqrt{(x_2-x_0)^2 + (y_2-y_0)^2} = \sqrt{(x'_2-x'_0)^2 + (y'_2-y'_0)^2} \qquad (2.67)$$
$$\sqrt{(\cos(\alpha+\beta)-1)^2 + (\sin(\alpha+\beta)-0)^2} = \sqrt{(\cos\beta-\cos\alpha)^2 + (\sin\beta-(-\sin\alpha))^2}$$

Square both sides of equation (2.67) and expand the squared terms to obtain

$$\cos^2(\alpha+\beta) - 2\cos(\alpha+\beta) + 1 + \sin^2(\alpha+\beta) = \cos^2\beta - 2\cos\alpha\cos\beta + \cos^2\alpha + \sin^2\beta + 2\sin\alpha\cos\beta + \sin^2\alpha$$

Observe that $\cos^2(\alpha+\beta) + \sin^2(\alpha+\beta) = 1$ as well as $\cos^2\beta + \sin^2\beta = 1$ and $\cos^2\alpha + \sin^2\alpha = 1$. Use these identities to simplify the above result and obtain the addition formula

$$\cos(\alpha+\beta) = \cos\alpha\cos\beta - \sin\alpha\sin\beta \qquad (2.68)$$

In equation (2.68) replace the angle $\beta$ by $-\beta$ to obtain

$$\cos(\alpha+-\beta) = \cos\alpha\cos(-\beta) - \sin\alpha\sin(-\beta)$$

and then use the even-odd identities to obtain

$$\cos(\alpha-\beta) = \cos\alpha\cos\beta + \sin\alpha\sin\beta \qquad (2.69)$$

Observe that a utilization of the cofunction formulas gives

$$\sin(\alpha+\beta) = \cos\left[\frac{\pi}{2} - (\alpha+\beta)\right] = \cos\left[(\frac{\pi}{2} - \alpha) - \beta\right]$$

In equation (2.69) replace $\alpha$ by $\frac{\pi}{2} - \alpha$ to show

$$\sin(\alpha+\beta) = \cos(\frac{\pi}{2} - \alpha)\cos\beta + \sin(\frac{\pi}{2} - \alpha)\sin\beta$$

which simplifies using the cofunction formulas. This gives the addition rule

$$\sin(\alpha+\beta) = \sin\alpha\cos\beta + \cos\alpha\sin\beta \qquad (2.70)$$

It is an easy exercise to replace $\beta$ by $-\beta$ and then use the even-odd properties to show

$$\sin(\alpha-\beta) = \sin\alpha\cos\beta - \cos\alpha\sin\beta \qquad (2.71)$$

Now if $\cos\alpha \neq 0$ and $\cos(\alpha + \beta) \neq 0$, then

$$\tan(\alpha + \beta) = \frac{\sin(\alpha + \beta)}{\cos(\alpha + \beta)} = \frac{\sin\alpha\cos\beta + \cos\alpha\sin\beta}{\cos\alpha\cos\beta - \sin\alpha\sin\beta}$$

Divide each of the terms in the numerator and denominator by $\cos\alpha\cos\beta$ and then simplify to obtain the result

$$\tan(\alpha + \beta) = \frac{\tan\alpha + \tan\beta}{1 - \tan\alpha\tan\beta} \qquad (2.72)$$

Replacing $\beta$ by $-\beta$ in equation (2.72) and simplifying the result produces the difference formula

$$\tan(\alpha - \beta) = \frac{\tan\alpha - \tan\beta}{1 + \tan\alpha\tan\beta} \qquad (2.73)$$

## Double-angle Formulas

In the addition formulas (2.68), (2.70), and (2.72) make the substitution $\alpha = \beta$ to obtain the double-angle formula

$$\sin 2\beta = 2\sin\beta\cos\beta \qquad (2.74)$$

$$\cos 2\beta = \cos^2\beta - \sin^2\beta \qquad (2.75)$$

$$\tan 2\beta = \frac{2\tan\beta}{1 - \tan^2\beta} \qquad (2.76)$$

Using the identity $\cos^2\beta + \sin^2\beta = 1$ the equation (2.75) can be expressed in one of the alternative forms

$$\cos 2\beta = 2\cos^2\beta - 1 \quad \text{or} \quad \cos 2\beta = 1 - 2\sin^2\beta \qquad (2.77)$$

## Half-angle Formulas

In the equations (2.77) replace the angle $\beta$ by $\alpha/2$ to obtain the half-angle formulas

$$\sin^2\frac{\alpha}{2} = \frac{1 - \cos\alpha}{2}, \quad \sin\frac{\alpha}{2} = (-1)^n\sqrt{\frac{1-\cos\alpha}{2}}, \quad n = \text{int}\left(\frac{\pi + |x|}{2\pi}\right) \qquad (2.78)$$

$$\cos^2\frac{\alpha}{2} = \frac{1 + \cos\alpha}{2}, \quad \cos\frac{\alpha}{2} = (-1)^n\sqrt{\frac{1+\cos\alpha}{2}}, \quad n = \text{int}\left(\frac{|x|}{2\pi}\right) \qquad (2.79)$$

Taking the ratio of the above results gives

$$\tan^2\frac{\alpha}{2} = \frac{\sin^2\frac{\alpha}{2}}{\cos^2\frac{\alpha}{2}} = \frac{1 - \cos\alpha}{1 + \cos\alpha} \quad \text{provided } \cos\alpha \neq -1 \qquad (2.80)$$

The equations (2.78), (2.79), and (2.80) can be expressed in the alternative forms

$$\sin\frac{\alpha}{2} = \pm\sqrt{\frac{1-\cos\alpha}{2}}, \quad \cos\frac{\alpha}{2} = \pm\sqrt{\frac{1+\cos\alpha}{2}}, \quad \tan\frac{\alpha}{2} = \pm\sqrt{\frac{1-\cos\alpha}{1+\cos\alpha}} \qquad (2.81)$$

The correct algebraic sign, + or −, is determined by the quadrant the angle $\alpha/2$ lies in.

## Product, Sum and Difference Formula

Products of the sine and cosine functions can be replaced by certain sums or differences and conversely, sums and differences of sine and cosine functions can be replaced by certain products. For example, adding the equations (2.70) and (2.71) gives the sum formula

$$\sin(\alpha + \beta) + \sin(\alpha - \beta) = 2\sin\alpha\cos\beta \qquad (2.82)$$

and subtracting the equations (2.70) and (2.71) gives the difference formula

$$\sin(\alpha + \beta) - \sin(\alpha - \beta) = 2\sin\beta\cos\alpha \qquad (2.83)$$

Similarly, by adding the equations (2.68) and (2.69) one obtains the sum formula

$$\cos(\alpha - \beta) + \cos(\alpha + \beta) = 2\cos\alpha\cos\beta \qquad (2.84)$$

and subtracting these equations gives

$$\cos(\alpha - \beta) - \cos(\alpha + \beta) = 2\sin\alpha\sin\beta \qquad (2.85)$$

The equations (2.82), (2.83), (2.84), and (2.85) allow one to express certain sine-cosine products as sums or differences. For example, make the following substitutions into the equations (2.82), (2.83), (2.84), and (2.85),

$$\alpha + \beta = A \quad \text{and} \quad \alpha - \beta = B \qquad (2.86)$$

Adding and subtracting the equations (2.86) implies that

$$\alpha = \frac{A+B}{2}, \quad \text{and} \quad \beta = \frac{A-B}{2} \qquad (2.87)$$

The equations (2.82), (2.83), (2.84), and (2.85) then take on the forms

$$\sin A + \sin B = 2\sin\left(\frac{A+B}{2}\right)\cos\left(\frac{A-B}{2}\right) \qquad (2.88)$$

$$\sin A - \sin B = 2\cos\left(\frac{A+B}{2}\right)\sin\left(\frac{A-B}{2}\right) \qquad (2.89)$$

$$\cos A + \cos B = 2\cos\left(\frac{A+B}{2}\right)\cos\left(\frac{A-B}{2}\right) \qquad (2.90)$$

$$\cos B - \cos A = 2\sin\left(\frac{A+B}{2}\right)\sin\left(\frac{A-B}{2}\right) \qquad (2.91)$$

## Simple Harmonic Motion

The equation
$$y = y_0 + A\cos\omega t + B\sin\omega t \tag{2.92}$$
where $y_0, A, B$ and $\omega$ are constants and $t$ represents time, can be written in the alternative form
$$y = y_0 + \sqrt{A^2+B^2}\left(\frac{A}{\sqrt{A^2+B^2}}\cos\omega t + \frac{B}{\sqrt{A^2+B^2}}\sin\omega t\right) \tag{2.93}$$

which is obtained by multiplying $\sqrt{A^2+B^2}$ to both the numerator and denominator of the sine and cosine terms. Consider the right triangle with sides $A$ and $B$ with angle $\phi$ illustrated in equation (2.93). This triangle has the property that
$$\sin\phi = \frac{B}{\sqrt{A^2+B^2}} \quad \text{and} \quad \cos\phi = \frac{A}{\sqrt{A^2+B^2}} \tag{2.94}$$
so that equation (2.92) can be written in the form
$$y = y_0 + \sqrt{A^2+B^2}(\cos\phi\cos\omega t + \sin\phi\sin\omega t) \quad \text{or} \quad y = y_0 + \sqrt{A^2+B^2}\cos(\omega t - \phi) \tag{2.95}$$

A graph of this function is illustrated in the figure 2-10.

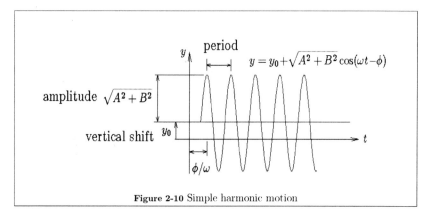

**Figure 2-10** Simple harmonic motion

The cosine function is periodic with period $2\pi$ and so the substitution $\theta = \omega t - \phi$ allows one to determine for what values of time $t$ does $\theta = 0$ and $\theta = 2\pi$, because during this time change the cosine function moves through one period. Note that when $\omega t - \phi = 0$, then $t = t_0 = \phi/\omega$ and when $\omega t - \phi = 2\pi$, $t = t_{2\pi} = \phi/\omega + 2\pi/\omega$. The time difference $t_{2\pi} - t_0 = \text{period} = 2\pi/\omega$ is the period of the oscillation. The quantity $f = \dfrac{1}{period} = \dfrac{\omega}{2\pi}$ is called the frequency of the oscillation. Associated with the graph in figure 2-10 one uses the following terminology and definitions

$y_0$ is called the vertical shift

$\sqrt{A^2 + B^2}$ is called the amplitude of the oscillation

$\omega$ is called the angular frequency with units [radians/second]

$\phi$ is called the phase angle or phase shift with units [radians]

$t$ is the time in units [seconds]

$P = 2\pi/\omega$ is the period of the oscillation in units [seconds]

$f = \dfrac{1}{p}$ is called the frequency of the oscillation with units called hertz [Hz=(seconds)$^{-1}$]

## Inverse Functions

Two functions $f(x)$ and $g(x)$ are said to be inverse functions of one another if $f(x)$ and $g(x)$ have the properties that

$$g(f(x)) = x \qquad \text{and} \qquad f(g(x)) = x \tag{2.96}$$

If $g(x)$ is an inverse function of $f(x)$, the notation $f^{-1}$, read "f-inverse", is used to denote the function $g$. That is, an inverse function of $f(x)$ is denoted $f^{-1}(x)$ and has the properties

$$f(f^{-1}(x)) = x \qquad \text{and} \qquad f^{-1}(f(x)) = x \tag{2.97}$$

Given a function $y = f(x)$, then by interchanging the symbols $x$ and $y$ there results $x = f(y)$. This is an equation which defines the inverse function. If the equation $x = f(y)$ can be solved for $y$ in terms of $x$, to obtain a single valued function, then this function is called the inverse function of $f(x)$. There then results the equivalent statements

$$x = f(y) \qquad \Longleftrightarrow \qquad y = f^{-1}(x) \tag{2.98}$$

The process of interchanging $x$ and $y$ in the representation $y = f(x)$ to obtain $x = f(y)$ implies that geometrically the graphs of $f$ and $f^{-1}$ are mirror images of each other about the line $y = x$. In order that the inverse function be single valued it is necessary that there are no horizontal lines, $y = constant$, which intersect the graph $y = f(x)$ more than once. An example of a function and its inverse is given in the figure 2-11.

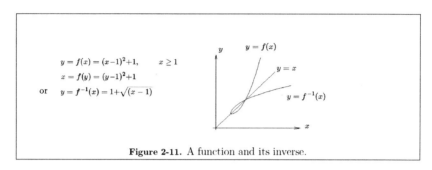

**Figure 2-11.** A function and its inverse.

### Inverse Trigonometric Functions

The inverse trigonometric functions are defined in the following table.

| | | Inverse Trigonometric Functions | |
|---|---|---|---|
| Function | Alternate notation | Definition | Interval for single-valuedness |
| arcsin $x$ | $\sin^{-1} x$ | $\sin^{-1} x = y$ if and only if $x = \sin y$ | $-\frac{\pi}{2} \leq y \leq \frac{\pi}{2}$ |
| arccos $x$ | $\cos^{-1} x$ | $\cos^{-1} x = y$ if and only if $x = \cos y$ | $0 \leq y \leq \pi$ |
| arctan $x$ | $\tan^{-1} x$ | $\tan^{-1} x = y$ if and only if $x = \tan y$ | $-\frac{\pi}{2} < y < \frac{\pi}{2}$ |
| arccot $x$ | $\cot^{-1} x$ | $\cot^{-1} x = y$ if and only if $x = \cot y$ | $0 < y < \pi$ |
| arcsec $x$ | $\sec^{-1} x$ | $\sec^{-1} x = y$ if and only if $x = \sec y$ | $0 \leq y \leq \pi,\ y \neq \frac{\pi}{2}$ |
| arccsc $x$ | $\csc^{-1} x$ | $\csc^{-1} x = y$ if and only if $x = \csc y$ | $-\frac{\pi}{2} \leq y \leq \frac{\pi}{2},\ y \neq 0$ |

The inverse trigonometric functions are multi-valued functions and so one must defined an interval over which single-valuedness occurs. The defined interval is called a branch of the inverse trigonometric function. Whenever a particular branch is required for certain problems, then these branches are called principal branches. The following table gives principal values for the inverse trigonometric functions.

| Principal Values for Regions Indicated | |
|---|---|
| $x < 0$ | $x \geq 0$ |
| $-\frac{\pi}{2} \leq \sin^{-1} x < 0$ | $0 \leq \sin^{-1} x \leq \frac{\pi}{2}$ |
| $\frac{\pi}{2} \leq \cos^{-1} x \leq \pi$ | $0 \leq \cos^{-1} x \leq \frac{\pi}{2}$ |
| $-\frac{\pi}{2} \leq \tan^{-1} x < 0$ | $0 \leq \tan^{-1} x < \frac{\pi}{2}$ |
| $\frac{\pi}{2} < \cot^{-1} x < \pi$ | $0 < \cot^{-1} x \leq \frac{\pi}{2}$ |
| $\frac{\pi}{2} \leq \sec^{-1} x \leq \pi$ | $0 \leq \sec^{-1} x < \frac{\pi}{2}$ |
| $-\frac{\pi}{2} \leq \csc^{-1} x < 0$ | $0 < \csc^{-1} x \leq \frac{\pi}{2}$ |

The inverse trigonometric functions are obtained by taking the mirror image of the trigonometric functions about the line $y = x$. The trigonometric functions and the inverse trigonometric functions are illustrated in the figures 2-12 and 2-13.

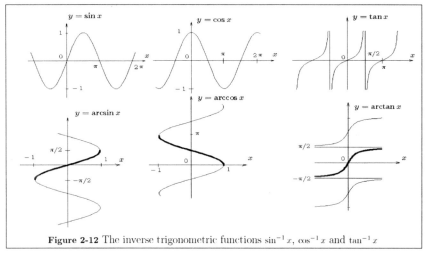

**Figure 2-12** The inverse trigonometric functions $\sin^{-1} x$, $\cos^{-1} x$ and $\tan^{-1} x$

Unless stated otherwise it is implicitly implied that principal values will be used in dealing with inverse trigonometric functions.

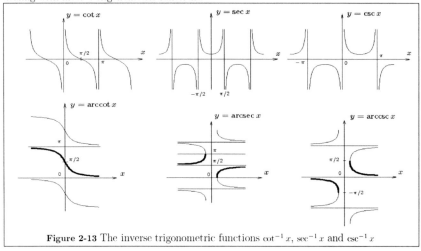

**Figure 2-13** The inverse trigonometric functions $\cot^{-1} x$, $\sec^{-1} x$ and $\csc^{-1} x$

## Principal Value Properties

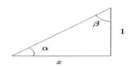

In reference to the right triangle illustrated, observe that if $\alpha = \cot^{-1} x$, then $\cot\alpha = x$ and $\tan\alpha = \frac{1}{x}$. Consequently, $\alpha = \cot^{-1} x = \tan^{-1}\left(\frac{1}{x}\right)$. Also observe that $\alpha + \beta = \frac{\pi}{2}$ and since $\tan\beta = x$ and $\cot\alpha = x$, there results

$$\cot^{-1} x + \tan^{-1} x = \frac{\pi}{2} \tag{2.99}$$

Other inverse trigonometric relationships can be determined in a similar manner and are listed below for $x > 0$.

$$\begin{aligned}
\cot^{-1} x &= \tan^{-1}\left(\frac{1}{x}\right) & \cot^{-1} x + \tan^{-1} x &= \frac{\pi}{2} \\
\sec^{-1} x &= \cos^{-1}\left(\frac{1}{x}\right) & \cos^{-1} x + \sin^{-1} x &= \frac{\pi}{2} \\
\csc^{-1} x &= \sin^{-1}\left(\frac{1}{x}\right) & \csc^{-1} x + \sec^{-1} x &= \frac{\pi}{2}
\end{aligned} \tag{2.100}$$

If $\theta = \sin^{-1}(-x)$, then $\sin\theta = -x$, and using the result $\sin(-\theta) = -\sin\theta = -(-x) = x$, one finds $\theta = -\sin^{-1} x$, and consequently, $\sin^{-1}(-x) = -\sin^{-1} x$. Similarly, if $\theta = \cos^{-1}(-x)$, then $\cos\theta = -x$, and using $\cos(\pi - \theta) = -\cos\theta = -(-x) = x$ there results $\cos^{-1} x = \pi - \theta = \pi - \cos^{-1}(-x)$. Similar relationships can be derived and are summarized below.

$$\begin{aligned}
\sin^{-1}(-x) &= -\sin^{-1} x & \cot^{-1}(-x) &= \pi - \cot^{-1} x \\
\cos^{-1}(-x) &= \pi - \cos^{-1} x & \sec^{-1}(-x) &= \pi - \sec^{-1} x \\
\tan^{-1}(-x) &= -\tan^{-1} x & \csc^{-1}(-x) &= -\csc^{-1} x
\end{aligned} \tag{2.101}$$

## Hyperbolic Functions

The hyperbolic functions are defined in terms of the exponential functions $e^t$ and $e^{-t}$ as given in the following table.

| Hyperbolic Functions | | |
|---|---|---|
| Function | Representation | Definition |
| Hyperbolic sine | $\sinh t$ | $\sinh t = \dfrac{e^t - e^{-t}}{2}$ |
| Hyperbolic cosine | $\cosh t$ | $\cosh t = \dfrac{e^t + e^{-t}}{2}$ |
| Hyperbolic tangent | $\tanh t$ | $\tanh t = \dfrac{e^t - e^{-t}}{e^t + e^{-t}} = \dfrac{\sinh t}{\cosh t}$ |
| Hyperbolic cotangent | $\coth t$ | $\coth t = \dfrac{e^t + e^{-t}}{e^t - e^{-t}} = \dfrac{\cosh t}{\sinh t}$ |
| Hyperbolic secant | $\operatorname{sech} t$ | $\operatorname{sech} t = \dfrac{2}{e^t + e^{-t}} = \dfrac{1}{\cosh t}$ |
| Hyperbolic cosecant | $\operatorname{csch} t$ | $\operatorname{csch} t = \dfrac{2}{e^t - e^{-t}} = \dfrac{1}{\sinh t}$ |

Graphs of the hyperbolic functions are given in the figure 2-14.

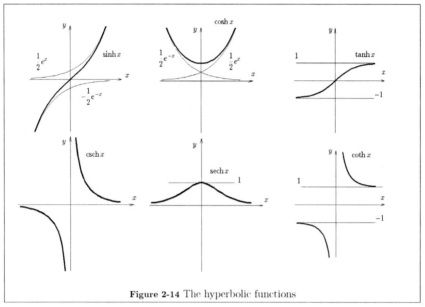

**Figure 2-14** The hyperbolic functions

## Hyperbolic Identities

The hyperbolic functions satisfy the identities

$$\cosh^2 t - \sinh^2 t = 1, \qquad 1 - \tanh^2 t = \operatorname{sech}^2 t, \qquad \coth^2 t - 1 = \operatorname{csch}^2 t \qquad (2.102)$$

Using the definitions for the hyperbolic functions, it is readily verified that

$$\cosh^2 t = \left[\frac{e^t + e^{-t}}{2}\right]^2 = \frac{1}{4}\left[e^{2t} + 2 + e^{-2t}\right], \qquad \sinh^2 t = \left[\frac{e^t - e^{-t}}{2}\right]^2 = \frac{1}{4}\left[e^{2t} - 2 + e^{-2t}\right]$$

Subtracting $\sinh^2 t$ from $\cosh^2 t$ gives the first result above. The other identities are derived in a similar manner.

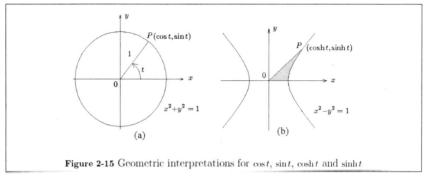

**Figure 2-15** Geometric interpretations for $\cos t$, $\sin t$, $\cosh t$ and $\sinh t$

The figure 2-15(a) illustrates the circle $x^2+y^2=1$ with unit radius. The point $P(\cos t,\sin t)$ is a point on the circumference of this circle lying where the ray $\theta = t$ intersects the circle. The figure 2-15(b) has the point $P(\cosh t,\sinh t)$ lying on the right branch of the hyperbola $x^2-y^2=1$. Note that $\cosh^2 t - \sinh^2 = 1$ with $\cosh t \geq 1$. Here the parameter $t$ represents twice the shaded area indicated. This last statement can be verified by calculating the shaded area, using methods from calculus, to show (Area=$t/2$).

### Properties of Hyperbolic Functions

Using the definition of the hyperbolic functions the following relationships can be verified

$$\sinh(-x) = -\sinh x, \qquad \cosh(-x) = \cosh x, \qquad \tanh(-x) = -\tanh x \qquad (2.103)$$

Hyperbolic functions representing sums and differences also follow directly from the definitions. For example,

$$\sinh x \cosh y = \left(\frac{e^x - e^{-x}}{2}\right)\left(\frac{e^y + e^{-y}}{2}\right) = \frac{1}{4}\left(e^{x+y} + e^{x-y} - e^{-(x-y)} - e^{-(x+y)}\right) \qquad (2.104)$$

$$\cosh x \sinh y = \left(\frac{e^x + e^{-x}}{2}\right)\left(\frac{e^y - e^{-y}}{2}\right) = \frac{1}{4}\left(e^{x+y} - e^{x-y} + e^{-(x-y)} - e^{-(x+y)}\right) \qquad (2.105)$$

Adding the equations (2.104) and (2.105) there results the addition formula

$$\sinh(x+y) = \sinh x \cosh y + \cosh x \sinh y$$

In a similar manner the following addition subtraction formulas are derived.

$$\sinh(x \pm y) = \sinh x \cosh y \pm \cosh x \sinh y$$
$$\cosh(x \pm y) = \cosh x \cosh y \pm \sinh x \sinh y \qquad (2.106)$$
$$\tanh(x \pm y) = \frac{\tanh x \pm \tanh y}{1 \pm \tanh x \tanh y}$$

The double argument identities are a special case of the above results in the special case $y = x$. This produces

$$\sinh 2x = 2\sinh x \cosh x$$
$$\cosh 2x = \cosh^2 x + \sinh^2 y = 2\cosh^2 x - 1 = 1 + 2\sinh^2 x \qquad (2.107)$$
$$\tanh 2x = \frac{2\tanh x}{1 + \tanh^2 x}$$

The results for $\cosh 2x$ from equation (2.107) can be used to construct the relations

$$\sinh^2 x = \frac{\cosh 2x - 1}{2}, \quad \cosh^2 x = \frac{\cosh 2x + 1}{2}, \quad \tanh^2 2x = \frac{\sinh^2 2x}{\cosh^2 2x} = \frac{\cosh 2x - 1}{\cosh 2x + 1} \qquad (2.108)$$

In the equations (2.108) replace $x$ by $x/2$ to obtain the half argument formulas

$$\sinh \frac{x}{2} = \begin{cases} +\sqrt{\frac{\cosh x - 1}{2}}, & \text{if } x > 0 \\ -\sqrt{\frac{\cosh x - 1}{2}}, & \text{if } x < 0 \end{cases}, \quad \cosh \frac{x}{2} = \sqrt{\frac{\cosh x + 1}{2}}, \qquad (2.109)$$

The hyperbolic tangent of a half argument can be written in several alternative forms since

$$\tanh \frac{x}{2} = \begin{cases} +\sqrt{\frac{\cosh x - 1}{\cosh x + 1}}, & \text{if } x > 0 \\ -\sqrt{\frac{\cosh x - 1}{\cosh x + 1}}, & \text{if } x < 0 \end{cases} = \sqrt{\frac{\cosh^2 x - 1}{(\cosh x + 1)^2}} = \frac{\sinh x}{\cosh x + 1} = \frac{\cosh x - 1}{\sinh x} \qquad (2.110)$$

Sums, differences and products of hyperbolic functions are derived using the definitions. Sum and difference formulas are given by

$$\begin{aligned} \sinh x + \sinh y &= 2\sinh \frac{x+y}{2} \cosh \frac{x-y}{2} \\ \sinh x - \sinh y &= 2\cosh \frac{x+y}{2} \sinh \frac{x-y}{2} \\ \cosh x + \cosh y &= 2\cosh \frac{x+y}{2} \cosh \frac{x-y}{2} \\ \cosh x - \cosh y &= 2\sinh \frac{x+y}{2} \sinh \frac{x-y}{2} \end{aligned} \qquad (2.111)$$

Products of hyperbolic functions produce the following relationships

$$\begin{aligned} \sinh x \sinh y &= \frac{1}{2}[\cosh(x+y) - \cosh(x-y)] \\ \cosh x \cosh y &= \frac{1}{2}[\cosh(x+y) + \cosh(x-y)] \\ \sinh x \cosh y &= \frac{1}{2}[\sinh(x+y) + \sinh(x-y)] \end{aligned} \qquad (2.112)$$

## Inverse Hyperbolic Functions

Using the trigonometric functions for guidance, the inverse hyperbolic functions are defined as follows.

$$\begin{aligned}
&\text{If } x = \sinh y &\iff& \quad y = \sinh^{-1} x = \text{Inverse hyperbolic sine of } x \\
&\text{If } x = \cosh y &\iff& \quad y = \cosh^{-1} x = \text{Inverse hyperbolic cosine of } x \\
&\text{If } x = \tanh y &\iff& \quad y = \tanh^{-1} x = \text{Inverse hyperbolic tangent of } x \\
&\text{If } x = \coth y &\iff& \quad y = \coth^{-1} x = \text{Inverse hyperbolic cotangent of } x \\
&\text{If } x = \operatorname{sech} y &\iff& \quad y = \operatorname{sech}^{-1} x = \text{Inverse hyperbolic secant of } x \\
&\text{If } x = \operatorname{csch} y &\iff& \quad y = \operatorname{csch}^{-1} x = \text{Inverse hyperbolic cosecant of } x
\end{aligned} \qquad (2.113)$$

Graphs of these functions are obtained by constructing the mirror image of the hyperbolic functions about the line $y = x$. These graphs are illustrated in the figure 2-16.

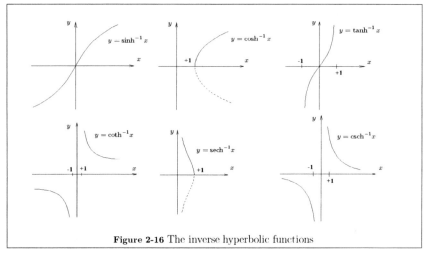

**Figure 2-16** The inverse hyperbolic functions

Examine the graph for $\sinh^{-1} x$ and note that this function is uniquely defined for all values of $x$. This is because the function $\sinh x$ is a monotone increasing function for $-\infty < x < \infty$. However, the function $\cosh^{-1} x$ is multiple valued because there is a positive and negative value associated with a given value of $x \geq 1$. Select the positive value for the upper branch of the $\cosh^{-1} x$ function.

Observe that if $y = \sinh^{-1} x$, then $\sinh y = -x$, and using $\sinh(-y) = -\sinh(y) = -(-x) = x$, there results $y = -\sinh^{-1}(x) = \sinh^{-1}(-x)$. Similarly,

$$\sinh^{-1}(-x) = -\sinh^{-1} x, \quad \cosh^{-1}(-x) = \cosh^{-1}(x), \quad \tanh^{-1}(-x) = -\tanh^{-1}(x)$$
$$\operatorname{csch}^{-1} x = \sinh^{-1}(1/x) \quad \operatorname{sech}^{-1} x = \cosh^{-1}(1/x) \quad \coth^{-1} x = \tanh^{-1}(1/x)$$
(2.114)

The inverse hyperbolic functions can be represented in terms of the logarithm function. If $x = \sinh y = \frac{e^y - e^{-y}}{2}$, then this equation can be written as the quadratic equation $e^{2y} - 2xe^y - 1 = 0$ in the variable $e^y$. Solve this equation for $e^y$ and show

$$e^y = \frac{2x \pm \sqrt{4x^2 + 4}}{2} = x \pm \sqrt{x^2 + 1} \tag{2.115}$$

Select the + sign because $e^y$ is always positive. Solving the equation (2.115) for $y$ gives the result

$$y = \sinh^{-1} x = \log(x + \sqrt{x^2 + 1}) \tag{2.116}$$

If $y = \cosh^{-1} x$, then $x = \cosh y = \frac{e^y + e^{-y}}{2}$ and multiplying this equation by $e^y$ produces the quadratic equation

$$e^{2y} - 2xe^y + 1 = 0$$

Solving for $e^y$ gives

$$e^y = x \pm \sqrt{x^2 - 1}$$

and solving for $y$ there results $y = \log(x \pm \sqrt{x^2 - 1})$. Here there are two values for the logarithm function, namely $\log(x + \sqrt{x^2 - 1})$ and $\log(x - \sqrt{x^2 - 1})$. These two values correspond to the upper and lower branches of the inverse hyperbolic cosine function illustrated in the figure 2-16. Note that

$$(x + \sqrt{x^2 - 1})(x - \sqrt{x^2 - 1}) = 1$$
$$\text{and} \quad \log(x + \sqrt{x^2 - 1}) + \log(x - \sqrt{x^2 - 1}) = \log 1 = 0$$
$$\text{and consequently} \quad \log(x - \sqrt{x^2 - 1}) = -\log(x + \sqrt{x^2 - 1})$$

which shows the symmetry of the $\cosh^{-1} x$ function. In a similar fashion it is possible to express the other hyperbolic functions in terms of logarithms to obtain

$$\sinh^{-1} x = \log(x + \sqrt{x^2 + 1}), \quad -\infty < x < \infty$$
$$\cosh^{-1} x = \log(x + \sqrt{x^2 - 1}), \quad x \geq 1$$
$$\tanh^{-1} x = \frac{1}{2} \log\left(\frac{1+x}{1-x}\right), \quad -1 < x < 1$$
$$\coth^{-1} x = \frac{1}{2} \log\left(\frac{x+1}{x-1}\right), \quad x > 1 \text{ or } x < -1$$
$$\operatorname{sech}^{-1} x = \log\left(\frac{1}{x} + \sqrt{\frac{1}{x^2} - 1}\right), \quad 0 < x \leq 1$$
$$\operatorname{csch}^{-1} x = \log\left(\frac{1}{x} + \sqrt{\frac{1}{x^2} + 1}\right), \quad x \neq 0$$
(2.117)

## Complex Numbers

Complex numbers are written in the form $z_0 = \alpha + i\beta$ where $\alpha$ and $\beta$ are real numbers and $i$ is called an imaginary unit having the property that $i^2 = -1$. Complex variables are written in the form $z = x + iy$ where $x$ and $y$ are real numbers, where $x$ is called the real part of $z$, written $x = \text{Re}\{z\}$ and $y$ is called the imaginary part of $z$ and is written $y = \text{Im}\{z\}$. Complex numbers are represented graphical by plotting number pairs $(x, y)$ in a plane called the $z$-plane. The polar form of a complex number is obtained by replacing $x$ and $y$ in terms of $r$ and $\theta$. The distance of $z$ from the origin is called the modulus of the complex number and is written $|z| = r$. The angle $\theta$ is called the argument or amplitude of the complex number. Note that $\theta$ is not unique as multiples of $2\pi$ can be added or subtracted from the angle $\theta$ to obtain $\theta \pm k2\pi$, for $k$ some integer. Complex numbers have the following properties.

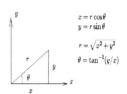

| Property | Operations |
|---|---|
| Equality | $z_1 = a + ib = z_2 = c + id$ if and only if $a = c$ and $b = d$ |
| Addition | $(a + ib) + (c + id) = (a + c) + i(b + d)$ |
| Subtraction | $(a + ib) - (c + id) = (a - c) + i(b - d)$ |
| Multiplication | $(a + ib)(c + id) = (ac - bd) + i(ad + bc)$ |
| Conjugate | If $z_1 = a + ib$, the conjugate of $z_1$ is denoted $\overline{z}_1 = a - ib$ |
| Division | $\dfrac{z_1}{z_2} = \dfrac{a + ib}{c + id} = \dfrac{(a + ib)(c - id)}{(c + id)(c - id)} = \dfrac{ac + bd}{c^2 + d^2} + i\dfrac{bc - ad}{c^2 + d^2}, \; z_2 \neq 0$ <br> Note division is multiplication of the numerator and denominator by the conjugate of denominator $c - id$ |
| Polar representation | $z = x + iy = r(\cos\theta + i\sin\theta) = re^{i\theta}, \quad e^{i\theta} = \cos\theta + i\sin\theta$ |
| Polar multiplication | $r_1(\cos\theta_1 + i\sin\theta_1)r_2(\cos\theta_2 + i\sin\theta_2) = r_1 r_2 (\cos(\theta_1 + \theta_2) + i\sin(\theta_1 + \theta_2))$ |
| Polar division | $\dfrac{r_1(\cos\theta_1 + i\sin\theta_1))}{r_2(\cos\theta_2 + i\sin\theta_2)} = \dfrac{r_1}{r_2}[\cos(\theta_1 - \theta_2) + i\sin(\theta_1 - \theta_2)]$ |
| DeMoivre's Theorem | $[r(\cos\theta + i\sin\theta)]^n = r^n(\cos n\theta + i\sin n\theta)$ |
| Roots of complex numbers | $[r(\cos\theta + i\sin\theta)]^{1/n} = r^{1/n}\left[\cos\left(\frac{\theta + 2m\pi}{n}\right) + i\sin\left(\frac{\theta + 2m\pi}{n}\right)\right]$, <br> where $m$ can have any of the integer values $\quad m = 0, 1, 2, 3, \ldots, n-1$ |

**Riemannian Sphere** A sphere, called the Riemannian[1] sphere, with diameter of unit length is placed on the $(x, y)$-plane with the South pole of the sphere tangent at the origin. Construct a straight line from a point $z = x + iy$ in the $(x, y)$-plane to the North pole of the sphere. This line intersects the sphere at the point $(\xi, \eta, \zeta)$. The coordinates $(x, y)$ and $(\xi, \eta, \zeta)$ are related as follows.

---

[1] Georg Friedrich Bernhard Riemann (1826-1866) German mathematician.

$(a)$ $\xi^2 + \eta^2 + (\zeta - 1/2)^2 = (1/2)^2$

$(b)$ $x = \dfrac{\xi}{1-\zeta}, \quad y = \dfrac{\eta}{1-\zeta}$

$(c)$ $\zeta = \dfrac{x^2 + y^2}{x^2 + y^2 + 1}$

$(d)$ $\xi = \dfrac{x}{x^2 + y^2 + 1}$

$(e)$ $\eta = \dfrac{y}{x^2 + y^2 + 1}$

$(f)$ $|z|^2 = z\bar{z} = \dfrac{\xi^2 + \eta^2}{(1-\zeta)^2}$

$(g)$ $|z|^2 = \dfrac{\zeta}{1-\zeta}$

$(h)$ $\zeta = \dfrac{|z|^2}{1+|z|^2}$

$(i)$ $\xi = \dfrac{x}{1+|z|^2}$

$(j)$ $\eta = \dfrac{y}{1+|z|^2}$

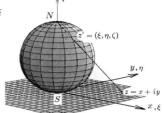

## Summary of Properties of Trigonometric and Hyperbolic Functions

### Pythagorean identities

$x^2 + y^2 = r^2$

$\left(\dfrac{x}{r}\right)^2 + \left(\dfrac{y}{r}\right)^2 = 1, \quad 1 + \left(\dfrac{y}{x}\right)^2 = \left(\dfrac{r}{x}\right)^2, \quad \left(\dfrac{x}{y}\right)^2 + 1 = \left(\dfrac{r}{y}\right)^2,$

$\cos^2\theta + \sin^2\theta = 1, \quad 1 + \tan^2\theta = \sec^2\theta, \quad \cot^2\theta + 1 = \csc^2\theta,$

### Addition Formulas

$\sin(A+B) = \sin A \cos B + \cos A \sin B \qquad \sin(A-B) = \sin A \cos B - \cos A \sin B$

$\cos(A+B) = \cos A \cos B - \sin A \sin B \qquad \cos(A-B) = \cos A \cos B + \sin A \sin B$

$\tan(A+B) = \dfrac{\tan A + \tan B}{1 - \tan A \tan B} \qquad \tan(A-B) = \dfrac{\tan A - \tan B}{1 + \tan A \tan B}$

### Double angle formulas

$\sin 2A = 2\sin A \cos A = \dfrac{2\tan A}{1 + \tan^2 A}$

$\cos 2A = \cos^2 A - \sin^2 A = 1 - 2\sin^2 A = 2\cos^2 A - 1 = \dfrac{1 - \tan^2 A}{1 + \tan^2 A}$

$\tan 2A = \dfrac{2\tan A}{1 - \tan^2 A} = \dfrac{2\cot A}{\cot^2 A - 1}$

### Half angle formulas

Sign depends upon quadrant $A/2$ lies in

$\sin\dfrac{A}{2} = \pm\sqrt{\dfrac{1-\cos A}{2}}$

$\cos\dfrac{A}{2} = \pm\sqrt{\dfrac{1+\cos A}{2}}$

$\tan\dfrac{A}{2} = \pm\sqrt{\dfrac{1-\cos A}{1+\cos A}} = \dfrac{\sin A}{1+\cos A} = \dfrac{1-\cos A}{\sin A}$

## Multiple angle formulas

$\sin 3A = 3\sin A - 4\sin^3 A$

$\cos 3A = 4\cos^3 A - 3\cos A$

$\tan 3A = \dfrac{3\tan A - \tan^3 A}{1 - 3\tan^2 A}$

$\sin 4A = 4\sin A \cos A - 8\sin^3 A \cos A$

$\cos 4A = 8\cos^4 A - 8\cos^2 A + 1$

$\tan 4A = \dfrac{4\tan A - 4\tan^3 A}{1 - 6\tan^2 A + \tan^4 A}$

$\sin 5A = 5\sin A - 20\sin^3 A + 16\sin^5 A$

$\cos 5A = 16\cos^5 A - 20\cos^3 A + 5\cos A$

$\tan 5A = \dfrac{\tan^5 A - 10\tan^3 A + 5\tan A}{1 - 10\tan^2 A + 5\tan^4 A}$

$\sin 6A = 6\cos^5 A \sin A - 20\cos^3 A \sin^3 A + 6\cos A \sin^5 A$

$\cos 6A = \cos^6 A - 15\cos^4 A \sin^2 A + 15\cos^2 A \sin^4 A - \sin^6 A$

$\tan 6A = \dfrac{6\tan A - 20\tan^3 A + 6\tan^5 A}{1 - 15\tan^2 A + 15\tan^4 A - \tan^6 A}$

## General sine and cosine multiple angle formula

$$\sin n\theta = \sum_{k=0}^{n} \binom{n}{k} \cos^k \theta \sin^{n-k} \theta \sin\left((n-k)\dfrac{\pi}{2}\right)$$

$$\cos n\theta = \sum_{k=0}^{n} \binom{n}{k} \cos^k \theta \sin^{n-k} \theta \cos\left((n-k)\dfrac{\pi}{2}\right)$$

Recurrence relation for the tangent function is given by

$$\tan(n+1)\theta = \dfrac{\tan n\theta + \tan\theta}{1 - \tan n\theta \tan\theta}$$

## Summation and difference formula

$\sin A + \sin B = 2\sin(\dfrac{A+B}{2})\cos(\dfrac{A-B}{2})$

$\cos A + \cos B = 2\cos(\dfrac{A+B}{2})\cos(\dfrac{A-B}{2})$

$\tan A + \tan B = \dfrac{\sin(A+B)}{\cos A \cos B}$

$\sin A - \sin B = 2\sin(\dfrac{A-B}{2})\cos(\dfrac{A+B}{2})$

$\cos A - \cos B = -2\sin(\dfrac{A-B}{2})\sin(\dfrac{A+B}{2})$

$\tan A - \tan B = \dfrac{\sin(A-B)}{\cos A \cos B}$

## Product formula

$\sin A \sin B = \dfrac{1}{2}\cos(A-B) - \dfrac{1}{2}\cos(A+B)$

$\cos A \cos B = \dfrac{1}{2}\cos(A-B) + \dfrac{1}{2}\cos(A+B)$

$\sin A \cos B = \dfrac{1}{2}\sin(A-B) + \dfrac{1}{2}\sin(A+B)$

## Additional relations

$\sin(A+B)\sin(A-B) = \sin^2 A - \sin^2 B$

$-\sin(A+B)\sin(A-B) = \cos^2 A - \cos^2 B$

$\cos(A+B)\cos(A-B) = \cos^2 A - \sin^2 B$

$\dfrac{\sin A \pm \sin B}{\cos A + \cos B} = \tan(\dfrac{A \pm B}{2})$

$\dfrac{\sin A \pm \sin B}{\cos A - \cos B} = -\cot(\dfrac{A \mp B}{2})$

$\dfrac{\sin A + \sin B}{\sin A - \sin B} = \dfrac{\tan(\dfrac{A+B}{2})}{\tan(\dfrac{A-B}{2})}$

## Powers of trigonometric functions

$$\sin^2 A = \frac{1}{2} - \frac{1}{2}\cos 2A \qquad \cos^2 A = \frac{1}{2} + \frac{1}{2}\cos 2A$$

$$\sin^3 A = \frac{3}{4}\sin A - \frac{1}{4}\sin 3A \qquad \cos^3 A = \frac{3}{4}\cos A + \frac{1}{4}\cos 3A$$

$$\sin^4 A = \frac{3}{8} - \frac{1}{2}\cos 2A + \frac{1}{8}\cos 4A \qquad \cos^4 A = \frac{3}{8} + \frac{1}{2}\cos 2A + \frac{1}{8}\cos 4A$$

## Principal values for inverse trigonometric functions

| $x < 0$ | $x \geq 0$ |
|---|---|
| $-\pi/2 \leq \sin^{-1} x < 0$ | $0 \leq \sin^{-1} x \leq \pi/2$ |
| $\pi/2 < \cos^{-1} x \leq \pi$ | $0 \leq \cos^{-1} x \leq \pi/2$ |
| $-\pi/2 < \tan^{-1} x < 0$ | $0 \leq \tan^{-1} x < \pi/2$ |
| $\pi/2 < \cot^{-1} x < \pi$ | $0 < \cot^{-1} x \leq \pi/2$ |
| $\pi/2 < \sec^{-1} x \leq \pi$ | $0 \leq \sec^{-1} x < \pi/2$ |
| $-\pi/2 \leq \csc^{-1} x < 0$ | $0 < \csc^{-1} x \leq \pi/2$ |

$x = 0 \longrightarrow x$

## Properties of inverse trigonometric functions

$$\sin^{-1}(-x) = -\sin^{-1} x \qquad \csc^{-1} x = \sin^{-1}(1/x)$$

$$\cos^{-1}(-x) = \pi - \cos^{-1} x \qquad \sec^{-1} x = \cos^{-1}(1/x)$$

$$\tan^{-1}(-x) = -\tan^{-1} x \qquad \cot^{-1} x = \tan^{-1}(1/x)$$

$$\cot^{-1}(-x) = \pi - \cot^{-1} x \qquad \sin^{-1} x + \cos^{-1} x = \pi/2$$

$$\sec^{-1}(-x) = \pi - \sec^{-1} x \qquad \tan^{-1} x + \cot^{-1} x = \pi/2$$

$$\csc^{-1}(-x) = -\csc^{-1} x \qquad \sec^{-1} x + \csc^{-1} x = \pi/2$$

## Symmetry properties of trigonometric functions

$$\sin\theta = -\sin(-\theta) = \cos(\pi/2 - \theta) = -\cos(\pi/2 + \theta) = +\sin(\pi - \theta) = -\sin(\pi + \theta)$$

$$\cos\theta = +\cos(-\theta) = \sin(\pi/2 - \theta) = +\sin(\pi/2 + \theta) = -\cos(\pi - \theta) = -\cos(\pi + \theta)$$

$$\tan\theta = -\tan(-\theta) = \cot(\pi/2 - \theta) = -\cot(\pi/2 + \theta) = -\tan(\pi - \theta) = +\tan(\pi + \theta)$$

$$\cot\theta = -\cot(-\theta) = \tan(\pi/2 - \theta) = -\tan(\pi/2 + \theta) = -\cot(\pi - \theta) = +\cot(\pi + \theta)$$

$$\sec\theta = +\sec(-\theta) = \csc(\pi/2 - \theta) = +\csc(\pi/2 + \theta) = -\sec(\pi - \theta) = -\sec(\pi + \theta)$$

$$\csc\theta = -\csc(-\theta) = \sec(\pi/2 - \theta) = +\sec(\pi/2 + \theta) = +\csc(\pi - \theta) = -\csc(\pi + \theta)$$

## Transformations

If $\tan\dfrac{u}{2} = A$, then

$$\sin u = \frac{2A}{1+A^2}, \qquad \cos u = \frac{1-A^2}{1+A^2}, \qquad \tan u = \frac{2A}{1-A^2}$$

The transformation $\sin A = u$, requires $\cos A = \sqrt{1 - u^2}$, and $\tan A = \dfrac{u}{\sqrt{1 - u^2}}$

The same idea holds for hyperbolic functions. The transformation $\sinh x = u$, requires $\cosh x = \sqrt{1 + u^2}$, and $\tanh x = \dfrac{u}{\sqrt{1 + u^2}}$.

## Euler identities

$$e^{i\theta} = \cos\theta + i\sin\theta, \qquad e^{-i\theta} = \cos\theta - i\sin\theta$$

## Exponential and trigonometric relations

$$\sin\theta = \frac{e^{i\theta} - e^{-i\theta}}{2i}$$

$$\cos\theta = \frac{e^{i\theta} + e^{-i\theta}}{2}$$

$$\tan\theta = \frac{e^{i\theta} - e^{-i\theta}}{i(e^{i\theta} + e^{-i\theta})} = -i\left(\frac{e^{i\theta} - e^{-i\theta}}{e^{i\theta} + e^{-i\theta}}\right)$$

$$\cot\theta = i\left(\frac{e^{i\theta} + e^{-i\theta}}{e^{i\theta} - e^{-i\theta}}\right)$$

$$\sec\theta = \frac{2}{e^{i\theta} + e^{-i\theta}}$$

$$\csc\theta = \frac{2i}{e^{i\theta} - e^{-i\theta}}$$

## Hyperbolic functions

$$\sinh A = \frac{e^A - e^{-A}}{2}$$

$$\cosh A = \frac{e^A + e^{-A}}{2}$$

$$\tanh A = \frac{e^A - e^{-A}}{e^A + e^{-A}}$$

$$\operatorname{csch} A = \frac{2}{e^A - e^{-A}}$$

$$\operatorname{sech} A = \frac{2}{e^A + e^{-A}}$$

$$\coth A = \frac{e^A + e^{-A}}{e^A - e^{-A}}$$

The above definitions satisfy the following properties.

$$\cosh^2 A - \sinh^2 A = 1$$
$$\operatorname{sech}^2 A + \tanh^2 A = 1$$
$$\coth^2 A - \operatorname{csch}^2 A = 1$$

$$\sinh A = \frac{1}{\operatorname{csch} A}$$
$$\cosh A = \frac{1}{\operatorname{sech} A}$$
$$\tanh A = \frac{1}{\coth A}$$
$$\tanh A = \frac{\sinh A}{\cosh A}$$

## Symmetry properties

$$\sinh(-A) = -\sinh A \qquad \operatorname{csch}(-A) = -\operatorname{csch} A$$
$$\cosh(-A) = \cosh A \qquad \operatorname{sech}(-A) = \operatorname{sech} A$$
$$\tanh(-A) = -\tanh A \qquad \coth(-A) = -\coth A$$

## Sum and difference relations

$$\sinh(A + B) = \sinh A \cosh B + \cosh A \sinh B$$
$$\cosh(A + B) = \cosh A \cosh B + \sinh A \sinh B$$
$$\tanh(A + B) = \frac{\tanh A + \tanh B}{1 + \tanh A \tanh B}$$

$$\sinh(A - B) = \sinh A \cosh B - \cosh A \sinh B$$
$$\cosh(A - B) = \cosh A \cosh B - \sinh A \sinh B$$
$$\tanh(A - B) = \frac{\tanh A - \tanh B}{1 - \tanh A \tanh B}$$

### Double argument formula

$$\sinh 2A = 2\sinh A \cosh A$$
$$\cosh 2A = \cosh^2 A + \sinh^2 A = 2\cosh^2 A - 1 = 1 + 2\sinh^2 A$$
$$\tanh 2A = \frac{2\tanh A}{1+\tanh^2 A}$$

### Half argument formulas

$$\sinh \frac{A}{2} = \pm\sqrt{\frac{\cosh A - 1}{2}} \quad \begin{cases} + \text{ for } A > 0 \\ - \text{ for } A < 0 \end{cases}$$

$$\cosh \frac{A}{2} = \sqrt{\frac{\cosh A + 1}{2}}$$

$$\tanh \frac{A}{2} = \pm\sqrt{\frac{\cosh A - 1}{\cosh A + 1}} = \frac{\sinh A}{\cosh A + 1} = \frac{\cosh A - 1}{\sinh A} \quad \begin{cases} + \text{ for } A > 0 \\ - \text{ for } A < 0 \end{cases}$$

### Multiple argument formula

$$\sinh 3A = 3\sinh A + 4\sinh^3 A \qquad \sinh 4A = 8\sinh^3 A \cosh A + 4\sinh A \cosh A$$
$$\cosh 3A = 4\cosh^3 A - 3\cosh A \qquad \cosh 4A = 8\cosh^4 A - 8\cosh^2 A + 1$$
$$\tanh 3A = \frac{3\tanh A + \tanh^3 A}{1 + 3\tanh^2 A} \qquad \tanh 4A = \frac{4\tanh A + 4\tanh^3 A}{1 + 6\tanh^2 A + \tanh^4 A}$$

### Sum and difference formula

$$\sinh A + \sinh B = 2\sinh(\tfrac{A+B}{2})\cosh(\tfrac{A-B}{2}) \qquad \sinh A - \sinh B = 2\cosh(\tfrac{A+B}{2})\sinh(\tfrac{A-B}{2})$$
$$\cosh A + \cosh B = 2\cosh(\tfrac{A+B}{2})\cosh(\tfrac{A-B}{2}) \qquad \cosh A - \cosh B = 2\sinh(\tfrac{A+B}{2})\sinh(\tfrac{A-B}{2})$$

### Product formula

$$\sinh A \sinh B = \tfrac{1}{2}\cosh(A+B) - \tfrac{1}{2}\cosh(A-B)$$
$$\sinh A \cosh B = \tfrac{1}{2}\sinh(A+B) + \tfrac{1}{2}\sinh(A-B)$$
$$\cosh A \cosh B = \tfrac{1}{2}\cosh(A+B) + \tfrac{1}{2}\cosh(A-B)$$

### Power formula

$$\sinh^2 A = \tfrac{1}{2}\cosh 2A - \tfrac{1}{2} \qquad \sinh^3 A = \tfrac{1}{4}\sinh 3A - \tfrac{3}{4}\sinh A$$
$$\cosh^2 A = \tfrac{1}{2}\cosh 2A + \tfrac{1}{2} \qquad \cosh^3 A = \tfrac{1}{4}\cosh 3A + \tfrac{3}{4}\cosh A$$

## Inverse hyperbolic functions

$$\sinh^{-1} A = \ln(A + \sqrt{A^2 + 1}), \quad -\infty < A < \infty$$

$$\cosh^{-1} A = \ln(A + \sqrt{A^2 - 1}), \quad A \geq 1 \text{ with } \cosh^{-1} A > 0$$

$$\tanh^{-1} A = \frac{1}{2}\ln\left(\frac{1+A}{1-A}\right), \quad -1 < A < 1$$

$$\operatorname{csch}^{-1} A = \ln\left(\frac{1}{x} + \sqrt{\frac{1}{A^2} + 1}\right), \quad A \neq 0$$

$$\operatorname{sech}^{-1} A = \ln\left(\frac{1}{A} + \sqrt{\frac{1}{A^2} - 1}\right), \quad 0 < A \leq 1, \; \operatorname{sech}^{-1} A > 0$$

$$\coth^{-1} A = \frac{1}{2}\ln\left(\frac{A+1}{A-1}\right), \quad A > 1 \text{ or } A < -1$$

## Relations between hyperbolic functions

$$\sinh^{-1} A = \operatorname{csch}^{-1}(1/A) \qquad \sinh^{-1}(-A) = -\sinh^{-1} A$$

$$\cosh^{-1} A = \operatorname{sech}^{-1}(1/A) \qquad \cosh^{-1}(-A) = \pi i - \cosh^{-1} A$$

$$\tanh^{-1} A = \coth^{-1}(1/A) \qquad \tanh^{-1}(-A) = -\tanh^{-1} A$$

## Periodic properties

$$\sinh(A + 2n\pi i) = \sinh A, \qquad \cosh(A + 2n\pi i) = \cosh A, \qquad \tanh(A + n\pi i) = \tanh A$$

where $n$ is an integer.

## Relations between hyperbolic and trigonometric functions

$$\sin(iA) = i\sinh A \qquad \sinh(iA) = i\sin A$$

$$\cos(iA) = \cosh A \qquad \cosh(iA) = \cos A \qquad i^2 = -1$$

$$\tan(iA) = i\tanh A \qquad \tanh(iA) = i\tan A$$

## Relations between inverse hyperbolic and inverse trigonometric functions

$$\sin^{-1}(iA) = i\sinh^{-1} A \qquad \sinh^{-1}(iA) = i\sin^{-1} A$$

$$\cos^{-1}(A) = \pm i\cosh^{-1} A \qquad \cosh^{-1} A = \pm i\cos A \qquad i^2 = -1$$

$$\tan^{-1}(ix) = i\tanh^{-1}(A) \qquad \tanh^{-1}(iA) = i\tan^{-1} A$$

## Multiple valued functions

The inverse functions $\arcsin x$, $\arccos x$ and $\arctan x$ are in general multiple-valued functions and so the solutions to the equations

$$\sin x = A, \qquad \cos x = A, \qquad \tan x = A$$

are given by the relations

$$x = (-1)^n \arcsin A + n\pi, \qquad x = \pm \arccos A + 2n\pi, \qquad x = \arctan A + n\pi, \quad A \text{ real}$$

where $n$ is an arbitrary integer. To obtain single-valued functions for the inverse trigonometric functions one must use the principal values associated with these functions.

# Chapter 3
# Geometry

The following is a review of selected material from the subject areas of plane, solid and analytic geometry.

**Classification of triangles**

A triangle is a three sided polygon or a polygon with three vertices and is usually called by one of the names given in the figure 3-1.

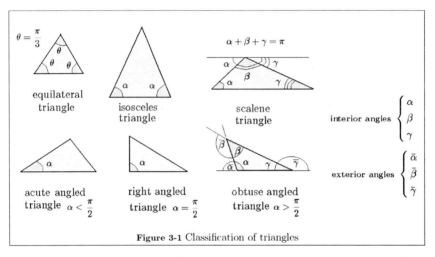

Figure 3-1 Classification of triangles

An equilateral triangle has equal angles and equal sides. An isosceles triangle has two equal angles and two equal sides. A scalene triangle has unequal angles and unequal sides. Triangles can also be called acute angled if all angles are less than $\pi/2$ radians, called right angled if one angle is $\pi/2$ radians, or called obtuse if one of the angles is greater than $\pi/2$ radians. From the scalene triangle in figure 3-1 it is an easy exercise to show the sum of the angles of a triangle is $\pi$ radians.

**Similar Triangles**

Many times angles are denoted by capital letters and the corresponding lower case letter is used for the length of the side opposite that angle. For example, in the figure 3-2 $A$ is used to denote the angle $\angle A$ with $a$ used to represent the length of the side opposite that angle. The same notation with primes (') is sometimes used to denote similar triangles.

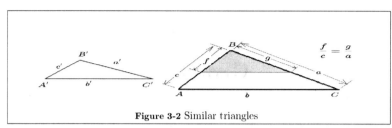

**Figure 3-2** Similar triangles

Recall that two triangles $ABC$ and $A'B'C'$ are called similar if their angles satisfy $\angle A = \angle A'$, $\angle B = \angle B'$, and $\angle C = \angle C'$. The mathematical notation $\triangle ABC \sim \triangle A'B'C'$ is used to denote similarity between two triangles. For any triangle the sum of the interior angles must equal 180° or $\pi$ radians. Consequently, if two angles of triangle $ABC$ are respectively equal to two angles from triangle $A'B'C'$, then the triangles must be similar. If the two triangles $ABC$ and $A'B'C'$ are similar, then the ratio of the sides of the two triangles are proportional to one another and consequently

$$\frac{a}{b} = \frac{a'}{b'}, \quad \frac{b}{c} = \frac{b'}{c'}, \quad \frac{a}{c} = \frac{a'}{c'} \tag{3.1}$$

Alternatively, $\frac{a}{a'} = \frac{b}{b'} = \frac{c}{c'}$.

The perimeter of both triangles are in the same ratio as their sides and this ratio is called the scale factor associated with the two triangles. That is,

$$\frac{a}{a'} = \frac{b}{b'} = \frac{c}{c'} = \frac{a+b+c}{a'+b'+c'} = \text{scale factor} \tag{3.2}$$

The ratio of the areas of two similar triangles equals the ratio of the squares of the corresponding sides, giving

$$\frac{\text{Area}\,ABC}{\text{Area}\,A'B'C'} = \frac{a^2}{a'^2} = \frac{b^2}{b'^2} = \frac{c^2}{c'^2} \tag{3.3}$$

## Congruent Triangles

Two triangles $\triangle ABC$ and $\triangle A'B'C'$ are congruent, written $\triangle ABC \cong \triangle A'B'C'$ if (i) three sides are equal (SSS) (iii) two sides and the included angle are equal (SAS) (iii) two sides and the angle opposite the longer side are equal (SSA) (iv) one side and two angles are equal (ASA) or (SAA). If two triangles are congruent, then one triangle can be superposed onto the other by (i) translation, (ii) a rotation, (iii) inversion (iv) reflection or (v) any combination of the previous operations.

## Inequalities

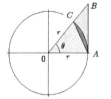

$$\text{area of triangle } 0AC < \text{area of sector } 0\widehat{AC} < \text{area of triangle } 0AB$$

$$\frac{1}{2}r^2 \sin\theta < \frac{1}{2}r^2\theta < \frac{1}{2}r^2 \tan\theta$$

$$\sin\theta < \theta < \tan\theta$$

## Golden Ratio

Divide a line segment into two parts labeled $a$ and $b$, with say $a$ the longer segment. The golden ratio $\frac{a}{b}$ occurs when the ratio of the total length to the longer side is in the same ratio as the longer side to the shorter side or

$$\frac{a+b}{a} = \frac{a}{b} \implies a^2 - ab - b^2 = 0 \implies \frac{a}{b} = \frac{1+\sqrt{5}}{2} \approx 1.61803$$

The golden ratio is called by many other names. For example, golden mean or golden proportion. This ratio can be found in many architectural designs, art work and has been used in the construction of numerous statues.

## Medians and Perpendicular Bisectors

Consider a general triangle ABC as illustrated below. Construct straight lines joining the midpoints of each side to the opposite vertices of the triangle. These lines are known as median lines for the triangle. The lengths of the medians of the triangle are

$$\ell_a = \frac{1}{2}\sqrt{2(b^2+c^2)-a^2}, \qquad \ell_b = \frac{1}{2}\sqrt{2(a^2+c^2)-b^2}, \qquad \ell_c = \frac{1}{2}\sqrt{2(b^2+a^2)-c^2}$$

The three medians intersect at a common point P known as the center of gravity of the triangle. The point P divides the medians in the ratio 2:1.

Take an arbitrary triangle ABC and move to the midpoint of each side and construct the perpendicular bisector. These perpendicular bisectors intersect at a point $P^*$. The distance of the point $P^*$ from each vertex of the triangle is the same and can be found from the relation $r = \frac{abc}{4A}$ which represents the product of the sides divided by four times the area A. This means it is possible to construct a circle centered at $P^*$ with radius $r$ which passes through each of the vertices of the triangle.

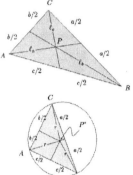

### Angle bisectors

The straight lines which bisect the interior angles of a triangle have the following lengths.

$$\overline{AA'} = \frac{\sqrt{bc[(b+c)^2 - a^2]}}{b+c}$$

$$\overline{BB'} = \frac{\sqrt{ca[(c+a)^2 - b^2]}}{c+a}$$

$$\overline{CC'} = \frac{\sqrt{ab[(a+b)^2 - c^2]}}{a+b}$$

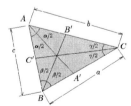

## Law of sines for triangle ABC

$$\frac{a}{\sin A} = \frac{b}{\sin B} = \frac{c}{\sin C} \quad (3.4)$$

## Law of cosines for triangle ABC

$$c^2 = a^2 + b^2 - 2ab\cos C$$
$$a^2 = b^2 + c^2 - 2bc\cos A \quad (3.5)$$
$$b^2 = c^2 + a^2 - 2ac\cos B$$

## Law of tangents for triangle ABC

$$\frac{a+b}{a-b} = \frac{\tan\frac{1}{2}(A+B)}{\tan\frac{1}{2}(A-B)}$$
$$\frac{b+c}{b-c} = \frac{\tan\frac{1}{2}(B+C)}{\tan\frac{1}{2}(B-C)} \quad (3.6)$$
$$\frac{c+a}{c-a} = \frac{\tan\frac{1}{2}(C+A)}{\tan\frac{1}{2}(C-A)}$$

## Area of triangle ABC

$$Area = \sqrt{s(s-a)(s-b)(s-c)}$$
where $s = \frac{1}{2}(a+b+c)$ is the semiperimeter $\quad (3.7)$
$$Area = \frac{1}{2}bh$$

### half-angle formulas for triangle

$$\sin^2\frac{A}{2} = \frac{(s-b)(s-c)}{bc}, \quad \sin^2\frac{B}{2} = \frac{(s-a)(s-c)}{ac}, \quad \sin^2\frac{C}{2} = \frac{(s-a)(s-b)}{ab},$$
$$\cos^2\frac{A}{2} = \frac{s(s-a)}{bc}, \quad \cos^2\frac{B}{2} = \frac{s(s-b)}{ac}, \quad \cos^2\frac{C}{2} = \frac{s(s-c)}{ab},$$
$$\tan^2\frac{A}{2} = \frac{(s-b)(s-c)}{s(s-a)}, \quad \tan^2\frac{B}{2} = \frac{(s-a)(s-c)}{s(s-b)}, \quad \tan^2\frac{C}{2} = \frac{(s-a)(s-b)}{s(s-c)}$$

where $s = \frac{1}{2}(a+b+c)$ is the semiperimeter of the triangle.

### Additional relations for above triangle

$$b = a\cos C + c\cos A, \quad \frac{a+b}{c} = \frac{\cos\frac{1}{2}(B-A)}{\cos\frac{1}{2}(B+A)}, \quad \frac{b-a}{c} = \frac{\sin\frac{1}{2}(B-A)}{\sin\frac{1}{2}(B+A)}$$

$$\sin A + \sin B + \sin C = 4\cos\frac{A}{2}\cos\frac{B}{2}\cos\frac{C}{2} \qquad \sin^2 A + \sin^2 B + \sin^2 C = 2(1 + \cos A\cos B\cos C)$$
$$\cos A + \cos B + \cos C = 1 + 4\sin\frac{A}{2}\sin\frac{B}{2}\sin\frac{C}{2} \qquad \cos^2 A + \cos^2 B + \cos^2 C = 1 - 2\cos A\cos B\cos C$$
$$\tan A + \tan B + \tan C = \tan A\tan B\tan C \qquad \cot A\cot B + \cot A\cot C + \cot B\cot C = 1$$

## Miscellaneous Properties of a Triangle

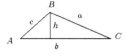

The sum of the angles of triangle is 180° or $\pi$ radians. The sum of the lengths of any two sides of a triangle is greater than the length of the third side. The difference between the lengths of any two sides of a triangle is smaller than the third side.

This gives the inequalities

$$|a-b| < c < a+b, \qquad |b-c| < a < b+c, \qquad |a-c| < b < a+c$$

## Two-dimensional rectangular coordinates $(x,y)$

Construct two straight lines $0x$ and $0y$ which intersect at right angles at the point 0 and then place distance scales on these lines, where distances are measured from the point 0 with positive in one direction and negative in the opposite direction for each line.

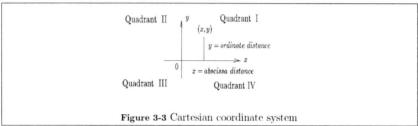

**Figure 3-3** Cartesian coordinate system

The line $0x$ is called the $x$-axis and the line $0y$ is called the $y$-axis. In plotting a point $P$, having coordinates $(x,y)$, the value of $x$ is called the abscissa or distance of point $P$ from the $y$-axis and the value $y$ is called the ordinate or distance of point $P$ from the $x$-axis. The abscissa and ordinate values are called the rectangular coordinates of the point $P$. Any system involving intersecting lines at right angles to specify a frame of reference is called a Cartesian coordinate system in honor of Rene Descartes (1596-1650).

## Translation and rotation of axes

Consider two sets of axes labeled $(x,y)$ and $(\bar{x},\bar{y})$ with origins 0 and $\bar{0}$ respectively as illustrated in the figures 3-4 (a),(b) and (c). A simple translation of axes occurs when the origin 0 of the $(x,y)$ coordinate system is moved to a point $(x_0, y_0)$ and then everything is relabeled with a bar over the symbols as illustrated in the figure 3-4(a). The relation between the $(x,y)$ and $(\bar{x},\bar{y})$ axes is then found to be

$$x = \bar{x} + x_0, \quad y = \bar{y} + y_0, \qquad \text{or} \qquad \bar{x} = x - x_0, \quad \bar{y} = y - y_0 \tag{3.8}$$

A rotation of axes occurs when the $(x, y)$ axes is rotated about the origin through an angle $\theta$ and then everything is re-labeled with a bar over the symbols as illustrated in the figure 3-4(b).

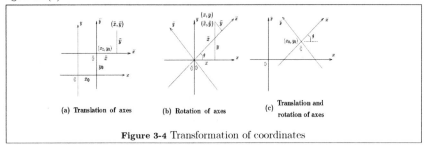

**Figure 3-4** Transformation of coordinates

The relation between the old set of rectangular axes and the new set of rotated axes is given by either set of equations

$$x = \bar{x}\cos\theta - \bar{y}\sin\theta \qquad \bar{x} = x\cos\theta + y\sin\theta$$
$$\text{or} \qquad (3.9)$$
$$y = \bar{x}\sin\theta + \bar{y}\cos\theta \qquad \bar{y} = y\cos\theta - \bar{x}\sin\theta$$

For the translation of axes to the point $(x_0, y_0)$ followed by a rotation of axes, as illustrated in the figure 3-4(c), the transformation of coordinates is given by either of the equations

$$x = \bar{x}\cos\theta - \bar{y}\sin\theta + x_0 \qquad \bar{x} = (x - x_0)\cos\theta + (y - y_0)\sin\theta$$
$$\text{or} \qquad (3.10)$$
$$y = \bar{x}\sin\theta + \bar{y}\cos\theta + y_0 \qquad \bar{y} = (y - y_0)\cos\theta - (x - x_0)\sin\theta$$

## Vectors

A vector can be thought of as a line segment with a direction specified by an arrowhead. A vector can be represented mathematically using boldface type, such as **V** or it can be represented by a symbol with an arrow over it such as $\vec{V}$. The length or magnitude of a vector $\vec{V}$ is denoted $|\vec{V}|$.

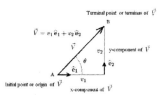

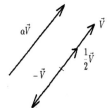

Vectors can be multiplied by a scalar. This has the effect of changing the length or magnitude of the vector. Multiplying a vector by $-1$ changes its direction by 180 degrees or $\pi$ radians. If $\alpha$ is a scalar, then the vector $\alpha\vec{V}$ is called a scalar multiple of $\vec{V}$. The length of the new vector is given by $|\alpha\vec{V}| = |\alpha||\vec{V}|$.

A vector starting at a point $A$ and ending with an arrowhead at the point $B$ has the following terminology associated with it. The point $A$ is called the origin or initial point of the vector $\vec{V}$ and the point $B$ is called the terminus or terminal point of the vector $V$. Vectors are called free vectors because it is possible to move a vector to some other position provided its length and direction are not changed. Movement of vector to some other point is usually done to emphasize some physical or geometric property of the vector system.

Two vectors $\vec{V}$ and $\vec{F}$ are added as follows. The vector sum $\vec{V}+\vec{F}$ is obtained by placing the origin of vector $\vec{F}$ at the terminus of the vector $\vec{V}$ and then joining the origin of $\vec{V}$ to the terminus point of $\vec{F}$ as illustrated. The vector sum $\vec{F}+\vec{V}$ is obtained by placing the origin of vector $\vec{V}$ at the terminus point of vector $\vec{F}$ and joining the origin of $\vec{F}$ to the terminus of vector $\vec{V}$ as illustrated.

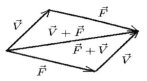

This is known as the parallelogram law for vector addition and shows that $\vec{V}+\vec{F}=\vec{F}+\vec{V}$. That is, vector addition is commutative under addition.

## Vector Components

A vector, with origin at 0 of a Cartesian coordinate system, can be resolved into its $x$-component and $y$-component by projecting the vector onto the $x$ and $y$ axes. The $x$-component is given by $v_1 = |\vec{V}|\cos\alpha$ and the $y$-component is given by $v_2 = |\vec{V}|\sin\alpha$ where $|\vec{V}|$ is the length or magnitude of the vector $\vec{V}$. By placing the origin of the vector $\vec{V}$ at the origin $(0,0)$ of the $x,y$ axes it is possible to determine the angle $\alpha$ between the vector and the $x$-axis. Note that,

$$\cos\alpha = \frac{v_1}{|\vec{V}|}, \qquad \sin\alpha = \frac{v_2}{|\vec{V}|}, \qquad \tan\alpha = \frac{v_2}{v_1}, \qquad |\vec{V}| = \sqrt{v_1^2 + v_2^2} \qquad (3.11)$$

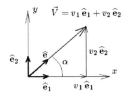

Define the unit vectors $\hat{e}_1$, $\hat{e}_2$ having unit length, which point in the directions of the $x$ and $y$ axes, then the vector $\vec{V}$ can be expressed in terms of its components as $v_1\hat{e}_1$ and $v_2\hat{e}_2$ by using vector addition to obtain $\vec{V} = v_1\hat{e}_1 + v_2\hat{e}_2$. Note $\vec{V} = |\vec{V}|\cos\alpha\,\hat{e}_1 + |\vec{V}|\sin\alpha\,\hat{e}_2 = |\vec{V}|(\cos\alpha\,\hat{e}_1 + \sin\alpha\,\hat{e}_2) = |\vec{V}|\hat{e}$ where $\hat{e}$ is a unit vector expressed in the direction of $\vec{V}$.

## Direction Cosines

Consider a vector $\vec{V} = v_1\hat{e}_1 + v_2\hat{e}_2$ and if necessary translate the origin of the vector $\vec{V}$ to the point $(0,0)$ of the $x,y$ coordinate system as illustrated above. A unit vector $\hat{e}$ in the direction of $\vec{V}$ is obtained by multiplying the vector $\vec{V}$ by the scale factor $\frac{1}{|\vec{V}|}$ where $|\vec{V}| = \sqrt{v_1^2 + v_2^2}$ is the magnitude or length of the vector $\vec{V}$. Scalar multiplication by the reciprocal of the vector magnitude produces the unit vector

$$\widehat{e} = \frac{1}{|\vec{V}|}\vec{V} = \frac{v_1}{\sqrt{v_1^2+v_2^2}}\,\widehat{e}_1 + \frac{v_2}{\sqrt{v_1^2+v_2^2}}\,\widehat{e}_2 \qquad (3.12)$$

Note that the angles $\alpha$ and $\beta$, that are formed between the vector $\vec{V}$ and the $x$ and $y$ axes, satisfy the relations

$$\cos\alpha = \frac{v_1}{\sqrt{v_1^2+v_2^2}}, \quad \text{and} \quad \cos\beta = \frac{v_2}{\sqrt{v_1^2+v_2^2}} = \sin\alpha \qquad (3.13)$$

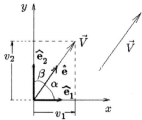

The angles $\alpha$ and $\beta$ are called the direction angles associated with the vector $\vec{V}$ and the quantities $\cos\alpha$ and $\cos\beta$ are called the direction cosines associated with the vector $\vec{V}$. The unit vector $\widehat{e}$ in the direction of $\vec{V}$ can then be expressed in terms of the direction cosines

$$\widehat{e} = \cos\alpha\,\widehat{e}_1 + \cos\beta\,\widehat{e}_2 = \cos\alpha\,\widehat{e}_1 + \sin\alpha\,\widehat{e}_2 \qquad (3.14)$$

This same terminology is used to define the direction angles and direction cosines of lines in both two-dimensional and three dimensional spaces. Note that the sum of the squares of the direction cosines must equal unity because $\widehat{e}$ was constructed as a unit vector.

In three dimensions, the vector components of the vector $\vec{V}$ are projections of the vector onto the $x, y$ and $z$-axes. These components are respectively,

$$|\vec{V}|\cos\alpha, \qquad |\vec{V}|\cos\beta, \qquad |\vec{V}|\cos\gamma$$

where $\alpha, \beta, \gamma$ are the direction angles that the vector makes with the $x, y$ and $z$-axes. The cosine of these angles $\cos\alpha, \cos\beta, \cos\gamma$ are called the direction cosines of the vector $\vec{V}$. In terms of its components or projections, the vector $\vec{V}$ can be expressed

$$\vec{V} = |\vec{V}|\,(\cos\alpha\,\widehat{e}_1 + \cos\beta\,\widehat{e}_2 + \cos\gamma\,\widehat{e}_3) = |\vec{V}|\,\widehat{e}$$

where $\widehat{e} = \cos\alpha\,\widehat{e}_1 + \cos\beta\,\widehat{e}_2 + \cos\gamma\,\widehat{e}_3$ is a unit vector in the direction of $\vec{V}$.

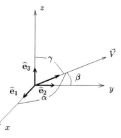

Any set of numbers $(a, b, c)$, not all zero, which are proportional to the direction cosines are called direction numbers of the vector $\vec{V}$. For example, given the vector $\vec{V} = a\,\widehat{e}_1 + b\,\widehat{e}_2 + c\,\widehat{e}_3$, the numbers $a, b, c$ determine the direction of the vector. By multiplying by $\frac{1}{\sqrt{a^2+b^2+c^2}} = \frac{1}{|\vec{V}|}$ there results the unit vector

$$\widehat{e} = \frac{1}{|\vec{V}|}\vec{V} = \cos\alpha\,\widehat{e}_1 + \cos\beta\,\widehat{e}_2 + \cos\gamma\,\widehat{e}_3$$

where
$$\cos\alpha = \frac{a}{\sqrt{a^2+b^2+c^2}}, \qquad \cos\beta = \frac{b}{\sqrt{a^2+b^2+c^2}}, \qquad \cos\gamma = \frac{c}{\sqrt{a^2+b^2+c^2}}$$
are the direction cosines associated with a unit vector in the direction $\vec{V}$.

## Properties of vectors

Let $\vec{A}, \vec{B}$ and $\vec{C}$ represent vectors and let $\alpha, \beta$ denote scalar quantities.

1. Two vectors $\vec{A}$ and $\vec{B}$ are equal if they have the same magnitude and direction. If $\vec{A} = a_1\,\hat{e}_1 + a_2\,\hat{e}_2 = \vec{B} = b_1\,\hat{e}_1 + b_2\,\hat{e}_2$, then their components must be equal and consequently, $a_1 = b_1$ and $a_2 = b_2$.

2. The sum of two vectors $\vec{A}$ and $\vec{B}$ is obtained using the parallelogram law for vector addition. That is, the vector sum $\vec{A} + \vec{B}$ is obtained by placing the origin of vector $\vec{B}$ at the terminus point of the vector $\vec{A}$ and then joining the origin of $\vec{A}$ to the terminus point of the vector $\vec{B}$. $\vec{A} + \vec{B} = [a_1\,\hat{e}_1 + a_2\,\hat{e}_2] + [b_1\,\hat{e}_1 + b_2\,\hat{e}_2] = (a_1 + b_1)\,\hat{e}_1 + (a_2 + b_2)\,\hat{e}_2$

3. The difference of two vectors $\vec{A}$ and $\vec{B}$ is written $\vec{A} - \vec{B}$ and is obtained by adding the vector $-\vec{B}$ to the vector $\vec{A}$. $\vec{A} - \vec{B} = [a_1\,\hat{e}_1 + a_2\,\hat{e}_2] - [b_1\,\hat{e}_1 + b_2\,\hat{e}_2] = (a_1 - b_1)\,\hat{e}_1 + (a_2 - b_2)\,\hat{e}_2$,

4. The following commutative laws hold

$$\vec{A} + \vec{B} = \vec{B} + \vec{A} \qquad\qquad \alpha\vec{A} = \vec{A}\alpha$$
Commutative law for addition  Commutative law for multiplication

5. The following associative laws hold

$$\vec{A} + (\vec{B} + \vec{C}) = (\vec{A} + \vec{B}) + \vec{C} \qquad\qquad \alpha(\beta\vec{A}) = (\alpha\beta)\vec{A}$$
Associative law for addition  Associative law for scalar multiplication

6. The following distributive laws hold

$$(\alpha + \beta)\vec{A} = \alpha\vec{A} + \beta\vec{A} \qquad \text{and} \qquad \alpha(\vec{A} + \vec{B}) = \alpha\vec{A} + \alpha\vec{B}$$

## Curve sketching

Given an equation of the form $f(x,y) = 0$, try to solve for $y$ in terms of $x$ or solve for $x$ in terms of $y$. Alternatively, make a substitution $x = x(t)$ and then solve for $y = y(t)$ to obtain a parametric representation of the function. Sometimes it is possible to make a change of variable to represent the curve in a different coordinate system. For example, the substitution $x = r\cos\theta$ and $y = r\sin\theta$ is used to represent the curve in polar coordinates. From either the explicit, implicit or parametric representation of the function, construct a table of $(x,y)$ values for selected $x$-values over some range. These points can be plotted and connected by a smooth curve to obtain a graphical representation of the given function.

In sketching curves take into account the following considerations.

1. Select the type of graph paper desired and label all axes.

2. The choice of scaling for the axes will affect the shape of the graph. The type of graph and scaling of axes depends upon the range of values being used and the choice of units being represented.
   The following are some sketches of curves easily recognized.

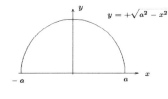
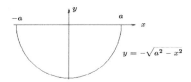

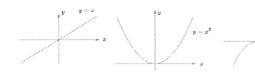

3. Look for symmetry in the graphs being sketched. There are five basic types of symmetry to be tested. Whenever it is possible to make a substitution in the original equation to obtain the same or an equivalent equation, then some type of symmetry is usually indicated. The following five substitutions illustrate this idea.

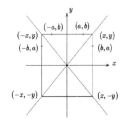

Substitute

$-x$ for $x$ to test for symmetry about the $y$-axis

$-y$ for $y$ to test for symmetry about the $x$-axis

$-x$ for $x$ and $-y$ for $y$ to test for symmetry about the origin

$x$ for $y$ and $y$ for $x$ to test for symmetry about the 45° line

$-x$ for $y$ and $-y$ for $x$ to test for symmetry about the 135° line

3. The axis intercepts occur where $x = 0$ and $y = 0$. That is, put $x = 0$ and solve for $y$ and then put $y = 0$ and solve for $x$.

4. Look for asymptotes. If after solving for $y$ in terms of $x$ there results a denominator or denominators which can go to zero, then equate each denominator equal to zero and solve for $x$. If after solving for $x$ in terms of $y$ there results a denominator or denominators which can go to zero, then equate each denominator equal to zero and

solve for $y$. Alternatively, take the limits $\lim_{x\to\infty}$ or $\lim_{y\to\infty}$ to determine if one of the variables approaches a definite value.

5. Look for excluded regions. That is, first solve for $x$ or $y$ and then examine the range of values for which real solutions will exist. If there is a range of values which produce complex numbers, then that range must be excluded from the graph.
6. For technical reports be sure to label axes with appropriate variables or names and state units of measurements. Illustrate the scale used on the axes and include a caption stating what the graph represents. In the body of the technical report explain any interesting features indicated by the graph. Note some technical journals require that graphs be done in specified formats, such as TIFF, JPG, BMP, Postscript, etc. Check with the journal editors for specified formats which must be applied to figures and graphs.

**Special Graph Paper**

The following is a presentation of special types of graph paper that occur frequently in science and engineering. The graph paper is presented in the following order.

1. Cartesian- equal spaced
2. Polar
3. Triangular or ternary
4. Hexagonal
5. Semi-log 2 cycles
6. Semi-log 3 cycles
7. Semi-log 6 cycles
8. Log-Log 2 by 2 cycle
9. Log-Log 4 by 2 cycles
10. Log-Log 6 by 3 cycles
11. Probability paper 0.1% to 99.9%

Note that semi-log and log-log paper comes in any number of cycles or fractions of a cycle. Logarithmic scales are used whenever numbers required for scaling an axis get large or the data covers a large range of values. Probability paper is used to test for a normal distribution. If data has a normal distribution, then the data plots up as a straight line. Cartesian and polar paper are self explanatory. The triangular or ternary paper is used whenever it is required to plot three related quantities. The axes of the ternary paper run from a selected vertex as a line perpendicular to the opposite side. Note that in planning the presentation of a graph, select the scaling of an axis in order that the graph be centered in the figure being presented. Hexagonal grids have applications in image processing, materials science, engineering and chemistry. Many other special kinds of graph paper can be constructed for specific purposes.

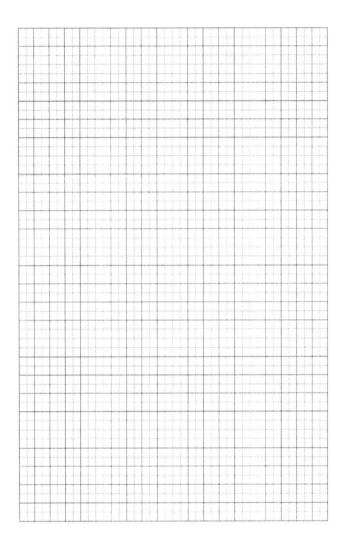

Cartesian equal spaced graph paper

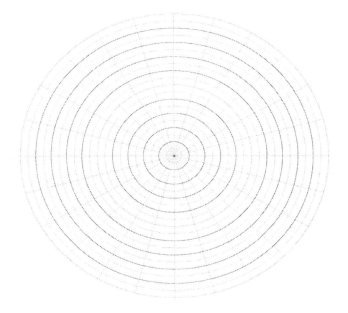

Polar graph paper

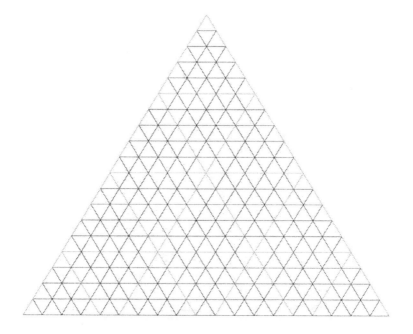

Triangular or ternary graph paper

Hexagonal graph paper

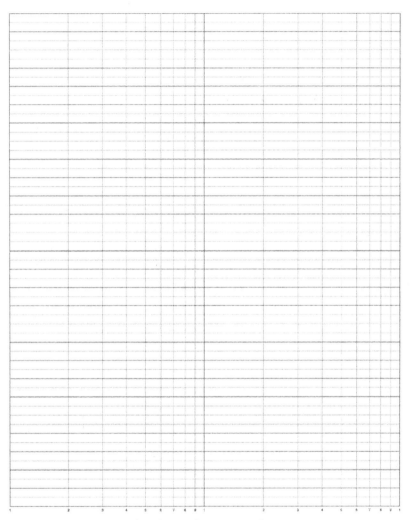

Semi-log graph paper, 2 cycles
Log scale on abscissa axis

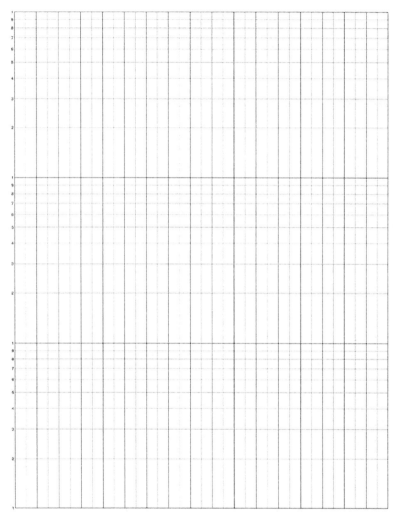

Semi-log graph paper, 3 cycles
Log scale on ordinate axis

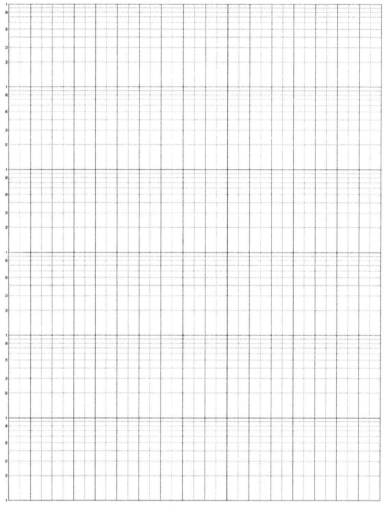

Semi-log graph paper, 6 cycles
Log scale on ordinate axis

Log-Log graph paper, 2 by 2 cycles

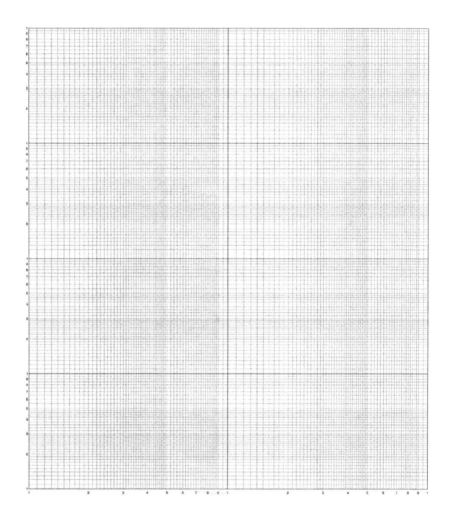

Log-Log graph paper, 4 by 2 cycles

Log-Log graph paper, 6 by 3 cycles

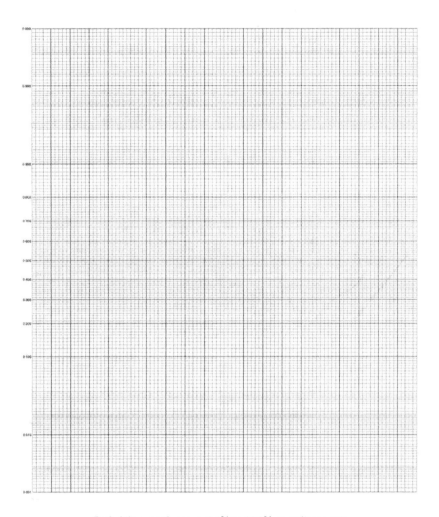

Probability graph paper, 0.1% to 99.9% on ordinate axis

Note that the equation $y = \alpha e^{\beta x}$ plots as a straight line on semi-log paper. Take the logarithm of both sides of the above equation to obtain $\ln y = \ln \alpha + \beta x$. Observe that the substitutions $Y = \ln y$ and $b = \ln \alpha$ gives the straight line $Y = \beta x + b$. Similarly, the equation $y = \alpha x^{\beta}$ plots as a straight line on log-log paper. Taking the logarithm of this equation gives $\ln y = \ln \alpha + \beta \ln x$ and the substitutions $Y = \ln y$, $b = \ln \alpha$ and $X = \ln x$ gives the straight line $Y = \beta X + b$.

## Straight lines

The equation of a straight line can be presented in many different ways.

**Point-slope formula** The equation of a line between two points $(x_1, y_1)$ and $(x_2, y_2)$ is given by either of the point-slope formulas

$$y - y_1 = m(x - x_1)$$
$$\text{or} \quad y - y_2 = m(x - x_2) \qquad \text{where } slope\ of\ line\ = m = \frac{y_2 - y_1}{x_2 - x_1} = \frac{\text{change in } y}{\text{change in } x} = \frac{\Delta y}{\Delta x} \qquad (3.15)$$

**Slope-intercept formula**

$$y = mx + b \quad \text{where } m \text{ is the slope and } b \text{ is the } y\text{-intercept.} \qquad (3.16)$$

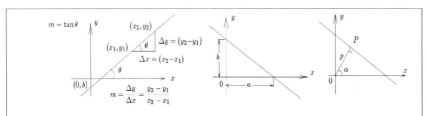

**Figure 3-5** Various parameters for representing straight lines

**Intercept formula**

$$\frac{x}{a} + \frac{y}{b} = 1 \quad \text{where } a \text{ is the } x\text{-intercept and } b \text{ is the } y\text{-intercept} \qquad (3.17)$$

**General formula for a straight line**

$$Ax + By + C = 0 \quad \text{where } A, B \text{ and } C \text{ are constants} \qquad (3.18)$$

**Normal form for equation of line**

$$x \cos \alpha + y \sin \alpha = p \qquad (3.19)$$

where, as illustrated in the figure 3-5, the segment $\overline{OP}$ is the perpendicular from the origin to a point $P$ on the line and $\alpha$ is the angle that the line segment $\overline{OP}$ makes with the positive $x$-axis and $p$ is the length of the segment $\overline{OP}$.

### Vector form for a straight line

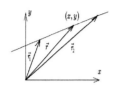

If $\vec{r}_1 = x_1\,\hat{e}_1 + y_1\,\hat{e}_2$ is the vector to the point $(x_1, y_1)$ and $\vec{r}_2 = x_2\,\hat{e}_1 + y_2\,\hat{e}_2$ is the vector to the point $(x_2, y_2)$, then a point $(x, y)$ on the straight line through the points $(x_1, y_1)$ and $(x_2, y_2)$ can be represented by the vector equation

$$\vec{r} = x\,\hat{e}_1 + y\,\hat{e}_2 = \vec{r}_1 + t(\vec{r}_2 - \vec{r}_1) \tag{3.20}$$

where $t$ is a scalar.

### Parametric equations for a straight line

Expanding the vector equation (3.20) and equating components, one obtains the parametric form for the equation of a straight line

$$x = x_1 + t(x_2 - x_1), \qquad y = y_1 + t(y_2 - y_1) \tag{3.27}$$

### Distance from point $P(x_0, y_0)$ to line $Ax + By + C = 0$

The perpendicular distance $d$ from a general point $P(x_0, y_0)$ to a given line $Ax + By + C = 0$ is given by

$$d = \frac{Ax_0 + By_0 + C}{\pm\sqrt{A^2 + B^2}} \tag{3.22}$$

where the sign is selected such that the distance is positive.

The equation of the straight line passing through the points $(x_1, y_1)$ and $(x_2, y_2)$ can be determined by evaluating the determinant $\begin{vmatrix} x & y & 1 \\ x_1 & y_1 & 1 \\ x_2 & y_2 & 1 \end{vmatrix} = 0$

### Equations for a circle

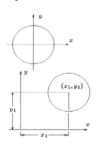

The equation of a circle having radius $b$ and centered at the origin is given by

$$x^2 + y^2 = b^2$$

The equation of a circle with radius $b$ centered at the point $(x_1, y_1)$ is obtained by translation of axes and has the form

$$(x - x_1)^2 + (y - y_1)^2 = b^2 \tag{3.23}$$

Equations of the form $x^2 + y^2 + Ax + By + C = 0$ can be placed in the form of equation (3.23) by completing the square on the $x$ and $y$ terms.

One finds $x_1 = -A/2$, $y_1 = -B/2$, $b^2 = \frac{1}{4}(A^2 + B^2 - 4C)$

Area of circle $= \pi b^2$, Perimeter of circle $= 2\pi b$

## Use of determinants

The equation of the circle which passes through the points $(x_1, y_1)$, $(x_2, y_2)$ and $(x_3, y_3)$ can be found by evaluating the determinant

$$\begin{vmatrix} x^2 + y^2 & x & y & 1 \\ x_1^2 + y_1^2 & x_1 & y_1 & 1 \\ x_2^2 + y_2^2 & x_2 & y_2 & 1 \\ x_3^2 + y_3^2 & x_3 & y_3 & 1 \end{vmatrix} = 0$$

### Triangle

A triangle with distinct vertices $(x_1, y_1), (x_2, y_2), (x_3, y_3)$, is illustrated in the figure 3-6. Denote the slope of the line joining the points $(x_1, y_1)$ to $(x_2, y_2)$ by $m_{21}$. In a similar fashion define the slope of the line connecting $(x_1, y_1)$ and $(x_3, y_3)$ by $m_{31}$ and the slope of the line connecting $(x_2, y_2)$ and $(x_3, y_3)$ by $m_{32}$.

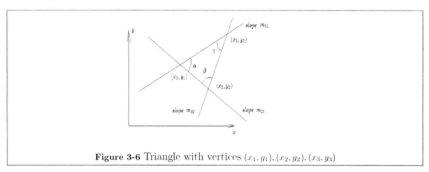

**Figure 3-6** Triangle with vertices $(x_1, y_1), (x_2, y_2), (x_3, y_3)$

In figure 3-6, verify the slopes of the lines are given by

$$m_{21} = \frac{y_2 - y_1}{x_2 - x_1}, \qquad m_{31} = \frac{y_3 - y_1}{x_3 - x_1}, \qquad m_{32} = \frac{y_3 - y_2}{x_3 - x_2} \qquad (3.24)$$

The interior angles $\alpha, \beta, \gamma$ of the resulting triangle are determined from the relations

$$\tan\alpha = \left|\frac{m_{21} - m_{31}}{1 + m_{21}m_{31}}\right|, \qquad \tan\beta = \left|\frac{m_{21} - m_{32}}{1 + m_{21}m_{32}}\right|, \qquad \tan\gamma = \left|\frac{m_{32} - m_{31}}{1 + m_{32}m_{31}}\right| \qquad (3.25)$$

Here the absolute values are used to insure that the tangents are positive. Note that whenever the product of the slopes of two sides of the triangle is equal to $-1$, then the sides form a right angle. Also if two sides have the same slope, then the sides are coincident.

The area of the triangle with vertices $(x_1, y_1), (x_2, y_2), (x_3, y_3)$ is given by the determinant

$$A = \pm\frac{1}{2}\begin{vmatrix} x_1 & y_1 & 1 \\ x_2 & y_2 & 1 \\ x_3 & y_3 & 1 \end{vmatrix} = \frac{1}{2}\left\{\begin{vmatrix} x_1 & y_1 \\ x_2 & y_2 \end{vmatrix} + \begin{vmatrix} x_2 & y_2 \\ x_3 & y_3 \end{vmatrix} + \begin{vmatrix} x_3 & y_3 \\ x_1 & y_1 \end{vmatrix}\right\} \qquad (3.26)$$

where the $\pm$ sign is selected so that the area is positive. Note that if the area is zero, then the three points will all lie on the same line.

**Area of a polygon**

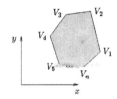

The area of polygon having vertices in the order
$\vec{V}_1 = x_1\hat{e}_1 + y_1\hat{e}_2, \vec{V}_2 = x_2\hat{e}_1 + y_2\hat{e}_2, \ldots, \vec{V}_n = x_n\hat{e}_1 + y_n\hat{e}_2$
can be obtained from the sum of n two by two determinants

$$Area = \frac{1}{2}\left\| \begin{matrix} x_1 & x_2 \\ y_1 & y_2 \end{matrix} \right| + \left| \begin{matrix} x_2 & x_3 \\ y_2 & y_3 \end{matrix} \right| + \left| \begin{matrix} x_3 & x_4 \\ y_3 & y_4 \end{matrix} \right| + \cdots + \left| \begin{matrix} x_n & x_1 \\ y_n & y_1 \end{matrix} \right| \right\| \quad (3.27)$$

which represents the area of $n-2$ triangles formed from the $n$ vertices.

## Polar coordinates

In contrast to rectangular coordinates where points are given in reference to distances from two lines, which intersect perpendicularly, a polar coordinate system is constructed as follows. Let 0 denote a fixed point called the pole or origin of the coordinate system and then construct a fixed line 0x called the polar axis. If the line 0x is rotated counterclockwise about the pole 0, this creates a ray from the origin lying at some positive angle $\theta$ with respect to the original polar axis. Points can be given coordinates $(r, \theta)$ where $r$ is a distance from 0 along the ray at angle $\theta$ with respect to the polar axis. The distance $r$ is called the radius vector and the angle $\theta$ is called the vectorial angle.

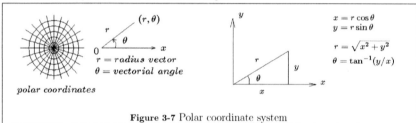

**Figure 3-7** Polar coordinate system

The vectorial distance $r$ can be positive or negative and to the vectorial angle $\theta$ it is possible to add or subtract multiples of $2\pi$. Consequently, points in polar coordinates can have many $(r, \theta)$ number pairs because the angle $\theta$ is not unique. That is, points $(r, \theta)$ can also have the coordinates $((-1)^n r, \theta + n\pi)$, for $n = 0, \pm 1, \pm 2, \ldots$

**Transformation equations**

The transformation equations relating rectangular and polar coordinates are given by

$$x = r\cos\theta, \quad y = r\sin\theta, \quad \text{with inverse} \quad r = \sqrt{x^2+y^2}, \quad \theta = \tan^{-1}(y/x) \quad (3.28)$$

## Curve sketching in polar coordinates

Given an equation in polar coordinates $(r, \theta)$ of the form $f(r, \theta) = 0$, one usually tries to solve for $r$ as a function of the angle $\theta$ to obtain $r = f(\theta)$. On can then form a table of $(r, \theta)$ values for selected values of $\theta$. These points can be plotted and then connected by a smooth curve to give a graphical representation of the function. If $f(r, -\theta) = f(r, \theta)$, then the graph will be symmetric with respect to the polar axis. If $f(r, \pi - \theta) = f(r, \theta)$, then the graph will be symmetric with respect to the $\pi/2$ ray or ninety-degree line. If $f(r, \theta) = f(-r, \theta)$, then the graph will be symmetric with respect to the pole. There are many tests for symmetries in polar coordinates as suggested by the following sketches.

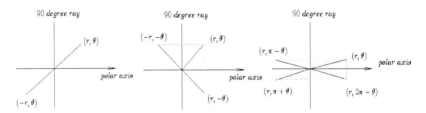

## Example 3-1. Curve sketching polar coordinates

To plot the polar curve $r = 10 \sin 2\theta$ make up a table of values such as the following

| $\theta$ deg | $2\theta$ deg | $2\theta$ radians | $\sin 2\theta$ | r |
|---|---|---|---|---|
| 0 | 0 | 0 | 0 | 0 |
| 15 | 30 | $\frac{\pi}{6}$ | 1/2 | 5 |
| 30 | 60 | $\frac{\pi}{3}$ | $\sqrt{3}/2$ | $5\sqrt{3}$ |
| 45 | 90 | $\frac{\pi}{2}$ | 1 | 10 |
| 60 | 120 | $\frac{2\pi}{3}$ | $\sqrt{3}/2$ | $5\sqrt{3}$ |
| 75 | 150 | $\frac{5\pi}{6}$ | 1/2 | 5 |
| 90 | 180 | $\pi$ | 0 | 0 |

$r = 10 \sin 2\theta$

Plot the above points and connect the points by a smooth curve. Then make use of symmetries, if available, to complete the curve. Alternatively, plot the parametric equations

$$x = r \cos \theta = 10 \sin 2\theta \cos \theta, \qquad y = r \sin \theta = 10 \sin 2\theta \sin \theta$$

to obtain an $(x, y)$ representation for the curve.

■

**Polar equation of line**

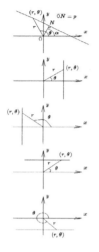

Denote by $(r,\theta)$ the coordinates of a general point on the line and denote by $\overline{ON} = p$ the perpendicular distance from the origin to the line, with $\alpha$ the angle of inclination of the segment $\overline{ON}$. From the geometry of the top figure on the left, show that the equation of the line can be represented in either of the forms

$$r\cos(\theta - \alpha) = p \quad \text{or} \quad r = p\sec(\theta - \alpha) \tag{3.29}$$

Note the following special cases.
If $\alpha = 0$, $p > 0$, equation of line is $r\cos\theta = p$
If $\alpha = 0$, $p < 0$, equation of line is $r\cos\theta = -p$
If $\alpha = \frac{\pi}{2}$, $p > 0$, equation of line is $r\sin\theta = p$
If $\alpha = \frac{\pi}{2}$, $p < 0$, equation of line is $r\sin\theta = -p$
Note also that if $p = 0$, then the line passes through the origin and so the equation of the line is written in the form $\theta = constant$.

**Polar equation of circle**

In polar coordinates construct a circle of radius $b$ centered at the point $(r_1, \theta_1)$ as illustrated in the figure 3-8.

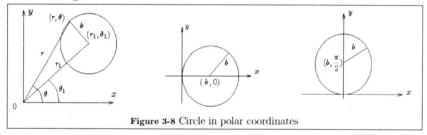

**Figure 3-8** Circle in polar coordinates

Apply the law of cosines to the triangle constructed in the figure 3-8 to obtain the equation of the circle in polar coordinates

$$r^2 + r_1^2 - 2rr_1\cos(\theta - \theta_1) = b^2 \tag{3.30}$$

Note the special case $\theta_1 = 0$, $r_1 = b$ gives the polar equation $r = 2b\cos\theta$, and the special case $\theta_1 = \pi/2$, $r_1 = b$, gives the polar equation $r = 2b\sin\theta$. When $r_1 = b$, equation (3.30) becomes $r = 2b\cos(\theta - \theta_1)$.

# 125

## Geometric shapes

The following is a selection of fundamental relationships from plane geometry, solid geometry and spherical trigonometry.

**Trapezoid**

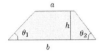

Consider a trapezoid with parallel sides a and b with height $h$ as illustrated.

Area $= \frac{1}{2}h(a+b)$

$s = a + b + h(\csc\theta_1 + \csc\theta_2) =$ perimeter

---

**Rectangle**

Area Rectangle $= bh$

$s = 2h + 2b =$ perimeter

Area Triangle ABC $= \frac{1}{2}bh$

---

**Circle**

Area $= \pi r^2$

$s = 2\pi r =$ perimeter

---

**Parallelogram**

Area $= bh = ab\sin\theta$

$s = 2a + 2b =$ perimeter

---

**Sector of circle**

Area $= \frac{1}{2}r\theta^2$

$s = r\theta =$ arc length subtended by sector

$\theta$ in radians

---

**Polar rectangle**

Area $= \frac{R+r}{2}(R-r)\theta$

Perimeter $= 2(R-r) + (R+r)\theta$

$\theta$ in radians

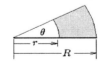

**Regular polygon of $n$ sides.**

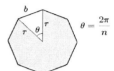

$$\text{Area} = \frac{1}{4} nb^2 \cot \frac{\pi}{n}$$
$s = nb = $ perimeter
$b^2 = 2r^2(1 - \cos\theta)$

$\theta = \frac{2\pi}{n}$

---

**Inscribed regular polygon**

$\theta = \frac{2\pi}{n}$

Given a circle with radius $r$, then inscribe a regular polygon of $n$ sides. One obtains
$$\text{Area} = \frac{1}{2} nr^2 \sin\theta$$
$s = 2nr \sin \frac{\theta}{2} = $ perimeter

---

**Circumscribed Regular polygon of $n$ sides.**

Circumscribe a polygon of $n$ sides about a circle of radius $r$ to obtain
$$\text{Area} = nr^2 \tan \frac{\theta}{2}$$
$s = 2nr \tan \frac{\theta}{2} = $ perimeter

$\theta = \frac{2\pi}{n}$

---

**Circle inscribed in triangle**

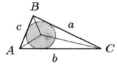

The center of the inscribed circle occurs where the angle bisectors meet.

$$r = \frac{c \sin \frac{A}{2} \sin \frac{B}{2}}{\cos \frac{C}{2}}$$

or $r = \frac{1}{s}\sqrt{s(s-a)(s-b)(s-c)} = $ radius of inscribed circle

$s = \frac{1}{2}(a+b+c) = $ semiperimeter of triangle

---

**Right circular cylinder**

The volume and total surface area of a cylinder of radius $r$ and height $h$ is given by

$Volume = \pi r^2 h$

$Total\ surface\ area = 2\pi rh + 2\pi r^2$

**Circle circumscribing a triangle**

The point $P$, which is equidistant from the vertices of the triangle, occurs at the intersection where the perpendicular bisectors of the sides meet. This point $P$ is called the point of concurrency of the perpendicular bisectors. The radius $r$ of the circumscribed circle is given by

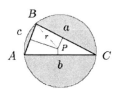

$$r = \frac{a}{2\sin A} = \frac{b}{2\sin B} = \frac{c}{2\sin C} \quad \text{or} \quad r = \frac{abc}{4\sqrt{s(s-a)(s-b)(s-c)}}$$

where $s = \frac{1}{2}(a+b+c)$ is the semiperimeter of the triangle.

---

**Sphere**

Sphere of radius $r$ has the following volume and surface area
$Volume = \frac{4}{3}\pi r^3$
$Surface\ area = 4\pi r^2$

---

**Rectangular parallelepiped**

The volume and surface area of a rectangular parallelepiped having sides a,b,c is given by
$Volume = abc$
$Surface\ area = 2(ab + ac + bc)$
$Diagonal = \sqrt{a^2 + b^2 + c^2}$

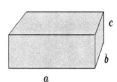

---

**Segment of a circle**

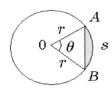

For the segment of circle of radius $r$ illustrated
$Area\ of\ shaded\ part = \frac{1}{2}r^2(\theta - \sin\theta)$
$Area\ of\ shaded\ part\ plus\ area\ of\ triangle\ 0AB = \frac{1}{2}r^2\theta$
$s = r\theta$ is the arc AB
$\theta$ is in radians

### Skewed parallelepiped

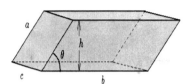

A parallelepiped with cross sectional area $bc$ and height $h$

$Volume = (bc)h = (bc)\,a\sin\theta$

---

### Slanted cylinder

A slanted cylinder with a circular cross section has a volume given by $V = \pi r^2 \ell = \dfrac{\pi r^2 h}{\sin\theta}$
The side surface area is $S = 2\pi r \ell = \dfrac{2\pi r \ell}{\sin\theta}$
The top +bottom area $= 2\pi r^2$

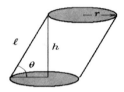

---

### Slanted cylinder with arbitrary cross section

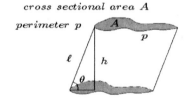

A slanted cylinder with an arbitrary cross sectional area $A$ having a cross section with perimeter $p$ has the volume
$V = A\ell = \dfrac{Ah}{\sin\theta}$
and side surface area $S = p\ell = \dfrac{ph}{\sin\theta}$

---

### Pyramid

The volume on any pyramid is one third the product of its base area times its altitude. $Volume = \dfrac{1}{3} A_B h$
If a plane cuts the pyramid at a height $H$, where $0 < H < h$, and $A_T$ is the top area, then volume of frustum of pyramid is given by
$Volume\ frustum = \dfrac{H}{3}(A_B + \sqrt{A_B A_T} + A_T)$

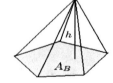

**Right circular cone**

A right circular cone has the volume
$$V = \frac{1}{3}\pi r^2 h$$
and surface area
$$S = \pi r \ell = \pi r \sqrt{r^2 + h^2}$$

---

**Frustum of right circular cone**

The volume and lateral surface area are given by the relations
$$V = \frac{1}{3}\pi h(r_1^2 + r_1 r_2 + r_2^2)$$
$$S = \pi(r_1 + r_2)\ell = \pi(r_1 + r_2)\sqrt{h^2 + (r_2 - r_1)^2}$$

---

**Spherical cap**

If a plane intersects a sphere to form a spherical cap of height $h$, then the volume $V$ and lateral surface area $S$ of the cap are given by $V = \frac{1}{3}\pi h^2(3r - h)$, $S = 2\pi r h$ where $r$ is the radius of the sphere.
Volume of spherical sector $V = \frac{2\pi}{3} r^2 h$
Surface area of spherical sector $S = \pi r(2h + a)$

---

**Torus**

A torus has the volume and surface area
$$V = \frac{1}{4}\pi^2(r_1 + r_2)(r_2 - r_1)^2$$
$$S = \pi^2(r_2^2 - r_1^2)^2 \text{ where } r_1 \text{ is the inner radius and } r_2 \text{ is the outer radius.}$$

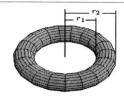

---

**Quadrilateral**

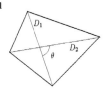

A quadrilateral is a polygon with four sides. The area is given by one-half the product of the diagonals times the sine of the angle of intersection of the diagonals.
$$Area = \frac{1}{2} D_1 D_2 \sin \theta$$

## Rhombus

A rhombus is an equilateral quadrilateral with area given by
$Area = \frac{1}{2}D_1 D_2$ where $D_1$ and $D_2$ are the diagonals.

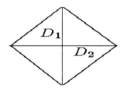

## Spherical trigonometry

Consider a sphere with origin 0 and unit radius with points A,B,C on the surface of the sphere. Construct three great circles on the sphere passing respectively through the points AB, BC and AC to form the minor arcs illustrated. These minor arcs define the spherical triangle ABC on the surface of the unit sphere.

Tangent lines to the great circles at the point A define the plane angle $\angle A$. Similarly, constructing tangent lines to the great circles at the points B and C to define the plane angles $\angle B$ and $\angle C$. The angles $\angle AOC = \beta$, $\angle AOB = \gamma$ and $\angle BOC = \alpha$ are respectively the same as the arc lengths $\widehat{AC}$, $\widehat{AB}$, and $\widehat{BC}$ on the surface of the unit sphere. Spherical triangles have the following properties.

### Law of sines for spherical triangles

$$\frac{\sin \alpha}{\sin A} = \frac{\sin \beta}{\sin B} = \frac{\sin \gamma}{\sin C}$$

### Law of cosines for spherical triangles

$$\cos \alpha = \cos \beta \cos \gamma + \sin \beta \sin \gamma \cos A$$
$$\cos \beta = \cos \gamma \cos \alpha + \sin \gamma \sin \alpha \cos B$$
$$\cos \gamma = \cos \alpha \cos \beta + \sin \alpha \sin \beta \cos C$$

### Half angle formulas for spherical triangles

$$\tan^2 \frac{A}{2} = \frac{\sin(s-\beta)\sin(s-\gamma)}{\sin(s)\sin(s-\alpha)}$$
$$\tan^2 \frac{B}{2} = \frac{\sin(s-\gamma)\sin(s-\alpha)}{\sin(s)\sin(s-\beta)}$$
$$\tan^2 \frac{C}{2} = \frac{\sin(s-\alpha)\sin(s-\beta)}{\sin(s)\sin(s-\gamma)}$$

where $s = \frac{1}{2}(\alpha + \beta + \gamma)$ is the semiperimeter of the spherical triangle on the unit sphere.

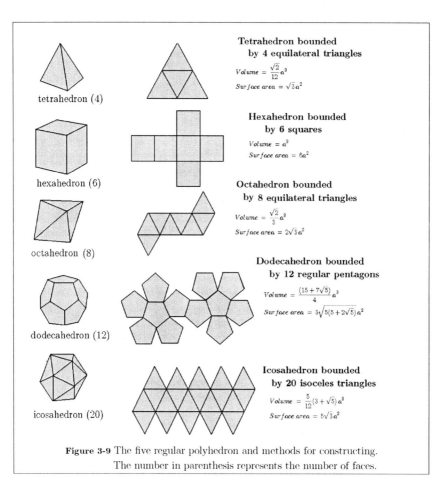

**Figure 3-9** The five regular polyhedron and methods for constructing. The number in parenthesis represents the number of faces.

## Polyhedron

A polyhedron is a solid having plane surfaces. If all the faces are congruent regular polygons and the polyhedral angles are congruent, then the solid is called a regular polyhedron.

Let $E$, $V$ and $F$ denote respectively the number of edges, the number of vertices and number of faces of a polyhedron, then Euler's theorem states that $E + 2 = V + F$.

Some well known regular polyhedron are:

A **tetahedron** is a polyhedron of four faces.
A **hexahedron** is a polyhedron of six faces.
A **octahedron** is a polyhedron of eight faces.
A **dodecahedron** is a polyhedron of twelve faces.
A **icosahedron** is a polyhedron of twenty faces.

It can be proven that there cannot be more than five regular polyhedron. The figure 3-9 illustrates the five regular polyhedron and methods for constructing them from cardboard.

**Conic sections**

A general equation of the second degree has the form

$$Ax^2 + Bxy + Cy^2 + Dx + Ey + F = 0 \qquad (3.31)$$

where $A, B, C, D, E, F$ are constants. All curves which have the form of equation (3.31) can be obtained by cutting a cone with a plane. The figure 3-10(a) illustrates a right circular cone obtained by constructing a circle in a horizontal plane and then moving perpendicular to the plane to a point $V$ above or below the center of the circle. The point $V$ is called the vertex of the cone. All the lines through the point $V$ and a point on the circumference of the circle are called generators of the cone. The set of all generators produces the right circular cone. The figure 3-10(b) illustrates a horizontal plane intersecting the cone in a circle. The figure 3-10(c) illustrates a nonhorizontal plane section which cuts two opposite generators. The resulting curve of intersection is called an ellipse. Figure 3-7(d) illustrates a plane parallel to a generator of the cone which also intersects the cone. The resulting curve of intersection is called a parabola. Any plane cutting both the upper and lower parts of a cone will intersect the cone in a curve called a hyperbola. This is illustrated in the figure 3-10(e).

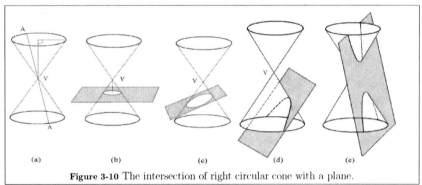

**Figure 3-10** The intersection of right circular cone with a plane.

**Rotation of axes**

Substitute the equations

$$x = \bar{x}\cos\theta - \bar{y}\sin\theta, \qquad y = \bar{x}\sin\theta + \bar{y}\cos\theta, \tag{3.32}$$

representing a rotation of axes, into the equation (3.31) and show there results the new equation equation

$$\bar{A}\bar{x}^2 + \bar{B}\bar{x}\bar{y} + \bar{C}\bar{y}^2 + \bar{D}\bar{x} + \bar{E}\bar{y} + \bar{F} = 0 \tag{3.33}$$

where $\bar{A}, \bar{B}, \bar{C}, \bar{D}, \bar{E}, \bar{F}$ are new coefficients given by

$$\begin{aligned}
\bar{A} &= A\cos^2\theta + B\cos\theta\sin\theta + C\sin^2\theta \\
\bar{B} &= B(\cos^2\theta - \sin^2\theta) + 2(C - A)\sin\theta\cos\theta \\
\bar{C} &= A\sin^2\theta - B\sin\theta\cos\theta + C\cos^2\theta \\
\bar{D} &= D\cos\theta + E\sin\theta \\
\bar{E} &= -D\sin\theta + E\cos\theta \\
\bar{F} &= F
\end{aligned} \tag{3.34}$$

The combination of terms $B^2 - 4AC$ is called the discriminant of equation (3.31). The discriminant is an invariant under the rotation of axes and is such that $B^2 - 4AC = \bar{B}^2 - 4\bar{A}\bar{C}$. Further, if $B^2 - 4AC = 0$, then a parabola results. If $B^2 - 4AC < 0$, then an ellipse results and if $B^2 - 4AC > 0$, then a hyperbola results. Note that the term $\bar{B}$ in equations (3.33) and (3.34) can be made zero by selecting $\theta$ to satisfy the relation

$$B(\cos^2\theta - \sin^2\theta) + (C-A)2\sin\theta\cos\theta = 0$$

$$\text{or} \qquad B\cos 2\theta + (C-A)\sin 2\theta = 0 \quad \Longrightarrow \quad \tan 2\theta = \frac{B}{A-C} \tag{3.35}$$

For this choice for $\theta$, the equation (3.33) will have the special form

$$\bar{A}\bar{x}^2 + \bar{C}\bar{y}^2 + \bar{D}\bar{x} + \bar{E}\bar{y} + \bar{F} = 0 \tag{3.36}$$

in the barred system of coordinates. The equation (3.36) has the following special cases.
(i) If $\bar{A} = \bar{C} = 0$, there results the straight line $\bar{D}\bar{x} + \bar{E}\bar{y} + \bar{F} = 0$
(ii) If either $\bar{A} = 0$ or $\bar{C} = 0$, then the equation (3.36) becomes a quadratic equation in one of the variables and linear in the other. By completing the square on the quadratic term there results the equation of a parabola in standard form.
(iii) If both $\bar{A} > 0, \bar{C} > 0$ (or $\bar{A} < 0$ and $\bar{C} < 0$), then by completing the square on the quadratic terms there results the equation of an ellipse.
(iv) IF $\bar{A}$ and $\bar{C}$ are of opposite sign and both different from zero, then by completing the square on the quadratic terms there results the equation of a hyperbola.

In addition to a rotation of axes it is sometimes necessary to also perform a translation of axes in order to reduce the equation (3.31) to a standard form. Note that it is possible that the above transformations of equation (3.31) could produce either a single point, intersecting lines or no real locus of points.

Conic sections can be defined as follows. In the $xy$-plane select a point $f$ and a line $\ell$ not through $f$. The set of points $P$ satisfying the condition that $r = \overline{Pf}$ is some multiple $e$ times the perpendicular distance $d = \overline{PP'}$ is called a conic section with eccentricity $e$, focus $f$ and directrix $\ell$. Here $r = ed$ and whenever $0 < e < 1$, the conic section is an ellipse, for $e = 1$, the conic section is a parabola, and for $e > 1$, the conic section is a hyperbola.

### Circle

A circle is the locus of points $(x, y)$ in a plane equidistant from a fixed point called the center of the circle. Note that no real locus occurs if the radius $r$ is imaginary.

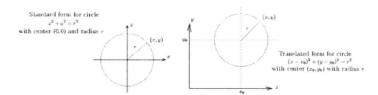

Parametric equations for a circle centered at $(x_0, y_0)$ are
$$x = x_0 + r\cos t, \qquad y = y_0 + r\sin t, \qquad 0 \le t \le 2\pi$$

When dealing with equations of the form $x^2 + y^2 + \alpha x + \beta y = \gamma$ it is customary to complete the square on the $x$ and $y$ terms to obtain

$$(x^2 + \alpha x + \frac{\alpha^2}{4}) + (y^2 + \beta y + \frac{\beta^2}{4}) = \gamma + \frac{\alpha^2}{4} + \frac{\beta^2}{4} \implies (x + \frac{\alpha}{2})^2 + (y + \frac{\beta}{2})^2 = r^2$$

where it is assumed that $r^2 = \gamma + \frac{\alpha^2}{4} + \frac{\beta^2}{4} > 0$.

### Ellipse

An ellipse is the locus of points $(x, y)$ in a plane where the sum of the distances $d_1$ and $d_2$ from two fixed points $(c, 0)$ and $(-c, 0)$ is a constant. The fixed points are called foci and $d_1 + d_2 = constant$.

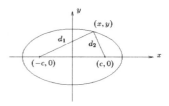

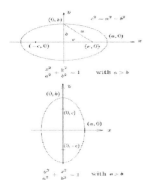

The standard form for an ellipse having foci on the $x$-axis is

$$b^2x^2 + a^2y^2 = a^2b^2 \quad \text{or} \quad \frac{x^2}{a^2} + \frac{y^2}{b^2} = 1 \quad \text{with } a > b$$

with the major axis $= 2a$, the minor axis $= 2b$ and eccentricity $e = \dfrac{c}{a} = \dfrac{\sqrt{a^2 - b^2}}{a}$. The center of the ellipse is $(0,0)$ and the distance from the center to the foci is $c = \sqrt{a^2 - b^2}$

The translated ellipse has the form

$$\frac{(x - x_0)^2}{a^2} + \frac{(y - y_0)^2}{b^2} = 1 \quad \text{with } a > b.$$

The standard form for the equation of an ellipse with the foci on the $y$-axis is given by

$$\frac{y^2}{a^2} + \frac{x^2}{b^2} = 1 \quad \text{with } a > b.$$

Note that the larger quantity is now under the $y^2$ term.

The area enclosed by an ellipse with foci on the $x$-axis is given by $Area = \pi ab$ and the perimeter of the ellipse is given by the relation

$$Perimeter = 4a \int_0^{\pi/2} \sqrt{1 - e^2 \sin^2 t}\, dt \quad \text{where } e \text{ is the eccentricity of the ellipse}$$

The parametric equations for the ellipse centered at $(x_0, y_0)$ is given by

$$x = x_0 + a\cos t, \qquad y = y_0 + b\sin t, \qquad 0 \leq t \leq 2\pi$$

## Parabola

The parabola is defined as the locus of points $(x, y)$ equidistant from a fixed point and a fixed line. The fixed point is called the focus of the parabola and the fixed line is called the directrix of the parabola. Standard forms for writing the parabola with vertex at the point $(0,0)$ are illustrated in the figures 3-11(a) and (b). Note that if $c > 0$ the parabola of figure 3-11(a) opens upward and the parabola of figure 3-11(b) opens to the right. If the sign of $c$ is reversed, then the parabola will open downward and to the left respectively and also the directrix switches sign.

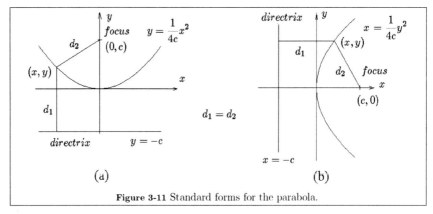

**Figure 3-11** Standard forms for the parabola.

If the vertex of the parabola is translated to the point $(x_0, y_0)$, then the equations of the parabola in figure 3-11(a) and (b) assume the form

$$(x - x_0)^2 = 4c(y - y_0), \quad \text{and} \quad (y - y_0)^2 = 4c(x - x_0)$$

**Use of determinants**

The equation of the parabola passing through the points $(x_1, y_1)$, $(x_2, y_2)$ and $(x_3, y_3)$ can be determined by evaluating the determinant

$$\begin{vmatrix} y & x^2 & x & 1 \\ y_1 & x_1^2 & x_1 & 1 \\ y_2 & x_2^2 & x_2 & 1 \\ y_3 & x_3^2 & x_3 & 1 \end{vmatrix} = 0$$

provided the following determinants are different from zero.

$$\begin{vmatrix} x_1^2 & x_1 & 1 \\ x_2^2 & x_2 & 1 \\ x_3^2 & x_3 & 1 \end{vmatrix} \neq 0, \quad \text{and} \quad \begin{vmatrix} x_1 & y_1 & 1 \\ x_2 & y_2 & 1 \\ x_3 & y_3 & 1 \end{vmatrix} \neq 0$$

**Hyperbola**

The hyperbola is defined as the locus of points $(x, y)$ which move such that the difference of its distances from two fixed points is a constant. The two fixed points are called the foci of the hyperbola. Standard forms for writing the hyperbola are illustrated in the figures 3-12(a) and (b).

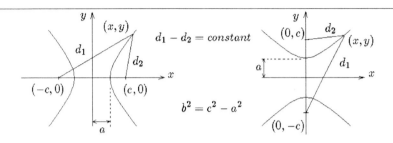

**Figure 3-12** Standard forms for the hyperbola.

Note that the hyperbola has two branches. The transverse axis has length $2a$ and the conjugate axis has length $2b$. These are the distances between points where the hyperbola crosses an axis. The ends of the transverse axis, the points $(a,0)$ and $(-a,0)$, are called vertices of the hyperbola. The origin is called the center of the hyperbola. The chord through either focus which is perpendicular to the transverse axis is called a latus rectum. Verify that the latus rectum intersects the hyperbola at the points $(c, b^2/a)$ and $(c, -b^2/a)$. The eccentricity of the hyperbola is defined as the ratio

$$\text{eccentricity} = e = \frac{c}{a} = \frac{\sqrt{a^2+b^2}}{a}$$

Note that the eccentricity of a hyperbola will always be greater than 1.

The hyperbola with transverse axis on the $x$-axis have the asymptotic lines $y = +\frac{b}{a}x$ and $y = -\frac{b}{a}x$. Any hyperbola with the property that the conjugate axis has the same length as the transverse axis is called a rectangular or equilateral hyperbola. Rectangular hyperbola are such that the asymptotic lines are perpendicular to each other. If two hyperbola are such that the transverse axis of either is the conjugate axis of the other, then they are called conjugate hyperbola. Conjugate hyperbola will have the same asymptotic lines as illustrated in the figure 3-13.

The equations for hyperbola with center $(x_0, y_0)$ are obtained by translation and one obtains

$$\frac{(x-x_0)^2}{a^2} - \frac{(y-y_0)^2}{b^2} = 1, \quad \text{or} \quad \frac{(y-y_0)^2}{a^2} - \frac{(x-x_0)^2}{b^2} = 1$$

depending upon whether the transverse axis is horizontal or vertical.

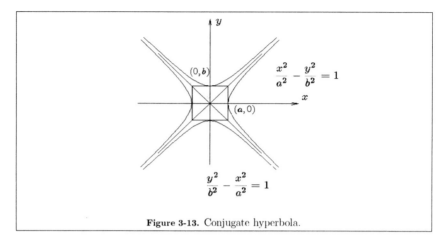

**Figure 3-13.** Conjugate hyperbola.

### Determinants and conic sections

The general equation of a conic section passing through the fixed points $(x_1, y_1)$, $(x_2, y_2)$, $(x_3, y_3)$, $(x_4, y_4)$, $(x_5, y_5)$ is obtained by evaluating the determinant

$$D = \begin{vmatrix} x^2 & xy & y^2 & x & y & 1 \\ x_1^2 & x_1 y_1 & y_1^2 & x_1 & y_1 & 1 \\ x_2^2 & x_2 y_2 & y_2^2 & x_2 & y_2 & 1 \\ x_3^2 & x_3 y_3 & y_3^2 & x_3 & y_3 & 1 \\ x_4^2 & x_4 y_4 & y_4^2 & x_4 & y_4 & 1 \\ x_5^2 & x_5 y_5 & y_5^2 & x_5 & y_5 & 1 \end{vmatrix} = 0$$

### Conic Sections in Polar Coordinates

Refer to the figure 3-14, where the ratio definition of an ellipse, hyperbola and parabola is the locus of a point $(x, y)$ which moves such that the ratio

$$\frac{OP}{QP} = e = \frac{\sqrt{x^2 + y^2}}{x + c} = constant \qquad (3.37)$$

The constant $e$ is called the eccentricity. In polar coordinates $x = r\cos\theta$ so that the distance $QP = c + r\cos\theta$, then equation (3.37) can be written as

$$e = \frac{r}{c + r\cos\theta} \qquad (3.38)$$

Solve for the radial distance $r$ from the equation (3.38) and show

$$r = \frac{ec}{1 - e\cos\theta} \qquad (3.39)$$

One then has the result that if

If $e < 1$, then equation (3.39) describes an ellipse
If $e > 1$, then equation (3.39) describes a hyperbola
If $e = 1$, then equation (3.39) describes a parabola

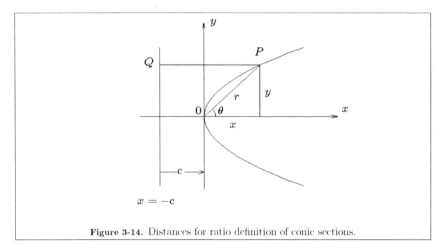

**Figure 3-14.** Distances for ratio definition of conic sections.

The equation of a conic section can also be expressed in the alternative forms

$$r = \frac{ec}{1 + e\cos\theta}, \qquad r = \frac{ec}{1 - e\sin\theta}, \qquad r = \frac{ec}{1 + e\sin\theta} \qquad (3.40)$$

## Functions and graphs

If a relation $y = f(x)$, called a functional relation, assigns a unique value for $y$ for each given $x$-value, then by plotting a set of $(x, y)$ values on rectangular coordinates there results a graphical representation of the function. If there exists constants $m$ and $M$ such that $m \le f(x) \le M$ for all values of $x$ over an interval, then the function is said to be bounded in the interval. Select any two points within the interval of definition $(a, b)$, say $x_1$ and $x_2$, if for all $x_1 < x_2$, the relation $f(x_1) \le f(x_2)$ is satisfied, then the function is called a monotonic increasing function over the interval $(a, b)$. If for all $x_1 < x_2$, one finds $f(x_1) \ge f(x_2)$, then $f(x)$ is called a monotonic decreasing function over the interval $(a, b)$.

## Plane curves

The following is a selection of special plane curves represented in both rectangular form and in parametric form. Many of the curves are named after early Greeks who studied geometry. In many instances the special curves were developed for some type of geometrical construction that the early Greeks were trying to perform.

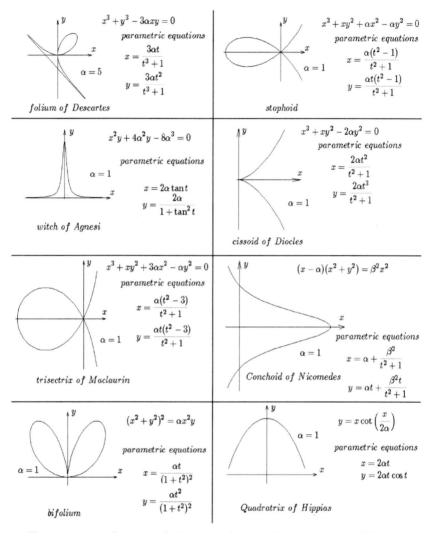

There are many other curves having unusual names that arise when studying ancient Greek mathematics. A sampling of some additional such curves are presented. However, it is left as an exercise to sketch graphs of these curves.

As a graphical exercise, plot graphs of the following functions.

| | | | |
|---|---|---|---|
| Limaçon of Pascal | $r = b - a\cos\theta$ | Astroid | $x^{2/3} + y^{2/3} = a^{2/3}$ |
| Conchoid of Diocles | $y^2(a-x) = x^3$ | Lituus | $r^2\theta = a^2$ |
| Neil's parabola | $y = ax^{3/2}$ | Cruciform | $(x^2 + y^2 - ax)^2 = a^2(x^2 + y^2)$ |
| Lemniscate of Bernoulli | $(x^2 + y^2)^2 - 2a^2(x^2 - y^2) = 0$ | Trisectrix | $y^2(a+x) = x^3(3a-x)$ |

## Plane curves in polar and parametric form

The following is a selection of plane curves represented in polar form.

### Rose curves

Rose curves have the general equation

$$r = \alpha \cos n\theta \quad \text{or} \quad r = \alpha \sin n\theta \qquad (3.41)$$

where $n$ is a positive integer. If $n$ is an odd integer, then the curves have $n$-loops called leaves. If $n$ is an even integer, then the curves have $2n$-leaves. Some examples are illustrated in the figure 3-15.

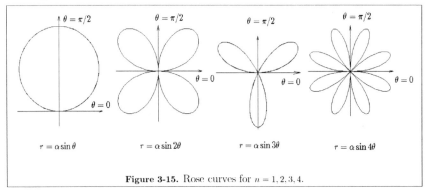

**Figure 3-15.** Rose curves for $n = 1, 2, 3, 4$.

## Spirals

The following are examples for the spirals known as the involute of a circle, the Archimedes spiral and the hyperbolic spiral.

### The involute of a circle

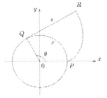

Consider a circle of radius $a$ and let $P$ denote a point on the circle, say for example $P$ is at the point $(a, 0)$ as illustrated in the accompanying figure. As the ray $0P$ moves through an angle $\theta$ to the position $0Q$, the arc length $s$ swept out along the circle is given by $s = a\theta$. Construct at the point $Q$ a tangent line to the circle and mark off the distance $s = a\theta$ along the tangent line so that $QR = s = a\theta$.

The locus of points $R$, as $Q$ moves around the circumference of the circle, is called the involute of the circle. Think of the length $QR$ as that of a string being unwound from the circle, where the string is being kept taut. The parametric equations describing this spiral are given by

$$x = a(\cos\theta + \theta\sin\theta)$$
$$y = a(\sin\theta - \theta\cos\theta)$$
(3.42)

An example spiral is illustrated in the figure 3-16.

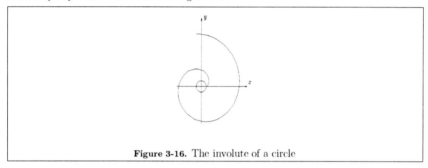

**Figure 3-16.** The involute of a circle

### The Archimedes Spiral

The polar equation $r = a\theta$, where $a$ is a constant, is known as the Archimedes spiral. The figure 3-17 illustrates the Archimedes spiral for positive and negative values of $a$.

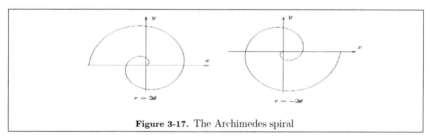

**Figure 3-17.** The Archimedes spiral

### The Hyperbolic Spiral

The hyperbolic spiral, sometimes referred to as the reciprocal spiral, has the form

$$r = \frac{a}{\theta}, \quad a \text{ is a constant} \tag{3.43}$$

The figure 3-18 illustrates the hyperbolic spiral for positive and negative values of $a$.

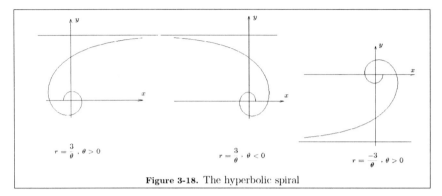

Figure 3-18. The hyperbolic spiral

## Cycloids

Consider a circle of radius $r_0$ rolling without slipping along the $x$-axis. Construct the line segments $0A < r_0$, $0B = r_0$ and $0C > r_0$ as illustrated below. The cycloids are obtained by tracing the motion of the points A,B and C, as the circle rolls without slipping along the $x$-axis.

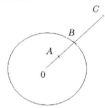

Following the motion of point $A$ as the circle rolls gives the curtate cycloid. Following the motion of the point $B$ on the rolling circle gives the cycloid and following the motion of the point $C$ gives the prolate cycloid. The equations for these cycloids are given by

$$x = r_0\theta - \beta\sin\theta, \quad y = r_0 - \beta\cos\theta \qquad (3.44)$$

where $\beta < r_0$ for a curtate cycloid, $\beta = r_0$ for a cycloid and $\beta > r_0$ for a prolate cycloid. Sketches of the curtate cycloid, cycloid and prolate cycloid are given in the figure 3-19.

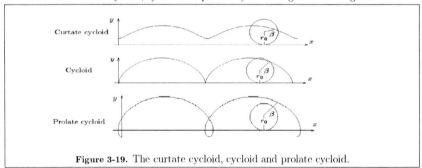

Figure 3-19. The curtate cycloid, cycloid and prolate cycloid.

## Epicycloids

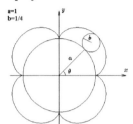

Following the motion of a point $P$ on the circumference of a small circle, which rolls without slipping on the outside of a larger circle, produces an epicycloid. The parametric equations describing an epicycloid are

$$x = (a+b)\cos\theta - b\cos\left[\left(\frac{a+b}{b}\right)\theta\right]$$
$$y = (a+b)\sin\theta - b\sin\left[\left(\frac{a+b}{b}\right)\theta\right] \quad (3.45)$$

If the ratio $a/b$ is an integer, then the tracing point $P$ will always return to its original position. If the ratio $a/b$ is a rational number, then $a$ and $b$ are said to be commensurable and the point $P$ will eventually return to its original position. If the ratio $a/b$ is an irrational number, then $a$ and $b$ are said to be incommensurable, and the point $P$ will never return to its original value.

## Hypocycloid

Following the motion of a point $P$ on the circumference of a small circle rolling, without slipping, on the inside of a larger circle produces a hypocycloid.

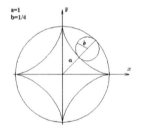

The parametric equations describing a hypocycloid are given by

$$x = (a-b)\cos\theta + b\cos\left[\left(\frac{a-b}{b}\right)\theta\right]$$
$$y = (a-b)\sin\theta - b\sin\left[\left(\frac{a-b}{b}\right)\theta\right] \quad (3.46)$$

Note that the above parametric equations are obtained from the equations (3.45) by replacing $b$ by the value $-b$.

## The Ovals of Cassini

The ovals of Cassini are illustrated in the figure 3-20. These curves describe the locus of a point $P$ such that the product of its distances from two fixed points is a constant. If the two fixed points are $(-a, 0)$ and $(a, 0)$ and the constant is $b^2$, then the ovals of Cassini are given by the equations

$$(x^2+y^2)^2 + a^4 - 2a^2(x^2-y^2) = b^2, \quad \text{or in polar form} \quad r^4 + a^4 - 2a^2 r^2 \cos 2\theta = b^4 \quad (3.47)$$

The polar form is a quadratic equation in $r^2$ and so it is possible to solve for $r$ to obtain

$$r = \pm\sqrt{a^2 \cos^2\theta \pm \sqrt{a^4 \cos^2 2\theta + b^4 - a^4}} \quad (3.48)$$

Now plot the parametric equations $x = r\cos\theta$, $y = r\sin\theta$ where $r$ is given by equation (3.48). One must consider the correct combinations of the $\pm$ signs to get the correct sections of the curves. By combining these sections the figure 3-20 results.

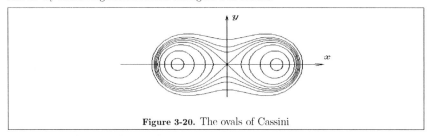

**Figure 3-20.** The ovals of Cassini

Note that the curve through the origin separates the inner curves from the outer curves. This curve through the origin results in the case $a = b$ in equation (3.47). The resulting curve is called a lemniscate. Selecting values of $b > a$ gives the curves exterior to the lemniscate and selecting values of $b < a$ gives the curves interior to the lemniscate.

**Solid Analytic Geometry**

The following are selected surfaces and their representations that occur frequently in the sciences and engineering. The general parametric equation of a surface is given by the parametric equations

$$x = x(u,v), \qquad y = y(u,v), \qquad z = z(u,v) \tag{3.56}$$

where $u,v$ are parameters which may or may not have a physical representation with respect to the coordinate system. The position vector to an arbitrary point on the surface can be represented in the parametric form

$$\vec{r} = \vec{r}(u,v) = x(u,v)\,\hat{e}_1 + y(u,v)\,\hat{e}_2 + z(u,v)\,\hat{e}_3 \qquad u,v \in \mathbb{R} \tag{3.50}$$

where $u,v$ are parameters called surface coordinates. The curves $\vec{r}(u,v_0)$ and $\vec{r}(u_0,v)$, where $u_0$ and $v_0$ are constants, are called coordinate curves. Tangent vectors to the coordinate curves are given by the partial derivatives

$$\frac{\partial \vec{r}}{\partial u} \quad \text{and} \quad \frac{\partial \vec{r}}{\partial v}$$

When these vectors are evaluated at a common point $(u,v)$ on the surface, it is possible to calculate the normal vector

$$\vec{N} = \frac{\partial \vec{r}}{\partial u} \times \frac{\partial \vec{r}}{\partial v}$$

to the surface at the point $(u,v)$. Note that $-\vec{N}$ is also normal to the surface at that point. The tangent vectors can also be used to calculate the equation of the tangent plane to the surface at the point $(u, v)$. The equation of the tangent plane to the surface at the point with surface coordinates $(u_0, v_0)$ is given by the parametric equation

$$\vec{r} = \vec{r}(s,t) = \vec{r}(u_0, v_0) + s\frac{\partial \vec{r}}{\partial u}\bigg|_{(u_0,v_0)} + t\frac{\partial \vec{r}}{\partial v}\bigg|_{(u_0,v_0)} \qquad s, t \in \mathbb{R}$$

where $s, t$ are parameters.

If $v$ is a function of $u$, write $v = v(u)$ and when this functional relation is substituted into the equation (3.50) one obtains a one parameter representation of a curve on the surface. This curve is traced out by the position vector

$$\vec{r} = \vec{r}(u) = \vec{r}(u, v(u)) = x(u, v(u))\,\hat{\mathbf{e}}_1 + y(u, v(u))\,\hat{\mathbf{e}}_2 + z(u, v(u))\,\hat{\mathbf{e}}_3 \qquad (3.51)$$

**Plane**

The parametric representation of the plane passing through the three points $(x_1, y_1, z_1)$, $(x_2, y_2, z_2)$, $(x_3, y_3, z_3)$, not on a line, can be represented using the position vector

$$\vec{r} = \vec{r}(u, v) = \vec{r}_1 + u(\vec{r}_2 - \vec{r}_1) + v(\vec{r}_3 - \vec{r}_1),$$

where $u, v \in \mathbb{R}$ are parameters and $\vec{r}_1, \vec{r}_2, \vec{r}_3$ are position vectors to the given points lying on the plane.

If $\hat{n} = n_1\,\hat{\mathbf{e}}_1 + n_2\,\hat{\mathbf{e}}_2 + n_3\,\hat{\mathbf{e}}_3$ is a nonzero normal vector to the plane, and $\vec{r}_1$ is a vector to a point on the plane, then the plane can be represented by either of the forms

$$\vec{n} \cdot (\vec{r} - \vec{r}_1) = 0 \qquad \text{or} \qquad n_1 x + n_2 y + n_3 z = d \qquad (3.52)$$

where $d$ is a constant and $\vec{r} = x\,\hat{\mathbf{e}}_1 + y\,\hat{\mathbf{e}}_2 + z\,\hat{\mathbf{e}}_3$ is a vector to a general point in the plane. In the special case $\hat{n}$ is a unit vector and $d \geq 0$, then $n_1^2 + n_2^2 + n_3^2 = 1$. In the equation (3.52), $d$ represents the distance from the origin to the plane and the unit vector $\hat{n}$, when placed at the origin, will point from the origin to the plane. This special form is called the normal form for the equation of the plane.

The general coordinate form for the representation of a plane is given by

$$Ax + By + Cz = D$$

where $A, B, C, D$ are constants. If one divides both sides of this equation by $\sqrt{A^2 + B^2 + C^2}$ one obtains the normal form for the equation of the plane

$$\frac{Ax + By + Cz}{\sqrt{A^2 + B^2 + C^2}} = \frac{D}{\sqrt{A^2 + B^2 + C^2}}$$

Here $d = \left|\frac{D}{\sqrt{A^2+B^2+C^2}}\right|$ is the distance of the plane from the origin and the direction cosines of the unit vector normal to the plane are given by

$$n_1 = \frac{A}{\sqrt{A^2 + B^2 + C^2}}, \qquad n_2 = \frac{B}{\sqrt{A^2 + B^2 + C^2}}, \qquad n_3 = \frac{C}{\sqrt{A^2 + B^2 + C^2}}$$

A line through the terminal points of the vectors $\vec{r}_1$ and $\vec{r}_2$ is given by $\vec{r} = \vec{r}(t) = \vec{r}_1 + t(\vec{r}_2 - \vec{r}_1)$ where $t$ is a parameter. This line intersects the plane $\vec{n} \cdot \vec{r} = d$ at the point $\vec{r}_0$ given by

$$\vec{r}_0 = \vec{r}_1 + t_0(\vec{r}_2 - \vec{r}_1) \quad \text{where} \quad t_0 = \frac{d - \vec{n} \cdot \vec{r}_1}{\vec{n} \cdot (\vec{r}_2 - \vec{r}_1)}$$

provided that the line is not parallel to the plane.

If the terminal points of the vectors $\vec{r}_1$, $\vec{r}_2$, $\vec{r}_3$ do not lie in a straight line, then the equation of the plane through these terminal points is given by the vector equation

$$\vec{n} \cdot (\vec{r} - \vec{r}_1) = 0. \quad \text{where} \quad \vec{n} = (\vec{r}_3 - \vec{r}_1) \times (\vec{r}_2 - \vec{r}_1) \text{ is normal to plane}$$

and $\vec{r} = x\,\hat{e}_1 + y\,\hat{e}_2 + z\,\hat{e}_3$ is a general position vector to an arbitrary point in the plane.

Consider two planes $\vec{n}_1 \cdot \vec{r} = d_1$ and $\vec{n}_2 \cdot \vec{r} = d_2$ which intersect. Along the line of intersection construct the normal vectors $\vec{n}_1$ and $\vec{n}_2$, then the angle $\alpha$ between these normal vectors, which represents the angle between the intersecting planes, is determined from the relation

$$\cos\alpha = \frac{\vec{n}_1 \cdot \vec{n}_2}{|\vec{n}_1| \cdot |\vec{n}_2|}$$

If $\vec{n}_1 \cdot (\vec{r} - \vec{r}_1) = 0$ and $\vec{n}_2 \cdot (\vec{r} - \vec{r}_2) = 0$ are two parallel planes, then it is necessary that $\vec{n}_1 \times \vec{n}_2 = \vec{0}$. In this case the distance between the parallel planes is given by

$$d = \frac{|\vec{n}_1 \cdot (\vec{r}_1 - \vec{r}_2)|}{|\vec{n}_1|} = \frac{|\vec{n}_2 \cdot (\vec{r}_1 - \vec{r}_2)|}{|\vec{n}_2|}$$

**Sphere**

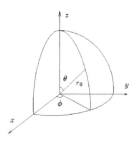

The equation of a sphere centered a $(x_0, y_0, z_0)$ with radius $r_0$ is given by the equation

$$(x - x_0)^2 + (y - y_0)^2 + (z - z_0)^2 = r_0^2 \qquad (3.53)$$

The parametric equations for representing this sphere are given by

$$\begin{aligned} x &= x_0 + r_0 \sin\theta \cos\phi \\ y &= y_0 + r_0 \sin\theta \sin\phi \\ z &= z_0 + r_0 \cos\theta \end{aligned} \quad \begin{aligned} 0 &\leq \theta \leq \pi \\ 0 &\leq \phi \leq 2\pi \end{aligned} \qquad (3.54)$$

where the parameters $\theta$, $\phi$ are illustrated in the special case the sphere is centered at the origin.

Volume $V = \dfrac{4\pi}{3}r^3$, Surface area $S = 4\pi r^2$

### Cylinder and Elliptic Cylinder

The equation of a cylinder with radius $r_0$ is given by $x^2 + y^2 = r_0^2$ for $a \leq z \leq b$. The equation of an elliptic cylinder with semi-axes $a,b$ is given by $\frac{x^2}{a^2} + \frac{y^2}{b^2} = 1$. The parametric equations of an elliptic cylinder are given by

$$\begin{aligned} x &= a\cos u \\ y &= b\sin u \\ z &= v \end{aligned} \qquad \begin{aligned} 0 &\leq u \leq 2\pi \\ a &\leq v \leq b \end{aligned} \qquad (3.55)$$

The position vector to a general point on the surface is given by

$$\vec{r} = \vec{r}(u,v) = a\cos u\, \hat{\mathbf{e}}_1 + b\sin u\, \hat{\mathbf{e}}_3 + v\, \hat{\mathbf{e}}_3 \qquad 0 \leq v \leq h,\ 0 \leq u \leq 2\pi$$

In the special case $a = b = r_0$ one obtains a right circular cylinder where the radius of the cylinder is $r_0$ and the parameters $u,\ v$ are as illustrated. The parameter $u$ can be interpreted as an angle and the parameter $v$ can be interpreted as a height.

### Hyperbolic and Parabolic cylinder

The equation of a hyperbolic cylinder symmetric about the $z$-axis is given by

$$\frac{x^2}{a^2} - \frac{y^2}{b^2} = 1$$

this surface can be represented in the parametric form

$$x = a\sinh u, \quad y = \pm b\cosh u, \quad z = v$$

for $-\alpha \leq u \leq \alpha$, $-\beta \leq v \leq \beta$ where $\alpha$ and $\beta$ are constants.

The parabolic cylinder is given by

$$y^2 = 2ax$$

and can be represented in the parametric form

$$x = 2au^2, \quad y = 2au, \quad z = v$$

for $-\alpha \leq u \leq \alpha$, $-\beta \leq v \leq \beta$ where $\alpha$ and $\beta$ are constants.

hyperbolic cylinder

parabolic cylinder

### Ellipsoid

An ellipsoid can be formed by rotating an ellipse centered at the origin about either the $x$ or $y$ axis. The equation of an ellipsoid centered at the point $(x_0, y_0, z_0)$ is given by

$$\frac{(x-x_0)^2}{a^2} + \frac{(y-y_0)^2}{b^2} + \frac{(z-z_0)^2}{c^2} = 1 \qquad (3.56)$$

where $a, b, c$ are constants. It can also be represented by the parametric equations

$$x = x_0 + a\cos u \sin v, \qquad y = y_0 + b\sin u \sin v, \qquad z = z_0 + c\cos v \qquad (3.57)$$

where $0 \leq u \leq 2\pi$ and $0 \leq v \leq \pi$. If $a = b > c$, then surface is called an oblate spheroid. If $a = b < c$, then the surface is called a prolate spheroid and with $a = b = c$ the surface reduces to a sphere with radius $a$. The volume is given by $V = \frac{4}{3}\pi abc$ and the surface area is calculated in terms of incomplete elliptic integrals.

### Elliptic cone

The equation of an elliptic cone having the $z$-axis as an axis of symmetry is given by

$$\frac{x^2}{a^2} + \frac{y^2}{b^2} = \frac{z^2}{c^2} \qquad (3.58)$$

where $a, b, c$ are constants. Here the plane $z = c = constant$, cuts the cone in an ellipse with semi-axes $a$ and $b$. The parametric equations of the elliptic cone with vertex at the origin is given by

$$x = a\cos u \sin v, \qquad y = b\sin u \sin v, \qquad z = c\sin v \qquad (3.59)$$

where $0 \leq u \leq 2\pi$ and $-\pi \leq v \leq \pi$.

### Hyperboloid of Two Sheets

The hyperboloid of two sheets centered at the origin with the $z$ axis an axis of symmetry is represented by the equation

$$\frac{x^2}{a^2} + \frac{y^2}{b^2} - \frac{z^2}{c^2} = -1 \qquad (3.60)$$

Here the plane $z = constant$, cuts the surface to form an ellipse. The plane $\alpha x + \beta y = \gamma$ cuts the surface in hyperbolas.

The surface can be represented by the parametric equations

$$x = a\cos u \cosh v, \qquad y = b\sin u \cosh v, \qquad z = c\sinh v \qquad (3.61)$$

where $0 \le u \le 2\pi$ and $-v_0 \le v \le v_0$.

**Hyperboloid of One Sheet**

The hyperboloid of one sheet centered at the origin, having the $z$-axis as an axis of symmetry, is given by the equation

$$\frac{x^2}{a^2} + \frac{y^2}{b^2} - \frac{z^2}{c^2} = 1 \qquad (3.62)$$

Here the planes $z = constant$, cut the surface in ellipses and the planes $\alpha x + \beta y = \gamma$ cut the surface in hyperbolas.

The parametric equations for the hyperboloid of one sheet are given by

$$x = a\cos u \cosh v, \qquad y = b\sin u \cosh v, \qquad z = c\sinh v \qquad (3.63)$$

where $0 \le u \le 2\pi$ and $-v_0 \le v \le v_0$.

**Elliptic Paraboloid**

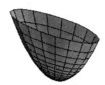

An elliptic paraboloid having the $z$-axis as an axis of symmetry is represented by the equation

$$\frac{x^2}{a^2} + \frac{y^2}{b^2} = \frac{z}{c} \qquad (3.64)$$

Here planes $z = constant$, cut the surface in an ellipse and planes $\alpha x + \beta y = \gamma$ cut the surface in a parabola.

The parametric equations representing an elliptic paraboloid are given by

$$x = a\cos u \sinh v, \qquad y = b\sin u \sinh v, \qquad z = c\sinh^2 v \qquad (3.65)$$

where $0 \le u \le v$ and $0 \le v \le v_0$.

**Hyperbolic Paraboloid**

A hyperbolic paraboloid having the $x$ axis as an axis of symmetry is represented by the equation

$$\frac{x^2}{a^2} - \frac{y^2}{b^2} = \frac{z}{c} \qquad (3.66)$$

Here planes $z = constant$, cut the surface in a hyperbola and the planes $x = constant$, $y = constant$, cut the surface in a parabola .

The parametric equations describing a hyperbolic paraboloid are given by

$$x = au, \qquad y = bv, \qquad z = c(u^2 - v^2) \qquad (3.67)$$

where $-u_0 \le u \le u_0$ and $-v_0 \le v \le v_0$.

The torus is symmetric about the $z$-axis and can be represented by the equation

$$(r_0 - \sqrt{x^2 + y^2})^2 + z^2 = r_1^2 \qquad (3.68)$$

where $r_1$ is the radius of a circular cross-section and $r_0$ is the radius of all the center points of the circular cross-sections. The volume is $V = 2\pi^2 r_1^2 r_0$ and the surface area is $S = 4\pi^2 r_1 r_0$.

The parametric equations representing a torus are given by

$$x = (r_0 + r_1 \cos v) \cos u, \qquad y = (r_0 + r_1 \cos v) \sin u, \qquad z = r_1 \sin v \qquad (3.69)$$

where $0 \leq u \leq 2\pi$ and $0 \leq v \leq 2\pi$.

**Möbius Strip**

The parametric equations representing a Möbius strip are given by

$$\begin{aligned} x &= \left(\ell - v \sin \frac{u}{2}\right) \sin u, \\ y &= \left(\ell - v \sin \frac{u}{2}\right) \cos u, \\ z &= v \cos \frac{u}{2} \end{aligned} \qquad (3.70)$$

where $0 \leq u \leq 2\pi$ and $-1 \leq v \leq 1$ where $\ell > v$. The parameter $\ell$ represents the radius of a circle. As you move through an angle $u$ around the circle, then ruled lines inclined at an angle $u/2$ are drawn. The $v$ parameter determines the length of the line drawn. A Möbius strip is a one sided surface obtained by twisting a rectangular piece of paper and gluing the ends.

**Klein bottle**

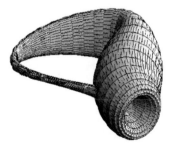

The Klein bottle is a two dimensional manifold where its inside is the same as its outside. The Klein bottle is sometimes viewed as the three-dimensional analog of the Möbius strip. The Klein bottle receives its name from the German mathematician Felix Klein (1849-1925).

The parametric equations for the Klein bottle, obtained from the Maclean reference, are

$$x = -\frac{2}{15} \cos u \, (3\cos v + 5\sin u \cos v \cos u - 30 \sin u - 60 \sin u \cos^6 u + 90 \sin u \cos^4 u)$$

$$y = \frac{-1}{15} \sin u \, (80 \cos v \cos^7 u \sin u + 48 \cos v \cos^6 u - 80 \cos v \cos^5 u \sin u - 48 \cos v \cos^4 u$$
$$- 5 \cos v \cos^3 u \sin u - 3 \cos v \cos^2 u + 5 \sin u \cos v \cos u + 3 \cos v - 60 \sin u)$$ \hfill (3.71)

$$z = \frac{2}{15} \sin v \, (3 + 5 \sin u \cos u)$$

### Geometry and Graphics

Historically, the ancient Greeks solved many algebraic equations graphically. For example, consider the cubic equation $x^3 - N = 0$ and its relation to the parabola $y = x^2$ and the circle $x^2 + y^2 - Nx - y = 0$. To find where the parabola and the circle intersect eliminate the variable $y$ from the equations defining the parabola and circle to obtain

$$x^2 + (x^2)^2 - Nx - x^2 = 0 \qquad \text{or} \qquad x^4 - Nx = x(x^3 - N) = 0$$

This shows that the parabola and circle intersect where $x = 0$ and $x = \sqrt[3]{N}$. The figure 3-21 illustrates the intersection of the parabola and the circle with the abscissa of the intersection representing the cube root of $N$ for $N = 2, 3, 4, 5, 6, 7, 8$ and $9$.

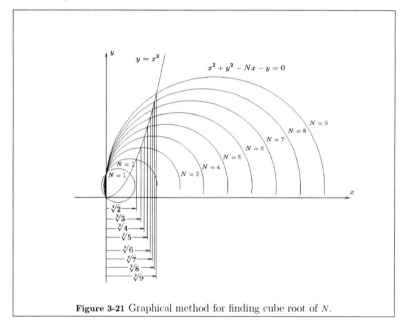

**Figure 3-21** Graphical method for finding cube root of $N$.

# Chapter 4
# Calculus

This chapter presents a review of selected material from the study area of calculus. Basic differentiation and integration formulas together with calculus concepts are developed and summarized for later reference. A short table of integrals is developed with a more extensive table of integrals presented in the Appendix B.

## Limits

The limit statement $\lim_{x \to x_0} f(x) = L$ is read, "the function $f(x)$ has the limit $L$ as $x$ approaches $x_0$". Such limit statements imply that $x$ can approach $x_0$ in any manner, but $x \neq x_0$.

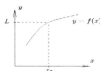

Sometimes the function $f(x)$ has a jump discontinuity at the point $x_0$ and the limiting value depends upon how $x$ approaches $x_0$. The right-hand limit of $x$ approaching $x_0$ is written

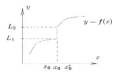

$$\lim_{x \to x_0^+} f(x) = L_2$$

and the left-hand limit of $x$ approaching $x_0$ is written

$$\lim_{x \to x_0^-} f(x) = L_1$$

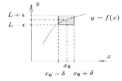

The limit statement $\lim_{x \to x_0} f(x) = L$ is to mean that for every $\epsilon > 0$ there exists a $\delta > 0$, which usually depends upon $\epsilon$, such that the difference

$$|f(x) - L| < \epsilon \quad \text{whenever} \quad 0 < |x - x_0| < \delta$$

This means the function $f(x)$ gets trapped inside the interval $(L-\epsilon, L+\epsilon)$ whenever $x$ satisfies $x_0 - \delta < x < x_0 + \delta$ and $x \neq x_0$. Here $\epsilon > 0$ can be made arbitrarily small and so the function $f(x)$ can be made as close to $L$ as desired.

The right-hand limit statement means $|f(x) - L| < \epsilon$ whenever $x_0 < x < x_0 + \delta$ and the left-hand limit statement means $|f(x) - L| < \epsilon$ whenever $x_0 - \delta < x < x_0$.

The limit statement $\lim_{x \to \infty} f(x) = L$ is to be read, "the limit of $f(x)$, as $x$ increases without bound, is $L$" and is taken to mean that for every $\epsilon > 0$, there exists and integer $N > 0$, such that for all $x > N$, the inequality $|f(x) - L| < \epsilon$ is satisfied.

## Properties of Limits

Assume the limits $\lim_{x \to x_0} f(x) = A$ and $\lim_{x \to x_0} g(x) = B$ exist, then

1. The limit of a sum or difference is the sum or difference of the limits

$$\lim_{x \to x_0} (f(x) \pm g(x)) = \lim_{x \to x_0} f(x) \pm \lim_{x \to x_0} g(x) = A \pm B$$

2. The limit of a product is the product of the limits

$$\lim_{x \to x_0} f(x)g(x) = \left(\lim_{x \to x_0} f(x)\right)\left(\lim_{x \to x_0} g(x)\right) = AB$$

3. The limit of a quotient is the quotient of the limits

$$\lim_{x \to x_0} \frac{f(x)}{g(x)} = \frac{\lim_{x \to x_0} f(x)}{\lim_{x \to x_0} g(x)} = \frac{A}{B}, \quad \text{provided } B \neq 0$$

## Continuity

A function $f(x)$ is said to be continuous at the point $x_0$ if

(i) $f(x_0)$ exists. The point $x_0$ is said to be in the domain of $f(x)$
(ii) $\lim_{x \to x_0} f(x)$ exists. That is, $f(x)$ must be defined everywhere in some interval about $x_0$
(iii) $\lim_{x \to x_0} f(x) = f(x_0)$

A continuous function is recognized as having the property that the curve representing the function can be drawn without lifting the pen or pencil from the paper.

If the function $f(x)$ is not continuous at a point $x_0$, then it is said to be discontinuous at that point. Note that if the function $f(x)$ is continuous at the point $x_0$, then both the left and right-hand limits have the same value with

$$\lim_{x \to x_0^-} f(x) = \lim_{x \to x_0^+} f(x)$$

A function $f(x)$ is said to be continuous for $a \leq x \leq b$ if $f(x)$ is continuous at every point $x$ satisfying $a \leq x \leq b$. When dealing with functions which are continuous over an interval $a \leq x \leq b$ it is to be understood that at the end points where $x = a$ it is implied that the function is continuous from the right and at the end point $x = b$ it is implied the function is continuous from the left.

If two functions $f(x)$ and $g(x)$ are continuous functions over an interval $a \leq x \leq b$, then the following functions are also continuous

$cf(x)$ where $c$ any constant

$f(x) + g(x)$ and $f(x) - g(x)$ the sum or difference of two functions

$f(x)g(x)$ the product of two functions

$\dfrac{f(x)}{g(x)}$ the quotient of two functions, provided $g(x) \neq 0$

Note 1: A continuous function of a continuous function is also continuous. That is, if $y = f(x)$ is a continuous function with the property that $y_0 = f(x_0)$ and $z = g(y)$ is a continuous function at the point $y_0$, then the function $z = g(f(x))$, called a composite function (or function of a function), is also continuous at $x = x_0$.

Note 2: The following are well known continuous functions over the domains where they are well defined.

| | | |
|---|---|---|
| *polynomials* | *exponential functions* | *inverse trigonometric functions* |
| *rational functions* | *roots of functions* | *logarithmic functions* |
| *trigonometric functions* | *inverse functions* | *composite functions* |

## $\epsilon - \delta$ Definition of Continuity

A function $f(x)$ is said to be continuous at the point $x = x_0$ if for every $\epsilon > 0$ there exists a $\delta > 0$ such that $|f(x) - f(x_0)| < \epsilon$, whenever $|x - x_0| < \delta$. Note that this is the same definition used to define the limit of a function with the exceptions that (i) the restriction $x \neq x_0$ has been removed and (ii) $L = f(x_0)$ is the limiting value.

## Intermediate Value Theorem

Let $y = f(x)$ denote a continuous function on the interval $a \leq x \leq b$ and construct the line $y = c = constant$ where $f(a) < c < f(b)$. The Intermediate Value Theorem states that there must exist at least one number $\xi$ satisfying $a < \xi < b$ such that $f(\xi) = c$.

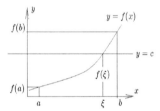

## Derivatives

If $y = f(x)$ is a continuous function of the single variable $x$, then this function can be represented as curve as illustrated in the figure 4-1. The line drawn connecting the points $(x, f(x))$ and $(x + \Delta x, f(x + \Delta x))$ on the curve is called the secant line. In the limit as $\Delta x \to 0$, the secant line becomes the tangent line. In figure 4-1 the symbol $f(x)$ or $y$ represents the height of the curve at the point $x$ and $f(x + \Delta x)$ represents the height of the curve at point $x + \Delta x$.

Consider the special case of finding the secant line between two fixed points $(x_0, f(x_0))$ and $(x_0 + \Delta x, f(x_0 + \Delta x))$. The equation of the secant line can be constructed using the point slope formula. The slope of the secant line through points $(x_0, f(x_0))$ and $(x_0 + \Delta x, f(x_0 + \Delta x))$ is given by

$$\text{Slope of secant line} = m_{secant} = \frac{\Delta y}{\Delta x} = \frac{f(x_0 + \Delta x) - f(x_0)}{\Delta x} = \frac{\text{change in y}}{\text{change in x}} \qquad (4.1)$$

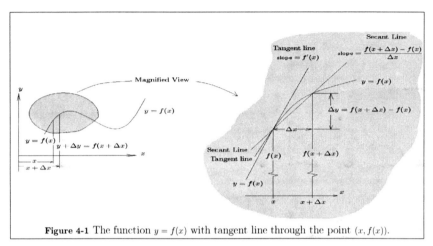

**Figure 4-1** The function $y = f(x)$ with tangent line through the point $(x, f(x))$.

and the equation of the secant line is obtained using the point-slope formula

$$y - f(x_0) = m_{secant}(x - x_0) \quad \text{or} \quad y - f(x_0) = \left[\frac{f(x_0 + \Delta x) - f(x_0)}{\Delta x}\right](x - x_0) \quad (4.2)$$

In the limit as $\Delta x$ tends toward zero, the secant line becomes the tangent line and the slope of the tangent line is represented

$$\text{Slope of tangent line at point } x_0 = m_{tangent} = \lim_{\Delta x \to 0} \frac{\Delta y}{\Delta x} = \lim_{\Delta x \to 0} \frac{f(x_0 + \Delta x) - f(x_0)}{\Delta x} = f'(x_0) \quad (4.3)$$

if this limit exists. The equation of the tangent line is then given by the point-slope formula

$$y - f(x_0) = m_{tangent}(x - x_0) \quad \text{or} \quad y - f(x_0) = f'(x_0)(x - x_0)$$

Whenever the limit defining the slope of the tangent line exists, the slope $m_{tangent}$ is called the derivative of $y$ with respect to $x$ and is denoted using the notation $m_{tangent} = f'(x)$. The derivative evaluated at a point $x_0$ is denoted $f'(x_0)$. The derivative function $f'(x)$ is a function derived from $f(x)$ and represents the instantaneous rate of change of the curve $y = f(x)$ with respect to $x$ or slope of the tangent line at point $(x, f(x))$ on the curve. The general situation is illustrated in the figure 4-1. Alternative representations for the derivative function are

$$\begin{pmatrix} \text{Derivative of } y \\ \text{with respect to } x \end{pmatrix} = m_{tangent} = \lim_{\Delta x \to 0} \frac{\Delta y}{\Delta x} = \frac{dy}{dx} = \frac{d}{dx}y = \frac{d}{dx}f(x) = y' = f'(x) = \frac{df(x)}{dx} \quad (4.4)$$

In general, the derivative of a function $y = f(x)$, at a general point $x$, is defined

$$\frac{dy}{dx} = f'(x) = \lim_{\Delta x \to 0} \frac{f(x + \Delta x) - f(x)}{\Delta x} = \lim_{\Delta x \to 0} \frac{\Delta y}{\Delta x} \quad \text{if this limit exits}$$

The derivative function represents the slope of the tangent line at the point $(x, f(x))$ on the curve.

The derivative function $f'(x)$ can be differentiated to obtain $\frac{d}{dx}f'(x) = f''(x)$ which is called a second derivative. The second derivative is related to the concavity of the curve. If $f''(x) > 0$, then the curve is concave upward and if $f''(x) < 0$, then the curve is concave downward. Higher derivatives are defined as derivatives of lower ordered derivatives.

## Basic Differentiation Rules

$\frac{d}{dx}c = 0$, $c$ constant

$\frac{d}{dx}x = 1$

$\frac{d}{dx}x^n = nx^{n-1}$

$\frac{d}{dx}e^x = e^x$

$\frac{d}{dx}\ln x = \frac{1}{x}$

$\frac{d}{dx}[u(x)v(x)] = u(x)\frac{dv}{dx} + \frac{du}{dx}v(x)$

$\frac{d}{dx}\left(\frac{u(x)}{v(x)}\right) = \frac{v(x)\frac{du}{dx} - u(x)\frac{dv}{dx}}{[v(x)]^2}$

$\frac{d}{dx}u(x)^n = nu(x)^{n-1}\frac{du}{dx}$

$\frac{d}{dx}e^{u(x)} = e^{u(x)}\frac{du}{dx}$

$\frac{d}{dx}\ln u(x) = \frac{1}{u(x)}\frac{du}{dx}$

$\frac{d}{dx}\sin u(x) = \cos u(x)\frac{du}{dx}$

$\frac{d}{dx}\cos u(x) = -\sin u(x)\frac{du}{dx}$

$\frac{d}{dx}\tan u(x) = \sec^2 u(x)\frac{du}{dx}$

$\frac{d}{dx}\sinh u(x) = \cosh u(x)\frac{du}{dx}$

$\frac{d}{dx}\cosh u(x) = \sinh u(x)\frac{du}{dx}$

$\frac{d}{dx}\tanh u(x) = \operatorname{sech}^2 u(x)\frac{du}{dx}$

## Differentials

For the curve $y = f(x)$, the change in $y$ in moving from $x$ to $x + \Delta x$ is given by

$$\Delta y = f(x + \Delta x) - f(x)$$

so that

$$\frac{\Delta y}{\Delta x} = \frac{f(x + \Delta x) - f(x)}{\Delta x} = f'(x) + \epsilon = \frac{dy}{dx} + \epsilon$$

where $\epsilon$ is a small quantity which tends to zero as $\Delta x$ tends toward zero. The change in $y$ is given by $\Delta y = f'(x)\Delta x + \epsilon \Delta x$. The differential of $x$ is defined $dx = \Delta x$ and the differential of $y$ is defined as $dy = f'(x)\,dx$.

## Properties of Differentials

1. $d\left(\frac{u}{v}\right) = \frac{v\,du - u\,dv}{v^2}$
2. $d(u^n) = nu^{n-1}\,du$
3. $d(u_1 \pm u_2 \pm \cdots \pm u_n) = du_1 \pm du_2 \pm \cdots \pm du_n$
4. $d(uv) = u\,dv + v\,du$

## Higher Derivatives

The derivative of a derivative gives the next higher ordered derivative. For example,

Second derivatives $\quad \frac{d}{dx}\left(\frac{dy}{dx}\right) = \frac{d^2y}{dx^2} = f''(x) = y''$

Third derivatives $\quad \frac{d}{dx}\left(\frac{d^2y}{dx^2}\right) = \frac{d^3y}{dx^3} = f'''(x) = y'''$

$\vdots$

$n$th derivatives $\quad \frac{d}{dx}\left(\frac{d^{n-1}y}{dx^{n-1}}\right) = \frac{d^ny}{dx^n} = f^{(n)}(x) = y^{(n)}$

## Parametric functions

If a function is defined parametrically, say $x = x(t)$ and $y = y(t)$, then it is possible to treat $y$ as a function of $x$ with the derivatives

$$y' = \frac{dy}{dx} = \frac{\frac{dy}{dt}}{\frac{dx}{dt}} \qquad y'' = \frac{d^2y}{dx^2} = \frac{d}{dx}\left(\frac{dy}{dx}\right) = \frac{d}{dt}\left(\frac{dy}{dx}\right)\frac{dt}{dx} = \frac{\frac{dx}{dt}\frac{d^2y}{dt^2} - \frac{dy}{dt}\frac{d^2x}{dt^2}}{\left(\frac{dx}{dt}\right)^3}$$

$$y''' = \frac{d^3y}{dx^3} = \frac{d}{dx}\left(\frac{d^2y}{dx^2}\right) = \frac{d}{dt}\left(\frac{d^2y}{dx^2}\right)\frac{dt}{dx} = \frac{(x')^2 y''' - x'y'x''' - 3x'x''y'' + 3y'(x'')^2}{(x')^5}$$

In general, $\frac{dy}{dx}\frac{dx}{dt} = \frac{dy}{dt}$. This is known as chain rule differentiation.

## Leibnitz Rule for Differentiating Products

Define the differential operators $D = \frac{d}{dx}, D^2 = \frac{d^2}{dx^2}, \ldots, D^n = \frac{d^n}{dx^n}$ such that

$$Dy = \frac{dy}{dx}, \quad D^2 y = \frac{d^2 y}{dx^2}, \quad \ldots, \quad D^n y = \frac{d^n y}{dx^n}$$

The Leibnitz rule for differentiating products can be written using the differential operator $D$ as follows.

$$D(uv) = uDv + vDu$$
$$D^2(uv) = uD^2v + 2DuDv + vD^2u$$
$$D^3(uv) = uD^3v + 3DuD^2v + 3D^2uDv + vD^3u$$
$$\vdots \qquad \vdots$$
$$D^n(uv) = uD^nv + \binom{n}{1}DuD^{n-1}v + \binom{n}{2}D^2uD^{n-2}v + \cdots + vD^nu$$

This can also be expressed using the summation notation as

$$D^n(uv) = \sum_{i=0}^{n}\binom{n}{i}D^iuD^{n-i}v \qquad \text{where} \qquad \binom{n}{r} = \frac{n!}{r!(n-r)!} \quad \text{are the binomial coefficients.}$$

## Partial Derivatives

If $u = u(x,y)$ is a function of two real variables $x$ and $y$, then the partial derivatives of $u$ with respect to $x$ and $y$ are defined

$$\frac{\partial u}{\partial x} = \lim_{\Delta x \to 0}\frac{u(x+\Delta x, y) - u(x,y)}{\Delta x}, \qquad \frac{\partial u}{\partial y} = \lim_{\Delta y \to 0}\frac{u(x, y+\Delta y) - u(x,y)}{\Delta y},$$

provided these limits exist. The partial derivative operator $\frac{\partial}{\partial x}$ is just like an ordinary derivative, except all other variables are held constant during the differentiation with respect to $x$. Similarly, the partial differential operator $\frac{\partial}{\partial y}$ is just like an ordinary derivative except all other variables are held constant during the differentiation with respect to $y$.

# 159

Higher partial derivatives are defined as a derivative of a lower ordered derivative. For example, The second partial derivatives of $u = u(x,y)$ are defined

$$\frac{\partial^2 u}{\partial x^2} = \frac{\partial}{\partial x}\left(\frac{\partial u}{\partial x}\right), \qquad \frac{\partial^2 u}{\partial y^2} = \frac{\partial}{\partial y}\left(\frac{\partial u}{\partial y}\right)$$

The second derivatives $\frac{\partial^2 u}{\partial x \partial y} = \frac{\partial}{\partial x}\left(\frac{\partial u}{\partial y}\right)$, $\frac{\partial^2 u}{\partial y \partial x} = \frac{\partial}{\partial y}\left(\frac{\partial u}{\partial x}\right)$ are called mixed partial derivatives. If both the function $u = u(x,y)$ and its first ordered partial derivatives are continuous functions, then the mixed partial derivatives are equal to one another, in which case it doesn't matter as to the order of the differentiation and consequently $\frac{\partial^2 u}{\partial x \partial y} = \frac{\partial^2 u}{\partial y \partial x}$.

The differential of the continuous function $u = u(x,y)$ is written $du = \frac{\partial u}{\partial x} dx + \frac{\partial u}{\partial y} dy$

## Implicit Differentiation

Implicit differentiation is a result of the following theorem.

### Implicit Function Theorem

*Let $U(x,y)$ denote a continuous function of two real variables $x$ and $y$ which possesses the continuous partial derivatives $\frac{\partial U(x,y)}{\partial x}$ and $\frac{\partial U(x,y)}{\partial y}$. Assume there exists a point $(x_0, y_0)$ such that $U(x_0, y_0) = 0$ is satisfied and $\frac{\partial U(x_0, y_0)}{\partial y} \neq 0$, then there exists a rectangle $a \leq x_0 \leq b$, $c \leq y_0 \leq d$ about the point $(x_0, y_0)$ such that for every $x$ in the interval $a \leq x \leq b$, the equation $U(x,y) = 0$ determines a single-valued function of $y$ in terms of $x$, denoted $y = f(x)$, which satisfies $c \leq y \leq d$. The function $y = f(x)$ satisfies the conditions*

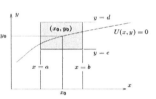

$$(i) \quad y_0 = f(x_0) \quad \text{and} \quad U(x,y) = U(x, f(x)) = 0$$

*for all $x$ within the rectangle. The function $y = f(x)$ is continuous and has a derivative which is also continuous.*

It is possible to use chain rule differentiation to calculate the derivative of the function $y = f(x)$ which is associated with the implicit function $U(x,y) = 0$. If $U(x,y) = 0$, then using the chain rule for differentiation gives

$$\frac{d}{dx} U(x,y) = \frac{\partial U}{\partial x} \frac{dx}{dx} + \frac{\partial U}{\partial y} \frac{dy}{dx} = \frac{d}{dx} 0$$

and solving for $\frac{dy}{dx}$ one finds

$$\frac{dy}{dx} = -\frac{\frac{\partial U}{\partial x}}{\frac{\partial U}{\partial y}} \qquad (4.5)$$

Continue to use the chain rule for differentiation to calculate higher ordered derivatives associated with $y = f(x)$. For example, a derivative with respect to $x$ gives the second derivative

$$\frac{d}{dx}\left(\frac{dy}{dx}\right) = \frac{d^2y}{dx^2} = -\frac{d}{dx}\left(\frac{\frac{\partial U}{\partial x}}{\frac{\partial U}{\partial y}}\right) = -\left[\frac{\frac{\partial U}{\partial y}\left(\frac{\partial^2 U}{\partial x^2} + \frac{\partial^2 U}{\partial x \partial y}\frac{dy}{dx}\right) - \frac{\partial U}{\partial x}\left(\frac{\partial^2 U}{\partial y \partial x} + \frac{\partial^2 U}{\partial y^2}\frac{dy}{dx}\right)}{\left(\frac{\partial U}{\partial y}\right)^2}\right] \quad (4.6)$$

where the derivative $\frac{dy}{dx}$ in equation (4.6) is obtained from the equation (4.5). Higher derivatives are calculated in a similar manner.

### Example 4-1.

Find $\frac{dy}{dx}$ if $x^3 y^2 + x \sin y = 0$.

**Solution:** Here $U(x, y) = x^3 y^2 + x \sin y$ has the partial derivatives

$$\frac{\partial U}{\partial x} = 3x^2 y^2 + \sin y \quad \text{and} \quad \frac{\partial U}{\partial y} = x^3 2y + x \cos y$$

If $U(x, y) = 0$, then $\frac{dU}{dx} = \frac{\partial U}{\partial x} + \frac{\partial U}{\partial y}\frac{dy}{dx} = 0$ or

$$\frac{dy}{dx} = -\frac{\frac{\partial U}{\partial x}}{\frac{\partial U}{\partial y}} = -\left(\frac{3x^2 y^2 + \sin y}{2x^3 y + x \cos y}\right) \quad (4.7)$$

Alternatively, by treating $y$ as a function of $x$

$$\frac{d}{dx}(x^3 y^2 + x \sin y) = \frac{d}{dx} 0 \implies x^3 2y \frac{dy}{dx} + 3x^2 y^2 + x \cos y \frac{dy}{dx} + \sin y = 0$$

and solving this last equation for $\frac{dy}{dx}$ there results the equation (4.7).

### Mean Value Theorem for Derivatives ■

The mean value theorem for derivatives is illustrated in figure 4-2 and can be described as follows. If $y = f(x)$ is a continuous curve over the interval $a \le x \le b$, then there exists at least one point $x = \xi$ where the slope of the tangent line $f'(\xi)$ is the same as the slope of the secant line through the points $(a, f(a))$ and $(b, f(b))$. That is, there must exist at least one point $x = \xi$ where

$$f'(\xi) = \frac{f(b) - f(a)}{b - a} \quad (4.48)$$

Figure 4-2 illustrates the secant line through the points $(a, f(a))$ and $(b, f(b))$ which has the slope $\frac{f(b) - f(a)}{b - a}$. Also illustrated is the tangent line through the point $(\xi_1, f(\xi_1))$ which is parallel to the secant line. The figure 4-2 illustrates a situation where more than one value of $\xi$ exists which satisfies the mean value theorem for derivatives and so through the point $(\xi_2, f(\xi_2))$ another tangent line has been constructed which has the same slope as the secant line.

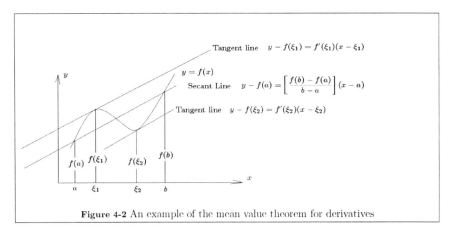

**Figure 4-2** An example of the mean value theorem for derivatives

### Rolle's Theorem

If $y = f(x)$ is a continuous and differentiable function everywhere over the interval $(a, b)$ and satisfies the conditions $f(a) = f(b)$, then there exists at least one point $\xi$, satisfying $a < \xi < b$ such that $f'(\xi) = 0$. Rolle's theorem does not apply to functions with corners or jump discontinuities in the interval $(a, b)$.

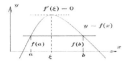

The Rolle's theorem states that there exists at least one point $\xi$ in the interval $(a, b)$ where the slope of the curve is zero. This occurs at a point where the function has a maximum or minimum value.

### Cauchy's Generalized Mean Value Theorem

Let $f(x)$ and $g(x)$ denote two functions which are both continuous and differentiable over the interval $(a, b)$. If $g(a) \neq g(b)$ and the derivative functions $f'(x)$ and $g'(x)$ are not zero simultaneously at some value $x$ within the interval $(a, b)$, then one can find a value $x = \xi$ such that

$$\frac{f(b) - f(a)}{g(b) - g(a)} = \frac{f'(\xi)}{g'(\xi)}, \qquad a < \xi < b \tag{4.9}$$

**Indeterminate Forms** $\quad \dfrac{0}{0}, \; \dfrac{\infty}{\infty}, \; 0 \cdot \infty, \; \infty - \infty, \; 0^0, \; \infty^0, \; 1^\infty$

A special case of the Cauchy generalized mean value theorem occurs when $b = x$ and the condition $f(a) = g(a) = 0$ is used in equation (4.9). This reduces equation (4.9) to the form

$$\frac{f(x)}{g(x)} = \frac{f'(\xi)}{g'(\xi)}, \qquad a < \xi < x$$

Note that if $x$ approaches $a$, then $\xi$ must also approach $a$ as it is sandwiched between the values $a$ and $x$. Consequently,

$$\lim_{x \to a} \frac{f(x)}{g(x)} = \lim_{\xi \to a} \frac{f'(\xi)}{g'(\xi)} = \lim_{x \to a} \frac{f'(x)}{g'(x)}, \qquad \text{provided the limit exists} \qquad (4.10)$$

This result is known as L'Hospital's Rule.

L'Hospital's rule is used whenever the indeterminate form 0/0 arises. In practice, it is possible to continue to apply L'Hospital's Rule until one finds a ratio involving some $n$th derivative, $\dfrac{f^{(n)}(x)}{g^{(n)}(x)}$, that is not indeterminate.

If the fraction $f(x)/g(x)$ takes on the indeterminate form $\infty/\infty$ as $x$ approaches some finite value or as $x$ becomes infinite, then it is possible to apply the L'Hospital's Rule given by equation (4.10).

If a difference of two functions $f(x) - g(x)$ takes on the indeterminate form $\infty - \infty$ as $x$ approaches a finite value or as $x$ becomes infinite, then one usually applies some algebraic manipulation in order to transform the difference into one of the indeterminate forms 0/0 or $\infty/\infty$. One suggested form is to write $f(x) - g(x) = \dfrac{\frac{1}{g(x)} - \frac{1}{f(x)}}{\frac{1}{f(x)} \frac{1}{g(x)}}$. Another suggested form is to write $f(x) - g(x) = \ln\left[\dfrac{e^{f(x)}}{e^{g(x)}}\right]$.

Whenever the product of two functions $f(x) \cdot g(x)$ takes on the form $0 \cdot \infty$ as $x$ approaches some finite value or as $x$ becomes infinite, then write the product in one of the forms

$$f(x) \cdot g(x) = \frac{f(x)}{\frac{1}{g(x)}} = \frac{g(x)}{\frac{1}{f(x)}}$$

so that the limiting value then assumes one of the forms 0/0 or $\infty/\infty$.

The indeterminate form $0^0$, $\infty^0$ or $1^\infty$ are usually associated with functions having the form $f(x)^{g(x)}$. In such cases, define

$$y = f(x)^{g(x)} \qquad \text{and} \qquad \ln y = g(x) \ln f(x)$$

Now if the original function becomes indeterminate as $x$ approaches some value $a$, the function $\ln y$ will take on the form $0 \cdot \infty$ and its limit can be determined by one of the above mentioned methods.

## Maximum and Minimum Values

Let $y = f(x)$ denote a continuous function which is not everywhere constant which has the property that both the first derivative $\dfrac{dy}{dx} = f'(x)$ and second derivative $\dfrac{d^2y}{dx^2} = f''(x)$ are continuous over some interval $a \le x \le b$. If there exists a point $x_0$ satisfying $a < x_0 < b$ such that $f'(x_0) = 0$, then the point $(x_0, f(x_0))$ is called a critical point of the function $y = f(x)$. Another way of saying this is that at a critical point of the function $y = f(x)$, the tangent to the curve is a horizontal line.

**Definition of relative maximum or local maximum**
If $x_0$ is a critical point and $f(x_0) \geq f(x)$ for all $x$ near $x_0$, then $f(x_0)$ is called a relative or local maximum value of $f(x)$.

**Definition of relative minimum or local minimum**
If $x_0$ is a critical point and $f(x_0) \leq f(x)$ for all $x$ near $x_0$, then $f(x_0)$ is called a relative or local minimum value of $f(x)$.

The existence of a local maximum or minimum value associated with a function $y = f(x)$ can be found by examining the shape of the curve in the vicinity of a critical point $(x_0, f(x_0))$. This shape can be determined by finding the slope of the tangent line to the curve at values of $x$ near $x_0$. The figure 4-3 illustrates various situations where $y = f(x)$ has a critical point. The figure 4-3(a) illustrates a local maximum value. The first derivative test for a local maximum is to test the slopes on either side of the critical point. Let $x$ move from left to right across the critical point $x_0$, then if the slope changes from positive to zero to negative, a local maximum value exists at the point $(x_0, f(x_0))$. The figure 4-3(b) illustrates a local minimum value. The first derivative test for a local minimum value is to examine the slope of the tangent line as $x$ moves from left to right across the critical value $x_0$. If the slope changes from negative to zero to positive, then a local minimum value exists for $y = f(x)$ at the point $(x_0, f(x_0))$.

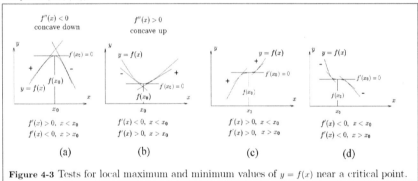

**Figure 4-3** Tests for local maximum and minimum values of $y = f(x)$ near a critical point.

The figures 4-3(c) and 4-3(d) illustrate situations where the first derivative test fails and the point $(x_0, f(x_0))$ corresponds to neither a maximum or minimum. In these situations the sign of the first derivative on either side of the critical point are the same and so the point $(x_0, f(x_0))$ is called a point of inflection.

The second derivative determines the concavity of the curve. If $f''(x) > 0$ in the vicinity of a critical point, then this implies the first derivative is increasing as $x$ increases or the slope of the tangent line is rotating counter clockwise as $x$ increases giving the curve a concave upward shape. Conversely, if $f''(x) < 0$ in the vicinity of a critical point, the slope of the tangent line decreases as $x$ increases, so that the tangent line rotates clockwise. When this happens the curve is said to be concave downward.

This gives the second derivative test for a relative maximum or minimum. At a local maximum, one finds $f''(x_0) < 0$ so the curve is concave downward and at a local minimum, one finds $f''(x_0) > 0$ so the curve is concave upward. A point where the curve changes the direction of concavity is called a point of inflection.

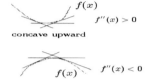

### Taylor Series for Functions of One Variable

The Taylor series expansion of a function $y = f(x)$ about the point $x_0$ is given by

$$y = f(x) = f(x_0) + f'(x_0)\frac{(x-x_0)}{1!} + f''(x_0)\frac{(x-x_0)^2}{2!} + \cdots + f^{(n)}(x_0)\frac{(x-x_0)^n}{n!} + R_n \qquad (4.11)$$

where $R_n$ is called a remainder term and can have one of the following forms.

$$\text{The Lagrange form} \qquad R_n = \frac{f^{(n+1)}(\xi_1)(x-a)^{n+1}}{(n+1)!}, \qquad x_0 < \xi_1 < x \qquad (4.12)$$

or

$$\text{The Cauchy form} \qquad R_n = \frac{f^{(n+1)}(\xi_2)(x-\xi_2)^n(x-x_0)}{n!}, \qquad x_0 < \xi_2 < x$$

where $\xi_1$ and $\xi_2$ represent some unknown points lying in the interval $(x_0, x)$. Note that in the special case $x_0 = 0$, then the above Taylor series reduces the well known Maclaurin series expansion of $y = f(x)$ about the origin.

### Examples of Series Expansions

$$\sin x = x - \frac{x^3}{3!} + \frac{x^5}{5!} - \frac{x^7}{7!} + \cdots + (-1)^n \frac{x^{2n+1}}{(2n+1)!} + \cdots$$

$$\cos x = 1 - \frac{x^2}{2!} + \frac{x^4}{4!} - \frac{x^6}{6!} + \cdots + (-1)^n \frac{x^{2n}}{(2n)!} + \cdots$$

$$\tan x = x + \frac{x^3}{3} + \frac{2x^5}{15} + \cdots + \frac{2^{2n}(2^{2n-1}-1)\mathfrak{B}_n x^{2n-1}}{(2n)!} + \cdots$$

$$e^x = 1 + x + \frac{x^2}{2!} + \frac{x^3}{3!} + \cdots + \frac{x^n}{n!} + \cdots$$

$$\ln(1+x) = x - \frac{x^2}{2} + \frac{x^3}{3} - \frac{x^4}{4} + \cdots + (-1)^{n+1}\frac{x^n}{n} + \cdots$$

$$\sinh x = x + \frac{x^3}{3!} + \frac{x^5}{5!} + \cdots + \frac{x^{2n+1}}{(2n+1)!} + \cdots$$

$$\cosh x = 1 + \frac{x^2}{2!} + \frac{x^4}{4!} + \cdots + \frac{x^{2n}}{(2n)!} + \cdots$$

$$\tanh x = x - \frac{x^3}{3} + \frac{2x^5}{15} + \cdots + (-1)^{n-1}\frac{2^{2n}(2^{2n}-1)\mathfrak{B}_n x^{2n-1}}{(2n)!} + \cdots$$

See chapter 7 for definition of the Bernoulli number $\mathfrak{B}_n$.

### Taylor Series for Functions of Two Variables

Using the result given by equation (4.11) it is possible to derive a Taylor series expansion associated with a function of two variables $f = f(x, y)$. Assume the function $f(x, y)$ is defined in a region about a fixed point $(x_0, y_0)$, where the points $(x_0, y_0)$ and $(x, y)$ can be connected by a straight line. Such regions are called connected regions. Further, let $f(x, y)$ possess $n$th-order partial derivatives which also exist in the region which surrounds the fixed point $(x_0, y_0)$. The Taylor's series expansion of $f(x, y)$ about the point $(x_0, y_0)$ is given by

$$f(x_0 + h, y_0 + k) = f(x_0, y_0) + \frac{\partial f(x_0, y_0)}{\partial x} h + \frac{\partial f(x_0, y_0)}{\partial y} k \\ + \frac{1}{2!} \left[ \frac{\partial^2 f(x_0, y_0)}{\partial x^2} h^2 + 2 \frac{\partial^2 f(x_0, y_0)}{\partial x \partial y} hk + \frac{\partial^2 f(x_0, y_0)}{\partial y^2} k^2 \right] + \cdots \quad (4.13)$$

This expansion can be represented in a simpler form by defining the differential operator

$$D = h \frac{\partial}{\partial x} + k \frac{\partial}{\partial y}, \qquad h \text{ and } k \text{ are constants.}$$

The Taylor series can then be represented in the form

$$f(x_0 + h, y_0 + k) = \sum_{j=0}^{n} \frac{1}{j!} D^j f(x, y) + R_{n+1}. \quad (4.14)$$

where all the derivatives are evaluated at the point $(x_0, y_0)$. Analogous to equation (4.12), the remainder term can be expressed as

$$R_{n+1} = \frac{1}{(n+1)!} D^{(n+1)} f(x, y), \qquad \text{to be evaluated at } (x, y) = (\xi, \eta) \quad (4.15)$$

where the point $(\xi, \eta)$, lies somewhere on the straight line connecting the points $(x_0 + h, y_0 + k)$ and $(x_0, y_0)$.

The equation (4.13) or (4.14) is derived by introducing a new independent variable $t$ which is the parameter for the straight line defined by the equations

$$x = x_0 + ht, \qquad y = y_0 + kt, \quad \text{with} \quad \frac{dx}{dt} = h, \quad \text{and} \quad \frac{dy}{dt} = k$$

where $h$ and $k$ are constants and $0 \le t \le 1$. Consider the function of the single variable $t$ defined by

$$F(t) = f(x, y) = f(x_0 + ht, y_0 + kt)$$

which is a composite function of the single variable $t$. The composite function can be expanded in a Maclaurin series about $t = 0$ to obtain

$$F(t) = F(0) + F'(0)t + F''(0)\frac{t^2}{2!} + \cdots + F^{(n)}(0)\frac{t^n}{n!} + F^{(n+1)}(\xi)\frac{t^{(n+1)}}{(n+1)!}, \quad 0 < \xi < t. \quad (4.16)$$

Evaluation of equation (4.16) at $t = 1$ gives $f(x_0 + h, y_0 + k)$.

The first $n$ derivatives of the function $F(t)$ are calculated using chain rule differentiation. The first derivative is

$$F'(t) = \frac{\partial f(x,y)}{\partial x}\frac{dx}{dt} + \frac{\partial f(x,y)}{\partial y}\frac{dy}{dt}$$
$$= \frac{\partial f(x,y)}{\partial x}h + \frac{\partial f(x,y)}{\partial y}k. \qquad (4.17)$$

By differentiating this expression, the second derivative can be determined as

$$F''(t) = \left[\frac{\partial^2 f(x,y)}{\partial x^2}h + \frac{\partial^2 f(x,y)}{\partial y\,\partial x}k\right]\frac{dx}{dt}$$
$$+ \left[\frac{\partial^2 f(x,y)}{\partial x\,\partial y}h + \frac{\partial^2 f(x,y)}{\partial y^2}k\right]\frac{dy}{dt}$$

or

$$F''(t) = \frac{\partial^2 f(x,y)}{\partial x^2}h^2 + 2\frac{\partial^2 f(x,y)}{\partial x\,\partial y}hk + \frac{\partial^2 f(x,y)}{\partial y^2}k^2. \qquad (4.19)$$

Continuing in this manner, higher derivatives of $F(t)$ can be calculated. For example, the third derivative is calculated from equation (4.49), to obtain

$$F'''(t) = \left[\frac{\partial^3 f}{\partial x^3}h^2 + 2\frac{\partial^3 f}{\partial y\,\partial x^2}hk + \frac{\partial^3 f}{\partial x\,\partial y^2}k^2\right]\frac{dx}{dt}$$
$$+ \left[\frac{\partial^3 f}{\partial y\,\partial x^2}h^2 + 2\frac{\partial^3 f}{\partial x\,\partial y^2}hk + \frac{\partial^3 f}{\partial y^3}k^2\right]\frac{dy}{dt}$$

or

$$F'''(t) = \frac{\partial^3 f}{\partial x^3}h^3 + 3\frac{\partial^3 f}{\partial y\,\partial x^2}h^2 k + 3\frac{\partial^3 f}{\partial x\,\partial y^2}hk^2 + \frac{\partial^3 f}{\partial y^3}k^3, \qquad (4.19)$$

Using the operator $D = h\frac{\partial}{\partial x} + k\frac{\partial}{\partial y}$ a pattern to these derivatives can be constructed

$$F'(t) = Df(x,y) = \left(h\frac{\partial}{\partial x} + k\frac{\partial}{\partial y}\right)f(x,y)$$
$$F''(t) = D^2 f(x,y) = \left(h\frac{\partial}{\partial x} + k\frac{\partial}{\partial y}\right)^2 f(x,y)$$
$$F'''(t) = D^3 f(x,y) = \left(h\frac{\partial}{\partial x} + k\frac{\partial}{\partial y}\right)^3 f(x,y)$$
$$\vdots \qquad \vdots$$
$$F^{(n)}(t) = D^n f(x,y) = \left(h\frac{\partial}{\partial x} + k\frac{\partial}{\partial y}\right)^n f(x,y).$$

Here the operator $D^n = \left(h\frac{\partial}{\partial x} + k\frac{\partial}{\partial y}\right)^n$ can be expanded just like the binomial expansion and

$$F^{(n)}(t) = D^n f(x,y) = h^n\frac{\partial^n f}{\partial x^n} + \binom{n}{1}h^{n-1}k\frac{\partial^n f}{\partial x^{n-1}\partial y} + \binom{n}{2}h^{n-2}k^2\frac{\partial^n f}{\partial x^{n-2}\partial y^2}$$
$$+ \cdots + \binom{n}{n-1}hk^{n-1}\frac{\partial^n f}{\partial x\,\partial y^{n-1}} + k^n\frac{\partial^n f}{\partial y^n}, \qquad (4.21)$$

where $\binom{n}{m} = \frac{n!}{m!(n-m)!}$ are the binomial coefficients.

In order to calculate the Maclaurin series about $t = 0$, each of the derivatives must be evaluated at the value $t = 0$ which corresponds to the point $(x_0, y_0)$ on the line. Substituting these derivatives into the Maclaurin series produces the result given by equation (4.13), where all derivatives are understood to be evaluated at the point $(x_0, y_0)$.

In order for the Taylor series to exist, all the partial derivatives of $f$ through the $n$th order must exist at the point $(x_0, y_0)$. In this case, write $f \in C^n$ over the connected region containing the points $(x_0, y_0)$ and $(x, y)$. The notation $f \in C^n$ is read, "$f$ belongs to the class of functions which have all partial derivatives through the $n$th order, and further, these partial derivatives are continuous functions in the connected region surrounding the point $(x_0, y_0)$."

In a similar fashion it is possible to derive the Taylor series expansion of a function $f = f(x, y, z)$ of three variables. Assume the Taylor series expansion is to be about the point $(x_0, y_0, z_0)$, then show the Taylor series expansion has the form

$$f(x_0 + h, y_0 + k, z_0 + \ell) = \sum_{j=0}^{n} \frac{1}{j!} D^j f(x, y, z) + R_{n+1} \qquad (4.21)$$

where

$$Df = \left( h\frac{\partial}{\partial x} + k\frac{\partial}{\partial y} + \ell\frac{\partial}{\partial z} \right) f = \left( h\frac{\partial f}{\partial x} + k\frac{\partial f}{\partial y} + \ell\frac{\partial f}{\partial z} \right) \qquad (4.22)$$

is a differential operator and $h = x - x_0$, $k = y - y_0$ and $\ell = z - z_0$. After expanding the derivative operator $D^j f$ for $j = 0, 1, 2, \ldots$, each of the derivatives are to be evaluated at the point $(x_0, y_0, z_0)$. The term $R_{n+1}$ is the remainder term given by

$$R_{n+1} = \frac{1}{(n+1)!} D^{(n+1)} f(x, y, z) \Big|_{(x,y,z)=(\xi,\eta,\zeta)} \qquad (4.23)$$

where the point $(\xi, \eta, \zeta)$ is some unknown point on the line connecting the points $(x_0, y_0, z_0)$ and $(x_0 + h, y_0 + k, z_0 + \ell)$.

Functions of $n$-variables $f = f(x_1, x_2, \ldots, x_n)$ have their Taylor series expansions derived in a manner similar to the above by employing a differential operator of the form

$$D = \left( h_1 \frac{\partial}{\partial x_1} + h_2 \frac{\partial}{\partial x_2} + \cdots + h_n \frac{\partial}{\partial x_n} \right) \qquad (4.24)$$

where $h_1 = x_1 - x_{10}, h_2 = x_2 - x_{20}, \ldots, h_n = x_n - x_{n0}$.

## Summary of Differentiation Rules

For $u, v, w$ functions of $x$ and $a, b, c$ constants, the rules for differentiation can be summarized as follows.

1. The derivative of a constant is zero. The curve $y = C = constant$, is a flat curve. As $x$ changes, there is no change in $y$ and consequently, the derivative is zero for all $x$.

$$\frac{d}{dx}C = 0$$

2. The curve $y = x$ has a constant slope of 1, hence the derivative of $y$ with respect to $x$ has the constant value of unity for all values of $x$.

$$\frac{d}{dx}x = 1$$

3. The derivative of $x$ raised to a power $m$ is the power $m$ times $x$ raised to the $(m-1)$ power.

$$\frac{d}{dx}x^m = mx^{m-1}$$

4. The derivative of a constant $C$ times a function is the constant times the derivative of the function.

$$\frac{d}{dx}Cf(x) = C\frac{df(x)}{dx}$$

5. The derivative of a sum (or difference) of functions equals the sum (or difference) of the derivatives. This is known as a distributive property of the derivative operator $\frac{d}{dx}$

$$\frac{d}{dx}(u \pm v \pm \cdots \pm w) = \frac{du}{dx} \pm \frac{dv}{dx} \pm \cdots \pm \frac{dw}{dx}$$

For example, the polynomial function $y = f(x) = 3x^7 - 4x^3 + 100x^2 - \pi x + 7$ has the derivative

$$\frac{d}{dx}(y) = \frac{d}{dx}(f(x)) = \frac{d}{dx}(3x^7 - 4x^3 + 100x^2 - \pi x + 7)$$

$$\frac{dy}{dx} = f'(x) = 3\frac{d}{dx}x^7 - 4\frac{d}{dx}x^3 + 100\frac{d}{dx}x^2 - \pi\frac{d}{dx}x + \frac{d}{dx}7$$

$$\frac{dy}{dx} = f'(x) = 3(7x^6) - 4(3x^2) + 100(2x) - \pi(1) + 0$$

$$\frac{dy}{dx} = f'(x) = 21x^6 - 12x^2 + 200x - \pi$$

6. The derivative of a product of two functions $u$ and $v$ equals the first function times the derivative of the second function plus the second function times the derivative of the first function.

$$\frac{d}{dx}(uv) = u\frac{dv}{dx} + v\frac{du}{dx} = uv\left(\frac{1}{u}\frac{du}{dx} + \frac{1}{v}\frac{dv}{dx}\right)$$

For example, if $y = f(x) = (x^2 + x)(x^3 - x^2)$, then

$$\frac{d}{dx}y = \frac{d}{dx}f(x) = \frac{d}{dx}[(x^2 + x)(x^3 - x^2)]$$

$$\frac{dy}{dx} = f'(x) = (x^2 + x)\frac{d}{dx}(x^3 - x^2) + (x^3 - x^2)\frac{d}{dx}(x^2 + x)$$

$$\frac{dy}{dx} = f'(x) = (x^2 + x)(3x^2 - 2x) + (x^3 - x^2)(2x + 1)$$

7. The derivative of a product of three functions is calculated using property 6 twice. For example,

$$\frac{d}{dx}(uvw) = uv\frac{dw}{dx} + w\frac{d}{dx}(uv)$$

$$\frac{d}{dx}(uvw) = uv\frac{dw}{dx} + wu\frac{dv}{dx} + wv\frac{du}{dx}$$

$$\frac{d}{dx}(uvw) = uvw\left(\frac{1}{u}\frac{du}{dx} + \frac{1}{v}\frac{dv}{dx} + \frac{1}{w}\frac{dw}{dx}\right)$$

8. The 'derivative of a product rule' can be generalized to differentiate a product of $n$-functions and one finds

$$\frac{d}{dx}(u_1 u_2 u_3 \cdots u_n) = (u_1 u_2 \cdots u_n)\left(\frac{1}{u_1}\frac{du_1}{dx} + \frac{1}{u_2}\frac{du_2}{dx} + \frac{1}{u_3}\frac{du_3}{dx} + \cdots + \frac{1}{u_n}\frac{du_n}{dx}\right)$$

9. If $y$ is a composite function defined by $y = y(u)$ where $u = u(x)$, then calculate the derivatives $\frac{dy}{du}$ and $\frac{du}{dx}$ with the derivative of $y = y(u(x))$ given by

$$\frac{dy}{dx} = \frac{dy}{du}\frac{du}{dx}$$

This result is known as chain rule differentiation.

10. The derivative of a function $u = u(x)$ raised to a power $m$ is given by

$$\frac{d}{dx}u^m = mu^{m-1}\frac{du}{dx}$$

The derivative of a function to a power equals the power times the function to the one less power times the derivative of the function with respect to $x$. This holds for both integer and noninteger values for $m$. A special case of property 8 when $m$ is an integer and also an example of property 9.

11. The above chain rule for differentiation can be generalized. If $y = y(u)$, $u = u(x)$ and $x = x(t)$, then $y$ can be treated as a function of $t$ with derivative

$$\frac{dy}{dt} = \frac{dy}{du}\frac{du}{dx}\frac{dx}{dt}$$

12. Derivative of an inverse function. If $y = y(x)$ has the inverse function of $x = x(y)$, which has the nonzero derivative $\frac{dx}{dy}$, then

$$\frac{dy}{dx} = \frac{1}{\frac{dx}{dy}} \quad \text{and} \quad \frac{d^2y}{dx^2} = -\frac{\frac{d^2x}{dy^2}}{\left(\frac{dx}{dy}\right)^3}$$

**13.** If the curve is defined by a set of parametric equations $x = x(u)$ and $y = y(u)$, with parameter $u$, then the derivative of $y$ with respect to $x$ is given by

$$\frac{dy}{dx} = \frac{\frac{dy}{du}}{\frac{dx}{du}}$$

**14.** The exponential function is the only function equal to its own derivative.

$$\frac{d}{dx}e^x = e^x$$

**15.** The derivative of the natural logarithm function is $1/x$.

$$\frac{d}{dx}\ln x = \frac{1}{x}$$

Note the exponential function and logarithm function are inverse functions with $\ln e^x = x$ and $e^{\ln x} = x$. Thus, if $y = e^x$, then $x = \ln y$, so by rule 12

$$\frac{dy}{dx} = e^x, \quad \text{and} \quad \frac{dx}{dy} = \frac{d}{dy}\ln y = \frac{1}{y}, \quad \text{with} \quad \frac{dy}{dx} = \frac{1}{\frac{dx}{dy}} = \frac{1}{\frac{1}{y}} = y = e^x$$

**16.** If $y = e^u$ and $u = u(x)$ is a function of $x$, then by the chain rule for differentiation

$$\frac{dy}{dx} = \frac{d}{dx}e^u = \frac{d}{du}e^u \frac{du}{dx} = e^u \frac{du}{dx} \quad \text{or} \quad \frac{d}{dx}e^u = e^u \frac{du}{dx}$$

A special case of this rule is $\frac{d}{dx}e^{ax} = ae^{ax}$

**17.** If $y = \ln u$ and $u = u(x)$ is a function of $x$, then

$$\frac{dy}{dx} = \frac{d}{dx}\ln u = \frac{1}{u}\frac{du}{dx} \quad \text{A special case of this rule is} \quad \frac{d}{dx}\ln(ax) = \frac{1}{x}$$

**18.** $\frac{d}{dx}a^x = \frac{d}{dx}e^{x \ln a} = e^{x \ln a}\ln a = a^x \ln a \quad \text{or} \quad \frac{d}{dx}a^x = a^x \ln a$

**19.** If $y = a^u$ and $u = u(x)$ if a function of $x$, then

$$\frac{dy}{dx} = \frac{d}{dx}a^u = \frac{d}{du}a^u \frac{du}{dx} = (a^u \ln a)\frac{du}{dx}$$

**20.** If $y = y(x)$ is defined implicitly by a function $f(x, y) = 0$, where $\frac{\partial f}{\partial y} \neq 0$, then

$$\frac{dy}{dx} = -\frac{\frac{\partial f}{\partial x}}{\frac{\partial f}{\partial y}}, \quad \text{and} \quad \frac{d^2 y}{dx^2} = -\frac{1}{\left(\frac{\partial f}{\partial y}\right)^3}\left[\frac{\partial^2 f}{\partial x^2}\left(\frac{\partial f}{\partial y}\right)^2 - 2\frac{\partial^2 f}{\partial x \partial y}\frac{\partial f}{\partial x}\frac{\partial f}{\partial y} + \frac{\partial^2 f}{\partial y^2}\left(\frac{\partial f}{\partial x}\right)^2\right]$$

**21.** If $y = f(u_1, u_2, \ldots, u_n)$ where $u_1 = u_1(x), u_2 = u_2(x), \ldots, u_n = u_n(x)$, then

$$\frac{dy}{dx} = \frac{\partial f}{\partial u_1}\frac{du_1}{dx} + \frac{\partial f}{\partial u_2}\frac{du_2}{dx} + \cdots + \frac{\partial f}{\partial u_n}\frac{du_n}{dx}$$

## The Indefinite Integral

The inverse operator associated with the differential operator $\frac{d}{dx}(\ )$ is called the antiderivative operator or integral operator and it is written $\int (\ )\, dx$. Some notation and nomenclature associated with the integrals

$$\int f(x)\, dx = F(x) + C \quad \text{and} \quad \int_a^b f(x)\, dx = F(x)\big|_a^b = F(b) - F(a) \qquad (4.25)$$

is as follows. The integral of the left-hand side of equation (4.25) is called an indefinite integral of $f(x)$ with respect to $x$ and the integral on the right-hand side of equation (4.25) is called a definite integral of $f(x)$ with respect to $x$ between the limits $a$ and $b$. The function $f(x)$ is called the integrand and the result $F(x)$ is termed the integral.

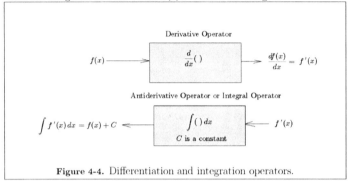

Figure 4-4. Differentiation and integration operators.

Picture the two operators $\frac{d}{dx}(\ )$ and $\int (\ )\,dx$ as illustrated in the figure 4-4. When the function $f(x)$ passes through the differential operator box there is produced the new function $f'(x)$ which is the derivative of $f(x)$. When the derivative function $f'(x)$ passes through the integral operator box, then the original function $f(x)$ is restored. Note that two functions $f(x)$ and $f(x) + C$, where $C$ is a constant independent of $x$, both have the same derivative. Consequently, the antiderivative operator or integral operator returns the original function to within an additive constant. The additive constant $C$ is termed the constant of integration.

### Rules for Integration

1. If $\dfrac{dF(x)}{dx} = f(x)$, then $\int f(x)\,dx = F(x) + C$

2. $\int Cu(x)\,dx = C \int u(x)\,dx$

    That is, the integral of a constant times a function equals the constant times the integral of the function.

3. $\int (u \pm v \pm \cdots \pm w)\,dx = \int u\,dx \pm \int v\,dx \pm \cdots \pm \int w\,dx$

    Here the integral of a sum (or difference) of functions equals the sum (or difference) of the integrals of each function. The integral operator is said to have the distributive property.

4. $\int u\,dv = uv - \int v\,du$

This result is known as integration by parts and it is sometimes written in the form $\int uv'\,dx = uv - \int u'v\,dx$ where $'$ always means differentiation with respect to the argument of the function.

5. $\int f(ax)\,dx = \dfrac{1}{a}\int f(\xi)\,d\xi$

Integrals can be scaled by making a change of variable. If $\xi = ax$, then $d\xi = a\,dx$.

6. $\int f(x)\,dx = \int f(u(\xi))\dfrac{du}{d\xi}\,d\xi$

Make the substitution $x = u(\xi)$ with $dx = u'(\xi)\,d\xi$ to place integrals in an alternative form.

7. $\int u^n\,du = \dfrac{1}{n+1}u^{n+1} + C, \quad n \neq -1$

The integral of a function raised to a power $n$ is the function raised to the power $n+1$ divided by $n+1$.

8. $\int \dfrac{du}{u} = \ln|u| + C$ where $\ln|u| = \begin{cases} \ln u, & u > 0 \\ \ln(-u), & u < 0 \end{cases}$

Here the logarithm function must have a positive argument.

9. $\int e^u\,du = e^u + C$

The integral of the exponential function produces the exponential function plus a constant of integration.

10. $\int a^u\,du = \int e^{u\ln a}\,du + C = \dfrac{1}{\ln a}e^{u\ln a} = \dfrac{1}{\ln a}a^u + C, \quad a > 0$

**Differentiation and Integration**

Observe that from a differentiation formula, there is immediately obtain an integration formula

$$\text{If } \dfrac{dF(x)}{dx} = f(x), \quad \text{then} \quad \int f(x)\,dx = F(x) + C$$

where $C$ is a constant of integration. For example, if

$$\dfrac{d}{dx}\left(\dfrac{x^{n+1}}{n+1}\right) = x^n, \quad \text{then} \quad \int x^n\,dx = \dfrac{x^{n+1}}{n+1} + C$$

The operators $\dfrac{d}{dx}$ and $\int (\ )\,dx$ are sometimes viewed as being inverse operators with the property that one operator undoes what the other operator does. The integral operator having the property of always adding a constant of integration. The inverse operators of differentiation and integration have the properties

$$\dfrac{d}{dx}\int f(x)\,dx = f(x) \quad \text{and} \quad \int \left[\dfrac{d}{dx}f(x)\right]dx = f(x) + C$$

The following are some well known examples of the above relationship between derivatives and indefinite integrals.

## Trigonometric Functions

| Derivative relationships $\frac{d}{dx}F(x) = f(x)$ | Integral relationships $\int f(x)\,dx = F(x) + C$ |
|---|---|
| $\frac{d}{dx}\sin u = \cos u \frac{du}{dx}$ | $\int \cos u\,du = \sin u + C$ |
| $\frac{d}{dx}\cos u = -\sin u \frac{du}{dx}$ | $\int \sin u\,du = -\cos u + C$ |
| $\frac{d}{dx}\tan u = \sec^2 u \frac{du}{dx}$ | $\int \sec^2 u\,du = \tan u + C$ |
| $\frac{d}{dx}\cot u = -\csc^2 u \frac{du}{dx}$ | $\int \csc^2 u\,du = -\cot u + C$ |
| $\frac{d}{dx}\sec u = \sec u \tan u \frac{du}{dx}$ | $\int \sec u \tan u\,du = \sec u + C$ |
| $\frac{d}{dx}\csc u = -\csc u \cot u \frac{du}{dx}$ | $\int \csc u \cot u\,du = -\csc u + C$ |

## Inverse Trigonometric Functions

| Derivative relationships $\frac{d}{dx}F(x) = f(x)$ | Integral relationships $\int f(x)\,dx = F(x) + C$ |
|---|---|
| $\frac{d}{dx}\sin^{-1} u = \frac{1}{\sqrt{1-u^2}}\frac{du}{dx},\quad -\frac{\pi}{2} < \sin^{-1} u < \frac{\pi}{2}$ | $\int \frac{du}{\sqrt{1-u^2}} = \sin^{-1} u + C$ |
| $\frac{d}{dx}\cos^{-1} u = \frac{-1}{\sqrt{1-u^2}}\frac{du}{dx},\quad 0 < \cos^{-1} u < \pi$ | $\int \frac{1}{\sqrt{1-u^2}}\,du = -\cos^{-1} u + C$ |
| $\frac{d}{dx}\tan^{-1} u = \frac{1}{1+u^2}\frac{du}{dx},\quad -\frac{\pi}{2} < \tan^{-1} u < \frac{\pi}{2}$ | $\int \frac{1}{1+u^2}\,du = \tan^{-1} u + C$ |
| $\frac{d}{dx}\cot^{-1} u = \frac{-1}{1+u^2}\frac{du}{dx},\quad 0 < \cot^{-1} u < \pi$ | $\int \frac{1}{1+u^2}\,du = -\cot^{-1} u + C$ |
| $\frac{d}{dx}\sec^{-1} u = \frac{\pm 1}{u\sqrt{u^2-1}}\frac{du}{dx},\quad \begin{cases} +1, & \text{if } 0 < \sec^{-1} u < \frac{\pi}{2} \\ -1, & \text{if } \frac{\pi}{2} < \sec^{-1} u < \pi \end{cases}$ | $\int \frac{1}{u\sqrt{u^2-1}}\,du = \sec^{-1}|u| + C$ |
| $\frac{d}{dx}\csc^{-1} u = \frac{\pm 1}{u\sqrt{u^2-1}}\frac{du}{dx},\quad \begin{cases} +1, & \text{if } -\frac{\pi}{2} < \csc^{-1} u < 0 \\ -1, & \text{if } 0 < \csc^{-1} u < \frac{\pi}{2} \end{cases}$ | $\int \frac{1}{u\sqrt{u^2-1}}\,du = \csc^{-1}|u| + C$ |

Note 1: Examine the above table of integrals and remember that when dealing with indefinite integrals one is confronted with an infinite number of solutions because of the arbitrary constant $C$ which is added to each integral. Observe that the inverse trigonometric function table gives the following integrals.

$$\int \frac{1}{\sqrt{1-x^2}}\,dx = \sin^{-1} x + C \quad \text{and} \quad \int \frac{1}{\sqrt{1-x^2}}\,dx = -\cos^{-1} x + C \qquad (4.26)$$

From these integrals one cannot draw the conclusion that $\sin^{-1} x$ and $-\cos^{-1} x$ are equal. The $C$'s, representing the constants of integration in the above formulas are not all the same. What is implied by the equations (4.26) is that the inverse functions $\sin^{-1} x$ and $-\cos^{-1} x$ differ by a constant. In fact $\sin^{-1} x = \dfrac{\pi}{2} - \cos^{-1} x$. These difficulties in notation can be overcome by using a symbol different from $C$ for the constant of integration. For example, after examining the above table of integrals observe that

$$\int \frac{du}{\sqrt{1-u^2}} = \sin^{-1} u + C = -\cos^{-1} u + k$$
$$\int \frac{du}{1+u^2} = \tan^{-1} u + C = -\cot^{-1} u + k \qquad (4.27)$$
$$\int \frac{du}{u\sqrt{u^2-1}} = \sec^{-1} |u| + C = -\csc^{-1} |u| + k$$

and then make note of the trigonometric identities

$$\cos^{-1} x + \sin^{-1} x = \frac{\pi}{2}, \qquad \cot^{-1} x + \tan^{-1} x = \frac{\pi}{2}, \qquad \csc^{-1} x + \sec^{-1} x = \frac{\pi}{2}$$

If the constants $C$ and $k$ in equations (4.27) are such that they differ by $\pi/2$, then the answers are the same. Take the first equation in (4.27) and observe that the substitution $\sin^{-1} u = \pi/2 - \cos^{-1} u$, gives the new constant $k = C + \pi/2$. A similar type of argument can be applied to the remaining equations in (4.27).

Note 2: To differentiate a function like $y = \sin^{-1} u$, where $u$ is a function of $x$, one uses implicit differentiation and writes

$$y = \sin^{-1} u \quad \Longleftrightarrow \quad \sin y = u$$

and then treats $y$ as a function of $x$ and differentiates to obtain

$$\frac{d}{dx}\sin y = \frac{du}{dx} \quad \text{or} \quad \cos y \frac{dy}{dx} = \frac{du}{dx} \implies \frac{dy}{dx} = \frac{d}{dx}\sin^{-1} u = \frac{1}{\cos y}\frac{du}{dx}$$
$$\frac{dy}{dx} = \frac{d}{dx}\sin^{-1} u = \frac{1}{\pm\sqrt{1-\sin^2 y}}\frac{du}{dx} \qquad (4.28)$$
$$\frac{dy}{dx} = \frac{d}{dx}\sin^{-1} u = \frac{1}{\pm\sqrt{1-u^2}}\frac{du}{dx}$$

Remember that the inverse sine function is multiple-valued and one usually selects the region from $-\pi/2$ to $+\pi/2$ where the arc sine function is a monotonic function as $u$ varies from $-1$ to $1$ (See figure 2-11). The value of $\arcsin u$ varying from $-\pi/2$ to $+\pi/2$ is called the principal value for the arc sine function. For this selection for the principal value one must use the $+$ sign for the square root function in equation (4.28) because the square root function is to represent the cosine function in equation (4.28). One should verify that the cosine function is positive over this region. Note that it is possible to define the arc sine function over some other interval where the inverse sine function is monotonic. For example, by selecting the interval $\pi/2$ to $3\pi/2$, then one would have to use the negative sign for the square root function in equation (4.28) because in this region the cosine function is negative.

The graphs of the hyperbolic functions and inverse hyperbolic functions must also be investigated in order to define regions where they are uniquely defined.

| Hyperbolic Functions | |
|---|---|
| Derivative relationships $\dfrac{d}{dx}F(x) = f(x)$ | Integral relationships $\int f(x)\,dx = F(x) + C$ |
| $\dfrac{d}{dx}\sinh u = \cosh u\, \dfrac{du}{dx}$ | $\int \cosh u\, du = \sinh u + C$ |
| $\dfrac{d}{dx}\cosh u = \sinh u\, \dfrac{du}{dx}$ | $\int \sinh u\, du = \cosh u + C$ |
| $\dfrac{d}{dx}\tanh u = \operatorname{sech}^2 u\, \dfrac{du}{dx}$ | $\int \operatorname{sech}^2 u\, du = \tanh u + C$ |
| $\dfrac{d}{dx}\coth u = -\operatorname{csch}^2 u\, \dfrac{du}{dx}$ | $\int \operatorname{csch}^2 u\, du = -\coth u + C$ |
| $\dfrac{d}{dx}\operatorname{sech} u = -\operatorname{sech} u \tanh u\, \dfrac{du}{dx}$ | $\int \operatorname{sech} u \tanh u\, du = -\operatorname{sech} u + C$ |
| $\dfrac{d}{dx}\operatorname{csch} u = -\operatorname{csch} u \coth u\, \dfrac{du}{dx}$ | $\int \operatorname{csch} u \coth u\, du = -\operatorname{csch} u + C$ |

| Inverse Hyperbolic Functions | |
|---|---|
| Derivative relationships $\dfrac{d}{dx}F(x) = f(x)$ | Integral relationships $\int f(x)\,dx = F(x) + C$ |
| $\dfrac{d}{dx}\sinh^{-1} u = \dfrac{1}{\sqrt{u^2+1}}\,\dfrac{du}{dx}$ | $\int \dfrac{du}{\sqrt{u^2+1}} = \sinh^{-1} u + C$ |
| $\dfrac{d}{dx}\cosh^{-1} u = \dfrac{\pm 1}{\sqrt{u^2-1}}\,\dfrac{du}{dx}, \begin{cases} +1 \text{ if } \cosh^{-1} u > 0, u > 1 \\ -1 \text{ if } \cosh^{-1} u < 0, u > 1 \end{cases}$ | $\int \dfrac{du}{\sqrt{u^2-1}} = \cosh^{-1} u + C$ |
| $\dfrac{d}{dx}\tanh^{-1} u = \dfrac{1}{1-u^2}\,\dfrac{du}{dx},\ -1 < u < 1$ | $\int \dfrac{du}{1-u^2} = \tanh^{-1} u + C$ |
| $\dfrac{d}{dx}\coth^{-1} u = \dfrac{1}{1-u^2}\,\dfrac{du}{dx},\ u > 1 \text{ or } u < -1$ | $\int \dfrac{du}{1-u^2} = \coth^{-1} u + C$ |
| $\dfrac{d}{dx}\operatorname{sech}^{-1} u = \dfrac{\pm 1}{u\sqrt{1-u^2}}\,\dfrac{du}{dx}, \begin{cases} -1 \text{ if } \operatorname{sech}^{-1} u > 0,\ 0 < u < 1 \\ +1 \text{ if } \operatorname{sech}^{-1} u < 0,\ 0 < u < 1 \end{cases}$ | $\int \dfrac{du}{u\sqrt{1-u^2}} = \operatorname{sech}^{-1} u + C$ |
| $\dfrac{d}{dx}\operatorname{csch}^{-1} u = \dfrac{\pm 1}{u\sqrt{1+u^2}}\,\dfrac{du}{dx}, \begin{cases} -1 \text{ if } u > 0 \\ +1 \text{ if } u < 0 \end{cases}$ | $\int \dfrac{du}{u\sqrt{1+u^2}} = \operatorname{csch}^{-1} u + C$ |

Make note of the following.

1. The hyperbolic sine function is monotone increasing for all values of $x$ so that the inverse hyperbolic sine function is well defined for all values of $x$.

2. The inverse hyperbolic cosine function is not well defined. Note that $\cosh x$ is greater than or equal to 1 for all values of $x$ and so the inverse hyperbolic cosine function is only defined for $x \geq 1$ and has two branches. Using the logarithmic expressions for the inverse hyperbolic functions it is customary to select the upper branch so that

$$\cosh^{-1} x = \ln(x + \sqrt{x^2 - 1}), \quad x \geq 1$$

and the lower branch is selected as $\quad \cosh^{-1} x = \ln(x - \sqrt{x^2 - 1}), \quad x \geq 1$
3. The inverse hyperbolic tangent function is well defined for $-1 < x < 1$.
4. The inverse hyperbolic cotangent function is well defined for $x < -1$ and $x > 1$.
5. The positive branch of the inverse hyperbolic secant function is given by

$$\operatorname{sech}^{-1} x = \ln\left(\frac{1}{x} + \sqrt{\frac{1}{x^2} - 1}\right), \quad 0 < x \leq 1$$

and the negative branch is given by $\quad \operatorname{sech}^{-1} x = \ln\left(\frac{1}{x} - \sqrt{\frac{1}{x^2} - 1}\right), \quad 0 < x \leq 1$
6. The inverse hyperbolic cosecant is well defined in the intervals $x < -1$ and $x > 1$.

## The Exponential and Logarithmic Functions

The exponential function $e^x = \exp x$ and the logarithmic function $\ln x$ are inverses of each other and consequently

$$e^{\ln x} = x \quad \text{and} \quad \ln e^x = x \tag{4.29}$$

Often times one is confronted with differentiating functions which are defined as functions raised to the power of another function, or functions represented as complicated products or quotients of functions. In such cases it is customary to take the logarithm before differentiating. This is known as logarithmic differentiation. For example, given the function $u(x)^{v(x)}$, define $y = u(x)^{v(x)}$ and then take the logarithm to obtain $\ln y = v(x) \ln u(x)$. The differentiation of this function produces the result

$$\frac{1}{y}\frac{dy}{dx} = v(x)\frac{1}{u(x)}\frac{du(x)}{dx} + \frac{dv(x)}{dx}\ln u(x)$$

and from this equation solve for the desired derivative $\frac{dy}{dx}$.

Note: If one did not know that $\frac{d}{dx}\ln x = \frac{1}{x}$ and was confronted with the integral $\int \frac{1}{x} dx$, then a way around this difficulty would be to use limits. For example, it is known that $\int x^\alpha \, dx = \frac{x^{\alpha+1}}{\alpha+1} + C$, for $\alpha \neq -1$ and so the above result can be expressed in the form

$$\int x^\alpha \, dx = \frac{x^{\alpha+1} - 1}{\alpha+1} + C', \quad \text{where} \quad C' = C + \frac{1}{\alpha+1}$$

is some new constant.

Examine this integral in the limit as $\alpha \to -1$ using L'Hospital's Rule to show

$$\lim_{\alpha \to -1} \frac{x^{\alpha+1}-1}{\alpha+1} = \lim_{\alpha \to -1} \frac{\frac{d}{d\alpha}(x^{\alpha+1}-1)}{\frac{d}{d\alpha}(\alpha+1)} = \lim_{\alpha \to -1} \frac{e^{(\alpha+1)\ln x} \ln x}{1} = \ln x$$

This integral can be found as the second entry in the following table.

| Exponential and Logarithmic Functions | |
|---|---|
| Derivative relationships $\frac{d}{dx}F(x) = f(x)$ | Integral relationships $\int f(x)\,dx = F(x) + C$ |
| $\frac{d}{dx}e^x = e^x$ | $\int e^x\,dx = e^x + C$ |
| $\frac{d}{dx}\ln x = \frac{1}{x}$ | $\int \frac{1}{x}\,dx = \ln x + C$ |
| $\frac{d}{dx}e^{\alpha x} = \alpha e^{\alpha x}$ | $\int \alpha e^{\alpha x}\,dx = e^{\alpha x} + C$ |
| $\frac{d}{dx}\ln \alpha x = \frac{1}{x}$ | $\int \frac{1}{x}\,dx = \ln x + C'$ <br> $\quad = \ln x + \ln \alpha + C$ <br> $\quad = \ln \alpha x + C$ <br> The constant of integration can be selected in terms of some other arbitrary constant if one chooses to modifiy the constant. |
| $\frac{d}{dx}e^{u(x)} = e^{u(x)}\frac{du}{dx}$ | $\int e^{u(x)}\,du = e^u + C$ |
| $\frac{d}{dx}\ln u(x) = \frac{1}{u(x)}\frac{du}{dx}$ | $\int \frac{du}{u(x)} = \ln u + C$ |

## The Definite Integral

A fundamental problem occurring in calculus is the area problem which can be described as follows. Find the area bounded by the curve $y = f(x)$, the $x$-axis, the lines $x = a$ and $x = b$. The situation is illustrated in the figure 4-5(a). To solve this problem divide the interval $(a, b)$ into $n$ equally spaced subintervals each having a length

$$\Delta x = \frac{b-a}{n} \tag{4.30}$$

and define the points $x_0 = a$, $x_1 = a + \Delta x$, $x_2 = a + 2\Delta x, \ldots, x_i = a + i\Delta x, \ldots, x_n = a + n\Delta x = b$. This is called partitioning the interval $(a, b)$ into $n$ subintervals. Interval partitioning is illustrated in the figure 4-5(b).

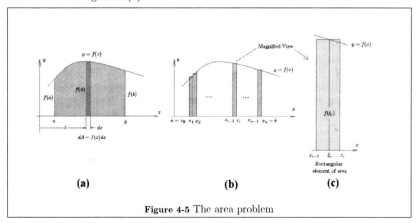

**Figure 4-5** The area problem

Define $\Delta x_i = x_i - x_{i-1}$ as the length of the $i$th subinterval $(x_{i-1}, x_i)$ and pick a point $\xi_i$ anywhere in the $i$th subinterval as illustrated in the figure 4-5(c). Now construct the height of the curve at the point $\xi_i$, $x_{i-1} \le \xi_i \le x_i$, and define the rectangular element of area associated with the $i$th subinterval as the rectangle with height $f(\xi_i)$ and base $\Delta x_i$ as illustrated in the figure 4-5(c). Perform this type of construction for each subinterval $i$, $i = 1, 2, \ldots, n$. The area associated with each rectangle can then be expressed

$$A_1 = f(\xi_1)\Delta x_1, \quad A_2 = f(\xi_2)\Delta x_2, \ldots, A_i = f(\xi_i)\Delta x_i, \ldots, A_n = f(\xi_n)\Delta x_n$$

Define $S_n$ as the sum of the rectangular elements of area that have just been constructed and write

$$S_n = f(\xi_1)\Delta x_1 + f(\xi_2)\Delta x_2 + \cdots + f(\xi_n)\Delta x_n = \sum_{j=1}^{n} f(\xi_j)\,\Delta x_j \qquad (4.31)$$

The sum $S_n$ represents an approximation to the area to be found. Let $n$ increase without bound, then each subinterval $\Delta x_i$ tends toward zero. If the limit $\lim_{n\to\infty} S_n$ exists, this limit is defined as the definite integral of $f(x)$ over the interval $(a, b)$ and is written as

$$\int_a^b f(x)\,dx = \lim_{n\to\infty} \sum_{i=1}^{n} f(\xi_i)\,\Delta x_i \qquad (4.32)$$

The integral sign $\int$ is an elongated $S$ to represent the summation process of rectangles over the interval $(a, b)$. Here $a$ is called the lower limit of integration and $b$ is called the upper limit of integration, $f(x)$ is called the integrand and $dA = f(x)\,dx$ is called a differential element of area and is representative of the rectangles being summed. This element of area is illustrated in the figure 4-5(a). Whenever $f(x)$ is a continuous function over the interval $(a,b)$, then it can be shown that the limit $\lim_{n\to\infty} S_n$ exists. Whenever this limit exists the function $f(x)$ is said to be Riemann integrable over the interval $(a,b)$. The definite integral $\int_a^b f(x)\,dx$ geometrically represents the area bounded by the curve $y = f(x)$, the $x$-axis, the line $x = a$ and the line $x = b$ as illustrated in the figure 4-5(a).

The definite integral satisfies the following properties.

1. $\int_a^b f(x)\,dx = -\int_b^a f(x)\,dx$

   That is, if the limits of integration are reversed, then the sign of the integral changes.

2. If $f(a)$ exists, then $\int_a^a f(x)\,dx = 0$

3. If $f(x) = C$, is a constant for all values of $x$ over the interval $(a,b)$, then

$$\int_a^b C\,dx = \lim_{n\to\infty} \sum_{j=1}^n C\Delta x_j = C \sum_{j=1}^n \Delta x_j = C(b-a)$$

Here the area is a rectangle of height $C$ and base $(b-a)$.

4. The definite integral is distributive

$$\int_a^b [u_1(x) \pm u_2(x) \pm \cdots \pm u_n(x)]\,dx = \int_a^b u_1(x)\,dx \pm \int_a^b u_2(x)\,dx \pm \cdots \pm \int_a^b u_n(x)\,dx$$

5. If $f(x)$ is a continuous function for $a \le x \le b$ and $\xi$ is a number satisfying $a < \xi < b$, then

$$\int_a^b f(x)\,dx = \int_a^\xi f(x)\,dx + \int_\xi^b f(x)\,dx$$

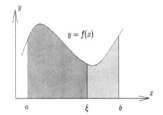

This result can also be written in one of the forms

$$\int_a^b f(x)\,dx - \int_a^\xi f(x)\,dx = \int_\xi^b f(x)\,dx \quad \text{or} \quad \int_a^b f(x)\,dx - \int_\xi^b f(x)\,dx = \int_a^\xi f(x)\,dx$$

6. **Mean value theorem for integrals**

If $y = f(x)$ is a continuous function over the interval $(a,b)$, then there exist a number $x = \xi$ satisfying $a \le \xi \le b$, such that $\int_a^b f(x)\,dx = f(\xi)(b-a)$. This states that the area bounded by the curve $y = f(x)$, the $x$-axis, the line $x = a$ and the line $x = b$

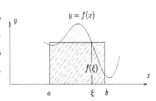

is the same as the area of a rectangle with base $(b-a)$ and height $f(\xi)$. Another way to express this result is to make the substitution $\xi = a + \epsilon(b-a)$ and express the mean value theorem as follows. There exists an value $\epsilon$, satisfying $0 \le \epsilon \le 1$ such that

$$\int_a^b f(x)\,dx = (b-a)f(a + \epsilon(b-a))$$

A graphical representation of the mean value theorem is illustrated above.

7. Assume the interval $(a, b)$ has been partitioned into $n$ equal subintervals $\Delta x_i$, $i = 1, \ldots, n$, with $x_0 = a$, $x_n = b$ and $\Delta x_i = \frac{b-a}{n}$, for $i = 1, \ldots, n$. If it is possible to find an indefinite integral of $f(x)$, say $\int f(x)\,dx = F(x) + C$, then $F'(x) = f(x)$, so that the following table of values can be calculated using the mean value theorem for derivatives.

| Interval | Mean value theorem for derivatives |
|---|---|
| $(x_0, x_1)$ | $F(x_1) + C - [F(x_0) + C] = F'(\xi_1)(x_1 - x_0) = f(\xi_1)(x_1 - x_0) = f(\xi_1)\Delta x_1$ |
| $(x_1, x_2)$ | $F(x_2) + C - [F(x_1) + C] = F'(\xi_2)(x_2 - x_1) = f(\xi_2)(x_2 - x_1) = f(\xi_2)\Delta x_2$ |
| $\vdots$ | $\vdots$ |
| $(x_{i-1}, x_i)$ | $F(x_i) + C - [F(x_{i-1}) + C] = F'(\xi_i)(x_i - x_{i-1}) = f(\xi_i)(x_i - x_{i-1}) = f(\xi_i)\Delta x_i$ |
| $\vdots$ | $\vdots$ |
| $(x_{n-1}, x_n)$ | $F(x_n) + C - [F(x_{n-1}) + C] = F'(\xi_n)(x_n - x_{n-1}) = f(\xi_n)(x_n - x_{n-1}) = f(\xi_n)\Delta x_n$ |

Addition of the terms on both the left and right-hand side of the above table produces the result

$$F(x_n) - F(x_0) = F(b) - F(a) = \sum_{j=1}^n f(\xi_j)\Delta x_j$$

and consequently, in the limit as $n$ increases without bound one finds

$$F(b) - F(a) = \int_a^b f(x)\,dx, \qquad \text{where } F(x) \text{ is any indefinite integral of } f(x)$$

This result is sometimes expressed in the form

$$\int_a^b f(x)\,dx = F(x)\Big|_a^b = F(b) - F(a)$$

and is known as the fundamental theorem of integral calculus.

8. The variable of integration in a definite integral is a dummy variable. For example, the following integrals are all the same

$$\int_a^b f(x)\,dx = \int_a^b f(t)\,dt = \int_a^b f(\xi)\,d\xi$$

The reason for this is that the definite integral depends only upon the limits $a$ and $b$, so no matter what symbol is used for the variable of integration one finds

$$\int_a^b f(t)\, dt = F(t)\Big|_a^b = F(b) - F(a) \quad \text{For example} \quad \int_2^3 2t\, dt = t^2\Big|_2^3 = 3^2 - 2^2 = 5$$

That is, the symbol used for the variable of integration does not appear in the final answer for the definite integral and consequently any dummy symbol can be used to evaluate the integral.

9.      If $F(x) = \int_a^x f(t)\, dt$, then $F'(x) = \frac{d}{dx} F(x) = f(x)$. This follows directly from the definition of a derivative.

$$F'(x) = \lim_{\Delta x \to 0} \frac{F(x + \Delta x) - F(x)}{\Delta x} = \lim_{\Delta x \to 0} \frac{\int_a^{x+\Delta x} f(t)\, dt - \int_a^x f(t)\, dt}{\Delta x} = \lim_{\Delta x \to 0} \frac{1}{\Delta x} \int_x^{x+\Delta x} f(t)\, dt$$

so that by the mean value theorem for integrals one obtains the result

$$F'(x) = \lim_{\Delta x \to 0} \frac{1}{\Delta x} f(x + \epsilon \Delta x)\, \Delta x = f(x) \qquad \text{where } 0 \leq \epsilon \leq 1$$

Consequently,

$$\frac{d}{dx} \int_a^x f(t)\, dt = f(x), \qquad \frac{d}{d\beta} \int_0^\beta f(t)\, dt = f(\beta), \qquad \frac{d}{d\alpha} \int_0^\alpha f(t)\, dt = f(\alpha)$$

10. 
$$\frac{d}{dx} \int_{\alpha(x)}^{\beta(x)} f(t)\, dt = f(\beta(x)) \frac{d\beta}{dx} - f(\alpha(x)) \frac{d\alpha}{dx}$$

If the limits of integration in a definite integral are functions of $x$, then using the chain rule for differentiation and the property 9, there results

$$\frac{d}{dx} \int_{\alpha(x)}^{\beta(x)} f(t)\, dt = \frac{d}{dx} \left[ \int_{\alpha(x)}^0 f(t)\, dt + \int_0^{\beta(x)} f(t)\, dt \right]$$

$$= \frac{d}{dx} \left[ \int_0^{\beta(x)} f(t)\, dt - \int_0^{\alpha(x)} f(t)\, dt \right]$$

$$= \frac{d}{d\beta} \left[ \int_0^\beta f(t)\, dt \right] \frac{d\beta}{dx} - \frac{d}{d\alpha} \left[ \int_0^\alpha f(t)\, dt \right] \frac{d\alpha}{dx}$$

$$= f(\beta(x)) \frac{d\beta}{dx} - f(\alpha(x)) \frac{d\alpha}{dx}$$

A more general result is the Leibnitz formula

$$\frac{d}{dx} \int_{\alpha(x)}^{\beta(x)} f(t, x)\, dt = \int_{\alpha(x)}^{\beta(x)} \frac{\partial f(t, x)}{\partial x}\, dt + f(\beta(x), x) \frac{d\beta}{dx} - f(\alpha(x), x) \frac{d\alpha}{dx}$$

11. **Integration by parts**

$$\int_a^b u(x)\, v'(x)\, dx = u(x) v(x)\Big|_a^b - \int_a^b v(x)\, u'(x)\, dx$$

Another way to express integration by parts is the following.
$$\int_a^b u\,dv = u\,v\Big]_a^b - \int_a^b v\,du$$

12. **Generalized first mean value theorem**

If both $f(x)$ and $g(x)$ are continuous over the interval $(a,b)$ and $g(x)$ does not change sign over the interval, then
$$\int_a^b f(x)g(x)\,dx = f(\xi)\int_a^b g(x)\,dx \quad \text{for some value } \xi \text{ satisfying } a < \xi < b$$

13. **Generalized second mean value theorem (Bonnet's theorem)**

If both $f(x)$ and $g(x)$ are continuous over the interval $(a,b)$ and $g(x) > 0$ is monotone decreasing function over the interval, then
$$\int_a^b f(x)g(x)\,dx = g(a)\int_a^\xi f(x)\,dx \quad \text{for some value } \xi \text{ satisfying } a < \xi < b$$

If $g(x) > 0$ is a monotone increasing function over the interval, then
$$\int_a^b f(x)g(x)\,dx = g(b)\int_\xi^b f(x)\,dx \quad \text{for some value } \xi \text{ satisfying } a < \xi < b$$

14. **Generalized mean value theorem**

If both $f(x)$ and $g(x)$ are continuous over the interval $(a,b)$ and $g(x) > 0$ is either a monotone decreasing or monotone increasing function over the interval, then
$$\int_a^b f(x)g(x)\,dx = g(a)\int_a^\xi f(x)\,dx + g(b)\int_\xi^b f(x)\,dx \quad \text{for some value } \xi \text{ satisfying } a < \xi < b$$

15. **Change of variable**

If $\xi = \xi(x)$ and its inverse function $x = x(\xi)$ are single-valued and continuous with continuous derivatives over some interval $(a,b)$, then
$$\int_a^b f(x)\,dx = \int_{\xi(a)}^{\xi(b)} f(x(\xi))\,\frac{dx}{d\xi}\,d\xi$$

Usually such substitutions are made in order to obtain an integration which is easier to perform. For example, to evaluate the integral $I = \int_0^\beta \frac{dx}{\sqrt{x^2+4}}$, the substitution $x = 2\sinh t$ with $dx = 2\cosh t\,dt$ can be used to obtain
$$I = \int_0^{\sinh^{-1}(\beta/2)} \frac{2\cosh t\,dt}{\sqrt{4\sinh^2 t + 4}} = \int_0^{\sinh^{-1}(\beta/2)} dt = t\,\Big|_0^{\sinh^{-1}(\beta/2)} = \sinh^{-1}(\beta/2)$$

## Improper Integrals

Consider a definite integral having the form $\int_a^b f(x)\,dx$. If one or more of the following conditions are satisfied

(i) $a$ is $\pm\infty$ or $b$ is $\pm\infty$ or $a = -\infty$ and $b = \infty$

(ii) If the integrand $f(x)$ becomes unbounded at one or more points within the integration interval $a \leq x \leq b$

(iii) Integrals where both conditions (i) and (ii) are satisfied.

then the integral is called an improper integral. If only condition (i) occurs, then the integral is termed an improper integral of the first kind. If only condition (ii) occurs, then the integral is termed an improper integral of the second kind. If condition (iii) is satisfied, then the integral is termed an improper integral of the third kind.

If $f(x)$ is a bounded function over every finite interval $a \leq x \leq b$, then an improper integral of the first kind with unbounded lower or upper limits can be defined by the limiting processes

$$\int_a^\infty f(x)\,dx \equiv \lim_{b \to \infty} \int_a^b f(x)\,dx, \quad \text{or} \quad \int_{-\infty}^b f(x)\,dx \equiv \lim_{a \to -\infty} \int_a^b f(x)\,dx \qquad (4.33)$$

Improper integrals of the first kind having the form $\int_{-\infty}^\infty f(x)\,dx$ can be written in the form

$$\int_{-\infty}^\infty f(x)\,dx = \int_{-\infty}^c f(x)\,dx + \int_c^\infty f(x)\,dx \qquad (4.34)$$

where $c$ is a finite real number. The integrals on the right-hand side of equation (4.34) can now be evaluated using limiting processes similar to those defined in equation (4.33). The above integrals are called convergent if the limits exist otherwise they are called divergent integrals.

If the integrand $f(x)$ becomes unbounded at the lower limit where $x = a$, it is usually expressed as the limiting process

$$\int_a^b f(x)\,dx = \lim_{\epsilon \to 0} \int_{a+\epsilon}^b f(x)\,dx \qquad (4.35)$$

Similarly, if $f(x)$ becomes unbounded at the upper limit where $x = b$, it is usually expressed using the limiting process

$$\int_a^b f(x)\,dx = \lim_{\epsilon \to 0} \int_a^{b-\epsilon} f(x)\,dx \qquad (4.36)$$

If there is a point $c$, satisfying $a < c < b$, where $f(x)$ is unbounded, then the integral is understood to be represented by the limiting process

$$\int_a^b f(x)\,dx = \lim_{\epsilon \to 0} \left[ \int_a^{c-\epsilon} f(x)\,dx + \int_{c+\epsilon}^b f(x)\,dx \right] \qquad (4.37)$$

If the limits in equations (4.35), (4.36), or (4.37) exist, then the integrals are called convergent, otherwise they are termed divergent integrals.

If the integrand $f(x)$ becomes unbounded at more than one point within the interval $a \leq x \leq b$, then the integral $\int_a^b f(x)\,dx$ must be broken up into a sum of integrals involving smaller intervals and using appropriate limiting processes similar to those defined above. Combinations of these limiting processes are also applied for the investigation of improper integrals of the third kind to determine if they converge or diverge.

**Bliss's Theorem**

Let $y = f(x)$ and $y = g(x)$ denote two continuous functions which are integrable over the interval $(a,b)$. Subdivide the interval $(a,b)$ into $n$ equally spaced points $x_i = a + i\Delta x_i$, for the index $i = 0, 1, 2, \ldots, n$, where $\Delta x_i = \frac{b-a}{n}$. This gives the partitions illustrated below. Bliss investigated the summation $\sum_{i=1}^{n} f(\xi_i) g(\zeta_i) \Delta x_i$ where $\xi_i$ and $\zeta_i$ represent arbitrary points within the $i$th sub-interval of the partitions.

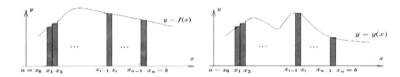

Here the arbitrary values $\xi_i$ and $\zeta_i$ satisfy $x_{i-1} \leq \xi_i \leq x_i$ and $x_{i-1} \leq \zeta_i \leq x_i$ for the index $i$ satisfying $i = 1, 2, \ldots, n$. Bliss's theorem then states that

$$\lim_{\Delta x_i \to 0} \sum_{i=1}^{n} f(\xi_i)\, g(\zeta_i) \Delta x_i = \int_a^b f(x) g(x)\, dx \qquad (4.38)$$

Note that this summation is not a Riemann summation because it involves two different functions and two arbitrary values $\xi_i$ and $\zeta_i$ in each of the subintervals $\Delta x_i$ for $i = 1, 2, \ldots, n$.

**Arc Length Formula**

The arc length $ds$ along a curve $y = f(x)$ from the point $(x, y)$ to $(x + dx, y + dy)$ is given by

$$ds = \sqrt{dx^2 + dy^2}$$

The total arc length along the curve between the points $(a, f(a))$ and $(b, f(b))$ is given by

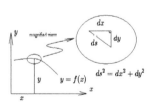

Element of arc length ds along curve

$$s = \int_a^b ds = \int_a^b \frac{ds}{dx}\, dx = \int_a^b \sqrt{1 + \left(\frac{dy}{dx}\right)^2}\, dx \qquad (4.39)$$

If the curve is defined by the parametric equations $x = x(t)$ and $y = y(t)$ for $t_0 \le t \le t_1$ with $x(t_0) = a$ and $x(t_1) = b$, then the arc length is obtained from the integral

$$s = \int_a^b ds = \int_{t_0}^{t_1} \frac{ds}{dt}\, dt = \int_{t_0}^{t_1} \sqrt{\left(\frac{dx}{dt}\right)^2 + \left(\frac{dy}{dt}\right)^2}\, dt \qquad (4.40)$$

If the curve is defined by the equation $x = g(y)$ for $\alpha \le y \le \beta$, then the arc length along the curve is given by

$$s = \int_\alpha^\beta ds = \int_\alpha^\beta \frac{ds}{dy}\, dy = \int_\alpha^\beta \sqrt{1 + \left(\frac{dx}{dy}\right)^2}\, dy \qquad (4.41)$$

**Area Formulas**

The area bounded by the curve $y = f(x)$, the $x$-axis, the lines $x = a$ and $x = b$, as illustrated in the figure 4-5(a), is given by

$$A = \int_a^b y\, dx = \int_a^b f(x)\, dx \qquad (4.42)$$

If the curve is given in the parametric form $x = x(t), y = y(t)$ for $t_0 \le t \le t_1$, where $x(t_0) = a$ and $x(t_1) = b$, then the area under the curve found from the integral

$$A = \int_a^b y\, dx = \int_{t_0}^{t_1} y(t) x'(t)\, dt \qquad (4.43)$$

The element of area $dA$ associated with an area element between two curves $y = f(x)$ and $y = g(x)$, for $a \le x \le b$, is written as

$$dA = \left| \begin{pmatrix} \text{height} \\ \text{curve 1} \end{pmatrix} - \begin{pmatrix} \text{height} \\ \text{curve 2} \end{pmatrix} \right| dx$$

or $\qquad dA = |f(x) - g(x)|\, dx$

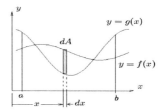

and the total area is obtained by summation of the elements of area to give

$$A = \int_a^b dA = \int_a^b |f(x) - g(x)|\, dx \qquad (4.44)$$

Note that absolute value signs are used because the given curves may or may not cross one another. If the curves do cross, then a positive area is required for the area. Alternatively, it is possible to solve for the points of intersection of the two curves $y = f(x)$ and $y = g(x)$ by solving the equations $f(x) = g(x)$ for the variable $x$. The integral can then be broken up into integrals over different sections where in each section it is possible to construct elements of area which, when summed, give positive results.

If the two curves are defined by equations of the form $x = g(y)$ and $x = f(y)$, for $\alpha \le y \le \beta$, then the element of area is defined $dA = |f(y) - g(y)|\,dy$ and the total area is determined from the evaluation of the integral

$$A = \int_\alpha^\beta dA = \int_\alpha^\beta |f(y) - g(y)|\,dy \qquad (4.45)$$

## Average Value of a Function

The average value of a set of n discrete values $y_1, y_2, \ldots, y_n$ is given by

$$y_{avg} = \frac{y_1 + y_2 + \cdots + y_n}{n}$$

If $y = f(x)$ is defined and continuous over the interval $a \le x \le b$, then it is possible to partition the interval $(a, b)$ into $n$ equal subintervals of length $\Delta x = \Delta x_i = \frac{b-a}{n}$, for $i = 1, 2, \ldots, n$. Select the points $x_1 = x_0 + \Delta x$, $x_2 = x_1 + \Delta x$, \ldots, $x_n = x_{n-1} + \Delta x$ and form the average of the numbers $f(x_1), f(x_2), \ldots, f(x_n)$ to obtain

$$y_{avg} = \frac{f(x_1) + f(x_2) + \cdots + f(x_n)}{n} = \frac{f(x_1) + f(x_2) + \cdots + f(x_n)}{\frac{b-a}{\Delta x}}$$

Observe that this average can be written as the Riemann summation

$$y_{avg} = \frac{1}{b-a} \sum_{i=1}^n f(x_i)\,\Delta x_i \qquad (4.46)$$

where each $\Delta x_i = \Delta x$. In the limit as $n$ increases without bound the Riemann sum becomes an integral which defines the average value of $y$ over the interval $[a, b]$ as

$$y_{avg} = \frac{1}{b-a} \int_a^b f(x)\,dx \qquad (4.47)$$

## Volumes

Consider a general curve $y = f(x)$ which is smooth and continuous and then imagine this curve rotated about the $x$-axis, as illustrated in the figure 4-6. At a general point $x$ satisfying $a \le x \le b$ construct an element of area $dA = y\,dx$. This element of area is representative of all the rectangles being summed to calculate the area under the curve. The rotation of this element of area about the $x$-axis creates a volume element $dV$ in the shape of a disc. The volume bounded by the surface of revolution and the planes $x = a$ and $x = b$ is a summation of the disc elements created by the rotation of the curve $y = f(x)$. The disc has a volume equal to its surface area times its thickness and can be represented $dV = \pi y^2\,dx$. A summation of the disc volume elements gives the total volume

$$V = \int_a^b dV = \int_a^b \pi y^2\,dx = \int_a^b \pi[f(x)]^2\,dx \qquad (4.48)$$

The figure 4-6 illustrates a cut-away view of the surface showing the volume element created by rotation of the curve $y = f(x)$. Sum these elements to obtain the total volume.

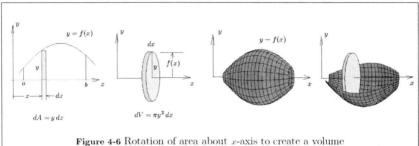

**Figure 4-6** Rotation of area about $x$-axis to create a volume

Another type of volume element is created if the curve $y = f(x)$, $a \le x \le b$, is rotated about the $y$-axis, as illustrated in the figures 4-7. In this case the element of area $dA$ becomes a cylindrical shell element. The element of volume of the cylindrical shell element of figure 4-7 is written

$$dV = (circumference)(height)(thickness) = 2\pi xy\,dx = 2\pi x f(x)\,dx$$

A summation of these cylindrical shell elements gives the total volume $V = \int_a^b 2\pi x f(x)\,dx$.

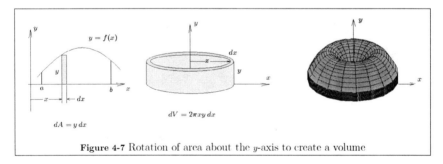

**Figure 4-7** Rotation of area about the $y$-axis to create a volume

Consider the element of volume formed when the element of area $dA$ is rotated about the line $y = -C$ as illustrated in the figure 4-8. The rotation generates a washer shaped element of volume as illustrated in the figure 4-9.

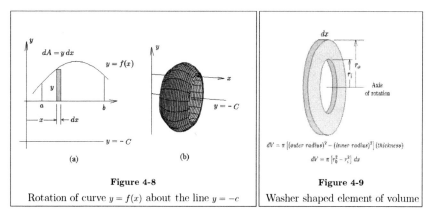

**Figure 4-8**
Rotation of curve $y = f(x)$ about the line $y = -c$

**Figure 4-9**
Washer shaped element of volume

The volume of the washer shaped element is given by

$$dV = \pi(r_o^2 - r_i^2)\,dx$$

where $r_o$ is the outer radius and $r_i$ is the inner radius of the washer element. A summation of these washer shaped volume elements gives the integration formula for volume

$$V = \int_a^b \pi[(y+c)^2 - c^2]\,dx = \int_a^b \pi[(f(x)+c)^2 - c^2]\,dx \tag{4.49}$$

Consider the area bounded by two smooth curves $y = f(x)$ and $y = g(x)$ which intersect at the points $x = a$ and $x = b$. This area is given by

$$A = \int_a^b dA = \int_a^b (f(x) - g(x))\,dx$$

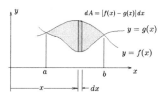

If this area is rotated about the $x$-axis, a surface of revolution is generated and the volume contained within this surface is determined by a summation of washer shaped elements of volume. This element of volume is given by

$$dV = \pi\left[f(x)^2 - g(x)^2\right]dx$$

and the total volume is obtained by summation of these volume elements

$$V = \int_a^b \pi\left[f(x)^2 - g(x)^2\right]dx$$

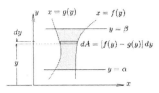

If the two curves are defined by the equations $x = f(y)$ and $x = g(y)$, for $\alpha \le y \le \beta$ and the curves do not cross, then the element of area $dA$ is a length, given by the difference in the $x$ values at position $y$, times a thickness $dy$ or $dA = [f(y) - g(y)]\,dy$.

When this element of area is rotated about the $x$-axis, then a cylindrical shell element of volume is produced. When the element of area is rotated about the $y$-axis, then a washer shaped element of volume is created.

Whenever the cross sectional area of a solid is known as a function of $x$, say $A = A(x)$, then a volume element can be written $dV = A(x)\,dx$ and a summation of these volume elements gives the total volume as

$$V = \int_a^b A(x)\,dx$$

Similarly, when the cross sectional area of a solid is a known function of position $y$, say $A = A(y)$, then the volume element can be expressed $dV = A(y)\,dy$. A summation of the volume elements gives the total volume

$$V = \int_\alpha^\beta A(y)\,dy$$

As an example, consider the problem to determine the total volume associated with a pyramid with cross sectional area in the shape of a square. Let $h$ denote the height of the pyramid and $b$ denote the length of one side of the base, then by using similar triangles one finds

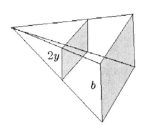

$$\frac{y}{x} = \frac{b/2}{h} \quad \text{or} \quad 2y = \frac{b}{h}x$$

This gives the cross sectional area

$$A = A(x) = (2y)^2 = \frac{b^2}{h^2}x^2$$

with element of volume $dV = A(x)\,dx$. Summation of the volume elements gives the total volume as

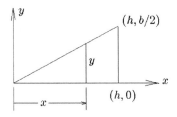

$$V = \int_0^h A(x)\,dx = \int_0^h \frac{b^2}{h^2}x^2\,dx = \frac{1}{3}b^2 h$$

As an exercise, change the notations in the above problem and derive the result given in the figure 1-2.

## Surface Area

When a curve $y = f(x)$ or $x = g(y)$ is rotated about a line, then a surface of revolution is formed. Consider the rotation of a trapezoidal element associated with a curve $y = f(x)$ as illustrated in the figure 4-10. The trapezoid element has corners $(x_{i-1}, f(x_{i-1}))$, $(x_i, f(x_i))$, $(x_{i-1}, 0)$ and $(x_i, 0)$. The rotation of this element produces the frustum of a cone which has a surface area given by

$$\Delta S_i = \pi[f(x_{i-1}) + f(x_i)]\sqrt{[x_i - x_{i-1}]^2 + [f(x_i)) - f(x_{i-1})]^2}$$

$$\text{or}\quad \Delta S_i = \pi[f(x_{i-1}) + f(x_i)]\sqrt{1 + \left[\frac{f(x_i) - f(x_{i-1})}{x_i - x_{i-1}}\right]^2} \Delta x_i \tag{4.50}$$

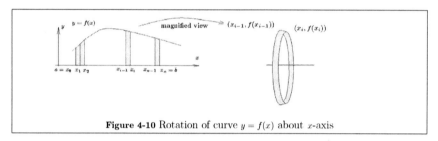

**Figure 4-10** Rotation of curve $y = f(x)$ about $x$-axis

The Intermediate Value Theorem says there exists a value $\xi_i$ satisfying $x_{i-1} \le \xi_i \le x_i$ such that

$$\frac{f(x_{i-1}) + f(x_i)}{2} = f(\xi_i) = \text{Average value of left and right side of trapezoid}$$

The mean value theorem for derivatives says there exists a value $\zeta_i$ such that

$$\frac{f(x_i) - f(x_{i-1})}{x_i - x_{i-1}} = f'(\zeta_i)$$

A summation of the elements of surface area associated with the surface generated by rotation of the curve $y = f(x)$ for $a \le x \le b$ is given by a summation of the elements from equation (4.50) to obtain

$$\sum_{i=1}^{n} \Delta S_i = \sum_{i=1}^{n} 2\pi f(\xi_i)\sqrt{1 + [f'(\zeta_i)]^2}\, \Delta x_i \tag{4.52}$$

The Bliss theorem allows one to write this surface area as

$$\sum_{i=1}^{n} 2\pi f(\xi_i)\sqrt{1 + [f'(\zeta_i)]^2}\, \Delta x_i = \int_{a}^{b} 2\pi f(x)\sqrt{1 + [f'(x)]^2}\, dx \tag{4.52}$$

provided the integral exists.

A heuristic method to remember this result is to go to a position $(x, y)$ on the curve $y = f(x)$ and sketch an element of arc length $ds$ and then rotate this element about an axis to form a ribbon shaped element of surface area. The area of this ribbon element is then

$$dS = 2\pi y\, ds = 2\pi y \sqrt{1 + \left(\frac{dy}{dx}\right)^2}\, dx \qquad (4.53)$$

and then substitute $y = f(x)$ and its derivative into this result and perform a summation of these elements between appropriate limits.

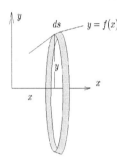

Note that the arc length can be determined from the arc length relation $ds^2 = dx^2 + dy^2$ which can be written in several different forms. The general heuristic method for constructing the area of the ribbon is

$$dS = 2\pi(\text{lever arm distance})(\text{element of arc length})$$

where $2\pi(\text{lever arm distance})$ is the circumference of the circle generated when the point $(x, y)$ is rotated about an axis. For example, if the element of arc length $ds$ is rotated about the $y$-axis, then

$$dS = 2\pi x\, ds \quad \text{with} \quad S = \int 2\pi x\, ds \qquad (4.54)$$

for the element of area and the total surface area.

If the element of arc length $ds$ is rotated about the $x$-axis, then the element of surface area $dS$ and the total surface area $S$ can be represented

$$dS = 2\pi y\, ds \quad \text{with} \quad S = \int 2\pi y\, ds \qquad (4.55)$$

In the equations (4.54) and (4.55) it is customary to substitute for the element of arc length using one of the forms

$$ds = \sqrt{1 + \left(\frac{dy}{dx}\right)^2}\, dx \qquad \text{or} \qquad ds = \sqrt{1 + \left(\frac{dx}{dy}\right)^2}\, dy \qquad (4.56)$$

depending upon (i) the form in which the curve is represented i.e. either $y = f(x)$ of $x = g(y)$ and (ii) the appropriate limits over which the surface area elements are summed.

Note also that if the surface is rotated about a line $y = constant$ or $x = constant$, then the lever arm distance must be adjusted accordingly. Also note that if the curve being rotated to form a surface of revolution is given in the parametric form $x = x(t)$ and $y = y(t)$, for $t_0 \le t \le t_1$, then the element of arc length can be represented

$$ds = \sqrt{\left(\frac{dx}{dt}\right)^2 + \left(\frac{dy}{dt}\right)^2}\, dt \qquad (4.57)$$

Appropriate substitution of these results into the equations (4.54) and (4.55) and the results integrated from $t_0$ to $t_1$ produces the surface area for the surface of revolution.

## Double Integrals

Consider a function $f(x,y)$ defined on a rectangle $R = \{(x,y)\,|\,a \le x \le b,\ c \le y \le d\}$ and assume the function $f(x_0, y)$, $x_0$ constant, is a continuous function of $y$ and the function $f(x, y_0)$, $y_0$ constant, is a continuous function of $x$. The integral $\int_c^d f(x_0, y)\, dy = \phi(x_0)$ exists for all values of $x_0$ selected in the interval $(a,b)$. Hence, the function $\phi(x) = \int_c^d f(x,y)\, dy$, $a \le x \le b$ is a continuous function of $x$. If the function $\phi(x)$ is integrated one obtains

$$\int_a^b \phi(x)\, dx = \int_a^b \left( \int_c^d f(x,y)\, dy \right) dx$$

Reversing the roles of $x$ and $y$ above, the function $\int_a^b f(x,y)\, dx = \psi(y)$ is a continuous function of $y$ and integration of $\psi(y)$ gives

$$\int_c^d \psi(y)\, dy = \int_c^d \left( \int_a^b f(x,y)\, dx \right) dy$$

This shows that double integrals can be treated as repeated first order integrals and if $f(x,y)$ is a continuous functions, then

$$\iint_R f(x,y)\, dxdy = \int_a^b \left( \int_c^d f(x,y)\, dy \right) dx = \int_c^d \left( \int_a^b f(x,y)\, dx \right) dy$$

Let $y = f(x)$ and $y = g(x)$ denote two different curves which are defined and continuous for $a \le x \le b$. Let $F(x,y)$ denote a function of $x$ and $y$ which is defined everywhere within the region $R$ bounded by the above curves and the lines $x = a$ and $x = b$ as illustrated in the figure 4-11. The notation $\iint_R F(x,y)\, dxdy$ is used to denote the indefinite double integral of the function $F(x,y)$ over the region $R$.

Integrals of the form

$$\int_a^b dx \int_{g(x)}^{f(x)} H(x,y)\, dy \quad \text{or} \quad \int_a^b \int_{g(x)}^{f(x)} H(x,y)\, dydx \quad \text{or} \quad \int_a^b \left[ \int_{g(x)}^{f(x)} H(x,y)\, dy \right] dx \qquad (4.58)$$

are called "definite double integrals" or "double integrals" of the function $F(x,y)$ over the region $R$.

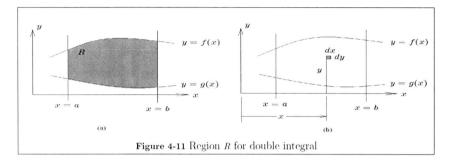

**Figure 4-11** Region $R$ for double integral

The integrals in equation (4.58) are sometimes referred to as iterated or repeated integrals. The above notation indicates that the inside integration is to be performed first. That is, in equation (4.58) hold $x$ constant and integrate with respect to $y$ and then substitute the limits for $y$. This is then followed by an integration with respect to $x$.

## Example 4-2.

Evaluate the double integral $I = \iint_R 5x^2 y\, dx dy$ where $R$ is the region between the line $y = \frac{1}{2}x$ and the parabola $x = 2y^2$.

**Solution:** First sketch the region $R$ to determine the limits of integration. The two curves intersect where the $y$-values are the same, or where $x = 2\left(\frac{x}{2}\right)^2$. This requires a solution to the equation $2x = x^2$ or $x(x-2) = 0$. From this equation verify the points of intersection $(0,0)$ and $(2,1)$. A sketch of the situation is illustrated in the figure 4-12.

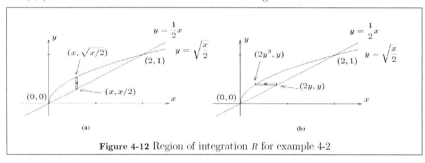

**Figure 4-12** Region of integration $R$ for example 4-2

The figure 4-12(a) illustrates the case where the element of area $dx dy$ is summed first in the $y$-direction. This gives

$$I_1 = \int_0^2 \left[\int_{x/2}^{\sqrt{x/2}} 5x^2 y\, dy\right] dx = \int_0^2 \left[5x^2 \frac{y^2}{2}\bigg|_{y=x/2}^{y=\sqrt{x/2}}\right] dx = \int_0^2 \left[\frac{5}{8}(2x^3 - x^4)\right] dx = \frac{5}{8}\left(\frac{x^4}{2} - \frac{x^5}{5}\right)\bigg|_0^2 = 1$$

The figure 4-12(b) illustrates the case where the element of area $dxdy$ is first summed in the $x$-direction. This gives the order of integration being reversed to give

$$I_2 = \int_0^1 \left[\int_{2y^2}^{2y} 5x^2 y\, dx\right] dy = \int_0^1 \left[\frac{5}{3}x^3 y\right]_{x=2y^2}^{x=2y} dy = \int_0^1 \frac{40}{3}(y^4 - y^7)\, dy = \frac{40}{3}\left(\frac{y^5}{5} - \frac{y^8}{8}\right)\Big|_0^1 = 1$$

Here $I_1 = I_2$ and in general, whenever $H(x,y)$ is continuous on the bounded region $R$, the iterated reversed double integrals will be the same. Note that this example introduces a situation where the region $R$ is such that one has an option as to which integration is performed first. Many double integration problems do not have this option because of the shape of the region $R$. Note also that once a summation direction is selected for the element $dxdy$, then it is required that (i) the limits of integration in either the $x$ or $y$ direction be calculated and (ii) the ability to solve for $x$ in terms of $y$ from the functions $y = f(x)$ and $y = g(x)$ to determine the limits of integration in the reversed direction.

■

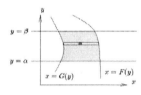

The order of integration is usually determined by how the region $R$ is defined. For example, if the area over which the double integral is to be performed is the region between two curves $x = G(y)$ and $x = F(y)$ and the lines $y = \alpha$, $y = \beta$, then because of the shape of the curves one must select an integration in the $x$-direction first.

The double integral can be expressed in one of the forms

$$\int_\alpha^\beta dy \int_{G(y)}^{F(y)} H(x,y)\, dx = \int_\alpha^\beta \int_{G(y)}^{F(y)} H(x,y)\, dxdy = \int_\alpha^\beta \left[\int_{G(y)}^{F(y)} H(x,y)\, dx\right] dy \quad (4.59)$$

The physical interpretation associated with the double integrals given by equations (4.58) is that $dA = dxdy$, located at the position $(x,y)$ within the region $R$, represents an element of area and the product $H(x,y)\, dA = H(x,y)\, dxdy$ denotes an element of volume in the shape of a parallelepiped with base $dA = dxdy$ and height $H$, which lies between the plane $z = 0$ and the surface $z = H(x,y)$ for $(x,y) \in R$. The double integral representing a summation of these volume elements. In the special case $H = 1$ for all values of $x,y$, then the double integrals are interpreted as representing the area of the region $R$.

## Double Integrals in Polar Coordinates

An element of area in polar coordinates is represented $dA = rdrd\theta$ and is illustrated in the figure 4-13(a). The area bounded by the curves $r = f(\theta)$, $r = g(\theta)$ and the rays $\theta = \alpha$ and $\theta = \beta$, represented in the figure 4-13(b), is represented by the double integral

$$\iint_{R_1} r\, drd\theta = \int_\alpha^\beta \int_{f(\theta)}^{g(\theta)} rdrd\theta, \quad R_1 = \{(r,\theta)\,|\, \alpha \le \theta \le \beta,\ f(\theta) \le r \le g(\theta)\,\} \quad (4.60)$$

The area bounded by the curves $\theta = F(r)$, $\theta = G(r)$ and the circles $r = a$ and $r = b$, represented in the figure 4-13(c), is represented by the double integral

$$\iint_{R_2} r\, dr d\theta = \int_a^b \int_{F(r)}^{G(r)} r\, d\theta dr, \quad R_2 = \{(r,\theta)\,|\, a \le r \le b,\ F(r) \le \theta \le G(r)\} \tag{4.61}$$

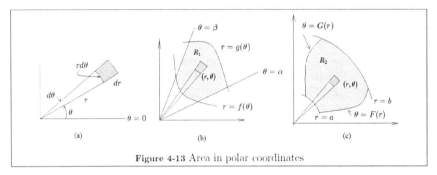

**Figure 4-13** Area in polar coordinates

Integrals having one of the forms

$$I_1 = \iint_{R_1} H(r,\theta)\, r dr d\theta = \int_\alpha^\beta \int_{f(\theta)}^{g(\theta)} H(r,\theta)\, r dr d\theta. \quad R_1 = \{(r,\theta)\,|\, \alpha \le \theta \le \beta,\ f(\theta) \le r \le g(\theta)\}$$

or

$$I_2 = \iint_{R_2} H(r,\theta)\, r\, d\theta dr = \int_a^b \int_{F(r)}^{G(r)} H(r,\theta)\, r\, d\theta dr, \quad R_2 = \{(r,\theta)\,|\, a \le r \le b,\ F(r) \le \theta \le G(r)\}$$

represent a double integration of the function $H(r,\theta)$ over a regions $R_1$ and $R_2$. Note that $I_1$ and $I_2$ represent double integrals where either $\theta$ is held constant for the first integral or $r$ is held constant for the first integral, the type of integral required being dependent upon how the region $R$ is defined. In both double integrals above, note that the integrand is $rH(r,\theta)$ this is because the element of area $dxdy$ in rectangular coordinates transforms to the element of area $rdrd\theta$ in polar coordinates.

## Triple Integrals

An element of volume in rectangular coordinates is given by $dV = dxdydz$ and integrals of the form

$$\iiint_R F(x,y,z)\, dxdydz$$

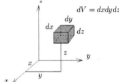

are called triple integrals of $F(x,y,z)$ over the region $R$. Integrals of the form

$$\int_a^b \left[ \int_{u_1}^{u_2} \left[ \int_{z_1}^{z_2} F(x,y,z)\,dz \right] dy \right] dx$$

are called definite integrals. Here three successive integrations must be performed.

If the first integration is in the $z$-direction, this integration is a summation of volume elements $dV$ in the $z$-direction to form a parallelepiped between the surfaces $z_1 = f_1(x,y)$ and $z_2 = f_2(x,y)$. If the second integration is in the $y$-direction, then sum the parallelepipeds in the $y$-direction to form a slab between the cylinders $u_2(x)$ and $u_1(x)$ The final integration in the $x$-direction is to sum the slabs in the $x$-direction between the planes $x = a$ and $x = b$.

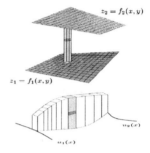

**Example 4-3.** Consider the region $R$ between the planes $6z = -18 - x$, $6z = 27 - x - 3y$, $y = 0$ and the cylinder $x^2 + y^2 - 9 = 0$ where $y \geq 0$. To evaluate an integral of the form

$$\iiint_R F(x,y,z)\,dxdydz$$

one must place limits on the integrals. Imagine an element of volume $dV = dxdydz$ placed at a point $(x,y,z)$ within the region $R$ illustrated in the figure 4-14. The element of volume can be summed in the $z$-direction between the planes $z_1 = (-18-x)/6$ and $z_2 = (27-x-3y)/6$. The element of volume then becomes a parallelepiped. Sum the parallelepiped elements in

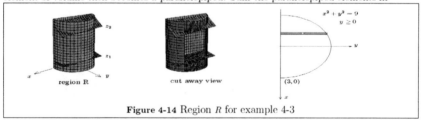

**Figure 4-14** Region $R$ for example 4-3

the $y$-direction between the limits $y = 0$ and $y = \sqrt{9-x^2}$ to form a slab element. The slab elements are then summed from $x = -3$ to $x = 3$. This gives the triple integral

$$\int_{-3}^{3} \int_{0}^{\sqrt{9-x^2}} \int_{(-18-x)/6}^{(27-x-3y)/6} F(x,y,z)\,dzdydx$$

■

In cylindrical coordinates $(r, \theta, z)$, illustrated in the figure 4-15, the element of volume $dV$ is given by $dV = rd\theta dz dr$.

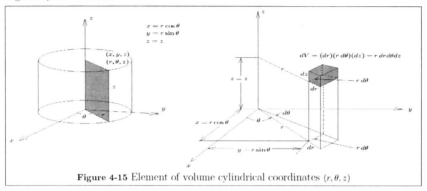

**Figure 4-15** Element of volume cylindrical coordinates $(r, \theta, z)$

In order to evaluate integrals of the form

$$\iiint_R F(r, \theta, z) dV = \iiint_R F(r, \theta, z) r\, dr d\theta dz$$

one must sum the volume element in the directions $z$, $\theta$ and $r$ or those directions which enable one to determine tractable limits of integration. In spherical coordinates $(\rho, \theta, \phi)$, illustrated in the figure 4-16, the element of volume $dV$ is given by $dV = \rho^2 \sin\theta\, d\rho d\theta d\phi$ and in order to evaluate integrals of the form

$$\iiint_R F(\rho, \theta, \phi)\, dV = \iint_R F(\rho, \theta, \phi)\, \rho^2 \sin\theta\, d\rho d\theta d\phi$$

one must perform summations in the directions $\rho$, $\theta$ and $\phi$ in some order. The order of summation depends upon how the region $R$ is specified.

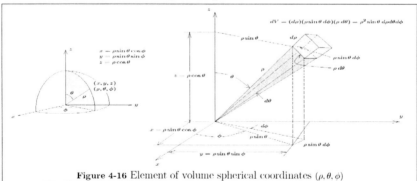

**Figure 4-16** Element of volume spherical coordinates $(\rho, \theta, \phi)$

## Evaluation of Integrals $\int f(x)\,dx$

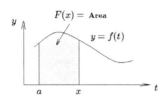

If the integral $\int f(x)\,dx$ cannot be evaluated, then define the function

$$F(x) = \int_a^x f(t)\,dt \quad \text{with derivative} \quad \frac{dF(x)}{dx} = f(x) \quad (4.62)$$

where $a$ is some convenient constant. The function $F(x)$ represents the area bounded by the $x$-axis, the lines at $t = a$ and $t = x$ and the curve $y = f(t)$ for $a \le t \le x$.

For example, if one did not know that $\frac{d}{dx}\ln x = \frac{1}{x}$, then it would be impossible to evaluate the integral $\int \frac{1}{x}\,dx$ in terms of known functions. Define the function

$$\ln x = \int_1^x \frac{1}{t}\,dt, \qquad 0 < x < \infty \qquad (4.63)$$

as the area under the curve $y = 1/t$ between $t=1$ and $t = x$. This function is called the natural logarithm function and it is possible to determine the properties of the natural logarithm function using the above integral definition. For example,

(i) $\ln 1 = \int_1^1 \frac{1}{t}\,dt = 0$

(ii) $\ln ax = \int_1^{ax} \frac{1}{t}\,dt = \int_1^x \frac{1}{t}\,dt + \int_x^{ax} \frac{1}{t}\,dt$ In the second integral make the substitution $t = xu$ with $dt = x\,du$ to show $\ln ax = \int_1^x \frac{1}{t}\,dt + \int_1^a \frac{1}{u}\,du = \ln x + \ln a$

which is called the addition theorem for the natural logarithm function.

(iii) Using properties (i) and (ii) consider the special case $x = \frac{1}{a}$ and show

$$\ln 1 = \ln\left(a\frac{1}{a}\right) = \ln a + \ln\frac{1}{a} = 0 \implies \ln\frac{1}{a} = -\ln a$$

(iv) Let $a_1, a_2, \ldots, a_n$ denote $n$ real values and use a repeated application of the addition theorem for logarithms to show $\ln(a_1 a_2 \cdots a_n) = \ln(a_1) + \ln(a_2) + \cdots + \ln(a_n)$

(v) A special case of property (iv) occurs when $a_i = a$ for $i = 1, 2, \ldots, n$. Then

$$\ln a^n = n \ln a$$

(vi) Consider the special case of property (v) when $a = b^{1/n} = \sqrt[n]{b}$, then

$$\ln b = n \ln b^{1/n} \implies \ln b^{1/n} = \frac{1}{n}\ln b$$

(vii) In the special case $0 < x < 1$, write

$$\ln x = \int_1^x \frac{1}{t}\,dt = -\int_x^1 \frac{1}{t}\,dt$$

which represents the negative of the area under the $1/t$ curve between $x$ and 1. Note also that the natural logarithm is not defined as a real quantity when $x < 0$.

## Substitutions

In general, chain rule for differentiation gives

$$\frac{d}{dx}f(u(x)) = \frac{df(u(x))}{du}\frac{du}{dx} \quad \text{and consequently} \quad \int \frac{df(u(x))}{du}\frac{du}{dx}dx = f(u(x)) + C$$

This suggests that sometimes it is possible to make a judicious transformation of the independent variable to transform the integral $\int f(x)\,dx$ into a different form.

A change of variable in an integral is usually made to (i) Change an elementary integral into a more complex one in order to gain knowledge of other integrable forms and (ii) Possibly reduce an integral to a much simpler form. There are no general rules for simplification of integrals or changing them to a recognizable form. However, experience has demonstrated that when an integral has certain special forms, then specific substitutions are known to produce simplifications. Some examples of the special forms and the substitutions used to simplify these forms are given in the table 4.1. Additional examples can be found in Appendix B.

Table 4.1  Substitutions for Evaluating Integrals

| Integral form | Suggested substitution |
|---|---|
| $\int f(ax)\,dx$ | $u = ax$ |
| $\int f(ax+b)\,dx$ | $u = ax+b$ |
| $\int f(\sqrt{a^2-x^2})\,dx$ | $u = a\sin\theta, \quad -\frac{\pi}{2} \leq \theta \leq \frac{\pi}{2}$ |
| $\int f(\sqrt{a^2+x^2})\,dx$ | $u = a\tan\theta, \quad \frac{-\pi}{2} < \theta < \frac{\pi}{2}$ |
| $\int f(\sqrt{x^2-a^2})\,dx$ | $u = a\sec\theta, \quad 0 \leq \theta < \frac{\pi}{2} \text{ or } \pi \leq \theta < \frac{3\pi}{2}$ |
| $\int f(\sqrt{ax+b})\,dx$ | $u = \sqrt{ax+b}$ |
| $\int f(\sqrt[n]{ax+b})\,dx$ | $u = \sqrt[n]{ax+b}$ |
| $\int f(e^{ax})\,dx$ | $u = e^{ax}$ |
| $\int f(\ln x)\,dx$ | $u = \ln x$ |
| $\int f(\sin^{-1}x)\,dx$ | $u = \sin^{-1}x$ Same idea for other inverse trigonometric functions |
| $\int f(\sin\theta,\cos\theta)\,d\theta$ | $u = \tan\frac{x}{2}, \quad \sin\theta = \frac{2u}{1+u^2}, \quad \cos\theta = \frac{1-u^2}{1+u^2}, \quad d\theta = \frac{2du}{1+u^2}$ |
| $\int f(\sqrt{a-x})\,dx$ | $x = a\cos 2\theta$ |
| $\int f(\sqrt{2ax-x^2})\,dx$ | $x = a(1-\cos\theta)$ |
| $\int f\left(\sqrt{\frac{a-x}{a+x}}\right)dx$ | $x = a\cos 2\theta$ |
| $\int f\left(\sqrt{a^{2n} \pm x^{2n}}\right)dx$ | $x^n = \frac{a^n}{z}$ |

**Partial Fractions to evaluate $\int f(x)\,dx$**

Another special case arises whenever the integrand $f(x)$ is a rational function having the form $f(x) = \dfrac{P_\ell(x)}{P_m(x)}$ where $P_\ell(x)$ and $P_m(x)$ are polynomials of degrees $\ell$ and $m$ respectively.

**Table 4.2** Partial Fraction Expansion for $\dfrac{P(x)}{Q(x)} = \dfrac{P_\ell(x)}{P_m(x)}$, $\ell < m$

| Factors of $Q(x)$ | Terms Required in Partial Fraction Expansion |
|---|---|
| Distinct linear factors | Products of distinct linear factors produce terms of the form |
| $(x-x_1)(x-x_2)\cdots(x-x_n)$ | $\dfrac{A_1}{x-x_1} + \dfrac{A_2}{x-x_2} + \cdots + \dfrac{A_n}{x-x_n}$ |
|  | where $A_1, A_2, \ldots, A_n$ are constants |
| Example | $\dfrac{5x-14}{(x-1)(x-4)} = \dfrac{3}{x-1} + \dfrac{2}{x-4}$ |
| Repeated linear factors | For each repeated factor of the form $(x-x_0)^n$ the partial fraction expansion must contain a sum on $n$ fractions having the form |
| $(x-x_0)^n$ | $\dfrac{B_1}{x-x_0} + \dfrac{B_2}{(x-x_0)^2} + \cdots + \dfrac{B_n}{(x-x_0)^n}$ |
|  | where $B_1, B_2, \ldots, B_n$ are constants. |
| Example | $\dfrac{x^2 - 8x + 18}{(x-5)^3} = \dfrac{1}{x-5} + \dfrac{2}{(x-5)^2} + \dfrac{3}{(x-5)^3}$ |
| Distinct quadratic factors | For each distinct quadratic factor, the partial fraction expansion must contain terms of the form |
| $(ax^2 + bx + c)$ | $\dfrac{C_1 x + D_1}{ax^2 + bx + c}$ |
|  | where $C_1, D_1$ are constants |
| Example | $\dfrac{3x^2 + 2x + 1}{(x-1)(x^2+x+4)} = \dfrac{1}{x-1} + \dfrac{2x+3}{x^2+x+4}$ |
| Repeated quadratic factors | For each repeated quadratic factor, the partial fraction expansion must include sums of the form |
| $(ax^2 + bx + c)^n$ | $\dfrac{C_1 x + D_1}{ax^2 + bx + c} + \dfrac{C_2 x + D_2}{(ax^2 + bx + c)^2} + \cdots + \dfrac{C_n x + D_n}{(ax^2 + bx + c)^n}$ |
|  | where $C_1, D_1, C_2, D_2, \ldots, C_n, D_n$ are constants |
| Example | $\dfrac{5x^4 + 8x^2 + 39x^2 + 25x + 50}{(x-2)(x^2+x+4)^2} = \dfrac{4}{x-2} + \dfrac{x+1}{x^2+x+4} + \dfrac{2x+3}{(x^2+x+4)^2}$ |

Consider the following cases.

**Case 1: $\ell \geq m$**   Here the degree of the polynomial in the numerator is greater than or equal to the degree of the polynomial in the denominator. In this case perform long division and rewrite the function $f(x)$ in the form

$$f(x) = F(x) + \frac{P(x)}{Q(x)}$$

where $F(x)$, $P(x)$ and $Q(x)$ are polynomials satisfying the property that the degree of $P(x)$ is less than the degree of $Q(x)$. One then obtains

$$\int f(x)\,dx = \int \frac{P_\ell(x)}{P_m(x)}\,dx = \int F(x)\,dx + \int \frac{P(x)}{Q(x)}\,dx$$

The function $F(x)$ can be immediately integrated since it is a polynomial.

**Case 2:** $\ell < m$

Whenever the degree of the polynomial in the numerator is less than the degree of the polynomial in the denominator the rational function is called proper and written

$$f(x) = \frac{P_\ell(x)}{P_m(x)} = \frac{P(x)}{Q(x)}, \qquad \ell < m$$

The integral of $f(x)$ has the form

$$\int f(x)\,dx = \int \frac{P(x)}{Q(x)}\,dx$$

Both cases one and two are now essentially the same problem of trying to reduce the rational function, $\frac{P(x)}{Q(x)}$ with degree of $P$ less than the degree of $Q$, to a form that is integrable. That is, the integrand must first be reduced to a form where the ratio $\frac{P(x)}{Q(x)}$ is a proper fraction. Then the proper fraction can be further simplified by factoring the denominator $Q(x)$ and examining the linear terms and quadratic terms after the factorization. The factored denominator $Q(x)$ determines the final form for the partial fraction. These different forms and the partial fraction expansions associated with these forms are listed in the table 4.2.

**Reduction Formula**

A reduction formula occurs whenever one integral can be expressed in a linear fashion in terms of another integral. This concept is illustrated by the following examples. Note in the following examples the constants of integration are ignored because it is always possible to collect all the constants of integration into one new constant which is then appended to the integral being calculated.

**Example 4-4.** Evaluate the integral $I_n = \int \cos^n x\,dx$.

**Solution:** Use integration by parts with

$$U = \cos^{n-1} x \qquad\qquad dV = \cos x\,dx$$
$$dU = (n-1)\cos^{n-2} x(-\sin x)\,dx \qquad V = \sin x$$

to obtain

$$I_n = \sin x \cos^{n-1} x + \int (n-1)\cos^{n-2} x \sin^2 x\,dx$$
$$I_n = \sin x \cos^{n-1} x + \int (n-1)\cos^{n-2} x(1 - \cos^2 x)\,dx$$
$$I_n = \sin x \cos^{n-1} x + (n-1)\int \cos^{n-2} x\,dx - (n-1)\int \cos^n x\,dx$$
$$I_n = \sin x \cos^{n-1} x + (n-1)I_{n-2} - (n-1)I_n$$

Solving this last equation for $I_n$ gives the reduction formula

$$I_n = \frac{1}{n}\sin x \cos^{n-1} x + \frac{n-1}{n}I_{n-2}$$

for $n = 2, 3, \ldots$. Use the known values $I_0 = \int dx = x$ and $I_1 = \int \cos x\, dx = \sin x$ and show

$$I_2 = \int \cos^2 x\, dx = \frac{1}{2}\sin x \cos x + \frac{1}{2}I_0 = \frac{1}{2}\sin x \cos x + \frac{1}{2}x$$

which produces the result

$$\int \cos^2 x\, dx = \frac{1}{2}x + \frac{1}{4}\sin 2x + C$$

∎

**Example 4-5.**

Define the integrals $S_m = \int x^m \sin nx\, dx$ and $C_m = \int x^m \cos nx\, dx$ where $m$ is an integer satisfying $m \geq 0$. Evaluate each integral using integration by parts and show

$$\begin{aligned} S_m &= \frac{-1}{n}x^m \cos nx + \frac{m}{n}C_{m-1} \\ C_m &= \frac{1}{n}x^m \sin nx - \frac{m}{n}S_{m-1} \end{aligned} \quad (4.64)$$

In the equations (4.64) replace $m$ by $m-1$ everywhere and show

$$\begin{aligned} S_{m-1} &= \frac{-1}{n}x^{m-1}\cos nx + \frac{m-1}{n}C_{m-2} \\ C_{m-1} &= \frac{1}{n}x^{m-1}\sin nx - \frac{m-1}{n}S_{m-2} \end{aligned} \quad (4.65)$$

Substituting the results from equations (4.65) into equations (4.64) produces the reduction formulas

$$\begin{aligned} S_m &= \frac{-1}{n}x^m \cos nx + \frac{m}{n^2}x^{m-1}\sin nx - \frac{m(m-1)}{n^2}S_{m-2} \\ C_m &= \frac{1}{n}x^m \sin nx + \frac{m}{n^2}x^{m-1}\cos nx - \frac{m(m-1)}{n^2}C_{m-2} \end{aligned} \quad (4.66)$$

Use the known values

$$S_0 = \int \sin nx\, dx = \frac{-1}{n}\cos nx \quad \text{and} \quad C_0 = \int \cos nx\, dx = \frac{1}{n}\sin nx$$

and substitute $m = 1$ into the equation (4.64) to show

$$\begin{aligned} S_1 &= \int x \sin nx\, dx = \frac{-1}{n}x \cos nx + \frac{1}{n^2}\sin nx \\ C_1 &= \int x \cos nx\, dx = \frac{1}{n}x \sin nx + \frac{1}{n^2}\cos nx \end{aligned} \quad (4.67)$$

The equations (4.66) are now a set of reduction formulas for calculating $S_m$ and $C_m$ for integer values of $m \geq 2$.

∎

## Differentiation and Integration of Arrays

The derivative and integral of the $n$-dimensional column vector $\overline{x}(t) = \begin{pmatrix} x_1(t) \\ x_2(t) \\ \vdots \\ x_n(t) \end{pmatrix}$ are defined

$$\frac{d\overline{x}(t)}{dt} = \begin{pmatrix} \frac{dx_1(t)}{dt} \\ \frac{dx_2(t)}{dt} \\ \vdots \\ \frac{dx_n(t)}{dt} \end{pmatrix} \quad \text{and} \quad \int_{t_1}^{t_2} \overline{x}(t)\,dt = \begin{pmatrix} \int_{t_1}^{t_2} x_1(t)\,dt \\ \int_{t_1}^{t_2} x_2(t)\,dt \\ \vdots \\ \int_{t_1}^{t_2} x_n(t)\,dt \end{pmatrix} \quad (4.68)$$

The derivative and integral of the $n \times n$ matrix $A(t) = \begin{pmatrix} a_{11}(t) & a_{12}(t) & \cdots & a_{1n}(t) \\ a_{21}(t) & a_{22}(t) & \cdots & a_{2n}(t) \\ \vdots & \vdots & \ddots & \vdots \\ a_{n1}(t) & a_{n2}(t) & \cdots & a_{nn}(t) \end{pmatrix}$ are given by

$$\frac{dA(t)}{dt} = \begin{pmatrix} \frac{da_{11}(t)}{dt} & \frac{da_{12}(t)}{dt} & \cdots & \frac{da_{1n}(t)}{dt} \\ \frac{da_{21}(t)}{dt} & \frac{da_{22}(t)}{dt} & \cdots & \frac{da_{2n}(t)}{dt} \\ \vdots & \vdots & \ddots & \vdots \\ \frac{da_{n1}(t)}{dt} & \frac{da_{n2}(t)}{dt} & \cdots & \frac{da_{nn}(t)}{dt} \end{pmatrix} \quad (4.69)$$

and

$$\int_{t_1}^{t_2} A(t)\,dt = \begin{pmatrix} \int_{t_1}^{t_2} a_{11}(t)\,dt & \int_{t_1}^{t_2} a_{12}(t)\,dt & \cdots & \int_{t_1}^{t_2} a_{1n}(t)\,dt \\ \int_{t_1}^{t_2} a_{21}(t)\,dt & \int_{t_1}^{t_2} a_{22}(t)\,dt & \cdots & \int_{t_1}^{t_2} a_{2n}(t)\,dt \\ \vdots & \vdots & \ddots & \vdots \\ \int_{t_1}^{t_2} a_{n1}(t)\,dt & \int_{t_1}^{t_2} a_{n2}(t)\,dt & \cdots & \int_{t_1}^{t_2} a_{nn}(t)\,dt \end{pmatrix} \quad (4.70)$$

The shorthand notation is to represent $A$ as $A = (a_{ij}(t))$, for $i, j = 1, 2, \ldots, n$ and then

$$\frac{dA(t)}{dt} = \left(\frac{da_{ij}(t)}{dt}\right) \quad \text{and} \quad \int_{t_1}^{t_2} A(t)\,dt = \left(\int_{t_1}^{t_2} a_{ij}(t)\,dt\right)$$

The derivative of the determinant of the $n \times n$ matrix $A$ is a summation of $n$-determinants where the $i$th determinant of the sum has the elements in the $i$th row differentiated. For example,

$$\frac{d}{dx}\begin{vmatrix} u_1 & v_1 \\ u_2 & v_2 \end{vmatrix} = \begin{vmatrix} u_1' & v_1' \\ u_2 & v_2 \end{vmatrix} + \begin{vmatrix} u_1 & v_1 \\ u_2' & v_2' \end{vmatrix}$$

and

$$\frac{d}{dx}\begin{vmatrix} u_1 & v_1 & w_1 \\ u_2 & v_2 & w_2 \\ u_3 & v_3 & w_3 \end{vmatrix} = \begin{vmatrix} u_1' & v_1' & w_1' \\ u_2 & v_2 & w_2 \\ u_3 & v_3 & w_3 \end{vmatrix} + \begin{vmatrix} u_1 & v_1 & w_1 \\ u_2' & v_2' & w_2' \\ u_3 & v_3 & w_3 \end{vmatrix} + \begin{vmatrix} u_1 & v_1 & w_1 \\ u_2 & v_2 & w_2 \\ u_3' & v_3' & w_3' \end{vmatrix}$$

Other differentiation properties for differentiation of a product of matrices $AB$, differentiation of a matrix times a column vector $A\overline{x}$ and differentiation of an inverse matrix $A^{-1}$ are

$$\frac{d}{dt}(AB) = \frac{dA}{dt}B + A\frac{dB}{dt}, \quad \frac{d}{dt}(A\overline{x}) = \frac{dA}{dt}\overline{x} + A\frac{d\overline{x}}{dt}, \quad \frac{d}{dt}A^{-1} = -A^{-1}\frac{dA}{dt}A^{-1} \quad (4.71)$$

where the left and right multiplication properties must be maintained.

## Inequalities Involving Integrals

**The Cauchy-Schwartz inequality**

$$\left| \int_a^b f(x)g(x)\,dx \right|^2 \le \left\{ \int_a^b |f(x)|^2\,dx \right\} \left\{ \int_a^b |g(x)|^2\,dx \right\} \qquad (4.72)$$

**The Holder inequality**

$$\int_a^b |f(x)g(x)|\,dx \le \left\{ \int_a^b |f(x)|^p\,dx \right\}^{1/p} \left\{ \int_a^b |g(x)|^q\,dx \right\}^{1/q} \qquad (4.73)$$

where $\frac{1}{p} + \frac{1}{q} = 1$ with $p > 1$ and $q > 1$.

**The Minkowski inequality**

$$\left\{ \int_a^b |f(x) + g(x)|^p\,dx \right\}^{1/p} \le \left\{ \int_a^b |f(x)|^p\,dx \right\}^{1/p} + \left\{ \int_a^b |g(x)|^p\,dx \right\}^{1/p} \qquad (4.74)$$

## General Series

A general infinite series $u_1 + u_2 + u_3 + \cdots$ is abbreviated $\sum_{j=1}^{\infty} u_j$. The sequence of partial sums is denoted $S_n = \sum_{j=1}^{n} u_j$. The infinite series $\sum_{j=1}^{\infty} u_j$ is said to be convergent if the sequence of partial sums $\{S_1, S_2, S_3, \ldots, S_m, \ldots, S_n, \ldots\}$ has the property that for every $\epsilon > 0$ an integer $N$ can be found such that $|S_m - S_n| < \epsilon$ whenever $m$ and $n$ satisfy the conditions $m > N$ and $n > N$. This is known as Cauchy's test for convergence of a sequence or series. If the sequence of partial sums does not converge, then the infinite series is said to be divergent. If $\lim_{n \to \infty} \sum_{j=1}^{n} u_j = \lim_{n \to \infty} S_n = S$, then $S$ is called the sum of the series. A necessary condition for the infinite series $\sum_{j=1}^{n} u_j$ to converge is for $\lim_{n \to \infty} u_n = 0$. The converse of this statement may or may not hold. That is, if $\lim_{n \to \infty} u_n = 0$, then the infinite series $\sum_{n=1}^{\infty} u_n$ may or may not converge. If the condition $\lim_{n \to \infty} u_n = 0$ is not satisfied, then the infinite series diverges.

A series $\sum_{n=1}^{\infty} u_n$ is said to be absolutely convergent if $\sum_{n=1}^{\infty} |u_n|$ converges. Whenever the series $\sum_{n=1}^{\infty} u_n$ converges, but the series of absolute values $\sum_{n=1}^{\infty} |u_n|$ diverges, then the series $\sum_{n=1}^{\infty} u_n$ is said to be conditionally convergent. If $\sum_{j=1}^{\infty} |u_j|$ converges, then $\sum_{j=1}^{\infty} u_j$ converges.

The following are some special tests to determine the convergence and divergence of an infinite series.

**Comparison test for convergence**

If $|a_n| \le b_n$ for all $n$ and $\sum_{j=1}^{\infty} b_j$ converges, then $\sum_{j=1}^{\infty} a_j$ is absolutely convergent.

**Comparison test for divergence**

If $a_n \ge b_n \ge 0$ for all $n$ and $\sum_{j=1}^{\infty} b_j$ diverges, then the infinite series $\sum_{j=1}^{\infty} a_j$ diverges.

**Cauchy's Integral test**

If there exists a function $f(x)$ which is continuous for $x \geq c$ and which decreases as $x$ increases and satisfies $\lim_{x \to \infty} f(x) = 0$ and is such that $a_n \leq f(n)$, then the convergence or divergence of the infinite series $\sum_{j=1}^{\infty} a_j$ is dependent upon whether the improper integral $\int_c^{\infty} f(x)\,dx$ converges or diverges.

**Ratio test (d'Alembert)**

If $a_n \neq 0$ for $n = 1, 2, 3, \ldots$, and $\lim_{n \to \infty} |\frac{a_{n+1}}{a_n}| = L$, then the infinite series is absolutely convergent if $L < 1$, divergent if $L > 1$ and the test fails if $L = 1$.

**Alternating series (Leibnitz)**

An alternating series has the form $\sum_{j=1}^{\infty}(-1)^{j+1} a_j$, where each $a_j > 0$. If $\lim_{j \to \infty} a_j = 0$ and $a_{j+1} \leq a_j$ for some value $j = N$ and all values of $j > N$, then the alternating series converges.

**Root test**

If $\lim_{n \to \infty} \sqrt[n]{|a_n|} = L$, then the infinite series $\sum_{j=1}^{\infty} a_j$ is absolutely convergent if $L < 1$, the series diverges if $L > 1$ and the test fails if $L = 1$.

**Raabe's test** If the limit given by $\lim_{n \to \infty} n\left(1 - \left|\frac{u_{n+1}}{u_n}\right|\right) = L$ exists, then the series $\sum_{n=1}^{\infty} u_n$ converges absolutely if $L > 1$, either diverges or converges conditionally if $L < 1$ and the test fails if $L = 1$.

**Gauss' test** If $\left|\frac{u_{n+1}}{u_n}\right| = 1 - \frac{L}{n} + \frac{c_n}{n^2}$, where $|c_n| < M$ for some constant $M$ and for all values $n$ larger than some value $N$, then the series $\sum_{n=1}^{\infty} u_n$ converges absolutely if $L > 1$, diverges or converges conditionally if $L \leq 1$.

**Geometric series**

The series $S = \sum_{n=1}^{\infty} ar^{n-1} = a + ar + ar^2 + ar^3 + \cdots$, for $a$ and $r$ constants, is called the geometric series. If $|r| < 1$, the series converges to $S = \frac{a}{1-r}$ and if $|r| \geq 1$, then the series diverges. The sum of the first n-terms of a geometric series is given by $S_n = \frac{a(1-r^n)}{1-r}$.

**P-series**

The series $\sum_{n=1}^{\infty} \frac{1}{n^p} = \frac{1}{1^p} + \frac{1}{2^p} + \frac{1}{3^p} + \cdots$ is called the p-series. This series converges if $p > 1$ and diverges for $p \leq 1$. This series, with $p = 1$, is called the harmonic series.

**Cauchy's Convergence Test**

The series $\sum u_n$ converges if and only if, for every $\epsilon > 0$, there exists a number $N$ such that
$$|u_n + u_{n+1} + \cdots + u_m| < \epsilon$$
whenever $m > n > N$.

### Uniform convergence

Consider a sequence of functions $\{u_n(x,y)\}$ defined everywhere within a region $R$ of the $xy$-plane. The infinite series $\sum_{j=1}^{\infty} u_j(x,y)$ is said to be uniformly convergent in $R$ to a function $f(x,y)$ if, for every $\epsilon > 0$, there exists a number $N$, independent of $x$ and $y$, such that $\left|\sum_{j=1}^{n} u_j(x,y) - f(x,y)\right| < \epsilon$, whenever $n > N$. Note that uniform convergence implies pointwise convergence.

### Weierstrass M-test for uniform convergence

Consider a sequence of functions $\{u_n(x,y)\}$ defined everywhere in a region $R$ of the $xy$-plane. If there exists a sequence of constant values $\{M_n\}$, such that for all values of (x,y) in $R$, $|u_n(x,y)| \leq M_n$, for $n = 1, 2, 3, \ldots$, then if the infinite series $\sum_{n=1}^{\infty} M_n$ converges, then the infinite series $\sum_{n=1}^{\infty} u_n(x,y)$ converges uniformly.

### Dirichlet's test

If the following three conditions are satisfied
(i) $a_n \geq a_{n+1} > 0$ for all $n$, the sequence $\{a_n\}$ is a monotone decreasing sequence.
(ii) $\lim_{n \to \infty} a_n = 0$
(iii) The infinite series $\sum_{n=1}^{\infty} u_n(x)$ converges for all $x \in [a,b]$,

then the series $\sum_{n=1}^{\infty} a_n u_n(x)$ is uniformly convergent for $x \in [a,b]$.

### Product of Series

If $\sum_{j=0}^{\infty} a_j$ and $\sum_{j=0}^{\infty} b_j$ are two absolutely convergent series, then Cauchy has shown that the product can be written

$$\left(\sum_{j=0}^{\infty} a_j\right)\left(\sum_{j=0}^{\infty} b_j\right) = \sum_{j=0}^{n}\left(\sum_{j=0}^{n} a_j b_{n-j}\right)$$

## Weierstrass Approximation Theorems

**For Polynomials** If $P(x) = \sum_{j=1}^{n} a_j x^j$ is a polynomial with real coefficients and $f(x)$ is a real continuous function defined over a closed interval $[a,b]$, then for every $\epsilon > 0$, there exists a polynomial such that

$$|P(x) - f(x)| < \epsilon \quad \text{for all } x \in [a,b]$$

**For Trigonometric Series** If $T(x) = \sum_{j=0}^{n}(a_j \cos j\omega x + b_j \sin j\omega x)$ is a trigonometric series and $f(x)$ is a real continuous function defined over a closed interval $[a,b]$, then for every $\epsilon > 0$, there exists a trigonometric series such that

$$|T(x) - f(x)| < \epsilon \quad \text{for all } x \in [a,b]$$

## Infinite Series of Functions

Assume that $\sum_{i=0}^{\infty} u_i(x)$ is a uniformly convergent series over some interval $(a,b)$.

(i) **Differentiation** If the derivatives $u_0'(x), u_1'(x), u_2'(x), \ldots$ exist and the series $\sum_{i=0}^{\infty} u_i'(x)$ converges uniformly for all $x \in (a,b)$, then it is possible to differentiate the series term by term, giving

$$\frac{d}{dx}\sum_{i=0}^{\infty} u_i(x) = \sum_{i=0}^{\infty} \frac{d}{dx} u_i(x) = \sum_{i=0}^{\infty} u_i'(x), \quad \text{for } x \in (a,b)$$

(ii) **Integration** If $\sum_{i=0}^{\infty} u_i(x)$ converges uniformly on the closed interval $[a,b]$ and each $u_i(x)$, $i = 0, 1, 2, \ldots$ is continuous throughout the open interval $(a,b)$, then

$$\int_{x_0}^{x_1} \sum_{i=0}^{\infty} u_i(x)\, dx = \sum_{i=0}^{\infty} \int_{x_0}^{x_1} u_i(x)\, dx \quad \text{for } a \le x_0 < x_1 \le b$$

As an example of the above consider the power series $f(x) = \sum_{n=0}^{\infty} c_n(x-x_0)^n$. This series converges for some values of $x$ and may diverge for other values of $x$. Usually the radius of convergence is the distance from the center of the series $x_0$ to the nearest singular point of the function $f(x)$, if such a singular point exists. The function $f(x)$ defined by the power series is continuous for all values of $x$ within the region of convergence and so it can be differentiated or integrated term by term.

### Transformations

On occasions a very slowly converging series can be replaced by a faster converging series by making a transformation. Let $S_n = \sum_{i=1}^{n} a_i$ denote the $n$th partial sum associated with the infinite series $\sum_{i=1}^{\infty} a_i$. Observe that the $n$th partial sum can also be written in the form

$$S_n = \sum_{i=1}^{n-1} i(a_i - a_{i+1}) + na_n = \sum_{i=1}^{n} a_i.$$

J.A. Shohat[1] has examined the two infinite series

$$A = \sum_{i=1}^{\infty} a_i \quad \text{and} \quad B = \sum_{i=1}^{\infty} i(a_i - a_{i+1})$$

and has shown that if $\lim_{n \to \infty} na_n = L$ exists, then the two series for $A$ and $B$ are either both convergent or both divergent, according as $L = 0$ or $L \ne 0$. In the case of convergence, $L = 0$, the two series $A$ and $B$ have the same sum and it turns out that the series $B$ converges much faster than the series for $A$.

## Numerical Integration

To evaluate the integral $\int_a^b f(x)\, dx$ partition the interval $(a,b)$ into $n$ equal parts by defining a step size $h = \dfrac{b-a}{n}$ together with the points

$$x_0 = a, \quad x_1 = a+h, \quad x_2 = a+2h, \ldots, \quad x_i = a+ih, \ldots, \quad x_n = a+nh = b$$

---

[1] J.A. Shohat, *On a Certain Transformation of Infinite Series*, American Mathematical Monthly, Vol 43, Pp 226-239, 1933.

Knowing $y = f(x)$, it is possible to calculate the heights of the curve at these $x$-values to obtain

$$y_0 = f(x_0), \ y_1 = f(x_1), \ y_2 = f(x_2), \ldots, y_i = f(x_i), \ldots, y_n = f(x_n)$$

The area bounded by the $x$-axis the lines $x = a$, $x = b$ and the curve $y = f(x)$ has now been divided up into a series of panels as illustrated in the figure 4-17.

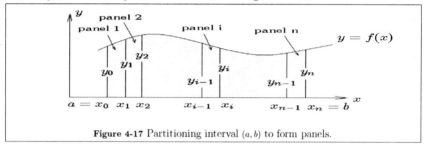

**Figure 4-17** Partitioning interval $(a, b)$ to form panels.

A single panel between $x_{i-1}$ and $x_i$ is defined by the lines $y = 0$, $x = x_{i-1}$, $x = x_i$, and some curve connecting the points $(x_{i-1}, y_{i-1})$ and $(x_i, y_i)$.

**Trapezoidal rule (1-panel formula)** Connecting the heights of the curve by straight line segments produces a series of panels in the shape of trapezoids. The area $A_1$ of a trapezoid is the average height times the base or $A_1 = \frac{h}{2}(y_{i-1} + y_i)$. Summing the area of all the trapezoids produced by the partitioning gives the trapezoidal rule formula for the total area under the curve

$$Total\,Area = \int_a^b f(x)\,dx = \sum_{i=1}^n \frac{h}{2}(y_{i-1} + y_i) = \frac{h}{2}[y_0 + 2y_1 + 2y_2 + \cdots + 2y_{n-1} + y_n] + Global\ Error$$

where the global error is found to be given by $Global\ Error = -\dfrac{(b-a)}{12}h^2 f''(\xi), \quad a < \xi < b$.

**Simpson's 1/3 rule (2-panel formula)**

After partitioning the interval $(a, b)$ into an even number of panels, pass a parabola through the points $(x_0, y_0), (x_1, y_1), (x_2, y_2)$ and then find the area under the parabola is given by

$$A_2 = \int_{x_0}^{x_2} f(x)\,dx = \frac{h}{3}[y_0 + 4y_1 + y_2] + local\ error$$

This is known as Simpson's 1/3 rule. Applying the 2-panel formula to all groups of 2-panels, produced by the partitioning, gives the integration formula for the total area under the curve as

$$Total\,Area = \int_a^b f(x)\,dx = \frac{h}{3}[y_0 + 4y_1 + 2y_2 + 4y_3 + 2y_4 + 4y_5 + \cdots + 2y_{n-2} + 4y_{n-1} + y_n] + Global\ Error$$

where the global error is given by $-\dfrac{(b-a)}{90}h^4 f^{(iv)}(\xi), \quad a < \xi < b$

# Chapter 5
# Vector Calculus

### Introduction
Recall that vectors and scalars are defined.

**Vector** *A vector is any quantity which possesses both magnitude and direction.*
**Scalar** *A scalar is a quantity which possesses a magnitude but does not possess a direction.*

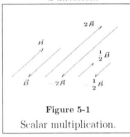

Figure 5-1
Scalar multiplication.

A vector can be represented by an arrow. The orientation of the arrow determines the direction of the vector, and the length of the arrow is associated with the magnitude of the vector. The tail end of the arrow is called the origin, and the arrowhead is called the terminus. Vectors are usually denoted by letters in bold face type. When a bold face type representation is inconvenient, then a letter with an arrow over it is employed, that is, $\vec{A}$, $\vec{B}$, $\vec{C}$. In this reference book the arrow notation is used in the following discussions.

### Properties of Vectors
Some important properties of vectors are
1. Two vectors $\vec{A}$ and $\vec{B}$ are equal if they have the same magnitude (length) and direction. Equality is denoted by $\vec{A} = \vec{B}$.
2. The magnitude of a vector is a nonnegative scalar quantity. The magnitude of a vector $\vec{B}$ is denoted by the symbols $B$ or $|\vec{B}|$.
3. A vector $\vec{B}$ is equal to zero only if its magnitude is zero. A vector whose magnitude is zero is called the zero or null vector and is denoted by the symbol $\vec{0}$.
4. Multiplication of a nonzero vector $\vec{B}$ by a positive scalar $m$ is denoted by $m\vec{B}$ and produces a new vector whose direction is the same as $\vec{B}$ but whose magnitude is $m$ times the magnitude of $\vec{B}$. Symbolically, $|m\vec{B}| = m|\vec{B}|$. If $m$ is a negative scalar the direction of $m\vec{B}$ is opposite to that of the direction of $\vec{B}$. In figure 5-1 several vectors obtained from $\vec{B}$ by scalar multiplication are exhibited.
5. Vectors are considered as "free vectors". The term "free vector" is used to mean the following. Any vector may be moved to a new position in space provided that in the new position it is parallel to and has the same direction as its original position. In many of the examples that follow, there are times when a given vector is moved to a convenient point in space in order to emphasize a special geometrical or physical concept. See for example figure 5-1.

## Vector Addition and Subtraction

Let $\vec{C} = \vec{A} + \vec{B}$ denote the sum of two vectors $\vec{A}$ and $\vec{B}$. To find the vector sum $\vec{A} + \vec{B}$, slide the origin of the vector $\vec{B}$ to the terminus point of the vector $\vec{A}$, then draw the line from the origin of $\vec{A}$ to the terminus of $\vec{B}$ to represent $\vec{C}$. Alternatively, start with the vector $\vec{B}$ and place the origin of the vector $\vec{A}$ at the terminus point of $\vec{B}$ to construct the vector $\vec{B} + \vec{A}$. Adding vectors in this way employs the parallelogram law for vector addition which is illustrated in the figure 5-2. Note that vector addition is commutative. That is, using the shifted vectors $\vec{A}$ and $\vec{B}$, denoted by the dashed positions in figure 5-2, the commutative law for vector addition $\vec{A} + \vec{B} = \vec{B} + \vec{A}$, is illustrated as the parallelogram law for vector addition.

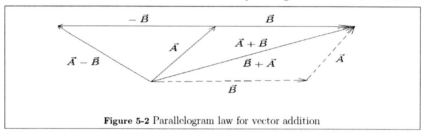

**Figure 5-2** Parallelogram law for vector addition

If $\vec{F} = \vec{A} - \vec{B}$ denotes the difference of two vectors $\vec{A}$ and $\vec{B}$, then $\vec{F}$ is determined by the above rule for vector addition by writing $\vec{F} = \vec{A} + (-\vec{B})$. Thus, subtraction is the addition of the vector $-\vec{B}$ to $\vec{A}$. In figure 5-2 observe that the vectors $\vec{A}$ and $\vec{B}$ are free vectors.

Vectors constitute a group under the operation of addition; that is, the following four properties are satisfied:

1. **Closure property:** If $\vec{A}$ and $\vec{B}$ belong to a set of vectors, then their sum $\vec{A} + \vec{B}$ must also belong to the same set.
2. **Associative property:** The insertion of parentheses or grouping of terms in vector summation is immaterial. That is,

$$(\vec{A} + \vec{B}) + \vec{C} = \vec{A} + (\vec{B} + \vec{C}) \tag{5.1}$$

3. **Identity element:** The zero or null vector when added to a vector does not produce a new vector. In symbols, $\vec{A} + \vec{0} = \vec{A}$. The null vector is called the identity element under addition.
4. **Inverse element:** If to each vector $\vec{A}$, there is associated a vector $\vec{E}$ such that under addition these two vectors produce the identity element, and $\vec{A} + \vec{E} = \vec{0}$, then the vector $\vec{E}$ is called the inverse of $\vec{A}$ under vector addition and is denoted by $\vec{E} = -\vec{A}$.

Additional properties satisfied by vectors include

5. **Commutative law:** If in addition all vectors of the group satisfy $\vec{A} + \vec{B} = \vec{B} + \vec{A}$, then the set of vectors is said to form a commutative group under vector addition.
6. **Distributive law:** The distributive law with respect to scalar multiplication is

$$m(\vec{A} + \vec{B}) = m\vec{A} + m\vec{B}, \qquad \text{where } m \text{ is a scalar.} \tag{5.2}$$

**Definition: (Linear dependence and independence of vectors)**
*Two nonzero vectors $\vec{A}$ and $\vec{B}$ are said to be linearly dependent if it is possible to find scalars $k_1$, $k_2$ not both zero, such that the equation*

$$k_1 \vec{A} + k_2 \vec{B} = \vec{0} \tag{5.3}$$

*is satisfied. If $k_1 = 0$ and $k_2 = 0$ are the only scalars for which the above equation is satisfied, then the vectors $\vec{A}$ and $\vec{B}$ are said to be linearly independent.*

This definition can be interpreted geometrically. If $k_1 \neq 0$, then equation (5.3) implies that $\vec{A} = -\frac{k_2}{k_1}\vec{B} = m\vec{B}$ showing that $\vec{A}$ is a scalar multiple of $\vec{B}$. That is, $\vec{A}$ and $\vec{B}$ have the same direction and therefore, they are called colinear vectors. If $\vec{A}$ and $\vec{B}$ are not colinear, then they are linearly independent (noncolinear). If two nonzero vectors $\vec{A}$ and $\vec{B}$ are linearly independent, then any vector $\vec{C}$ lying in the plane of $\vec{A}$ and $\vec{B}$ can be expressed as a linear combination of the vectors $\vec{A}$ and $\vec{B}$. Construct as in figure 5-3 a parallelogram with diagonal $\vec{C}$ and sides parallel to the vectors $\vec{A}$ and $\vec{B}$.

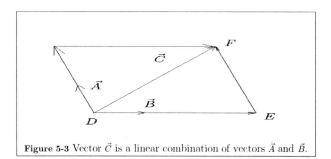

**Figure 5-3** Vector $\vec{C}$ is a linear combination of vectors $\vec{A}$ and $\vec{B}$.

Since the vector side $\overrightarrow{DE}$ is parallel to $\vec{B}$ and the vector side $\overrightarrow{EF}$ is parallel to $\vec{A}$, then there exists scalars $m$ and $n$ such that $\overrightarrow{DE} = m\vec{B}$ and $\overrightarrow{EF} = n\vec{A}$. With vector addition,

$$\vec{C} = \overrightarrow{DE} + \overrightarrow{EF} = m\vec{B} + n\vec{A} \tag{5.4}$$

which shows that $\vec{C}$ is a linear combination of the vectors $\vec{A}$ and $\vec{B}$.

**Example 5-1.** Show that the medians of a triangle meet at a trisection point.

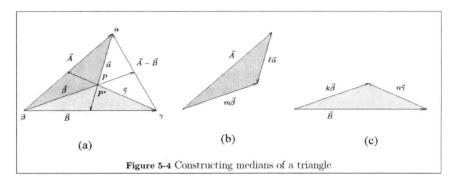

**Figure 5-4** Constructing medians of a triangle

**Solution:** Let the sides of a triangle with vertices $\alpha, \beta, \gamma$ be denoted by the vectors $\vec{A}$, $\vec{B}$, and $\vec{A} - \vec{B}$ as illustrated in the figure 5-4. Further, let $\vec{\alpha}$, $\vec{\beta}$, $\vec{\gamma}$ denote the vectors from the respective vertices of $\alpha$, $\beta$, $\gamma$ to the midpoints of the opposite sides. By construction, these vectors must satisfy the vector equations

$$\vec{A} + \vec{\alpha} = \frac{1}{2}\vec{B} \qquad \vec{B} + \frac{1}{2}(\vec{A} - \vec{B}) = \vec{\beta} \qquad \vec{B} + \vec{\gamma} = \frac{1}{2}\vec{A}. \tag{5.5}$$

Let the vectors $\vec{\alpha}$ and $\vec{\beta}$ intersect at a point designated by $P$. Similarly, let the vectors $\vec{\beta}$ and $\vec{\gamma}$ intersect at the point designated $P^*$. The problem is to show that the points $P$ and $P^*$ are the same. Figures 5-4(b) and 5-4(c) illustrate that for suitable scalars $k$, $\ell$, $m$, $n$, the points $P$ and $P^*$ determine the vectors equations

$$\vec{A} + \ell\vec{\alpha} = m\vec{\beta} \qquad \text{and} \qquad \vec{B} + n\vec{\gamma} = k\vec{\beta}. \tag{5.6}$$

In these equations the scalars $k$, $\ell$, $m$, $n$ are unknowns to be determined. Use the set of equations (5.5), to solve for the vectors $\vec{\alpha}$, $\vec{\beta}$, $\vec{\gamma}$ in terms of the vectors $\vec{A}$ and $\vec{B}$ and show

$$\vec{\alpha} = \frac{1}{2}\vec{B} - \vec{A} \qquad \vec{\beta} = \frac{1}{2}(\vec{A} + \vec{B}) \qquad \vec{\gamma} = \frac{1}{2}\vec{A} - \vec{B}. \tag{5.7}$$

These equations can now be substituted into the equations (5.6) to yield, after some simplification, the equations

$$(1 - \ell - \frac{m}{2})\vec{A} = (\frac{m}{2} - \frac{\ell}{2})\vec{B} \quad \text{and} \quad (\frac{k}{2} - \frac{n}{2})\vec{A} = (1 - n - \frac{k}{2})\vec{B}.$$

Since the vectors $\vec{A}$ and $\vec{B}$ are linearly independent (noncolinear), the scalar coefficients in the above equation must equal zero, because if these scalar coefficients were not zero, then the vectors $\vec{A}$ and $\vec{B}$ would be linearly dependent (colinear) and a triangle would not exist.

By equating to zero the scalar coefficients in these equations, there results the simultaneous scalar equations

$$(1 - \ell - \frac{m}{2}) = 0, \qquad (\frac{m}{2} - \frac{\ell}{2}) = 0, \qquad (\frac{k}{2} - \frac{n}{2}) = 0, \qquad (1 - n - \frac{k}{2}) = 0$$

The solution of these equations produces the fact that $k = \ell = m = n = \frac{2}{3}$ and hence the conclusion $P = P^*$ is a trisection point.

■

## Unit Vectors

A vector having length or magnitude of one is called a unit vector. If $\vec{A}$ is a nonzero vector of length $|\vec{A}|$, a unit vector in the direction of $\vec{A}$ is obtained by multiplying the vector $\vec{A}$ by the scalar $m = \frac{1}{|\vec{A}|}$. The unit vector so constructed is denoted

$$\hat{e}_A = \frac{\vec{A}}{|\vec{A}|} \qquad \text{and satisfies} \qquad |\hat{e}_A| = 1.$$

The symbol $\hat{e}$ is reserved for unit vectors and the notation $\hat{e}_A$ is to be read "a unit vector in the direction of $\vec{A}$." The hat or carat ($\hat{\phantom{x}}$) notation is used to represent a unit vector or normalized vector.

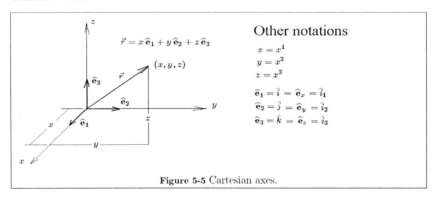

**Figure 5-5** Cartesian axes.

The figure 5-5 illustrates unit base vectors $\hat{e}_1$, $\hat{e}_2$, $\hat{e}_3$ in the directions of the positive $x, y, z$-coordinate axes in a rectangular Cartesian coordinate system. At times it is convenient to replace the symbols $x$, $y$, $z$, respectively by the symbols $x^1$, $x^2$, $x^3$. Here the superscript notation should not be confused with powers of $x$. The symbols $x^i$, $i = 1, 2, 3$, are to represent variables and not $x$ raised to a power. Whenever it is required to raise a variable like $x^2$ to a power use parentheses for this purpose and write $(x^2)^2$. The substitutions $x = x^1, y = x^2$ and $z = x^3$ are made in many textbooks to write mathematical quantities in a more compact form.

The unit base vectors in the direction of the $x, y, z$ or $x^1, x^2, x^3$ axes can be represented by a variety of notations. Some of the more common notations employed in various textbooks to denote rectangular unit base vectors are

$$\hat{i},\ \hat{j},\ \hat{k}, \qquad \hat{e}_x,\ \hat{e}_y,\ \hat{e}_z, \qquad \hat{i}_1,\ \hat{i}_2,\ \hat{i}_3, \qquad \bar{I}_x,\ \bar{I}_y,\ \bar{I}_z, \qquad \hat{e}_1,\ \hat{e}_2,\ \hat{e}_3$$

Historically the unit vectors $\hat{i}, \hat{j}, \hat{k}$ were employed to represent the Cartesian components of a vector. The notation $\hat{e}_1$, $\hat{e}_2$, $\hat{e}_3$ to represent the unit base vectors in the direction of the $x^1, x^2, x^3$ axes will be used in the discussions that follow. The subscript and superscript notation makes it easier to generalize to vectors in $n$-dimensions involving coordinates $(x^1, x^2, x^3, \ldots, x^n)$ and unit vectors in these directions being denoted $\hat{e}_1, \hat{e}_2, \hat{e}_3, \ldots, \hat{e}_n$.

Another important notation to be adhered to is that of using the symbol $\vec{r}$ to represent a general position vector of an arbitrary point $(x, y, z)$ in three-dimensional space. The vector $\vec{r}$ is understood to mean the vector

$$\vec{r} = x\,\hat{e}_1 + y\,\hat{e}_2 + z\,\hat{e}_3 = x^1\,\hat{e}_1 + x^2\,\hat{e}_2 + x^3\,\hat{e}_3 \tag{5.8}$$

which is a linear combination of the unit base vectors.

## Scalar or Dot Product (inner product)

The scalar or dot product of two vectors is sometimes referred to as an inner product of vectors.

**Definition: (Dot product)** *The scalar or dot product of two vectors $\vec{A}$ and $\vec{B}$ is denoted*

$$\vec{A} \cdot \vec{B} = |\vec{A}|\,|\vec{B}|\,\cos\theta, \tag{5.9}$$

*and represents the magnitude of $\vec{A}$ times the magnitude $\vec{B}$ times the cosine of $\theta$, where $\theta$ is the angle between the vectors $\vec{A}$ and $\vec{B}$ when their origins are made to coincide.*

The angle between any two of the orthogonal unit base vectors $\hat{e}_1, \hat{e}_2, \hat{e}_3$ in Cartesian coordinates is $90°$ or $\frac{\pi}{2}$ radians. Using the results $\cos\frac{\pi}{2} = 0$ and $\cos 0 = 1$, there results the following dot product relations for these unit vectors

$$\hat{e}_1 \cdot \hat{e}_1 = 1 \qquad \hat{e}_2 \cdot \hat{e}_1 = 0 \qquad \hat{e}_3 \cdot \hat{e}_1 = 0$$
$$\hat{e}_1 \cdot \hat{e}_2 = 0 \qquad \hat{e}_2 \cdot \hat{e}_2 = 1 \qquad \hat{e}_3 \cdot \hat{e}_2 = 0$$
$$\hat{e}_1 \cdot \hat{e}_3 = 0 \qquad \hat{e}_2 \cdot \hat{e}_3 = 0 \qquad \hat{e}_3 \cdot \hat{e}_3 = 1$$

Using an index notation the above dot products can be expressed $\hat{e}_i \cdot \hat{e}_j = \delta_{ij}$ where the subscripts $i$ and $j$ can take on any of the values $1, 2, 3$. Here $\delta_{ij}$ is the Kronecker delta symbol defined by $\delta_{ij} = \begin{cases} 1, & i = j \\ 0, & i \neq j \end{cases}$.

The dot product satisfies the following properties which are given without proof:

**Commutative law** $\quad \vec{A} \cdot \vec{B} = \vec{B} \cdot \vec{A}$
**Distributive law** $\quad \vec{A} \cdot (\vec{B} + \vec{C}) = \vec{A} \cdot \vec{B} + \vec{A} \cdot \vec{C}$
**Magnitude squared** $\quad \vec{A} \cdot \vec{A} = A^2 = |\vec{A}|^2$

The physical interpretation of projection can be assigned to the dot product as is illustrated in figure 5-6. In this figure $\vec{A}$ and $\vec{B}$ are nonzero vectors with $\hat{e}_A$ and $\hat{e}_B$ unit vectors in the directions of $\vec{A}$ and $\vec{B}$, respectively. The figure 5-6 illustrates the physical interpretation of the following equations:

$$\hat{e}_B \cdot \vec{A} = |\vec{A}| \cos\theta = \text{Projection of } \vec{A} \text{ onto direction of } \vec{B}$$
$$\hat{e}_A \cdot \vec{B} = |\vec{B}| \cos\theta = \text{Projection of } \vec{B} \text{ onto direction of } \vec{A}.$$

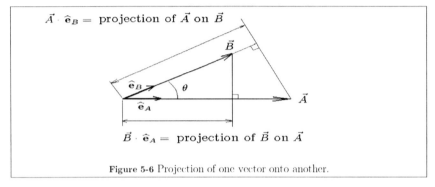

**Figure 5-6** Projection of one vector onto another.

In general, the dot product of a nonzero vector with a unit vector can be interpreted as the projection of the given vector onto the direction of the unit vector. The dot product of a vector with a unit vector is a basic fundamental concept which arises in a variety of science and engineering applications.

Observe that if the dot product of two vectors is zero, $\vec{A} \cdot \vec{B} = |\vec{A}||\vec{B}| \cos\theta = 0$, then this implies that either $\vec{A} = \vec{0}$, $\vec{B} = \vec{0}$, or $\theta = \frac{\pi}{2}$. If $\theta = \frac{\pi}{2}$, then $\vec{A}$ is perpendicular to $\vec{B}$ or the projection of $\vec{B}$ on $\vec{A}$ is zero.

A curve $C$ in Euclidean three space can be expressed $\vec{r} = x(t)\hat{e}_1 + y(t)\hat{e}_2 + z(t)\hat{e}_3$ where $t$ is a parameter. The derivative of the position vector with respect to the parameter $t$ is given by $\frac{d\vec{r}}{dt} = \frac{dx}{dt}\hat{e}_1 + \frac{dy}{dt}\hat{e}_2 + \frac{dz}{dt}\hat{e}_3$ and represents a tangent vector to the curve at the position corresponding to the parameter value $t$. Using $d\vec{r} = dx\,\hat{e}_1 + dy\,\hat{e}_2 + dz\,\hat{e}_3$ calculate the element of arc length squared given by the dot product $d\vec{r} \cdot d\vec{r} = dx^2 + dy^2 + dz^2 = ds^2$ and observe that the magnitude of the derivative $\frac{d\vec{r}}{dt}$ is given by $|\frac{d\vec{r}}{dt}| = \sqrt{\left(\frac{dx}{dt}\right)^2 + \left(\frac{dy}{dt}\right)^2 + \left(\frac{dz}{dt}\right)^2} = \frac{ds}{dt}$ where $s$ is arc length. Use the chain rule for differentiation to show

$$\frac{d\vec{r}}{dt} = \frac{d\vec{r}}{ds}\frac{ds}{dt} \quad \text{which gives} \quad \frac{d\vec{r}}{ds} = \hat{e}_t = \frac{1}{|\frac{d\vec{r}}{dt}|}\frac{d\vec{r}}{dt}$$

Here $\frac{d\vec{r}}{ds}$ represents a derivative of the position vector $\vec{r}$ with respect to arc length and has the physical interpretation of denoting a unit tangent vector to the curve.

### Direction Cosines Associated With Vectors

Let $\vec{A}$ be a nonzero vector having its origin at the origin of a rectangular Cartesian coordinate system. The dot products

$$\vec{A} \cdot \hat{e}_1 = A_1 \qquad \vec{A} \cdot \hat{e}_2 = A_2 \qquad \vec{A} \cdot \hat{e}_3 = A_3 \tag{5.10}$$

represent, respectively, the components or projections of the vector $\vec{A}$ onto the $x, y$ and $z$-axes. The projections $A_1, A_2, A_3$ of the vector $\vec{A}$ onto the coordinate axes are scalars which are called the components of the vector $\vec{A}$. From the definition of the dot product of two vectors, the scalar components of the vector $\vec{A}$ satisfy the equations

$$A_1 = \vec{A} \cdot \hat{e}_1 = |\vec{A}| \cos \alpha, \qquad A_2 = \vec{A} \cdot \hat{e}_2 = |\vec{A}| \cos \beta, \qquad A_3 = \vec{A} \cdot \hat{e}_3 = |\vec{A}| \cos \gamma, \tag{5.11}$$

where $\alpha, \beta, \gamma$ are, respectively, the smaller angles between the vector $\vec{A}$ and the $x, y, z$ coordinate axes and are illustrated in figure 5-7.

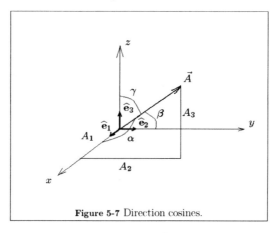

**Figure 5-7** Direction cosines.

The vector quantities

$$\vec{A}_1 = A_1 \hat{e}_1, \qquad \vec{A}_2 = A_2 \hat{e}_2, \qquad \vec{A}_3 = A_3 \hat{e}_3 \tag{5.12}$$

are called the vector components of the vector $\vec{A}$. From the addition property of vectors, the vector components of $\vec{A}$ may be added to obtain

$$\vec{A} = A_1 \hat{e}_1 + A_2 \hat{e}_2 + A_3 \hat{e}_3 = |\vec{A}|(\cos \alpha \, \hat{e}_1 + \cos \beta \, \hat{e}_2 + \cos \gamma \, \hat{e}_3). \tag{5.13}$$

This vector representation is called the component form of the vector $\vec{A}$.

If $\vec{A} = A_1\,\hat{e}_1 + A_2\,\hat{e}_2 + A_3\,\hat{e}_3$ and $\vec{B} = B_1\,\hat{e}_1 + B_2\,\hat{e}_2 + B_3\,\hat{e}_3$, are vectors in component form, then the dot product of these vectors is given by

$$\vec{A} \cdot \vec{B} = |\vec{A}| \cdot |\vec{B}| \cos\theta = (A_1\,\hat{e}_1 + A_2\,\hat{e}_2 + A_3\,\hat{e}_3) \cdot (B_1\,\hat{e}_1 + B_2\,\hat{e}_2 + B_3\,\hat{e}_3) = A_1 B_1 + A_2 B_2 + A_3 B_3$$

$$\vec{A} \cdot \vec{A} = A_1^2 + A_2^2 + A_3^2 = A \cdot A \cos 0 = A^2 = |\vec{A}|^2$$

From the component form of $\vec{A}$, it is possible to construct a unit vector in the direction of $\vec{A}$. This is accomplished by multiplying the vector $\vec{A}$ by the nonzero scalar $\frac{1}{|\vec{A}|}$. The unit vector in the direction of $\vec{A}$ that results is obtained from equation (5.13) and can be represented

$$\hat{e}_A = \frac{\vec{A}}{|\vec{A}|} = \cos\alpha\,\hat{e}_1 + \cos\beta\,\hat{e}_2 + \cos\gamma\,\hat{e}_3,$$

where the components of the unit vector $\hat{e}_A$ are called the direction cosines associated with the vector $\vec{A}$. Note that the direction cosines associated with the vector $\vec{A}$ must satisfy

$$\hat{e}_A \cdot \hat{e}_A = \cos^2\alpha + \cos^2\beta + \cos^2\gamma = 1$$

which states that the sum of squares of the direction cosines of the vector must equal unity. That is, if you find a unit vector in the direction of $\vec{A}$, then the components of the unit vector represent the direction cosines of the vector.

### Example 5-2.

Find the equation of the line which passes through the two distinct points $(x_1, y_1, z_1)$ and $(x_2, y_2, z_2)$.

**Solution:** Define the position vectors

$$\vec{r}_1 = x_1\,\hat{e}_1 + y_1\,\hat{e}_2 + z_1\,\hat{e}_3$$

$$\vec{r}_2 = x_2\,\hat{e}_1 + y_2\,\hat{e}_2 + z_2\,\hat{e}_3$$

from the origin to the given points and let

$$\vec{r} = x\,\hat{e}_1 + y\,\hat{e}_2 + z\,\hat{e}_3$$

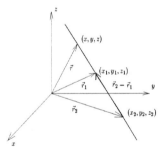

represent the vector to a variable point $(x, y, z)$ on the line as illustrated in the accompanying figure.

Observe that the vector $\vec{r}_2 - \vec{r}_1$ is a vector parallel to the line. Using vector addition any point on the line can be represented as

$$\vec{r} = \vec{r}_1 + t(\vec{r}_2 - \vec{r}_1) \tag{5.14}$$

where $t$ is a scalar parameter. An alternate form for the equation of the line is given by

$$\vec{r} = \vec{r}_2 + \lambda(\vec{r}_2 - \vec{r}_1)$$

where $\lambda$ is some scalar. By equating coefficients of unit vectors in equation (5.14) there results the parametric equations of the line

$$x = x_1 + t(x_2 - x_1), \qquad y = y_1 + t(y_2 - y_1), \qquad z = z_1 + t(z_2 - z_1)$$

If the quantities $(x_2 - x_1)$, $(y_2 - y_1)$ and $(z_2 - z_1)$ are different from zero, then the equation of the line can be represented in the symmetric form

$$\frac{x - x_1}{x_2 - x_1} = \frac{y - y_1}{y_2 - y_1} = \frac{z - z_1}{z_2 - z_1} = t$$

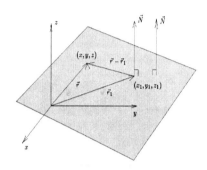

### Example 5-3.

Find the equation of the plane which passes through the point $(x_1, y_1, z_1)$ and is perpendicular to the vector $\vec{N} = N_1\,\hat{e}_1 + N_2\,\hat{e}_2 + N_3\,\hat{e}_3$.

**Solution:** Let $\vec{r}_1 = x_1\,\hat{e}_1 + y_1\,\hat{e}_2 + z_1\,\hat{e}_3$ denote a vector from the origin to the given point and let $\vec{r} = x\,\hat{e}_1 + y\,\hat{e}_2 + z\,\hat{e}_3$ denote a position vector from the origin to a variable point $(x, y, z)$ lying in the plane.

By construction the vector $\vec{r} - \vec{r}_1$ lies in the plane and is therefore perpendicular to the given vector $\vec{N}$. This requires that

$$\vec{N} \cdot (\vec{r} - \vec{r}_1) = 0$$

or in scalar form the equation of the plane is given by

$$N_1(x - x_1) + N_2(y - y_1) + N_3(z - z_1) = 0$$

∎

### The Cross Product or Outer Product

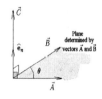

The cross or outer product of two vectors $\vec{A}$ and $\vec{B}$ is a new vector $\vec{C}$ given by

$$\vec{C} = \vec{A} \times \vec{B} = |\vec{A}||\vec{B}| \sin\theta\,\hat{e}_n, \qquad (5.15)$$

where $\theta$ is the smaller angle between the two vectors $\vec{A}$ and $\vec{B}$ when their origins coincide, and $\hat{e}_n$ is a unit vector perpendicular to the plane containing the vectors $\vec{A}$ and $\vec{B}$.

The direction of $\hat{e}_n$ is determined by the right-hand rule. Place the fingers of your right-hand in the direction of $\vec{A}$ and rotate the fingers toward the vector $\vec{B}$, then the thumb of the right-hand points in the direction $\vec{C}$. The vectors $\vec{A}$, $\vec{B}$, $\vec{C}$ then form a right-handed system.

A special case of the above definition is when $\vec{A}$ is parallel to $\vec{B}$. In this case $\theta = 0$ and consequently if the cross product of these two vectors is the zero vector, $\vec{A} \times \vec{B} = \vec{0}$, then either $\vec{A} = \vec{0}$ or $\vec{B} = \vec{0}$ or $\vec{A}$ is parallel to $\vec{B}$. The cross product possesses the following properties which are stated without proof.

**Properties of the Cross Product**
1. $\vec{A} \times \vec{B} = -\vec{B} \times \vec{A}$ (noncommutative)
2. $\vec{A} \times (\vec{B} + \vec{C}) = \vec{A} \times \vec{B} + \vec{A} \times \vec{C}$ (distributive law)
3. $m(\vec{A} \times \vec{B}) = (m\vec{A}) \times \vec{B} = \vec{A} \times (m\vec{B})$ $m$ a scalar
4. $\vec{A} \times \vec{A} = \vec{0}$

Note that the cross products of the orthogonal unit vectors $\hat{e}_1$, $\hat{e}_2$, $\hat{e}_3$ satisfy the relations

$$\hat{e}_1 \times \hat{e}_1 = \vec{0} \qquad \hat{e}_2 \times \hat{e}_1 = -\hat{e}_3 \qquad \hat{e}_3 \times \hat{e}_1 = \hat{e}_2$$
$$\hat{e}_1 \times \hat{e}_2 = \hat{e}_3 \qquad \hat{e}_2 \times \hat{e}_2 = \vec{0} \qquad \hat{e}_3 \times \hat{e}_2 = -\hat{e}_1 \qquad (5.16)$$
$$\hat{e}_1 \times \hat{e}_3 = -\hat{e}_2 \qquad \hat{e}_2 \times \hat{e}_3 = \hat{e}_1 \qquad \hat{e}_3 \times \hat{e}_3 = \vec{0}.$$

The above results can be expressed $\hat{e}_i \times \hat{e}_j = e_{1ij}\,\hat{e}_1 + e_{2ij}\,\hat{e}_2 + e_{3ij}\,\hat{e}_3$ where $i,j$ can take on any of the values $1, 2$ or $3$. The term $e_{ijk}$ is called the e-permutation symbol and defined as

$$e_{ijk} = \begin{cases} +1, & \text{if } ijk \text{ is an even permutation of the integers } 123 \\ -1, & \text{if } ijk \text{ is an odd permutation of the integers } 123 \\ 0, & \text{in all other cases} \end{cases}$$

Let $\vec{A} = A_1\,\hat{e}_1 + A_2\,\hat{e}_2 + A_3\,\hat{e}_3$ and $\vec{B} = B_1\,\hat{e}_1 + B_2\,\hat{e}_2 + B_3\,\hat{e}_3$ be two nonzero vectors and form the cross product

$$\vec{A} \times \vec{B} = (A_1\,\hat{e}_1 + A_2\,\hat{e}_2 + A_3\,\hat{e}_3) \times (B_1\,\hat{e}_1 + B_2\,\hat{e}_2 + B_3\,\hat{e}_3). \qquad (5.17)$$

The cross product can be expanded by using the distributive law to obtain

$$\begin{aligned}\vec{A} \times \vec{B} = &\ A_1 B_1\,\hat{e}_1 \times \hat{e}_1 + A_1 B_2\,\hat{e}_1 \times \hat{e}_2 + A_1 B_3\,\hat{e}_1 \times \hat{e}_3 \\ &+ A_2 B_1\,\hat{e}_2 \times \hat{e}_1 + A_2 B_2\,\hat{e}_2 \times \hat{e}_2 + A_2 B_3\,\hat{e}_2 \times \hat{e}_3 \\ &+ A_3 B_1\,\hat{e}_3 \times \hat{e}_1 + A_3 B_2\,\hat{e}_3 \times \hat{e}_2 + A_3 B_3\,\hat{e}_3 \times \hat{e}_3.\end{aligned} \qquad (5.18)$$

Simplification produces the important cross product formula

$$\vec{A} \times \vec{B} = (A_2 B_3 - A_3 B_2)\,\hat{e}_1 + (A_3 B_1 - A_1 B_3)\,\hat{e}_2 + (A_1 B_2 - A_2 B_1)\,\hat{e}_3, \qquad (5.19)$$

a result that can be expressed in the determinant form

$$\vec{A} \times \vec{B} = \begin{vmatrix} \hat{e}_1 & \hat{e}_2 & \hat{e}_3 \\ A_1 & A_2 & A_3 \\ B_1 & B_2 & B_3 \end{vmatrix} = \begin{vmatrix} A_2 & A_3 \\ B_2 & B_3 \end{vmatrix}\hat{e}_1 - \begin{vmatrix} A_1 & A_3 \\ B_1 & B_3 \end{vmatrix}\hat{e}_2 + \begin{vmatrix} A_1 & A_2 \\ B_1 & B_2 \end{vmatrix}\hat{e}_3. \qquad (5.20)$$

In summary, the cross product of two vectors $\vec{A}$ and $\vec{B}$ is a new vector $\vec{C}$, where

$$\vec{C} = \vec{A} \times \vec{B} = C_1\,\hat{e}_1 + C_2\,\hat{e}_2 + C_3\,\hat{e}_3$$

with components

$$C_1 = A_2 B_3 - A_3 B_2, \qquad C_2 = A_3 B_1 - A_1 B_3, \qquad C_3 = A_1 B_2 - A_2 B_1 \qquad (5.21)$$

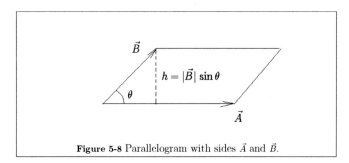

**Figure 5-8** Parallelogram with sides $\vec{A}$ and $\vec{B}$.

A geometric interpretation that can be assigned to the magnitude of the cross product of two vectors is illustrated in figure 5-8. The area of the parallelogram having the vectors $\vec{A}$ and $\vec{B}$ for its sides is given by

$$\text{Area} = |\vec{A}| \cdot h = |\vec{A}||\vec{B}|\sin\theta = |\vec{A} \times \vec{B}|. \tag{5.22}$$

Therefore, the magnitude of the cross product of two vectors represents the area of the parallelogram formed from these vectors when their origins are made to coincide.

**Vector Identities**

The following vector identities are often needed to simplify various equations in science and engineering. These identities are stated without proofs.

1. $\vec{A} \times \vec{B} = -\vec{B} \times \vec{A}$ (5.23)
2. $\vec{A} \cdot (\vec{B} \times \vec{C}) = \vec{B} \cdot (\vec{C} \times \vec{A}) = \vec{C} \cdot (\vec{A} \times \vec{B})$ (5.24)

   An identity known as the triple scalar product.
3. $(\vec{A} \times \vec{B}) \times (\vec{C} \times \vec{D}) = \vec{C}(\vec{D} \cdot \vec{A} \times \vec{B}) - \vec{D}(\vec{C} \cdot \vec{A} \times \vec{B})$
   $= \vec{B} \cdot (\vec{A} \cdot \vec{C} \times \vec{D}) - \vec{A}(\vec{B} \cdot \vec{C} \times \vec{D})$ (5.25)
4. $\vec{A} \times (\vec{B} \times \vec{C}) = \vec{B}(\vec{A} \cdot \vec{C}) - \vec{C}(\vec{A} \cdot \vec{B})$ (5.26)
5. $(\vec{A} \times \vec{B}) \cdot (\vec{C} \times \vec{D}) = (\vec{A} \cdot \vec{C})(\vec{B} \cdot \vec{D}) - (\vec{A} \cdot \vec{D})(\vec{B} \cdot \vec{C})$ (5.27)
6. $\vec{A} \times (\vec{B} \times \vec{C}) + \vec{B} \times (\vec{C} \times \vec{A}) + \vec{C} \times (\vec{A} \times \vec{B}) = \vec{0}$ (5.28)

In the second identity above, the quantity $\vec{A} \cdot (\vec{B} \times \vec{C})$ is called a triple scalar product and can be represented using determinants

$$\vec{A} \cdot (\vec{B} \times \vec{C}) = \begin{vmatrix} A_1 & A_2 & A_3 \\ B_1 & B_2 & B_3 \\ C_1 & C_2 & C_3 \end{vmatrix}.$$

A physical interpretation that can be assigned to the triple scalar product is that its absolute value represents the volume of the parallelepiped formed by the three noncoplaner vectors $\vec{A}, \vec{B}, \vec{C}$. The absolute value is needed because sometimes the triple scalar product is negative. This physical interpretation can be obtained from an analysis of figure 5-9.

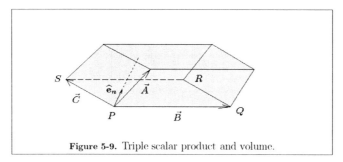

Figure 5-9. Triple scalar product and volume.

In figure 5-9 note the following.
(a) The magnitude $|\vec{B} \times \vec{C}|$ represents the area of the parallelogram $PQRS$.
(b) The unit vector $\hat{e}_n = \dfrac{\vec{B} \times \vec{C}}{|\vec{B} \times \vec{C}|}$ is normal to the plane containing the vectors $\vec{B}$ and $\vec{C}$.
(c) The dot product $\vec{A} \cdot \hat{e}_n = \vec{A} \cdot \dfrac{\vec{B} \times \vec{C}}{|\vec{B} \times \vec{C}|} = h$ represents the projection of $\vec{A}$ on $\hat{e}_n$ and produces the height of the parallelepiped. These results demonstrate that

$$\left|\vec{A} \cdot (\vec{B} \times \vec{C})\right| = |\vec{B} \times \vec{C}|\, h = (\text{Area of base})(\text{Height}) = \text{Volume}.$$

so that the magnitude of the triple scalar product is the volume of the parallelepiped formed when the origins of the three vectors are made to coincide.

**Example 5-4.** Derive the law of sines for the triangle illustrated in the figure 5-10.

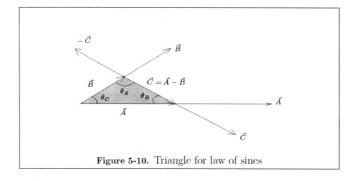

Figure 5-10. Triangle for law of sines

**Solution:** Since the vectors $\vec{A}, \vec{B}$ and $\vec{C}$ are free vectors, these vectors can be moved to the positions illustrated in figure 5-10. Also sketch the vector $-\vec{C}$ as illustrated. The new positions for the vectors $\vec{A}, \vec{B}, \vec{C}$ and $-\vec{C}$ are to better visualize certain vector cross products associated with the law of sines. Examine figure 5-10 and note the following cross products
$$\vec{C} \times \vec{A} = (\vec{A} - \vec{B}) \times \vec{A} = \vec{A} \times \vec{A} - \vec{B} \times \vec{A} = -\vec{B} \times \vec{A} = \vec{A} \times \vec{B}$$
and $\quad \vec{B} \times (-\vec{C}) = \vec{B} \times (-\vec{A} + \vec{B}) = \vec{B} \times (-\vec{A}) + \vec{B} \times \vec{B} = \vec{A} \times \vec{B}.$

Taking the magnitude of the above cross products gives
$$|\vec{C} \times \vec{A}| = |\vec{A} \times \vec{B}| = |\vec{B} \times (-\vec{C})|$$
or
$$AC \sin\theta_B = AB \sin\theta_C = BC \sin\theta_A.$$
Dividing by the product of the vector magnitudes $ABC$ produces the law of sines
$$\frac{\sin\theta_A}{A} = \frac{\sin\theta_B}{B} = \frac{\sin\theta_C}{C}.$$

■

## Scalar and Vector Fields

Often arising in the areas of science and engineering are the concepts of a scalar and vector field. A scalar field is a function from $R^n$ to $R$. That is, a scalar field $\phi$ is a one-to-one correspondence between points in space and the scalar quantity $\phi$. Scalar fields in one-dimension are written $\phi = \phi(x)$, in two-dimensions $\phi = \phi(x,y)$, in three-dimensions $\phi = \phi(x,y,z), \ldots$, in $n$-dimensions $\phi = \phi(x_1, x_2, \ldots, x_n)$. Examples of scalar fields are humidity, temperature, pressure, electric potential, etc. The scalar field $\phi$ is assumed to be continuous and differentiable. Scalar fields in two dimensions can be represented graphically by constructing the equi-level curves (contour plots) $\phi(x,y) = constant$ for various values of the constant. Scalar fields in three dimensions can be represented by graphing the equi-level surfaces $\phi(x,y,z) = constant$ for various values of the constant.

## The Gradient

The gradient function grad $\phi$ is a field characteristic that describes the spatial rate of change of a scalar field. Let $\phi = \phi(x,y,z)$ represent a scalar field, then the gradient of $\phi$ is a vector and is written
$$\operatorname{grad} \phi = \frac{\partial \phi}{\partial x} \hat{e}_1 + \frac{\partial \phi}{\partial y} \hat{e}_2 + \frac{\partial \phi}{\partial z} \hat{e}_3. \tag{5.29}$$
Assume that the scalar field $\phi = \phi(x,y,z)$ possesses first partial derivatives throughout some region of space in order that the gradient vector exists. The operator
$$\nabla = \frac{\partial}{\partial x} \hat{e}_1 + \frac{\partial}{\partial y} \hat{e}_2 + \frac{\partial}{\partial z} \hat{e}_3 \tag{5.30}$$
is called the "del" operator and can be used to express the gradient in the operator form
$$\operatorname{grad} \phi = \nabla \phi = \left( \frac{\partial}{\partial x} \hat{e}_1 + \frac{\partial}{\partial y} \hat{e}_2 + \frac{\partial}{\partial z} \hat{e}_3 \right) \phi. \tag{5.31}$$
Note that this operator is not commutative and $\nabla \phi \neq \phi \nabla$.

Let $\vec{r}(s)$ denote an arbitrary space curve which passes through the point $P(x,y,z)$ of the region $R$, where the scalar function $\phi$ exists and has all first-order partial derivatives which are continuous. Here the space curve is expressed in terms of the arc length parameter $s$, where $s$ is measured from some fixed point on the curve. In general, the scalar field $\phi = \phi(x,y,z)$ varies with position and has different values when evaluated at different points in space. Let us evaluate $\phi$ at points along the arbitrary curve $\vec{r}$ to determine how $\phi$ changes with position along the curve. The rate of change of $\phi$ with respect to arc length along the curve is given by

$$\frac{d\phi}{ds} = \frac{\partial \phi}{\partial x}\frac{dx}{ds} + \frac{\partial \phi}{\partial y}\frac{dy}{ds} + \frac{\partial \phi}{\partial z}\frac{dz}{ds}$$

$$\frac{d\phi}{ds} = \left(\frac{\partial \phi}{\partial x}\hat{e}_1 + \frac{\partial \phi}{\partial y}\hat{e}_2 + \frac{\partial \phi}{\partial z}\hat{e}_3\right) \cdot \left(\frac{dx}{ds}\hat{e}_1 + \frac{dy}{ds}\hat{e}_2 + \frac{dz}{ds}\hat{e}_3\right)$$

$$\frac{d\phi}{ds} = \operatorname{grad} \phi \cdot \frac{d\vec{r}}{ds} = \nabla \phi \cdot \hat{e}_t,$$

where the right-hand side is to be evaluated at a point $P$ on the curve $\vec{r}(s)$ in $R$. The right-hand side of this equation is the dot product of the gradient vector with the unit tangent vector to the curve at the point $P$ and physically represents the projection of the vector grad $\phi$ in the direction of the tangent vector. Note that the curve $\vec{r}(s)$ represents an arbitrary curve through the point $P$, and hence, the unit tangent vector represents an arbitrary direction. Therefore, the derivative $\frac{d\phi}{ds} = \operatorname{grad} \phi \cdot \vec{e}$ can be interpreted as representing the rate of change of $\phi$ as one moves in the direction $\vec{e}$. Here the derivative equals the projection of the vector grad $\phi$ in the direction $\vec{e}$. Such derivatives are called directional derivatives.

**Definition: (Directional derivative)** *The component of the gradient $\phi$ in the direction of a unit vector $\hat{e}$ is equal to $\nabla\phi \cdot \hat{e}$ and is called the directional derivative of $\phi$ in the direction $\hat{e}$. The directional derivative is written as*

$$\frac{d\phi}{ds} = \operatorname{grad} \phi \cdot \vec{e} = \nabla \phi \cdot \hat{e} \tag{5.32}$$

*where $s$ denotes distance in the direction $\vec{e}$.*

The directional derivative is a measure of how the scalar field $\phi$ changes as a point moves in a specified direction. Since the maximum projection of a vector is the magnitude of the vector itself, the gradient of $\phi$ is a vector which points in the direction of the greatest rate of change of $\phi$. The length of the vector is $|\operatorname{grad} \phi|$ and represents the magnitude of this greatest rate of change. In other words, the gradient of a scalar field is a vector field and represents the direction and magnitude of the greatest rate of change of the scalar field.

**Example 5-5.** Show the gradient of $\phi$ is a normal vector to the surface

$$\phi = \phi(x,y,z) = c = \text{constant}.$$

**Solution:** Let $\vec{r}(s) = x(s)\,\hat{e}_1 + y(s)\,\hat{e}_2 + z(s)\,\hat{e}_3$, where $s$ is arc length, represent any curve lying in the surface $\phi(x,y,z) = c$. Along this curve, we have the scalar $\phi = \phi(x(s), y(s), z(s)) = c$ and the rate of change of $\phi$ along this curve is given by

$$\frac{d\phi}{ds} = \frac{\partial \phi}{\partial x}\frac{dx}{ds} + \frac{\partial \phi}{\partial y}\frac{dy}{ds} + \frac{\partial \phi}{\partial z}\frac{dz}{ds} = \frac{dc}{ds} = 0 \quad \text{or} \quad \frac{d\phi}{ds} = \text{grad}\,\phi \cdot \frac{d\vec{r}}{ds} = \text{grad}\,\phi \cdot \hat{e}_t = 0.$$

The resulting equation tells us that the vector grad $\phi$ is perpendicular to the unit tangent vector to the curve on the surface. But this unit tangent vector lies in the tangent plane to the surface at the point of evaluation for the gradient. Thus, grad $\phi$ is normal to the surface $\phi(x,y,z) = c$. The family of surfaces $\phi = \phi(x,y,z) = c$, for various values of $c$, are called level surfaces. In two-dimensions, the curves $\phi = \phi(x,y) = c$ are called level curves. The gradient of $\phi$ is a vector perpendicular to these level surfaces or level curves. Note that the vector $-\text{grad}\,\phi$ is also a normal vector. ∎

**Example 5-6.** Find the unit tangent vector at a point on the curve defined by the intersection of the two surfaces $F(x,y,z) = c_1$ and $G(x,y,z) = c_2$, where $c_1$ and $c_2$ are constants.
**Solution:** If two surfaces $F = c_1$ and $G = c_2$ intersect in a curve, then at a point $(x_0, y_0, z_0)$ common to both surfaces and on the curve, it is possible to calculate the normal vectors to both surfaces. These normal vectors are $\nabla F = \text{grad}\,F$ and $\nabla G = \text{grad}\,G$ which are evaluated at the point $(x_0, y_0, z_0)$ common to both surfaces and on the curve of intersection of the surfaces. The cross product $(\nabla F) \times (\nabla G)$, evaluated at $(x_0, y_0, z_0)$ is a vector tangent to the curve of intersection and perpendicular to both of the normal vectors $\nabla F$ and $\nabla G$. A unit tangent vector to the curve of intersection is constructed by dividing this cross product by its magnitude to obtain

$$\hat{e}_t = \frac{\nabla F \times \nabla G}{|\nabla F \times \nabla G|}.$$

∎

**Example 5-7.** In two-dimensions a curve $y = f(x)$ can be represented in the implicit form $\phi = \phi(x,y) = y - f(x) = 0$ so that

$$\text{grad}\,\phi = \frac{\partial \phi}{\partial x}\hat{e}_1 + \frac{\partial \phi}{\partial y}\hat{e}_2 = -f'(x)\,\hat{e}_1 + \hat{e}_2 = \vec{N}$$

is a vector normal to the curve at the point $(x, f(x))$. A unit normal vector to the curve is given by

$$\hat{e}_n = \frac{-f'(x)\,\hat{e}_1 + \hat{e}_2}{\sqrt{1 + (f'(x))^2}}$$

Another way to construct this normal vector is as follows. The position vector $\vec{r}$ describing the curve $y = f(x)$ is given by $\vec{r} = x\,\hat{e}_1 + f(x)\,\hat{e}_2$ with tangent $\frac{d\vec{r}}{dx} = \hat{e}_1 + f'(x)\,\hat{e}_2$. The unit tangent vector to the curve is given by $\hat{e}_t = \frac{\hat{e}_1 + f'(x)\,\hat{e}_2}{\sqrt{1 + (f'(x))^2}}$. The vector $\hat{e}_3$ is perpendicular to the planar surface containing the curve and consequently the vector $\hat{e}_3 \times \hat{e}_t$ is normal to the curve. This cross product is given by

$$\hat{e}_3 \times \hat{e}_t = \frac{-f'(x)\hat{e}_1 + \hat{e}_2}{\sqrt{1 + (f'(x))^2}} = \hat{e}_n$$

and produces a unit normal vector to the curve. Note that there are always two normals to every curve or surface. It is important to observe that if $\vec{N}$ is normal to a point on the surface, then the vector $-\vec{N}$ is also a normal to the same point on the surface. If the surface is a closed surface, then one normal is called an inward normal and the other an outward normal. ∎

A vector field is a vector-valued function representing a mapping from $R^n$ to a vector $\vec{V}$. Any vector which varies as a function of position is space is said to represent a vector field. The vector field is a one-to-one correspondence between points in space and a vector quantity $\vec{V}$ which is assumed to be continuous and differentiable within some region $R$. Examples of vector fields are velocity, electric force, mechanical force, etc. Vector fields can be represented graphically by plotting vectors at selected points within a region. These kind of graphical representations are called vector field plots. Alternative to plotting many vectors at selected points to visualize a vector field, it is sometimes easier to use the concept of field lines associated with a vector field. A field line is a curve where at each point $(x, y, z)$ of the curve, the tangent vector to the curve has the same direction as the vector field at that point. If $\vec{r} = x(t)\hat{e}_1 + y(t)\hat{e}_2 + z(t)\hat{e}_3$ is the position vector describing a field line, then the vector $\frac{d\vec{r}}{dt} = \frac{dx}{dt}\hat{e}_1 + \frac{dy}{dt}\hat{e}_2 + \frac{dz}{dt}\hat{e}_3$ is in the direction of the tangent to the field line and this tangent vector must be proportional to $\vec{V}$ at the point of tangency if it has the same direction as $\vec{V}$. In other words $\frac{d\vec{r}}{dt}$ and $\vec{V}$ must be colinear at each point on the curve representing the field line. This requires

$$\frac{d\vec{r}}{dt} = \frac{dx}{dt}\hat{e}_1 + \frac{dy}{dt}\hat{e}_2 + \frac{dz}{dt}\hat{e}_3 = k\left[V_1(x,y,z)\hat{e}_1 + V_2(x,y,z)\hat{e}_2 + V_3(x,y,z)\hat{e}_3\right]$$

where $k$ is some proportionality constant. Equating like components in the above equation one obtains the system of differential equations

$$\frac{dx}{dt} = kV_1(x,y,z), \qquad \frac{dy}{dt} = kV_2(x,y,z), \qquad \frac{dz}{dt} = kV_3(x,y,z)$$

which must be solved to obtain the field lines.

## Integration of Vectors

Let $\vec{u}(s) = u_1(s)\hat{e}_1 + u_2(s)\hat{e}_2 + u_3(s)\hat{e}_3$ denote a vector function of arc length, where the components $u_i(s)$, $i = 1, 2, 3$ are continuous functions. The indefinite integral of $\vec{u}(s)$ is defined as the indefinite integral of each component of the vector. This is expressed

$$\int \vec{u}(s)\,ds = \int u_1(s)\,ds\,\hat{e}_1 + \int u_2(s)\,ds\,\hat{e}_2 + \int u_3(s)\,ds\,\hat{e}_3 + \vec{C},$$
$$= \vec{U}(s) + \vec{C}. \tag{5.33}$$

where $\vec{U}(s)$ is a vector such that $\frac{d\vec{U}}{ds} = \vec{u}(s)$ and $\vec{C}$ is a vector constant of integration.

The definite integral of $\vec{u}$ is defined as

$$\int_a^b \vec{u}(s)\,ds = \vec{U}(b) - \vec{U}(a), \quad \text{where} \quad \frac{d\vec{U}(s)}{ds} = \vec{u}(s).$$

Some properties n the integration of vector functions are stated without proof:

1. $\int [\vec{c}_1 \cdot \vec{u}(s) + \vec{c}_2 \cdot \vec{v}(s)]\,ds = \vec{c}_1 \cdot \int \vec{u}(s)\,ds + \vec{c}_2 \cdot \int \vec{v}(s)\,ds,$ where $\vec{c}_1$ and $\vec{c}_2$ are constant vectors.

2. $\int \vec{c} \times \vec{u}(s)\,ds = \vec{c} \times \int \vec{u}(s)\,ds,$ where $\vec{c}$ is a constant vector.

3. $\int_a^b f(s)\vec{u}(s)\,ds = f(s)\vec{U}(s)\Big|_a^b - \int_a^b f'(s)\vec{U}(s)\,ds,$ where $f(s)$ is a scalar function and $\frac{d\vec{U}(s)}{ds} = \vec{u}(s).$

**Example 5-8.** The acceleration of a particle is given by $\vec{a} = \sin t\,\hat{e}_1 + \cos t\,\hat{e}_2.$ If at time $t = 0$ the position and velocity of the particle are given by

$$\vec{r}(0) = 6\,\hat{e}_1 - 3\,\hat{e}_2 + 4\,\hat{e}_3 \quad \text{and} \quad \vec{V}(0) = 7\,\hat{e}_1 - 6\,\hat{e}_2 - 5\,\hat{e}_3,$$

find the position and velocity as a function of time.

**Solution:** An integration of the acceleration with respect to time produces the velocity and

$$\int \vec{a}(t)\,dt = \vec{V} = \vec{V}(t) = -\cos t\,\hat{e}_1 + \sin t\,\hat{e}_2 + \vec{c}_1,$$

where $\vec{c}_1$ is a vector constant of integration. From the above initial condition for the velocity, the constant $\vec{c}_1$ is calculated from

$$\vec{V}(0) = -\hat{e}_1 + \vec{c}_1 = 7\,\hat{e}_1 - 6\,\hat{e}_2 - 5\,\hat{e}_3 \quad \text{or} \quad \vec{c}_1 = 8\,\hat{e}_1 - 6\,\hat{e}_2 - 5\,\hat{e}_3.$$

Consequently, the velocity can be expressed as a function of time in the form

$$\vec{V} = \vec{V}(t) = \frac{d\vec{r}}{dt} = (-\cos t + 8)\,\hat{e}_1 + (\sin t - 6)\,\hat{e}_2 - 5\,\hat{e}_3.$$

An integration of the velocity with respect to time produces the position vector as a function of time and

$$\int \vec{V}(t)\,dt = \int \frac{d\vec{r}}{dt}\,dt = \int (-\cos t + 8)\,dt\,\hat{e}_1 + \int (\sin t - 6)\,dt\,\hat{e}_2 - 5\int dt\,\hat{e}_3 + \vec{c}_2$$

$$\vec{r}(t) = (-\sin t + 8t)\,\hat{e}_1 + (-\cos t - 6t)\,\hat{e}_2 - 5t\,\hat{e}_3 + \vec{c}_2,$$

where $\vec{c}_2$ is a vector constant of integration. The vector constant of integration is determined from the above initial conditions. One finds

$$\vec{r}(0) = -\hat{e}_2 + \vec{c}_2 = 6\,\hat{e}_1 - 3\,\hat{e}_2 + 4\,\hat{e}_3 \quad \text{or} \quad \vec{c}_2 = 6\,\hat{e}_1 - 2\,\hat{e}_2 + 4\,\hat{e}_3.$$

The position vector as a function of time can be expressed as

$$\vec{r} = \vec{r}(t) = (-\sin t + 8t + 6)\,\hat{e}_1 + (-\cos t - 6t - 2)\,\hat{e}_2 + (-5t + 4)\,\hat{e}_3.$$

∎

**Example 5-9.** A particle in a force field $\vec{F} = \vec{F}(x,y,z)$ having a position vector $\vec{r} = x\,\hat{e}_1 + y\,\hat{e}_2 + z\,\hat{e}_3$ moves according to Newton's second law such that

$$\vec{F} = m\vec{a} = m\frac{d\vec{V}}{dt} \quad \text{or} \quad \vec{F}\,dt = m\,d\vec{V}.$$

An integration over the time interval $t_1$ to $t_2$ produces

$$\int_{t_1}^{t_2} \vec{F}\,dt = m\vec{V}(t_2) - m\vec{V}(t_1).$$

The quantity $\int_{t_1}^{t_2} \vec{F}\,dt$ is called the linear impulse on the particle over the time interval $(t_1, t_2)$. The quantity $m\vec{V}$ is called the linear momentum of the particle. The above equation tells us that the linear impulse equals the change in linear momentum. ∎

### Line Integrals of Scalar and Vector Functions

An important type of vector integration is integration by line integrals. Let $C$ be a curve defined by a position vector

$$\vec{r} = x\,\hat{e}_1 + y\,\hat{e}_2 + z\,\hat{e}_3,$$

where the coordinates $x, y, z$ are defined by some parametric set of equations. The element of arc length along the curve, when squared, is given by

$$ds^2 = d\vec{r}\cdot d\vec{r} = dx^2 + dy^2 + dz^2.$$

An integration (summation) produces the following formulas for the arc length $s$.

1. If $y = y(x)$ and $z = z(x)$ are known in terms of the parameter $x$, the arc length between two points $P_0(x_0, y_0, z_0)$ and $P_1(x_1, y_1, z_1)$ on the curve can be represented in the form

$$s = \int_{x_0}^{x_1} \sqrt{1 + \left(\frac{dy}{dx}\right)^2 + \left(\frac{dz}{dx}\right)^2}\,dx. \tag{5.34}$$

2. If the parametric equations of the curve are given by $x = x(t)$, $y = y(t)$ and $z = z(t)$, the arc length between two points $P_0$ and $P_1$ on the curve is given by

$$s = \int_{t_0}^{t_1} \sqrt{\left(\frac{dx}{dt}\right)^2 + \left(\frac{dy}{dt}\right)^2 + \left(\frac{dz}{dt}\right)^2}\,dt, \tag{5.35}$$

where the parametric values $t = t_0$ and $t = t_1$ correspond to the points $P_0$ and $P_1$ and

$$x(t_0) = x_0, \quad y(t_0) = y_0, \quad z(t_0) = z_0$$
$$x(t_1) = x_1, \quad y(t_1) = y_1, \quad z(t_1) = z_1.$$

The above formulas result indirectly from the following limiting process. On that part of the curve between the given points $P_0(x_0, y_0, z_0)$ and $P_1(x_1, y_1, z_1)$, the arc length is divided into $n$ segments by a set of numbers $s_0 < s_1 < \ldots < s_n$, where corresponding to each $s_i$ there is a position vector $\vec{r}(s_i)$ as illustrated in figure 5-11.

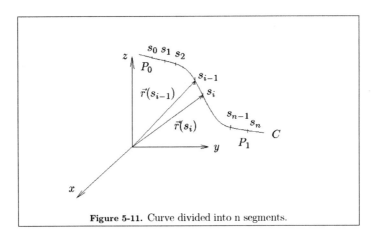

**Figure 5-11.** Curve divided into n segments.

The element of arc length $\Delta s_i$ is defined as $\Delta s_i = |\vec{r}(s_i) - \vec{r}(s_{i-1})| = |\Delta \vec{r}_i|$. The total arc length is obtained from the sum of these elements of arc length as the number of these lengths increase without bound. In symbols, this limit is denoted as

$$s = \lim_{n \to \infty} \sum_{i=1}^{n} \Delta s_i = \int_{s_0}^{s_n} ds.$$

The above definition for arc length suggests how values of a scalar field can be summed as a point moves through the scalar field along a curve $C$.

**Definition:** (Line integral of scalar function.) Let $f = f(x, y, z)$ denote a scalar function of position. The line integral of $f$ along a curve $C$ is defined as

$$\int_C f(x, y, z) \, ds = \lim_{n \to \infty} \sum_{i=1}^{n} f(x_i^*, y_i^*, z_i^*) \Delta s_i, \qquad (5.36)$$

where $(x_i^*, y_i^*, z_i^*)$ is a point on the curve in the ith subinterval $\Delta s_i$ and where the symbol $\int_C$ denotes an integral taken along the curve $C$.

Similarly, define the summation of a vector field along a curve $C$ as a line integral of a vector function.

**Definition: (Line integral involving a dot product.)**
Let $\vec{F} = \vec{F}(x,y,z) = F_1\hat{e}_1 + F_2\hat{e}_2 + F_3\hat{e}_3$ denote a vector function of position. The line integral of $\vec{F}$ along the curve $C$, characterized by a position vector $\vec{r} = x\hat{e}_1 + y\hat{e}_2 + z\hat{e}_3$, is defined as

$$\int_C \vec{F} \cdot d\vec{r} = \lim_{n \to \infty} \sum_{i=1}^{n} \vec{F}(x_i^*, y_i^*, z_i^*) \cdot \frac{\Delta \vec{r}_i}{\Delta s_i} \Delta s_i$$

$$\int_C \vec{F} \cdot d\vec{r} = \int_C \vec{F} \cdot \frac{d\vec{r}}{ds} ds = \int_C \left( F_1 \frac{dx}{ds} + F_2 \frac{dy}{ds} + F_3 \frac{dz}{ds} \right) ds, \quad (5.37)$$

where $(x_i^*, y_i^*, z_i^*)$ is a point on the ith subinterval with arc length $\Delta s_i$.

In the above definition the dot product $\vec{F} \cdot \frac{d\vec{r}}{ds}$ represents the projection of the vector $\vec{F}$ or component of $\vec{F}$ in the direction of the tangent vector to the curve $C$. The line integral of the vector function may be thought of as representing a summation of the tangential components of $\vec{F}$ along the curve $C$. Line integrals of this type arise in the calculation of the work done in moving along a curve within a force field.

In particular, the above line integral can be expressed in the form

$$\int_C \vec{F} \cdot d\vec{r} = \int_C \vec{F} \cdot \frac{d\vec{r}}{ds} ds = \int_C \vec{F} \cdot \hat{e}_t \, ds = \int_C F_1 dx + F_2 dy + F_3 dz, \quad (5.38)$$

where at each point on the curve $C$, the dot product $\vec{F} \cdot \hat{e}_t$ is a scalar function of position and represents the projection of $\vec{F}$ on the unit tangent vector to the curve.

Summations of cross products along a curve produce another type of line integral.

**Definition: (Line integral involving cross products.)** The line integral $\int_C \vec{F} \times d\vec{r}$ is defined by the limiting process

$$\int_C \vec{F} \times d\vec{r} = \int_C \vec{f} \times \frac{d\vec{r}}{ds} ds = \lim_{n \to \infty} \sum_{i=1}^{n} \vec{F}(x_i^*, y_i^*, z_i^*) \times \Delta \vec{r}_i, \quad (5.39)$$

where $\vec{F} = \vec{F}(x_i^*, y_i^*, z_i^*)$ is the value of $\vec{F}$ at the point $(x_i^*, y_i^*, z_i^*)$ in the ith subinterval of arc length on the curve $C$.

Integrals of this type arise in the calculation of magnetic dipole moments associated with current loops.

**Work Done.** Consider a particle moving from a point $A$ to a point $B$ along a curve $C$ which lies in a force field $\vec{F}(x,y,z)$. To obtain a physical representation of the above line integral construct the following vectors: (i) the position vector $\vec{r}$, (ii) the force vector $\vec{F}$, and (iii) the tangent vector $d\vec{r}$, where all vectors are evaluated at some general point $(x,y,z)$ on the given curve.

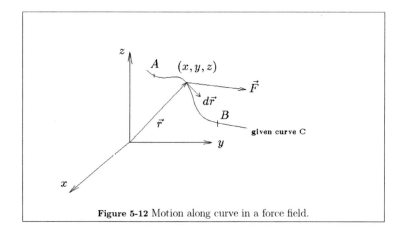

**Figure 5-12** Motion along curve in a force field.

The dot product $\vec{F} \cdot \frac{d\vec{r}}{ds}$ represents the projection of $F$ onto the unit tangent vector and the integration sums these projections along the curve. This produces the line integral

$$W_{AB} = \int_C \vec{F} \cdot d\vec{r} = \int_A^B \vec{F} \cdot \frac{d\vec{r}}{ds} ds = \int_A^B \vec{F} \cdot \hat{e}_t \, ds$$

representing a summation of the force times distance traveled along the curve $C$. Consequently, the above integral represents the work done in moving through the force field from $A$ to $B$ along the curve $C$.

**Example 5-10.** Let a particle with constant mass $m$ move along a curve $C$ which lies in a vector force field $\vec{F} = \vec{F}(x, y, z)$. Also, let $\vec{r}$ denote the position vector of the particle in the force field and on the curve $C$. As the particle moves along the curve, at each point $(x, y, z)$ of the curve, the particle experiences a force $\vec{F}(x, y, z)$ which is determined by the vector force field. Using Newton's second law of motion, write

$$\vec{F} = m\vec{a} = m\frac{d^2\vec{r}}{dt^2} = m\frac{d\vec{V}}{dt}.$$

The work done in moving along the curve $C$ between two points $A$ and $B$ can then be expressed as

$$W_{AB} = \int_A^B \vec{F} \cdot d\vec{r} = \int_A^B \vec{F} \cdot \frac{d\vec{r}}{dt} dt = \int_A^B \vec{F} \cdot \vec{V} \, dt = \int_A^B m\frac{d\vec{V}}{dt} \cdot \vec{V} \, dt = \int_A^B m\vec{V} \cdot \frac{d\vec{V}}{dt} dt.$$

Employ the vector identity

$$\frac{1}{2}\frac{d}{dt}(V^2) = \frac{1}{2}\frac{d}{dt}\left(\vec{V}\cdot\vec{V}\right) = \vec{V}\cdot\frac{d\vec{V}}{dt},$$

to show the line integral can be expressed in the form

$$\int_A^B \vec{F}\cdot\frac{d\vec{r}}{dt}\,dt = \int_A^B \vec{F}\cdot\vec{V}\,dt = \int_A^B \frac{m}{2}\frac{d}{dt}(V^2)\,dt,$$

which is easily integrated. Thus, the work done in moving from point $A$ to $B$ can be expressed

$$W_{AB} = \int_A^B \vec{F}\cdot d\vec{r} = \frac{m}{2}V^2\Big|_A^B = \frac{m}{2}(V_B^2 - V_A^2) = T_B - T_A.$$

In this equation the line integral $W_{AB} = \int_A^B \vec{F}\cdot d\vec{r}$ is called the work done in moving the particle from $A$ to $B$ through the force field $\vec{F}$. The quantity $T = \frac{m}{2}V^2$ is called the kinetic energy of the particle. The above equation tells us that the work done in moving a particle from $A$ to $B$ in a force field $\vec{F}$ must equal the change in the kinetic energy of the particle between the points $A$ and $B$.

■

## Representation of Line Integrals

The line integral $\int \vec{F}\cdot d\vec{r}$ can be expressed in many different forms:

1. 
$$\int_A^B \vec{F}\cdot d\vec{r} = \int_{t_A}^{t_B} \vec{F}\cdot\frac{d\vec{r}}{dt}\,dt = \int_{t_A}^{t_B} \vec{F}\cdot\vec{V}\,dt$$

   Integrals of this form are used if $\vec{F} = \vec{F}(t)$ and $\vec{V} = \vec{V}(t)$ are known functions of the parameter $t$.

2. 
$$\int_A^B \vec{F}\cdot d\vec{r} = \int_A^B \vec{F}\cdot\frac{d\vec{r}}{ds}\,ds = \int_{s_A}^{s_B} \vec{F}\cdot\hat{e}_t\,ds$$

   Here $\vec{F}\cdot\hat{e}_t$ is the tangential component of the force $\vec{F}$ along the curve $C$. This form of the line integral is used if $\vec{F} = \vec{F}(s)$ and $\hat{e}_t$ are known functions of the arc length $s$.

3. For a force field given by

$$\vec{F} = \vec{F}(x,y,z) = F_1(x,y,z)\hat{e}_1 + F_2(x,y,z)\hat{e}_2 + F_3(x,y,z)\hat{e}_3$$

   and the position vector of a point $(x,y,z)$ on a curve $C$ represented by

$$\vec{r} = x\hat{e}_1 + y\hat{e}_2 + z\hat{e}_3 \quad\text{with}\quad d\vec{r} = dx\,\hat{e}_1 + dy\,\hat{e}_2 + dz\,\hat{e}_3,$$

   the work done can be represented in the form

$$\int_A^B \vec{F}\cdot d\vec{r} = \int_A^B F_1\,dx + F_2\,dy + F_3\,dz.$$

Line integrals are written in this form when a parametric representation of the curve is known. In the special case where $\vec{r} = x\,\hat{e}_1 + 0\,\hat{e}_2 + 0\,\hat{e}_3$, the above line integral reduces to an ordinary integral.

4. The line integral $\int_C \vec{F} \cdot d\vec{r}$ may be broken up into a sum of line integrals along different portions of the curve $C$. If the curve $C$ is comprised of $n$ separate curves $C_1, C_2, \ldots, C_n$, then write

$$\int_C \vec{F} \cdot d\vec{r} = \int_{C_1} \vec{F} \cdot d\vec{r} + \int_{C_2} \vec{F} \cdot d\vec{r} + \cdots + \int_{C_n} \vec{F} \cdot d\vec{r}.$$

5. When the curve $C$ is a simple closed curve (i.e., the curve does not intersect itself), the line integral is represented in the form $\oint \vec{F} \cdot d\vec{r}$, where the direction of integration is in the positive sense unless specified otherwise. Note that when the curve is a simple closed curve, there is no need to specify a beginning and end point for the integration. One need only specify a direction to the integration. The integration is said to be in the positive sense or in the negative sense. The sense of integration is the same as that for angular measure. The situation is illustrated in figure 5-13.

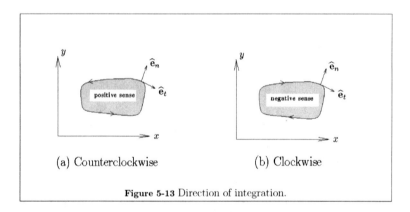

**Figure 5-13** Direction of integration.

The direction of integration can be referenced with respect to the unit outward normal $\hat{e}_n$ and to the unit tangent vector $\hat{e}_t$ to the simple close curve. By the right-hand rule, $\hat{e}_b = \hat{e}_t \times \hat{e}_n$ is the binormal to the closed curve. The direction of integration is either clockwise or counterclockwise as viewed from the terminus of the binormal vector. A counterclockwise direction is the positive sense and a clockwise direction is the negative sense. If the direction of integration is reversed, then the sign of the line integral changes.

## Surface and Volume Integrals

In this section various types of surface integrals are introduced. In particular, surface integrals of the form

$$\iint_S f(x,y,z)\,dS, \qquad \iint_S \vec{F}(x,y,z)\cdot d\vec{S}, \qquad \iiint_S \vec{F}(x,y,z)\times d\vec{S}$$

where $d\vec{S}$ is a vector element of surface area, are defined and illustrated. Throughout the following discussion all surfaces $S$ are considered to be orientated (two-sided) surfaces.

Consider a surface in space with a scalar element of surface area $dS$ constructed at some general point on the surface as is illustrated in figure 5-14.

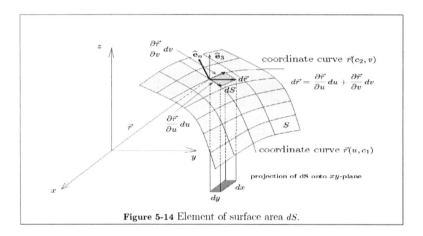

**Figure 5-14** Element of surface area $dS$.

In the representation of various surface integrals, it is convenient to define a vector element of surface area $d\vec{S}$ whose magnitude is $dS$ and whose direction is the same as the unit outward normal $\hat{e}_n$ to the surface. This vector element of area is defined

$$d\vec{S} = \hat{e}_n\,dS \qquad \text{or} \qquad \vec{\Delta}S = \hat{e}_n\,\Delta S$$

where $dS$ is a scalar element of surface area and $\hat{e}_n$ is a unit normal to the surface. The element of surface area $dS$ is formed from the differential of the position vector $\vec{r}$ to a general point on the surface. If $\vec{r} = \vec{r}(u,v)$, then the differential $d\vec{r} = \frac{\partial \vec{r}}{\partial u}du + \frac{\partial \vec{r}}{\partial v}dv$ lies in the tangent plane to the surface and then the vectors $\frac{\partial \vec{r}}{\partial u}du$ and $\frac{\partial \vec{r}}{\partial v}dv$ are used to construct the element of surface area $dS$. The following discussions show how to calculate the normal to a surface and how to construct the element of surface area.

**Normal to a Surface**

Note that if $\hat{e}_n$ is a normal to a smooth surface, then $-\hat{e}_n$ is also normal to the surface. That is, all smooth orientated surfaces possess two normals. If the surface is a closed surface, there is an inside surface and an outside surface. The outside surface is called the positive side of the surface. The unit normal to the positive side of a surface is called the positive normal or outward normal. If the surface is not closed, then arbitrarily select one side of the surface and call it the positive side. The normal drawn to this positive side is called the outward normal.

If the surface is expressed in an implicit form $F(x,y,z) = 0$, then a unit normal to the surface can be obtained from the gradient vector using the relation:

$$\hat{e}_n = \frac{\text{grad } F}{|\text{grad } F|}.$$

If the surface is expressed in the explicit form $z = z(x,y)$, then a unit normal to the surface can be found using the above gradient relation with $F = z(x,y) - z = 0$. This gives the unit normal

$$\hat{e}_n = \frac{\text{grad }[z(x,y)-z]}{|\text{grad }[z(x,y)-z]|} = \frac{\frac{\partial z}{\partial x}\hat{e}_1 + \frac{\partial z}{\partial y}\hat{e}_2 - \hat{e}_3}{\sqrt{1 + \left(\frac{\partial z}{\partial x}\right)^2 + \left(\frac{\partial z}{\partial y}\right)^2}}.$$

Surfaces can also be expressed in the parametric form

$$x = x(u,v), \qquad y = y(u,v), \qquad z = z(u,v),$$

where $u$ and $v$ are parameters. The functions $x(u,v)$, $y(u,v)$, and $z(u,v)$ must be such that one and only one point $(u,v)$ maps to any given point on the surface. These functions are also assumed to be continuous and differentiable. In this case, the position vector to a point on the surface can be represented as

$$\vec{r} = \vec{r}(u,v) = x(u,v)\,\hat{e}_1 + y(u,v)\,\hat{e}_2 + z(u,v)\,\hat{e}_3.$$

The number pair $(u,v)$ is called the surface coordinates of a point on the surface. The curves

$$\vec{r}(u,v)|_{v=\text{Constant}} \qquad \text{and} \qquad \vec{r}(u,v)|_{u=\text{Constant}}$$

sweep out coordinate curves on the surface and the vectors $\frac{\partial \vec{r}}{\partial u}, \frac{\partial \vec{r}}{\partial v}$ are tangent vectors to these coordinate curves. A unit normal to the surface can then be calculated from the cross product of these tangent vectors and

$$\hat{e}_n = \frac{\frac{\partial \vec{r}}{\partial u} \times \frac{\partial \vec{r}}{\partial v}}{\left|\frac{\partial \vec{r}}{\partial u} \times \frac{\partial \vec{r}}{\partial v}\right|}.$$

Whenever the cross product $\dfrac{\partial \vec{r}}{\partial u} \times \dfrac{\partial \vec{r}}{\partial v} \neq \vec{0}$, the surface is called a smooth surface. If at a point with surface coordinates $(u_0, v_0)$ this cross product equals the zero vector, then the point is called a singular point of the surface.

**Element of Surface Area**

Consider the case where the surface is given in the explicit form $z = z(x, y)$. In this case, the position vector of a point on the surface is given by

$$\vec{r} = \vec{r}(x, y) = x\,\hat{e}_1 + y\,\hat{e}_2 + z(x, y)\,\hat{e}_3. \tag{5.40}$$

The curves

$$\vec{r}(x, y)|_{y=\text{Constant}} \qquad \text{and} \qquad \vec{r}(x, y)|_{x=\text{Constant}}$$

are coordinate curves lying in the surface. When these curves intersect at a common point $(x, y, z)$ the vectors

$$\frac{\partial \vec{r}}{\partial x} = \hat{e}_1 + \frac{\partial z}{\partial x}\hat{e}_3 \qquad \text{and} \qquad \frac{\partial \vec{r}}{\partial y} = \hat{e}_2 + \frac{\partial z}{\partial y}\hat{e}_3$$

are tangent to the coordinate curves, and consequently the differential of the position vector

$$d\vec{r} = \frac{\partial \vec{r}}{\partial x}dx + \frac{\partial \vec{r}}{\partial y}dy$$

lies in the tangent plane to the surface at the common point of intersection of the coordinate curves. This differential is illustrated in figure 5-14 where the components of the differential $d\vec{r}$ form the sides of an elemental parallelogram which defines the element of area $dS$ on the surface. The area of this elemental parallelogram is calculated from the cross product relation

$$\left(\frac{\partial \vec{r}}{\partial x}dx\right) \times \left(\frac{\partial \vec{r}}{\partial y}dy\right) = \begin{vmatrix} \hat{e}_1 & \hat{e}_2 & \hat{e}_3 \\ dx & 0 & \frac{\partial z}{\partial x}dx \\ 0 & dy & \frac{\partial z}{\partial y}dy \end{vmatrix} = \left(-\frac{\partial z}{\partial x}\hat{e}_1 - \frac{\partial z}{\partial y}\hat{e}_2 + \hat{e}_3\right)dx\,dy. \tag{5.41}$$

The area $dS$ of the elemental parallelogram is the magnitude of the above cross product, and defined as the element of surface area

$$dS = \sqrt{1 + \left(\frac{\partial z}{\partial x}\right)^2 + \left(\frac{\partial z}{\partial y}\right)^2}\,dx\,dy. \tag{5.42}$$

Given a surface defined in the explicit form $z = z(x, y)$, use the gradient operation on the function $\phi(x, y, z) = z - z(x, y) = 0$ to obtain the unit normal

$$\hat{e}_n = \frac{\text{grad } \phi}{|\text{grad } \phi|} = \frac{-\frac{\partial z}{\partial x}\hat{e}_1 - \frac{\partial z}{\partial y}\hat{e}_2 + \hat{e}_3}{\sqrt{1 + (\frac{\partial z}{\partial x})^2 + (\frac{\partial z}{\partial y})^2}}. \tag{5.43}$$

The vector element of surface area is then defined

$$d\vec{S} = \hat{e}_n\,dS = \left(-\frac{\partial z}{\partial x}\hat{e}_1 - \frac{\partial z}{\partial y}\hat{e}_2 + \hat{e}_3\right)dx\,dy.$$

Taking the dot product of both sides of the above equation with the unit vector $\hat{e}_3$ gives

$$|\hat{e}_3 \cdot \hat{e}_n| dS = dx\, dy \quad \text{or} \quad dS = \frac{dx\, dy}{|\hat{e}_3 \cdot \hat{e}_n|} = \frac{dx\, dy}{\cos \gamma} \tag{5.44}$$

with an absolute value placed upon the dot product to ensure that the surface area is positive (i.e., recall that there are two normals to the surface which differ in sign). The equation (5.44) is interpreted as an element of surface area in terms of its projection onto the $xy$ plane. The angle $\gamma = \gamma(x,y)$ is the angle between the outward normal to the surface and the unit vector $\hat{e}_3$. This representation of the element of surface area is valid provided that $\cos \gamma \neq 0$. That is, it is assumed that the surface is such that the normal to the surface is nowhere parallel to the $xy$ plane.

For surfaces which have a normal parallel to the $xy$ plane, the element of surface area can be projected onto either of the planes $x = 0$ or $y = 0$. If the surface element is projected onto the plane $x = 0$, then it takes the form

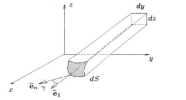

$$dS = \frac{dy\, dz}{|\hat{e}_1 \cdot \hat{e}_n|} \tag{5.45}$$

and if projected onto the plane $y = 0$ it has the form

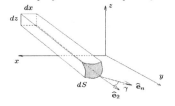

$$dS = \frac{dx\, dz}{|\hat{e}_2 \cdot \hat{e}_n|}. \tag{5.46}$$

The limits of integration are with respect to one of the projected areas $dxdy$, $dydz$ or $dxdz$ depending upon the projection selected. The integration limits range over the plane region $R$ determined by where the surface area $S$ is projected.

**Surface Placed in a Scalar Field**

If a surface is placed in a region of a scalar field $f(x,y,z)$, divide the surface $S$ into $n$ small areas

$$\Delta S_1, \; \Delta S_2, \ldots, \Delta S_n.$$

For $n$ large, define $f_i = f_i(x_i, y_i, z_i)$ as the value of the scalar field associated with the $i$th surface element $\vec{\Delta} S_i$ and let $i$ range from 1 to $n$.

The summation of the elements $f_i \Delta \vec{S}_i$ over all $i$ as $n$ increases without bound defines the surface integral

$$\iint_R f(x,y,z)\, \hat{e}_n\, dS = \iint_R f(x,y,z)\, d\vec{S} = \lim_{n \to \infty} \sum_{i=1}^{n} f_i(x_i, y_i, z_i) \Delta \vec{S}_i, \qquad (5.47)$$

where the integration limits are determined by the way one represents the element of surface area $d\vec{S}$. The integral can be represented in different forms depending upon how the given surface is specified and projection of the element of surface area $dS$.

**Surface Placed in a Vector Field**

For a surface $S$ in a region of a vector field $\vec{F} = \vec{F}(x,y,z)$, the integral

$$\iint_R \vec{F} \cdot d\vec{S} = \iint_R \vec{F} \cdot \hat{e}_n\, dS \qquad (5.48)$$

represents a scalar which is the sum of the projections of $\vec{F}$ onto the normals to the surface elements. Like the previous surface integrals, divided the surface into $n$ small surface elements $\Delta S_i$, $i = 1, \ldots, n$, and let $\vec{F}_i = \vec{F}(x_i, y_i, z_i)$ represent the value of the vector field over the $i$th surface element. The summation of the elements

$$\vec{F}_i \cdot \Delta \vec{S}_i = \vec{F}_i \cdot \hat{e}_{n_i} \Delta S_i$$

over all surface elements represents the sum of the normal components of $\vec{F}_i$ multiplied by $\Delta S_i$ as $i$ varies from 1 to $n$. A summation gives the surface integral

$$\lim_{n \to \infty} \sum_{i=1}^{n} \vec{F}_i \cdot \Delta \vec{S}_i = \iint_R \vec{F} \cdot d\vec{S}. \qquad (5.49)$$

Again, the limits of integration on this integral depends upon how the given surface is represented. Integrals of this type arise when calculating the volume rate of change associated with velocity fields. It is called a flux integral and represents the amount of a substance moving across an imaginary surface placed within the vector field.

The vector integral

$$\iint_R \vec{F} \times d\vec{S}$$

represents a vector which is obtained by summing the vector elements $\vec{F}_i \times \Delta \vec{S}_i$ over the given surface. This produces

$$\lim_{n \to \infty} \sum_{i=1}^{n} \vec{F}_i \times \Delta \vec{S}_i = \iint_R \vec{F} \times d\vec{S}. \qquad (5.50)$$

Integrals of this type arise in some integral theorems.

Each of the above surface integrals can be represented in different forms depending upon for how the element of surface area is represented. The form in which the given surface is represented usually dictates the method used to calculate the surface area element. Sometimes the representation of a surface in a different form is helpful in determining the limits of integration to certain surface integrals.

### Surface Area from Parametric Form

When a surface is represented in parametric form, the position vector of a point on the surface is given by

$$\vec{r} = \vec{r}(u,v) = x(u,v)\,\hat{e}_1 + y(u,v)\,\hat{e}_2 + z(u,v)\,\hat{e}_3.$$

A unit normal vector to the surface, at a point having the surface coordinates $(u,v)$, can be found from the relation

$$\hat{e}_n = \frac{\dfrac{\partial \vec{r}}{\partial u} \times \dfrac{\partial \vec{r}}{\partial v}}{\left|\dfrac{\partial \vec{r}}{\partial u} \times \dfrac{\partial \vec{r}}{\partial v}\right|}.$$

That is, the differential of the position vector $\vec{r} = \vec{r}(u,v)$ is

$$d\vec{r} = \frac{\partial \vec{r}}{\partial u}\,du + \frac{\partial \vec{r}}{\partial v}\,dv \tag{5.51}$$

lies in the tangent plane to the surface and defines an element of surface area determined by the area $dS$ of the elemental parallelogram having sides $\frac{\partial \vec{r}}{\partial u} du$ and $\frac{\partial \vec{r}}{\partial v} dv$ as illustrated in the figure 5-14. The area $dS$ of the elemental parallelogram is given by the magnitude of the cross product of the two vectors defining its sides and can be expressed

$$dS = \left|\left(\frac{\partial \vec{r}}{\partial u}\,du\right) \times \left(\frac{\partial \vec{r}}{\partial v}\,dv\right)\right| = \left|\frac{\partial \vec{r}}{\partial u} \times \frac{\partial \vec{r}}{\partial v}\right| du\,dv. \tag{5.52}$$

By using the dot product relation

$$(\vec{A} \times \vec{B}) \cdot (\vec{C} \times \vec{D}) = (\vec{A} \cdot \vec{C})(\vec{B} \cdot \vec{D}) - (\vec{A} \cdot \vec{D})(\vec{B} \cdot \vec{C})$$

verify that

$$\left|\frac{\partial \vec{r}}{\partial u} \times \frac{\partial \vec{r}}{\partial v}\right| = \sqrt{EG - F^2}, \tag{5.53}$$

where

$$E = \frac{\partial \vec{r}}{\partial u} \cdot \frac{\partial \vec{r}}{\partial u} = \left(\frac{\partial x}{\partial u}\right)^2 + \left(\frac{\partial y}{\partial u}\right)^2 + \left(\frac{\partial z}{\partial u}\right)^2$$

$$F = \frac{\partial \vec{r}}{\partial u} \cdot \frac{\partial \vec{r}}{\partial v} = \frac{\partial x}{\partial u}\frac{\partial x}{\partial v} + \frac{\partial y}{\partial u}\frac{\partial y}{\partial v} + \frac{\partial z}{\partial u}\frac{\partial z}{\partial v}$$

$$G = \frac{\partial \vec{r}}{\partial v} \cdot \frac{\partial \vec{r}}{\partial v} = \left(\frac{\partial x}{\partial v}\right)^2 + \left(\frac{\partial y}{\partial v}\right)^2 + \left(\frac{\partial z}{\partial v}\right)^2.$$

The element of surface area $dS$, given by equation (5.52), can then be represented in the form

$$dS = \sqrt{EG - F^2}\,du\,dv, \quad \text{and} \quad S = \iint_{R_{uv}} \sqrt{EG - F^2}\,du\,dv, \tag{5.54}$$

where the integration is over those parameter values $u$ and $v$ which define the surface.

The surface integrals given by equations (5.47) and (5.48) can also be represented in terms of the parameters $u$ and $v$. These integrals have the forms

$$\iint_R f(x,y,z)\, d\vec{S} = \iint_{R_{uv}} f(x(u,v), y(u,v), z(u,v))\sqrt{EG - F^2}\, \hat{e}_n\, du\, dv$$

and

$$\iint_R \vec{F} \cdot d\vec{S} = \iint_{R_{uv}} \vec{F}(x(u,v), y(u,v), z(u,v)) \cdot \hat{e}_n \sqrt{EG - F^2}\, du\, dv.$$

(5.55)

## Volume Integrals

The summation of scalar and vector fields over a region of space can be expressed by the volume integrals of the form

$$\iiint_V f(x,y,z)\, d\tau \quad \text{and} \quad \iiint_V \vec{F}(x,y,z)\, d\tau,$$

where $d\tau = dx\, dy\, dz$ is an element of volume and $V$ is the region over which the integrations are to extend.

The integral of the scalar field is an ordinary triple integral like those occurring in calculus. The triple integral of the vector function $\vec{F} = \vec{F}(x, y, z)$ can be expressed as

$$\iiint_V \vec{F}\, d\tau = \hat{e}_1 \iiint_V F_1(x,y,z)\, d\tau + \hat{e}_2 \iiint_V F_2(x,y,z)\, d\tau + \hat{e}_3 \iiint_V F_3(x,y,z)\, d\tau, \qquad (5.56)$$

where each component is a scalar triple integral.

There are times when it may be appropriate to make a change of variables

$$x = x(u,v,w), \quad y = y(u,v,w), \quad z = z(u,v,w) \qquad (5.57)$$

in order to evaluate a triple integral. To express the volume element $d\tau$ in terms of the new variables $u, v,$ and $w$, let $\vec{r}$ be the position vector to a general point $(x, y, z)$ expressed in terms of the new variables $(u, v, w)$. The position vector to a general point in the $(u, v, w)$ space is represented

$$\vec{r} = \vec{r}(u,v,w) = x(u,v,w)\,\hat{e}_1 + y(u,v,w)\,\hat{e}_2 + z(u,v,w)\,\hat{e}_3$$

The differential $d\vec{r}$ is then expressed as the vector

$$d\vec{r} = \frac{\partial \vec{r}}{\partial u}\, du + \frac{\partial \vec{r}}{\partial v}\, dv + \frac{\partial \vec{r}}{\partial w}\, dw$$

which represents a small change within the region of $(u, v, w)$ space. The differential $d\vec{r}$ is interpreted as the diagonal of a parallelepiped having the vector sides

$$\frac{\partial \vec{r}}{\partial u}\, du, \quad \frac{\partial \vec{r}}{\partial v}\, dv, \quad \text{and} \quad \frac{\partial \vec{r}}{\partial w}\, dw.$$

The volume of this parallelepiped is given by the triple scalar product relation

$$d\tau = \left| \frac{\partial \vec{r}}{\partial u} \cdot \left( \frac{\partial \vec{r}}{\partial v} \times \frac{\partial \vec{r}}{\partial w} \right) \right| du\, dv\, dw.$$

Recall that the triple scalar product can be expressed as a determinant and so this result can be represented in the alternate form

$$d\tau = \left| J\left( \frac{x, y, z}{u, v, w} \right) \right| du\, dv\, dw,$$

where

$$J\left( \frac{x, y, z}{u, v, w} \right) = \begin{vmatrix} \frac{\partial x}{\partial u} & \frac{\partial x}{\partial v} & \frac{\partial x}{\partial w} \\ \frac{\partial y}{\partial u} & \frac{\partial y}{\partial v} & \frac{\partial y}{\partial w} \\ \frac{\partial z}{\partial u} & \frac{\partial z}{\partial v} & \frac{\partial z}{\partial w} \end{vmatrix}$$

is called the Jacobian determinant of the transformation given by equations (5.57).

As an example, the volume element $d\tau = dx\, dy\, dz$ under the change of variable to cylindrical coordinates $(r, \theta, z)$, where

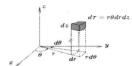

$$x = r\cos\theta, \quad y = r\sin\theta, \quad z = z$$

has the Jacobian $\begin{vmatrix} r\cos\theta & -r\sin\theta & 0 \\ \sin\theta & r\cos\theta & 0 \\ 0 & 0 & 1 \end{vmatrix} = r$

and volume element $d\tau = r\, dr\, d\theta\, dz$.

As another example, the volume element $d\tau = dx\, dy\, dz$ under the change of variable to spherical coordinates $(\rho, \theta, \phi)$, where

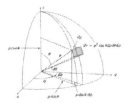

$$x = \rho\sin\theta\cos\phi, \quad y = \rho\sin\theta\sin\phi, \quad z = \rho\cos\theta$$

has the Jacobian $\begin{vmatrix} \sin\theta & \rho\cos\theta\cos\phi & -\rho\sin\theta\sin\phi \\ \sin\theta\sin\phi & \rho\cos\theta\sin\phi & \rho\sin\theta\cos\phi \\ \cos\theta & -\rho\sin\theta & 0 \end{vmatrix} = \rho^2 \sin\theta$

and the volume element $d\tau = \rho^2 \sin\theta\, d\rho\, d\theta\, d\phi$

### Divergence

Let $\vec{F}(x, y, z)$ denote a continuous vector field with continuous derivatives in some region $R$ of space. Visualize a surface placed in this vector field, such that at each point $(x, y, z)$ on the surface there is associated a vector $\vec{F}(x, y, z)$. This image provides a way of visualizing the vector field. Think of the surface as being punctured by arrows of different lengths. The arrows representing the direction and magnitude of the vectors in the vector field evaluated at a point on the surface. The situation is illustrated in figure 5-15.

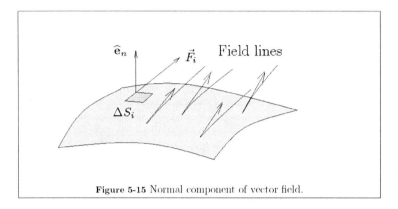

**Figure 5-15** Normal component of vector field.

Another way to visualize a vector field is to create a bundle of curves in space, where each curve in the bundle has the property that at every point $(x,y,z)$ on the curve, the direction of the tangent vector to the curve is the same as the direction of the vector field $\vec{F}(x,y,z)$ at that point. Curves with this property are called field lines associated with the vector field $\vec{F}$. Let $\vec{r} = x\,\hat{e}_1 + y\,\hat{e}_2 + z\,\hat{e}_3$ be a position vector to a point on a curve, then $d\vec{r} = dx\,\hat{e}_1 + dy\,\hat{e}_2 + dz\,\hat{e}_3$ is in the direction of the tangent to the curve. If the curve is also a field line, then when $d\vec{r}$ and $\vec{F}$ are evaluated at the same point on the field line, these vectors must have the same direction and consequently $d\vec{r} = \alpha \vec{F}$, where $\alpha$ is a proportionality constant. From this equation there results the differential equations defining the field lines.

$$d\vec{r} = \alpha \vec{F} \implies \frac{dx}{F_1} = \frac{dy}{F_2} = \frac{dz}{F_3} = \alpha$$

**Divergence of a Vector Field**

The study of field lines leads to the concept of intensity of a vector field or the density of the field lines in a region. To visualize this, place a surface in a vector field and try to determine how the vector field punctures this surface. Imagine the surface is divided into $n$ small areas $\Delta S_i$ and let $\vec{F}_i = \vec{F}(x_i, y_i, z_i)$ denote the value of the vector field associated with each surface element. The dot product $\vec{F}_i \cdot \Delta \vec{S}_i = \vec{F}_i \cdot \hat{e}_n \Delta S_i$ represents the projection of the vector $\vec{F}_i$ onto the normal to the element $\Delta S_i$ multiplied by the area of the element. Such a product is a measure of the number of field lines which pass through the area $\Delta S_i$ and is called a flux across the surface boundary. The total flux across the surface is denoted by

$$\varphi = \lim_{\substack{n \to \infty \\ \Delta S_i \to 0}} \sum_{i=1}^{n} \vec{F}_i \cdot \Delta \vec{S}_i = \iint_R \vec{F} \cdot d\vec{S}. \tag{5.58}$$

The surface area over which the integration is performed can be part of a surface or it can be over all points of a closed surface. The evaluation of a flux integral over a closed surface is a measure of the total contribution of the normal component of the vector field over the surface.

The term flux in some applications can mean flow. For example, let an imaginary plane surface of area one square centimeter be placed perpendicular to a uniform velocity flow of magnitude $V_0$, such that the velocity is the same at all points over the surface. In one second there results a column of fluid $V_0$ units long which passes through the unit of surface area. The dimension of the flux integral is volume per unit of time and can be interpreted as the rate of flow or flux of the velocity across the surface.

In the above example, the flux was a quantity which is recognized as volume rate of flow. In many other problems the flux is only a definition and does not readily have any physical meaning. For example, the electric flux over the surface of a sphere due to a point charge at its center is given by the integral $\iint_R \vec{E} \cdot d\vec{S}$, where $\vec{E}$ is the electrostatic intensity. The flux cannot be interpreted as flow because nothing is flowing. In this case the flux is considered as a measure of the density of the field lines that pass through the surface of the sphere.

The flux depends upon the size of the surface that is placed in the vector field under consideration and therefore cannot be used to describe a characteristic of the vector field. Imagine an arbitrary closed surface placed in a vector field and the flux integral over this surface is evaluated. Next calculate the volume enclosed by the surface and form the ratio of $\frac{\text{Flux}}{\text{Volume}}$. In the limit as the volume and surface area of our arbitrary closed surface approach zero, the ratio of $\frac{\text{Flux}}{\text{Volume}}$ is the measure of a point characteristic of the vector field called the divergence. Symbolically, the divergence is a scalar quantity and is defined as

$$\text{div } \vec{F} = \lim_{\substack{\Delta V \to 0 \\ \Delta S \to 0}} \frac{\iint_R \vec{F} \cdot d\vec{S}}{\Delta V}. \tag{5.59}$$

The following example is the evaluation of the above divergence limit in the special case where the closed surface is a sphere. Consider a sphere of radius $\epsilon > 0$ centered at some selected point $(x_0, y_0, z_0)$ in a vector field

$$\vec{F} = \vec{F}(x,y,z) = F_1(x,y,z)\,\hat{e}_1 + F_2(x,y,z)\,\hat{e}_2 + F_3(x,y,z)\,\hat{e}_3.$$

A sphere, centered at $(x_0, y_0, z_0)$, can be represented in the parametric form

$$x = x_0 + \epsilon \sin\theta \cos\phi, \qquad 0 \le \phi \le 2\pi$$
$$y = y_0 + \epsilon \sin\theta \sin\phi, \qquad 0 \le \theta \le \pi$$
$$z = z_0 + \epsilon \cos\theta$$

where $\theta$ and $\phi$ are parameters. Alternatively, points on the sphere can be represented by the position vector

$$\vec{r} = \vec{r}(\theta, \phi) = (x_0 + \epsilon \sin\theta \cos\phi)\,\hat{\mathbf{e}}_1 + (y_0 + \epsilon \sin\theta \sin\phi)\,\hat{\mathbf{e}}_2 + (z_0 + \epsilon \cos\theta)\,\hat{\mathbf{e}}_3 \qquad (5.60)$$

The element of surface area on the sphere is represented $dS = \sqrt{EG - F^2} = \epsilon^2 \sin\theta\, d\phi\, d\theta$. The unit normal to the surface of the sphere is given by

$$\hat{\mathbf{e}}_n = \frac{\frac{\partial \vec{r}}{\partial \theta} \times \frac{\partial \vec{r}}{\partial \phi}}{\left|\frac{\partial \vec{r}}{\partial \theta} \times \frac{\partial \vec{r}}{\partial \phi}\right|} = \sin\theta \cos\phi\,\hat{\mathbf{e}}_1 + \sin\theta \sin\phi\,\hat{\mathbf{e}}_2 + \cos\theta\,\hat{\mathbf{e}}_3.$$

The flux integral over this surface can be expressed

$$\varphi = \iint_R \vec{F}(x,y,z) \cdot d\vec{S} = \iint_R \vec{F}(x,y,z) \cdot \hat{\mathbf{e}}_n\, dS$$

$$\varphi = \int_{\phi=0}^{\phi=2\pi} \int_{\theta=0}^{\theta=\pi} \vec{F}(x_0 + \epsilon \sin\theta \cos\phi, y_0 + \epsilon \sin\theta \sin\phi, z_0 + \epsilon \cos\theta) \cdot \hat{\mathbf{e}}_n \epsilon^2 \sin\theta\, d\theta\, d\phi.$$

By expanding $\vec{F}$ in a Taylor's series in powers of $\epsilon$ about $\epsilon = 0$, there results

$$\vec{F} = \vec{F}(x_0, y_0, z_0) + \epsilon \frac{d\vec{F}}{d\epsilon} + \frac{\epsilon^2}{2!}\frac{d^2\vec{F}}{d\epsilon^2} + \frac{\epsilon^3}{3!}\frac{d^3\vec{F}}{d\epsilon^3} + \cdots. \qquad (5.61)$$

where all derivatives with respect to $\epsilon$ are to be evaluated at $\epsilon = 0$. Substituting the expressions for the unit normal and the Taylor's series into the flux integral produces

$$\varphi = \epsilon^2 \mu_0 + \epsilon^3 \mu_1 + \epsilon^4 \mu_2 + \cdots,$$

where

$$\mu_0 = \int_0^\pi \int_0^{2\pi} \vec{F}(x_0, y_0, z_0) \cdot \hat{\mathbf{e}}_n \sin\theta\, d\phi\, d\theta$$

$$\mu_1 = \int_0^\pi \int_0^{2\pi} \frac{d\vec{F}}{d\epsilon} \cdot \hat{\mathbf{e}}_n \sin\theta\, d\phi\, d\theta$$

$$\mu_2 = \int_0^\pi \int_0^{2\pi} \frac{d^2\vec{F}}{d\epsilon^2} \cdot \hat{\mathbf{e}}_n \sin\theta\, d\phi\, d\theta,$$

plus higher order terms in $\epsilon$. The vector $\vec{F}(x_0, y_0, z_0)$ is a constant and an evaluation of the integral defining $\mu_0$ produces $\mu_0 = 0$. To evaluate $\mu_1$ calculate

$$\frac{d\vec{F}}{d\epsilon} = \frac{\partial \vec{F}}{\partial x} \sin\theta \cos\phi + \frac{\partial \vec{F}}{\partial y} \sin\theta \sin\phi + \frac{\partial \vec{F}}{\partial z} \cos\theta$$

and by expanding the terms in this expression one obtains the result

$$\begin{aligned}\frac{d\vec{F}}{d\epsilon} =\ & \left(\frac{\partial F_1}{\partial x} \sin\theta \cos\phi + \frac{\partial F_1}{\partial y} \sin\theta \sin\phi + \frac{\partial F_1}{\partial z} \cos\theta\right) \hat{\mathbf{e}}_1 \\ & + \left(\frac{\partial F_2}{\partial x} \sin\theta \cos\phi + \frac{\partial F_2}{\partial y} \sin\theta \sin\phi + \frac{\partial F_2}{\partial z} \cos\theta\right) \hat{\mathbf{e}}_2 \\ & + \left(\frac{\partial F_3}{\partial x} \sin\theta \cos\phi + \frac{\partial F_3}{\partial y} \sin\theta \sin\phi + \frac{\partial F_3}{\partial z} \cos\theta\right) \hat{\mathbf{e}}_3,\end{aligned}$$

where all the derivatives are to be evaluated at $\epsilon = 0$. The evaluation of the integral $\mu_1$, gives

$$\mu_1 = \frac{4}{3}\pi \left(\frac{\partial F_1}{\partial x} + \frac{\partial F_2}{\partial y} + \frac{\partial F_3}{\partial z}\right).$$

The flux then has the form

$$\varphi = \frac{4}{3}\pi\epsilon^3 \left( \frac{\partial F_1}{\partial x} + \frac{\partial F_2}{\partial y} + \frac{\partial F_3}{\partial z} \right) + \epsilon^4 \mu_2 + \epsilon^5 \mu_3 + \cdots.$$

The volume of the sphere is given by $\frac{4}{3}\pi\epsilon^3$ and consequently the limit of the ratio of $\frac{\text{Flux}}{\text{Volume}}$ as $\epsilon$ tends toward zero produces the scalar relation

$$\text{div}\,\vec{F} = \lim_{\substack{\Delta V \to 0 \\ \Delta S \to 0}} \frac{\iint_R \vec{F} \cdot d\vec{S}}{\Delta V} = \frac{\partial F_1}{\partial x} + \frac{\partial F_2}{\partial y} + \frac{\partial F_3}{\partial z}. \tag{5.62}$$

Recalling the definition of the operator $\nabla$, the mathematical expression of the divergence may be represented

$$\text{div}\,\vec{F} = \nabla \cdot \vec{F} = \left( \frac{\partial}{\partial x}\hat{e}_1 + \frac{\partial}{\partial y}\hat{e}_2 + \frac{\partial}{\partial z}\hat{e}_3 \right) \cdot (F_1\hat{e}_1 + F_2\hat{e}_2 + F_3\hat{e}_3)$$

$$\text{div}\,\vec{F} = \lim_{\substack{\Delta V \to 0 \\ \Delta S \to 0}} \frac{\iint_R \vec{F} \cdot d\vec{S}}{\Delta V} = \nabla \vec{F} = \frac{\partial F_1}{\partial x} + \frac{\partial F_2}{\partial y} + \frac{\partial F_3}{\partial z}. \tag{5.63}$$

The above result holds for an arbitrary simple closed surface enclosing a volume in the limit as the surface area and volume tend toward zero.

### Gauss Divergence Theorem

A relation known as the Gauss divergence theorem exists between the flux and divergence of a vector field. If $V$ is a volume bounded by a closed surface $S$ and $\vec{F}$ is a continuous vector function of position which has continuous derivatives, then the Gauss divergence theorem states

$$\iiint_V \text{div}\,\vec{F}\,d\tau = \iiint_V \nabla \cdot \vec{F}\,d\tau = \iint_S \vec{F} \cdot d\vec{S} = \iint_S \vec{F} \cdot \hat{e}_n\,dS. \tag{5.64}$$

This theorem can also be represented in the expanded form as

$$\iiint_V \left( \frac{\partial F_1}{\partial x} + \frac{\partial F_2}{\partial y} + \frac{\partial F_3}{\partial z} \right) dx\,dy\,dz = \iint_S (F_1\hat{e}_1 + F_2\hat{e}_2 + F_3\hat{e}_3) \cdot \hat{e}_n\,dS, \tag{5.65}$$

where $\hat{e}_n$ is the exterior or positive normal to the closed surface $S$.

To prove Gauss' divergence theorem, one usually verifies the integrals

$$\iiint_V \frac{\partial F_1}{\partial x}\,d\tau = \iint_S F_1\hat{e}_1 \cdot \hat{e}_n\,dS$$

$$\iiint_V \frac{\partial F_2}{\partial y}\,d\tau = \iint_S F_2\hat{e}_2 \cdot \hat{e}_n\,dS \tag{5.66}$$

$$\iiint_V \frac{\partial F_3}{\partial z}\,d\tau = \iint_S F_3\hat{e}_3 \cdot \hat{e}_n\,dS.$$

245

The addition of these integrals produces the desired result. However, the arguments used in proving each of the above integrals are essentially the same. For this reason only the last integral is evaluated and the verification of the other integrals is left as an exercise. Let the closed surface $S$ be composed of an upper half $S_2$ defined by $z = z_2(x, y)$ and a lower half $S_1$ defined by $z = z_1(x, y)$ as illustrated in figure 5-16. Also illustrated in the figure 5-16 is an element $dx dy$ projected upward and intersecting the surface in the surface elements $dS_1$ and $dS_2$.

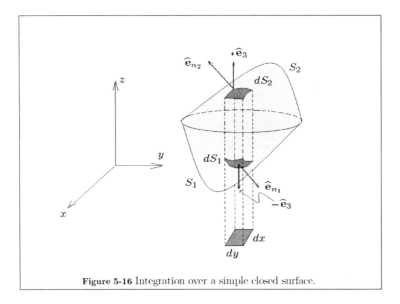

Figure 5-16 Integration over a simple closed surface.

Integrate the last integral of equation (5.66) in the $z$-direction to obtain

$$\iiint\limits_{V} \frac{\partial F_3}{\partial z} dz dx\, dy = \iint F_3(x,y,z)\big|_{z_1(x,y)}^{z_2(x,y)} dx\, dy$$

$$= \iint\limits_{S_2} F_3(x,y,z_2(x,y))\, dx\, dy - \iint\limits_{S_1} F_3(x,y,z_1(x,y))\, dx\, dy.$$

The elements of surface area can be represented by

$$dS_2 = \frac{dx\, dy}{\hat{\mathbf{e}}_3 \cdot \hat{\mathbf{e}}_{n_2}} \quad \text{on surface } S_2, \qquad dS_1 = \frac{dx\, dy}{-\hat{\mathbf{e}}_3 \cdot \hat{\mathbf{e}}_{n_1}} \quad \text{on surface } S_1.$$

so that the above integral can be expressed as

$$\iint_V \frac{\partial F_3}{\partial z} d\tau = \iint_{S_2} F_3 \hat{\mathbf{e}}_3 \cdot \hat{\mathbf{e}}_{n_2} dS_2 + \iint_{S_1} F_3 \hat{\mathbf{e}}_3 \cdot \hat{\mathbf{e}}_{n_1} dS_1 = \iint_S F_3 \hat{\mathbf{e}}_3 \cdot \hat{\mathbf{e}}_n dS,$$

which establishes the desired result.

Similarly by dividing the surface into appropriate sections and projecting the surface elements of these sections onto appropriated planes, the remaining integrals may be verified.

### Physical Interpretation of Divergence

The divergence of a vector field is a scalar field which is interpreted as representing the flux per unit volume diverging from a small neighborhood of a point. In the limit as the volume of the neighborhood tends toward zero, the limit of the ratio of flux divided by volume is called the instantaneous flux per unit volume at a point or the instantaneous flux density at a point.

Let $\vec{F}(x,y,z)$ define a vector field which is continuous with continuous derivatives in a region $R$ such that the divergence exists at point $P_0$ of $R$. The following terminology is associated with the divergence.

If div $\vec{F} > 0$, then a source is said to exist at point $P_0$.

If div $\vec{F} < 0$, then a sink is said to exist at point $P_0$.

If div $\vec{F} = 0$, then $\vec{F}$ is called solenoidal and no sources or sinks exist.

Gauss' divergence theorem states that if div $\vec{F} = 0$, then the flux $\varphi = \iint_S \vec{F} \cdot d\vec{S}$ over the closed surface vanishes. When the flux vanishes the vector field is called solenoidal, and in this case, the flux of the vector field $\vec{F}$ into a volume exactly equals the flux of the field $\vec{F}$ out of the volume. Consider the field lines discussed earlier and visualize a bundle of these field lines forming a tube. Cut the tube by two plane areas $S_1$ and $S_2$ normal to the field lines as in figure 5-17.

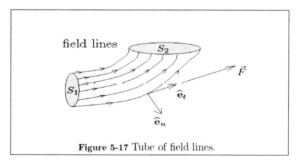

**Figure 5-17** Tube of field lines.

The sides of the tube are composed of field lines, and consequently the direction of the tangents to the field lines are in the same direction as the vector field $\vec{F}$. Therefore, $\vec{F} \cdot \hat{e}_n = 0$ everywhere on the sides of the tube. The sides of the tube consist of field lines, and therefore there is no flux of the vector field across the sides of the tube and all the flux enters, through $S_1$, and leaves through $S_2$. In particular, if $\vec{F}$ is solenoidal and div $\vec{F} = 0$, then

$$\iint_{S_1} \vec{F} \cdot d\vec{S}_1 = - \iint_{S_2} \vec{F} \cdot d\vec{S}_2$$

which implies the flux or flow into $S_1$ must equal the flow leaving $S_2$.

Physically, the divergence assigns a number to each point of space where the vector field exists. The number assigned by the divergence is a scalar and represents the rate per unit volume at which the field issues (or enters) from (or toward) a point. In terms of figure 5-17, if more flux lines enter $S_1$ than leaves $S_2$, the divergence is negative and a sink is said to exist. If more flux lines leave $S_2$ than enter $S_1$, a source is said to exist.

### Green's Theorem in the Plane

Let $C$ denote a simple closed curve enclosing a region $R$ of the $xy$ plane. If $M(x,y)$ and $N(x,y)$ are continuous functions with continuous derivatives in the region $R$, then Green's theorem in the plane can be written as

$$\oint_C M(x,y)\,dx + N(x,y)\,dy = \iint_R \left( \frac{\partial N}{\partial x} - \frac{\partial M}{\partial y} \right) dx\,dy, \tag{5.67}$$

where the line integral is taken in a counterclockwise direction around the simple closed curve $C$ which encloses the region $R$.

To prove this theorem, let $y = y_2(x)$ and $y = y_1(x)$ be single-valued continuous functions which describe the upper and lower portions $C_2$ and $C_1$ of the simple closed curve $C$ in the interval $x_1 \leq x \leq x_2$ as illustrated in figure 5-18.

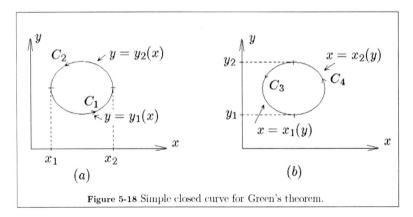

**Figure 5-18** Simple closed curve for Green's theorem.

The last double integral in equation (5.67) can be expressed

$$-\iint_R \frac{\partial M}{\partial y} dx\, dy = -\int_{x_1}^{x_2} \int_{y_1(x)}^{y_2(x)} \frac{\partial M}{\partial y} dy\, dx = -\int_{x_1}^{x_2} M(x,y)\Big|_{y_1(x)}^{y_2(x)} dx$$

$$= \int_{x_1}^{x_2} [M(x, y_1(x)) - M(x, y_2(x))]\, dx \qquad (5.68)$$

$$= \int_{x_1}^{x_2} M(x, y_1(x))\, dx + \int_{x_2}^{x_1} M(x, y_2(x))\, dx$$

$$= \int_{C_1} M(x, y_1(x))\, dx + \int_{C_2} M(x, y_2(x))\, dx = \oint_C M(x,y)\, dx.$$

Now let $x = x_1(y)$ and $x = x_2(y)$ be single-valued continuous functions which describe the left and right sections $C_3$ and $C_4$ of the curve $C$ in the interval $y_1 \le y \le y_2$. The first double integral in equation (5.67) can be expressed

$$\iint_R \frac{\partial N}{\partial x} dx\, dy = \int_{y_1}^{y_2} \int_{x_1(y)}^{x_2(y)} \frac{\partial N}{\partial x} dx\, dy = \int_{y_1}^{y_2} N(x,y)\Big|_{x_1(y)}^{x_2(y)} dy$$

$$= \int_{y_1}^{y_2} [N(x_2(y), y) - N(x_1(y), y)]\, dy \qquad (5.69)$$

$$= \int_{y_1}^{y_2} N(x_2(y), y)\, dy + \int_{y_2}^{y_1} N(x_1(y), y)\, dy$$

$$= \int_{C_3} N(x_1(y), y)\, dy + \int_{C_4} N(x_2(y), y)\, dy = \oint_C N(x,y)\, dy.$$

Adding the results of equations (5.68) and (5.69) produces the Green's theorem in the plane.

**Solution of Differential Equations by Line Integrals**

The total differential of a function $\phi = \phi(x, y)$ is

$$d\phi = \frac{\partial \phi}{\partial x} dx + \frac{\partial \phi}{\partial y} dy. \qquad (5.70)$$

When the right-hand side of this equation is set equal to zero, the resulting equation is called an exact differential equation, and $\phi = \phi(x, y) = k = Constant$ is called a primitive or integral of this equation.

A differential equation of the form

$$M(x,y)\, dx + N(x,y)\, dy = 0 \qquad (5.71)$$

is an exact differential equation if there exists a function $\phi = \phi(x, y)$ such that

$$\frac{\partial \phi}{\partial x} = M(x,y) \quad \text{and} \quad \frac{\partial \phi}{\partial y} = N(x,y).$$

If such a function $\phi$ exists, then the mixed second partial derivatives must be equal and

$$\frac{\partial^2 \phi}{\partial x\, \partial y} = \frac{\partial M}{\partial y} = \frac{\partial^2 \phi}{\partial y\, \partial x} = \frac{\partial N}{\partial x}. \qquad (5.72)$$

Hence a necessary condition that the differential equation be exact is that $\frac{\partial M}{\partial y} = \frac{\partial N}{\partial x}$. If the differential equation is exact, then Green's theorem tells us that the line integral of $M\,dx + N\,dy$ around a closed curve equals zero, since

$$\oint_C M\,dx + N\,dy = \iint_R \left(\frac{\partial N}{\partial x} - \frac{\partial M}{\partial y}\right) dx\,dy = 0. \tag{5.73}$$

For an arbitrary path of integration, such as the path illustrated in figure 5-19.

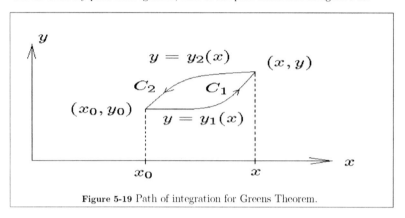

**Figure 5-19** Path of integration for Greens Theorem.

The line integral around the closed path $C$ in figure 5-19 can be written

$$\oint_C M\,dx + N\,dy = \int_{x_0}^{x} M(x, y_1(x))\,dx + N(x, y_1(x))\,dy$$
$$+ \int_{x}^{x_0} M(x, y_2(x))\,dx + N(x, y_2(x))\,dy = 0.$$

where $y = y_2(x)$ denotes the upper path $C_2$ and $y = y_1(x)$ denotes the lower path $C_1$. The above integral can be expressed in the form

$$\int_{x_0}^{x} M(x, y_1(x))\,dx + N(x, y_1(x))\,dy = \int_{x_0}^{x} M(x, y_2(x))\,dx + N(x, y_2(x))\,dy. \tag{5.74}$$

Equation (5.74) tells us that the line integral of $M\,dx + N\,dy$ from $(x_0, y_0)$ to $(x, y)$ is independent of the path joining these two points.

It is now demonstrated that the line integral

$$\int_{(x_0, y_0)}^{(x, y)} M(x, y)\,dx + N(x, y)\,dy$$

is a function of $x$ and $y$ which is related to the solution of the exact differential equation $M\,dx + N\,dy = 0$. Observe that if $M\,dx + N\,dy$ is an exact differential, there exists a function $\phi = \phi(x,y)$ such that $\frac{\partial \phi}{\partial x} = M$ and $\frac{\partial \phi}{\partial y} = N$, and the above line integral reduces to

$$\int_{(x_0,y_0)}^{(x,y)} \frac{\partial \phi}{\partial x}\,dx + \frac{\partial \phi}{\partial y}\,dy = \int_{(x_0,y_0)}^{(x,y)} d\phi = \phi(x,y) - \phi(x_0,y_0). \tag{5.75}$$

Thus the solution of the differential equation $M\,dx + N\,dy = 0$ can be represented in the form $\phi(x,y) = $ Constant, where the function $\phi$ can be obtained from the integral

$$\phi(x,y) - \phi(x_0,y_0) = \int_{(x_0,y_0)}^{(x,y)} M(x,y)\,dx + N(x,y)\,dy. \tag{5.76}$$

Since the line integral is independent of the path of integration, any convenient path of integration from $(x_0,y_0)$ to $(x,y)$ can be selected.

**Area Inside a Simple Closed Curve.**

A very interesting special case of Green's theorem concerns the area enclosed by a simple closed curve. Consider the simple closed curve illustrated in figure 5-20. Green's theorem states that the area inside the simple closed curve can be determined if one knows the values of $x, y$ on the boundary of the curve.

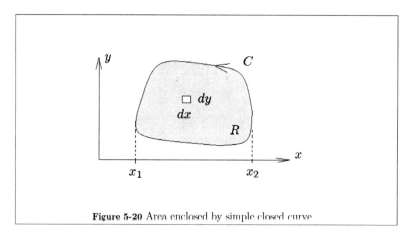

**Figure 5-20** Area enclosed by simple closed curve

In Green's theorem the functions $M$ and $N$ are arbitrary; therefore, consider the special case $M = -y$ and $N = 0$ to obtain

$$\oint_C -y\,dx = \iint_R dx\,dy = A = \text{Area enclosed by C}. \tag{5.77}$$

Similarly, the special case $M = 0$ and $N = x$ in Green's theorem produces

$$\oint_C x\,dy = \iint_R dx\,dy = A = \text{Area enclosed by C}. \tag{5.78}$$

Adding the results from equations (5.77) and (5.78) produces

$$2A = \oint_C x\,dy - y\,dx \quad \text{or} \quad A = \frac{1}{2}\oint_C x\,dy - y\,dx = \iint_R dx\,dy. \tag{5.79}$$

Therefore the area enclosed by a simple closed curve $C$ can be expressed as a line integral around the boundary of the region $R$ enclosed by $C$. This result shows that by knowing the values of $x$ and $y$ on the boundary $C$, the area $R$ enclosed by the boundary can be determined. This is the concept behind the device known as a planimeter, which is a mechanical instrument used for measuring the area of a plane figure by moving a pointer around the surrounding boundary curve.

## Change of Variable in Integration

Often it is convenient to change variables in an integration in order to make the integrals more tractable. If $x, y$ are variables which are related to another set of variables $u, v$ by a set of transformation equations

$$x = x(u, v) \qquad y = y(u, v) \tag{5.80}$$

and if these equations are continuous and have partial derivatives, then it is possible to calculate

$$dx = \frac{\partial x}{\partial u}du + \frac{\partial x}{\partial v}dv \qquad dy = \frac{\partial y}{\partial u}du + \frac{\partial y}{\partial v}dv. \tag{5.81}$$

Using Green's theorem, the area integral (5.79) can be expressed in the form

$$\iint_r dx\,dy = \frac{1}{2}\oint_C x(u,v)\left[\frac{\partial y}{\partial u}du + \frac{\partial y}{\partial v}dv\right] - y(u,v)\left[\frac{\partial x}{\partial u}du + \frac{\partial x}{\partial v}dv\right]$$

$$= \frac{1}{2}\oint_C \left[x\frac{\partial y}{\partial u} - y\frac{\partial x}{\partial u}\right]du + \left[x\frac{\partial y}{\partial v} - y\frac{\partial x}{\partial v}\right]dv. \tag{5.82}$$

Let $M(u,v) = x\frac{\partial y}{\partial u} - y\frac{\partial x}{\partial u}$ and $N(u,v) = x\frac{\partial y}{\partial v} - y\frac{\partial x}{\partial v}$ and show

$$\frac{1}{2}\left(\frac{\partial N}{\partial u} - \frac{\partial M}{\partial v}\right) = \begin{vmatrix} \frac{\partial x}{\partial u} & \frac{\partial y}{\partial u} \\ \frac{\partial x}{\partial v} & \frac{\partial y}{\partial v} \end{vmatrix} = \frac{\partial x}{\partial u}\frac{\partial y}{\partial v} - \frac{\partial x}{\partial v}\frac{\partial y}{\partial u} = J\left(\frac{x,y}{u,v}\right). \tag{5.83}$$

where the determinant $J$ is called the Jacobian determinant of the transformation from $(x,y)$ to $(u,v)$. The area integral can then be expressed in the form

$$A = \iint_R dx\,dy = \iint_{u,v} J\left(\frac{x,y}{u,v}\right) du\,dv, \tag{5.84}$$

where the limits of integration must be changed to range over the variables $u, v$ which define the region $R$.

In general, an integral of the form

$$\iint_{R_{xy}} f(x, y)\, dx\, dy$$

under a change of variables $x = x(u, v)$, $y = y(u, v)$ becomes

$$\iint_{R_{uv}} f(x(u, v), y(u, v))\, J\left(\frac{x, y}{u, v}\right) du\, dv,$$

where the integrand is expressed in terms of $u$ and $v$ and $dx\, dy$ is replaced by $J\left(\frac{x,y}{u,v}\right) du\, dv$.

## The Curl of a Vector Field

Let $\vec{F} = \vec{F}(x, y, z)$ denote a continuous vector field possessing continuous derivatives, and let $P_0$ denote a point in this vector field having coordinates $(x_0, y_0, z_0)$. Consider an arbitrary surface $S$ which contains the point $P_0$ and construct a unit normal $\hat{e}_n$ to the surface at point $P_0$. On the surface construct a simple closed curve $C$ which encircles the point $P_0$. The work done in moving around this closed curve is called the circulation at the point $P_0$. The circulation is a scalar quantity and is expressed as

$$\oint_C \vec{F} \cdot d\vec{r} = \text{Circulation of } \vec{F} \text{ around } C \text{ on the surface } S,$$

where the integration is taken counterclockwise. If the circulation is divided by the area $\Delta S$ enclosed by the simple closed curve $C$ on $S$, then the limit of the ratio $\frac{\text{Circulation}}{\text{Area}}$ as the area $\Delta S$ tends toward zero, is called the component of the curl of $\vec{F}$ in the direction $\hat{e}_n$ and is written as

$$(\operatorname{curl} \vec{F}) \cdot \hat{e}_n = \lim_{\Delta S \to 0} \frac{\oint_C \vec{F} \cdot d\vec{r}}{\Delta S}. \tag{5.85}$$

Construct a simple surface and evaluate the curl of a vector field $\vec{F}$ at a point $P_0(x_0, y_0, z_0)$, on the surface. For example, consider a plane which passes through $P_0$ which is parallel to the $xy$ plane. This plane has the unit normal $\hat{e}_n = \hat{e}_3$ at all points on the plane. In this plane, let us consider the circulation at $P_0$ due to a circle of radius $\epsilon$ centered at $P_0$. The equation of this circle in parametric form is

$$x = x_0 + \epsilon \cos\theta, \quad y = y_0 + \epsilon \sin\theta, \quad z = z_0$$

and the circulation can be expressed as

$$I = \oint_C \vec{F} \cdot d\vec{r} = \int_0^{2\pi} \vec{F}(x_0 + \epsilon \cos\theta, y_0 + \epsilon \sin\theta, z_0)\left(-\epsilon \sin\theta\, \hat{e}_1 + \epsilon \cos\theta\, \hat{e}_2\right) d\theta.$$

By expanding $\vec{F}$ in a Taylor series about $\epsilon = 0$, there results

$$\vec{F}(x_0 + \epsilon\cos\theta, y_0 + \epsilon\sin\theta, z_0) = \vec{F}(x_0, y_0, z_0) + \epsilon\frac{d\vec{F}}{d\epsilon} + \frac{\epsilon^2}{2!}\frac{d^2\vec{F}}{d\epsilon^2} + \cdots,$$

where all derivatives are evaluated at $\epsilon = 0$. The circulation can now be written as

$$I = \epsilon\mu_0 + \epsilon^2\mu_1 + \epsilon^3\mu_2 + \cdots,$$

where

$$\mu_0 = \int_0^{2\pi} \vec{F}(x_0, y_0, z_0)\, d\vec{\xi}$$

$$\mu_1 = \int_0^{2\pi} \frac{d\vec{F}}{d\epsilon}\, d\vec{\xi}$$

$$\mu_2 = \int_0^{2\pi} \frac{1}{2!}\frac{d^2\vec{F}}{d\epsilon^2}\, d\vec{\xi}$$

$$\cdots$$

with $d\vec{\xi} = (-\sin\theta\,\hat{e}_1 + \cos\theta\,\hat{e}_2)\,d\theta$. The vector $\vec{F}(x_0, y_0, z_0)$ is a constant and the integral $\mu_0$ is easily shown to be zero. The vector $\dfrac{d\vec{F}}{d\epsilon}$ when expanded is given by

$$\frac{d\vec{F}}{d\epsilon} = \frac{\partial\vec{F}}{\partial x}\cos\theta + \frac{\partial\vec{F}}{\partial y}\sin\theta = \left(\frac{\partial F_1}{\partial x}\cos\theta + \frac{\partial F_1}{\partial y}\sin\theta\right)\hat{e}_1$$
$$+ \left(\frac{\partial F_2}{\partial x}\cos\theta + \frac{\partial F_2}{\partial y}\sin\theta\right)\hat{e}_2$$
$$+ \left(\frac{\partial F_3}{\partial x}\cos\theta + \frac{\partial F_3}{\partial y}\sin\theta\right)\hat{e}_3,$$

where the partial derivatives are all evaluated at $\epsilon = 0$. It is readily verified that the integral $\mu_1$ reduces to

$$\mu_1 = \pi\left(\frac{\partial F_2}{\partial x} - \frac{\partial F_1}{\partial y}\right).$$

The area of the circle surrounding the point $P_0$ is $\pi\epsilon^2$, and consequently the ratio of the circulation divided by the area, in the limit as $\epsilon$ tends toward zero, produces

$$(\operatorname{curl}\vec{F})\cdot\hat{e}_3 = \frac{\partial F_2}{\partial x} - \frac{\partial F_1}{\partial y}. \tag{5.86}$$

Similarly, by considering other planes through the point $P_0$ which are parallel to the $xz$ and $yz$ planes, arguments similar to those above produce the relations

$$(\operatorname{curl}\vec{F})\cdot\hat{e}_2 = \frac{\partial F_1}{\partial z} - \frac{\partial F_3}{\partial x} \quad\text{and}\quad (\operatorname{curl}\vec{F})\cdot\hat{e}_1 = \frac{\partial F_3}{\partial y} - \frac{\partial F_2}{\partial z}. \tag{5.87}$$

These relations show that the mathematical expression for curl $\vec{F}$ may be written as

$$\operatorname{curl}\vec{F} = \left(\frac{\partial F_3}{\partial y} - \frac{\partial F_2}{\partial z}\right)\hat{e}_1 + \left(\frac{\partial F_1}{\partial z} - \frac{\partial F_3}{\partial x}\right)\hat{e}_2 + \left(\frac{\partial F_2}{\partial x} - \frac{\partial F_1}{\partial y}\right)\hat{e}_3. \tag{5.88}$$

The curl $\vec{F}$ can be expressed by using the operator $\nabla$ in the determinant form

$$\text{curl } \vec{F} = \nabla \times \vec{F} = \begin{vmatrix} \hat{e}_1 & \hat{e}_2 & \hat{e}_3 \\ \frac{\partial}{\partial x} & \frac{\partial}{\partial y} & \frac{\partial}{\partial z} \\ F_1 & F_2 & F_3 \end{vmatrix}. \tag{5.89}$$

**Physical Interpretation of Curl**

The curl of a vector field is itself a vector field. If curl $\vec{F} = \vec{0}$ at all points of a region $R$ where $\vec{F}$ is defined, then the vector field $\vec{F}$ is called an irrotational vector field, otherwise the vector field is called rotational.

The circulation $\oint_C \vec{F} \cdot d\vec{r}$ about a point $P_0$ can be written as

$$\oint_C \vec{F} \cdot \frac{d\vec{r}}{ds} ds = \oint_C \vec{F} \cdot \hat{e}_t ds.$$

The integration of the quantity $\vec{F} \cdot \hat{e}_t$ represents the summation of the projections of the vector $\vec{F}(x, y, z)$ onto the unit tangent vector to the curve $C$ at every point $P$ on the curve $C$. If the summation of these tangential components around the simple closed curve is positive or negative, then there is a moment about the point $P_0$ which causes a rotation. The circulation is thus a measure of the forces tending to produce a rotation about point $P_0$. The curl is the limit of the circulation divided by an area as the area tends toward zero. The curl is thought of as a measure of the circulation density of the field or as a measure of the angular velocity produced by the vector field.

Consider the two-dimensional velocity field $\vec{V} = V_0 \hat{e}_1$, $0 \leq y \leq h$, where $V_0$ is constant, which is illustrated in figure 5-21(a). The velocity field is uniform, and to each point $(x, y)$ there corresponds a constant velocity vector in the $\hat{e}_1$ direction. The curl of this velocity field is zero since the derivative of a constant is zero. The given velocity field is an example of an irrotational vector field.

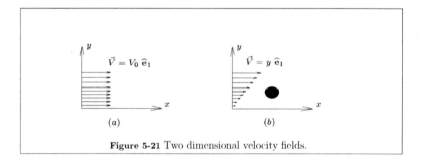

**Figure 5-21** Two dimensional velocity fields.

In comparison, consider the two-dimensional velocity field $\vec{V} = y\,\hat{e}_1$, $0 \le y \le h$, which is illustrated in figure 5-21(b). Here the velocity field may be thought of as representing the flow of fluid in a river. The curl of this velocity field is

$$\text{curl } \vec{V} = \nabla \times \vec{V} = \begin{vmatrix} \hat{e}_1 & \hat{e}_2 & \hat{e}_3 \\ \frac{\partial}{\partial x} & \frac{\partial}{\partial y} & \frac{\partial}{\partial z} \\ y & 0 & 0 \end{vmatrix} = -\hat{e}_3.$$

In this example, the velocity field is rotational. Consider a spherical ball dropped into this velocity field. The curl $\vec{V}$ tells us that the ball rotates in a clockwise direction about an axis normal to the $xy$ plane. Observe the difference in velocities of the water particles acting upon the upper and lower surfaces of the sphere which cause the clockwise rotation.

Using the right-hand rule, let the fingers of the right hand move in the direction of the rotation. The thumb then points in the $-\hat{e}_3$ direction.

The curl tells us the direction of rotation, but it does not tell us the angular velocity associated with a point as the following example illustrates. Consider a basin of water in which the water is rotating with a constant angular velocity $\vec{\omega} = \omega_0 \hat{e}_3$. The velocity of a particle of fluid at a position vector $\vec{r} = x\hat{e}_1 + y\hat{e}_2$ is given by

$$\vec{V} = \vec{\omega} \times \vec{r} = \begin{vmatrix} \hat{e}_1 & \hat{e}_2 & \hat{e}_3 \\ 0 & 0 & \omega_0 \\ x & y & 0 \end{vmatrix} = -\omega_0 y\,\hat{e}_1 + \omega_0 x\,\hat{e}_2.$$

The curl of this velocity field is

$$\text{curl } \vec{V} = \nabla \times \vec{V} = \begin{vmatrix} \hat{e}_1 & \hat{e}_2 & \hat{e}_3 \\ \frac{\partial}{\partial x} & \frac{\partial}{\partial y} & \frac{\partial}{\partial z} \\ -\omega_0 y & \omega_0 x & 0 \end{vmatrix} = 2\omega_0 \hat{e}_3.$$

The curl tells us the direction of the angular velocity but not its magnitude.

## Stokes' Theorem

Let $\vec{F} = \vec{F}(x,y,z)$ denote a continuous vector field having continuous derivatives in a region of space. Let $S$ denote an open two-sided surface in the region of the vector field. For any simple closed curve $C$ lying on the surface $S$, the following integral relation holds

$$\iint_R \text{curl } \vec{F} \cdot d\vec{S} = \iint_R \left(\nabla \times \vec{F}\right) \cdot \hat{e}_n\, dS = \oint_C \vec{F} \cdot d\vec{r}, \tag{5.90}$$

where the surface integrations are understood to be over the portion of the surface $S$ enclosed by the simple closed curve $C$ and the line integral around $C$ is in the positive sense. The above integral relation is known as Stokes' theorem. In scalar form, the line and surface integrals in Stokes' theorem can be expressed as

$$\iint_R (\text{curl } \vec{F}) \cdot d\vec{S} = \iint_R \left[\left(\frac{\partial F_3}{\partial y} - \frac{\partial F_2}{\partial z}\right)\hat{e}_1 + \left(\frac{\partial F_1}{\partial z} - \frac{\partial F_3}{\partial x}\right)\hat{e}_2 + \left(\frac{\partial F_2}{\partial x} - \frac{\partial F_1}{\partial y}\right)\hat{e}_3\right] \cdot \hat{e}_n\, dS$$

and $\oint_C \vec{F} \cdot d\vec{r} = \oint_C F_1\,dx + F_2\,dy + F_3\,dz,$ where $\hat{\mathbf{e}}_n$ is a unit normal to the surface $S$ and the path of integration is counterclockwise around the normal.

To prove Stokes' theorem it is customary to verify each of the following integral relations

$$\iint_R \left( \frac{\partial F_1}{\partial z} \hat{\mathbf{e}}_2 \cdot \hat{\mathbf{e}}_n - \frac{\partial F_1}{\partial y} \hat{\mathbf{e}}_3 \cdot \hat{\mathbf{e}}_n \right) dS = \oint_C F_1\,dx$$

$$\iint_R \left( \frac{\partial F_2}{\partial x} \hat{\mathbf{e}}_3 \cdot \hat{\mathbf{e}}_n - \frac{\partial F_2}{\partial z} \hat{\mathbf{e}}_1 \cdot \hat{\mathbf{e}}_n \right) dS = \oint_C F_2\,dy \qquad (5.91)$$

$$\iint_R \left( \frac{\partial F_3}{\partial y} \hat{\mathbf{e}}_1 \cdot \hat{\mathbf{e}}_n - \frac{\partial F_3}{\partial x} \hat{\mathbf{e}}_2 \cdot \hat{\mathbf{e}}_n \right) dS = \oint_C F_3\,dz.$$

Then an addition of these integrals would produce the Stokes' theorem. However, the arguments used in proving the above integrals are repetitious, and so only the first integral will be verified leaving a verification of the other integrals as an exercise.

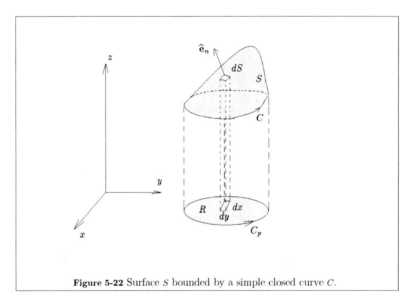

**Figure 5-22** Surface $S$ bounded by a simple closed curve $C$.

Let $z = z(x,y)$ define the surface $S$ and consider the projections of the surface $S$ and the curve $C$ onto the plane $z = 0$ as illustrated in figure 5-22. Call these projections $R$ and $C_p$ as illustrated. The unit normal to the surface has been shown to be of the form

$$\hat{\mathbf{e}}_n = \frac{-\frac{\partial z}{\partial x}\hat{\mathbf{e}}_1 - \frac{\partial z}{\partial y}\hat{\mathbf{e}}_2 + \hat{\mathbf{e}}_3}{\sqrt{1 + (\frac{\partial z}{\partial x})^2 + (\frac{\partial z}{\partial y})^2}}. \qquad (5.92)$$

Consequently,

$$\hat{e}_n \cdot \hat{e}_2 = \frac{-\frac{\partial z}{\partial y}}{\sqrt{1+(\frac{\partial z}{\partial x})^2 + (\frac{\partial z}{\partial y})^2}}, \qquad \hat{e}_n \cdot \hat{e}_3 = \frac{1}{\sqrt{1+(\frac{\partial z}{\partial x})^2 + (\frac{\partial z}{\partial y})^2}}. \qquad (5.93)$$

Also note that the element of surface area can be expressed as

$$dS = \sqrt{1+\left(\frac{\partial z}{\partial x}\right)^2 + \left(\frac{\partial z}{\partial y}\right)^2}\, dx\, dy$$

and the first integral of equation (5.91) can be simplified to the form

$$\iint_R \left(\frac{\partial F_1}{\partial z}\hat{e}_2 \cdot \hat{e}_n - \frac{\partial F_1}{\partial y}\hat{e}_3 \cdot \hat{e}_n\right) dS = \iint_R \left(-\frac{\partial F_1}{\partial z}\frac{\partial z}{\partial y} - \frac{\partial F_1}{\partial y}\right) dx\, dy. \qquad (5.94)$$

Thus, on the surface $S$ where $z = z(x,y)$ and $F_1 = F_1(x,y,z) = F_1(x,y,z(x,y))$, a differentiation of the composite function $F_1$ produces

$$\frac{\partial F_1(x,y,z(x,y))}{\partial y} = \frac{\partial F_1}{\partial y} + \frac{\partial F_1}{\partial z}\frac{\partial z}{\partial y} \qquad (5.95)$$

which is the integrand in the integral (5.94) with the sign changed. Therefore,

$$\iint_R \left(-\frac{\partial F_1}{\partial z}\frac{\partial z}{\partial y} - \frac{\partial F_1}{\partial y}\right) dx\, dy = -\iint_R \frac{\partial F_1(x,y,z(x,y))}{\partial y}\, dx\, dy. \qquad (5.96)$$

Now by using Greens theorem with $M(x,y) = F_1(x,y,z(x,y))$ and $N(x,y) = 0$, the integral (5.96) can be expressed as

$$-\iint_R \frac{\partial F_1(x,y,z(x,y))}{\partial y}\, dx\, dy = \int_{C_p} F_1(x,y,z(x,y))\, dx = \int_C F_1(x,y,z)\, dx \qquad (5.97)$$

which verifies the first integral of (5.91). The remaining integrals in equation (5.91) may be verified in a similar manner.

### Related Integral Theorems

Let $\phi$ denote a scalar field and $\vec{F}$ a vector field. These fields are assumed to be continuous with continuous derivatives. For the volumes, surfaces, and simple closed curves of Stokes' theorem and the divergence theorem, there exist the additional integral relationships

$$\iiint_V \operatorname{curl} \vec{F}\, d\tau = \iiint_V \nabla \times \vec{F}\, d\tau = \iint_R \hat{e}_n \times \vec{F}\, dS \qquad (5.98)$$

$$\iiint_V \operatorname{grad} \phi\, d\tau = \iiint_V \nabla \phi\, d\tau = \iint_R \phi\, \hat{e}_n\, dS \qquad (5.99)$$

$$\iint_R \hat{e}_n \times \operatorname{grad} \phi\, dS = \iint_R \hat{e}_n \times \operatorname{grad} \phi\, dS = -\iint_R \operatorname{grad} \phi \times d\vec{S} = \oint_C \phi\, d\vec{r}. \qquad (5.100)$$

The integral relation (5.98) follows from the divergence theorem. In the divergence theorem, substitute $\vec{F} = \vec{H} \times \vec{C}$, where $\vec{C}$ is an arbitrary constant vector. By using the vector relations

$$\text{div } \vec{F} = \nabla \cdot \vec{F} = \nabla(\vec{H} \times \vec{C}) = \vec{C} \cdot (\nabla \times \vec{H}) \quad \text{and}$$
$$\vec{F} \cdot \hat{e}_n = (\vec{H} \times \vec{C}) \cdot \hat{e}_n = \vec{H}(\vec{C} \times \hat{e}_n) = \vec{C}(\hat{e}_n \times \vec{H}) \quad \text{(triple scalar product)},$$
(5.101)

the divergence theorem can be written as

$$\iiint_V \text{div } (\vec{H} \times \vec{C}) \, d\tau = \iiint_V \vec{C} \cdot (\nabla \times \vec{H}) \, d\tau = \iint_R (\vec{H} \times \vec{C}) \cdot \hat{e}_n \, dS = \iint_R \vec{C} \cdot (\hat{e}_n \times \vec{H}) \, dS \quad (5.102)$$

Since $\vec{C}$ is a constant vector

$$\vec{C} \cdot \iiint_V \nabla \times \vec{H} \, d\tau = \vec{C} \cdot \iint_R \hat{e}_n \times \vec{H} \, dS. \quad (5.103)$$

For arbitrary $\vec{C}$ this relation implies

$$\iiint_V \nabla \times \vec{H} \, d\tau = \iint_R \hat{e}_n \times \vec{H} \, dS. \quad (5.104)$$

In this integral replace $\vec{H}$ by $\vec{F}$ ($\vec{H}$ is arbitrary) to obtain the relation (5.98).

The integral (5.99) also is a special case of the divergence theorem. In the divergence theorem substitute $\vec{F} = \phi \vec{C}$, where $\phi$ is a scalar function of position and $\vec{C}$ is an arbitrary constant vector and show

$$\iint_V \text{div } \vec{F} \, d\tau = \iiint_V \nabla(\phi \vec{C}) \, d\tau = \iiint_V \vec{C} \cdot \nabla \phi \, d\tau = \iint_R \vec{C} \phi \, d\vec{S}. \quad (5.105)$$

This relation, for arbitrary $\vec{C}$, produces the integral relation (5.99).

The integral (5.100) is another special case of Stokes' theorem. In Stokes' theorem substitute $\vec{F} = \phi \vec{C}$, where $\vec{C}$ is a constant vector, to show

$$\iint_R (\text{curl } \vec{F}) \cdot d\vec{S} = \iint_R \nabla \times (\phi \vec{C}) \cdot \hat{e}_n \, dS = \iint_R (\nabla \phi \times \vec{C}) \cdot \hat{e}_n \, dS$$
$$= \iint_R (\hat{e}_n \times \nabla \phi) \cdot \vec{C} \, dS = \oint_C \vec{C} \phi \, d\vec{r}. \quad (5.106)$$

For arbitrary $\vec{C}$, this integral implies the relation (5.100). That is, it is possible to factor out the constant vector $\vec{C}$ as long as the constant vector is different from zero. Under these conditions the integral relation (5.100) must hold.

### Region of Integration

Green's, Gauss' and Stokes' theorems are valid only if certain conditions are satisfied. In these theorems it has been assumed that the integrands are continuous inside the region and on the boundary where the integrations occur. Also assumed is that all necessary derivatives of these integrands exist and are continuous over the regions or boundaries of the integration. In the study of the various vector and scalar fields arising in engineering and physics, there are times when discontinuities occur at points inside the regions or on the boundaries of the integration. Under these circumstances the above theorems are still valid but they must be modified slightly. Modification is done by using superposition of the integrals over each side of a discontinuity and under these circumstances there usually results some kind of a jump condition involving the value of the field on either side of the discontinuity.

If a region of space has the property that every simple closed curve within the region can be deformed or shrunk in a continuous manner to a single point within the region, without intersecting a boundary of the region, then the region is said to be simply connected. If in order to shrink or reduce a simply closed curve to a point, the curve must leave the region under consideration, then the region is said to be a multiply connected region. An example of a multiply connected region is the surface of a torus. Here a circle which encloses the hole of this doughnut-shaped region cannot be shrunk to a single point without leaving the surface, and so the region is called a multiply connected region.

If a region is multiply connected it usually can be modified by introducing imaginary cuts or lines within the region and requiring that these lines cannot be crossed. By introducing appropriate cuts, it is possible to modify a multiply connected region into a simply connected region. The theorems of Gauss, Green, and Stokes are applicable to simply connected regions or multiply connected regions which can be reduced to simply connected regions by introducing suitable cuts.

### Green's First and Second Identities

Two special cases of the divergence theorem are known as Green's first and second identities. These identities have many uses in studying scalar and vector fields arising in science and engineering.

In the divergence theorem, make the substitution $\vec{F} = \psi \nabla \phi$ to obtain

$$\iiint_V \nabla \vec{F} \, d\tau = \iiint_V \nabla \cdot (\psi \nabla \phi) \, d\tau = \iint_R \psi \nabla \phi \, d\vec{S}. \tag{5.107}$$

Using the relation

$$\nabla \cdot (\psi \nabla \phi) = \psi \nabla^2 \phi + \nabla \psi \cdot \nabla \phi$$

simplifies equation (5.107) to

$$\iiint_V \left( \psi \nabla^2 \phi + \nabla \psi \cdot \nabla \phi \right) d\tau = \iint_R \psi \nabla \phi \, d\vec{S}. \tag{5.108}$$

This result is known as Green's first identity.

In Green's first identity interchange $\psi$ and $\phi$ to obtain

$$\iiint_V \left(\phi\nabla^2\psi + \nabla\phi\cdot\nabla\psi\right) d\tau = \iint_R \phi\nabla\psi\, d\vec{S}. \qquad (5.109)$$

Subtracting equation (5.108) from equation (5.109) produces Green's second identity

$$\iiint_V \left(\phi\nabla^2\psi - \psi\nabla^2\phi\right) d\tau = \iint_R \left(\phi\nabla\psi - \psi\nabla\phi\right) d\vec{S}. \qquad (5.110)$$

## Additional Operators

The del operator in Cartesian coordinates

$$\nabla = \frac{\partial}{\partial x}\hat{e}_1 + \frac{\partial}{\partial y}\hat{e}_2 + \frac{\partial}{\partial z}\hat{e}_3 \qquad (5.111)$$

has been used to express the gradient of a scalar field and the divergence and curl of a vector field. Other operators involving $\nabla$ which prove to be useful are the following.

In the following list of operators let $\vec{A}$ denote a continuous vector function of position.

1. The operator $\vec{A}\cdot\nabla$ is defined as

$$\vec{A}\cdot\nabla = (A_1\hat{e}_1 + A_2\hat{e}_2 + A_3\hat{e}_3)\cdot\left(\frac{\partial}{\partial x}\hat{e}_1 + \frac{\partial}{\partial y}\hat{e}_2 + \frac{\partial}{\partial z}\hat{e}_3\right)$$
$$= A_1\frac{\partial}{\partial x} + A_2\frac{\partial}{\partial y} + A_3\frac{\partial}{\partial z}. \qquad (5.112)$$

Note that $\vec{A}\cdot\nabla$ is an operator which can operate on vector or scalar quantities.

2. The operator $\vec{A}\times\nabla$ is defined as

$$\vec{A}\times\nabla = \begin{vmatrix} \hat{e}_1 & \hat{e}_2 & \hat{e}_3 \\ A_1 & A_2 & A_3 \\ \frac{\partial}{\partial x} & \frac{\partial}{\partial y} & \frac{\partial}{\partial z} \end{vmatrix}$$
$$= \left(A_2\frac{\partial}{\partial z} - A_3\frac{\partial}{\partial y}\right)\hat{e}_1 + \left(A_3\frac{\partial}{\partial x} - A_1\frac{\partial}{\partial z}\right)\hat{e}_2 + \left(A_1\frac{\partial}{\partial y} - A_2\frac{\partial}{\partial x}\right)\hat{e}_3. \qquad (5.113)$$

This operator is a vector operator.

3. The Laplacian operator $\nabla^2 = \nabla\cdot\nabla$ in rectangular Cartesian coordinates is given by

$$\nabla^2 = \frac{\partial^2}{\partial x^2} + \frac{\partial^2}{\partial y^2} + \frac{\partial}{\partial z^2}. \qquad (5.114)$$

This operator can operate on vector or scalar quantities.

One must be careful in the use of operators because in general, they are not commutative. They operate only on the quantities to their immediate right.

## Properties of the Del Operator

For $f, g$ scalar functions of position and $\vec{A}, \vec{B}$ vector functions of position, the del operator has the following properties which are stated without proof:

| | |
|---|---|
| 1. | $\nabla(f+g) = \nabla f + \nabla g$ |
| 2. | $\nabla(f\vec{A}) = (\nabla f) \cdot \vec{A} + f(\nabla \cdot \vec{A})$ |
| 3. | $\nabla \times (f\vec{A}) = (\nabla f) \times \vec{A} + f(\nabla \times \vec{A})$ |
| 4. | $\nabla \cdot (\vec{A} \times \vec{B}) = \vec{B}(\nabla \times \vec{A}) - \vec{A}(\nabla \times \vec{B})$ |
| 5. | $\nabla(\vec{A} + \vec{B}) = \nabla \vec{A} + \nabla \vec{B}$ |
| 6. | $\nabla \times (\vec{A} + \vec{B}) = \nabla \times \vec{A} + \nabla \times \vec{B}$ |
| 7. | $(\vec{A} \cdot \nabla)f = \vec{A} \cdot \nabla f$ |
| 8. | $(\vec{A} \times \nabla)f = \vec{A} \times \nabla f$ |
| 9. | For $f = f(u)$ and $u = u(x, y, z)$, then $\nabla f = \dfrac{df}{du} \cdot \nabla u$ |
| 10. | For $f = f(u_1, u_2, \ldots, u_n)$ and $u_i = u_i(x, y, z)$, $i = 1, 2, \ldots n$, $\nabla f = \dfrac{\partial f}{\partial u_1}\nabla u_1 + \dfrac{\partial f}{\partial u_2}\nabla u_2 + \cdots + \dfrac{\partial f}{\partial u_n}\nabla u_n$ |
| 11. | $\nabla \times (\vec{A} \times \vec{B}) = \vec{A}(\nabla \cdot \vec{B}) - \vec{B}(\nabla \cdot \vec{A}) + (\vec{B} \cdot \nabla)\vec{A} - (\vec{A} \cdot \nabla)\vec{B}$ |
| 12. | $\nabla(\vec{A} \cdot \vec{B}) = \vec{A} \times (\nabla \times \vec{B}) + \vec{B} \times (\nabla \times \vec{A}) + (\vec{B} \cdot \nabla)\vec{A} + (\vec{A} \cdot \nabla)\vec{B}$ |
| 13. | $\nabla \times (\nabla \times \vec{A}) = \nabla(\nabla \cdot \vec{A}) - \nabla^2 \vec{A}$ |

## Curvilinear Coordinates

In this section the concept of curvilinear coordinates is introduced and the representation of scalars and vectors in these new coordinates are studied.

If associated with each point $(x, y, z)$ of a rectangular coordinate system there is a set of variables $(u, v, w)$ such that $x, y, z$ can be expressed in terms of $u, v, w$ by a set of functional relationships or transformations equations, then $(u, v, w)$ are called the curvilinear coordinates of the point $(x, y, z)$. Such transformation equations are expressible in the form

$$x = x(u, v, w), \qquad y = y(u, v, w), \qquad z = z(u, v, w) \tag{5.115}$$

and the inverse transformation can be expressed as

$$u = u(x, y, z), \qquad v = v(x, y, z), \qquad w = w(x, y, z) \tag{5.116}$$

It is assumed that the transformation equations (5.115) and (5.116) are single valued and continuous functions with continuous derivatives. It is also assumed that the transformation equations (5.115) are such that the inverse transformation (5.116) exists, because this condition assures us that the correspondence between the variables $(x, y, z)$ and $(u, v, w)$ is a one-to-one correspondence.

The position vector

$$\vec{r} = x\,\hat{e}_1 + y\,\hat{e}_2 + z\,\hat{e}_3 \tag{5.117}$$

of a general point $(x, y, z)$ can be expressed in terms of the curvilinear coordinates $(u, v, w)$ by utilizing the transformation equations (5.115). The position vector $\vec{r}$, when expressed in terms of the curvilinear coordinates, becomes

$$\vec{r} = \vec{r}(u, v, w) = x(u, v, w)\,\hat{e}_1 + y(u, v, w)\,\hat{e}_2 + z(u, v, w)\,\hat{e}_3 \tag{5.118}$$

and an element of arc length squared is $ds^2 = d\vec{r} \cdot d\vec{r}$. In the curvilinear coordinates where $\vec{r} = \vec{r}(u, v, w)$ is a function of the curvilinear coordinates there results

$$d\vec{r} = \frac{\partial \vec{r}}{\partial u} du + \frac{\partial \vec{r}}{\partial v} dv + \frac{\partial \vec{r}}{\partial w} dw. \tag{5.119}$$

The differential element $d\vec{r}$ is used to produce the element of arc length squared

$$\begin{aligned} d\vec{r} \cdot d\vec{r} = ds^2 = & \frac{\partial \vec{r}}{\partial u} \cdot \frac{\partial \vec{r}}{\partial u} du\,du + \frac{\partial \vec{r}}{\partial u} \cdot \frac{\partial \vec{r}}{\partial v} du\,dv + \frac{\partial \vec{r}}{\partial u} \cdot \frac{\partial \vec{r}}{\partial w} du\,dw \\ & + \frac{\partial \vec{r}}{\partial v} \cdot \frac{\partial \vec{r}}{\partial u} dv\,du + \frac{\partial \vec{r}}{\partial v} \cdot \frac{\partial \vec{r}}{\partial v} dv\,dv + \frac{\partial \vec{r}}{\partial v} \cdot \frac{\partial \vec{r}}{\partial w} dv\,dw \\ & + \frac{\partial \vec{r}}{\partial w} \cdot \frac{\partial \vec{r}}{\partial u} dw\,du + \frac{\partial \vec{r}}{\partial w} \cdot \frac{\partial \vec{r}}{\partial v} dw\,dv + \frac{\partial \vec{r}}{\partial w} \cdot \frac{\partial \vec{r}}{\partial w} dw\,dw. \end{aligned} \tag{5.120}$$

The quantities

$$\begin{aligned} g_{11} &= \frac{\partial \vec{r}}{\partial u} \cdot \frac{\partial \vec{r}}{\partial u} & g_{12} &= \frac{\partial \vec{r}}{\partial u} \cdot \frac{\partial \vec{r}}{\partial v} & g_{13} &= \frac{\partial \vec{r}}{\partial u} \cdot \frac{\partial \vec{r}}{\partial w} \\ g_{21} &= \frac{\partial \vec{r}}{\partial v} \cdot \frac{\partial \vec{r}}{\partial u} & g_{22} &= \frac{\partial \vec{r}}{\partial v} \cdot \frac{\partial \vec{r}}{\partial v} & g_{23} &= \frac{\partial \vec{r}}{\partial v} \cdot \frac{\partial \vec{r}}{\partial w} \\ g_{31} &= \frac{\partial \vec{r}}{\partial w} \cdot \frac{\partial \vec{r}}{\partial u} & g_{32} &= \frac{\partial \vec{r}}{\partial w} \cdot \frac{\partial \vec{r}}{\partial v} & g_{33} &= \frac{\partial \vec{r}}{\partial w} \cdot \frac{\partial \vec{r}}{\partial w} \end{aligned} \tag{5.121}$$

are called the metric components of the curvilinear coordinate system. The metric components may be thought of as the elements of a symmetric matrix, since $g_{ij} = g_{ji}$, $i, j = 1, 2, 3$. These metrices play an important role in the subject area of tensor calculus.

The vectors $\frac{\partial \vec{r}}{\partial u}$, $\frac{\partial \vec{r}}{\partial v}$, $\frac{\partial \vec{r}}{\partial w}$, used to calculate the metric components $g_{ij}$ have the following physical interpretation. The vector $\vec{r} = \vec{r}(u, c_2, c_3)$, where $u$ is a variable and $v = c_2$, $w = c_3$ are constants, traces out a curve in space called a coordinate curve. Families of these curves create a coordinate system. Coordinate curves can also be viewed as being generated by the intersection of the coordinate surfaces $v(x, y, z) = c_2$ and $w(x, y, z) = c_3$. The tangent vector to the coordinate curve is calculated with the partial derivative $\frac{\partial \vec{r}}{\partial u}$. Similarly, the curves

$\vec{r} = \vec{r}(c_1, v, c_3)$ and $\vec{r} = \vec{r}(c_1, c_2, w)$ are coordinate curves and have the respective tangent vectors $\frac{\partial \vec{r}}{\partial v}$ and $\frac{\partial \vec{r}}{\partial w}$. The magnitude of these tangent vectors is used to define the scalar magnitudes

$$h_1 = h_u = |\frac{\partial \vec{r}}{\partial u}|, \qquad h_2 = h_v = |\frac{\partial \vec{r}}{\partial v}|, \qquad h_3 = h_w = |\frac{\partial \vec{r}}{\partial w}|. \tag{5.122}$$

The unit tangent vectors to the coordinate curves are given by the relations

$$\hat{e}_u = \frac{1}{h_1}\frac{\partial \vec{r}}{\partial u}, \qquad \hat{e}_v = \frac{1}{h_2}\frac{\partial \vec{r}}{\partial v}, \qquad \hat{e}_w = \frac{1}{h_3}\frac{\partial \vec{r}}{\partial w}. \tag{5.123}$$

The coordinate surfaces and coordinate curves may be formed from the equations (5.115) and (5.116) are illustrated in figure 5-23.

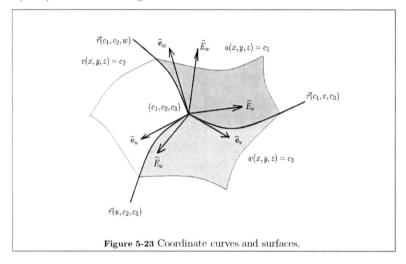

**Figure 5-23** Coordinate curves and surfaces.

Consider the point $u = c_1$, $v = c_2$, $w = c_3$ in the curvilinear coordinate system. This point can be viewed as being created by the intersection of the three surfaces

$$u = u(x, y, z) = c_1, \qquad v = v(x, y, z) = c_2, \qquad w = w(x, y, z) = c_3$$

obtained from the inverse transformation equations (5.116).

Figure 5-23 illustrates the surfaces $u(x, y, z) = c_1$ and $v(x, y, z) = c_2$ which intersect in the curve $\vec{r} = \vec{r}(c_1, c_2, w)$. The point where this curve intersects the surface $w(x, y, z) = c_3$, is $(c_1, c_2, c_3)$. The vector grad $u(x, y, z)$ is a vector normal to the surface $u(x, y, z) = c_1$. A unit normal to the $u(x, y, z) = c_1$ surface has the form $\widehat{E}_u = \frac{\text{grad } u}{|\text{grad } u|}$. Similarly, calculate the vectors $\widehat{E}_v = \frac{\text{grad } v}{|\text{grad } v|}$, and $\widehat{E}_w = \frac{\text{grad } w}{|\text{grad } w|}$ representing unit normal vectors to the coordinate surfaces $v(x, y, z) = c_2$ and $w(x, y, z) = c_3$.

The unit tangent vectors $\hat{\mathbf{e}}_u$, $\hat{\mathbf{e}}_v$, $\hat{\mathbf{e}}_w$ and the unit normal vectors $\widehat{\mathbf{E}}_u$, $\widehat{\mathbf{E}}_v$, $\widehat{\mathbf{E}}_w$ are identical if and only if $g_{ij} = 0$ for $i \neq j$ and under these conditions the curvilinear coordinate system is called an orthogonal coordinate system since the coordinate curves and coordinate surfaces intersect perpendicularly at the point $(c_1, c_2, c_3)$.

### Example 5-11.

Consider the identity transformation between the coordinates $(x, y, z)$ and $(u, v, w)$ given by $u = x$, $v = y$, and $w = z$. The position vector is

$$\vec{r}(x, y, z) = x\,\hat{\mathbf{e}}_1 + y\,\hat{\mathbf{e}}_2 + z\,\hat{\mathbf{e}}_3,$$

and in this rectangular coordinate system, the element of arc length squared is given by $ds^2 = dx^2 + dy^2 + dz^2$. In this space the metric components are

$$g_{ij} = \begin{pmatrix} h_1^2 & 0 & 0 \\ 0 & h_2^2 & 0 \\ 0 & 0 & h_3^2 \end{pmatrix} = \begin{pmatrix} 1 & 0 & 0 \\ 0 & 1 & 0 \\ 0 & 0 & 1 \end{pmatrix},$$

and the coordinate system is orthogonal.

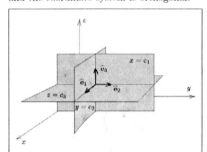

**Figure 5-24** Cartesian coordinate system.

In rectangular coordinates consider the family of surfaces

$$x = c_1, \qquad y = c_2, \qquad z = c_3,$$

where $c_1, c_2, c_3$ take on the integer values $1, 2, 3, \ldots$. These surfaces intersect in lines which are the coordinate curves. The vectors grad $x = \hat{\mathbf{e}}_1$, grad $y = \hat{\mathbf{e}}_2$, and grad $z = \hat{\mathbf{e}}_3$ are the unit vectors which are normal to the coordinate surfaces.

The vectors

$$\frac{\partial \vec{r}}{\partial x} = \hat{\mathbf{e}}_1, \qquad \frac{\partial \vec{r}}{\partial y} = \hat{\mathbf{e}}_2, \qquad \frac{\partial \vec{r}}{\partial z} = \hat{\mathbf{e}}_3$$

can also be viewed as being tangent to the coordinate curves. The situation is illustrated in figure 5-24.

∎

## Example 5-12.

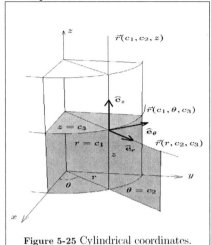

**Figure 5-25** Cylindrical coordinates.

In cylindrical coordinates $(r, \theta, z)$, the transformation equations (5.115) become

$$x = x(r, \theta, z) = r\cos\theta$$
$$y = y(r, \theta, z) = r\sin\theta$$
$$z = z(r, \theta, z) = z$$

and the inverse transformation (5.116) can be written

$$r = r(x, y, z) = \sqrt{x^2 + y^2}$$
$$\theta = \theta(x, y, z) = \arctan\frac{y}{x}$$
$$z = z(x, y, z) = z$$

In cylindrical coordinates $u = r$, $v = \theta, w = z$ the position vector (5.117) to a general point is represented $\vec{r} = \vec{r}(r, \theta, z) = r\cos\theta\,\hat{\mathbf{e}}_1 + r\sin\theta\,\hat{\mathbf{e}}_2 + z\,\hat{\mathbf{e}}_3$. Consequently, the coordinate curve $\vec{r} = \vec{r}(c_1, \theta, c_3) = c_1\cos\theta\,\hat{\mathbf{e}}_1 + c_1\sin\theta\,\hat{\mathbf{e}}_2 + c_3\,\hat{\mathbf{e}}_3$, where $c_1$ and $c_3$ are constants, represents the circle $x^2 + y^2 = c_1^2$ in the plane $z = c_3$. The curve $\vec{r} = \vec{r}(c_1, c_2, z) = c_1\cos c_2\,\hat{\mathbf{e}}_1 + c_1\sin c_2\,\hat{\mathbf{e}}_2 + z\,\hat{\mathbf{e}}_3$ represents a straight line parallel to the $z$-axis which is normal to the $xy$ plane at the point $r = c_1$, $\theta = c_2$. The curve $\vec{r} = \vec{r}(r, c_2, c_3) = r\cos c_2\,\hat{\mathbf{e}}_1 + r\sin c_2\,\hat{\mathbf{e}}_2 + c_3\,\hat{\mathbf{e}}_3$ represents a straight line in the plane $z = c_3$, which extends in the direction $\theta = c_2$. These coordinate curves are illustrated in the figure 5-25.

The tangent vectors to the coordinate curves are obtained by differentiation

$$\frac{\partial\vec{r}}{\partial r} = \cos\theta\,\hat{\mathbf{e}}_1 + \sin\theta\,\hat{\mathbf{e}}_2, \qquad \frac{\partial\vec{r}}{\partial\theta} = -r\sin\theta\,\hat{\mathbf{e}}_1 + r\cos\theta\,\hat{\mathbf{e}}_2, \qquad \frac{\partial\vec{r}}{\partial z} = \hat{\mathbf{e}}_3$$

and the unit vectors in the direction of these tangents are given by

$$\hat{\mathbf{e}}_r = \cos\theta\,\hat{\mathbf{e}}_1 + \sin\theta\,\hat{\mathbf{e}}_2, \qquad \hat{\mathbf{e}}_\theta = -\sin\theta\,\hat{\mathbf{e}}_1 + \cos\theta\,\hat{\mathbf{e}}_2, \qquad \hat{\mathbf{e}}_z = \hat{\mathbf{e}}_3$$

and are illustrated in figure 5-25. The element of arc length squared is

$$ds^2 = d\vec{r}\cdot d\vec{r} = dr^2 + r^2 d\theta^2 + dz^2$$

and the metric components of the space are

$$g_{ij} = \begin{pmatrix} h_1^2 & 0 & 0 \\ 0 & h_2^2 & 0 \\ 0 & 0 & h_3^2 \end{pmatrix} = \begin{pmatrix} 1 & 0 & 0 \\ 0 & r^2 & 0 \\ 0 & 0 & 1 \end{pmatrix}.$$

Observe that this is an orthogonal system where $g_{ij} = 0$ for $i \neq j$. The surface $r = c_1$ is a cylinder, whereas the surface $\theta = c_2$ is a plane perpendicular to the $xy$ plane and passing through the $z$-axis. The surface $z = c_3$ is a plane parallel to the $xy$ plane.

■

### Example 5-13.

The spherical coordinates $(r, \theta, \phi)$ are related to the rectangular coordinates through the transformation equations

$$x = x(\rho, \theta, \phi) = \rho \sin\theta \cos\phi, \qquad y = y(\rho, \theta, \phi) = \rho \sin\theta \sin\phi, \qquad z = z(\rho, \theta, \phi) = \rho \cos\theta$$

which can be obtained from the geometry of figure 5-26.

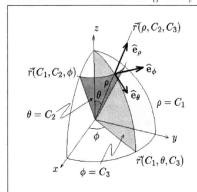

**Figure 5-26** Spherical coordinates.

The position vector to a point in spherical coordinates is

$$\vec{r} = \vec{r}(\rho, \theta, \phi) = \rho \sin\theta \cos\phi\, \hat{\mathbf{e}}_1 + \rho \sin\theta \sin\phi\, \hat{\mathbf{e}}_2 + \rho \cos\theta\, \hat{\mathbf{e}}_3.$$

This position vector generates the coordinate curves

$$\vec{r} = \vec{r}(c_1, c_2, \phi), \qquad \vec{r} = \vec{r}(c_1, \theta, c_3), \qquad \vec{r} = \vec{r}(\rho, c_2, c_3),$$

where $c_1, c_2, c_3$ are constants. These coordinate curves are, respectively, circles of radius $c_1 \sin c_2$, meridian lines on the surface of the sphere, and a line normal to the sphere.

The surfaces $\rho = c_1$, $\theta = c_2$, and $\phi = c_3$ are, respectively, spheres, circular cones, and planes passing through the $z$-axis. These coordinate surfaces and coordinate curves are illustrated in the figure 5-26.

The unit tangent vectors to the coordinate curves are obtained from normalization of the derivatives $\dfrac{\partial \vec{r}}{\partial \rho}, \dfrac{\partial \vec{r}}{\partial \theta}, \dfrac{\partial \vec{r}}{\partial \phi}$ to obtain

$$\hat{\mathbf{e}}_\rho = \sin\theta \cos\phi\, \hat{\mathbf{e}}_1 + \sin\theta \sin\phi\, \hat{\mathbf{e}}_2 + \cos\theta\, \hat{\mathbf{e}}_3, \qquad h_1 = h_\rho = 1$$

$$\hat{\mathbf{e}}_\theta = \cos\theta \cos\phi\, \hat{\mathbf{e}}_1 + \cos\theta \sin\phi\, \hat{\mathbf{e}}_2 - \sin\theta\, \hat{\mathbf{e}}_3, \qquad h_2 = h_\theta = \rho$$

$$\hat{\mathbf{e}}_\phi = -\sin\phi\, \hat{\mathbf{e}}_1 + \cos\phi\, \hat{\mathbf{e}}_2, \qquad h_3 = h_\phi = \rho \sin\theta.$$

The element of arc length squared is

$$ds^2 = d\rho^2 + \rho^2\, d\theta^2 + \rho^2 \sin^2\theta\, d\phi^2,$$

and the metric components of this space are given by

$$g_{ij} = \begin{pmatrix} h_1^2 & 0 & 0 \\ 0 & h_2^2 & 0 \\ 0 & 0 & h_3^2 \end{pmatrix} = \begin{pmatrix} 1 & 0 & 0 \\ 0 & \rho^2 & 0 \\ 0 & 0 & \rho^2 \sin^2\theta \end{pmatrix}.$$

Note that the curvilinear coordinate system is an orthogonal system.

## Left and Right-handed Coordinate Systems

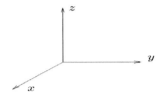
right-handed Cartesian

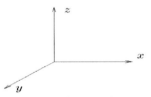
left-handed Cartesian

In a right-handed Cartesian coordinate system, if you place the fingers of your right hand in the direction of the $x$-axis and rotate these fingers 90° to the $y$-axis, then your thumb will point in the direction of the positive $z$ axis. In a similar fashion, you can recognize a left-handed Cartesian coordinate system by placing the fingers of the left hand in the direction of the $x$ axis and then rotate these fingers 90° to the $y$ axis, then the thumb of the left hand will point in the direction of the positive $z$ axis.

Note in spherical coordinates $(\rho, \theta, \phi)$, the transformation equations

$$x = \rho \cos\theta \cos\phi, \qquad y = \rho \sin\theta \sin\phi, \qquad z = \rho \cos\theta$$

are used to define the coordinate curves and the unit tangent vectors $\hat{e}_\rho$, $\hat{e}_\theta$, $\hat{e}_\phi$ to these coordinate curves form a right-handed system. That is, if you place the fingers of the right hand in the direction $\hat{e}_\rho$ and rotate the fingers 90° to $\hat{e}_\theta$, then the thumb of the right hand points in the direction $\hat{e}_\phi$.

Observe that if the coordinates are listed as $(\rho, \phi, \theta)$, then by placing the fingers of the left-hand in the direction $\hat{e}_\rho$ and rotating the fingers 90° to the second direction $\hat{e}_\phi$ you find that your thumb is pointing in the direction $\hat{e}_\theta$. This, then becomes a left-handed coordinate system. This shows that when one is confronted with a coordinate system $(x_1, x_2, x_3)$, the positioning of the variables in the notation $(x_1, x_2, x_3)$ determines if it is a left or right-handed coordinate system. Check this out with the previous cylindrical coordinate system $(r, \theta, z)$. Also note that many European textbooks use left-handed coordinate systems.

## Gradient, Divergence, Curl and Laplacian

In applied mathematics one must know how to represent the gradient, divergence, curl and Laplacian in a general orthogonal curvilinear coordinate system. Recall that for the representations $\phi = \phi(x,y,z)$ and $\vec{F} = F_1(x,y,z)\hat{e}_1 + F_2(x,y,z)\hat{e}_3 + F_3(x,y,z)\hat{e}_3$ the gradient, divergence, curl and Laplacian in rectangular coordinates are given by

$$\operatorname{grad}\phi = \frac{\partial \phi}{\partial x}\hat{e}_1 + \frac{\partial \phi}{\partial y}\hat{e}_2 + \frac{\partial \phi}{\partial z}\hat{e}_3$$

$$\operatorname{div}\vec{F} = \frac{\partial F_1}{\partial x} + \frac{\partial F_2}{\partial y} + \frac{\partial F_3}{\partial z}$$

$$\operatorname{curl}\vec{F} = \left(\frac{\partial F_3}{\partial y} - \frac{\partial F_2}{\partial z}\right)\hat{e}_1 + \left(\frac{\partial F_1}{\partial z} - \frac{\partial F_3}{\partial x}\right)\hat{e}_2 + \left(\frac{\partial F_2}{\partial x} - \frac{\partial F_1}{\partial y}\right)\hat{e}_3$$

$$\nabla^2 \phi = \frac{\partial^2 \phi}{\partial x^2} + \frac{\partial^2 \phi}{\partial y^2} + \frac{\partial^2 \phi}{\partial z^2}$$

For

$$\phi = \phi(u,v,w) \quad \text{and} \quad \vec{F} = F_1(u,v,w)\hat{e}_u + F_2(u,v,w)\hat{e}_v + F_3(u,v,w)\hat{e}_w$$

expressed in terms of a set of general orthogonal curvilinear coordinates, $(u,v,w)$, the gradient, divergence, curl and Laplacian can be shown to have the forms

gradient $\quad \nabla \phi = \operatorname{grad} \phi = \dfrac{1}{h_1}\dfrac{\partial \phi}{\partial u}\hat{e}_u + \dfrac{1}{h_2}\dfrac{\partial \phi}{\partial v}\hat{e}_v + \dfrac{1}{h_3}\dfrac{\partial \phi}{\partial w}\hat{e}_w$ \hfill (5.124)

divergence $\quad \operatorname{div}\vec{F} = \nabla \cdot \vec{F} = \dfrac{1}{h_1 h_2 h_3}\left[\dfrac{\partial (h_2 h_3 F_1)}{\partial u} + \dfrac{\partial (h_1 h_3 F_2)}{\partial v} + \dfrac{\partial (h_1 h_2 F_3)}{\partial w}\right]$ \hfill (5.125)

curl $\quad \nabla \times \vec{F} = \dfrac{1}{h_1 h_2 h_3}\begin{vmatrix} h_1 \hat{e}_u & h_2 \hat{e}_v & h_3 \hat{e}_w \\ \frac{\partial}{\partial u} & \frac{\partial}{\partial v} & \frac{\partial}{\partial w} \\ h_1 F_1 & h_2 F_2 & h_3 F_3 \end{vmatrix}$ \hfill (5.126)

Laplacian $\quad \nabla^2 \phi = \dfrac{1}{h_1 h_2 h_3}\left[\dfrac{\partial}{\partial u}\left(\dfrac{h_2 h_3}{h_1}\dfrac{\partial \phi}{\partial u}\right) = \dfrac{\partial}{\partial v}\left(\dfrac{h_1 h_3}{h_2}\dfrac{\partial \phi}{\partial v}\right) + \dfrac{\partial}{\partial w}\left(\dfrac{h_1 h_2}{h_3}\dfrac{\partial \phi}{\partial w}\right)\right]$ \hfill (5.127)

Expanding the determinant in equation (5.126) illustrates that the curl can also be represented in the form

$$\nabla \times \vec{F} = \frac{1}{h_2 h_3}\left[\frac{\partial (F_3 h_3)}{\partial v} - \frac{\partial (F_2 h_2)}{\partial w}\right]\hat{e}_u + \frac{1}{h_1 h_3}\left[\frac{\partial (F_1 h_1)}{\partial w} - \frac{\partial (F_3 h_3)}{\partial u}\right]\hat{e}_v + \frac{1}{h_1 h_2}\left[\frac{\partial (F_2 h_2)}{\partial u} - \frac{\partial (F_1 h_1)}{\partial v}\right]\hat{e}_w.$$

For example, in cylindrical coordinates $(r,\theta,z)$ one finds $h_1 = 1, h_2 = r, h_3 = 1$ and so the gradient, divergence, curl and Laplacian in cylindrical coordinates have the forms

gradient of $u = u(r,\theta,z)$ is written $\quad \operatorname{grad} u = \nabla u = \dfrac{\partial u}{\partial r}\hat{e}_r + \dfrac{1}{r}\dfrac{\partial u}{\partial \theta}\hat{e}_\theta + \dfrac{\partial u}{\partial z}\hat{e}_z$

divergence of $\vec{V} = \vec{V}(r,\theta,z) = V_r \hat{e}_r + V_\theta \hat{e}_\theta + V_z \hat{e}_z$ is $\quad \operatorname{div}\vec{V} = \nabla \cdot \vec{V} = \dfrac{1}{r}\dfrac{\partial (rV_r)}{\partial r} + \dfrac{1}{r}\dfrac{\partial V_\theta}{\partial \theta} + \dfrac{\partial V_z}{\partial z}$

curl of $\vec{V} = \vec{V}(r,\theta,z) = V_r \hat{e}_r + V_\theta \hat{e}_\theta + V_z \hat{e}_z$ is

$$\nabla \times \vec{V} = \left(\frac{1}{r}\frac{\partial V_z}{\partial \theta} - \frac{\partial V_\theta}{\partial z}\right)\hat{e}_r + \left(\frac{\partial V_r}{\partial z} - \frac{\partial V_z}{\partial r}\right)\hat{e}_\theta + \frac{1}{r}\left(\frac{\partial}{\partial r}(rV_\theta) - \frac{\partial V_r}{\partial \theta}\right)\hat{e}_z$$

Laplacian of $u = u(r,\theta,z)$ is $\quad \Delta u = \nabla^2 u = \dfrac{\partial^2 u}{\partial r^2} + \dfrac{1}{r}\dfrac{\partial u}{\partial r} + \dfrac{1}{r^2}\dfrac{\partial^2 u}{\partial \theta^2} + \dfrac{\partial^2 u}{\partial z^2}$

# Chapter 6
# Ordinary Differential Equations

Treating $u$ as a dependent variable which is a function of one or more independent variables, a differential equation is defined as an equation relating values of the dependent variable $u$, the derivatives of $u$, and possibly some other known functions.

If the derivatives in the above definition are ordinary derivatives, the equation is called an **ordinary differential equation.** If the derivatives in the above definition are partial derivatives, the equation is called a **partial differential equation.** An ordinary differential equation can be expressed in the implicit form

$$F(x, u, \frac{du}{dx}, \ldots, \frac{d^n u}{dx^n}) = 0,$$

where $F$ is a function of the independent variable $x$, the dependent variable $u = u(x)$ and the derivatives of $u$. A partial differential equation can be expressed in the implicit form

$$F(x, y, u, \frac{\partial u}{\partial x}, \frac{\partial u}{\partial y}, \frac{\partial^2 u}{\partial x^2}, \frac{\partial^2 u}{\partial x \partial y}, \frac{\partial^2 u}{\partial y^2}, \ldots) = 0$$

where $x$ and $y$ are the independent variables and $u = u(x, y)$ is the dependent variable with $F$ some prescribed function. If the highest ordered derivative in the differential equation is of the $n$th order, then the equation is called an $n$th order differential equation. It should be noted that identities such as $\frac{du}{dx} = \frac{1}{\frac{dx}{du}}$ are not considered as differential equations. The following is a review of selected topics concerning ordinary differential equations

**Linear Differential Operators**

The operator concept proves to be a convenient notation in representing linear ordinary differential equations. A linear operator is an operator $L(\ )$ which transforms a function $u$ into a function $L(u)$ such that the following properties are satisfied

- $L(cu) = cL(u)$ for all constants $c$
- $L(u_1 + u_2) = L(u_1) + L(u_2)$ for all functions $u_1$ and $u_2$ in the domain of $L(\ )$.

Operators which are not linear are called nonlinear.

An important consequence of the above two properties is that

$$L(0) = 0 \quad \text{and} \quad L(u_1 + u_2 + \cdots + u_n) = L(u_1) + L(u_2) + \cdots + L(u_n)$$

for all functions $u_1, \ldots, u_n$ in the domain of $L(\ )$. An example of a linear differential operator involving the function $y = y(x)$ is

$$L(y) = a_0(x)\frac{d^n y}{dx^n} + a_1(x)\frac{d^{n-1} y}{dx^{n-1}} + \cdots + a_{n-1}(x)\frac{dy}{dx} + a_n(x) y$$

where $a_0(x), \ldots, a_n(x)$ are continuous functions of $x$ over some interval $I$.

Every differential equation involving one dependent variable $y$ and one independent variable, $x$, can be written in the form $L(y) = f(x)$, where $L(\ )$ is either a linear or nonlinear differential operator. If $f(x) = 0$ the differential equation is said to be a **homogeneous differential equation**. If $f(x)$ is different from zero, the equation is said to be a **nonhomogeneous differential equation**. If L is a linear differential operator, the equation is called a **linear differential equation**, otherwise it is called a **nonlinear differential equation**. The operator representation is just one way of writing a differential equation.

The following are some additional forms for representing ordinary differential equations. A differential equation expressed in the functional form

$$F(x, y, \frac{dy}{dx}, \frac{d^2y}{dx^2}, \ldots, \frac{d^ny}{dx^n}) = 0$$

is called an implicit form. If it is possible to solve for the highest derivative in the implicit form, then

$$\frac{d^ny}{dx^n} = f(x, y, \frac{dy}{dx}, \ldots, \frac{d^{n-1}y}{dx^{n-1}}).$$

This form of the differential equation is called an explicit form. Both these forms are relations between an independent variable $x$, a dependent variable $y$, and its derivatives $\frac{dy}{dx}, \ldots, \frac{d^ny}{dx^n}$.

## Solutions to Differential Equations

One type of solution to a given differential equation is the following. The general solution to a differential equation is any family of continuous functions (curves), where for each curve $y = y(x)$ from the family of curves, the curve $y$ together with its derivatives can be substituted into the differential equation to produce an identity which is satisfied for all values of $x$ over some interval. Whenever every member of the solution family satisfies the differential equation over the given interval, the solution family is called a general solution of the differential equation. Solution families can be represented in an explicit, implicit, parametric, integral or a series form. If the differential equation is an $n$th ordered differential equation, then the general solution family will usually have $n$ arbitrary independent constants and the solution family of curves is called an $n$-parameter family of curves.

For example, an equation of the form

$$F(x, y, c_1) = 0, \tag{6.1}$$

where $c_1$ is a constant, represents a family of curves dependent upon the parameter $c_1$. This family of curves is called a one parameter family of curves. Differentiating this equation with respect to $x$ gives

$$\frac{\partial F}{\partial x} + \frac{\partial F}{\partial y}\frac{dy}{dx} = g(x, y, y', c_1) = 0 \tag{6.2}$$

where $g$ is some new function of the variables $x$, $y$, $y'$ and the parameter $c_1$. If $\frac{\partial F}{\partial y}$ is different from zero and if somehow it is possible to eliminate the parameter $c_1$ between the equations (6.1) and (6.2), then there results a first order differential equation

$$y' = f(x,y). \tag{6.3}$$

The differential equation (6.3) is the differential equation of lowest order satisfied by the family of curves (6.1) and the family of curves (6.1) is called the general solution of the differential equation (6.3). The essential point to note is that the parameter $c_1$ of the family of curves does not appear in the differential equation of the family of curves. Also observe every curve in the family of curves satisfies the differential equation.

The above concept can be generalized to families of curves having the form

$$f(x, y, c_1, c_2, \ldots, c_n) = 0 \tag{6.4}$$

with $n$-independent parameters. This equation is called a $n$-parameter family of curves with $c_1, \ldots, c_n$ as the parameters. If one calculates $n$ successive derivatives of equation (6.4) there results n derived equations involving $x$, $y$, the derivatives $y'$, $y''$, ..., $y^{(n)}$, and the constants $c_1, c_2, \ldots, c_n$. If by some algebraic process it is possible to eliminate the $n$ independent constants from these $(n+1)$ equations, there results the differential equation associated with the family of curves defined by equation (6.4). The derived equation is an $n$th order differential equation and can be expressed in the implicit form

$$F(x, y, y', y'', \ldots, y^{(n)}) = 0 \tag{6.5}$$

The equation (6.4) is termed the general solution of the $n$th order differential equation (6.5). Observe that the general solution to a $n$th order differential equation contains $n$ arbitrary constants.

A solution of the first order differential equations $\frac{dy}{dx} = f(x,y)$ which passes through a specified point $(x_0, \alpha_0)$ is said to be a solution of the initial value problem (IVP)

$$\frac{dy}{dx} = f(x,y), \quad y(x_0) = \alpha_0 \tag{6.6}$$

Similarly, a solution of the $n$th order differential $\frac{d^n y}{dx^n} = f(x, y, y', \ldots, y^{(n-1)})$ which satisfies initial conditions at the point $x_0$ of the form

$$y(x_0) = \alpha_0, \ y'(x_0) = \alpha_1, \ y''(x_0) = \alpha_2, \cdots, y^{(n-1)}(x_0) = \alpha_{n-1}$$

where $\alpha_0, \alpha_1, \ldots, \alpha_{n-1}$ are constants, is said to be a solution of the initial value problem (IVP)

$$\begin{aligned}\frac{d^n y}{dx^n} &= f(x, y, y', \ldots, y^{(n-1)}), \\ y(x_0) &= \alpha_0, \ y'(x_0) = \alpha_1, \ y''(x_0) = \alpha_2, \cdots, y^{(n-1)}(x_0) = \alpha_{n-1}\end{aligned} \tag{6.7}$$

The solution to the above initial value problems represents the processes of selecting one curve, from the infinite family of solutions, which satisfies the assign initial conditions at the single point $x_0$. This requires that specific values be assigned to the $n$-parameters in a general solution family. Note that the values assigned to constants or variables in a family of solution curves to a differential equation are not restricted to real numbers. Many times it is advantageous to include complex numbers in the representation of a solution.

### Existence and Uniqueness of Solutions

Consider the problem of determining a solution family to a first order differential equation of the form $\frac{dy}{dx} = f(x,y)$ where $f(x,y)$ is a specified function. It would be nice to know (a) a solution family can be found where all curves in the family are continuous functions with derivatives that are also continuous (Existence of a solution) and (b) there is one and only one member from the solution family which satisfies the prescribed initial conditions (Uniqueness of a solution).

Under certain rather general conditions a solution to the initial value problem

$$\frac{dy}{dx} = f(x,y), \qquad y(x_0) = y_0$$

can be shown to exist and be unique in a region $R$ which contains the initial point $(x_0, y_0)$. The region $R$, where a solution exists, can be represented by a rectangle centered at $(x_0, y_0)$ and defined by

$$R = \{\,(x,y)\,|\,|x - x_0| < a,\ |y - y_0| < b\,\},$$

for some positive values of the constants $a$ and $b$. Usually, the region $R$ is confined to a small rectangle about the initial point $(x_0, y_0)$, and in this case, the existence and uniqueness of solutions to the initial value problem is restricted to this local region. Within this local region the following terminology is applied. The function $f(x,y)$ on the right-hand side of the first order differential equation is said to be continuous on $R$ if $f(x,y)$ is continuous at all points of $R$. If there are points $(x^*, y^*)$, where $f(x^*, y^*)$ is not defined, then such points are called singular points, while all other points of $R$ are called regular points.

THEOREM. (Cauchy[1] existence theorem.) When $f(x,y)$ and $\frac{\partial f}{\partial y}(x,y)$ are continuous functions over a region $R = \{\,(x,y)\,|\,|x - x_0| < a,\ |y - y_0| < b\,\}$ about the initial point $(x_0, y_0)$, then there exists a unique function $y(x)$ continuous on the interval $|x - x_0| < a$ such that the curve $(x, y(x))$ lies in $R$ and satisfies the initial value problem

$$\frac{dy}{dx} = f(x,y), \quad y(x_0) = y_0.$$

The Cauchy theorem gives conditions for determining if a solution to a differential equation exists and is unique.

---

[1] Augustin-Louis Cauchy (1789–1857) French Mathematician

(1) For some problems, the interval over which the solution exists is some small interval about the initial point. In special problems the region where the solution exists is determined by the distance from the initial point $(x_0, y_0)$ to the nearest singular point $(x^*, y^*)$, and for some other problems the interval of existence is determined by the distance from the initial point to the boundary of the region $R$.

(2) The Cauchy theorem implies that if a solution exists and is continuous, when the region $R$ is the entire $x$-$y$ plane, then the solution must remain bounded over any finite interval.

(3) The conditions of continuity of $f(x, y)$ and $\frac{\partial f}{\partial y}$ are sufficient conditions for the existence and uniqueness of a solution. If these conditions are not satisfied, then the initial value problem might still have a unique solution or it may have more than one solution or a solution might not even exist.

(4) In the special case where $f(x, y) = f(x)$ is a continuous function of $x$ alone, the fundamental theorem of integral calculus insures us that the initial value problem has a solution in the form of a definite integral.

(5) Existence and uniqueness theorems are not restricted to first-order equations. For a linear differential equation of order $n$ having the form

$$L(y) = a_0(x)\frac{d^n y}{dx^n} + a_1(x)\frac{d^{n-1} y}{dx^{n-1}} + \cdots + a_{n-1}(x)\frac{dy}{dx} + a_n(x)y = f(x),$$

there is a theorem concerning the existence and uniqueness of solutions. In essence this theorem states that if the coefficients $a_0(x), a_1(x), \ldots, a_n(x)$ and the right-hand side $f(x)$ are all continuous functions on some interval $I = \{x \mid a \leq x \leq b\}$, with $a_0(x) \neq 0$ for all $x \in I$, then the initial value problem to solve $L(y) = f(x)$ subject to the conditions

$$y(x_0) = \alpha_0, \quad y'(x_0) = \alpha_1, \quad \ldots, \quad y^{(n-1)}(x_0) = \alpha_{n-1}$$

has a unique solution. If for some value $x^* \in I$, one finds that $a_0(x^*) = 0$, then $x^*$ is a singular point and in this case the initial value problem may or may not have a unique solution over the interval I.

## Solutions Containing a Complex Number

Complex numbers are written in the form $z = x + iy$ where $i$ is called an imaginary unit with the property $i^2 = -1$. Complex numbers often are used in the representation of differential equations and their solutions. In many application areas, it is desirable to have complex solutions to differential equations. For example, complex solutions are used in elementary circuit courses taken by electrical engineering students. (Note in electrical engineering the symbol $i$ is reserved to represent current, therefore many electrical engineering texts denote complex numbers using the notation $z = x + jy$ where $j^2 = -1$.) The following is an example of a differential equation where complex numbers are used.

## Example 6-1.

The Cauchy theorem for existence and uniqueness of solutions to ordinary differential equations can be used to prove various identities containing complex numbers. The following example illustrates this idea. Consider the solution of the initial value problem

$$\frac{dy}{dx} = iy, \quad y(0) = 1, \tag{6.8}$$

where $i^2 = -1$. It is easy to verify directly by differentiation that both the functions

$$y_1(x) = e^{ix} \quad \text{and} \quad y_2(x) = \cos x + i \sin x$$

are solutions of the initial value problem (6.8). The Cauchy theorem states that the solution of the initial value problem must be unique. Consequently, the solutions given above must be different forms of the same solution and so the conclusion $y_1(x) = y_2(x)$ for all $x$. This produces the identity

$$e^{ix} = \cos x + i \sin x \tag{6.9}$$

known as Euler's[2] identity. It is a very important identity that is used quite often in various fields of mathematics.

■

### Differential Equations Easily Solved

The following is a review of some methods for finding solutions to certain easily recognized types of differential equations. Many differential equations can be solved by using the methods studied in calculus. Direct integration of an equation is perhaps the easiest solution technique to recognize.

#### Integration Methods

Nonhomogeneous differential equations of the form $\frac{d^n y}{dx^n} = f(x)$ or homogeneous differential equations of the form $\frac{d^n y}{dx^n} = 0$ can be solved by performing n-successive integrations. For example,

If $\dfrac{dy}{dx} = f(x)$, then $y = \displaystyle\int f(x)\,dx + c_1$

If $\dfrac{d^2 y}{dx^2} = f(x)$, then $\dfrac{dy}{dx} = \displaystyle\int f(x)\,dx + c_1$, and $y = \displaystyle\int \left[\int f(x)\,dx\right] dx + c_1 x + c_2$

If $\dfrac{d^3 y}{dx^3} = f(x)$, then $y = \displaystyle\int \left[\int \left[\int f(x)\,dx\right] dx\right] dx + c_1 \frac{x^2}{2} + c_2 x + c_3$

Note the order of the differential equation indicates how many integrations must be performed, and consequently, the order of the equation tells us the number of arbitrary constants appearing in the general solution. That is, after each integration, add a constant

---
[2] Leonard Euler (1707–1783), Swiss Mathematician.

of integration. Each of the above examples illustrates that a family of curves, called the solution family, is obtained by direct integration methods.

A solution to the $n$th order differential equation $\frac{d^n y}{dx^n} = f(x)$ can be constructed by performing $n$-integrations to obtain

$$y(x) = c_0 + c_1 x + c_2 \frac{x^2}{2!} + \cdots + c_{n-1} \frac{x^{n-1}}{(n-1)!} + \int_0^x \int_0^{x_1} \int_0^{x_2} \cdots \int_0^{x_{n-1}} f(x_n) \, dx_n \, dx_{n-1} \cdots dx_3 \, dx_2 \, dx_1$$

This solution satisfies the $n$ initial conditions

$$y(0) = c_0, \quad y'(0) = c_1, \quad y''(0) = c_2, \ldots, y^{(n-1)}(0) = c_{n-1} \qquad (6.10)$$

Related to the above solution is the function

$$F(x) = \int_0^x \frac{(x-t)^{n-1}}{(n-1)!} f(t) \, dt$$

By the Leibniz' rule for differentiating an integral, calculate the derivatives

$$F'(x), \quad F''(x), \ldots, F^{(n-1)}(x), \quad F^{(n)}(x)$$

to verify that the function

$$Y(x) = c_0 + c_1 x + c_2 \frac{x^2}{2!} + \cdots + c_{n-1} \frac{x^{n-1}}{(n-1)!} + \int_0^x \frac{(x-t)^{n-1}}{(n-1)!} f(t) \, dt$$

is a solution to the $n$th order differential equation $\frac{d^n Y}{dx^n} = f(x)$ and that this solution satisfies the initial conditions given in equation (6.10). Because the solution must be unique, the two solutions must equal one another or $Y(x) = y(x)$. That is, $y(x)$ and $Y(x)$ are just two different ways of representing the same solution.

**Separation of Variables**

Any first-order differential equation which can be reduced algebraically to the form

$$F(y) \, dy = G(x) \, dx \qquad (6.11)$$

is called a separable equation. Here the dependent and independent variables have been separated. An integration of both sides of equation (6.11) produces

$$\int F(y) \, dy = \int G(x) \, dx + c, \qquad (6.12)$$

where $c$ is a constant of integration. If in equation (6.12), the integration cannot be performed, as sometimes happens, then as an alternative measure form an initial value problem with the condition $y(x_0) = y_0$. Then the solution can be expressed in integral form as

$$\int_{y_0}^y F(s) \, ds = \int_{x_0}^x G(s) \, ds \qquad (6.13)$$

provided the integrands are defined over the intervals of integration. The definite integration form given by equation (6.13) can often be used even when the integrals are known.

Any differential equations having the form

$$F(x)G(y)\,dx + f(x)g(y)\,dy = 0 \implies \int \frac{F(x)}{f(x)}\,dx + \int \frac{g(y)}{G(y)}\,dy = C$$

then the variables can be separated and the resulting equation integrated. Differential equations having the form

$$\frac{dy}{dx} = \frac{y}{x} + g(x)f\left(\frac{y}{x}\right)y^m \implies \frac{dv}{f(v)v^m} = g(x)x^{m-1}\,dx$$

can be reduced to a differential equation where the variables are separable by first making the substitution $y = vx$, where $v = v(x)$ is a function to be determined.

Any function $f(x, y)$ which satisfies the property $f(tx, ty) = t^n f(x, y)$, for all values of $x$ and $y$, is called a homogeneous function of degree $n$. Differential equations of the form

$$M(x, y)\,dx + N(x, y)\,dy = 0,$$

where both $M(x, y)$ and $N(x, y)$ are homogeneous functions of the same degree, can be reduced to a form where the variables are separable by making the substitution $y = vx$, where $v = v(x)$ is a function to be determined.

First order differential equations of the form $\frac{dy}{dx} = f\left(\frac{y}{x}\right)$ are called homogeneous equations and using the substitution $y = vx$ the equation becomes separable so that an integration gives the solution

$$\ln x = \int \frac{dv}{f(v) - v} + C$$

In the special case where $f(v) = v$, the solution is found to be the family of lines $y = Cx$, where $C$ is a constant. All members of the solution family pass through the origin.

Many special differential equations can be reduced to a form where the variables are separable by making an appropriate substitution. For example, the differential equation

$$xf(xy)\,dx + yg(xy)\,dy = 0$$

can be transformed into a separable form by making the substitution $xy = v$ with differential $x\,dy + y\,dx = dv$.

## Exact Differential Equations

If a function $\phi = \phi(x,y)$ is continuous and differentiable over some region $R$, then the exact or total differential of this function is given by

$$d\phi = \frac{\partial \phi}{\partial x} dx + \frac{\partial \phi}{\partial y} dy. \qquad (6.14)$$

The equation $d\phi = 0$ is called an exact differential equation and the family of curves $\phi(x,y) = c$, with parameter $c$, represents the solution family of this exact differential equation. The family of curves $\phi(x,y) = c$ are called level curves associated with the solution family.

A differential equation of the form

$$M(x,y)\,dx + N(x,y)\,dy = 0 \qquad (6.15)$$

is said to be an exact differential equation if there exists a scalar function $\phi(x,y)$ such that

$$\frac{\partial \phi}{\partial x} = M(x,y) \quad \text{and} \quad \frac{\partial \phi}{\partial y} = N(x,y) \qquad (6.16)$$

If $\phi(x,y)$ has continuous second order partial derivatives, then the mixed partial derivative $\frac{\partial^2 \phi}{\partial x\,\partial y} = \frac{\partial M}{\partial y}$ must equal $\frac{\partial^2 \phi}{\partial y\,\partial x} = \frac{\partial N}{\partial x}$. Note that differentiating the equations (6.16) with respect to $x$ and $y$, respectively, one finds that the requirement

$$\frac{\partial M}{\partial y} = \frac{\partial N}{\partial x} \quad \Longrightarrow \quad \frac{\partial^2 \phi}{\partial x\,\partial y} = \frac{\partial^2 \phi}{\partial y\,\partial x} \qquad (6.17)$$

is a necessary condition for the differential equation (6.15) to be exact. If equation (6.15) is an exact differential equation, then it possess a family of solution curves having the form

$$\phi = \phi(x,y) = \int M(x,y)\,\partial x + \int \left( N(x,y) - \frac{\partial}{\partial y} \int M(x,y)\,\partial x \right) dy = c = \text{constant} \qquad (6.18)$$

where $\partial x$ is an integration with respect to $x$ holding $y$ constant. That is, if the differential equation (6.15) is an exact differential equation, then there exists a function $\phi = \phi(x,y)$ such that

$$\frac{\partial \phi}{\partial x} = M(x,y) \quad \text{and} \quad \frac{\partial \phi}{\partial y} = N(x,y) \qquad (6.19)$$

with $\frac{\partial M}{\partial y} = \frac{\partial N}{\partial y}$. Integrating the first equation of (6.19) with respect to $x$ says the solution family must have the form $\phi = \int M(x,y)\,\partial x + f(y) = c$ where $f(y)$ is to be determined. Using the second equation from (6.19) produces

$$\frac{\partial \phi}{\partial y} = \frac{\partial}{\partial y} \int M(x,y)\,\partial x + \frac{df(y)}{dy} = N(x,y) \qquad (6.20)$$

which implies $f(y) = \int \left( N(x,y) - \frac{\partial}{\partial y} \int M(x,y)\,\partial x \right) dy$ giving the solution family (6.18).

**Integrating Factors $\mu(x, y)$**

If the differential equation

$$m(x, y)\, dx + n(x, y)\, dy = 0$$

is not an exact differential equation and there exists a function $\mu = \mu(x, y)$ such that

$$\mu\, m(x, y)\, dx + \mu\, n(x, y)\, dy = 0$$

is an exact differential equation, then the function $\mu$ is called an integrating factor of the differential equation.

It can be shown that if the ratio $\dfrac{\frac{\partial m}{\partial y} - \frac{\partial n}{\partial x}}{n}$ is a function of $x$ only, then there exists an integrating factor having the form $\mu = \mu(x)$ which is obtained by solving the differential equation

$$\frac{d\mu}{\mu} = \frac{\frac{\partial m}{\partial y} - \frac{\partial n}{\partial x}}{n}\, dx$$

Similarly, if the ratio $\dfrac{\frac{\partial n}{\partial x} - \frac{\partial m}{\partial y}}{m}$ is a function of $y$ only, then there exists an integrating factor $\mu = \mu(y)$ which is obtained by solving the differential equation

$$\frac{d\mu}{\mu} = \frac{\frac{\partial n}{\partial x} - \frac{\partial m}{\partial y}}{m}\, dy$$

Construction of integrating factors in other special cases can be found in many texts on differential equations.

## Linear First Order Differential Equations

Differential equations having the form

$$\frac{dy}{dx} + P(x) y = Q(x) \tag{6.21}$$

have the integrating factor $\mu = \exp\left(\int P(x)\, dx\right)$. Multiplication of equation (6.21) by the integrating factor $\mu$ will always produce the exact differential equation

$$d[\mu y] = Q(x)\, \mu\, dx$$

which can be integrated to produce the solution

$$y = \frac{1}{\mu} \int Q(x)\, \mu\, dx + \frac{C}{\mu}, \qquad \mu = \exp\left(\int P(x)\, dx\right) = e^{\int P(x)\, dx}$$

where $C$ is a constant of integration.

**Bernoulli's Equation**

Any equation of the form
$$\frac{dy}{dx} + P(x)y = Q(x)y^n, \qquad n \neq 0, 1 \qquad (6.22)$$
is called a Bernoulli[3] equation. The substitution
$$u = y^{1-n} \quad \text{with derivative} \quad \frac{du}{dx} = (1-n)y^{-n}\frac{dy}{dx}$$
reduces equation (6.22) to the form
$$\frac{1}{1-n}\frac{du}{dx} + P(x)u = Q(x), \qquad n \neq 1$$
which is a linear first-order differential equation in the variable $u$ with solution
$$u \exp\left[(1-n)\int P(x)\,dx\right] = (1-n)\int Q(x)\exp\left[(1-n)\int P(x)\,dx\right]\,dx + C$$
where $u = y^{1-n}$ with $n \neq 1$. The case $n = 1$ gives a linear first order equation for $y$ with solution
$$y = C^* \exp\left[(Q(x) - P(x))\,dx\right]$$
where $C^*$ is a constant.

**Shifting of axes**

A differential equation having the form
$$\frac{dy}{dx} = f\left(\frac{a_1 x + b_1 y + c_1}{a_2 x + b_2 y + c_2}\right), \qquad (6.23)$$
where $a_i$, $b_i$, $c_i$, $i = 1$, and 2, are constants, may be solved by a translation of axes. By making the change of variables $y = Y + k$ and $x = X + h$, where $k$ and $h$ are constants to be determined, the differential equation can be expressed as
$$\frac{dY}{dX} = \frac{dy}{dx}\frac{dx}{dX} = f\left[\frac{a_1(X+h) + b_1(Y+k) + c_1}{a_2(X+h) + b_2(Y+k) + c_2}\right]. \qquad (6.24)$$
By selecting $h$ and $k$ such that
$$a_1 h + b_1 k + c_1 = 0 \quad \text{and} \quad a_2 h + b_2 k + c_2 = 0 \qquad (6.25)$$
equation (6.24) is reduced to the form
$$\frac{dY}{dX} = f\left(\frac{a_1 X + b_1 Y}{a_2 X + b_2 Y}\right). \qquad (6.26))$$
Differential equations having this form are homogeneous equations that can be further simplified by making the substitution $Y = VX$.

The conditions (6.25) can be thought of as representing the point of intersection $(h, k)$ of the straight lines
$$a_1 x + b_1 y + c_1 = 0 \quad \text{and} \quad a_2 x + b_2 y + c_2 = 0. \qquad (6.27)$$
If these lines are parallel, then of course there cannot be a solution for $(h, k)$. However, if the lines are parallel, equation (6.23) can be simplified by making the substitution $u = a_1 x + b_1 y$. This produces the equation $a_2 x + b_2 y = ku$ for some constant $k$.

---

[3] James Bernoulli (1654–1705), Swiss mathematician.

## Dependent Variable Absent

Differential equations of the form $F(x, y', y'') = 0$ are second-order differential equations with the dependent variable $y$ absent. By introducing the change of variables

$$v = y' = \frac{dy}{dx} \quad \text{and} \quad \frac{dv}{dx} = y'' = \frac{d^2y}{dx^2} \tag{6.28}$$

the above equation becomes $F(x, v, v') = 0$, which is a first-order equation with respect to the variable $v = y'$. In general, higher order equations with the dependent variable absent can be reduced to an equation of one lower order in the new variable $v = y'$.

## Independent Variable Absent

Differential equations of the form $F(y, y', y'') = 0$ are second-order equations which have the independent variable $x$ absent. Introduce a new variable $v$ and assume $v = v(y)$ is a function of $y$ and make the change of variable

$$\frac{dy}{dx} = y' = v = v(y). \tag{6.29}$$

The chain rule for differentiating a composite function produces the derivatives

$$\frac{d^2y}{dx^2} = y'' = \frac{d}{dx}(y') = \frac{d}{dx}(v) = \frac{dv}{dy}\frac{dy}{dx} = v\frac{dv}{dy}$$

and the given differential equation can be written in the form $F(y, v, v\frac{dv}{dy}) = 0$ which is a first-order equation with dependent variable $v$ and independent variable $y$. If one can solve this first-order equation for $v = v(y)$, then it is possible to separate the variables in equation (6.29) and then an integration will produce the solution $y$ as a function of $x$.

## Parametric Solutions to Differential Equations

The parametric equations

$$x = x(t), \quad \text{and} \quad y = y(t), \quad a \le t \le b, \tag{6.30}$$

with parameter $t$, represent a curve in the $x, y$ plane. If the parameter $t$ is eliminated from these equations, one finds $y$ is a function of $x$. All functions $y = y(x)$ can be represented in some type of parametric form. The parametric representation of a function is not unique and so one must be careful to define the domain of the parameter and the range of the function, as this dictates the portion of the curve $y = y(x)$ represented by the parametric equations.

To obtain parametric solutions to first-order differential equations make note of the following. If a first-order differential equation is represented in the implicit form $F(x, y, y') = 0$, then make the change of variables $y' = \frac{dy}{dx} = p$, to produce the new equation

$$F(x, y, p) = 0. \tag{6.31}$$

Representing first order differential equations in this form presents the following possibilities.

**Case I** Solve equation (6.31) for the variable $y$, to obtain $y$ as a function of $x$ and $p$, say $y = H(x,p)$. Differentiating this equation with respect to $x$ gives

$$\frac{dy}{dx} = y' = p = \frac{\partial H}{\partial x} + \frac{\partial H}{\partial p}\frac{dp}{dx},$$

which is a first-order differential equation in the variable $p = p(x)$. If it is possible to solve this first-order differential equation to obtain $p$, then there results the parametric set of equations

$$y = H(x,p), \quad p = p(x)$$

which represents the solution of the differential equation (6.31) in a parametric form with $x$ as a parameter.

**Case II** If it is possible to solve equation (6.31) for the variable $x$ as a function of $y$ and $p$, say $x = G(y,p)$, then differentiating this expression with respect to $y$ gives

$$\frac{dx}{dy} = \frac{1}{p} = \frac{\partial G}{\partial y} + \frac{\partial G}{\partial p}\frac{dp}{dy},$$

which is a first-order differential equation in the variable $p = p(y)$. If it is possible to solve this first-order differential equation for $p$ as a function of $y$, then the solution can be represented in the parametric form

$$x = G(y,p), \quad p = p(y).$$

This parametric representation of the solution has $y$ for the parameter.

**Case III** If it is possible to solve equation (6.31) for the variable $y' = p$, there results the explicit representation of the first-order differential equation as $y' = f(x,y)$. The equation is now in a form where the previous solution methods can be investigated.

Some differential equations can be solved using trial and error substitutions to construct functions easily integrated, with the solution in a parametric form. That is, select a trial function and see if by substitution of this function into the differential equation, there results integrals which are easy to obtain. For example, to obtain a parametric solution to differential equations having the form $f(y, y') = 0$, where the independent variable is absent, make the substitution $y' = G(t)$, where $G(t)$ is a trial solution. Upon substituting this trial solution into the differential equation there results $f[y, G(t)] = 0$. If the partial derivative of $f$ with respect to $y$ is different from zero, then theoretically it is possible to solve this equation for $y$ as a function of $t$. If $y = y(t)$, then by using the relation

$$dx = \frac{dy}{y'} = \frac{\frac{dy}{dt}\,dt}{G(t)}, \tag{6.32}$$

there results a separable equation which is our test function. If it is possible to integrate this test function, one obtains $x$ as a function of $t$ and consequently there results the parametric solution

$$x = x(t) = c + \int \frac{\frac{dy}{dt} dt}{G(t)} \tag{6.34}$$
$$y = y(t) \quad \text{(where } y(t) \text{ is obtained from the solution of } f(y, G(t)) = 0 \text{ )}$$

In a similar manner, it is possible to investigate differential equations having the form $f(x, y') = 0$, where the dependent variable $y$ is absent, by substituting in a trial solution having the form $y' = \frac{dy}{dx} = G(t)$. This gives the equation $f(x, G(t)) = 0$, which determines the variable $x$ as a function of $t$, say $x = x(t)$. Using the chain rule for differentiation and the relation

$$\frac{dy}{dt} = \frac{dy}{dx}\frac{dx}{dt} = G(t)x'(t),$$

one finds the integral expression

$$y = c + \int G(t)\, x'(t)\, dt.$$

This produces a parametric solution having the form

$$x = x(t), \quad \text{(from } f(x, G(t)) = 0 \text{ )} \quad \text{and} \quad y = y(t) = c + \int G(t)\, x'(t)\, dt$$

### Linear nth Order Differential Equations

Let $L(y)$ denote the $n$th order linear differential equation

$$L(y) = a_0 D^n y + a_1 D^{n-1} y + \cdots + a_{n-2} D^2 y + a_{n-1} Dy + a_n y = 0, \qquad D = \frac{d}{dx} \tag{6.34}$$

where $a_0, a_1, \ldots, a_{n-1}, a_n$ are coefficients which are functions of $x$ or constants and $a_0 \neq 0$. Recall $L(\ )$ is a linear operator if the conditions

$$L(cy) = cL(y) \quad \text{with } c \text{ a constant.}$$
$$L(y_1 + y_2) = L(y_1) + L(y_2) \tag{6.35}$$

are satisfied.

In order to develop solutions to $n$th order linear homogeneous and nonhomogeneous differential equations having the forms

$$L(y) = 0, \qquad x \in I = \{x \,|\, a \le x \le b\} \quad \text{and} \quad L(y) = r(x), \quad x \in I$$

where a solution $y$ is subject to initial conditions

$$y(a) = \alpha_0, \quad y'(a) = \alpha_1, \ldots, y^{(n-1)}(a) = \alpha_{n-1}$$

at some point $x = a$ belonging to the interval $I$, the following two theorems are employed.

**THEOREM 1.** *If $y_1(x)$ is a solution of a linear homogeneous differential equation $L(y) = 0$, then $y = cy_1(x)$ is also a solution for all constants $c$.*

**Proof:** By hypothesis $L(y_1) = 0$, and hence for $y = cy_1$, where $c$ is constant, there results

$$L(y) = L(cy_1) = cL(y_1) = c(0) = 0,$$

and thus $y = cy_1$ is also a solution. Theorem 1 is a special case of the following more general theorem.

**THEOREM 2.** *(Superposition principle) If $y_1, y_2, \ldots, y_m$ are each solutions of an nth-order linear homogeneous differential equation $L(y) = 0$, then any linear combination of these solutions*

$$c_1 y_1 + c_2 y_2 + \cdots + c_m y_m,$$

*where $c_1, c_2, \ldots, c_m$ are constants, is also a solution.*

**Proof:** By hypothesis $L(y_i) = 0$ for $i = 1, 2, \ldots, m$. Form the linear combination

$$y = \sum_{i=1}^{m} c_i y_i = c_1 y_1 + \cdots + c_m y_m,$$

where $c_1, c_2, \ldots, c_m$ are constants. Then by the linearity properties of $L(\ )$,

$$\begin{aligned} L(y) &= L(c_1 y_1 + c_2 y_2 + \cdots + c_m y_m) \\ &= L(c_1 y_1) + L(c_2 y_2) + \cdots + L(c_m y_m) \\ &= c_1 L(y_1) + c_2 L(y_2) + \cdots + c_m L(y_m) \\ &= c_1(0) + c_2(0) + \cdots + c_m(0) = 0 \end{aligned}$$

and hence $y$ is a solution.

**Definition: (Linear dependence)** Given a set of functions $\{y_1(x), y_2(x), \ldots, y_m(x)\}$ each well defined over some interval $I = \{x \mid a \leq x \leq b\}$. This set of functions is said to be linearly dependent over the interval $I$ if there exists constants $k_1, k_2, \ldots, k_m$ not all zero such that the relation

$$k_1 y_1(x) + k_2 y_2(x) + \cdots + k_m y_m(x) = 0 \tag{6.36}$$

holds for all $x \in I$.

If for some $i$ the condition $k_i \neq 0$, then from the relation (6.36) it is possible to solve for $y_i$ as

$$y_i(x) = -\frac{k_1}{k_i} y_1 - \frac{k_2}{k_i} y_2 - \cdots - \frac{k_m}{k_i} y_m.$$

This equation shows that $y_i$ can be expressed as some linear combination of the remaining $y_k$, $k \neq i$ functions. That is, it can be made dependent upon the remaining functions of the set.

**Definition: (Linear independence)** *If the only constants $k_i$, $i = 1, 2, \ldots, m$, for which relation (6.36) holds are the constants $k_1 = 0$, $k_2 = 0, \ldots, k_m = 0$, then the set of functions $\{y_1(x), y_2(x), \ldots, y_m(x)\}$ is said to be a linearly independent set of functions on I.*

## Solutions to $n$th Order Linear Homogeneous Differential Equations

Consider the $n$th order linear homogeneous differential equation

$$L(y) = a_0 D^n y + a_1 D^{n-1} y + \cdots + a_{n-2} D^2 y + a_{n-1} Dy + a_n y = 0, \qquad D = \frac{d}{dx} \qquad (6.37)$$

where $a_0, a_1, \ldots, a_{n-1}, a_n$ are coefficients of the differential equation which are functions of $x$ or constants and $a_0 \neq 0$. If it is possible to obtain a set of $n$ solutions $\{y_1, \ldots, y_n\}$ from which all other solutions of the differential equation can be constructed, then the set of $n$ solutions is called a fundamental system, basis, or fundamental set of solutions to the differential equation (6.37).

**Definition: (Fundamental system)** *A fundamental system or basis for the solution of an nth order linear homogeneous differential equation $L(y) = 0$ is a set of $n$ linearly independent solutions of the differential equation over an interval I.*

A general solution of a $n$th-order linear homogeneous differential equation must contain $n$ distinct arbitrary constants. By obtaining a fundamental system of $n$ independent solutions $\{y_1, y_2, \ldots, y_n\}$ associated with the $n$th-order linear homogeneous differential equation $L(y) = 0$, the theorem 2 can be used to construct the general solution. That is, the general solution can be represented as a linear combination of functions from the fundamental set of solutions and written as

$$y = y(x) = c_1 y_1 + c_2 y_2 + \cdots + c_n y_n, \qquad (6.38)$$

where $c_1, \ldots, c_n$ are arbitrary constants. This solution contains $n$ arbitrary constants and hence it can be called the general solution.

The coefficients of $y$ and the derivatives of $y$ occurring in the given differential equation $L(y) = 0$ play an important part in determining the type of solution that exists. When each of these coefficients are continuous over the interval $I$, it is possible to verify that the constants in the general solution, given by equation (6.38) can be chosen such that the initial value problem has a unique solution. That is, the solution of the initial value problem to solve $L(y) = 0$ subject to the $n$ initial conditions

$$y(x_0) = \alpha_0, \quad y'(x_0) = \alpha_1, \ldots, y^{(n-1)}(x_0) = \alpha_{n-1}, \qquad (6.39)$$

with $\alpha_0, \alpha_1, \ldots, \alpha_{n-1}$ constants, has a unique solution over I. Consequently, one and only one solution function, which satisfies the prescribed initial conditions, can be constructed.

Note that after one obtains a fundamental set of solutions $\{y_1,\ldots,y_n\}$, then these functions have the property that each function from the set must satisfy the differential equation, so that $L(y_i) = 0$, for $i = 1, 2, \ldots, n$. Further, this fundamental set must contain $n$ linearly independent functions defined and continuous over the given interval I, otherwise the linear combination given by equation (6.36) would not contain $n$ arbitrary distinct constants. Therefore, it would be nice to have a test to determine if our set of functions is linearly dependent or linearly independent. Such a test involves the Wronskian[4] determinant.

**Wronskian Determinant**

Assume it is possible to find $n$ solutions $\{y_1,\ldots,y_n\}$ of the linear homogeneous differential equation

$$L(y) = a_0 D^n y + a_1 D^{n-1} y + \cdots + a_{n-2} D^2 y + a_{n-1} Dy + a_n y = 0, \qquad D = \frac{d}{dx} \qquad (6.40)$$

If these $n$ solutions are not linearly independent, then there exists nonzero constants $K_1, K_2, \ldots, K_n$ such that

$$G(x) = K_1 y_1 + K_2 y_2 + \cdots + K_n y_n = 0 \qquad (6.41)$$

for all $x \in I$. Differentiate this relation $(n\text{-}1)$ times to obtain the equations

$$G'(x) = K_1 y_1' + K_2 y_2' + \cdots + K_n y_n' = 0$$
$$G''(x) = K_1 y_1'' + k_2 y_2'' + \cdots + K_n y_n'' = 0$$
$$\vdots \qquad\qquad\qquad\qquad\qquad\qquad (6.42)$$
$$G^{(n-1)}(x) = K_1 y_1^{(n-1)} + K_2 y_2^{(n-1)} + \cdots + k_n y_n^{(n-1)} = 0.$$

Here the $n$th derivative is some linear combination of the lower $(n\text{-}1)$ derivatives since by hypothesis each $y_i$, $i = 1, \ldots, n$ satisfies the $n$th-order linear differential equation $L(y_i) = 0$, $i = 1, 2, \ldots, n$. Equations (6.42) and (6.41) are $n$ equations from which one must determine the $n$ constants $K_1, K_2, \ldots, K_n$, not all zero. By Cramer's[5] rule if these equations are to have a nonzero solution, then the determinant

$$W = \begin{vmatrix} y_1 & y_2 & y_3 & \cdots & y_n \\ y_1' & y_2' & y_3' & \cdots & y_n' \\ y_1'' & y_2'' & y_3'' & \cdots & y_n'' \\ \vdots & \vdots & \vdots & \ddots & \vdots \\ y_1^{(n-1)} & y_2^{(n-1)} & y_3^{(n-1)} & \cdots & y_n^{(n-1)} \end{vmatrix}$$

must equal zero. The determinant $W$ is called the Wronskian associated with the solution set $\{y_1, y_2, \ldots, y_n\}$ defined on I. A necessary condition for the set $\{y_1, y_2, \ldots, y_n\}$ to be a linearly dependent set is for the Wronskian to equal zero for all $x$. If the Wronskian determinant is not identically zero for all $x \in I$, then the set of functions is said to be linearly independent on I.

---
[4] Hoëne Wronski (1778-1853), Polish mathematician.
[5] Gabriel Cramer (1704-1752), Swiss mathematician.

**Example 6-2.**
If two functions $y_1(x)$ and $y_2(x)$ are linearly dependent, then one function must be some constant multiple of the other function. One can show this directly from the definition. Another way to show this relation is to examine the Wronskian relation. The assumption that $y_1$ and $y_2$ are dependent implies

$$W = \begin{vmatrix} y_1(x) & y_2(x) \\ y_1'(x) & y_2'(x) \end{vmatrix} = 0$$

for all $x$. By evaluating the Wronskian determinant there results the relation

$$\frac{y_2'}{y_2} = \frac{y_1'}{y_1}$$

which can be integrated to show $y_2(x) = c y_1(x)$, where $c$ is a constant of integration. ∎

Note that when given a $n$th-order linear differential equation, it would be nice if it were possible to integrate it $n$ times to solve for $y$. If this is possible, then after each integration add an arbitrary constant to produce a total of $n$ constants. In solving a differential equation, be sure to reduce the constants in an expression to their simplest form. For example, to solve the differential equation $y'''' = 0$, perform four integrations to obtain

$$y'''' = 0$$
$$y''' = c_1$$
$$y'' = c_1 x + c_2$$
$$y' = c_1 \frac{x^2}{2} + c_2 x + c_3$$
$$y = c_1 \frac{x^3}{3!} + c_2 \frac{x^2}{2!} + c_3 x + c_4$$

Here the constants $c_1/3!$ and $c_2/2!$ can be relabeled since $c_1$ and $c_2$ are arbitrary. The solution can be written in the form $y = k_1 x^3 + k_2 x^2 + k_3 x + k_4$ by relabeling the constants. The fundamental set of solutions is given by $\{1, x, x^2, x^3\}$.

As another example of the reduction of a constant to its simplest form consider the function $y = x + c_1 \sin c_2$. This function can be written as $y = x + c_3$, where $c_3 = c_1 \sin c_2$ is a new arbitrary constant. That is, independent constants are referred to as essential arbitrary constants and usually mean constants that have been reduced to their simplest form.

A fundamental set of solutions is an important concept because any solution of a $n$th-order linear homogeneous differential equation $L(y) = 0$ can be expressed as some linear combination of these basis solutions. Knowing a fundamental set of solutions, it is possible to construct a linear combination of these solution to represents the general solution. The construction process guarantees the general solution possesses $n$ independent constants.

A fundamental set allows one to construct a general solution to a homogeneous linear $n$th-order differential equation, and from this general solution it is possible to construct the unique solution, over the interval I, which satisfies the initial value problem.

**Characteristic Equation**

Consider the $n$th order linear homogeneous differential equation

$$L(y) = (a_0 D^n + a_1 D^{n-1} + \cdots + a_{n-1} D + a_n)y = 0, \quad D = \frac{d}{dx} \tag{6.43}$$

with coefficients $a_i$, $i = 0, \ldots, n$ which are real constants, and $a_0 \neq 0$. Assume a solution to equation (6.43) having the form of an exponential function, $y = e^{mx}$, with $m$ a constant to be determined. Observe

$$Dy = me^{mx}, \quad D^2 y = m^2 e^{mx}, \quad D^3 y = m^3 e^{mx}, \quad \ldots \quad D^n y = m^n e^{mx}$$

so that when the above exponential function and derivatives are substituted into equation (6.43), the following equation results for determining the constant $m$

$$(a_0 m^n + a_1 m^{n-1} + \cdots + a_{n-1} m + a_n)e^{mx} = 0. \tag{6.44}$$

If $y = e^{mx}$ is to be a solution, then equation (6.44) must equal zero for all values of the independent variable $x$. The exponential function is never zero for finite values of $x$ and the only way for $y = e^{mx}$ to be a solution of equation (6.44) is for $m$ to satisfy the equation

$$(a_0 m^n + a_1 m^{n-1} + \cdots + a_{n-1} m + a_n) = 0. \tag{6.45}$$

This equation is a polynomial of degree $n$ in the variable $m$ and is called the characteristic equation associated with the homogeneous equation (6.43). The roots of the characteristic equation are called characteristic roots. In general, if the characteristic equation is a polynomial of degree $n$, then there will be $n$ roots. Consider the following cases where the characteristic equation has distinct real roots, complex roots, and repeated roots.

**Distinct Real Roots of Characteristic Equation** If $m_1, m_2, \ldots, m_n$ are the $n$ distinct roots of the characteristic equation, then a fundamental system associated with the differential equation (6.43) can be written as $\{y_1, y_2, \ldots, y_n\}$ where

$$y_1 = e^{m_1 x}, \quad y_2 = e^{m_2 x}, \quad \ldots \quad y_n = e^{m_n x}$$

and the general solution can be written as any linear combination of these solutions

$$y = \sum_{i=1}^{n} c_i y_i = c_1 y_1 + c_2 y_2 + \cdots + c_n y_n$$

where $c_1, c_2, \ldots, y_n$ are arbitrary constants.

**Complex or Imaginary Roots of Characteristic Equation**

Consider the problem of finding the general solution to the differential equation

$$\frac{d^2y}{dx^2} + y = y'' + y = (D^2 + 1)y = 0, \qquad D = \frac{d}{dx}$$

Assume an exponential solution $y = e^{mx}$ with $y' = me^{mx}$ and $y'' = m^2 e^{mx}$. Now substitute $y$ and $y''$ into the differential equation to obtain the characteristic equation

$$m^2 + 1 = 0$$

with characteristic roots $m_1 = i$ and $m_2 = -i$, where $i^2 = -1$. A fundamental system or basis of solutions can be written as

$$y_1 = e^{ix} \qquad \text{and} \qquad y_2 = e^{-ix}.$$

Employ the theorem 2 together with the Euler's identity, to form the linear combinations

$$y_3 = \frac{1}{2}y_1 + \frac{1}{2}y_2 = \frac{e^{ix} + e^{-ix}}{2} = \cos x$$

and

$$y_4 = \frac{1}{2i}y_1 - \frac{1}{2i}y_2 = \frac{e^{ix} - e^{-ix}}{2i} = \sin x,$$

(see page 94), which are real solutions of the given differential equation. From the set of solutions $\{y_1, y_2, y_3, y_4\}$, select any two linearly independent solutions to form a fundamental system. The general solution to the given differential equation can then be expressed as some linear combination of this fundamental set. Selecting the solutions $\{\cos x, \sin x\}$, as the fundamental set of solutions, produces a general solution of the form

$$y = c_1 \sin x + c_2 \cos x$$

which represent an oscillatory motion. This solution can be written in the alternative forms

$$y = A\sin(x \pm \phi) \quad \text{or} \quad y = A\cos(x \pm \psi), \tag{6.46}$$

where the angles $\psi$ and $\phi$ are called phase angles and the quantity $A$ is called the amplitude of the oscillatory motion. The $\pm$ signs chosen depend upon the signs of the constants $c_1$ and $c_2$. It is possible to write the real solutions in the form

$$y = \sqrt{c_1^2 + c_2^2}\left(\frac{c_1}{\sqrt{c_1^2 + c_2^2}}\sin x + \frac{c_2}{\sqrt{c_1^2 + c_2^2}}\cos x\right) \tag{6.47}$$

The geometry of figure 6-1 shows that by treating $c_1$ and $c_2$ as the sides of the right triangles illustrated, then it is possible to employ the identities

$$\sin(A \pm B) = \sin A \cos B \pm \cos A \sin B$$

$$\cos(A \pm B) = \cos A \cos B \mp \sin A \sin B.$$

to convert equation (6.47) to one of the forms in equation (6.46).

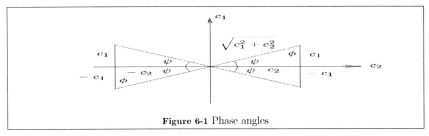

**Figure 6-1** Phase angles

In general, if the characteristic equation has a characteristic root $m_1 = a + ib$, then the conjugate root $m_2 = \overline{m}_1 = a - ib$ must also be a root of the characteristic equation, because complex roots occur in conjugate pairs whenever the polynomial equation has real coefficients. If $a + ib$ and $a - ib$ are complex roots of the characteristic equation, then

$$y_1 = e^{(a+ib)x} = e^{ax}e^{ibx} = e^{ax}(\cos bx + i \sin bx)$$
$$y_2 = e^{(a-ib)x} = e^{ax}e^{-ibx} = e^{ax}(\cos bx - i \sin bx)$$

are complex solutions to the differential equations. If real solutions are desired, then it is customary to employ the Euler identity to convert the complex solutions to real solutions. The previous theorem 2 states that other forms for the solutions can be obtained by taking linear combinations of the complex solutions. For example, the functions

$$y_3 = \frac{1}{2}(y_1 + y_2) = e^{ax} \cos bx$$
$$y_4 = \frac{1}{2i}(y_1 - y_2) = e^{ax} \sin bx$$

are linear combinations of $y_1$ and $y_2$ and represent real solutions to the given linear homogeneous differential equation which are to be associated with the complex roots $a + ib$ and $a - ib$.

**Repeated Roots for the Characteristic Equation** Assume the characteristic equation has a repeated root. As a specific example, let us determine the general solution to the differential equation

$$L(y) = y'' - 2y' + y = 0.$$

This equation is recognized as a differential equation with constant coefficients and so it is customary to assume an exponential solution in the form $y = e^{mx}$ where $m$ is a constant to be determined. The substitution of $y = e^{mx}$ and its derivatives into the differential equation produces the characteristic equation $(m-1)^2 = 0$. This equation has the repeated root $m_1 = 1, m_2 = 1$, which is called a root of multiplicity two, and hence only one independent solution $y_1(x) = e^x$ is obtained. Whenever a root repeats itself, then there exists a second linearly independent solution having the form $y_2 = u(x)y_1(x) = u(x)e^x$, where $u(x)$ is an unknown function to be determined. Substitute the assumed second solution and its derivatives

$$y_2' = ue^x + u'e^x \quad \text{and} \quad y_2'' = ue^x + 2u'e^x + u''e^x.$$

into the differential equation and show there results, after simplification, the equation $u''e^x = 0$, which is a new differential equation for determining the unknown function $u = u(x)$. The above example indicates that $y_2$ is a solution of the given equation provided that $u = u(x)$ is a solution of the differential equation $u'' = 0$. Integrating this equation twice gives $u' = c_1$ and $u = u(x) = c_1x + c_2$. Consequently, a second linearly independent solution can be represented in the form

$$y_2 = y_2(x) = uy_1 = (c_1x + c_2)e^x$$

provided $c_1 \neq 0$. Let $c_1 = 1$ and $c_2 = 0$, then $y_2(x) = xe^x$ is an independent solution which can be thought of as being obtained from $y_1(x)$ by multiplication by $x$. As an exercise, verify that the general solution is obtained from the linear combination of functions in the fundamental set $\{e^x, xe^x\}$. Also verify that the general solution has the form $y(x) = (c_1 + c_2x)e^x$. This modification rule of multiplication by $x$ whenever a characteristic root repeats itself can be shown to hold in general.

**THEOREM 3.** *Let $L(y) = 0$ denote the linear differential equation with constant coefficients given by equation (6.43). If $m = \alpha$ is a root (real or complex) of multiplicity $k$ of the characteristic equation associated with the differential equation (6.43), then the $k$ functions of the set*

$$e^{\alpha x}, \; xe^{\alpha x}, \; x^2e^{\alpha x}, \ldots, x^{k-1}e^{\alpha x}$$

*are $k$- independent solutions of the differential equation $L(y) = 0$.*

**Proof:** It is an easy exercise to show the set of functions $e^{\alpha x}, xe^{\alpha x}, \ldots, x^{k-1}e^{\alpha x}$ are linearly independent functions. The linear operator $L(y)$, defined by equation (6.43), is a polynomial operator in $D$ which can be factored and so the differential equation can be written in the alternative form

$$L(y) = a_0(D - \alpha_1)^{k_1}(D - \alpha_2)^{k_2} \ldots (D - \alpha)^k y = 0, \qquad (6.48)$$

where $\alpha_1, \alpha_2, \ldots, \alpha$ are the characteristic roots of multiplicity $k_1, k_2, \ldots, k$. To prove the theorem it is only necessary to show that $y_n = x^n e^{\alpha x}$ is a solution of equation (6.48) for $n = 0, 1, 2, \ldots, k - 1$, or alternatively show that $(D - \alpha)^k y_n = 0$ for values of $n$ satisfying $n = 0, 1, 2, \ldots, k - 1$. Toward this purpose let $y_n = x^n e^{\alpha x}$ and calculate the derivative

$$Dy_n = nx^{n-1}e^{\alpha x} + \alpha x^n e^{\alpha x}.$$

From these relations it is easy to show
$$(D-\alpha)y_n = nx^{n-1}e^{\alpha x}.$$

Continuing in this manner, show
$$(D-\alpha)^2 y_n = n(n-1)x^{n-2}e^{\alpha x}$$
$$(D-\alpha)^3 y_n = n(n-1)(n-2)x^{n-3}e^{\alpha x}$$
$$(D-\alpha)^4 y_n = n(n-1)(n-2)(n-3)x^{n-4}e^{\alpha x}$$
$$\vdots$$
$$(D-\alpha)^k y_n = n(n-1)\cdots(n-(k-1))x^{n-k}e^{\alpha x}.$$

In this last equation, observe that $(D-\alpha)^k y_n = 0$ for the $k$ values $n = 0, 1, 2, \ldots, k-1$ and the theorem is proven.

## The Phase Plane

Consider solutions $y_1(t)$, $y_2(t)$ of the two dimensional system of differential equations

$$\begin{aligned}\frac{dy_1}{dt} &= \alpha y_1 + \beta y_2\\ \frac{dy_2}{dt} &= \gamma y_1 + \delta y_2,\end{aligned} \quad \text{or} \quad \frac{d}{dt}\begin{bmatrix} y_1 \\ y_2 \end{bmatrix} = \begin{bmatrix} \alpha & \beta \\ \gamma & \delta \end{bmatrix}\begin{bmatrix} y_1 \\ y_2 \end{bmatrix} \quad \text{or} \quad \frac{d\vec{y}}{dt} = A\vec{y} \qquad (6.49)$$

where $\vec{y} = \text{col}(y_1, y_2)$, is a column vector and $A = \begin{bmatrix} \alpha & \beta \\ \gamma & \delta \end{bmatrix}$ is a coefficient matrix with elements $\alpha, \beta, \gamma, \delta$ which are constants. Construct a set of $y_1$, $y_2$ perpendicular axes called the phase plane associated with the solution set $y_1(t)$, $y_2(t)$ to the system of equations (6.49). The parametric curves $y_1(t)$, $y_2(t)$, with parameter $t$, when plotted in the phase plane are called trajectories of the differential system (6.49).

If $\vec{y}_0$ is a constant vector which is a solution of equation (6.49) then $\vec{y}_0$ is called a critical point of the system. Critical points are sometimes referred to as rest points or stationary points of the system. Critical points and phase plane trajectories are important because they describe all possible solutions to a system of differential equations (6.49).

The point $(y_1, y_2) = (0, 0)$ is a critical point or stationary point of the above system of equations (6.49). The phase plane trajectories of the system (6.49) are determined by assuming a solution of the form $\vec{y} = \vec{u}e^{\lambda t}$ where $\lambda$ is a constant to be determined and $\vec{u}$ is a constant vector to be determined. Differentiate the assumed solution to obtain $\frac{d\vec{y}}{dt} = \lambda \vec{u}e^{\lambda t}$ and the substitute both $\vec{y}$ and $\frac{d\vec{y}}{dt}$ into the differential equation to obtain

$$\lambda \vec{u}e^{\lambda t} = A\vec{u}e^{\lambda t} \quad \Longrightarrow \quad (A - \lambda I)\vec{u} = \vec{0} \qquad (6.50)$$

This system of equations has a nonzero solution provided

$$\det(A - \lambda I) = |A - \lambda I| = \begin{vmatrix} \alpha - \lambda & \beta \\ \gamma & \delta - \lambda \end{vmatrix} = 0 \qquad (6.51).$$

This gives the characteristic equation

$$\lambda^2 - (\alpha + \delta)\lambda + (\alpha\delta - \beta\gamma) = 0.$$

The solutions $\lambda_1$, $\lambda_2$ of the characteristic equation are called the eigenvalues of the matrix $A$. Corresponding to each distinct eigenvalue $\lambda_1, \lambda_2$, there are eigenvectors $\vec{u}_1, \vec{u}_2$ determined by the equation (6.50). In the case of distinct eigenvalues the solution of the vector system can be represented as

$$\vec{y} = c_1 \vec{u}_1 e^{\lambda_1 t} + c_2 \vec{u}_2 e^{\lambda_2 t}, \tag{6.52}$$

where $c_1$, $c_2$ are arbitrary constants and $\vec{u}_1$ and $\vec{u}_2$ are the eigenvectors corresponding to the eigenvalues $\lambda_1$ and $\lambda_2$ respectively. In the case of repeated characteristic roots, one must use other techniques to obtain the solution. Various types of trajectories can arise depending upon the values of $\lambda_1$ and $\lambda_2$. The following special cases are presented and are representative of the types of solutions possible.

**CASE I** ($\lambda_1 > 0, \lambda_2 > 0, \lambda_1 \ne \lambda_2$)  The system of differential equations

$$\frac{dy_1}{dt} = y_1 \quad \text{and} \quad \frac{dy_2}{dt} = 2y_2 \tag{6.53}$$

is a special case of equation (6.49) and this system of differential equations has the solutions

$$y_1 = c_1 e^{\lambda_1 t}, \qquad y_2 = c_2 e^{\lambda_2 t}, \quad \text{or} \quad \vec{y} = c_1 \begin{pmatrix} 1 \\ 0 \end{pmatrix} e^t + c_2 \begin{pmatrix} 0 \\ 1 \end{pmatrix} e^{2t} \tag{6.54}$$

where $c_1, c_2$ are arbitrary constants and $\lambda_1 = 1$ and $\lambda_2 = 2$. The phase plane trajectories can be expressed as the parametric equations

$$y_1 = c_1 e^t, \qquad y_2 = c_2 e^{2t}$$

which start at time $t = 0$ at an arbitrary point $(c_1, c_2) \ne (0, 0)$ in the phase plane. These types of trajectories appear as in figure 6-2. The critical point $(0, 0)$ is called an unstable node because any solution trajectory beginning near the critical point moves away from it.

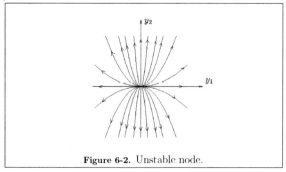

**Figure 6-2.** Unstable node.

The direction field of the solution trajectories is determined from the differential equation
$$\frac{dy_2}{dy_1} = \frac{2y_2}{y_1}$$
and the trajectories (solution curves) are the solutions of this equation. This equation is easily solved to produce the solution family $y_2 = c(y_1)^2$, where $c$ is a constant.

**CASE II** ( $\lambda_1 < 0$, $\lambda_2 < 0$, $\lambda_1 \neq \lambda_2$ ) If both eigenvalues are negative, then in this special case the trajectories appear as in figure 6-2 with the exception that all trajectories move toward the origin as $t \to \infty$. The origin (critical point) is called a node or stable node under these conditions.

**CASE III** ( $\lambda_1 \neq \lambda_2$, and $\lambda_1$ and $\lambda_2$ are of opposite sign) The system
$$\frac{dy_1}{dt} = -y_1, \qquad \frac{dy_2}{dt} = 2y_2$$
with eigenvalues $\lambda_1 = -1$ and $\lambda_2 = 2$ is a special case of equation (6.54) and has the solutions
$$y_1 = c_1 e^{-t}, \qquad y_2 = c_2 e^{2t} \quad \text{or} \quad \vec{y} = c_1 \begin{pmatrix} 1 \\ 0 \end{pmatrix} e^{-t} + c_2 \begin{pmatrix} 0 \\ 1 \end{pmatrix} e^{2t} \qquad (6.55)$$
with $c_1, c_2$ arbitrary constants. The phase plane trajectories corresponding to these solutions, defined by the parametric equations from equation (6.55), start at an arbitrary point $(c_1, c_2) \neq (0,0)$ and as $t \to \infty$ the point $y_1 \to 0$ and $y_2$ increases without bound. These trajectories are illustrated in figure 6-3.

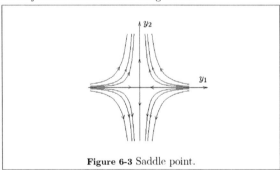

**Figure 6-3** Saddle point.

The direction field of these trajectories is determined by the differential equation
$$\frac{dy_2}{dy_1} = -2\frac{y_2}{y_1}$$
and the trajectories are the phase plane family of solutions is given by $y_2(y_1)^2 = c$, where $c$ is a constant. In this case the critical point $(0,0)$ is called a saddle point.

**CASE IV** ($\lambda_1 = \lambda_2$) The system of differential equations

$$\frac{d}{dt}\begin{bmatrix} y_1 \\ y_2 \end{bmatrix} = \begin{bmatrix} 0 & 1 \\ -1 & 2 \end{bmatrix}\begin{bmatrix} y_1 \\ y_2 \end{bmatrix}$$

is a special case of equation (6.49). One finds that the characteristic equations associated with the matrix $A = \begin{bmatrix} 0 & 1 \\ -1 & 2 \end{bmatrix}$ has the repeated roots $\lambda_1 = \lambda_2 = -1$. The matrix differential equation is equivalent to the system of scalar differential equations

$$\frac{dy_1}{dt} = y_2, \qquad \frac{dy_2}{dt} = -y_1 - 2y_2$$

Making the substitutions $y = y_1$ and $\frac{dy}{dt} = y_2$, produces the second order ordinary differential equation

$$\frac{d^2y}{dt^2} + 2\frac{dy}{dt} + y = 0$$

with solution given by $y = c_1 e^{-t} + c_2 t e^{-t}$, where $c_1$ and $c_2$ are arbitrary constants and from this solution, the solutions to the scalar differential system can be represented

$$y_1 = y = c_1 e^{-t} + c_2 t e^{-t}, \quad \text{and} \quad y_2 = \frac{dy}{dt} = (c_2 - c_1)e^{-t} - c_2 t e^{-t} \qquad (6.56)$$

The substitution $c_3 = c_1 - c_2$ can be implemented to write the solution to the matrix differential equation in the form

$$\vec{y} = \begin{bmatrix} y_1 \\ y_2 \end{bmatrix} = c_3 \begin{bmatrix} 1 \\ -1 \end{bmatrix} e^{-t} + c_2 \begin{bmatrix} 1+t \\ -t \end{bmatrix} e^{-t}.$$

The direction field determined by these trajectories is given by the differential equation

$$\frac{dy_2}{dy_1} = \frac{-y_1 - y_2}{y_2}$$

which has the solution trajectories

$$\frac{1}{2}\ln\left|1 + \frac{y_2}{y_1} + \left(\frac{y_2}{y_1}\right)^2\right| - \frac{1}{\sqrt{3}}\arctan\left[\frac{2y_2 + y_1}{\sqrt{3}y_1}\right] + \ln|y_1| = c,$$

where $c$ is a constant. These trajectories are illustrated in figure 6-4. For this example the eigenvalues were all negative and consequently the trajectories approach the critical point as $t$ increases without bound. The critical point in this case is called a stable node.

A special case where the eigenvalues are positive produces trajectories as in figure 6-4 with the exception that they move away from the origin as $t \to \infty$. In this case the node is called unstable.

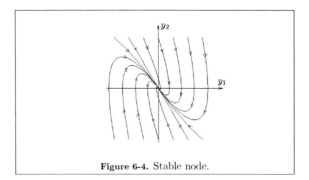

**Figure 6-4.** Stable node.

**CASE V (Complex eigenvalues)** If the eigenvalues are complex numbers and $\lambda_1 = a+ib$ and $\lambda_2 = a-ib$, then the general solution to equation (6.49) takes the form

$$\begin{bmatrix} y_1 \\ y_2 \end{bmatrix} = e^{at} \begin{bmatrix} \alpha_1 \\ \alpha_2 \end{bmatrix} \sin bt + e^{at} \begin{bmatrix} \beta_1 \\ \beta_2 \end{bmatrix} \cos bt$$

where $\alpha_1, \alpha_2, \beta_1, \beta_2$ are constants. The phase plane trajectories for this system are spirals of the type illustrated in figure 6-5. If $a$ is positive, the trajectories move away from the origin, and if $a$ is negative, the trajectories approach the origin as $t \to \infty$. In this case the critical point is called a focus.

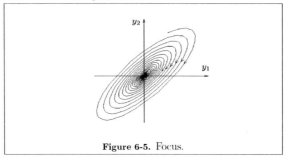

**Figure 6-5.** Focus.

**CASE VI (Pure imaginary roots)** For eigenvalues $\lambda_1 = ib$ and $\lambda_2 = -ib$, the general solution of equation (6.49) has the form

$$\vec{y} = \begin{bmatrix} y_1 \\ y_2 \end{bmatrix} = \begin{bmatrix} \alpha_1 \\ \alpha_2 \end{bmatrix} \sin bt + \begin{bmatrix} \beta_1 \\ \beta_2 \end{bmatrix} \cos bt, \quad \alpha_1, \alpha_2, \beta_1, \beta_2 \text{ constants.}$$

A special case of these equations are the trajectories $y_1 = \alpha_1 \sin bt$ and $y_2 = \beta_2 \cos bt$. These trajectories are part of the family

$$\left( \frac{y_1}{\alpha_1} \right)^2 + \left( \frac{y_2}{\beta_2} \right)^2 = 1.$$

and are ellipses centered at the origin. In this case the critical point is called a center.

## Boundary Value Problems

Consider a general solution to the linear second-order differential equation

$$L(y) = a_0(x)\frac{d^2y}{dx^2} + a_1(x)\frac{dy}{dx} + a_2(x)y = 0, \qquad a \leq x \leq b$$

which has the form $y = c_1 y_1(x) + c_2 y_2(x)$ where $c_1, c_2$ are constants and $\{y_1(x), y_2(x)\}$ is a fundamental set of solutions. What type of conditions can be imposed upon the general solution? An initial value problem (IVP) imposes two conditions at a single point $a$ having the form $y(a) = \alpha$ and $y'(a) = \beta$, $\alpha, \beta$ constants. One must then proceed further to determine whether these imposed conditions are reasonable for a solution to exist.

If some linear combination of $y(x)$ and $y'(x)$ are to satisfy prescribed conditions at two different points, say at the end points $x = a$ and $x = b$, then this type of condition on the general solution is known as a boundary value problem (BVP). Here conditions are imposed on the general solution at two different points $x = a$ and $x = b$. Again, one must determine if the imposed conditions are reasonable or unreasonable in order for a solution to exist.

The following are some examples of boundary value problems for a specific linear second-order differential equation.

## Example 6-3.

Let us examine three different boundary value problems (BVP) associated with the general solution $y = c_1 \sin x + c_2 \cos x$ of the ordinary differential equation $y'' + y = 0$ for $0 \leq x \leq \pi$.

**BVP I** Find the solution satisfying the boundary conditions:

$$B_1(y) = y(0) = 0 \quad \text{and} \quad B_2(y) = y(\pi) = 0$$

where the notation $B(\ )$ denotes a boundary operator. The first boundary condition requires $y(0) = c_1 \sin 0 + c_2 \cos 0 = 0$ and implies $c_2 = 0$. This reduces the solution family to solutions of the form $y(x) = c_1 \sin x$. The second boundary condition requires $y(\pi) = c_1 \sin \pi = 0$. Note that $\sin \pi = 0$, so that the second condition is satisfied for all values of the constant $c_1$. The solution of the boundary value problem

$$y'' + y = 0, \quad y(0) = 0, \quad y(\pi) = 0$$

can then be expressed as the one parameter family of curves $y = c_1 \sin x$, where $c_1$ is arbitrary. This BVP has an infinite number of solutions.

**BVP II** Find the solution satisfying the boundary conditions

$$B_1(y) = y(0) = 0 \quad \text{and} \quad B_2(y) = y(\pi) = 7$$

Here the first boundary condition remains the same, so that solutions must be of the form $y = c_1 \sin x$. The second boundary condition requires $y(\pi) = c_1 \sin \pi = 7$. Notice $c_1$ cannot be chosen to satisfy this condition since $\sin \pi = 0$. Consequently, this type of boundary value problem has no solution.

**BVP III** Find the solution of the boundary value problem

$$B_1(y) = y(0) = 0 \quad \text{and} \quad B_2(y) = y(\pi) - y'(\pi) = 1$$

Again, the first boundary condition requires solutions have the form $y = c_1 \sin x$. The second boundary condition requires that $c_1 \sin \pi - c_1 \cos \pi = 1$ which implies $c_1 = 1$, so that the solution of the boundary value problem $y'' + y = 0$, with boundary conditions $B_1(y) = y(0) = 0$, and $B_2(y) = y(\pi) - y'(\pi) = 1$ is the single function $y = \sin x$. In this case a unique solution exists.

The above three boundary value problems illustrate that the solution of a boundary value problem for an ordinary differential equation can have no solution, a unique solution, or an infinite number of solutions.

In the above example the notation $B(\ )$ is used to denote a boundary operator just like $L(\ )$ was used to denote a differential operator. Consider various types of boundary operators associated with solutions over the interval $a \leq x \leq b$. Dirichlet boundary conditions require that the values of the dependent variable be specified at the end points of the interval,

$$B_1(y) = y(a) = \text{A specified quantity.}$$
$$B_2(y) = y(b) = \text{A specified quantity.}$$

Neumann[6] boundary conditions require values of the first derivatives be specified at the end points of the interval.

$$B_1(y) = y'(a) = \text{A specified quantity.}$$
$$B_2(y) = y'(b) = \text{A specified quantity.}$$

A mixed boundary value problem requires boundary conditions of the type

$$B_1(y) = \alpha y(a) + \beta y'(a) = \text{A specified quantity.}$$
$$B_2(y) = \gamma y(b) + \delta y'(b) = \text{A specified quantity.}$$

---

[6] Carl Gottfried Neumann (1832-1925), German mathematician.

where $\alpha, \beta, \gamma$ and $\delta$ are constants not all zero. Mixed boundary value problems can be thought of as some linear combination of Dirichlet and Neumann boundary conditions.

## Solution of Nonhomogeneous Linear Differential Equations

Consider the nonhomogeneous differential equation $L(y) = r(x)$, with $r(x) \ne 0$ and where $L(y)$ is a linear differential operator with constant coefficients defined by equation (6.43). General solutions of the homogeneous equation $L(y) = 0$ are called complementary solutions and are represented using the notation $y_c$. Here $y_c$ satisfies the equation $L(y_c) = 0$ and is understood to represent a general solution to the homogeneous equation. Any solution of the nonhomogeneous equation $L(y) = r(x)$ is called a particular solution and is usually represented using the notation $y_p$.

**THEOREM 4.** *For $L(y)$, a linear differential operator, and $y_p(x)$ any particular solution of the nonhomogeneous differential equation satisfying $L(y_p) = r(x)$ and $y_c(x)$ the complementary solution of the homogeneous differential equation $L(y_c) = 0$, then the general solution $y = y(x)$ of the nonhomogeneous differential equation $L(y(x)) = r(x)$ can be expressed in the form*

$$y = y(x) = y_c(x) + y_p(x)$$

**Proof:** For $y_p(x)$ a particular solution satisfying $L(y_p) = r(x)$ and $y(x)$ any solution of the nonhomogeneous equation, then the function $z(x) = y(x) - y_p(x)$ must satisfy the homogeneous equation $L(z) = 0$, since

$$L(z) = L(y(x) - y_p(x)) = L(y(x)) - L(y_p(x)) = r(x) - r(x) = 0.$$

The most general solution to the homogeneous equation $L(z) = 0$ can be expressed as a linear combination of $n$-linearly independent functions $y_1, y_2, ..., y_n$, called the fundamental set or solution basis, and consequently

$$z = y_c = y_c(x) = c_1 y_1 + c_2 y_2 + \ldots + c_n y_n,$$

where $c_1, c_2, ..., c_n$ are constants. The linear combination $y = y(x) = y_c(x) + y_p(x)$ is referred to as the general solution associated with the nonhomogeneous differential equation $L(y) = r(x)$.

The above theorem tells us that to find the general solution to a linear ordinary nonhomogeneous differential equation $L(y) = r(x)$ one must first find the general solution $y_c$, the complementary solution, to the corresponding homogeneous equation $L(y) = 0$ and then find any particular solution $y_p$ to the nonhomogeneous differential equation $L(y_p) = r(x)$.

## Method of Undetermined Coefficients

The method of undetermined coefficients is applicable when the differential operator $L(y)$ is linear and has constant coefficients and the right-hand side of the nonhomogeneous equation $L(y) = r(x)$ contains:

(a) Polynomial terms $x^m$

(b) Exponential terms $e^{\lambda x}$

(c) Special trigonometric functions $\sin(\lambda x)$ and $\cos(\lambda x)$

(d) Linear combinations of the above functions.

In order to find a particular solution, one must find a function $y_p$, which when operated upon by the linear operator $L$, produces the right-hand side $r(x)$. Since $L$ is a linear differential operator with constant coefficients, the right-hand side $r(x)$ must be some linear combination of the function $y_p$ and its derivatives. Consider the special cases where the right-hand side is a polynomial in x, an exponential term $e^{\lambda x}$ or one of the trigonometric functions $\sin \lambda x$ or $\cos \lambda x$. Observe that if the right-hand side $r(x)$ contains polynomials of highest degree $n$, then such terms can be construct from linear combinations of the derivatives of $x^n$. If the right-hand side contains exponential terms $e^{\lambda x}$, then such terms can be constructed from linear combinations of the derivatives of $e^{\lambda x}$. Also note that if the right-hand side contains either of the trigonometric functions $\sin \lambda x$ or $\cos \lambda x$, then the derivatives of these terms also produce the same sine and cosine terms multiplied by some constants. Consequently, if the right-hand side is composed of polynomial terms, exponential terms or sine, cosine terms, one must examine these functions and their derivatives and try to construct from these functions, a linear independent set of functions

$$S = \{f_1(x), f_2(x), \ldots, f_m(x)\},$$

where no function in the set $S$ also occurs in the complementary solution. If S is a linearly independent set of functions, obtained from the derivatives of the right-hand side, then one would expect that the right-hand side $r(x)$ is built up as some linear combination of the functions in $S$. Consequently, assume a particular solution to the differential equation having the form

$$y_p(x) = A f_1(x) + B f_2(x) + \cdots + C f_m(x),$$

where $A, B, \ldots, C$ are undetermined coefficients which must be chosen to force this linear combination to be a particular solution of the given homogeneous equation.

## Example 6-4.

Find the general solution to the differential equation

$$L(y) = y'' - y = \sin 3x \tag{6.57}$$

**Solution:** First solve the homogeneous equation $L(y) = 0$ and obtain the complementary solution

$$y_c = c_1 e^x + c_2 e^{-x} \quad \text{or} \quad y_c = k_1 \sinh x + k_2 \cosh x.$$

In order to find a particular solution one must find a function $y_p$, which when operated upon by the operator $L$, produces the right-hand side $r(x) = \sin 3x$. In this example, all the derivatives of the right hand side contain at least one of the terms in the set

$$S = \{\sin 3x, \cos 3x\}$$

multiplied by some constant. In fact every derivative of $r(x) = \sin 3x$ is some linear combination of the functions in the set $S$ and none of the functions in $S$ is also a complementary solution. Therefore, assume a particular solution which is a linear combination of the functions in $S$ and write

$$y_p = A \sin 3x + B \cos 3x \tag{6.58}$$

where $A$ and $B$ are undetermined constant coefficients. If $y_p$ and its derivatives are combined in some linear combination as specified by the differential equation, then select the constants $A$ and $B$ which produce the right-hand side of the given equation. Substituting equation (6.58) into equation (6.57) produces the equation

$$L(y_p) = -10A \sin 3x - 10B \cos 3x = \sin 3x. \tag{6.59}$$

In order for equation (6.59) to be an identity for all $x$, it is necessary to force $A$ and $B$ to satisfy the equations $-10A = 1$ and $-10B = 0$. From these equations the coefficients $A = -1/10$ and $B = 0$ result. The assumed particular solution, represented by equation (6.58), then becomes

$$y_p = -\frac{1}{10} \sin 3x$$

and the general solution to equation (6.57) can be expressed as

$$y = y_c + y_p = k_1 \sinh x + k_2 \cosh x - \frac{1}{10} \sin 3x$$

■

## Example 6-5.

Find the general solution to the differential equation

$$L(y) = y'' + 9y = \sin 3x \tag{6.60}$$

**Solution:** First solve the homogeneous equation $L(y) = 0$ and obtain the complementary solution

$$y_c = c_1 \sin 3x + c_2 \cos 3x$$

Next examine the right-hand side $r(x) = \sin 3x$ and its derivatives. Note that every derivative of $r(x)$ is some linear combination of members from the set of independent functions

$$S = \{\sin 3x, \cos 3x\}.$$

If one assumes a solution

$$y_p = A \sin 3x + B \cos 3x$$

one finds $L(y_p) = 0$, since the terms in $S$ also occur in the complementary solution. To overcome this difficulty, examine the derivatives of two functions $u(x)$ and $xu(x)$ and list these derivatives in the sets $S_1$ and $S_2$. There results after taking the derivatives of these functions

$$S_1 = \begin{bmatrix} u \\ u' \\ u'' \\ \vdots \\ u^{(n)} \end{bmatrix}, \quad S_2 = \begin{bmatrix} xu \\ xu' + u \\ xu'' + 2u' \\ \vdots \\ xu^{(n)} + nu^{(n-1)} \end{bmatrix}.$$

Observe that the functions in $S_1$, when multiplied by $x$, occur in $S_2$. Also note that if $L$ is a linear differential operator and $L(u) = 0$, then $L(u)$ is just some linear combination of elements from the set $S_1$ which produces zero and that this same linear combination of the elements from $S_2$ would produce

$$xL(u) + M(u),$$

where $M(u)$ is some new linear differential operator. By hypothesis, $L(u) = 0$ and since $M(u)$ is an operator different from $L(u)$, one finds $y = xu$ is a solution of $L(y) = r(x)$ if the conditions $L(u) = 0$ and $M(u) = r(x)$ are satisfied.

For the above problem, the set $S$ contains the functions $\sin 3x$ and $\cos 3x$ which also occur in the complementary solution and so elements from $S$ could not be used to construct a particular solution. The previous discussion suggests that the set $S$ be modified by multiplying each term in the set by $x$. This produces the modified set

$$S_x = \{x \sin 3x, x \cos 3x\}.$$

Taking a linear combination of the functions in the modified set $S_x$, assume a particular solution of the form

$$y_p = Ax\sin 3x + Bx\cos 3x. \tag{6.61}$$

When equation (6.61) is substituted into the differential equation (6.60) there results

$$L(y_p) = xL(A\sin 3x + B\cos 3x) \\ + 6A\cos 3x - 6B\sin 3x = \sin 3x. \tag{6.62}$$

The first term in this equation is zero because $L$ operating upon the complementary solution produces zero. The remaining terms are the results of the operator $M$ and tell us that, in order for $y_p$ to be a solution, one must choose the coefficients $A$ and $B$ to satisfy the equations $6A = 0$ and $-6B = 1$. This gives the particular solution

$$y_p = -\frac{x}{6}\cos 3x$$

and the general solution

$$y = y_c + y_p = c_1\sin 3x + c_2\cos 3x - \frac{x}{6}\cos 3x.$$

∎

### Example 6-6.

Solve the differential equation

$$(D^2 + D)y = 4xe^{-x}$$

**Solution:** When given a nonhomogeneous linear equation, it is mandatory to first solve the homogeneous equation. Assume an exponential solution $y = e^{mx}$ to the homogeneous equation $(D^2 + D)y = 0$ and obtain the characteristic equation $m^2 + m = 0$ with characteristic roots $m = 0$ and $m = -1$. These roots produce the fundamental set $\{1, e^{-x}\}$. The complementary solution can then be written as

$$y_c = c_1 + c_2 e^{-x},$$

where $c_1$, $c_2$ are arbitrary constants. To find a particular solution, examine the right-hand side of the given equation and also the derivatives of this right hand side, and then form the independent set

$$S = \{xe^{-x}, e^{-x}\}.$$

which contains the essential terms occurring in these derivatives. Note that $e^{-x}$ in $S$ also occurs in the complementary solution, consequently, one must modify the set $S$ by multiplying every member of the set by $x$ to obtain the set

$$S_x = \{x^2 e^{-x}, \, x e^{-x}\}.$$

No member in this modified set occurs in the complementary solution, and therefore for $A$ and $B$ undetermined coefficients, one may assume that a particular solution is

$$y_p = Ax^2 e^{-x} + Bx e^{-x}.$$

Substitute this solution into the nonhomogeneous equation to obtain

$$(D^2 + D)y_p = -2Axe^{-x} + (2A - B)e^{-x} = 4x e^{-x}.$$

Compare like terms and show $A = -2$ and $B = -4$. Hence, the particular solution can be written as

$$y_p = -(2x^2 + 4x)e^{-x}$$

and the general solution becomes

$$y = y_c + y_p = c_1 + c_2 e^{-x} - (2x^2 + 4x) e^{-x}.$$

∎

To summarize, one method for solving a constant-coefficient, nonhomogeneous linear differential equations $L(y) = r(x)$ is as follows.
1. First solve the homogeneous equation $L(y) = 0$ and obtain the complementary solution $y_c$.
2. Next, examine the basic terms occurring in the right-hand side of the differential equation. If $r(x)$ is a sum of terms $r(x) = u_1(x) + u_2(x) + \cdots + u_n(x)$ then one must examine all derivatives which can be listed in sets as

$$S_1 = \begin{bmatrix} u_1 \\ u_1' \\ u_1'' \\ \vdots \end{bmatrix}, \quad S_2 = \begin{bmatrix} u_2 \\ u_2' \\ u_2'' \\ \vdots \end{bmatrix}, \quad \ldots \quad S_n = \begin{bmatrix} u_n \\ u_n' \\ u_n'' \\ \vdots \end{bmatrix}.$$

If the functions and their derivatives in the sets $S_i$, $i = 1, \ldots, n$ produce finite sets, such that the functions in each set can be represented as a linear combination of a finite number of independent terms, then the method of undetermined coefficients is applicable.
3. If one or more of the sets $S_i$, $i = 1, \ldots, n$, have a term which also occurs in the complementary solution, then modify only those sets by multiplying every

member of the set by $x$. Then examine again all the sets $S_i$, including the modified set, for solutions which occur in the complementary function. If there are any of the complementary functions in a set then modify the whole set by multiplying every member of the set by $x$. Continue this process until no further modifications are necessary.

4. Multiply the elements of the sets $S_i$, $i = 1, 2, \ldots, n$, by undetermined coefficients and add them and use this as your assumed particular solution $y_p$.

5. Substitute $y_p$ into the differential equation and choose the undetermined coefficients to force $y_p$ to be a solution. Observe that if one assumes the wrong form for the particular solution to begin with, then step 5 becomes an impossible task, and consequently, a re-examination the sets $S_i$ is required in order to construct an assumed solution of a different form.

### Example 6-7.

This is an example of the method of undetermined coefficients where many modifications are necessary before an assumed solution is constructed. Solve the differential equation

$$L(y) = \frac{d^4y}{dx^4} = 3x + 2\sin 3x \qquad (6.63)$$

**Solution:** First solve the homogeneous equation $L(y) = \frac{d^4y}{dx^4} = 0$ to obtain the fundamental set $\{1, x, x^2, x^3\}$ and the complementary solution

$$y_c = c_0 + c_1 x + c_2 x^2 + c_3 x^3. \qquad (6.64)$$

Next, examine the right hand side $r(x)$ and observe that the basic terms associated with the derivatives of $r(x)$ are some linear combination of the elements in the sets

$$S_1 = \{1, x\}, \qquad S_2 = \{\sin 3x, \cos 3x\}.$$

Note that the members in $S_1$ also occur in the complementary solution, and hence this set is modified to $S_x = \{x, x^2\}$ by multiplying every member of the set by $x$. Since the elements in the modified set also occur in the complementary solution, this set must be modified again by multiplying every term by $x$. This produces the set $S_{x^2} = \{x^2, x^3\}$. This set must be modified again, until the results in the modified set do not occur in the complementary solution. After multiplication by $x$, twice in succession, the final version of the modified set is $S_{x^4} = \{x^4, x^5\}$. From the set $S = S_{x^4} \cup S_2$ there results the assumed particular solution

$$y_p = Ax^4 + Bx^5 + E\sin 3x + F\cos 3x, \qquad (6.65)$$

where $A, B, E$ and $F$ are undetermined coefficients. Substituting equation (6.65) into equation (6.63) gives, after some algebra,

$$L(y_p) = 24A + 120Bx + 81E\sin 3x + 81F\cos 3x = 3x + 2\sin 3x \qquad (6.66)$$

which implies

$$A = 0, \quad 120B = 3, \quad 81E = 2, \quad 81F = 0$$

if $y_p$ is to be a solution.

Solve for the constants $A, B, E, F$ and obtain the particular solution

$$y_p = \frac{1}{40}x^5 + \frac{2}{81}\sin 3x.$$

One can then construct the general solution $y = y_c + y_p$.

∎

## Variation of Parameters

Assume that $L(y) = 0$ is a linear, $n$th-order homogeneous differential equation where

$$L(y) = \sum_{i=0}^{n} a_i D^{n-i} y = a_0 \frac{d^n y}{dx^n} + a_{n-1}\frac{d^{n-1}y}{dx^{n-1}} + \cdots + a_{n-2}\frac{d^2 y}{dx^2} + a_{n-1}\frac{dy}{dx} + a_n y, \qquad D = \frac{d}{dx}$$

where $a_0(x) \neq 0$. Further assume that $\{y_1, y_2, \ldots, y_n\}$ is a known set of fundamental solutions to the homogeneous differential equation $L(y) = 0$. The general solution to the homogeneous equation can then be represented as a linear combination of these functions and written as

$$y_c = c_1 y_1 + c_2 y_2 + \cdots + c_n y_n, \qquad (6.67)$$

where $c_1, c_2, \ldots, c_n$ are arbitrary constants.

The method of variation of parameters is used for obtaining particular solutions to the nonhomogeneous differential equation $L(y) = r(x)$. It should be noted that for a differential operator with coefficients $a_i$, it is not necessary for the coefficients $a_i$ to be constants. The method of variation of parameters is applicable even in the cases where the coefficients $a_i$, $i = 1, \ldots, n$, are functions of $x$. The method of variation of parameters is more general than the method of undetermined coefficients for finding particular solutions to linear differential equations.

The method of variation of parameters assumes that a particular solution exists of the form

$$y_p = v_1(x)y_1 + v_2(x)y_2 + \cdots + v_n(x)y_n \qquad (6.68)$$

This form is obtained from equation (6.67) by replacing the constants by unknown functions $v_1, v_2, \ldots, v_n$ which are functions of $x$ to be determined. When equation (6.68) is substituted into the nonhomogeneous differential equation $L(y) = r(x)$, there results a single equation involving $n$ unknown functions. Select $(n\text{-}1)$ other independent and consistent relations between the functions $(v_1, v_2, \ldots, v_n)$, and in this way, one can obtain $n$ equations in $n$ unknowns to determine the functions $v_1, \ldots, v_n$.

If one requires that the functions $(v_1, v_2, \ldots, v_n)$ satisfy the additional $(n\text{-}1)$ conditions

$$\begin{aligned} v'_1 y_1 + v'_2 y_2 + \cdots + v'_n y_n &= 0 \\ v'_1 y'_1 + v'_2 y'_2 + \cdots + v'_n y'_n &= 0 \\ v'_1 y''_1 + v'_2 y''_2 + \cdots + v'_n y''_n &= 0 \\ &\vdots \\ v'_1 y_1^{(n-2)} + v'_2 y_2^{(n-2)} + \cdots + v'_n y_n^{(n-2)} &= 0, \end{aligned} \qquad (6.69)$$

then these conditions together with equation (6.68) being substituted into the given differential equation, produces the final equation

$$v'_1 y_1^{(n-1)} + v'_2 y_2^{(n-1)} + \cdots + v'_n y_n^{(n-1)} = \frac{r(x)}{a_0(x)}, \qquad \text{for } a_0(x) \neq 0$$

to be added to the previous set of equations (6.69). This produces a particularly convenient set of conditions where the determinant of the coefficients of the unknown functions is the determinant

$$W = \begin{vmatrix} y_1 & y_2 & y_3 & \cdots & y_n \\ y'_1 & y'_2 & y'_3 & \cdots & y'_n \\ y''_1 & y''_2 & y''_3 & \cdots & y''_n \\ \vdots & \vdots & \vdots & \ddots & \vdots \\ y_1^{(n-1)} & y_2^{(n-1)} & y_3^{(n-1)} & \cdots & y_n^{(n-1)} \end{vmatrix}. \qquad (6.70)$$

The determinant $W$ is the Wronskian determinant constructed from the fundamental set of solutions associated with the homogeneous equation. Now solve for the unknowns $v'_1, v'_2, \ldots, v'_n$ followed by an integration to construct the functions $v_1, v_2, \ldots, v_n$.

The following examples illustrate the method of variation of parameters applied to first- and second-order equations.

### Example 6-8.

Use variation of parameters to solve the linear first-order differential equation

$$L(y) = \frac{dy}{dx} + P(x)y = Q(x) \qquad (6.71)$$

**Solution:** Here the fundamental set associated with the homogeneous differential equation $L(y) = \frac{dy}{dx} + P(x)y = 0$ is the single function

$$y_1 = y_1(x) = \exp\left[-\int_0^x P(x)\,dx\right]$$

which is a solution of the homogeneous equation $L(y) = 0$. The complementary solution is $y_c = c_1 y_1(x)$. Now assume a particular solution, to the nonhomogeneous equation $L(y) = Q(x)$, of the form $y_p = v(x) y_1(x)$. Here there is only one condition needed in order to determine $v = v(x)$, and this condition is for $y_p$ to be a solution of equation (6.71). Substituting $y_p = v(x) y_1(x)$ into equation (6.71) results in the condition

$$v y_1'' + v' y_1' + P(x) v y_1' = Q(x)$$

for determining $v = v(x)$. Rearranging terms, gives

$$v[y_1'' + P(x) y_1'] + v' y_1' = Q(x) \quad \text{or} \quad v L(y_1) + v' y_1' = Q(x). \tag{6.72}$$

The condition $L(y_1) = 0$ is satisfied, since $y_1$ is a fundamental solution to the homogeneous equation and hence the condition (6.72) simplifies to

$$v' = \frac{dv}{dx} = \frac{Q(x)}{y_1(x)} = Q(x) \exp\left[\int_0^x P(x)\, dx\right]. \tag{6.73}$$

Integrating this equation produces

$$v = v(x) = \int_0^x Q(s) \exp\left[\int_0^s P(\xi)\, d\xi\right] ds$$

and consequently $y_p = v(x) y_1(x)$ is a particular solution. The general solution can then be written as $y = y_c + y_p$.

∎

### Example 6-9.
Apply the method of variation of parameters to find the solution of

$$L(y) = y'' + y = f(x), \tag{6.74}$$

where $x \in I = \{x \mid a \leq x \leq b\}$, and $f(x)$ is any continuous function over the interval I.
**Solution:** A fundamental system associated with the homogeneous differential equation $L(y) = 0$ is $\{y_1, y_2\}$ where

$$y_1 = \sin x \quad \text{and} \quad y_2 = \cos x. \tag{6.75}$$

The complementary solution is therefore

$$y_c = c_1 \sin x + c_2 \cos x.$$

Using the method of variation of parameters one assumes a particular solution $y_p$ of the form

$$y_p = v_1(x) y_1 + v_2(x) y_2 = v_1(x) \sin x + v_2(x) \cos x. \tag{6.76}$$

Here there are two unknown functions $v_1$ and $v_2$ to be determined. Using equation (6.70) as a guide, require that $v_1$ and $v_2$ satisfy the first condition

$$v'_1 y_1 + v'_2 y_2 = 0. \tag{6.77}$$

Differentiating equation (6.76) produces the result

$$y'_p = v_1 y'_1 + v_2 y'_2 + (v'_1 y_1 + v'_2 y_2),$$

which is simplified by equation (6.77) to the form

$$y'_p = v_1 y'_1 + v_2 y'_2. \tag{6.78}$$

Differentiating equation (6.78) gives

$$y''_p = v_1 y''_1 + v_2 y''_2 + v'_1 y'_1 + v'_2 y'_2. \tag{6.79}$$

When equations (6.79) and (6.76) are substituted into the differential equation (6.74), and terms are rearranged, there results a second condition being imposed upon $v_1$ and $v_2$, namely

$$L(y_p) = v_1 L(y_1) + v_2 L(y_2) + v'_1 y'_1 + v'_2 y'_2 = f(x). \tag{6.80}$$

In equation (6.80) note that $L(y_1) = 0$ and $L(y_2) = 0$, since $\{y_1, y_2\}$ are functions from the fundamental set associated with the homogeneous equation. To summarize, the two conditions that must be satisfied by the functions $v_1$ and $v_2$ are

$$\begin{aligned} v'_1 y_1 + v'_2 y_2 &= 0 \\ v'_1 y'_1 + v'_2 y'_2 &= f(x) \end{aligned} \tag{6.81}$$

which were obtained from equations (6.77) and (6.80). Solving the system of equations (6.81) we have, using Cramer's rule, the following equations for determining the derivatives $v'_1$ and $v'_2$

$$v'_1 = \frac{dv_1}{dx} = \frac{\begin{vmatrix} 0 & y_2 \\ f(x) & y'_2 \end{vmatrix}}{\begin{vmatrix} y_1 & y_2 \\ y'_1 & y'_2 \end{vmatrix}} \quad \text{and} \quad v'_2 = \frac{dv_2}{dx} = \frac{\begin{vmatrix} y_1 & 0 \\ y'_1 & f(x) \end{vmatrix}}{\begin{vmatrix} y_1 & y_2 \\ y'_1 & y'_2 \end{vmatrix}}$$

or

$$\frac{dv_1}{dx} = \frac{-f(x)y_2(x)}{W} \quad \text{and} \quad \frac{dv_2}{dx} = \frac{f(x)y_1(x)}{W}. \tag{6.82}$$

where $W$ is the Wronskian determinant

$$W = \begin{vmatrix} y_1 & y_2 \\ y'_1 & y'_2 \end{vmatrix} = \begin{vmatrix} \sin x & \cos x \\ \cos x & -\sin x \end{vmatrix} = -1.$$

Observe that in equation (6.82) it is required to divide by the Wronskian $W$. One should ask the question, "Can the Wronskian ever be zero?" The answer to this question is that under rather general conditions on linear second order differential operators $L$, it can be demonstrated that if $y_1, y_2$ are linearly independent solutions of $L(y) = 0$, for $x \in I$, then the Wronskian is never zero. Integrating equations (6.82) produces the desired solutions for $v_1$ and $v_2$. These solutions are given by

$$v_1 = v_1(x) = \int_a^x f(\xi) \cos \xi \, d\xi \quad \text{and}$$
$$v_2 = v_2(x) = \int_a^x -f(\xi) \sin \xi \, d\xi. \tag{6.83}$$

Substituting $v_1$ and $v_2$ from equation (6.83) into equation (6.76) gives the particular solution

$$y_p = \sin x \int_a^x f(\xi) \cos \xi \, d\xi - \cos x \int_a^x f(\xi) \sin \xi \, d\xi. \tag{6.84}$$

Here a dummy variable of integration has been placed in equation (6.84). This change of variable has been introduced so that when the integrals were substituted into equation (6.76) there would be no danger of bringing the functions $\sin x$ and $\cos x$ under the integral sign and integrating the wrong functions. The particular solution given by equation (6.84) can also be written in the alternate form

$$y_p = \int_a^x f(\xi) \sin(x - \xi) \, d\xi \tag{6.85}$$

and the general solution can be written $y = y_c + y_p$. The Leibniz rule for differentiating an integral can be used to verify directly that the function defined by equation (6.85) is a particular solution of equation (6.74). Verify the derivatives

$$\frac{dy_p}{dx} = \int_a^x f(\xi) \cos(x - \xi) \, d\xi$$
$$\frac{d^2 y_p}{dx^2} = -\int_a^x f(\xi) \sin(x - \xi) \, d\xi + f(x)$$

and then show $y_p'' + y_p = f(x)$, and hence $y_p$ is a particular solution.

The functions defined by the integrals in equations (6.83) are solutions of the differential equations (6.82) and these solutions can be represented in the alternate forms

$$v_1 = v_1(x) = \int_{k_1}^x f(\xi) \cos \xi \, d\xi = c_1 + \int_a^x f(\xi) \cos \xi \, d\xi$$
$$v_2 = v_2(x) = \int_{k_2}^x f(\xi) \sin \xi \, d\xi = c_2 + \int_a^x f(\xi) \sin \xi \, d\xi,$$

where $k_1$ and $k_2$ are constants. The general solution to equation (6.74) can then be written in the form

$$y(x) = y_1(x) \int_{k_1}^x f(\xi) \cos \xi \, d\xi + y_2(x) \int_{k_2}^x f(\xi) \sin \xi \, d\xi.$$

The form of the new solution is an alternate form of $y = y_c + y_p$ because one can write for any integrable function $F(\xi)$

$$\int_k^x F(\xi)\,d\xi = \int_k^a F(\xi)\,d\xi + \int_a^x F(\xi)\,d\xi = c + \int_a^x F(\xi)\,d\xi,$$

where $c$ is a constant.

For general second order linear differential operators let $\{y_1(x),\ y_2(x)\}$ denote a fundamental set associated with the differential equation

$$L(y) = y'' + \alpha(x)y' + \beta(x)y = 0,$$

for $a \le x \le b$. This set of solutions has the Wronskian $W = \begin{vmatrix} y_1 & y_2 \\ y_1' & y_2' \end{vmatrix} = y_1 y_2' - y_2 y_1'$, and a particular solution of the nonhomogeneous differential equation $L(y) = r(x)$ is given by $y_p = v_1(x)y_1(x) + v_2(x)y_2(x)$, where $v_1, v_2$ are determined by solving the system of equations $v_1' y_1 + v_2' y_2 = 0$ and $v_1' y_1' + v_2' y_2' = r(x)$. Solving this system gives the solutions

$$v_1 = -\int_a^x \frac{r(\xi)y_2(\xi)}{W(\xi)}\,d\xi \quad \text{and} \quad v_2 = \int_a^x \frac{r(\xi)y_1(\xi)}{W(\xi)}\,d\xi.$$

A general solution can then be expressed as

$$y = y_c + y_p = c_1 y_1(x) + c_2 y_2(x) + \int_a^x [y_2(x)y_1(\xi) - y_1(x)y_2(\xi)] \frac{r(\xi)}{W(\xi)}\,d\xi$$

The above result can be generalized to $n$th order linear differential operators $L(y) = a_0(x)D^n y + a_1(x)D^{n-1}y + \cdots + a_{n-1}(x)Dy + a_n(x)y$, with coefficients $a_i(x)$, $i = 0, 1, \ldots, n$ which are continuous over the interval $I = \{x \mid a \le x \le b\}$ where it is assumed that $a_0(x) > 0$ for all $x \in I$. If one knows a fundamental set of solutions $\{y_1(x), y_2(x), \ldots, y_n(x)\}$ associated with the homogeneous differential equation $L(y) = 0$, then the general solution of the nonhomogeneous differential equation $L(y) = f(x)$ can be written in a form

$$y = c_1 y_1 + c_2 y_2 + \cdots + c_n y_n + \int_a^x g(x;\xi)f(\xi)\,d\xi,$$

where $c_1, c_2, \ldots, c_n$ are constants and the function $g(x;\xi)$ is defined in terms of a $n$th order determinant

$$g(x;\xi) = \frac{(-1)^{n-1}}{a_0(\xi)W(\xi)} \begin{vmatrix} y_1(x) & y_2(x) & \cdots & y_n(x) \\ y_1(\xi) & y_2(\xi) & \cdots & y_n(\xi) \\ y_1'(\xi) & y_2'(\xi) & \cdots & y_n'(\xi) \\ \vdots & \vdots & \ddots & \vdots \\ y_1^{(n-2)}(\xi) & y_2^{(n-2)}(\xi) & \cdots & y_n^{(n-2)}(\xi) \end{vmatrix}$$

where $W(\xi)$ is the Wronskian determinant associated with the fundamental set. The function $g(x;\xi)$ is called a one-point Green's function.

## Differential Equations with Variable Coefficients

A linear $n$th order differential equation with variable coefficients can be represented in the form where $y$ and all derivatives of $y$ are represented on the left-hand side of the equation and then denoting the left-hand side by a linear differential operator symbol $L(\ )$. This produces a differential equation having the form

$$L(y) = a_0(x)\frac{d^n y}{dx^n} + a_1(x)\frac{d^{n-1}y}{dx^{n-1}} + a_2(x)\frac{d^{n-2}y}{dx^{n-2}} + \cdots$$
$$+ a_{n-2}(x)\frac{d^2 y}{dx^2} + a_{n-1}(x)\frac{dy}{dx} + a_n(x)y = r(x), \qquad a_0(x) \neq 0 \tag{6.86}$$

If the right-hand side of the equation $r(x)$ is different from zero, the equation is said to be nonhomogeneous, and if $r(x) = 0$, the equation is termed homogeneous. The coefficients of the differential equation (6.86) are denoted $a_0(x), a_1(x), \ldots, a_n(x)$. These coefficient functions and the right hand-side $r(x)$ are assumed to be continuous real functions with continuous derivatives over some interval $I = \{x \mid a \leq x \leq b\}$. Let $x_0$ be a point in the interval $I$ and let $\gamma_0, \gamma_1, \ldots, \gamma_{n-1}$ be constants, then the initial value problem (IVP) to solve

$$L(y) = r(x), \quad x \in I$$
$$y(x_0) = \gamma_0, \ y'(x_0) = \gamma_1, \ldots, y^{(n-1)}(x_0) = \gamma_{n-1}$$

has a unique solution in I provided $a_0(x_0) \neq 0$. If for some value $x = x_0 \in I$ the coefficient $a_0(x_0)$ is zero, then the point $x_0$ is called a singular point of the differential equation.

## Cauchy or Euler Equations

The Cauchy type equations (sometimes referred to as Euler equations) are a special case of equation (6.86) and have the form

$$L(y) = (x^n D^n + \alpha_1 x^{n-1} D^{n-1} + \cdots + \alpha_{n-2} x^2 D^2 + \alpha_{n-1} x D + \alpha_n)y = 0, \qquad D = \frac{d}{dx} \tag{6.87}$$

where $\alpha_1, \alpha_2, \ldots, \alpha_n$ are constants. These types of equations are easy to recognize and solve since in each term of the equation, the power to which $x$ is raised, is the same as the order of the derivative in that term. Cauchy equations can be reduced to differential equations with constant coefficients by making the change of variable

$$x = e^t = \exp(t) \tag{6.88}$$

This change of variable reduces the equation (6.87) to an equation with constant coefficients which has the general form

$$L(y) = \frac{d^n y}{dt^n} + \beta_1 \frac{d^{n-1}y}{dt^{n-1}} + \cdots + \beta_{n-1}\frac{dy}{dt} + \beta_n y = 0, \tag{6.89}$$

where $\beta_1, \beta_2, \ldots, \beta_n$ are constants and $t$ denotes the new independent variable. There are two methods for determining the solution to the Cauchy-Euler equation (6.87).

**Method 1**

Make the substitution $x = e^t$ and produce an equation having the form of equation (6.89). Then assume an exponential solution $y = e^{\lambda t}$ to the resulting differential equation with constant coefficients. This produces a characteristic equation which is a polynomial equation in $\lambda$ with characteristic roots $\lambda_1, \ldots, \lambda_n$. Once the characteristic roots are determined, then a fundamental set of solutions associated with the transformed equation (6.89) can be constructed. The inverse transformation $t = \ln x$, from equation (6.88) converts the fundamental set of solutions, to produce a fundamental set of solutions to the original Cauchy-Euler equation (6.87).

**Method 2**

Instead of making the change of variable given by equation (6.88), note that if equation (6.89) has a solution of the form $y = \exp(\lambda t)$, then the substitution $x = e^t$ specified by equation (6.88) implies that the solution to the Cauchy-Euler equation (6.87) must be of the form $y = e^{\lambda t} = (e^t)^\lambda = x^\lambda$. The characteristic equation can therefore be determined by assuming a solution $y = x^\lambda$ to the Cauchy-Euler equation. From the resulting characteristic equation determine the characteristic roots $\lambda_1, \lambda_2, \ldots, \lambda_n$. There are three cases to consider.

**CASE I (Distinct characteristic roots)** If the characteristic equation associated with equation (6.89) has $n$ distinct roots ($\lambda_1, \lambda_2, \ldots, \lambda_n$), the fundamental set of solutions to equation (6.89) can be represented as

$$y_1 = e^{\lambda_1 t}, \; y_2 = e^{\lambda_2 t}, \ldots, y_n = e^{\lambda_n t}. \tag{6.90}$$

These solutions are given in terms of the independent variable $t$. From equation (6.88) $t$ is related to $x$ by $t = \ln x$, and consequently the fundamental set of solutions to equation (6.87) has the form

$$y_1 = x^{\lambda_1}, \; y_2 = x^{\lambda_2}, \ldots, y_n = x^{\lambda_n} \tag{6.91}$$

**CASE II (Characteristic equation has repeated roots)** If the characteristic equation associated with equation (6.89) has a root $\lambda = r$ of multiplicity $m$, a subset from the set of fundamental solutions to equation (6.89) corresponding to these $m$ repeated roots produces solutions having the form

$$y_1 = e^{rt}, \; y_2 = te^{rt}, \; y_3 = t^2 e^{rt}, \ldots, y_m = t^{m-1} e^{rt}. \tag{6.92}$$

The relation $t = \ln x$ enables us to find the corresponding solutions in the $x$-domain as

$$y_1 = x^r, \; y_2 = x^r \ln x, \; y_3 = x^r (\ln x)^2, \ldots, y_m = x^r (\ln x)^{m-1}. \tag{6.93}$$

That is, each time a characteristic root repeats itself, multiply by $\ln x$.

**CASE III (Characteristic equation has imaginary roots)** If $\lambda = \alpha \pm i\beta$ are imaginary roots of the characteristic equation associated with equation (6.89), then these type of roots produce solutions in the $t$-domain of the form

$$y_1 = e^{\alpha t} \sin \beta t \quad \text{and} \quad y_2 = e^{\alpha t} \cos \beta t.$$

Using the transformation equation $t = \ln x$, the corresponding solutions to equation (6.87) can be expressed in the form

$$y_1 = x^\alpha \sin(\beta \ln x) \quad \text{and} \quad y_2 = x^\alpha \cos(\beta \ln x). \tag{6.94}$$

## Second Order Exact Differential Equations

The differential equation

$$L(y) = a_0(x)y'' + a_1(x)y' + a_2(x)y = 0, \quad ' = \frac{d}{dx}, \quad '' = \frac{d^2}{dx^2} \tag{6.95}$$

is said to be an exact differential equation if it can be written in the form

$$\frac{d}{dx}[A(x)y' + B(x)y] = 0. \tag{6.96}$$

Clearly, if equation (6.95) is to be represented in the form of equation (6.96), then it is necessary that $A(x)$ and $B(x)$ satisfy

$$A = a_0(x), \quad \frac{dA}{dx} + B = a_1(x), \quad \frac{dB}{dx} = a_2(x). \tag{6.97}$$

The equations in (6.97) can be combined, by differentiating the middle expression, to obtain the following result. If the differential equation (6.95) is an exact differential equation, it is required that the coefficients satisfy

$$a_0'' - a_1' + a_2 = 0. \tag{6.98}$$

Whenever a linear second order differential equation is an exact differential equation, one integral is

$$A(x)y' + B(x)y = c_1, \quad A = a_0(x), \quad B = a_1(x) - a_0'(x) \tag{6.99}$$

where $c_1$ is a constant of integration.

## Adjoint Operators

If equation (6.95) is not an exact differential equation and the condition given by equation (6.97) is not satisfied by the coefficients, then let us ask the question "Does there exist a function $\mu = \mu(x)$ such that $\mu(x)L(y) = 0$ is an exact differential equation?" If such a function $\mu(x)$ exists, it is called an integrating factor for the differential equation. If an integrating factor exists, then the differential equation (6.95) can be written in the form

$$\mu a_0 y'' + \mu a_1 y' + \mu a_2 y = \frac{d}{dx}[A(x)y' + B(x)y] = 0. \tag{6.100}$$

The identities

$$\frac{d}{dx}[(a_0\mu)y'] = (a_0\mu)y'' + (a_0\mu)'y' \tag{6.101}$$

and

$$\frac{d}{dx}\{[(a_1\mu) - (a_0\mu)']y\} = [(a_1\mu) - (a_0\mu)']y + [(a_1\mu)' - (a_0\mu)'']y', \tag{6.102}$$

provide a way to represent equation (6.100) in the form

$$\frac{d}{dx}\{(a_0\mu)y' + [(a_1\mu) - (a_0\mu)']y\} + [(a_0\mu)'' - (a_1\mu)' + a_2\mu]y = \frac{d}{dx}[A(x)y' + B(x)y] = 0 \tag{6.103}$$

This equation implies that $\mu(x)L(y) = 0$ is an exact differential equation provided that $\mu = \mu(x)$ satisfies the equation

$$L^*(\mu) = (a_0\mu)'' - (a_1\mu)' + a_2\mu = 0. \tag{6.104}$$

The operator $L^*$, defined by equation (6.104), is called the adjoint operator associated with the second order differential operator $L$ of equation (6.95) and the equation $L^*(\mu) = 0$ is called the adjoint equation associated with equation (6.95). If one multiplies equation (6.95) by the integrating factor $\mu$ and then integrates each term by parts, one also obtains the same result given by equation (6.104). There are many special cases of equation (6.95), where integrating factors can be constructed.

It should be noted that adjoint operators and adjoint equations play an important role in the study of boundary value problems and Green's functions associated with both ordinary and partial differential equations.

The adjoint operator $L^*(\ )$ associated with a general $n$th order linear differential operator $L(\ )$ is defined as follows.

**Definition: (Adjoint operator)** *The adjoint operator $L^*(y)$ associated with the linear operator*

$$L(y) = a_0(x)\frac{d^n y}{dx^n} + a_1(x)\frac{d^{n-1} y}{dx^{n-1}} + \cdots + a_{n-2}(x)\frac{d^2 y}{dx^2} + a_{n-1}(x)\frac{dy}{dx} + a_n(x)y \tag{6.105}$$

*is given by*

$$L^*(y) = (-1)^n \frac{d^n}{dx^n}[a_0(x)y] + (-1)^{n-1}\frac{d^{n-1}}{dx^{n-1}}[a_1(x)y] + \cdots \\ + \frac{d^2}{dx^2}[a_{n-2}(x)y] - \frac{d}{dx}[a_{n-1}(x)y] + a_n(x)y. \tag{6.106}$$

*If $L = L^*$, then $L$ is said to be self-adjoint.*

For example, if
$$L(y) = a_0(x)\frac{d^2 y}{dx^2} + a_1(x)\frac{dy}{dx} + a_2(x)y \qquad (6.107)$$
is a linear second order differential operator, then the adjoint operator associated with $L(\ )$ is given by
$$L^*(y) = \frac{d^2}{dx^2}[a_0(x)y] - \frac{d}{dx}[a_1(x)y] + a_2(x)y \qquad (6.108)$$
The following relation exists between the adjoint operator $L^*$ and the linear second order differential operator $L$
$$vL(u) - uL^*(v) = \frac{d}{dx}[(a_0 v)u' - (a_0 v)'u + a_1 uv]. \qquad (6.109)$$
The relation given by equation (6.109) is known as Lagrange's[7] identity for the second order differential operator $L$. This identity is valid at all points $x$ over some interval $I$, where the functions $L(u), L(v), L^*(u)$, and $L^*(v)$ exist and are continuous. Otherwise the functions $u = u(x)$ and $v = v(x)$ can be arbitrary.

The Lagrange identity for the general $n$th order linear differential operators $L$ and $L^*$ of equations (6.105) and (6.106) is given by
$$vL(u) - uL^*(v) = \frac{d}{dx}[P(u, v)]. \qquad (6.110)$$
where
$$\begin{aligned}
P(u,v) = &u[(-1)^{n-1}D^{n-1}(a_0 v) + (-1)^{n-2}D^{n-2}(a_1 v) + \cdots + (-1)D(a_{n-2}v) + (a_{n-1}v)] \\
&+ (Du)[(-1)^{n-2}D^{n-2}(a_0 v) + (-1)^{n-3}D^{n-3}(a_1 v) + \cdots - D(a_{n-3}v) + a_{n-2}v] \\
&+ (D^2 u)[(-1)^{n-3}D^{n-3}(a_0 v) + \cdots - D(a_{n-4}v) + a_{n-3}v] \\
&+ \cdots \\
&+ (D^{n-2}u)[-D(a_0 v) + a_1 v] \\
&+ (D^{n-1}u)[a_0 v].
\end{aligned} \qquad (6.111)$$
is called the bilinear concomitant, or conjunct of the functions $u$ and $v$ and $D = \frac{d}{dx}$ is a differential operator. The conjunct of the functions $u$ and $v$ can also be represented in the form of a double summation
$$P(u, v) = \sum_{i=0}^{n}\sum_{j=0}^{i-1}(-1)^j D^j[a_{n-i}(x)v(x)]D^{i-j-1}u. \qquad 1$$

The integral of the Lagrange identity, given by equation (6.110) produces the Green's identity
$$\int_a^b v(x)L(u)\,dx - \int_a^b u(x)L^*(v)\,dx = P[u(b), v(b)] - P[u(a), v(a)] \qquad (6.113)$$

---
[7] Joseph Louis Lagrange (1736-1813), French Mathematician.

which involves the conjunct of $u$ and $v$ evaluated at the boundary points $x = a$ and $x = b$. Green's identity can be used to obtain solutions of certain boundary value problems (BVP).

## Forms Associated with Second Order Differential Equations

The homogeneous differential equation

$$L_1(y) = a_0(x)y'' + a_1(x)y' + a_2(x)y = 0, \qquad a_0(x) \neq 0, \qquad ' = \frac{d}{dx} \qquad (6.114)$$

can be represented in many different forms by making appropriate substitutions. The **standard form** for representing equation (6.114) is to divide by the leading coefficient $a_0(x)$ and write

$$L_2(y) = y'' + \beta(x)y' + \alpha(x)y = 0, \qquad (6.115)$$

where

$$\beta(x) = \frac{a_1(x)}{a_0(x)} \quad \text{and} \quad \alpha(x) = \frac{a_2(x)}{a_0(x)}, \qquad a_0(x) \neq 0$$

This division is valid provided $a_0(x) \neq 0$ for all $x$ on the interval $I$ over which the solution is desired. If for some value or values of $x$, $a_0(x) = 0$, then these values of $x$ are called singular points of the differential equation. Singular points are classified as follows: Assume $a_0(x_0) = 0$, then if both the limits

$$\lim_{x \to x_0} (x - x_0) \frac{a_1(x)}{a_0(x)} \quad \text{and} \quad \lim_{x \to x_0} (x - x_0)^2 \frac{a_2(x)}{a_0(x)} \qquad (6.116)$$

exist, then $x_0$ is referred to as a **regular singular point** of the differential equation. If these limits do not exist, then $x_0$ is called an **irregular singular point** of the differential equation. All points $x$ for which $a_0(x)$ is different from zero are termed **ordinary points** of the differential equation.

Any function $f(x)$ which can be represented by a series having sums involving both positive and negative powers of $(x - x_0)$ of the form

$$f(x) = \sum_{i=1}^{M} \frac{b_i}{(x - x_0)^i} + \sum_{i=0}^{\infty} c_i (x - x_0)^i, \qquad (6.117)$$

is said to have a Laurent[8] series expansion about the point $x_0$. The terms $b_i$ for $i = 1, \ldots, M$ and $c_i$, for $i = 0, 1, 2, \ldots$ are constants which are called the coefficients of the Laurent series. If the index $M$ is a finite positive integer and the term $b_M$ is different from zero, then the singularity at $x_0$ is called a pole of order $M$. In the Laurent series (6.117) the summation term going from 1 to $M$ is called the principal part of the Laurent series expansion. The conditions given in equation (6.116) show that if the differential equation (6.115) has a regular singular point at $x = x_0$, then

---

[8] Pierre Alphonse Laurent (1813-1854) French mathematician and physicist.

at worst the coefficient function $\beta(x)$ has a pole of order one and the coefficient function $\alpha(x)$ has a pole of order two. These statements imply the coefficient $\alpha(x)$ has a Laurent series expansion with a principal part containing two terms and the coefficient $\beta(x)$ has a Laurent series expansion with the principal part containing only one term. In this case, the singular point is called a regular singular point. Laurent series are used extensively in the theory of complex variables.

A function $f(x)$ is called analytic at a point $x_0$ if it can be expressed as a power series having the form

$$f(x) = \sum_{n=0}^{\infty} a_n (x - x_0)^n \qquad (6.118)$$

which has a positive radius of convergence. Note that a point $x_0$ is called an ordinary point of the differential equation if both the functions $a_1(x)/a_0(x)$ and $a_2(x)/a_0(x)$ are analytic functions at $x_0$. If this condition is not satisfied, then $x_0$ is called a singular point of the differential equation. A singular point at $x_0$ is classified as being a regular singular point if the conditions of equations (6.116) are satisfied, otherwise it is called an irregular singular point.

Another form into which the second-order differential equation (6.114) or (6.115) can be changed is the **normal form.** Make the change of variable

$$y = y(x) = u(x)v(x) \qquad (6.119)$$

in equation (6.114) and select $u(x)$ such that the resulting differential equation in $v = v(x)$ has no term which involves the first derivative of $v$, then the equation in $v$ is called the **normal form** of equation (6.114). Differentiating equation (6.119) and substituting the derivatives into equation (6.114) one obtains the derived equation

$$v'' + \left(\frac{2u'}{u} + \beta(x)\right) v' + \left[\frac{u'' + \beta(x)u' + \alpha(x)u}{u}\right] v = 0. \qquad (6.120)$$

To remove the first derivative term involving $v'$, impose the condition

$$\frac{2u'}{u} + \beta(x) = 0.$$

The value

$$u = u(x) = \exp\left[-\int \frac{1}{2}\beta(x)\,dx\right]$$

then simplifies equation (6.120) to the normal form

$$v'' + \left[\alpha(x) - \frac{\beta^2(x)}{4} - \frac{\beta'(x)}{2}\right] v = 0. \qquad (6.121)$$

or form which does not contain the first derivative of $v$.

## Abel's Formula

In equation (6.120) note that if $u(x)$ were chosen as a solution to equation (6.115), then $L_2(u) = 0$, and so equation (6.120) would reduce to the equation

$$v'' + \left[\frac{2u'}{u} + \beta(x)\right] v' = 0. \tag{6.122}$$

This implies that if one solution, $u(x)$, to equation (6.115), is known, a second linearly independent solution can be assumed of the form $y = u(x) \cdot v(x)$, where $v = v(x)$ satisfies equation (6.122). Hence, if one solution $u = u(x)$ of equation (6.115) is known, the change of variable $y = u(x)v(x)$ produces a first-order differential equation in the new variable $v'$. This procedure is called the **method of reduction of order**. The method of reduction of order produces a second linearly independent solution provided one solution $u(x)$ of equation (6.115) is known and the differential equation (6.122) can be solved for $v = v(x)$. In equation (6.122), let $v' = z$ and $v'' = z'$ to reduce the equation to the first-order equation

$$\frac{dz}{dx} + \left[\frac{2u'}{u} + \beta(x)\right] z = 0.$$

This equation is easily solved by separation of the variables and then integrating. One integration gives

$$z = z(x) = \frac{dv}{dx} = \frac{1}{u^2(x)} \exp\left[-\int_{x_0}^{x} \beta(x)\,dx\right]$$

and another integration gives $v = v(x)$. Consequently, the second linearly independent solution can be represented as

$$y = y(x) = u(x) \int_{x_0}^{x} \frac{1}{u^2(s)} \exp\left[-\int_{x_0}^{s} \beta(t)\,dt\right] ds$$

a result known as Abel's[9] formula.

To summarize, if $y_1(x)$ is a known solution of

$$L_2(y) = y'' + \beta(x)y' + \alpha(x)y = 0,$$

then a second linearly independent solution can be obtained from Abel's formula

$$y_2(x) = y_1(x) \int_{x_0}^{x} \frac{1}{y_1^2(s)} \exp\left[-\int_{x_0}^{s} \beta(t)\,dt\right] ds. \tag{6.123}$$

The general solution is then any linear combination of the functions $y_1$ and $y_2$.

---

[9] Niels Henrik Abel (1802-1829) Norwegian Mathematician.

## Self Adjoint Form

Still another form associated with differential equations (6.114) and (6.115) is the **self-adjoint form**.

$$L(y) = \frac{d}{dx}\left[p(x)\frac{dy}{dx}\right] + q(x)y = 0. \tag{6.124}$$

To transform equation (6.114) into this self-adjoint form multiply the differential equation (6.114) by a function $h(x)$ and compare the result with the expanded form of equation (6.124). Note that if $h(x)$ satisfies the equations

$$h(x)a_0(x) = p(x), \qquad h(x)a_1(x) = p'(x), \qquad h(x)a_2(x) = q(x) \tag{6.125}$$

then it is possible to transform equation (6.114) to the self-adjoint form given by equation (6.124). In order for the relations given by equations (6.125) to be consistent, it is necessary to select $h(x)$ as the solution to the differential equation

$$p'(x) = h(x)a_0'(x) + h'(x)a_0(x) = h(x)a_1(x).$$

This produces the function

$$h = h(x) = \frac{1}{a_0(x)} \exp\left[\int \frac{a_1(x)}{a_0(x)} dx\right], \qquad a_0(x) \neq 0.$$

which transforms the equation (6.114) into the self-adjoint form of equation (6.124). Note that the operator $L^*$ of equation (6.124) equals the operator $L$ and hence $L$ is called a self-adjoint operator.

## Nonhomogeneous Equations

For the nonhomogeneous equation

$$L(y) = y'' + \beta(x)y' + \alpha(x)y = r(x), \qquad x \in I = \{x \mid a \leq x \leq b\} \tag{6.126}$$

assume $\alpha(x)$ and $\beta(x)$ are defined and continuous for all $x \in I$. After obtaining two linearly independent solutions of the homogeneous equation $L(y) = 0$, it is then necessary to obtain a particular solution to the nonhomogeneous equation. The method of variation of parameters is applicable for constructing particular solutions to linear differential equations with variable coefficients. Let $y_1(x)$, $y_2(x)$ be a fundamental set associated with the homogeneous equation $L(y) = 0$. By the method of variation of parameters construct the general solution in the form

$$y(x) = K_1 y_1(x) + K_2 y_2(x) + \int_a^x Y_2(x;\xi) r(\xi)\, d\xi$$

where $K_1, K_2$ are constants and

$$Y_2(x;\xi) = \frac{y_1(\xi)y_2(x) - y_1(x)y_2(\xi)}{y_1(\xi)y_2'(\xi) - y_1'(\xi)y_2(\xi)}$$

is called a one-point Green's function or influence function. The one-point Green's function satisfies the conditions

$$Y_2(\xi;\xi) = 0 \quad \text{and} \quad Y_2'(\xi;\xi) = 1.$$

The function $Y_2(x;\xi)$ can be constructed directly from the fundamental set of solutions associated with the homogeneous equation. If in addition the requirement that the solution to equation (6.126) satisfy the initial value problem $y(a) = A$, $y'(a) = B$, then $K_1$ and $K_2$ must be selected such that the equations

$$y(a) = K_1 y_1(a) + K_2 y_2(a) = A$$
$$y'(a) = K_1 y_1'(a) + K_2 y_2'(a) = B$$

are satisfied. Solving for the constants $K_1$ and $K_2$, enables the solution to be written in the form

$$y(x) = AY_1(x;a) + BY_2(x;a) + \int_a^x Y_2(x;\xi) r(\xi)\, d\xi. \tag{6.127}$$

where

$$Y_1(x;a) = \frac{y_1(x) y_2'(a) - y_1'(a) y_2(x)}{y_1(a) y_2'(a) - y_1'(a) y_2(a)}$$

Observe that $Y_1(x;a)$ satisfies the initial conditions

$$Y_1(a;a) = 1 \quad \text{and} \quad Y_1'(a;a) = 0.$$

The functions $Y_1$ and $Y_2$ are linear combinations of functions from the fundamental set and hence they both satisfy the homogeneous equation. The functions $Y_1$ and $Y_2$ can be easily constructed using algebraic methods. These results can be summarized as follows.

Let $\{y_1(x), y_2(x)\}$ be a fundamental system associated with the homogeneous equation $L(y) = 0$ and construct the functions

$$Y_1(x;\xi) = c_1 y_1(x) + c_2 y_2(x)$$
$$Y_2(x;\xi) = c_3 y_1(x) + c_4 y_2(x) \tag{6.128}$$

by appropriately choosing the constants $c_1$, $c_2$, $c_3$ and $c_4$ such that the following conditions are satisfied

$$Y_1(\xi;\xi) = 1 \qquad Y_1'(\xi;\xi) = 0$$
$$Y_2(\xi;\xi) = 0 \qquad Y_2'(\xi;\xi) = 1.$$

The functions $Y_1$ and $Y_2$ are called a normalized fundamental set associated with the homogeneous equation (6.126). The solution to the nonhomogeneous equation (6.126) can be expressed as

$$y(x) = AY_1(x;a) + BY_2(x;a) + \int_a^x Y_2(x;u) r(u)\, du. \tag{6.129}$$

The function $Y_2(x;u)$ is called the Green's function for the one point boundary value problem. To prove the above result, observe that the function $y = AY_1 + BY_2$ satisfies the homogeneous equation and has two arbitrary constants, so that one need only show the integral term is a particular solution of the nonhomogeneous equation. Using the Leibniz's rule one obtains

$$y_p = \int_a^x Y_2(x;u)r(u)\,du$$
$$y_p' = \int_a^x Y_2'(x;u)r(u)\,du + Y_2(x;x)r(x)$$
$$y_p'' = \int_a^x Y_2''(x;u)r(u)\,du + Y_2'(x;x)r(x).$$

Observe that the function $Y_2$ satisfies $Y_2(x;x) = 0$ and $Y_2'(x;x) = 1$. Upon substituting the above integrals into equation (6.126), there results

$$L(y_p) = \int_a^x L(Y_2)r(u)\,du + r(x) = r(x)$$

Observe that $L(Y_2) = 0$ because $Y_2$ is a linear combination of solutions of the homogeneous equation. Note also that the above form of the solution requires equation (6.126) be written in standard form.

### Series Representation

The solution of differential equations with variable coefficients are in general extremely difficult to construct. However, in certain special cases, the solution in the form of an infinite series can be obtained. The following is a review of some fundamental concepts concerning infinite series.

**Properties of Infinite Series**

1. $n$th term test: If the series $\sum_{n=1}^{\infty} u_n$ converges, then the $n$th term must approach zero as $n$ increases without bound and $\lim_{n \to \infty} u_n = 0$.

2. $n$th term test: If the limit $\lim_{n \to \infty} u_n$ does not exist or is a number different from zero, then the series $\sum_{n=1}^{\infty} u_n$ diverges.

3. Absolute value: If $\sum_{n=1}^{\infty} |u_n|$ is convergent, then $\sum_{n=1}^{\infty} u_n$ is called an absolutely convergent series.

4. Ratio test: If $\lim_{n \to \infty} \frac{u_{n+1}}{u_n} = \rho$ exists, then the series $\sum_{n=1}^{\infty} u_n$, $u_n \geq 0$ converges if $\rho < 1$, diverges if $\rho > 1$, and the test fails if $\rho = 1$.

5. Root test: If $\lim_{n \to \infty} \sqrt[n]{|u_n|} = \rho$ exists, then the series $\sum_{n=1}^{\infty} u_n$ is absolutely convergent if $\rho < 1$, diverges if $\rho > 1$, and the test fails if $\rho = 1$.

6. Comparison test for convergence: If $|u_n| \leq v_n$ for $n = 1, 2, 3, \ldots$ and $\sum_{n=1}^{\infty} v_n$ converges, then $\sum_{n=1}^{\infty} u_n$ is absolutely convergent.

7. Comparison test for divergence: If $u_n \geq v_n \geq 0$ for $n = 1, 2, 3, \ldots$ and $\sum_{n=1}^{\infty} v_n$ diverges, then $\sum_{n=1}^{\infty} u_n$ diverges.

8. Integral test: If $u_n = f(n)$ and $f(x)$ is a positive continuous function on the interval $m \leq x < \infty$ such that $f(x)$ decreases as $x$ increases and $\lim_{x \to \infty} f(x) = 0$, then the series $\sum_{n=1}^{\infty} u_n$ converges if and only if the improper integral $\int_m^{\infty} f(x)\, dx$ exists.

9. Alternating series test: If $u_{n+1} \leq u_n$ for $n = 1, 2, 3, \ldots$ and $\lim_{n \to \infty} u_n = 0$, then the alternating series $\sum_{n=1}^{\infty} (-1)^n u_n$ converges.

An infinite series of the form

$$\sum_{n=0}^{\infty} a_n (z - \alpha)^n = a_0 + a_1 (z - \alpha) + a_2 (z - \alpha)^2 + \cdots + a_m (z - \alpha)^m + \cdots \qquad (6.130)$$

is called a power series in $(z - \alpha)$. In this series, $a_0, a_1, a_2, \ldots$ are constants and are called the coefficients of the power series, and the constant $\alpha$ is called the center of the power series. This power series converges for all values of $z$ satisfying $|z - \alpha| < R$, where $R$ is called the radius of convergence of the power series and can be calculated from the limits

$$\frac{1}{R} = \lim_{m \to \infty} \left| \frac{a_{m+1}}{a_m} \right| \quad \text{or} \quad \frac{1}{R} = \lim_{m \to \infty} \sqrt[m]{|a_m|} \qquad (6.131)$$

if these limits exist. In the above power series, $z, \alpha, a_i, i = 0, 1, 2, \ldots$, can be complex numbers or real numbers.

If the power series given by equation (6.130) converges for each value of $z$ within the circular region (or interval) $I = \{z \mid |z - \alpha| < R\}$ of radius $R$ centered at $\alpha$, then the function $f(z)$ defined by

$$f(z) = \sum_{n=0}^{\infty} a_n (z - \alpha)^n, \quad z \in I \qquad (6.132)$$

has the following properties:

1. $f(z)$ is a continuous function in I.

2. $f(z)$ has a derivative given by $f'(z) = \sum_{n=0}^{\infty} n a_n (z - \alpha)^{n-1}$, $z \in I$ and this series representing the derivative has the same radius of convergence as equation (6.132) (i.e. term wise differentiation is permissible.)

3. $f(z)$ has an integral given by $\int_{\alpha}^{z} f(z)\, dz = \sum_{n=0}^{\infty} \frac{a_n}{n+1} (z - \alpha)^{n+1}$, $z \in I$ and the series representing the integral has the same radius of convergence as equation (6.132). (i.e., term wise integration is permissible .)

4. In working with series it should be observed that the summation index is a dummy index and at times it becomes convenient to shift the summation index. For example,

$$\sum_{n=m}^{\infty} a_n (z - \alpha)^n = \sum_{j=m}^{\infty} a_j (z - \alpha)^j = \sum_{n=0}^{\infty} a_{n+m} (z - \alpha)^{n+m}.$$

Here $n$ was replaced by $n+m$ and the initial value of the index $n$ was shifted from $n=m$ to $n=0$.

5. If $\sum_{n=0}^{\infty} a_n x^n = 0$ for all $x$ in some interval $I$ and the coefficients $a_n$ of the power series are constants, then $a_n = 0$ for all $n$. For if the function

$$F(x) = a_0 + a_1 x + a_2 x^2 + a_3 x^3 + \cdots$$

is identically zero for all $x$, then when $x = 0$ there results $a_0 = 0$. If $F(x)$ is identically zero, then the derivative $F'(x) = 0$ and, consequently, $F'(0) = a_1 = 0$. Continuing in this manner

$$F''(0) = 2! a_2 = 0, \quad F'''(0) = 3! a_3 = 0, \cdots$$

which implies $a_n = 0$ for all $n$.

### Some Well Known Power Series

If $f(x)$ has derivatives of all orders and can be represented in some interval of convergence about $x = \alpha$ by the power series

$$f(x) = f(\alpha) + f'(\alpha)(x-\alpha) + \frac{f''(\alpha)}{2!}(x-\alpha)^2 + \cdots + \frac{f^{(n-1)}(\alpha)}{(n-1)!}(x-\alpha)^{n-1} + \cdots$$
$$f(x) = \sum_{n=0}^{\infty} \frac{f^{(n)}(\alpha)}{n!}(x-\alpha)^n \tag{6.133}$$

then $f(x)$ is said to have a **Taylor**[10] series expansion about the point $x = \alpha$. The $N$th partial sum of the above Taylor series

$$S_N = \sum_{n=0}^{N} \frac{f^{(n)}(\alpha)}{n!}(x-\alpha)^n \tag{6.134}$$

is called the $N$th degree Taylor polynomial of $f(x)$ at $x = \alpha$. It is often convenient to make the substitution $x = \alpha + h$ and represent the Taylor series expansion in the form

$$f(\alpha + h) = f(\alpha) + h f'(\alpha) + \frac{h^2}{2!} f''(\alpha) + \cdots + \frac{h^{n-1}}{(n-1)!} f^{(n-1)}(\alpha) + \cdots \tag{6.135}$$

**Taylor's Theorem** states that if a function $f(x)$ and its first $n$ derivatives are single-valued and continuous in the interval $\alpha \leq x \leq \alpha + h$, then $f(x)$ can be represented in the form

$$f(x) = f(\alpha) + f'(\alpha)(x-\alpha) + \frac{f''(\alpha)}{2!}(x-\alpha)^2 + \cdots + \frac{f^{(n-1)}(\alpha)}{(n-1)!}(x-\alpha)^{n-1} + R_n$$
$$\text{or} \quad f(\alpha + h) = f(\alpha) + h f'(\alpha) + \frac{h^2}{2!} f''(\alpha) + \cdots + \frac{h^{n-1}}{(n-1)!} f^{(n-1)}(\alpha) + R_n \tag{6.136}$$

---
[10] Brook Taylor (1685-1731)

where $R_n$ is a remainder term which can be represented in the Lagrange form

$$R_n = \frac{(x-\alpha)^n}{n!} f^{(n)}(\alpha + \theta(x-\alpha)), \quad \text{or} \quad R_n = \frac{h^n}{n!} f^{(n)}(\alpha + \theta h), \quad 0 < \theta < 1 \quad (6.137)$$

A **Maclaurin**[11] series is a special case of a Taylor series with center $\alpha = 0$, and

$$f(x) = \sum_{n=0}^{\infty} \frac{f^{(n)}(0)}{n!} x^n = f(0) + \frac{f'(0)}{1!} x + \frac{f''(0)}{2!} x^2 + \cdots + \frac{f^{(m)}(0)}{m!} x^m + \cdots$$

The following are some easily recognizable Taylor series which converge for all values of $x$

$$e^x = 1 + x + \frac{x^2}{2!} + \frac{x^3}{3!} + \cdots + \frac{x^n}{n!} + \cdots$$

$$\sin x = x - \frac{x^3}{3!} + \frac{x^5}{5!} - \cdots + (-1)^n \frac{x^{2n+1}}{(2n+1)!} + \cdots$$

$$\cos x = 1 - \frac{x^2}{2!} + \frac{x^4}{4!} - \cdots + (-1)^n \frac{x^{2n}}{(2n)!} + \cdots$$

$$\sinh x = x + \frac{x^3}{3!} + \frac{x^5}{5!} + \cdots + \frac{x^{2n+1}}{(2n+1)!} + \cdots$$

$$\cosh x = 1 + \frac{x^2}{2!} + \frac{x^4}{4!} + \cdots + \frac{x^{2n}}{(2n)!} + \cdots$$

Some other series of interest are:

$$\ln(1+x) = x - \frac{x^2}{2} + \frac{x^3}{3} - \cdots + (-1)^n \frac{x^{n+1}}{(n+1)} + \cdots \quad \text{which converges for } -1 < x \leq 1.$$

The binomial series

$$(1+x)^n = 1 + \binom{n}{1} x + \binom{n}{2} x^2 + \binom{n}{3} x^3 + \cdots \quad \text{which converges for } |x| < 1.$$

A special case of the binomial series occurs for $n = -1$

$$(1 \pm x)^{-1} = 1 \mp x + x^2 \mp x^3 + x^4 \mp \cdots, \quad |x| < 1$$

## Solution of Differential Equations by Series Methods

In constructing a series solution to the second order linear differential equation

$$\frac{d^2 y}{dx^2} + \frac{a_1(x)}{a_0(x)} \frac{dy}{dx} + \frac{a_2(x)}{a_0(x)} y = 0 \quad (6.138)$$

the type of series one works with depends upon the point chosen as the center of the series. In particular, if a point $x = x_0$ is an ordinary point of the linear ordinary differential equation, then there exists a power series solution, centered at $x = x_0$, of the form

$$y(x) = \sum_{n=0}^{\infty} C_n (x - x_0)^n = C_0 + C_1 (x - x_0) + C_2 (x - x_0)^2 + \cdots \quad (6.139)$$

---

[11] Colin Maclaurin (1698-1746) Scottish mathematician

In general, the radius of convergence of this series is the distance from the center of the series $x = x_0$ to the nearest singular point of the differential equation, if a singular point exists.

If a point $x = x_0$ is a regular singular point of a linear ordinary differential equation, then a Frobenius[12] type series solution exists. A Frobenius type series solution centered at $x = x_0$, has the form

$$y(x) = \sum_{n=0}^{\infty} C_n(r)(x-x_0)^{n+r} = (x-x_0)^r \left[ C_0 + C_1(x-x_0) + C_2(x-x_0)^2 + \cdots \right], \qquad (6.140)$$

where $C_0$ must be different from zero and $r$ is a constant to be determined. Note the notation $C_n = C_n(r)$ is introduced to emphasize that the coefficients in a Frobenius series solution depend upon the value selected for the constant $r$. The radius of convergence of this series is the annular region $0 < |x - x_0| < R$, where $R$ is the distance from the center $x_0$ to the nearest other singular point of the differential equation, if such a singular point exists.

If the point $x = x_0$ is an irregular singular point of the ordinary differential equation, then a Frobenius type solution may or may not exist.

**The point $x_0$ is an ordinary point**

If $x_0$ is an ordinary point of the differential equation (6.138) and the coefficients in the differential equation satisfy the conditions

$$\frac{a_1(x)}{a_0(x)} = \sum_{n=0}^{\infty} A_n(x-x_0)^n, \quad \text{for} \quad |x-x_0| < R_1 \qquad (6.141)$$

and

$$\frac{a_2(x)}{a_0(x)} = \sum_{n=0}^{\infty} B_n(x-x_0)^n, \quad \text{for} \quad |x-x_0| < R_2 \qquad (6.142)$$

then the differential equation (6.138) has a power series solution

$$y = y(x) = \sum_{n=0}^{\infty} C_n(x-x_0)^n, \quad \text{for} \quad |x-x_0| < R = \min(R_1, R_2) \qquad (6.143)$$

This assumed solution has the derivatives

$$\frac{dy}{dx} = \sum_{n=0}^{\infty} nC_n(x-x_0)^{n-1}, \quad \text{and} \quad \frac{d^2y}{dx^2} = \sum_{n=0}^{\infty} n(n-1)C_n(x-x_0)^{n-2} \qquad (6.144)$$

When the equation (6.141), (6.142), (6.143), and (6.144) are substituted into the differential equation (6.138) and like terms are collected there results the equation

$$\sum_{n=0}^{\infty} E_n x^n = E_0 + E_1(x-x_0) + E_2(x-x_0)^2 + \cdots + E_n(x-x_0)^n + \cdots = 0 \qquad (6.145)$$

---

[12] Ferdinand George Frobenius (1849–1917), German mathematician.

where $E_0, E_1, E_2, \ldots, E_n, \ldots$ are each some function of the coefficients from the sets

$$A = \{A_0, A_1, \ldots, A_n, \ldots\}, \qquad B = \{B_0, B_1, \ldots, B_n, \ldots\}, \qquad C = \{C_0, C_1, \ldots, C_n, \ldots\}$$

which represent sets containing the coefficients from the series in equations (6.141), (6.142), and (6.143).

In order for the equation (6.145) to be satisfied for all values of $x$, require

$$E_0 = 0, \quad E_1 = 0, \quad E_2 = 0, \quad \ldots \quad E_n = 0, \quad \ldots \tag{6.146}$$

This produces a set of equations where elements from the set $C$ can be determined. The equation $E_n = 0$ usually produces a recurrence relation to be satisfied by the coefficients from the set $C$. The conditions $E_0 = 0$, $E_1 = 0, \ldots$, together with the condition $E_n = 0$ enable one to express the coefficients $C_2, C_3, C_4, \ldots$ in terms of the coefficients $C_0$ and $C_1$.

### Example 6-10.

Find a power series solution about $x = 0$ which satisfies the differential equation

$$y'' + y = 0$$

**Solution:** Here $x_0 = 0$ is an ordinary point of the given differential equation and so a power series solution exists having the form $y = \sum_{n=0}^{\infty} C_n x^n$, where $C_n, n = 0, 1, 2, \ldots$ are constants to be determined. Differentiate the assumed series solution

$$y' = \sum_{n=0}^{\infty} n C_n x^{n-1}, \qquad y'' = \sum_{n=0}^{\infty} n(n-1) C_n x^{n-2}$$

and substitute these derivatives into the given differential equation to obtain

$$\sum_{n=0}^{\infty} n(n-1) C_n x^{n-2} + \sum_{n=0}^{\infty} C_n x^n = 0. \tag{6.147}$$

Examine the sums in equation (6.147) and observe that the first sum is zero for $n = 0$ and $n = 1$. Hence, one can write

$$\sum_{n=2}^{\infty} n(n-1) C_n x^{n-2} + \sum_{n=0}^{\infty} C_n x^n = 0. \tag{6.148}$$

By shifting the summation index on the first sum in equation (6.148) it is possible to express this equation in the form

$$\sum_{n=0}^{\infty} E_n x^n = \sum_{n=0}^{\infty} \left[(n+2)(n+1) C_{n+2} + C_n\right] x^n = 0. \tag{6.149}$$

For equation (6.149) to be identically zero for all $x$, the coefficients must satisfy

$$E_n = (n+2)(n+1)C_{n+2} + C_n = 0 \qquad (6.150)$$

for $n = 0, 1, 2, \ldots$. Equation (6.150) is called a recurrence relation between the coefficients of the power series. From this recurrence relation solve for the higher coefficients in terms of the lower numbered coefficients. This produces the equation

$$C_{n+2} = \frac{-C_n}{(n+2)(n+1)}$$

which is valid for $n = 0, 1, 2, \ldots$. When substituting in the values $n = 0, 1, 2, \ldots$, there results a set of equations which must be satisfied if the series is to be a solution of the given differential equation. These equations can be written as

$$n=0, \quad C_2 = \frac{-C_0}{2 \cdot 1} \qquad\qquad n=1, \quad C_3 = \frac{-C_1}{3 \cdot 2}$$

$$n=2, \quad C_4 = \frac{-C_2}{4 \cdot 3} \qquad\qquad n=3, \quad C_5 = \frac{-C_3}{5 \cdot 4}$$

$$\vdots \qquad\qquad\qquad\qquad \vdots$$

$$n=2m, \quad C_{2m} = \frac{-C_{2m-2}}{(2m)(2m-1)} \qquad n=2m-1, \quad C_{2m+1} = \frac{-C_{2m-1}}{(2m+1)(2m)}$$

Note that multiplying the even coefficients together there results

$$C_2 C_4 C_6 \cdots C_{2m-2} C_{2m} = \frac{(-1)^m C_0 C_2 C_4 C_6 \ldots C_{2m-2}}{(1 \cdot 2)(3 \cdot 4) \cdot (2m-1)(2m)}.$$

This simplifies to

$$C_{2m} = \frac{(-1)^m C_0}{(2m)!}, \quad m = 1, 2, 3, \ldots \qquad (6.151)$$

Similarly, multiplying the odd coefficients together and simplifying gives

$$C_{2m+1} = \frac{(-1)^m C_1}{(2m+1)!}, \quad m = 1, 2, 3, \ldots \qquad (6.152)$$

Substituting these constants into the assumed power series solution produces

$$y = C_0 + C_1 x - \frac{C_0 x^2}{2!} - \frac{C_1 x^3}{3!} + \frac{C_0 x^4}{4!} + \frac{C_1 x^5}{5!} - \frac{C_0 x^6}{6!} - \frac{C_1 x^7}{7!} + \cdots,$$

where $C_0$ and $C_1$ are arbitrary constants. After rearranging terms, the solution can be represented as

$$\begin{aligned} y = &C_0 \left( 1 - \frac{x^2}{2!} + \frac{x^4}{4!} - \frac{x^6}{6!} + \cdots + \frac{(-1)^m x^{2m}}{(2m)!} + \cdots \right) \\ &+ C_1 \left( x - \frac{x^3}{3!} + \frac{x^5}{5!} - \cdots - \frac{(-1)^m x^{2m+1}}{(2m+1)!} + \cdots \right). \end{aligned} \qquad (6.153)$$

where $C_0$ and $C_1$ are arbitrary constants.

Observe that the series in equation (6.153) are the series for $\cos x$ and $\sin x$. The equation (6.153) is the series form of the general solution $y = C_0 \cos x + C_1 \sin x$ where $C_0, C_1$ are arbitrary constants.

■

**The point $x_0$ is a regular singular point**

Let the point $x_0$ be a singular point of the differential equation (6.138). This singular point is called an isolated singularity of the differential equation if and only if it is possible to find a neighborhood of $x_0$ containing no other singular points. This isolated singular point is called a regular singular point if the following conditions are satisfied

$$(x - x_0)\frac{a_1(x)}{a_0(x)} = \sum_{n=0}^{\infty} A_n(x - x_0)^n, \quad \text{for} \quad |x - x_0| < R_1 \qquad (6.154)$$

and

$$(x - x_0)^2 \frac{a_2(x)}{a_0(x)} = \sum_{n=0}^{\infty} B_n(x - x_0)^n, \quad \text{for} \quad |x - x_0| < R_2 \qquad (6.155)$$

where $A_n$ and $B_n$ are constants. If equations (6.154) and (6.155) are satisfied, then the differential equation (6.138) can be expressed in the form

$$(x - x_0)^2 \frac{d^2y}{dx^2} + \left[(x - x_0)\frac{a_1(x)}{a_0(x)}\right](x - x_0)\frac{dy}{dx} + \left[(x - x_0)^2 \frac{a_2(x)}{a_0(x)}\right] y = 0 \qquad (6.156)$$

Assume a Frobenius type solution

$$y = y(x) = (x - x_0)^r \sum_{n=0}^{\infty} C_n(x - x_0)^n = \sum_{n=0}^{\infty} C_n(x - x_0)^{n+r} \qquad (6.157)$$

where $C_n$, $n = 0, 1, 2, \ldots$ are constants to be determined with the requirement $C_0$ can be arbitrary, but $C_0 \neq 0$. The exponent $r$, called the exponent belonging to the singularity at $x_0$, is also a constant to be determined.

The derivatives of the assumed solution have the forms

$$\frac{dy}{dx} = \sum_{n=0}^{\infty}(n+r)C_n(x - x_0)^{n+r-1}, \quad \text{and} \quad \frac{d^2y}{dx^2} = \sum_{n=0}^{\infty}(n+r)(n+r-1)C_n(x - x_0)^{n+r-2} \qquad (6.158)$$

Substituting the results from equations (6.158), (6.157), (6.154) and (6.155) into the differential equation (6.156) and then rearranging terms one obtains an equation of the form

$$(x - x_0)^r \left[E_0 + E_1(x - x_0) + E_2(x - x_0)^2 + \cdots + E_n(x - x_0)^n + \cdots\right] = 0 \qquad (6.159)$$

where again the terms $E_0, E_1, \ldots$ are functions of the coefficients from the equations (6.154), (6.155) and (6.157). In order for equation (6.159) to be satisfied for all values of $x$ it is required that

$$E_0 = 0, \quad E_1 = 0, \quad \ldots, \quad E_n = 0, \quad \ldots \qquad (6.160)$$

The equation $E_0$ will be a quadratic equation of the form

$$E_0 = C_0 \left[ r^2 + (A_0 + 1)r + B_0 \right] = 0 \tag{6.161}$$

which is called the indicial equation associated with the differential equation (6.156) and the assumed solution (6.157). Note the requirement that $C_0 \neq 0$ requires the solution of a quadratic equation in $r$ have two roots $r_1$ and $r_2$, called the indicial roots associated with the Frobenius solution. The equation $E_n = 0$ produces a recurrence relation between the coefficients $C_n$ which is dependent upon the value of $r$ selected. An analysis of the equations $E_0 = 0$, $E_1 = 0$ and $E_n = 0$ produces the following cases to be considered.

**CASE I** If the indicial roots $r_1, r_2$ are distinct and do not differ by an integer, then there exists independent solutions having the form

$$y_1(x) = (x - x_0)^{r_1} \sum_{n=0}^{\infty} C_n(r_1)(x - x_0)^n, \quad C_0(r_1) \neq 0,$$

$$y_2(x) = (x - x_0)^{r_2} \sum_{n=0}^{\infty} C_n(r_2)(x - x_0)^n, \quad C_0(r_2) \neq 0$$

corresponding to the exponents $r_1$ and $r_2$. Here the recurrence relation for $C_n(r)$ is dependent upon which of the indicial roots is used in the equation $E_n = 0$.

**CASE II** If the indicial roots are equal, $r_1 = r_2$, then two linearly independent solutions can be found having the forms

$$y_1(x) = (x - x_0)^{r_1} \sum_{n=0}^{\infty} C_n(r_1)(x - x_0)^n, \quad C_0(r_1) \neq 0,$$

$$y_2(x) = y_1(x) \ln(x - x_0) + (x - x_0)^{r_1} \sum_{n=1}^{\infty} C_n^*(r_1)(x - x_0)^n,$$

where the coefficients $C_n^*(r_1)$ may or may not equal zero.

**CASE III** If the indicial roots $r_1$ and $r_2$ differ by an integer and $r_1 - r_2 = N$, then two linearly independent solutions can be found having the forms

$$y_1(x) = (x - x_0)^{r_1} \sum_{n=0}^{\infty} C_n(r_1)(x - x_0)^n, \quad C_0(r_1) \neq 0,$$

$$y_2(x) = K y_1(x) \ln(x - x_0) + (x - x_0)^{r_2} \sum_{n=0}^{\infty} C_n^*(r_2)(x - x_0)^n, \quad C_0^*(r_2) \neq 0$$

the coefficient $K$ may or may not equal zero.

## Numerical Solution to First Order Differential Equations

There are numerous numerical methods available for solving first order ordinary differential equations. The fourth order classic Runge-Kutta method to solve the initial value problem

$$\frac{dy}{dx} = f(x, y), \quad y(x_0) = y_0, \quad x_0 \leq x \leq x_n \tag{6.162}$$

is one of the more popular methods because the method is rugged enough to handle a wide range of reasonably well behaved differential equations and the method can be extended to handle systems of differential equations.

### Runge-Kutta fourth order method (classic form)

This method is a single step marching method to solve the initial value problem given by equation (6.162). The algorithm begins at the initial point $(x_0, y_0)$ and uses a step size $h$ to advance the solution to the next point $(x_1, y_1)$ where $x_1 = x_0 + h$ and $y_1$ is determined from the calculations

$$k_1 = hf(x_0, y_0)$$
$$k_2 = hf(x_0 + \frac{1}{2}h, y_0 + \frac{1}{2}k_1)$$
$$k_3 = hf(x_0 + \frac{1}{2}h, y_0 + \frac{1}{2}k_2)$$
$$k_4 = hf(x_0 + y, y_0 + k_3)$$
$$y_1 = y_0 + \frac{1}{6}(k_1 + 2k_2 + 2k_3 + k_4)$$

The point $(x_0, y_0)$ in equation (6.162) is now replaced by $(x_1, y_1)$ and the algorithm is repeated. The repeating stops when the stepping process gets to the end point $x_n$. The accuracy of the algorithm is controlled by the step size $h$. Using a very small step size $h$ takes many calculations to achieve the final point $(x_n, y_n)$, but the final global error is usually small. Here the global error is on the order $O(h^4)$. A rough rule of thumb is to solve the differential equation with a step size $h$ and make note of the final value $(x_n, y_n)$, then solve the problem again using a step size $h/2$ to see how halving the step size effects the final answer $(x_n, y_n)$. If there is a large difference in the final result, then halve the step size again and repeat the process of examining the error. When there is a very small change in the final result you can have confidence that the algorithm is providing proper answers. Plotting the points $(x_0, y_0), (x_1, y_1), (x_2, y_2), \ldots (x_n, y_n)$ and connecting the points by straight line segments gives one form a graphical representation of the solution.

Higher ordered differential equations can be reduced to a system of first order differential equations by making appropriate substitutions. The above Runge-Kutta method can then be applied to each first order differential equation in the system of first order equations.

# Chapter 7
# Special Functions

## Integer Function

The integer function Int($x$) or $[x]$ is defined

$[x] = \text{Int}(x) = n, \quad n \leq x < n+1, \quad \text{for} \quad n = 0, \pm 1, \pm 2, \ldots$

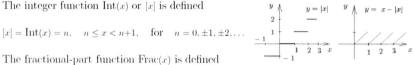

The fractional-part function Frac($x$) is defined

$\text{Frac}(x) = x - [x]$

## Heaviside Step Function

The Heaviside step function, sometimes called the unit step function, is defined

$$H(x - x_0) = \begin{cases} 0, & x < x_0 \\ \frac{1}{2}, & x = x_0 \\ 1, & x > x_0 \end{cases}$$

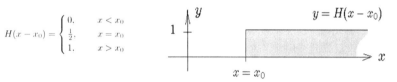

## Impulse Function and Dirac Delta Function

Impulse function
$$\delta_\epsilon(x - x_0) = \frac{1}{\epsilon}[H(x - x_0) - H(x - (x_0 + \epsilon))]$$

Dirac delta function
$$\delta(x - x_0) = \lim_{\epsilon \to 0} \delta_\epsilon(x - x_0)$$

**Other ways to define the Dirac delta function**

$$\delta(x - x_0) = \lim_{T \to \infty} \sqrt{\frac{T}{\pi}} e^{-T(x-x_0)^2}, \qquad \delta(x - x_0) = \lim_{T \to \infty} \frac{1}{2T} \text{sech}\left(\frac{x - x_0}{T}\right)$$

**Properties of the Dirac delta function**

($i$) $\quad \int_a^x f(x)\delta(x - x_0)\, dx = f(x_0)H(x - x_0), \quad a \leq x_0$

($ii$) $\quad \delta(x - x_0) = \dfrac{d}{dx} H(x - x_0)$

($iii$) $\quad \delta(x^2 - x_0^2) = \dfrac{1}{2|x_0|}[\delta(x + x_0) + \delta(x - x_0)]$

($iv$) $\quad \delta(cx) = \dfrac{1}{|c|}\delta(x)$

($v$) $\quad \int_{-\infty}^{\infty} f(t)\delta(t - x_0)\, dt = f(x_0)$

## The Inverse Trigonometric Functions

The inverse trigonometric functions can be defined by integrals as follows.

$$\arcsin(x) = \int_0^x \frac{dt}{\sqrt{1-t^2}}, \qquad \arctan(x) = \int_0^x \frac{dt}{1+t^2}, \qquad \mathrm{arcsec}\,(x) = \int_1^x \frac{dt}{t\sqrt{t^2-1}}$$

$$\arccos(x) = \int_x^1 \frac{dt}{\sqrt{1-t^2}}, \qquad \mathrm{arccot}\,(x) = \int_x^\infty \frac{dt}{1+t^2}, \qquad \mathrm{arccsc}\,(x) = \begin{cases} \int_{-\infty}^x \frac{dt}{t\sqrt{t^2-1}}, & -\infty < x \le -1 \\ \int_x^\infty \frac{dt}{t\sqrt{t^2-1}}, & 1 \le x < \infty \end{cases}$$

## The Gamma Function

The Gamma function is defined

$$\Gamma(x) = \int_0^\infty e^{-t} t^{x-1}\,dt \qquad (7.1)$$

and is illustrated in the figure 7-1.

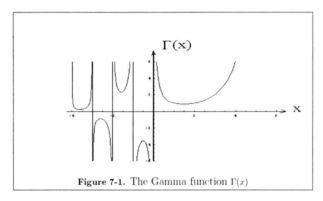

**Figure 7-1.** The Gamma function $\Gamma(x)$

For the special values $n = 0, -1, -2, -3, \ldots$ the Gamma function $\Gamma(n)$ is undefined and for these special values the function $1/\Gamma(n)$ is defined to be zero. Other special values are

$$\Gamma(1) = 1, \qquad \Gamma(\tfrac{1}{2}) = \sqrt{\pi}, \qquad \Gamma(-\tfrac{1}{2}) = -2\sqrt{\pi} \qquad (7.2)$$

Using integration by parts, the equation (7.1) can be integrated to produce the relation

$$\Gamma(\alpha+1) = \alpha\Gamma(\alpha) \qquad (7.3)$$

and using this relation it follows that

$$\Gamma(2) = \Gamma(1) = 1, \quad \Gamma(3) = 2\Gamma(2) = 2!, \quad \Gamma(4) = 3\Gamma(3) = 3!, \quad \ldots, \quad \Gamma(n+1) = n! \qquad (7.4)$$

The relation (7.2) together with the equation (7.3) allows one to calculate additional special values for the Gamma function. For example,

$$\Gamma(\tfrac{3}{2}) = \Gamma(\tfrac{1}{2}+1) = \tfrac{1}{2}\sqrt{\pi}, \qquad \Gamma(-\tfrac{3}{2}) = \frac{\Gamma(-\tfrac{3}{2}+1)}{-\tfrac{3}{2}} = \tfrac{4}{3}\sqrt{\pi}$$

Some other important relationships involving the Gamma function are

$$\Gamma(1-x)\Gamma(x) = \frac{\pi}{\sin(\pi x)} \qquad (7.5)$$

$$\Gamma(x)\Gamma(x+\tfrac{1}{2}) = 2^{1-2x}\sqrt{\pi}\,\Gamma(2x) \qquad (7.6)$$

$$\Gamma(x)\Gamma(x+\tfrac{1}{m})\Gamma(x+\tfrac{2}{m})\cdots\Gamma(x+\tfrac{m-1}{m}) = (2\pi)^{(m-1)/2} m^{1/2-mx}\Gamma(mx) \qquad (7.7)$$

Associated with the Gamma function are the following functions.

The upper incomplete gamma function $\quad \Gamma(a,x) = \int_x^\infty t^{a-1}e^{-t}\,dt \qquad (7.8)$

The lower incomplete gamma function $\quad \gamma(a,x) = \int_0^x t^{a-1}e^{-t}\,dt \qquad (7.9)$

Note that $\Gamma(a) = \gamma(a,x) + \Gamma(a,x)$.

### The Beta Function

The beta function $B(x,y)$ is defined

$$B(x,y) = \int_0^1 t^{x-1}(1-t)^{y-1}\,dt = 2\int_0^{\pi/2} \sin^{2x-1}u\,\cos^{2y-1}u\,du \qquad (7.10)$$

The integral (7.10) can be evaluated in terms of the Gamma function to obtain

$$B(x,y) = \frac{\Gamma(x)\Gamma(y)}{\Gamma(x+y)} = B(y,x), \qquad x>0, \quad y>0 \qquad (7.11)$$

The incomplete beta function is defined

$$B(x;a,b) = \int_0^x t^{a-1}(1-t)^{b-1}\,dt \qquad (7.12)$$

### Bessel Functions

The differential equation

$$t^2 \frac{d^2z}{dt^2} + t\frac{dz}{dt} + (t^2 - \nu^2)z = 0 \qquad (7.13)$$

is known as Bessel's differential equation. This equation arises in many applied problems involving cylindrical coordinates. The quantity $\nu$ in equation (7.13) is called a parameter and can be any real number. The Bessel equation (7.13) has a regular singular point at $t=0$ and so there exists a Frobenius type solution $z = \sum_{n=0}^\infty c_n t^{n+r}$. There exists solutions $J_\nu(t)$ and $J_{-\nu}(t)$, corresponding to $r=\nu$ and $r=-\nu$, for $\nu$ not an integer, where

$$J_\nu(t) = \sum_{m=0}^\infty \frac{(-1)^m}{m!\,\Gamma(\nu+m+1)} \left(\frac{t}{2}\right)^{2m+\nu} \qquad (7.14)$$

is defined as a Bessel function of the first kind of order $\nu$. In equation (7.14) the function $\Gamma$ is the Gamma function. The function $J_{-\nu}(t)$ is obtained from equation (7.14) by replacing $\nu$ by $-\nu$. For $\nu$ different from an integer the functions $J_\nu(t)$ and $J_{-\nu}(t)$ represent two linearly independent solutions to the Bessel differential equation (7.13).

In the special case $\nu = m$ is an integer, the Bessel functions $J_m(x)$ and $J_{-m}(x)$ satisfy the relation $J_{-m}(x) = (-1)^m J_m(x)$ and so these functions are no longer linearly independent functions. For any value of $\nu$ the Weber and Schlafi ratio[1]

$$Y_\nu(t) = \frac{J_\nu(t)\cos\nu\pi - J_{-\nu}(t)}{\sin\nu\pi} \tag{7.15}$$

is a linear combination of $J_\nu(t)$ and $J_{-\nu}(t)$ which can be used to define Bessel functions of the second kind of order $\nu$. The limiting process $Y_n(t) = \lim_{\nu \to n} Y_\nu(t)$ is used to define Bessel functions of the second kind with integer values $n$. Some texts refer to Bessel functions of the second kind of order $\nu$ as Neumann functions of order $\nu$ and are denoted using the notation $N_\nu(x)$. Graphs of selected Bessel functions are given in the figures 7-2 and 7-3. Note the zeros of the Bessel functions are not equally spaced like the sine and cosine functions. Denote by $\xi_{kn}$ the $n$th zero of the $k$th order Bessel function, then $J_k(\xi_{kn}) = 0$, for $n = 1, 2, 3, \ldots$.

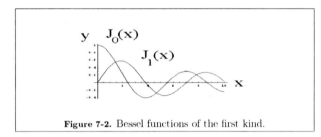

Figure 7-2. Bessel functions of the first kind.

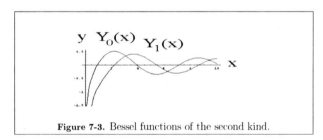

Figure 7-3. Bessel functions of the second kind.

The zeros of the Bessel functions can be found in the many handbooks.[2] The first five zeros for the $J_0, J_1, Y_0, Y_1$ Bessel functions are given in the following table.

---

[1] Wilhelm Eduard Weber (1804-1891), German mathematician, Ludwig Schlafli (1814-1895), Swiss mathematician.
[2] See for example: M. Abramowitz and I.A. Stegun, Handbook of Mathematical Functions, Dover Publications, 1972.

| First five zero's of the Bessel functions |
| --- |
| $J_0(x), J_1(x), Y_0(x), Y_1(x)$ |
| $J_0(x) = 0$ for x= 2.4048, 5.5201, 8.6537, 11.7915, 14.9309, ... |
| $J_1(x) = 0$ for x= 3.8317, 7.0156, 10.1735, 13.3237, 16.4706 ... |
| $Y_0(x) = 0$ for x= 0.8936, 3.9577, 7.0861, 10.2223, 13.3611, ... |
| $Y_1(x) = 0$ for x= 2.1971, 5.4297, 8.5960, 11.7492, 14.8974, ... |

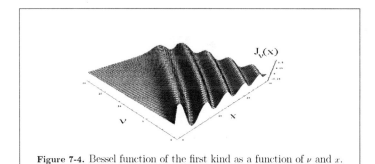

**Figure 7-4.** Bessel function of the first kind as a function of $\nu$ and $x$.

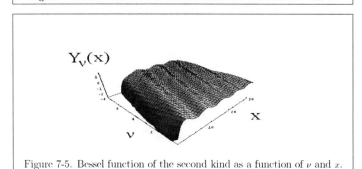

Figure 7-5. Bessel function of the second kind as a function of $\nu$ and $x$.

The Bessel functions can be of integer order, fractional order or noninteger nonfractional orders. The figure 7-4 illustrates the behavior of the Bessel function $J_\nu(x)$ when plotted to form the surface $z = J_\nu(x)$ as a function of $\nu$ and $x$. Note the behavior of the surface and

## Table 7.1 Bessel Function Properties

| | |
|---|---|
| 1. | $\frac{d}{dx}\left[x^{-\nu}J_\nu(x)\right] = -x^{-\nu}J_{\nu+1}(x)$    or    $J'_\nu(x) = \frac{\nu}{x}J_\nu(x) - J_{\nu+1}(x)$ |
| 2. | $\frac{d}{dx}[x^\nu J_\nu(x)] = x^\nu J_{\nu-1}(x)$    or    $J'_\nu(x) = J_{\nu-1}(x) - \frac{\nu}{x}J_\nu(x)$ |
| 3. | $J_{\nu-1}(x) + J_{\nu+1}(x) = \frac{2\nu}{x}J_\nu(x), \qquad Y_{\nu-1}(x) + Y_{\nu+1}(x) = \frac{2\nu}{x}Y_\nu(x)$ |
| 4. | $J_{\nu-1}(x) - J_{\nu+1}(x) = 2J'_\nu(x)$ |
| 5. | $\int x^\nu J_{\nu-1}(x)\,dx = x^\nu J_\nu(x) + C$ |
| 6. | $\int x^{-\nu} J_{\nu+1}(x)\,dx = -x^{-\nu}J_\nu(x) + C$ |
| 7. | $\int J_{\nu+1}(x)\,dx = \int J_{\nu-1}(x)\,dx - 2J_\nu(x) + C$ |
| 8. | $\frac{d}{dx}J_0(x) = -J_1(x), \qquad \frac{d}{dx}[J_0(\lambda x)] = -\lambda J_1(\lambda x)$ |
| 9. | $(\beta^2 - \alpha^2)\int_a^b x J_n(\alpha x) J_n(\beta x)\,dx = \left[x\left(\alpha J_n(\beta x) J'_n(\alpha x) - \beta J_n(\alpha x) J'_n(\beta x)\right)\right]_a^b$ |
| 10. | $J_{\frac{1}{2}}(x) = \sqrt{\frac{2}{\pi x}}\sin x, \qquad J_{\frac{3}{2}}(x) = \sqrt{\frac{2}{\pi x}}\left(\frac{\sin x}{x} - \cos x\right), \qquad Y_{\frac{1}{2}}(x) = -J_{-\frac{1}{2}}(x)$ |
| 11. | $J_{-\frac{1}{2}}(x) = \sqrt{\frac{2}{\pi x}}\cos x, \qquad J_{-\frac{3}{2}}(x) = -\sqrt{\frac{2}{\pi x}}\left(\frac{\cos x}{x} + \sin x\right), \qquad Y_{\frac{3}{2}}(x) = J_{-\frac{3}{2}}(x)$ |
| 12. | $\int_a^b x J_k^2(\lambda x)\,dx = \frac{b^2}{2}\left\{|J'_k(\lambda b)|^2 + J_k^2(\lambda b) - \frac{k^2}{\lambda^2 b^2}J_k^2(\lambda b)\right\} - \frac{a^2}{2}\left\{|J'_k(\lambda a)|^2 + J_k^2(\lambda a) - \frac{k^2}{\lambda^2 a^2}J_k^2(\lambda a)\right\}$ |
| 13. | $J_0(x) = 1 - \frac{x^2}{2^2} + \frac{x^4}{2^4 2!} - \frac{x^6}{2^6 3!} + \cdots$ |
| 14. | $x\frac{d}{dx}J_\nu(\lambda_n x) = \nu J_\nu(\lambda_n x) - \lambda_n x J_{\nu+1}(\lambda_n x)$ |
| 15. | $2\int x J_\nu^2(x)\,dx = x^2|J'_\nu(x)|^2 + (x^2 - \nu^2)|J_\nu(x)|^2 + c$ |
| 16. | $\exp\left(\frac{x}{2}[t - \frac{1}{t}]\right) = \sum_{n=-\infty}^{\infty} J_n(x)t^n$    Generating function |
| 17. | $\int_0^b x J_k(\lambda_n x) J_k(\lambda_m x)\,dx = \frac{b^2}{2}J_{k+1}^2(\lambda_n b)\delta_{mn}$    for $m$, $n$ integers and $\lambda_n$ satisfying $J_k(\lambda_n b) = 0$ for $n = 1, 2, 3, \ldots$ |

observe the function is bounded for large positive arguments $x$ and it approaches zero for large values of the order $\nu$. The figure 7-5 illustrates the surface $z = Y_\nu(x)$. The surface plot illustrates that the function $Y_\nu(x)$ becomes unbounded as $x$ approaches zero.

The table 7.1 lists several important properties of the Bessel functions. In table 7.1 observe that in all of the differentiation and integration properties the variable $x$ is a dummy variable and can be replaced by some other symbol. For example in entry number 4, replace $x$ by $\beta x$ and write

$$J'_\nu(\beta x) = \frac{1}{2\beta}[J_{\nu-1}(\beta x) - J_{\nu+1}(\beta x)].$$

Here the prime notation always means differentiation with respect to the argument of the function. For example, $J'_\nu(\xi) = \frac{d}{d\xi}J_\nu(\xi)$. Let $\xi = \beta x$, then by chain rule differentiation

$$\frac{d}{dx}[J_\nu(\beta x)] = \frac{d}{d\xi}J_\nu(\xi)\frac{d\xi}{dx} = J'_\nu(\beta x)\beta.$$

Note in particular the implied differential relations from entry 8 of table 7.1

$$J'_0(x) = -J_1(x) \quad \text{and} \quad \frac{d}{dx}[J_0(\lambda x)] = -\lambda J_1(\lambda x). \tag{7.16}$$

Both Bessel functions $J_n(x)$ and $Y_n(x)$ satisfy the same differential equation, and so they possess many of the same properties. Hence in many of the properties listed in table 7.1 it is possible to replace $J_n(x)$ by $Y_n(x)$.

### Example 7-1. (Bessel function properties.)

Show $\quad 2\int xJ_\nu^2(x)\,dx = x^2[J'_\nu(x)]^2 + (x^2 - \nu^2)[J_\nu(x)]^2 + c$

where $c$ is a constant of integration.

**Solution:** Multiply Bessel's equation by $2y'$ and obtain

$$2x^2y''y' + 2x(y')^2 + 2x^2yy' - 2\nu^2 yy' = 0. \tag{7.17}$$

Use of the identity

$$\frac{d}{dx}\left[x^2(y')^2\right] = 2x^2y'y'' + 2x(y')^2$$

enables us to express equation (7.17) in the form

$$d[x^2(y')^2] = -2x^2yy'\,dx + 2\nu^2 yy'\,dx.$$

This form is easily integrated to produce

$$x^2(y')^2 = -\int 2x^2yy'\,dx + \nu^2 y^2 + c, \tag{7.18}$$

where $c$ is a constant of integration. Integrate the remaining integral in equation (7.18) by parts to obtain the equation

$$2\int xy^2\,dx = x^2(y')^2 + (x^2 - \nu^2)y^2 + c$$

which for $y = J_\nu(x)$ reduces to

$$2\int xJ_\nu^2(x)\,dx = x^2\left[J'_\nu(x)\right]^2 + (x^2 - \nu^2)\left[J_\nu(x)\right]^2 + c.$$

which is entry number 15 in the table 7.1.

■

### Summary of solutions to Bessel's differential equation.

1. For $\nu$ different from an integer, the Bessel differential equation

$$t^2\frac{d^2z}{dt^2} + t\frac{dz}{dt} + (t^2 - \nu^2)z = 0 \tag{7.19}$$

has the general solution $\quad z = c_1 J_\nu(t) + c_2 J_{-\nu}(t) \tag{7.20}$

For all values of $\nu$ the general solution can be written

$$z = c_1 J_\nu(t) + c_2 Y_\nu(t) \tag{7.21}$$

where $c_1$ and $c_2$ denote arbitrary constants.

2. The substitutions $t = \lambda x$ and $y = z$ transforms the Bessel differential equation (7.19) into the form

$$x^2\frac{d^2y}{dx^2} + x\frac{dy}{dx} + (\lambda^2 x^2 - \nu^2)y = 0 \tag{7.22}$$

which for all values of $\nu$ has the general solution

$$y = c_1 J_\nu(\lambda x) + c_2 Y_\nu(\lambda x) \tag{7.23}$$

where $c_1$ and $c_2$ denote arbitrary constants.

3. The substitutions $t = \beta x^\gamma$ and $y = x^\alpha z$ transforms the equations (7.19) and (7.21). The new differential equation is

$$x^2\frac{d^2y}{dx^2} + (1 - 2\alpha)x\frac{dy}{dx} + [(\beta\gamma x^\gamma)^2 + \alpha^2 - \nu^2\gamma^2]y = 0 \tag{7.24}$$

with general solution

$$y = x^\alpha [c_1 J_\nu(\beta x^\gamma) + c_2 Y_\nu(\beta x^\gamma)] \tag{7.25}$$

where $c_1$ and $c_2$ denote arbitrary constants.

4. A more general form of Bessel's equation results using the substitutions

$$z = t^{(1-a)/2} e^{-(b/r)t^r} y \text{ and } t = \left(\frac{sx}{\sqrt{d}}\right)^{1/s}.$$

It can then be shown that the equation

$$x^2\frac{d^2y}{dx^2} + x(a + 2bx^r)\frac{dy}{dx} + [c + dx^{2s} + b(a + r - 1)x^r + b^2 x^{2r}]y = 0$$

has the solution

$$y = x^\alpha e^{-\beta x^r}[c_1 J_\nu(\lambda x^s) + c_2 Y_\nu(\lambda x^s)]$$

where $\alpha = \dfrac{1-a}{2}$, $\beta = \dfrac{b}{r}$, $\lambda = \dfrac{\sqrt{d}}{s}$, $\nu = \dfrac{\sqrt{(1-a)^2 - 4c}}{2s}$ with the restrictions that $(1-a)^2 - 4c \geq 0 \quad d > 0 \quad r \neq 0 \quad s \neq 0$.

## Modified Bessel's Equation

In Bessel's equation (7.13) replace $t$ by $it$, where $i^2 = -1$, to obtain the modified Bessel equation

$$t^2 \frac{d^2 z}{dt^2} + t \frac{dz}{dt} - (t^2 + \nu^2)z = 0. \tag{7.26}$$

The functions

$$I_\nu(x) = i^{-\nu} J_\nu(ix) \quad \text{and} \quad I_{-\nu}(x) = i^\nu J_{-\nu}(ix) \tag{7.27}$$

are solutions of the modified Bessel equation and are called modified Bessel functions of the first kind of order $\nu$. The functions

$$\begin{aligned} K_\nu(x) &= \frac{\pi}{2\sin(\nu\pi)}[I_{-\nu}(x) - I_\nu(x)] \quad n \neq 0, 1, 2, 3, \dots \\ K_n(x) &= \lim_{\nu \to n} K_\nu(x) \quad n = 0, 1, 2, 3, \dots \end{aligned} \tag{7.28}$$

are also solutions of the modified Bessel equation and are called modified Bessel functions of the second kind of order $\nu$. The figures 7-6 and 7-7 illustrate these modified Bessel functions for selected values of $\nu$.

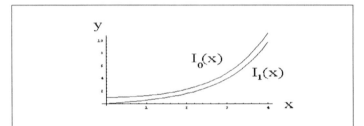

Figure 7-6. Modified Bessel functions of the first kind of orders 0 and 1.

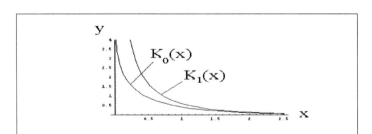

Figure 7-7. Modified Bessel functions of the second kind of orders 0 and 1.

For all values of $\nu$ the general solution to the modified Bessel differential equation

$$x^2\frac{d^2y}{dx^2} + x\frac{dy}{dx} - (x^2+\nu^2)y = 0 \qquad (7.29)$$

can be written

$$y = c_1 I_\nu(x) + c_2 K_\nu(x) \qquad (7.30)$$

where $c_1$ and $c_2$ are arbitrary constants.

A change of variables can be made in the differential equation (7.29) to obtain the following more general result involving a parameter $\lambda$.

For all values of $\nu$ the modified Bessel differential equation

$$x^2\frac{d^2y}{dx^2} + x\frac{dy}{dx} - (\lambda^2 x^2 + \nu^2)y = 0 \qquad (7.31)$$

has the general solution

$$y = c_1 I_\nu(\lambda x) + c_2 K_\nu(\lambda x) \qquad (7.32)$$

where $c_1$ and $c_2$ are arbitrary constants.

## The hypergeometric function

The hypergeometric or Gauss differential equation has the form

$$x(1-x)\frac{d^2y}{dx^2} + [c - (a+b+1)x]\frac{dy}{dx} - ab\, y = 0 \qquad (7.33)$$

where $a, b, c$ are constants which can be real or complex. This differential equation has singular points at $x = 0$, $x = 1$ and $x = \infty$. To deal with the singular point at infinity it is customary to make the substitution $z = 1/x$, and map the singular point at $x = \infty$ to the point $z = 0$. Assume a Frobenius type solution $y = y(x) = \sum_{n=0}^{\infty} k_n x^{n+r}$ about the singular point at $x = 0$, and verify that there are two indicial roots $r = 0$ and $r = 1 - c$. Corresponding to the indicial root $r = 0$ there exists a solution to the hypergeometric equation having the form

$$y = y(x) = k_0 \left[ 1 + \frac{ab}{1!c}x + \frac{a(a+1)b(b+1)}{2!c(c+1)}x^2 + \cdots \right.$$
$$\left. + \frac{a(a+1)(a+2)\cdots(a+m-1)b(b+1)(b+2)\cdots(b+m-1)}{m!c(c+1)(c+2)\cdots(c+m-1)}x^m + \cdots \right] \qquad (7.34)$$

where $k_0$ is a constant. Introducing the factorial function

$$(\alpha)_n = \alpha(\alpha+1)(\alpha+2)\cdots(\alpha+n-1) = \prod_{k=1}^{n}(\alpha+k-1) \qquad (7.35)$$

with $(\alpha)_0 = 1$ and it being understood that $\alpha \neq 0$. The solution given by equation (7.34) can then be expressed in terms of the factorial functions $(a)_n$, $(b)_n$, $(c)_n$ as

$$y = y(x) = k_0 \left[ 1 + \sum_{n=1}^{\infty} \frac{(a)_n (b)_n}{n!\,(c)_n} x^n \right] \qquad (7.36)$$

The solution given by equation (7.34), in the case $k_0 = 1$, is denoted using the notation

$$F(a,b;c;x) \text{ or } {}_2F_1(a,b;c;x) \text{ which is short for } {}_2F_1\begin{bmatrix} a,b; \\ c; \end{bmatrix} x \quad (7.37)$$

and is called the hypergeometric function. Thus,

$$F(a,b;c;x) = {}_2F_1(a,b;c;x) = 1 + \sum_{n=1}^{\infty} \frac{(a)_n (b)_n}{(c)_n} \frac{x^n}{n!} \quad (7.38)$$

A second independent solution to the hypergeometric differential equation, relative to the singularity at $x = 0$ and the second indicial root $r = 1 - c$, can be shown to have the form

$$y = x^{1-c} F(a - c + 1, b - c + 1; 2 - c; x) \quad (7.39)$$

where $c$ is a constant and $c \neq 1, 2, 3, \ldots$. This second independent solution, given by equation (7.39), is well defined as long as $c$ is different from a positive integer. The equations (7.38) and (7.39) represent two linearly independent solutions of the hypergeometric equation provided $c \neq 0, 1, 2, \ldots$.

Hence, for $c$ not zero or a positive integer, the general solution of the hypergeometric equation, associated with the singularity at $x = 0$, can be represented in the form

$$y = y(x) = k_0 F(a,b;c;x) + k_0^* x^{1-c} F(a - c + 1, b - c + 1; 2 - c; x) \quad (7.40)$$

where $k_0$ and $k_0^*$ are arbitrary constants. Using the ratio test on the series given by equation (7.38) one finds that $F(a,b;c;x)$ converges for $|x| < 1$.

The hypergeometric function can also be expressed in the form

$$F(a,b;c;z) = \sum_{n=0}^{\infty} \frac{(a)_n (b)_n}{(c)_n} \frac{z^n}{n!} = \frac{\Gamma(c)}{\Gamma(a)\Gamma(b)} \sum_{n=0}^{\infty} \frac{\Gamma(a+n)\Gamma(b+n)}{\Gamma(c+n)} \frac{z^n}{n!} \quad (7.41)$$

which converges for $|z| < 1$.

A Frobenius type solutions associated with the singular point at $x = 1$ are obtained by assuming a solution of the form $y = y(x) = \sum_{n=0}^{\infty} k_n (1-x)^{n+r}$, where the $k_n$ are constants to be determined and $k_0 \neq 0$. This form for a solution leads to the two indicial roots $r = 0$ and $r = c - a - b$ relative to the singular point at $x = 1$. Consequently, two independent solutions associated with the singularity at $x = 1$ can be expressed the form

$$\begin{aligned} y = y(x) &= F(a,b;a+b-c+1;1-x) \\ \text{and} \quad y = y(x) &= (1-x)^{c-a-b} F(c-a, c-b; c-a-b+1; 1-x) \end{aligned} \quad (7.42)$$

The radius of convergence for the power series in powers of $1 - x$ is given by $|x - 1| < 1$ or $0 < x < 2$.

For the singular point at $x = \infty$, assume a Frobenius type solution having the form $y = y(x) = \sum_{n=0}^{\infty} K_n (\frac{1}{x})^{n+r}$, where $K_n$ are constants to be determined with $K_0 \neq 0$. It can then be verified that $r = a$ and $r = b$ are the indicial roots associated with the singularity at $x = \infty$. This leads to two independent solutions associated with the singularity at $x = \infty$ which have the form

$$y = y(x) = x^{-a} F(a, a-c+1; a-b+1; \frac{1}{x})$$
and
$$y = y(x) = x^{-b} F(b, b-c+1; b-a+1; \frac{1}{x})$$
(7.43)

These series in powers of $x^{-1}$ are found to have the radius of convergence $|x| > 1$.

The hypergeometric differential equation is a linear second order differential equation with Frobenius type series representation for the independent solutions in the vicinity of the singular points at $x = 0$, $x = 1$ and $x = \infty$. These six solutions are given by the equations (7.40), (7.42), and (7.43). The hypergeometric differential equation is of second order and so only two independent solutions from the six solutions constructed are needed to obtain a general solution. Since all solutions of the hypergeometric differential equation can be represented as some linear combination of two linearly independent solutions, it should not be surprising that linear relationships must exists between the six solutions constructed.

**Recursion Formula**

$$cF(a, b-1; c; z) - cF(a-1, b; c; z) + (a-b)zF(a, b; c+1; z) = 0$$
$$cF(a, b; c; z) - cF(a, b+1; c; z) + azF(a+1, b+1; c+1; z) = 0$$
$$cF(a, b; c; z) - cF(a+1, b; c; z) + bzF(a+1, b+1; c+1; z) = 0$$
$$cF(a, b; c; z) - (c-b)F(a, b; c+1; z) - bF(a, b+1; c+1; z) = 0$$
$$cF(a, b; c; z) - (c-a)F(a, b; c+1; z) - aF(a+1, b; c+1; z) = 0$$
$$cF(a, b; c; z) - (c-a)F(a, b+1; c+1; z) - a(1-z)F(a+1, b+1; c+1; z) = 0$$
$$cF(a, b; c; z) + (b-c)F(a+1, b; c+1; z) - b(1-z)F(a+1, b+1; c+1; z) = 0$$
$$c(a-b)F(a, b; c; z) - a(c-b)F(a+1, b; c+1; z) + b(c-a)F(a, b+1; c+1; z) = 0$$

**Special Values for the Hypergeometric Function**

Observe that if one of the coefficients $a$ or $b$ is a negative integer, then the series defining the hypergeometric function terminates giving a polynomial result. Some examples are,

$$F(-n, b; b; z) = (1+z)^n$$
$$P_n(z) = F(n+1, -n; \frac{1-z}{2}) \quad \text{The Legendre polynomials}$$
$$T_n(1-2z) = F(-n, n; \frac{1}{2}; z) \quad \text{The Chebyshev polynomials}$$
$$P_n(1-2z) = F(-n, n+1; 1, z) \quad \text{The Legendre polynomials}$$
(7.44)

Some other special values associated with the hypergeometric function are

$$F(\tfrac{1}{2},\tfrac{1}{2};\tfrac{3}{2};z^2) = \tfrac{1}{x}\arcsin z \qquad F(a,b;b;z) = (1-z)^{-a}$$

$$F(1,1;2,z) = \tfrac{-1}{z}\log(1-z) \qquad \gamma(\alpha,z) = \alpha^{-1} z^\alpha \,{}_1F_1(\alpha;\alpha+1;-z) \qquad (7.45)$$

$$F(\tfrac{-1}{2},\tfrac{1}{2};\tfrac{1}{2};\sin^2 z) = \cos z \qquad B_z(p,q) = p^{-1} z^p F(p,1-q;p+1;z)$$

The reference Abramowitz and Stegun has a multitude of similar relations involving the hypergeometric function.

## Generalized hypergeometric function

The generalized hypergeometric function is defined

$$_pF_q \begin{bmatrix} \alpha_1,\alpha_2,\cdots,\alpha_p; \\ \beta_1,\beta_2,\cdots,\beta_q; \end{bmatrix} z \Bigg] = \sum_{k=0}^{\infty} \frac{(\alpha_1)_k (\alpha_2)_k \cdots (\alpha_p)_k}{(\beta_1)_k (\beta_2)_k \cdots (\beta_q)_k} \frac{z^k}{k!} \qquad (7.46)$$

and the hypergeometric function $_2F_1(a,b;c;z) = F(a,b;c;z) = {}_2F_1\begin{bmatrix} a,b; \\ c; \end{bmatrix} x \Bigg]$ is a special case of the above when $\alpha_1 = a$, $\alpha_2 = b$ and $\beta_1 = c$. Other special cases are

$$_0F_0(-;-;z) = \sum_{n=0}^{\infty} \frac{z^n}{n!} = e^z$$

$$_1F_0(a;-;z) = \sum_{n=0}^{\infty} \frac{(a)_n z^n}{n!},$$

$$_1F_1(a;b;z) = \sum_{n=0}^{\infty} \frac{(a)_n z^n}{(b)_n n!}$$

## The Riemann Differential Equation

A differential equation with three distinct regular singular points a,b,c, with $a \neq b \neq c$, is the Riemann differential equation

$$\frac{d^2 u}{dz^2} + p(z)\frac{du}{dx} + q(z) = 0 \qquad (7.47)$$

where

$$p(z) = \frac{1-\alpha-\alpha'}{z-a} + \frac{1-\beta-\beta'}{z-b} + \frac{1-\gamma-\gamma'}{z-c}$$

$$q(z) = \left[\frac{\alpha\alpha'(a-b)(a-c)}{z-a} + \frac{\beta\beta'(b-c)(b-a)}{z-b} + \frac{\gamma\gamma'(c-a)(c-b)}{z-c}\right]\frac{1}{(z-a)(z-b)(z-c)} \qquad (7.48)$$

where $\alpha,\alpha'$ are the exponents belonging to $z = a$, $\beta,\beta'$ are the exponents belonging to $z = b$ and $\gamma,\gamma'$ are the exponents belonging to $z = c$. The constants $\alpha,\alpha',\beta,\beta',\gamma,\gamma'$ are such that

$$\alpha + \alpha' + \beta + \beta' + \gamma + \gamma' = 1 \qquad (7.49)$$

A change in notation enables one to write the Riemann differential equation in a more compact form. Define the regular singular points as $z_1 = a$, $z_2 = b$, $z_3 = c$ and add the additional points $z_4 = z_2$ and $z_5 = z_3$. Use the subscript notation $\alpha_1 = \alpha$, $\alpha_2 = \beta$, $\alpha_3 = \gamma$, $\alpha_1' = \alpha'$, $\alpha_2' = \beta'$, $\alpha_3' = \gamma'$, to denote the exponents, then the equation (7.47) can be expressed in the form

$$\frac{d^2 u}{dz^2} + \left( \sum_{n=1}^{3} \frac{1 - \alpha_n - \alpha_n'}{z - z_n} \right) \frac{du}{dz} + \left( \sum_{n=1}^{3} \frac{\alpha_n \alpha_n' (z_n - z_{n+1})(z_n - z_{n+2})}{z - z_n} \right) \frac{u}{(z - z_1)(z - z_2)(z - z_3)} = 0 \quad (7.50)$$

with

$$\sum_{n=1}^{3} (\alpha_n + \alpha_n') = 1 \qquad (7.51)$$

Examine the exponent differences $\alpha_n - \alpha_n'$ for $n = 1, 2, 3$. If none of these differences is an integer, then equation (7.50) has two linearly independent solutions $u_{1,n}$ and $u_{2,n}$ in the neighborhood of the singular point $z = z_n$ given by

$$u_{1,n}(z) = (z - z_n)^{\alpha_n} \sum_{m=0}^{\infty} K_m (z - z_n)^m$$
$$u_{2,n}(z) = (z - z_n)^{\alpha_n'} \sum_{m=0}^{\infty} K_m' (z - z_n)^m \qquad (7.52)$$

for $n = 1, 2, 3$, where $K_m$ and $K_m'$ are constants with $K_0 \neq 0$ and $K_0' \neq 0$. If one or more of the exponent differences equals an integer, then case III of the Frobenius solution requires that one or more of the series solutions have a logarithm term.

Using the Papperitz P-notation, the complete set of solutions to the Riemann differential equation is denoted

$$u(z) = P \left\{ \begin{array}{cccc} z_1 & z_2 & z_3 & \\ \alpha_1 & \alpha_2 & \alpha_3 & z \\ \alpha_1' & \alpha_2' & \alpha_3' & \end{array} \right\} \qquad (7.53)$$

where the regular singular points of the Riemann differential equation are placed in the first row and directly under the singular points are placed the corresponding exponents. The fourth column contains the independent variable z.

Note that the hypergeometric equation is a special case of Riemann's differential equation and so using the P-notation, the set of solutions to the hypergeometric equation can be expressed

$$u(z) = P \left\{ \begin{array}{cccc} 0 & \infty & 1 & \\ 0 & a & 0 & z \\ 1-c & b & c-a-b & \end{array} \right\} \qquad (7.54)$$

**Generating Functions**

A generating function for a set of functions $\{\phi_n(x)\}$, $n = 0, 1, 2, \ldots$ is a function $g(x,t)$ which has some kind of a power series expansion in the variable $t$ such that the coefficient of the

$m$th term, or some scaling of the coefficient, defines the $m$th function $\phi_m(x)$. One possible form for a generating function is

$$g(x,t) = \sum_{n=0}^{\infty} c_n \phi_n(x) t^n = c_0 \phi_0(x) + c_1 \phi_1(x) t + c_2 \phi_2(x) t^2 + \cdots + c_m \phi_m(x) t^m + \cdots \qquad (7.55)$$

where $c_i$, $i = 0, 1, 2, \ldots$ are constants. The table 7.1 contains a generating function for the Bessel function of the first kind $J_n(x)$ whose form differs from that above in that the summation index for the given generating function varies over all values of the index $n$. There can be many different forms associated with generating functions. The equation (7.55) is just one of the more basic forms. Note that for equations of the form given by equation (7.55), for a fixed integer $m > 0$, the function $\phi_m(x)$ can be obtained from the generating function by differentiation and scaling

$$\phi_m(x) = \frac{1}{c_m\, m!} \frac{\partial^m}{\partial x^m} g(x,t) \bigg|_{t=0} \qquad (7.56)$$

A recurrence relation for a set of functions $\{\phi_n(x)\}$ is any equation which relates two or more functions from the set. Sometimes derivatives occur in a recurrence formula relating members of the set. Recurrence relations can sometimes be constructed from derivatives of the generating function or by comparing coefficients of powers of $t$ on both sides of an equation involving the generating function.

**Example 7-2.** (Recurrence relation.)
The Fibonacci numbers $\{\phi_n\}$ where $\phi_0 = 1$, $\phi_1 = 1$, $\phi_2 = 2$, $\phi_3 = 3$, $\phi_4 = 5$, $\phi_5 = 8$, ... can be obtained from the generating function

$$\frac{1}{1-t-t^2} = 1 + t + 2t^2 + 3t^3 + 5t^4 + 8t^5 + 13t^6 + 21t^7 + 34t^8 + \cdots$$

and members of the set $\{\phi_n\}$ satisfy the recurrence relation

$$\phi_{n+1} = \phi_n + \phi_{n-1}, \qquad n = 0, 1, 2, 3, \ldots$$

where $\phi_n = 0$ for $n < 0$ and $\phi_0 = 1$.

■

Note that the series associated with a generating function doesn't even have to converge. Use the series expansion to define the set of functions $\{\phi_n(x)\}$. The following are generating functions for some special functions of applied mathematics.

**The Polynomial Set $\{x^n\}$**
Generating function

$$g(x,t) = \frac{1}{1-xt} = \sum_{n=0}^{\infty} x^n t^n = 1 + xt + x^2 t^2 + x^3 t^3 + \cdots \qquad |xt| < 1$$

**The Sine Functions** $\{\sin nx\}$

Generating function

$$g(x,t) = \frac{t \sin x}{1 - 2t \cos x + t^2} = \sum_{n=0}^{\infty} (\sin nx)\, t^n = (\sin x)\, t + (\sin 2x)\, t^2 + (\sin 3x)\, t^3 + \cdots$$

Recurrence relation $\quad \sin nx = 2\cos x \sin(n-1)x - \sin(n-2)x, \quad n > 1$

**The Cosine functions** $\{\cos nx\}$

Generating function

$$g(x,t) = \frac{1 - t \cos x}{1 - 2t \cos x + t^2} = \sum_{n=0}^{\infty} (\cos nx)\, t^n = 1 + (\cos x)\, t + (\cos 2x)\, t^2 + (\cos 3x)\, t^3 + \cdots$$

Recurrence relation $\quad \cos nx = 2\cos x \cos(n-1)x - \cos(n-2)x, \quad n > 1$

**The Exponential Functions** $\{e^{nx}\}$

Generating function

$$g(x,t) = \frac{1}{1 - te^x} = \sum_{n=0}^{\infty} (e^{nx})\, t^n = 1 + (e^x)\, t + (e^{2x})\, t^2 + \cdots$$

**The Bernoulli Polynomials** $\{B_n(x)\}$

Generating function

$$g(x,t) = \frac{te^{xt}}{e^t - 1} = \sum_{n=0}^{\infty} B_n(x) \frac{t^n}{n!} = B_0(x) + B_1(x)\frac{t}{1!} + B_2(x)\frac{t^2}{2!} + B_3(x)\frac{t^3}{3!} + \cdots$$

$$g(x,t) = 1 + \left(x - \frac{1}{2}\right)\frac{t}{1!} + \left(x^2 - x + \frac{1}{6}\right)\frac{t^2}{2!} + \left(x^3 - \frac{3}{2}x^2 + \frac{x}{2}\right)\frac{t^3}{3!} + \left(x^4 - 2x^3 + x^2 - \frac{1}{30}\right)\frac{t^4}{4!} + \cdots$$

$$B_0(x) = 1 \qquad\qquad B'_n(x) = nB_{n-1}(x)$$

$$B_1(x) = x - \frac{1}{2} \qquad\qquad B_n(x) - B_n(a) = n \int_a^x B_{n-1}(x)\, dx$$

$$B_2(x) = x^2 - x + \frac{1}{6} \qquad\qquad B_n(1-x) = (-1)^n B_n(x)$$

$$B_3(x) = x^3 - \frac{3}{2}x^2 + \frac{1}{2}x \qquad\qquad B_n(1+x) = B_n(x) + nx^{n-1}$$

The Bernoulli numbers $B_n$ are obtained from the Bernoulli polynomials evaluated at $x = 0$. The first few Bernoulli numbers are

$$B_0 = B_0(0) = 1, \qquad B_1 = B_1(0) = -\frac{1}{2}, \qquad B_2 = B_2(0) = \frac{1}{6}, \qquad B_3 = B_3(0) = 0, \qquad B_4 = B_4(0) = -\frac{1}{30}$$

and in general $B_{2n+1} = 0$, for $n = 1, 2, 3, \ldots$ and $B_n = B_n(0) = (-1)^n B_n(1)$, for $n = 0, 1, 2, \ldots$
The Bernoulli polynomials are related to the summation $\sum_{k=1}^{m} k^n$ where $m$ and $n$ are positive integers. It can be shown that

$$\sum_{k=1}^{m} k^n = \frac{B_{n+1}(m+1) - B_{n+1}}{n+1}$$

Some texts define Bernoulli numbers $\mathfrak{B}_n$ as those numbers obtained from the generating function

$$\frac{x}{e^x-1} + \frac{x}{2} - 1 = \sum_{m=1}^{\infty} (-1)^{m-1} \mathfrak{B}_m \frac{x^{2m}}{(2m)!} = \mathfrak{B}_1 \frac{x^2}{2!} - \mathfrak{B}_2 \frac{x^4}{4!} + \mathfrak{B}_3 \frac{x^6}{6!} + \cdots + (-1)^{n-1} \mathfrak{B}_n \frac{x^{2n}}{(2n)!} + \cdots, \quad |x| < 2\pi$$

The first few Bernoulli numbers defined in this alternative fashion are given by

$$\mathfrak{B}_1 = 1/6, \quad \mathfrak{B}_2 = 1/30, \quad \mathfrak{B}_3 = 1/42, \quad \mathfrak{B}_4 = 1/30, \quad \mathfrak{B}_5 = 5/66, \quad \mathfrak{B}_6 = 691/2730, \ldots$$

### The Euler Polynomials $\{E_n(x)\}$

Generating function

$$g(x,t) = \frac{2e^{xt}}{e^t + 1} = \sum_{n=0}^{\infty} E_n(x) \frac{t^n}{n!} = E_0(x) + E_1(x)\frac{t}{1!} + E_2(x)\frac{t^2}{2!} + E_3(x)\frac{t^3}{3!} + \cdots$$

$$g(x,t) = 1 + (x - \frac{1}{2})\frac{t}{1!} + (x^2 - x)\frac{t^2}{2!} + (x^3 - \frac{3}{2}x^2 + \frac{1}{4})\frac{t^3}{3!} + (x^4 - 2x^3 + x)\frac{t^4}{4!} + \cdots$$

$$E_0(x) = 1$$
$$E_1(x) = x - \frac{1}{2}$$
$$E_2(x) = x^2 - x$$
$$E_3(x) = x^3 - \frac{3}{2}x^2 + \frac{1}{4}$$
$$E_4(x) = x^4 - 2x^3 + x$$

$$E_n'(x) = n E_{n-1}(x)$$
$$E_n(x) - E_n(a) = n \int_a^x E_{n-1}(x)\,dx$$
$$E_n(1 - x) = (-1)^n E_n(x)$$
$$E_n(1 + x) = 2x^n - E_n(x)$$

The Euler numbers $E_n = 2^n E_n(\frac{1}{2})$, for $n = 0, 1, 2, \ldots$, are obtained from a scaling of $2^n$ of the evaluation of the Euler polynomials at $1/2$. The first few Euler numbers are

$$E_0 = E_0(\tfrac{1}{2}) = 1, \quad E_1 = 2 E_1(\tfrac{1}{2}) = 0, \quad E_2 = 2^2 E_2(\tfrac{1}{2}) = -1, \quad E_4 = 2^4 E_4(\tfrac{1}{2}) = 5, \quad E_6 = 2^6 E_6(\tfrac{1}{2}) = -61$$

and in general

$$E_{2n+1} = 0 \text{ for } n = 0, 1, 2, \ldots \quad \text{and} \quad E_n(0) = -E_n(1)$$

The Euler polynomials and Bernoulli polynomials are related with

$$E_n(x) = \frac{2}{n+1}\left[B_{n+1}(x) - 2^{n+1} B_{n+1}(\tfrac{x}{2})\right]$$

$$B_n(x) = 2^{-n} \sum_{k=0}^{n} \binom{n}{k} B_{n-k} E_k(2x), \quad n = 0, 2, 3, \ldots$$

Note that some text define Euler numbers in a different way. The Euler numbers $\mathfrak{E}_n$ can be defined using the generating functions

$$\operatorname{sech} x = 1 - \mathfrak{E}_1 \frac{x^2}{2!} + \mathfrak{E}_2 \frac{x^4}{4!} - \mathfrak{E}_3 \frac{x^6}{6!} + \cdots, \quad |x| < \frac{\pi}{2}$$

$$\sec x = 1 + \mathfrak{E}_1 \frac{x^2}{2!} + \mathfrak{E}_2 \frac{x^4}{4!} + \mathfrak{E}_3 \frac{x^6}{6!} + \cdots, \quad |x| < \frac{\pi}{2}$$

These alternative definitions give the first few Euler numbers $\mathfrak{E}_n$ having the values

$$\mathfrak{E}_1 = 1, \quad \mathfrak{E}_2 = 5, \quad \mathfrak{E}_3 = 61, \quad \mathfrak{E}_4 = 1385, \quad \mathfrak{E}_5 = 50,521$$

**The Legendre Functions of the First Kind** $\{P_n(x)\}$

Generating function

$$g(x,t) = \frac{1}{\sqrt{1-2tx+t^2}} = \sum_{n=0}^{\infty} P_n(x) t^n = P_0(x) + P_1(x) t + P_2(x) t^2 + \cdots$$

$$g(x,t) = 1 + xt + \frac{1}{2}(-1+3x^2) t^2 + \frac{1}{2}(-3x+5x^3) t^3 + \frac{1}{8}(3-30x^2+35x^4) t^4 + \cdots$$

$$P_0(x) = 1, \qquad P_1(x) = x, \qquad P_2(x) = \frac{1}{2}(3x^2 - 1), \qquad P_3(x) = \frac{1}{2}(5x^3 - 3x)$$

Recurrence formula $\qquad (n+1)P_{n+1}(x) = (2n+1)xP_n(x) - nP_{n-1}(x)$

The Legendre functions of the first kind are polynomials which can be calculated using the Rodrigue's's formula

$$P_n(x) = \frac{1}{2^n n!} \frac{d^n}{dx^n} (x^2 - 1)^n$$

The functions $P_n(x)$ satisfy the Legendre differential equation

$$(1-x^2)\frac{d^2 y}{dx^2} - 2x\frac{dy}{dx} + n(n+1)y = 0$$

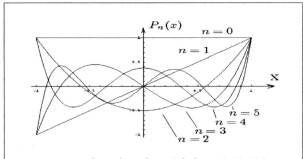

**Figure 7-8** Legendre polynomials for n=0,1,2,3,4,5

**The Associated Legendre Functions of the First Kind** $\{P_n^m(x)\}$

Generating function

$$g(x,t;m) = \frac{(2m)!(1-x^2)^{m/2} t^m}{2^m m!(1-2tx+t^2)^{m+1/2}} = \sum_{n=m}^{\infty} P_n^m(x) t^n = P_m^m(x) t^m + P_{m+1}^m(x) t^{m+1} + P_{m+2}^m(x) t^{m+3} + \cdots$$

$$g(x,t;1) = \sqrt{1-x^2}\, t + 3x\sqrt{1-x^2}\, t^2 + \frac{3}{2}\sqrt{1-x^2}\,(-1+5x^2) t^3 + \frac{5}{2}x\sqrt{1-x^2}\,(-3+7x^2) t^4 + \cdots$$

$$g(x,t;2) = 3(1-x^2) t^2 + 15x(1-x^2) t^3 + \frac{3}{2}(1-x^2)(-5+35x^2) t^4 + \cdots$$

$$g(x,t;3) = 15(1-x^2)^{3/2} t^3 + 105x(1-x^2)^{3/2} t^4 + \frac{15}{2}(1-x^2)^{3/2}(-7+63x^2) t^5 + \cdots$$

(7.57)

$$P_1^1(x) = \sqrt{1-x^2}$$
$$P_2^1(x) = 3x\sqrt{1-x^2}$$
$$P_3^1(x) = \frac{3}{2}(5x^2-1)\sqrt{1-x^2}$$

$$P_2^2(x) = 3(1-x^2)$$
$$P_3^2(x) = 15x(1-x^2)$$
$$P_4^2(x) = \frac{15}{2}(1-x^2)(7x^2-1)$$

$$P_3^3(x) = 15(1-x^2)^{3/2}$$
$$P_4^3(x) = 105x(1-x^2)^{3/2}$$
$$P_5^3(x) = \frac{105}{2}(1-x^2)^{3/2}(9x^2-1)$$

**Recurrence relation**

$$(n+1-m)P_{n+1}^m(x) = (2n+1)xP_n^m(x) - (n+m)P_{n-1}^m(x)$$

The associated Legendre functions of the first kind can also be calculated by differentiating the Legendre functions of the first kind using the relation

$$P_n^m(x) = (1-x^2)^{m/2}\frac{d^m}{dx^m}P_n(x)$$

Note that

$$P_n^0(x) = P_n(x) \quad \text{and} \quad P_n^m(x) = 0 \quad \text{if } m > n$$

**The Hermite Polynomials $\{H_n(x)\}$**

Generating function

$$g(x,t) = e^{2tx-t^2} = \sum_{n=0}^{\infty} H_n(x)\frac{t^n}{n!} = H_0(x) + H_1(x)\frac{t}{1!} + H_2(x)\frac{t^2}{2!} + H_3(x)\frac{t^3}{3!} + \cdots$$

$$g(x,t) = 1 + 2xt + (4x^2-2)\frac{t^2}{2!} + (+8x^3-12x)\frac{t^3}{3!} + (16x^4-48x^2+12)\frac{t^4}{4!} + \cdots$$

$$H_0(x) = 1$$
$$H_1(x) = 2x$$
$$H_2(x) = 4x^2 - 2$$

$$H_3(x) = 8x^3 - 12x$$
$$H_4(x) = 16x^4 - 48x^2 + 12$$
$$H_5(x) = 32x^5 - 160x^3 + 120x$$

$$H_n(x) = (-1)^n e^{x^2}\frac{d^n}{dx^n}(e^{-x^2})$$

Recurrence relations

$$H_{n+1}(x) = 2xH_n(x) - 2nH_{n-1}(x) \quad \text{and} \quad \frac{dH_n(x)}{dx} = 2nH_{n-1}(x)$$

An alternative definition of the Hermite polynomials is given by

$$He_n(x) = e^{x^2/2}\frac{d^n}{dx^n}\left(e^{-x^2/2}\right)$$

which can also be obtained from the generating function $e^{xt-t^2/2} = \sum_{n=0}^{\infty} He_n(x)\frac{t^n}{n!}$. This definition is preferred by statisticians because the weight function $e^{-x^2/2}$ is related to the normal probability curve. The function $y = He_n(x)$ is a solution of the differential equation $y'' - 2xy' + 2ny = 0$ and satisfies the recurrence relation

$$He_{n+1}(x) = xHe_n(x) - nHe_{n-1}(x)$$

where

$$He_0(x) = 1, \quad He_1(x) = x, \quad He_2(x) = x^2 - 1, \quad He_3(x) = x^3 - 3x, \quad He_4(x) = x^4 - 6x^2 + 3, \ldots$$

A scaling factor can be used to relate the two definitions giving $H_n(x) = 2^{n/2} He_n(\sqrt{2}x)$.

**The Laguerre Polynomials** $\{L_n(x)\}$

Generating function

$$g(x,t) = \frac{e^{-xt/(1-t)}}{1-t} = \sum_{n=0}^{\infty} L_n(x)\frac{t^n}{n!} = L_0(x) + L_1(x)\frac{t}{1!} + L_2(x)\frac{t^2}{2!} + L_3(x)\frac{t^3}{3!} + \cdots$$

$$g(x,t) = 1 - (x-1)t + (x^2 - 4x + 2)\frac{t^2}{2!} - (x^3 - 9x^2 + 18x - 6)\frac{t^3}{3!} + (x^4 - 16x^3 + 72x^2 - 96x + 24)\frac{t^4}{4!} + \cdots$$

$$L_0(x) = 1 \qquad L_3(x) = -(x^3 - 9x^2 + 18x - 6)$$
$$L_1(x) = -(x-1) \qquad L_4(x) = x^4 - 16x^3 + 72x^2 - 96x + 24$$
$$L_2(x) = x^2 - 4x + 2 \qquad L_5(x) = -(x^5 - 25x^4 + 200x^3 + 600x^2 - 600x + 120)$$

$$L_n(x) = e^x \frac{d^n y}{dx^n}(x^n e^{-x})$$

Recurrence relations

$$L_{n+1}(x) = (2n+1-x)L_n(x) - n^2 L_{n-1}(x) \qquad \text{and} \qquad \frac{dL_n(x)}{dx} = n\frac{dL_{n-1}(x)}{dx} - nL_{n-1}(x)$$

**The Associated Laguerre Polynomials** $\{L_n^m(x)\}$[3]

Generating function

$$g(x,t;m) = (-1)^m \frac{t^m}{(1-t)^{m+1}} e^{-xt/(1-t)} = \sum_{n=m}^{\infty} L_n^m(x)\frac{t^n}{n!} = L_m^m(x)\frac{t^m}{m!} + L_{m+1}^m \frac{t^{m+1}}{(m+1)!} + L_{m+2}^m(x)\frac{t^{m+2}}{(m+2)!} + \cdots$$

$$g(x,t;1) = -t + (-4 + 2x)\frac{t^2}{2!} + (-18 + 18x - 3x^2)\frac{t^3}{3!} + (-96 + 144x - 48x^2 + 4x^3)\frac{t^4}{4!} + \cdots$$

$$g(x,t;2) = 2\frac{t^2}{2!} + (18 - 6x)\frac{t^3}{3!} + (144 - 96x + 12x^2)\frac{t^4}{4!} + \cdots$$

$$g(x,t;3) = -6\frac{t^3}{3!} + (-96 + 24x)\frac{t^4}{4!} + (-1200 + 600x - 120x^2)\frac{t^5}{5!} + \cdots$$

$$L_1^1(x) = -1 \qquad\qquad L_2^2(x) = 2 \qquad\qquad L_3^3(x) = -6$$
$$L_2^1(x) = 2x - 4 \qquad\qquad L_3^2(x) = -6x + 18 \qquad\qquad L_4^3(x) = 24x - 96$$
$$L_3^1(x) = -3x^2 + 18x - 18 \qquad L_4^2(x) = 12x^2 - 96x + 144 \qquad L_5^3(x) = -120x^2 + 600x - 1200$$

The associated Laguerre polynomials can be obtained by differentiating the Laguerre polynomials

$$\frac{d^m}{dx^m}L_n(x) = L_n^m(x), \qquad L_n^0(x) = L_n(x), \qquad L_n^m(x) = 0 \quad \text{if } m > n$$

---

[3] Several alternative definitions of the associated Laguerre polynomials can be found in the literature.

Recurrence relations

$$\frac{n-m+1}{n+1}L_{n+1}^m(x) = -(x+m-2n-1)L_n^m(x) - n^2 L_{n-1}^m(x) \quad \text{and} \quad \frac{d}{dx}L_n^m(x) = L_n^{m+1}(x)$$

### The Chebyshev Polynomials of the First Kind $\{T_n(x)\}$

Generating function

$$g(x,t) = \frac{1-tx}{1-2tx+t^2} = \sum_{n=0}^{\infty} T_n(x) t^n = T_0(x) + T_1(x)t + T_2(x)t^2 + T_3(x)t^3 + \cdots$$

$$g(x,t) = 1 + xt + (-1 + 2x^2)t^2 + (-3x + 4x^3)t^3 + (1 - 8x^2 + 8x^4)t^4 + \cdots$$

$T_0(x) = 1$         $T_3(x) = 4x^3 - 3x$

$T_1(x) = x$         $T_4(x) = 8x^4 - 8x^2 + 1$

$T_2(x) = 2x^2 - 1$  $T_5(x) = 16x^5 - 20x^3 + 5x$

The Chebyshev polynomials of the first kind have the properties

$T_n(x) = \cos(n \cos^{-1} x)$    $T_{2n+1}(0) = 0$    $T_n(-1) = (-1)^n$

$T_n(-x) = (-1)^n T_n(x)$    $T_{2n}(0) = (-1)^n$    $T_n(1) = 1$

**Rodrigue's formula**    $T_n(x) = \frac{(-2)^n n!}{(2n)!}\sqrt{1-x^2}\,\frac{d^n}{dx^n}(1-x^2)^{n-1/2}$

Recurrence relation    $T_{n+1}(x) = 2x T_n(x) - T_{n-1}(x)$

### The Chebyshev Polynomials of the Second Kind $\{U_n(x)\}$

Generating function

$$g(x,t) = \frac{1}{1-2tx+t^2} = \sum_{n=0}^{\infty} U_n(x) t^n = U_0(x) + U_1(x)t + U_2(x)t^2 + U_3(x)t^3 + \cdots$$

$$g(x,t) = 1 + 2xt + (-1 + 4x^2)t^2 + (-4x + 8x^3)t^3 + (1 - 12x^2 + 16x^4)t^4 + (6x - 32x^3 + 32x^5)t^5 + \cdots$$

$U_0(x) = 1$         $U_3(x) = 8x^3 - 4x$

$U_1(x) = 2x$        $U_4(x) = 16x^4 - 12x^2 + 1$

$U_2(x) = 4x^2 - 1$  $U_5(x) = 32x^5 - 32x^3 + 6x$

The Chebyshev polynomials of the second kind have the properties

$U_n(x) = \frac{\sin(n+1)\cos^{-1} x}{\sin \cos^{-1} x}$    $U_{2n+1}(0) = 0$    $U_n(-1) = (-1)^n(n+1)$

$U_n(-x) = (-1)^n U_n(x)$    $U_{2n}(0) = (-1)^n$    $U_n(1) = n+1$

**Rodrigue's formula**    $U_n(x) = \frac{(-2)^n (n+1)!}{(2n+1)!\sqrt{1-x^2}}\frac{d^n}{dx^n}(1-x^2)^{n+1/2}$

Recurrence relation    $U_{n+1}(x) = 2x U_n(x) - U_{n-1}(x)$

The Chebyshev polynomials of the first kind and second kind are related and satisfy

$T_n(x) = U_n(x) - xU_{n-1}(x)$    and    $(1-x^2)U_n(n-1)(x) = xT_n(x) - T_{n+1}(x)$

## Orthogonal Functions

Consider a set of functions $\{\phi_n(x)\}$ for $n = 0, 1, 2, \ldots$ where each function is real valued and sectionally continuous over some interval $a \leq x \leq b$. The set of function $\{\phi_n(x)\}$ is called orthogonal over the interval $(a, b)$ with respect to a weight function $r(x) > 0$ if two arbitrary functions $\phi_j(x)$ and $\phi_i(x)$, from the set satisfy the inner product relation

$$(\phi_i(x), \phi_j(x)) = \int_a^b r(x)\phi_i(x)\phi_j(x)\, dx = \begin{cases} 0, & i \neq j \\ \|\phi_i\|^2, & j = i \end{cases}$$

where

$$\|\phi_i(x)\|^2 = (\phi_i, \phi_i) = \int_a^b r(x)\phi_i(x)^2\, dx$$

Here the notation $(u, v) = \int_a^b r(x)u(x)v(x)\, dx$ is called an inner product of the functions $u = u(x)$ and $v = v(x)$. The notation $(u, u) = \|u\|^2 = \int_a^b r(x)u^2(x)\, dx$ denotes an inner product of a function with itself and is called a norm squared. If the norm squared of a set of orthogonal functions $\{\phi_n(x)\}$ is unity, then the functions are said to be orthonormal.

## Sturm-Liouville Systems

A regular Sturm-Liouville system consists of a linear homogeneous differential equation

$$L(y) = \frac{d}{dx}\left(p(x)\frac{dy}{dx}\right) + q(x)y = -\lambda r(x)y \tag{7.58}$$

over an interval $a \leq x \leq b$ containing a parameter $\lambda$ and subject to boundary conditions at each end point of the form

$$\beta_1 y(a) + \beta_2 y'(a) = 0 \quad \text{and} \quad \beta_3 y(b) + \beta_4 y'(b) = 0 \tag{7.59}$$

where $\beta_1, \beta_2, \beta_3, \beta_4$ are real constants independent of $\lambda$ which are such that $\beta_1^2 + \beta_2^2 \neq 0$ and $\beta_3^2 + \beta_4^2 \neq 0$ cannot both be zero simultaneously. The operator $L(y)$ is a self-adjoint differential operator and the coefficients in the Sturm-Liouville differential equation (7.58) must be such that the functions $p(x), p'(x), q(x), r(x)$ are real and continuous with the additional requirement $p(x) > 0$ and $r(x) > 0$ over the solution interval $a \leq x \leq b$. Whenever the boundary conditions given by equation (7.59) are replaced by periodic boundary conditions of the form

$$y(a) = y(b) \qquad y'(a) = y'(b), \tag{7.60}$$

where the prime $'$ denotes differentiation with respect to the argument of the function, then the set of equations (7.58) and (7.60) is called a periodic Sturm-Liouville system.

Note the parameter $\lambda$ occurring in equation (7.58) can be written in different forms. For example replace $\lambda$ by $\sqrt{\lambda}, \lambda^2, -\lambda^2, -2\lambda$ or some other representation of a constant. The form selected is usually made in order that the differential equation take on an easier form to solve or the choice of a particular form makes some resulting algebra easier. Also note that the differential operator $L(y)$ is self-adjoint.

There can be no solutions, a unique solution or an infinite number of solutions to the Sturm-Liouville system given by the equation (7.58). The number and type solutions depends upon the value selected for $\lambda$. The parameter $\lambda$ is to be selected to obtain nonzero continuous solutions.

Values of $\lambda$ for which nonzero solutions exist are called eigenvalues. The set of eigenvalues associated with a Sturm-Liouville problem is called the spectrum of the problem. If all eigenvalues are real and there exists an infinite number of them, they can be labeled $\lambda_1, \lambda_2, \ldots, \lambda_n, \ldots$ where $\lambda_1$ is the smallest eigenvalue and $\lambda_n \to \infty$ as $n \to \infty$. The corresponding nonzero solution functions are called eigenfunctions. The German word 'eigen' means peculiar or specific. (The German word for eigenvalue is 'eigenwert'.) To each eigenvalue $\lambda_n$ there corresponds an eigenfunction solution to the Sturm-Liouville system which is denoted using a subscript notation as $y_n(x) = y(x; \lambda_n)$. Observe that if $y_n$ is an eigenfunction, then $cy_n$ is also an eigenfunction for nonzero constants $c$. Sometimes it is convenient to label the lowest eigenvalue as $\lambda_0$ and start the indexing at zero rather than one.

The eigenfunction solutions $y_n(x)$, $n = 1, 2, 3, \ldots$ to the Sturm-Liouville system have exactly $n - 1$ zeros on the interval $a \leq x \leq b$ and the set of eigenfunctions are orthogonal over the interval $(a, b)$ with respect to the weight function $r(x)$. That is, the inner product associated with two different eigenfunctions satisfies

$$(y_n, y_m) = \int_a^b r(x) y_n(x) y_m(x)\, dx = 0 \quad \text{for } m \neq n \tag{7.61}$$

and the norm squared is nonzero for each value of $n$

$$(y_n, y_n) = ||y_n||^2 = \int_a^b r(x) y_n^2(x)\, dx \neq 0. \tag{7.62}$$

If $f(x)$ is any function which is piecewise continuous[4] over the interval $(a, b)$ and possess a derivative $f'(x)$ which is also piecewise continuous, then it can be shown that under certain conditions, the function $f(x)$ can be represented by a series involving the eigenfunctions $y_n(x)$. This series is called a generalized Fourier series and has the form

$$f(x) = \sum_{n=n_0}^{\infty} c_n y_n(x). \tag{7.63}$$

where $c_n$ are constants called the Fourier coefficients. The Fourier coefficients can be represented as an inner product divided by a norm squared. That is, for $n = 1, 2, 3, \ldots$, write

$$c_n = \frac{(f, y_n)}{||y_n||^2} = \frac{\int_a^b r(x) f(x) y_n(x)\, dx}{\int_a^b r(x) y_n^2(x)\, dx} \tag{7.64}$$

---

[4] A function $f(x)$ is piecewise continuous over an interval if the interval can be subdivided into a finite number of subintervals and inside each subinterval $f(x)$ is continuous with finite limits at the end points of each subinterval.

The Sturm-Liouville system is called a regular system if the coefficients in the differential equation satisfy (i) $p(x) > 0$, $r(x) > 0$, (ii) $p(x), q(x), r(x)$ are real and continuous everywhere, and (iii) boundary conditions of the type given by equations (7.59) exist.

If any one of the regularity conditions is not satisfied, then the Sturm-Liouville problem is called singular. For example, if $p(x)$ vanishes at an end point or if one of the functions $p(x), q(x), r(x)$ becomes infinite at an end point, or all the boundary condition constants are zero, or one or both of the end points $a, b$ becomes infinite, then the Sturm-Liouville system is said to be singular.

**Orthogonality**

Assume that for different eigenvalues $\lambda_n$ and $\lambda_m$ there are two different nonzero solutions $y_n(x)$ and $y_m(x)$ of the differential equation $L(y) = -\lambda r(x) y$, where $L(y)$ is the differential operator defined by equation (7.58). That is,

$$\begin{aligned} \text{for } \lambda = \lambda_m & \qquad L(y_m(x)) = -\lambda_m r(x) y_m(x) \\ \text{and for } \lambda = \lambda_n & \qquad L(y_n(x)) = -\lambda_n r(x) y_n(x). \end{aligned} \qquad (7.65)$$

Employ the Green's identity

$$\begin{aligned} \int_a^b [u L(v) - v L(u)] \, dx &= p(x) \left[ u(x) v'(x) - v(x) u'(x) \right]_a^b \\ &= p(b) \left[ u(b) v'(b) - v(b) u'(b) \right] \\ &\quad - p(a) \left[ u(a) v'(a) - v(a) u'(a) \right]. \end{aligned} \qquad (7.66)$$

and make the substitutions $u = y_n(x)$ and $v = y_m(x)$, then the equations (7.65) simplify the resulting Green's identity to the relation

$$\begin{aligned} \int_a^b [y_n L(y_m) - y_m L(y_n)] \, dx &= p(x) \left[ y_n(x) y_m'(x) - y_m(x) y_n'(x) \right]_a^b \\ \int_a^b [y_n (-\lambda_m r(x) y_m) - y_m (-\lambda_n r(x) y_n)] \, dx &= p(x) \left[ y_n(x) y_m'(x) - y_m(x) y_n'(x) \right]_a^b \\ (\lambda_n - \lambda_m) \int_a^b r(x) y_n(x) y_m(x) \, dx &= p(b) \left[ y_n(b) y_m'(b) - y_m(b) y_n'(b) \right] \\ &\quad - p(a) \left[ y_n(a) y_m'(a) - y_m(a) y_n'(a) \right]. \end{aligned} \qquad (7.67)$$

Observe that the left hand side of the equation (7.67) has an integral which represents the inner product of the two eigenfunctions. To show this inner product is zero, so that the eigenfunctions are orthogonal for $n \neq m$, it must be demonstrated that the right-hand side of equation (7.67) is zero. Toward this end consider the following cases:

**Case 1.** If $p(a) = 0$ and $p(b) = 0$, then the right-hand side is zero and consequently the inner product satisfies $(y_n, y_m) = 0$ and so the eigenfunctions must be orthogonal. Note that this requires a singular Sturm-Liouville problem and further, no boundary conditions are required.

**Case 2.** If $p(a) = 0$ but $p(b) \neq 0$ then it must be required that both solutions $y_n$ and $y_m$ satisfy the boundary conditions

$$\beta_3 y_n(b) + \beta_4 y_n'(b) = 0 \quad \text{and} \quad \beta_3 y_m(b) + \beta_4 y_m'(b) = 0. \tag{7.68}$$

This is again a singular Sturm-Liouville problem. In order to have nonzero boundary conditions it is necessary that the determinant of the coefficients in equations (7.68) be equal to zero. This implies for $\beta_3, \beta_4$ different from zero that,

$$y_n(b) y_m'(b) - y_n'(b) y_m(b) = 0.$$

This condition together with the condition $p(a) = 0$ makes the right-hand side of equation (7.67) zero, which shows the inner product of the eigenfunctions is zero and hence the functions $y_n(x)$ and $y_m(x)$ are orthogonal.

**Case 3.** If $p(b) = 0$ but $p(a) \neq 0$ then it is necessary that both $y_n$ and $y_m$ satisfy boundary conditions of the type

$$\beta_1 y_n(a) + \beta_2 y_n'(a) = 0 \quad \text{and} \quad \beta_1 y_m(a) + \beta_2 y_m'(a) = 0. \tag{7.69}$$

Again, for $\beta_1, \beta_2$ nonzero the determinant of the coefficients requires

$$y_n(a) y_m'(a) - y_m(a) y_n'(a) = 0$$

and hence the right-hand side of equation (7.67) is zero and the eigenfunctions are orthogonal. Note these conditions require a singular Sturm-Liouville problem.

**Case 4.** The assumption that both $p(a) \neq 0$ and $p(b) \neq 0$ requires both $y_n$ and $y_m$ satisfy boundary conditions at the end points such that

$$\begin{aligned} \beta_1 y_n(a) + \beta_2 y_n'(a) &= 0 & \beta_3 y_n(b) + \beta_4 y_n'(b) &= 0 \\ \beta_1 y_m(a) + \beta_2 y_m'(a) &= 0 & \beta_3 y_m(b) + \beta_4 y_m'(b) &= 0. \end{aligned} \tag{7.70}$$

These conditions imply both

$$y_n(a) y_m'(a) - y_m(a) y_n'(a) = 0 \quad \text{and} \quad y_n(b) y_m'(b) - y_m(b) y_n'(b) = 0$$

which in turn make the right-hand side of equation (7.67) zero and consequently the eigenfunctions are orthogonal. Note this requires a regular Sturm-Liouville problem.

The generalized Fourier series representation of a function $f(x)$ given by equation (7.63) with Fourier coefficients given by equation (7.64) follows directly from the orthogonality properties of the eigenfunctions. Observe that if a piecewise smooth function $f(x)$ is represented as a series of eigenfunctions, then

$$f(x) = \sum_{n=1}^{\infty} c_n y_n(x) = c_1 y_1(x) + c_2 y_2(x) + c_3 y_3(x) + \cdots \tag{7.71}$$

with constants $c_n$ to be determined. (Here it is assumed the indexing of the summation begins with n=1.) Now multiply both sides of equation (7.71) by $r(x)y_m(x)\,dx$ and integrate both sides of the resulting equation from $a$ to $b$. There results the series of inner products

$$(f, y_m) = c_1(y_1, y_m) + c_2(y_2, y_m) + \cdots + c_m(y_m, y_m) + \cdots. \tag{7.72}$$

The set of functions $\{y_n\}$ are orthogonal over the interval $(a, b)$ so the only nonzero term on the right-hand side of equation (7.72) is the term with the index $m$. Hence the equation (7.72) simplifies to

$$(f, y_m) = c_m(y_m, y_m) = c_m ||y_m||^2 \tag{7.73}$$

which shows the coefficients of the series expansion are given by an inner product divided by a norm squared or

$$c_m = \frac{(f, y_m)}{||y_m||^2} = \frac{\int_a^b r(x) f(x) y_m(x)\,dx}{\int_a^b r(x) y_m^2(x)\,dx} \quad \text{for} \quad m = 1, 2, 3, \ldots. \tag{7.74}$$

**Example 7-3.** (**Orthogonal Trigonometric Functions**) Solve the Sturm-Liouville problem

$$y'' + \lambda y = 0 \qquad 0 \le x \le L \qquad y(0) = 0 \quad \text{and} \quad y(L) = 0$$

**Solution:** This is a regular Sturm-Liouville problem with coefficients $p(x) = 1$, $q(x) = 0$ and $r(x) = 1$. The differential equation is already in self-adjoint form. The problem is to find values of $\lambda$ for which there exists nonzero solutions. Examine the cases $\lambda = -\omega^2$ of negative eigenvalues, $\lambda = 0$ of a zero eigenvalue, and $\lambda = \omega^2$ of positive eigenvalues. The quantity $\omega$ is always assumed to be a positive constant.

**Case 1** $\lambda = -\omega^2$. The solution of the differential equation $y'' - \omega^2 y = 0$ is represented in the form $y = C_1 \sinh(\omega x) + C_2 \cosh(\omega x)$, with $C_1, C_2$ constants, as the algebra will be easier for this form of the solution. The condition

$y(0) = 0$ requires $C_1 \sinh(0) + C_2 \cosh(0) = 0$ or $C_2 = 0$

$y(L) = 0$ requires $C_1 \sinh(L) = 0$ or $C_1 = 0.$

This shows that only the trivial solution exists. The trivial solution is not wanted.

**Case 2** $\lambda = 0$. To solve the differential equation $y'' = 0$ integrate the differential equation twice to obtain $y = y(x) = C_1 x + C_2$, with $C_1, C_2$ constants. The boundary conditions require

$y(0) = 0$ or $y(0) = C_2 = 0.$ At $x = L$ In addition, require that

$y(L) = 0$ or $y(L) = C_1 L = 0$ which requires $C_1 = 0.$

This gives $y = 0$ as the solution. This is the trivial solution and so go on to the next case.

**Case 3** $\lambda = \omega^2$. The solutions to the differential equation $y'' + \omega^2 y = 0$ are written in the form $y = y(x) = C_1 \sin \omega x + C_2 \cos \omega x$ where $C_1, C_2$ are arbitrary constants. The boundary conditions require

$$y(0) = 0 \quad \text{or} \quad y(0) = C_2 = 0$$
$$y(L) = 0 \quad \text{or} \quad y(L) = C_1 \sin \omega L = 0.$$

If $C_1 = 0$ there results the trivial solution which is not desired and so let $C_1 = 1$ for convenience. This produces the requirement $\sin \omega L = 0$. The figure 7-9 illustrates a graph of the sine curve which oscillates between $+1$ and $-1$.

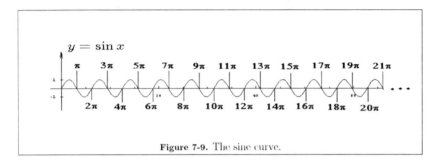

**Figure 7-9.** The sine curve.

Observe the sine curve has an infinite number of zeros that are equally spaced at multiples of $\pi$. Hence the equation $\sin \omega L = 0$ has an infinite number of solutions. The solutions of $\sin \omega L = 0$ require that

$$\omega L = n\pi \quad \text{for} \quad n = 1, 2, 3, \ldots, \tag{7.75}$$

To show there are an infinite number of possible solutions, write $\omega$ using an index notation,

$$\omega = \omega_n = \frac{n\pi}{L} \quad \text{for} \quad n = 1, 2, 3, \ldots$$

These values or $\omega_n$ are selected to satisfy the required boundary conditions. The corresponding eigenvalues are also written using an index notation

$$\lambda = \lambda_n = \omega_n^2 = \left(\frac{n\pi}{L}\right)^2 \quad \text{for} \quad n = 1, 2, 3, \ldots$$

**Remark 1.** Here the notations for $\omega$ and $\lambda$ have been changed by placing a subscript $n$ on them to emphasize that there is more than one value which gives a nonzero solution.

**Remark 2.** The value $n = 0$ is not included as this value gives $\omega = 0$ which in turn gives $y = 0$ which is the trivial solution. Recall that nonzero solutions are required to the Sturm-Liouville problem. Also the case for $\lambda = 0$ was previously discussed.

Associated with each eigenvalue there is an eigenfunction which we write as

$$y_n(x) = y(x;\lambda_n) = \sin\omega_n x = \sin\frac{n\pi x}{L} \quad \text{for} \quad n = 1,2,3,\ldots \tag{7.76}$$

These eigenfunctions are also written using an index notation to emphasize that there is more than one. The set of functions $\{\sin\frac{n\pi x}{L}\}$ represents an orthogonal set over the interval $(0,L)$ with respect to the weight function $r(x) = 1$.

■

### Example 7-4. (Trigonometric functions)

Solve the periodic Sturm-Liouville problem

$$\frac{d^2 F}{dx^2} + \lambda F = 0, \quad -L \le x \le L$$

subject to the periodic boundary conditions $F(-L) = F(L)$  $F'(-L) = F'(L)$.

**Solution:** Consider the following cases.

**Case 1:** $\lambda = -\omega^2$, $\omega > 0$. The solution of $\frac{d^2 F}{dx^2} - \omega^2 F = 0$ is written in terms of hyperbolic functions $F = F(x) = c_1\sinh\omega x + c_2\cosh\omega x$. Select this form for the solution because the subsequent algebra will be easier. Applying the boundary conditions show

$$F(-L) = c_1\sinh(-\omega L) + c_2\cosh(-\omega L) = F(L) = c_1\sinh\omega L + c_2\cosh\omega L$$

$$F'(-L) = c_1\omega\cosh(-\omega L) + c_2\omega\sinh(-\omega L) = F'(L) = c_1\omega\cosh\omega L + c_2\omega\sinh\omega L.$$

The solution of this system gives $c_1 = c_2 = 0$ and so $F$ turns out to produce the trivial solution. Discard this case.

**Case 2:** $\lambda = 0$. The solution to $\frac{d^2 F}{dx^2} = 0$ is $F = F(x) = c_1 x + c_2$. The periodic boundary conditions require

$$F'(-L) = c_1 = F'(L) = c_1 \quad \text{and}$$

$$F(-L) = c_1(-L) + c_2 = F(L) = c_1 L + c_2.$$

This requires $c_1 = 0$, and the constant $c_2$ be arbitrary. Therefore, $\lambda = 0$ is an eigenvalue with corresponding eigenfunction $F(x) = c_2$, with $c_2$ arbitrary and different from zero. Set $c_2 = 1$ and label the eigenfunction $F_0(x) = 1$, because if $F_0(x)$ is an eigenfunction then any constant times $F_0(x)$ is still an eigenfunction.

**Case 3:** $\lambda = \omega^2$, $\omega > 0$. The solution to the equation $\frac{d^2 F}{dx^2} + \omega^2 F = 0$ is given by

$$F = F(x) = c_1\cos\omega x + c_2\sin\omega x$$

with

$$F' = F'(x) = -c_1\omega\sin\omega x + c_2\omega\cos\omega x.$$

The periodic boundary conditions require

$$F(-L) = c_1\cos(-\omega L) + c_2\sin(-\omega L) = F(L) = c_1\cos\omega L + c_2\sin\omega L$$

which requires $2c_2\sin\omega L = 0$ with $c_1$ arbitrary. The derivative condition requires

$F'(-L) = -c_1\omega\sin(-\omega L) + c_2\omega\cos(-\omega L) = F(L) = -c_1\omega\sin\omega L + c_2\omega\cos\omega L$

which implies $2c_1\sin\omega L = 0$, with $c_2$ arbitrary. Thus, $c_1$ and $c_2$ can both be arbitrary with the requirement $\sin\omega L = 0$. This equation has an infinite number of solutions which can be expressed

$$\omega L = n\pi \quad \text{for} \quad n = 1, 2, 3, \ldots \quad \text{or} \quad \omega = \omega_n = \frac{n\pi}{L} \quad \text{for} \quad n = 1, 2, 3, \ldots$$

Note the relabeling of the eigenvalue solutions to reflect the fact that there are an infinite number of them. The eigenvalues are therefore given by

$$\lambda = \lambda_n = \omega_n^2 = \frac{n^2\pi^2}{L^2} \quad \text{for} \quad n = 1, 2, 3, \ldots$$

The corresponding eigenfunctions are $\{\sin\frac{n\pi x}{L}, \cos\frac{n\pi x}{L}\}$ for $n = 1, 2, 3, \ldots$. Note the function $F(x) = c_1\cos\frac{n\pi x}{L} + c_2\sin\frac{n\pi x}{L}$ is a solution for arbitrary values of $c_1$ and $c_2$. Let $c_1 = 1$ and $c_2 = 0$ giving only cosine terms and let $c_1 = 0$ and $c_2 = 1$ to obtain only sine terms.

In summary the periodic Sturm-Liouville problem

$$\frac{d^2F}{dx^2} + \lambda F = 0, \quad -L \leq x \leq L \tag{7.77}$$

with periodic boundary conditions $F(-L) = F(L)$ and $F'(-L) = F(L)$ has the eigenvalues $\lambda_0 = 0$, $\lambda_n = \frac{n^2\pi^2}{L^2}$ for $n = 1, 2, 3, \ldots$ with corresponding eigenfunctions $\{1, \sin\frac{n\pi x}{L}, \cos\frac{n\pi x}{L}\}$. These eigenfunctions satisfy the orthogonality conditions

$$(1, \sin\frac{n\pi x}{L}) = \int_{-L}^{L} \sin\frac{n\pi x}{L} dx = 0 \quad n = 1, 2, 3, \ldots$$

$$(1, \cos\frac{n\pi x}{L}) = \int_{-L}^{L} \cos\frac{n\pi x}{L} dx = 0$$

$$(\sin\frac{n\pi x}{L}, \sin\frac{m\pi x}{L}) = \int_{-L}^{L} \sin\frac{n\pi x}{L}\sin\frac{m\pi x}{L} dx = 0 \quad n \neq m \tag{7.77}(a)$$

$$(\cos\frac{n\pi x}{L}, \cos\frac{m\pi x}{L}) = \int_{-L}^{L} \cos\frac{n\pi x}{L}\cos\frac{m\pi x}{L} dx = 0 \quad n \neq m$$

$$(\cos\frac{n\pi x}{L}, \sin\frac{m\pi x}{L}) = \int_{-L}^{L} \cos\frac{n\pi x}{L}\sin\frac{m\pi x}{L} dx = 0 \quad \text{for all } n, m \text{ values.}$$

with norm squared given by

$$(1, 1) = ||1||^2 = \int_{-L}^{L} dx = 2L$$

$$(\sin\frac{n\pi x}{L}, \sin\frac{n\pi x}{L}) = ||\sin\frac{n\pi x}{L}||^2 = \int_{-L}^{L} \sin^2\frac{n\pi x}{L} dx = L \tag{7.77}(b)$$

$$(\cos\frac{n\pi x}{L}, \cos\frac{n\pi x}{L}) = ||\cos\frac{n\pi x}{L}||^2 = \int_{-L}^{L} \cos^2\frac{n\pi m}{L} dx = L$$

## Fourier Trigonometric Series

If $f(x)$ and $f'(x)$ are piecewise continuous functions over an interval $0 < \alpha \leq x \leq \alpha + 2L$, then $f(x)$ can be represented by the Fourier trigonometric series

$$\widetilde{f}(x) = \frac{a_0}{2} + \sum_{n=1}^{\infty} \left( a_n \cos \frac{n\pi x}{L} + b_n \sin \frac{n\pi x}{L} \right)$$

where

$$a_n = \frac{1}{L} \int_{\alpha}^{\alpha+2L} f(x) \cos \frac{n\pi x}{L} \, dx, \quad n = 0, 1, 2, \ldots, \qquad b_n = \frac{1}{L} \int_{\alpha}^{\alpha+2L} f(x) \sin \frac{n\pi x}{L} \, dx, \quad n = 1, 2, \ldots$$

are the Fourier coefficients. The Fourier trigonometric series $\widetilde{f}(x)$ is a periodic function with period $2L$ which represents $f(x)$ over the interval $\alpha \leq x \leq \alpha + 2L$ and converges to $f(x)$ at points $x$ were the function is continuous and converges to points $\frac{1}{2}[f(x^+) - f(x^-)]$ at points $x$ where the function $f(x)$ has a jump discontinuity.

The Fourier series can be represented using complex variables as follows.

$$\widetilde{f}(x) = \sum_{n=-\infty}^{\infty} c_n e^{i n\pi x/L}, \qquad \text{where} \qquad c_n = \frac{1}{L} \int_{\alpha}^{\alpha+2L} f(x) e^{-i n\pi x/L} \, dx = \begin{cases} \frac{1}{2}(a_n - i b_n), & n > 0 \\ \frac{1}{2}(a_{-n} + i b_{-n}), & n < 0 \\ \frac{1}{2}a_0, & n = 0 \end{cases}$$

**Selected Fourier Trigonometric Series**

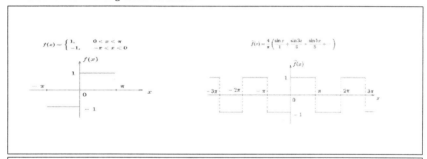

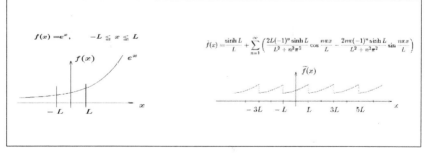

Note that the Fourier trigonometric series representation $\tilde{f}(x)$ of the exponential function, represents $e^x$ only on the interval $-L < x < L$ and at those points where a jump discontinuity occurs the Fourier trigonometric series converges to one-half the sum of the left and right limits.

## Example 7-5.   (Bessel function boundary condition y(b) = 0)

Consider the Sturm-Liouville problem involving Bessel's differential equation

$$L_1(y) = x^2 \frac{d^2y}{dx^2} + x\frac{dy}{dx} + (\lambda^2 x^2 - k^2)y = 0 \quad 0 \le x \le b \tag{7.78}$$

where $k$ and $\lambda$ are constants. This equation is subject to the boundary condition $y(b) = 0$. The given equation is Bessel's equation which is not in self-adjoint form. Here $a_0(x) = x^2$, $a_1(x) = x$ and $a_2(x) = \lambda^2 x^2 - k^2$. By multiplying the given equation by $\mu = \frac{1}{x}$ there results the self-adjoint form

$$L(y) = x\frac{d^2y}{dx^2} + \frac{dy}{dx} + \left(\lambda^2 x - \frac{k^2}{x}\right)y = 0$$

or

$$L(y) = \frac{d}{dx}\left(x\frac{dy}{dx}\right) - \frac{k^2}{x}y = -\lambda^2 xy. \quad 0 < x < b. \tag{7.79}$$

This is a singular Sturm-Liouville problem with $\lambda^2$ replacing $\lambda$ and weight function $r(x) = x$. Note that this equation is singular because $p(0) = 0$. In this case only one boundary condition is needed in order to insure orthogonality of the solution set. The coefficients are given by $p(x) = x$, $q(x) = -\frac{k^2}{x}$ and weight function $r(x) = x$. The only bounded solution of this equation on the interval $0 \le x \le b$ is given by some constant times $y = y(x;\lambda) = J_k(\lambda x)$. Let $\xi_{kn}$, for $n = 1, 2, 3, \ldots$, denote the $n$th zero of the $k$th order Bessel function. That is, define $\xi_{kn}$ as those values which satisfy $J_k(\xi_{kn}) = 0$ for $n = 1, 2, 3, \ldots$. The first couple of zero's $\xi_{k1}, \xi_{k2}, \ldots$ are illustrated in the figure 7-10.

If $y(b) = J_k(\lambda b) = 0$, then $\lambda b = \xi_{kn}$ or $\lambda = \lambda_n = \frac{\xi_{kn}}{b}$ for $n = 1, 2, 3, \ldots$. These are the unequally spaced eigenvalues for the above Sturm-Liouville problem. The corresponding eigenfunctions are

$$y_n(x) = y(x;\lambda_n) = J_k(\lambda_n x) = J_k(\frac{\xi_{kn}}{b}x) \quad n = 1, 2, 3, \ldots$$

These are orthogonal functions over the interval $(0, b)$ with respect to the weight function $x$ and have the inner product

$$(y_n, y_m) = \int_0^b x y_n(x) y_m(x)\, dx = \begin{cases} 0 & m \ne n \\ ||y_n||^2 & m = n \end{cases} \tag{7.80}$$

and norm squared

$$(y_n, y_n) = ||y_n||^2 = \int_0^b x J_k^2(\lambda_n x)\, dx = \frac{b^2}{2}J_{k+1}^2(\lambda_n b). \tag{7.81}$$

Note the weight function $r(x) = x$ is required for representing the inner product integral.

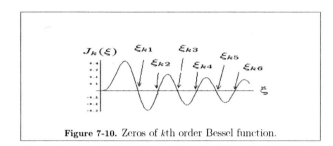

**Figure 7-10.** Zeros of $k$th order Bessel function.

The norm squared result is calculated as follows. Examine the entry 15 from table 7.1 and replace $x$ by $\lambda_n x$ and $\nu$ by $k$ to obtain

$$2\lambda_n^2 \int_0^b x J_k^2(\lambda_n x)\,dx = \left[\lambda_n^2 x^2 [J_k'(\lambda_n x)]^2 + (\lambda_n^2 x^2 - k^2)[J_k(\lambda_n x)]^2\right]_0^b. \tag{7.82}$$

The boundary condition requires $J_k(\lambda_n b) = 0$, so the integral (7.82) simplifies to

$$2 \int_0^b x J_k^2(\lambda_n x)\,dx = b^2 [J_k'(\lambda_n b)]^2. \tag{7.83}$$

Now use entry number 14 of table 7.1 and remember that the prime notation means differentiation with respect to the argument of the function and show

$$x \frac{d}{dx} J_k(\lambda_n x) = x J_k'(\lambda_n x)\lambda_n = k J_k(\lambda_n x) - \lambda_n x J_{k+1}(\lambda_n x). \tag{7.84}$$

Evaluate the equation (7.84) at $x = b$, using the given boundary condition to obtain

$$J_k'(\lambda_n b) = -J_{k+1}(\lambda_n b). \tag{7.85}$$

Substitute this result into the equation (7.83) to obtain the norm squared given by equation (7.81). The series expansion

$$f(x) = \sum_{n=1}^\infty c_n y_n(x) = \sum_{n=1}^\infty c_n J_k(\lambda_n x) \tag{7.86}$$

is called a Fourier-Bessel series with coefficients given by an inner product divided by a norm squared

$$c_n = \frac{(f, y_n)}{\|y_n\|^2} = \frac{\int_0^b x f(x) J_k(\lambda_n x)\,dx}{\int_0^b x J_k^2(\lambda_n x)\,dx}.$$

∎

**Example 7-6.** (Bessel function with different boundary condition)
Solve the singular Sturm-Liouville problem

$$\frac{d}{dx}\left(x\frac{dy}{dx}\right) - \frac{k^2}{x}y = -\lambda^2 xy \qquad 0 \leq x \leq b \tag{7.87}$$

subject to the boundary condition

$$by'(b) + \beta y(b) = 0. \tag{7.88}$$

where $\beta$ is a constant. Note that any boundary condition having the form $c_1 y'(b) + c_2 y(b) = 0$, with $c_1, c_2$ constants, can be converted to the form of equation (7.88) if one multiplies the boundary condition by $b/c_1$ and defines $bc_2/c_1 = \beta$ as a new constant. The only bounded solutions of the Bessel equation over the interval $0 \leq x \leq b$ is given by some constant times $y = y(x) = J_k(\lambda x)$. The boundary condition given by equation (7.88) requires $\lambda$ be chosen such that the equation

$$(\lambda b) J'_k(\lambda b) + \beta J_k(\lambda b) = 0. \tag{7.89}$$

Let the roots of this equation be denoted by the eigenvalues

$$\lambda = \lambda_n = \frac{\alpha_{kn}}{b} \qquad \text{for} \quad n = 1, 2, 3, \ldots \tag{7.90}$$

where $\alpha_{kn}$ are the roots satisfying

$$\alpha_{kn} J'_k(\alpha_{kn}) + \beta J_k(\alpha_{kn}) = 0 \quad \text{for} \quad n = 1, 2, 3, \ldots. \tag{7.91}$$

The Table 7.1 aids in writing the equation (7.91) in the alternative form

$$(k + \beta) J_k(\alpha_{kn}) - \alpha_{kn} J_{k+1}(\alpha_{kn}) = 0. \tag{7.92}$$

The corresponding eigenvalues are then $y_n(x) = J_k(\lambda_n x)$. These functions satisfy the orthogonality condition

$$(y_n, y_m) = \int_0^b x J_k(\lambda_n x) J_k(\lambda_m x)\, dx = \begin{cases} 0 & m \neq n \\ ||y_n||^2 & m = n \end{cases} \tag{7.93}$$

where the norm squared is given by

$$(y_n, y_n) = ||y_n||^2 = \int_0^b x J_k^2(\lambda_n x)\, dx = \left[\frac{\beta^2 + \lambda_n^2 b^2 - k^2}{2\lambda_n^2}\right] J_k^2(\lambda_n b). \tag{7.94}$$

To obtain this result for the norm squared we used entry number 12, from the table 7.1 with $a = 0$ to obtain

$$\int_0^b x J_k^2(\lambda_n x)\, dx = \frac{b^2}{2}\left\{[J'_k(\lambda b)]^2 + J_k^2(\lambda b) - \frac{k^2}{\lambda^2 b^2} J_k^2(\lambda b)\right\}. \tag{7.95}$$

Use the boundary condition equation from equation (7.89) to eliminate the derivative term in equation (7.95) to obtain

$$\int_0^b x J_k^2(\lambda_n x)\, dx = \frac{b^2}{2}\left\{[-\frac{\beta J_k(\lambda_n b)}{\lambda_n b}]^2 + J_k^2(\lambda b) - \frac{k^2}{\lambda^2 b^2} J_k^2(\lambda b)\right\}$$
$$= \left\{\frac{\beta^2 + \lambda_n^2 b^2 - k^2}{2\lambda_n^2}\right\} J_k^2(\lambda_n b). \qquad (7.96)$$

Remark: Here we have assumed that $k \neq 0$. If $k = 0$ and $\beta = 0$, then $\lambda_0 = 0$ is an eigenvalue corresponding to the eigenfunction $y_0(x) = J_0(0) = 1$. In this case the eigenvalues are the roots of the equation $J_0'(\lambda b) = 0$ or $J_1(\lambda b) = 0$ and the norm squared is

$$||y_0||^2 = ||1||^2 = \int_0^b x\, dx = \frac{b^2}{2}. \qquad (7.97)$$

This is the only case where $\lambda = 0$ can be an eigenvalue of the given Sturm-Liouville system. A series expansion of the form

$$f(x) = \sum_{n=1}^{\infty} c_n y_n(x) = \sum_{n=1}^{\infty} c_n J_k(\lambda_n x) \qquad (7.98)$$

is similar in form the equation (7.86), however, the eigenvalues are different.

∎

**Example 7-7.** (Bessel function on interval $a \leq x \leq b$)
Solve the Sturm-Liouville system

$$\frac{d}{dx}\left(x \frac{dy}{dx}\right) - \frac{k^2}{x} y = -\lambda^2 x y \qquad 0 < a < x < b \qquad (7.99)$$

subject to the boundary conditions $y(a) = 0$ and $y(b) = 0$. The domain for the solution of the Bessel equation is away from the origin so that the general solution can be written $y(x) = c_1 J_k(\lambda x) + c_2 Y_k(\lambda x)$ where $c_1$ and $c_2$ are arbitrary constants. Select the constants $c_1$ and $c_2$ such that the boundary condition $y(b) = 0$ is automatically satisfied and then select $c_1 = Y_k(\lambda b)$ and $c_2 = -J_k(\lambda b)$ to obtain the solution

$$y(x) = y(x; \lambda) = U_k(\lambda_n x) = Y_k(\lambda b) J_k(\lambda x) - J_k(\lambda b) Y_k(\lambda x). \qquad (7.100)$$

Here the eigenvalues $\lambda = \lambda_n$ are chosen as the roots of the equation

$$y(a) = U_k(\lambda_n a) = Y_k(\lambda_n b) J_k(\lambda_n a) - J_k(\lambda_n b) Y_k(\lambda_n a) = 0. \qquad (7.101)$$

The corresponding eigenfunctions are $y_n(x) = U_k(\lambda_n x)$ which satisfy the orthogonality condition

$$(y_n, y_m) = \int_a^b x U_k(\lambda_n x) U_k(\lambda_m x)\, dx = \begin{cases} 0 & m \neq n \\ ||y_n||^2 & m = n \end{cases} \qquad (7.102)$$

where the norm squared is

$$(y_n, y_n) = ||y_n||^2 = \int_a^b x U_k^2(\lambda_n x)\,dx = \frac{1}{2}\left\{b^2[U_k'(\lambda_n b)]^2 - a^2[U_k'(\lambda_n a)]^2.\right\} \tag{7.103}$$

To show how this last integral originates consider the Bessel equation

$$\frac{d}{dx}\left(x\frac{dy}{dx}\right) + (\lambda^2 x - \frac{m^2}{x})y = 0.$$

Multiply this equation by $2xy'$ to obtain

$$2xy'\frac{d}{dx}(xy') + (2\lambda^2 x^2 - 2m^2)yy' = 0.$$

Now integrate from $a$ to $b$ giving

$$\int_a^b \left[\frac{d}{dx}[x^2(y')^2] + (\lambda^2 x^2 - m^2)\frac{d}{dx}(y^2)\right]dx = 0.$$

Now integrate all the terms. The middle term is to be integrated by parts. The result is

$$x^2(y')^2\Big|_a^b + \lambda^2\left[x^2 y^2\Big|_a^b - 2\int_a^b xy^2\,dx\right] - m^2 y^2\Big|_a^b = 0$$

which can be expressed in the form

$$\begin{aligned}\int_a^b xy^2\,dx &= \frac{1}{2\lambda^2}\left[x^2(y')^2 + \lambda^2 x^2 y^2 - m^2 y^2\right]_a^b \\ &= \frac{1}{2\lambda^2}\left[b^2(y'(b))^2 + \lambda^2 b^2 y^2(b) - m^2 y^2(b)\right] \\ &\quad - \frac{1}{2\lambda^2}\left[a^2(y'(a))^2 + \lambda^2 a^2 y^2(a) - m^2 y^2(a)\right].\end{aligned} \tag{7.104}$$

Here $y(x)$ is any solution of Bessel's equation. Our solution satisfies $y(a) = 0$ and $y(b) = 0$ with

$$y'(b) = \lambda_n U_k'(\lambda_n b) \quad \text{and} \quad y'(a) = \lambda_n U_k'(\lambda_n a) \tag{7.105}$$

which simplifies the equation (7.104) to the norm squared result cited earlier. This result can also be written in the alternate form

$$||U_k(\lambda_n x)||^2 = \frac{2\left(J_k^2(\lambda_n a) - J_k^2(\lambda_n b)\right)}{\pi^2 \lambda_n^2 J_k^2(\lambda_n a)}. \tag{7.106}$$

Series expansions of the form

$$f(x) = \sum_{n=1}^{\infty} c_n y_n(x) = \sum_{n=1}^{\infty} c_n U_k(\lambda_n x) \tag{7.107}$$

generate another type of Fourier-Bessel expansion. ∎

**Example 7-8.** (Legendre equation)

Find the eigenvalues and eigenfunctions associated with the singular Sturm-Liouville problem
$$(1-x^2)\frac{d^2y}{dx^2} - 2x\frac{dy}{dx} + \lambda y = 0 \qquad -1 \leq x \leq 1$$
where there are no boundary conditions. Note this equation is already in self-adjoint form and can be written
$$\frac{d}{dx}\left[(1-x^2)\frac{dy}{dx}\right] + \lambda y = 0$$
which is now in the form of equation (7.58). Here $p(x) = 1 - x^2$ is zero at both the boundary end points, $q(x) = 0$ and the weight function is $r(x) = 1$. For $\lambda = \lambda_n = n(n+1)$ there results the Legendre equation. The only bounded solutions on the interval $-1 \leq x \leq 1$ are the Legendre polynomials $P_n(x)$ for $n = 0, 1, 2, \ldots$. These functions are orthogonal over the interval $(-1, 1)$ with respect to the weight function $r(x) = 1$. The norm squared is given by
$$||P_k||^2 = (P_k, P_k) = \frac{2}{2k+1}, \qquad k = 0, 1, 2, \ldots. \tag{7.108}$$

Series expansions of the form $f(x) = \sum_{n=0}^{\infty} c_n P_n(x)$, $-1 \leq x \leq 1$ are called Fourier-Legendre series with Fourier coefficients given by
$$c_n = \frac{(f, P_n)}{||P_n||^2} = \frac{2n+1}{2} \int_{-1}^{1} f(x) P_n(x)\, dx$$

∎

**Sturm-Liouville Theorem**

*Whenever the Sturm-Liouville system*
$$\frac{d}{dx}[p(x)y'] + [q(x) + \lambda r(x)]y = 0, \qquad a \leq x \leq b$$
*subject to the boundary conditions*
$$\beta_1 y(a) + \beta_2 y'(a) = 0, \qquad \beta_3 y(b) + \beta_4 y'(b) = 0, \qquad \beta_1^2 + \beta_2^2 \neq 0, \qquad \beta_3^2 + \beta_4^2 \neq 0$$
*is such that the coefficients $p(x)$, $p'(x)$, $q(x)$, $r(x)$, and $[p(x)r(x)]''$ are real and continuous on the interval $(a, b)$ with coefficients $r(x) > 0$ and $p(x) > 0$ and the constants $\beta_1, \beta_2, \beta_3, \beta_4$ are real and independent of $\lambda$, then there exists an infinite set of real eigenvalues $\lambda_n$ and corresponding real eigenfunctions $\{y_n(x)\}$ which are orthogonal on $(a, b)$ with respect to the weight function $r(x)$ and satisfy*
$$(y_n, y_m) = \int_a^b r(x) y_n(x) y_m(x)\, dx = \begin{cases} 0, & m \neq n \\ ||y_n||^2, & m = n \end{cases}$$
*where* $\quad ||y_n||^2 = \int_a^b r(x) y_n^2(x)\, dx$

For $f(x)$ and $f'(x)$ piecewise smooth functions over the interval $(a,b)$ the generalized Fourier series expansion of $f(x)$ has the form

$$f(x) = \sum_{n=0}^{\infty} c_n y_n(x),$$

where $c_n$ are the Fourier coefficients, given by

$$c_n = \frac{(f, y_n)}{\| y_n \|^2} = \frac{1}{\| y_n \|^2} \int_a^b r(x) f(x) y_n(x) \, dx = \frac{\int_a^b r(x) f(x) y_n(x) \, dx}{\int_a^b r(x) y_n^2(x) \, dx}$$

The generalized Fourier series converges to $f(x)$ at each point $x$ where $f(x)$ is continuous. The series converges to $\frac{1}{2}[f(x^+) + f(x^-)]$ at points $x$ where $f(x)$ has a jump discontinuity.

The table 7.2 contains a listing of orthogonal functions frequently encountered when solving partial differential equations in Cartesian, cylindrical and spherical coordinates. Series expansions using the trigonometric functions from table 7.2 produce the familiar Fourier trigonometric series expansions.

The associated Legendre functions, Hermite polynomials, Laguerre polynomials, associated Laguerre polynomials, Chebyshev polynomials are additional examples of orthogonal functions which arise in many areas of applied mathematics.

### The Associated Legendre Functions

The functions $P_n^m(x)$ satisfy the Legendre associated differential equation

$$(1 - x^2)\frac{d^2 y}{dx^2} - 2x\frac{dy}{dx} + \left\{ n(n+1) - \frac{m^2}{1 - x^2} \right\} y = 0, \qquad -1 \leq x \leq 1$$

where $m$ and $n$ are nonnegative integers. These functions satisfy the orthogonality property

$$\int_{-1}^{1} P_m^j(x) P_n^j(x) \, dx = \begin{cases} 0, & m \neq n \\ \| P_n^j(x) \|^2, & m = n \end{cases}$$

where $\quad \| P_n^j(x) \|^2 = (P_n^j, P_n^j) = \int_{-1}^{1} \left[P_n^j(x)\right]^2 dx = \frac{2}{2n+1} \frac{(n+j)!}{(n-j)!}$

The Fourier associated Legendre series has the form

$$f(x) = \sum_{n=m}^{\infty} c_n P_n^m(x) = c_m P_m^m(x) + c_{m+1} P_{m+1}^m(x) + c_{m+2} P_{m+2}^m(x) + \cdots$$

with Fourier coefficients

$$c_k = \frac{(f, P_k^m)}{\| P_k^m \|^2} = \frac{2k+1}{2} \frac{(k-m)!}{(k+m)!} \int_{-1}^{1} f(x) P_k^m(x) \, dx$$

## Table 7.2 Orthogonal Functions

| Name | Sturm-Liouville Differential Equation | Boundary Conditions | Weight Function | Eigenvalues | Orthogonal Functions | Norm Squared |
|---|---|---|---|---|---|---|
| Sine | $y'' + \lambda y = 0$, $0 \leq x \leq L$ | $y(0) = 0$, $y(L) = 0$ | $1$ | $\lambda_n = \left(\frac{n\pi}{L}\right)^2$, $n = 1, 2, \ldots$ | $\{y_n\} = \left\{\sin\frac{n\pi x}{L}\right\}$ | $\left\|\sin\frac{n\pi x}{L}\right\|^2 = \frac{L}{2}$ |
| Cosine | $y'' + \lambda y = 0$, $0 \leq x \leq L$ | $y'(0) = 0$, $y'(L) = 0$ | $1$ | $\lambda_0 = 0$, $\lambda_n = \left(\frac{n\pi}{L}\right)^2$, $n = 1, 2, \ldots$ | $\left\{1, \cos\frac{n\pi x}{L}\right\}$ | $\|1\|^2 = L$, $\left\|\cos\frac{n\pi x}{L}\right\|^2 = \frac{L}{2}$ |
| Fourier Trigonometric | $y'' + \lambda y = 0$, $-L \leq x \leq L$ | $y(-L) = y(L)$, $y'(-L) = y'(L)$ | $1$ | $\lambda_0 = 0$, $\lambda_n = \left(\frac{n\pi}{L}\right)^2$, $n = 1, 2, 3, \ldots$ | $\left\{1, \cos\frac{n\pi x}{L}, \sin\frac{n\pi x}{L}\right\}$ | $\|1\|^2 = 2L$, $\left\|\sin\frac{n\pi x}{L}\right\|^2 = L$, $\left\|\cos\frac{n\pi x}{L}\right\|^2 = L$ |
| Bessel$_1$ | $\frac{d}{dx}(xy') - \frac{k^2}{x}y + \lambda^2 xy = 0$, $0 \leq x \leq b$ | $y(b) = 0$ | $x$ | $\lambda_n = \frac{\alpha_{kn}}{b}$ for $n = 1, 2, 3, \ldots$, $J_k(\alpha_{kn}) = 0$ | $\{y_n\} = \{J_k(\lambda_n x)\}$ | $\|y_n\|^2 = \frac{b^2}{2}J_{k+1}^2(\lambda_n b)$ |
| Bessel$_2$ | $\frac{d}{dx}(xy') - \frac{k^2}{x}y + \lambda^2 xy = 0$, $0 \leq x \leq b$ | $by'(b) + hy(b) = 0$ | $x$ | $\lambda_n = \frac{\alpha_{kn}}{b}$ for $n = 1, 2, \ldots$, $\alpha_{kn}J_k'(\alpha_{kn}) + hJ_k(\alpha_{kn}) = 0$, $n = 1, 2, \ldots$ | $\{y_n\} = \{J_k(\lambda_n x)\}$ | $\|y_n\|^2 = \left[\frac{b^2 - k^2/\lambda_n^2}{2}\right]J_k^2(\lambda_n b)$, $k \neq 0$ |
| Bessel$_3$ | $\frac{d}{dx}(xy') - \frac{k^2}{x}y + \lambda^2 xy = 0$, $0 \leq x \leq b$ | $y'(a) = 0$, $y'(b) = 0$ | $x$ | $J_k'(\lambda_n a) = 0$ | $\{y_n\} = \{J_k(\lambda_n x)\}$ | |
| Legendre$_1$ | $(1-x^2)y'' - 2xy' + \lambda y = 0$, $-1 \leq x \leq 1$ | | $1$ | $\lambda_n = n(n+1)$ | $\{y_n\} = \{P_n(x)\}$ | $\|y_n\|^2 = \frac{2}{2n+1}$, $n = 0, 1, 2, \ldots$ |
| Legendre$_2$ | $\frac{d}{d\theta}(\sin\theta\frac{dy}{d\theta}) + \lambda\sin\theta\,y = 0$, $0 \leq \theta \leq \pi$ | | $\sin\theta$ | $\lambda_n = n(n+1)$ | $\{y_n\} = \{P_n(\cos\theta)\}$ | $\|P_n(\cos\theta)\|^2 = \frac{2}{2n+1}$, $n = 1, 2, \ldots$ |

**The Hermite Polynomials**

The Hermite polynomials $H_n(x)$ satisfy the differential equation

$$\frac{d^2y}{dx^2} - 2x\frac{dy}{dx} + 2n\,y = 0, \qquad -\infty < x\infty$$

and satisfy the orthogonality condition

$$(H_m, H_n) = \int_{-\infty}^{\infty} e^{-x^2} H_m(x) H_n(x)\,dx = \begin{cases} 0, & m \neq n \\ \|H_n\|^2, & m = n \end{cases}$$

where $\quad \|H_n\|^2 = (H_n, H_n) = \int_{-\infty}^{\infty} e^{-x^2}[H_n(x)]^2\,dx = 2^n\, n!\sqrt{\pi}$

Here the weight function is $e^{-x^2}$ is obtained by placing the differential equation in self-adjoint form.

The Fourier-Hermite series has the form

$$f(x) = \sum_{n=0}^{\infty} c_n H_n(x) = c_0 H_0(x) + c_1 H_1(x) + c_2 H_2(x) + \cdots$$

with Fourier coefficients

$$c_n = \frac{(f, H_n)}{\|H_n\|^2} = \frac{1}{2^n n!\sqrt{\pi}} \int_{-\infty}^{\infty} e^{-x^2} f(x) H_n(x)\,dx$$

**The Laguerre Polynomials**

The Laguerre polynomials $L_n(x)$ satisfy the Laguerre differential equation

$$x\frac{d^2y}{dx^2} + (1-x)\frac{d^2y}{dx^2} + n\,y = 0, \qquad 0 \leq x < \infty$$

and satisfy the orthogonality condition

$$(L_m, L_n) = \int_0^{\infty} e^{-x} L_m(x) L_n(x)\,dx = \begin{cases} 0, & m \neq n \\ \|L_n\|^2, & m = n \end{cases}$$

where $\quad \|L_n\|^2 = (L_n, L_n) = \int_0^{\infty} e^{-x}[L_n(x)]^2\,dx = (n!)^2$

The weight function $e^{-x}$ is obtained by placing the Laguerre differential equation in self adjoint form.

The Fourier-Laguerre series has the form

$$f(x) = \sum_{n=0}^{\infty} c_n L_n(x) = c_0 L_0(x) + c_1 L_1(x) + c_2 L_2(x) + \cdots$$

with Fourier coefficients

$$c_n = \frac{(f, L_n)}{\|L_n\|^2} = \frac{1}{(k!)^2} \int_0^{\infty} e^{-x} f(x) L_k(x)\,dx$$

**The Associated Laguerre Polynomials**

Associated Laguerre polynomials $L_n^m(x)$ are solutions of the associated Laguerre differential equation

$$x\frac{d^2y}{dx^2} + (m+1-x)\frac{dy}{dx} + (n-m)y = 0, \qquad 0 \leq x < \infty$$

and satisfy the orthogonality condition

$$(L_n^m(x), L_j^m(x)) = \int_0^\infty x^m e^{-x} L_n^m(x) L_j^m(x)\,dx = \begin{cases} 0, & j \neq m \\ \|L_n^m\|^2, & j = n \end{cases}$$

where $\quad \|L_n^m\|^2 = (L_n^m, L_n^m) = \int_0^\infty x^m e^{-x} [L_n^m(x)]^2\,dx = \dfrac{(n!)^3}{(n-m)!}$

The weight function $x^m e^{-x}$ is obtained by placing the associated Laguerre differential equation in self adjoint form.

The Fourier-associated Laguerre series has the form

$$f(x) = \sum_{n=m}^\infty c_n L_n^m(x) = c_m L_m^m(x) + c_{m+1} L_{m+1}^m(x) + c_{m+2} L_{m+1}^m(x) + \cdots$$

with Fourier coefficients

$$c_k = \frac{(f, L_k^m)}{\|L_k^m\|^2} = \frac{(k-m)!}{(k!)^3} \int_0^\infty x^m e^{-x} L_k^m(x) f(x)\,dx$$

**The Chebyshev Polynomials of the First Kind**

The Chebyshev polynomials of the first kind $T_n(x)$ are solutions of the Chebyshev differential equation

$$(1-x^2)\frac{d^2y}{dx^2} = x\frac{dy}{dx} + n^2 y = 0, \qquad n = 0, 1, 2, \ldots, \qquad -1 < x < 1$$

and satisfy the orthogonality condition

$$(T_n, T_m) = \int_{-1}^1 \frac{T_n(x) T_m(x)}{\sqrt{1-x^2}}\,dx = \begin{cases} 0, & m \neq n \\ \|T_n\|^2, & m = n \end{cases}$$

where $\quad \|T_n\|^2 = (T_n, T_n) = \int_{-1}^1 \frac{[T_n(x)]^2}{\sqrt{1-x^2}}\,dx = \begin{cases} \pi, & n = 0 \\ \frac{\pi}{2}, & n = 1, 2, 3 \ldots \end{cases}$

The weight function $1/\sqrt{1-x^2}$ is obtained by placing the Chebyshev differential equation in self adjoint form.

The Fourier-Chebyshev series has the form

$$f(x) = \sum_{n=0}^\infty c_n T_n(x) = c_0 T_0(x) + c_1 T_1(x) + c_2 T_2(x) + \cdots$$

where

$$c_k = \frac{(f, T_k)}{\|T_k\|^2} = \begin{cases} \frac{1}{\pi} \int_{-1}^1 \frac{f(x) T_0(x)}{\sqrt{1-x^2}}\,dx, & k = 0 \\ \frac{2}{\pi} \int_{-1}^1 \frac{f(x) T_k(x)}{\sqrt{1-x^2}}\,dx, & k > 0 \end{cases}$$

**The Chebyshev Polynomials of the Second Kind**

The Chebyshev polynomials of the second kind $U_n(x)$ satisfy the differential equation

$$(1-x^2)y'' - 3xy' + n(n+2)y = 0$$

obtained by differentiating the Chebyshev differential equation. The function $U_n(x)$ satisfy the orthogonality relation

$$(U_n, U_m) = \int_{-1}^{1} \sqrt{1-x^2}\, U_m(x) U_n(x)\, dx = \begin{cases} 0, & m \neq n \\ \|U_n\|^2, & m = n \end{cases}$$

where $\quad \|U_n\|^2 = (U_n, U_n) = \int_{-1}^{1} \sqrt{1-x^2}\,[U_n(x)]^2\, dx = \dfrac{\pi}{2} \quad$ for all values of $n$

The Chebyshev Fourier series has the form

$$f(x) = \sum_{n=0}^{\infty} c_n U_n(x) = c_0 U_0(x) + c_1 U_1(x) + c_2 U_2(x) + \cdots$$

where the Fourier coefficients are given by

$$c_k = \frac{2}{\pi} \int_{-1}^{1} \sqrt{1-x^2}\, f(x) U_k(x)\, dx, \qquad k = 0, 1, 2, \ldots$$

**Spherical Harmonics**

In terms of spherical coordinates $(\rho, \theta, \phi)$ where $\theta$ is the polar angle $0 \leq \theta \leq \pi$ and $\phi$ is the azimuthal angle $0 \leq \phi \leq 2\pi$, the function $\Phi_m(\phi) = \frac{1}{\sqrt{2\pi}} e^{i m \phi}$ is orthonormal with respect to the azimuthal angle and the associated Legendre polynomial $P_n^m(\cos\theta)$ is orthonormal with respect to the polar angle. The following scaled products of the functions $\Phi_m$ and $P_n^m$ defined by

$$Y_n^m(\theta, \phi) = (-1)^m \sqrt{\frac{2n+1}{4\pi} \frac{(n-m)!}{(n+m)!}}\, P_n^m(\cos\theta)\, e^{i m \phi}$$

are called spherical harmonics. These functions are orthonormal over the surface of a sphere.

## Fourier's Integral Theorem

If $f(x)$ and $f'(x)$ are piecewise continuous functions over every interval $(-L, L)$ and $\int_{-\infty}^{\infty} |f(x)|\, dx$ converges, then $f(x)$ can be represented in the integral form

$$f(x) = \int_0^{\infty} [A(\omega) \cos \omega x + B(\omega) \sin \omega x]\, d\omega \tag{7.109}$$

where

$$A(\omega) = \frac{1}{\pi} \int_{-\infty}^{\infty} f(x) \cos \omega x\, dx \quad \text{and} \quad B(\omega) = \frac{1}{\pi} \int_{-\infty}^{\infty} f(x) \sin \omega x\, dx \tag{7.110}$$

If $x_0$ is a point of discontinuity, then $f(x)$ converges to $\frac{1}{2}[f(x_0^+) + f(x_0^-)]$ which is the average of the left and right limits as $x$ approaches $x_0$.

**Equivalent Forms of the Fourier Integral Theorem**

(i) $\quad f(x) = \dfrac{1}{2\pi}\int_{-\infty}^{\infty} d\omega \int_{-\infty}^{\infty} du\, f(u)\cos\omega(x-u) = \dfrac{1}{2\pi}\int_{-\infty}^{\infty}\int_{-\infty}^{\infty} f(u)\cos\omega(x-u)\,du\,d\omega$

(ii) $\quad f(x) = \dfrac{2}{\pi}\int_{0}^{\infty} \sin\omega x\, d\omega \int_{0}^{\infty} f(u)\sin\omega u\, du \qquad$ if $f(x)$ is an odd function of $x$.

(iii) $\quad f(x) = \dfrac{2}{\pi}\int_{0}^{\infty} \cos\omega x\, d\omega \int_{0}^{\infty} f(u)\cos\omega u\, du, \qquad$ if $f(x)$ is an even function of $x$.

The Fourier integral theorem can also be expressed in the exponential form

$$f(x) = \dfrac{1}{2\pi}\int_{-\infty}^{\infty}\int_{-\infty}^{\infty} f(\xi) e^{i\omega(\xi-x)}\,d\xi\,d\omega \tag{7.111}$$

## The Fresnel Integrals

The Fresnel integrals are defined

$$C(x) = \int_{0}^{x} \cos(t^2)\,dt \qquad \text{and} \qquad S(x) = \int_{0}^{x} \sin(t^2)\,dt$$

and the complementary Fresnel functions

$$c(x) = \int_{x}^{\infty} \cos(t^2)\,dt \qquad \text{and} \qquad s(x) = \int_{x}^{\infty} \sin(t^2)\,dt$$

These functions are used in certain applications of optics and antenna theory. Their graphs are illustrated in the figure 7-11. The Fresnel integrals have the property that

$$\lim_{x\to\infty} C(x) = \dfrac{1}{2}\sqrt{\dfrac{\pi}{2}} \qquad \text{and} \qquad \lim_{x\to\infty} S(x) = \dfrac{1}{2}\sqrt{\dfrac{\pi}{2}}$$

so that the complementary Fresnel functions can be written

$$c(x) = \dfrac{1}{2}\sqrt{\dfrac{\pi}{2}} - C(x) \qquad \text{and} \qquad s(x) = \dfrac{1}{2}\sqrt{\dfrac{\pi}{2}} - S(x)$$

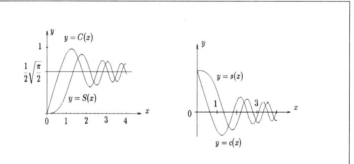

**Figure 7-11.** The Fresnel integrals $C(x), S(x)$ and complementary functions $c(x), s(x)$

## Integral Equations

A Fredholm integral equation of the first kind has the form

$$f(x) = \int_a^b k(x,t)\phi(t)\,dt, \quad a \leq x \leq b \tag{7.112}$$

Here $\phi(t)$, the unknown to be solved for, is under the integral sign and the limits of integration are over the finite interval $(a,b)$. The quantities $a, b, f(x), k(x,t)$ are known quantities with $k(x,t)$ called the kernel of the equation.

A Fredholm integral equation of the second kind with a parameter has the form

$$\phi(x) = f(x) + \lambda \int_a^b k(x,t)\phi(t)\,dt, \quad a \leq x \leq b \tag{7.113}$$

where the limits of integration are given along with known functions $f(x)$ and kernel $k(x,t)$. Here $\lambda$ is called the parameter of the integral equation. Sometimes the Fredholm integral equation of the second kind is written as

$$\mu\phi(x) = f(x) + \int_a^b k(x,t)\phi(t)\,dt, \quad a \leq x \leq b \tag{7.114}$$

where $\mu$ is the parameter and $\phi(x)$ is the unknown to be determined.

The Fredholm integral equation of the third kind has the form

$$\mu\psi(x)\phi(x) = f(x) + \int_a^b k(x,t)\phi(t)\,dt, \quad a \leq x \leq b \tag{7.115}$$

where $\psi(x)$ is a known function.

The Volterra integral equation of the first kind has the form

$$f(x) = \int_a^x k(x,t)\phi(t)\,dt, \tag{7.116}$$

where $\phi(t)$ is to be determined.

The Volterra integral equation of the second kind has one of the forms

$$\phi(x) = f(x) + \lambda \int_a^x k(x,t)\phi(t)\,dt, \tag{7.117}$$

$$\text{or} \quad \mu\phi(x) = f(x) + \int_a^x k(x,t)\phi(t)\,dt, \tag{7.118}$$

The Fredholm integral equations have fixed finite limits of integration while the Volterra integral equations have indefinite integrals.

The adjective singular is placed before the name whenever one or both of the following conditions are satisfied.

  (i) The interval of integration is infinite.

  (ii) The integrand becomes unbounded within the interval of integration.

## The Error Function

The error function and complementary error function are sometimes defined[5]

$$\text{erf}(x) = \frac{2}{\sqrt{\pi}} \int_0^x e^{-t^2}\,dt, \qquad \text{erfc}(x) = \frac{2}{\sqrt{\pi}} \int_x^\infty e^{-t^2}\,dt = 1 - \text{erf}(x)$$

These functions represent the area under the bell shaped curve $y = \frac{2}{\sqrt{\pi}} e^{-t^2}$ illustrated in the figure 7-12, where the factor $2/\sqrt{\pi}$ is used as a normalization factor so that the total area under the right half of the curve is 1. That is $\int_0^\infty e^{-t^2}\,dt = \frac{1}{2}\Gamma(\frac{1}{2}) = \frac{\sqrt{\pi}}{2}$. The error function satisfies the properties

$$\lim_{x \to \infty} \text{erf}(x) = 1, \qquad \text{erf}(0) = 0, \qquad \text{erf}(-x) = -\text{erf}(x)$$

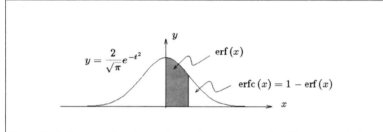

**Figure 7-12.** The error function and complementary error function represented as areas.

An alternative definition of the error function is given by $\text{erf}(x) = \int_0^x e^{-t^2}\,dt$ where the scale factor of $2/\sqrt{\pi}$ has been omitted.

A generalized error function can be defined

$$\text{erf}_n(x) = \frac{n!}{\sqrt{\pi}} \int_0^x e^{-t^n}\,dt$$

which has the special cases

$$\text{erf}_0(x) = \frac{1}{e\sqrt{\pi}} x, \qquad \text{erf}_1(x) = \frac{1}{\sqrt{\pi}}(1 - e^{-x}), \qquad \text{erf}_2(x) = \text{erf}(x)$$

## The Sine, Cosine and Exponential Integrals

The functions

$$Si(x) = \int_0^x \frac{\sin t}{t}\,dt \qquad \text{and} \qquad si(x) = \int_x^\infty \frac{\sin t}{t}\,dt$$

are called the sine integral and complementary sine integral. Note that

$$\lim_{x \to \infty} Si(x) = \frac{\pi}{2}, \qquad \text{and} \qquad si(x) = \frac{\pi}{2} - Si(x)$$

---

[5] Alternative definitions can be found in the literature.

The function
$$ci(x) = \int_x^\infty \frac{\cos t}{t} dt$$
is called the cosine integral. The integral
$$\text{Ei}(x) = \int_x^\infty \frac{e^{-t}}{t} dt$$
is called the exponential integral. Note that the exponential integral is a special case of the incomplete gamma function. The function
$$\text{li}(x) = \int_0^x \frac{dt}{\ln t}$$
called the logarithmic integral, is related to the exponential integral using the relation
$$\text{li}(e^{-x}) = -\text{Ei}(x)$$

## Elliptic Integral of the First Kind

The integral
$$u = F(k, \phi) = \int_0^\phi \frac{dt}{\sqrt{1 - k^2 \sin^2 t}}, \qquad 0 < k < 1, \quad \phi = \text{am}\, u$$
$$= \int_0^x \frac{d\xi}{\sqrt{(1-\xi^2)(1-k^2\xi^2)}}, \qquad x = \sin \phi$$
is called the incomplete elliptic integral of the first kind, where $\phi$ is called the amplitude of $u$ and denoted $\phi = \text{am}\, u$. The integral
$$K = F(k, \pi/2) = \int_0^{\pi/2} \frac{dt}{\sqrt{1 - k^2 \sin^2 t}} = \int_0^1 \frac{d\xi}{\sqrt{(1-\xi^2)(1-k^2\xi^2)}}$$
is called the complete elliptic integral of the first kind.

The incomplete elliptic integral of the first kind is used to define the Jacobi elliptic functions
$$x = \sin(\text{am}\, u) = \text{sn}\, u$$
$$\sqrt{1 - x^2} = \cos(\text{am}\, u) = \text{cn}\, u$$
$$\sqrt{1 - k^2 x^2} = \sqrt{1 - k^2 \text{sn}^2 u} = \text{dn}\, u$$
These functions are pronounce by saying each letter in the name. For example, the $\text{sn}\, u$ function is called the "s","n" of "u" function. Related to the above definitions are the additional Jacobi elliptic functions

$$\text{ns}\, u = \frac{1}{\text{sn}\, u} \qquad \text{sc}\, u = \frac{\text{sn}\, u}{\text{cn}\, u} \qquad \text{cs}\, u = \frac{\text{cn}\, u}{\text{sn}\, u}$$
$$\text{nc}\, u = \frac{1}{\text{cn}\, u} \qquad \text{sd}\, u = \frac{\text{sn}\, u}{\text{dn}\, u} \qquad \text{dc}\, u = \frac{\text{dn}\, u}{\text{cn}\, u}$$
$$\text{nd}\, u = \frac{1}{\text{dn}\, u} \qquad \text{cd}\, u = \frac{\text{cn}\, u}{\text{dn}\, u} \qquad \text{ds}\, u = \frac{\text{dn}\, u}{\text{sn}\, u}$$

The three basic functions $\operatorname{sn} u$, $\operatorname{cn} u$ and $\operatorname{dn} u$ have the derivatives

$$\frac{d}{du}\operatorname{sn} u = \operatorname{cn} u \operatorname{dn} u, \qquad \frac{d}{du}\operatorname{cn} u = -\operatorname{sn} u \operatorname{dn} u, \qquad \frac{d}{du}\operatorname{dn} u = -k^2 \operatorname{sn} u \operatorname{cn} u$$

from which there results the integrals

$$\int \operatorname{cn} u \operatorname{dn} u \, du = \operatorname{sn} u + C, \qquad \int \operatorname{sn} u \operatorname{dn} u \, du = -\operatorname{cn} u + C, \qquad \int k^2 \operatorname{sn} u \operatorname{cn} u = -\operatorname{dn} u + C$$

The functions $\operatorname{sn} u$, $\operatorname{cn} u$ and $\operatorname{sc} u$ are similar in behavior to the sine, cosine and tangent trigonometric functions. They frequently arise in mechanics and physics problems.

**Addition Theorems**

$$\operatorname{cn}^2(u,k) + \operatorname{sn}^2(u,k) = 1, \qquad \operatorname{dn}^2(u,k) + k^2 \operatorname{sn}^2(u,k) = 1$$

$$\operatorname{sn}(u+v,k) = \frac{\operatorname{sn}(u,k)\operatorname{cn}(v,k)\operatorname{dn}(v,k) + \operatorname{sn}(v,k)\operatorname{cn}(u,k)\operatorname{dn}(u,k)}{1 - k^2 \operatorname{sn}^2(u,k)\operatorname{sn}^2(v,k)}$$

$$\operatorname{cn}(u+v,k) = \frac{\operatorname{cn}(u,k)\operatorname{cn}(v,k) - \operatorname{sn}(u,k)\operatorname{sn}(v,k)\operatorname{dn}(u,k)\operatorname{dn}(v,k)}{1 - k^2 \operatorname{sn}^2(u,k)\operatorname{sn}^2(v,k)}$$

$$\operatorname{dn}(u+v,k) = \frac{\operatorname{dn}(u,k)\operatorname{dn}(v,k) - k^2 \operatorname{sn}(u,k)\operatorname{sn}(v,k)\operatorname{cn}(u,k)\operatorname{cn}(v,k)}{1 - k^2 \operatorname{sn}^2(u,k)\operatorname{sn}^2(v,k)}$$

## Elliptic Integral of the Second Kind

The integral

$$E(k,\phi) = \int_0^\phi \sqrt{1 - k^2 \sin^2 t}\, dt, \qquad 0 < k < 1$$

$$= \int_0^x \frac{\sqrt{1 - k^2 \xi^2}}{\sqrt{1 - \xi^2}}\, d\xi, \qquad x = \sin\phi$$

is called the incomplete elliptic integral of the second kind. The integral

$$E = E(k, \pi/2) = \int_0^{\pi/2} \sqrt{1 - k^2 \sin^2 t}\, dt = \int_0^1 \frac{\sqrt{1 - k^2 \xi^2}}{\sqrt{1 - \xi^2}}\, d\xi, \qquad 0 < k < 1$$

is called the complete elliptic integral of the second kind.

## Elliptic Integral of the Third Kind

The integral

$$\Pi(k, n, \phi) = \int_0^\phi \frac{dt}{(1 + n\sin^2 t)\sqrt{1 - k^2 \sin^2 t}} = \int_0^x \frac{d\xi}{(1 + n\xi^2)\sqrt{(1 - \xi^2)(1 - k^2 \xi^2)}}$$

where $x = \sin\phi$ and $0 < k < 1$ is called the incomplete elliptic integral of the third kind and the integral

$$\Pi(k, n, \pi/2) = \int_0^{\pi/2} \frac{dt}{(1 + n\sin^2 t)\sqrt{1 - k^2 \sin^2 t}} = \int_0^1 \frac{d\xi}{(1 + n\xi^2)\sqrt{(1 - \xi^2)(1 - k^2 \xi^2)}}$$

is called the complete elliptic integral of the third kind.

## Riemann Zeta Function

The Riemann zeta function is defined

$$\zeta(z) = \sum_{k=1}^{\infty} \frac{1}{k^z} = \frac{1}{1^z} + \frac{1}{2^z} + \frac{1}{3^z} + \frac{1}{4^z} + \cdots + \frac{1}{n^z} + \cdots, \qquad \text{Re}\{z\} > 1$$

Some special values associated with the Riemann zeta function are

$$\zeta(2) = \frac{\pi^2}{6}, \qquad \zeta(6) = \frac{\pi^6}{945} \qquad \zeta(z) = 2^z \pi^{z-1} \sin\left(\frac{\pi z}{2}\right) \Gamma(1-z) \zeta(1-z)$$

$$\zeta(4) = \frac{\pi^4}{90}, \qquad \zeta(8) = \frac{\pi^8}{9450} \qquad \sum_{n=1}^{\infty} \frac{(-1)^{n+1}}{n^z} = \zeta(z) - \frac{1}{2^{z-1}} \zeta(z)$$

The zeta function is related to the Euler numbers and Bernoulli numbers and is an important function with applications in number theory and the study of prime numbers.

A generalization of the Riemann zeta function is the Hurwitz zeta function

$$\zeta(x,q) = \sum_{k=1}^{\infty} \frac{1}{(k+q)^x}$$

## The Laplace Transform

The Laplace transform is used to solve linear ordinary differential equations and linear partial differential equations and is used quite frequently in electrical engineering. The Laplace transform of a real function $f(t)$ is defined

$$F(s) = \mathcal{L}\{f(t);\ t \to s\} = \int_0^{\infty} e^{-st} f(t)\, dt \qquad (7.119)$$

provided the parameter $s$ can be selected such that the integral exists.

The inverse Laplace transform can be calculated using table look up or by using methods from the theory of complex variables where it is shown that

$$f(t) = \mathcal{L}^{-1}\{F(s);\ s \to t\} = \frac{1}{2\pi i} \int_{\gamma-i\infty}^{\gamma+i\infty} e^{st} F(s)\, ds \qquad (7.120)$$

where $\gamma$ is a constant selected to the right of all singularities of $F(s)$. Properties of the Laplace transform and some examples are given in the tables 7.3 and 7.4.

**Example 7-9.** The Laplace transform of $f(t) = \sin \omega t$ is given by

$$F(s) = \mathcal{L}\{\sin \omega t;\ t \to s\} = \int_0^{\infty} e^{-st} \sin \omega t\, dt = \left. \frac{e^{-st}(-s \sin \omega t - \omega \cos \omega t)}{s^2 + \omega^2} \right|_0^{\infty} = \frac{\omega}{s^2 + \omega^2}$$

and so the inverse Laplace transform is

$$\mathcal{L}^{-1}\{\frac{\omega}{s^2 + \omega^2};\ s \to t\} = \sin \omega t$$

■

| | Table 7.3 | | |
|---|---|---|---|
| | **Laplace Transform Properties** | | |
| 1. | $f(t) = \mathcal{L}^{-1}\{F(s)\}$ | $F(s) = \mathcal{L}\{f(t)\}$ | Comments |
| 2. | $c_1 f(t) + c_2 g(t)$ | $c_1 F(s) + c_2 G(s)$ | Linearity property |
| 3. | $f'(t)$ | $sF(s) - f(0^+)$ | Derivative property |
| 4. | $f''(t)$ | $s^2 F(s) - sf(0^+) - f'(0^+)$ | Second derivative |
| 5. | $f^{(n)}(t)$ | $s^n F(s) - s^{n-1} f(0^+) - \ldots -sf^{(n-2)}(0^+) - f^{(n-1)}(0^+)$ | $n$th derivative |
| 6. | $\int_0^t f(\tau)\,d\tau$ | $\dfrac{F(s)}{s}$ | Integration property |
| 7. | $tf(t)$ | $-F'(s)$ | Multiplication by $t$ |
| 8. | $t^n f(t)$ | $(-1)^n F^{(n)}(s)$ | Repeated multiplication by $t$ |
| 9. | $\dfrac{f(t)}{t}$ | $\int_s^\infty F(s)\,ds$ | Division by $t$ |
| 10. | $e^{at} f(t)$ | $F(s-a)$ | First shift property |
| 11. | $f(t-a)H(t-a)$ | $e^{-as} F(s),\quad a > 0$ | Second shift property |
| 12. | $H(t-a)$ | $\dfrac{e^{-as}}{s},\quad a > 0$ | Heaviside function |
| 13. | $\delta(t-a)$ | $e^{-as},\quad a > 0$ | Dirac delta function |
| 14. | $\dfrac{1}{a} f\!\left(\dfrac{t}{a}\right)$ | $F(as),\quad a > 0$ | Scaling property |
| 15. | $f(t)^* g(t)$ | $F(s) G(s)$ | Convolution property $f(t)^* g(t) = \int_0^t f(\tau) g(t-\tau)\,d\tau$ $g(t)^* f(t) = \int_0^t g(\tau) f(t-\tau)\,d\tau$ |
| 16. | $\lvert f(t) \rvert \le M e^{\alpha t}$ | $\lim_{s \to \infty} F(s) = 0$ | $f(t)$ of exponential order. |

| Table 7.4 |||
|---|---|---|
| **Short Table of Laplace Transforms** |||
| | $f(t) = \mathcal{L}^{-1}\{F(s)\}$ | $F(s) = \mathcal{L}\{f(t)\}$ |
| 1. | $1$ | $\dfrac{1}{s}$ |
| 2. | $t$ | $\dfrac{1}{s^2}$ |
| 3. | $t^n$ | $\dfrac{n!}{s^{n+1}}$    $n$ integer, $s > 0$ |
| 4. | $t^\alpha$ | $\dfrac{\Gamma(\alpha+1)}{s^{\alpha+1}}$    $\alpha > -1$, $s > 0$ |
| 5. | $e^{\omega t}$ | $\dfrac{1}{s-\omega}$ |
| 6. | $\sin \omega t$ | $\dfrac{\omega}{s^2+\omega^2}$ |
| 7. | $\cos \omega t$ | $\dfrac{s}{s^2+\omega^2}$ |
| 8. | $\sinh \omega t$ | $\dfrac{\omega}{s^2-\omega^2}$ |
| 9. | $\cosh \omega t$ | $\dfrac{s}{s^2-\omega^2}$ |
| 10. | $\left(\dfrac{t}{\beta}\right)^{n/2} J_n(2\sqrt{\beta t})$ | $\dfrac{1}{s^{n+1}}e^{-\beta/s}$    $n > -1$ |
| 11. | $J_0(\omega t)$ | $\dfrac{1}{\sqrt{s^2+\omega^2}}$ |
| 12. | $\dfrac{1}{\sqrt{\omega}}e^{\omega t}\operatorname{erf}(\sqrt{\omega t})$ | $\dfrac{1}{\sqrt{s}(s-\omega)}$ |
| 13. | $\dfrac{1}{\sqrt{\pi t}}e^{-\omega^2/4t}$ | $\dfrac{1}{\sqrt{s}}e^{-\omega\sqrt{s}}$ |
| 14. | $\dfrac{\omega}{2\sqrt{\pi t^3}}e^{-\omega^2/4t}$ | $e^{-\omega\sqrt{s}}$ |
| 15. | $\operatorname{erf}\left(\dfrac{\omega}{2\sqrt{t}}\right)$ | $\dfrac{1}{s}\left(1-e^{-\omega\sqrt{s}}\right)$ |
| 16. | $\operatorname{erfc}\left(\dfrac{\omega}{2\sqrt{t}}\right)$ | $\dfrac{1}{s}e^{-\omega\sqrt{s}}$ |
| 17. | $e^{\beta^2 t+\alpha\beta}\operatorname{erfc}\left(\beta\sqrt{t}+\dfrac{\alpha}{2\sqrt{t}}\right)$ | $\dfrac{1}{\sqrt{s}(\beta+\sqrt{s})}e^{-\alpha\sqrt{s}}$ |
| 18. | $H(t-a)$ | $\dfrac{1}{s}e^{-as}$,    $a > 0$ |
| 19. | $\delta(t-a)$ | $e^{-as}$,    $a > 0$ |
| 20. | $\dfrac{x}{b}+\dfrac{2}{\pi}\sum_{n=1}^{\infty}\dfrac{(-1)^n}{n}e^{-n^2\pi^2 t/b^2}\sin\dfrac{n\pi x}{b}$ | $\dfrac{\sinh x\sqrt{s}}{s\sinh b\sqrt{s}}$ |

## The Fourier Transforms

The Fourier exponential transform pair can be defined[6] by splitting the exponential form of the Fourier integral theorem

$$f(x) = \frac{1}{2\pi} \int_{-\infty}^{\infty} \int_{-\infty}^{\infty} f(\xi) e^{i\omega(\xi-x)} \, d\xi \, d\omega \qquad (7.121)$$

into an inner and outer integral by defining

$$\mathcal{F}_e\{f(x); x \to \omega\} = F(\omega) = \frac{1}{2\pi} \int_{-\infty}^{\infty} f(\xi) e^{i\omega\xi} \, d\xi$$

$$\mathcal{F}_e^{-1}\{F(\omega); \omega \to x\} = f(x) = \int_{-\infty}^{\infty} F(\omega) e^{-i\omega x} \, d\omega \qquad (7.122)$$

For the Fourier transform of a function $f(x)$ to exist, the function $f(x)$ must be sectionally continuous and the absolute integral $\int_{-\infty}^{\infty} |f(x)| \, dx$ must exist. The function $F(\omega)$ is called the frequency domain representation of the function $f(t)$.

### Example 7-10. Symmetry factor
If $\mathcal{F}_e\{f(x); x \to \omega\} = \frac{1}{2\pi} \int_{-\infty}^{\infty} f(\xi) e^{i\omega\xi} \, d\xi = F(\omega)$ with $f(x) = \int_{-\infty}^{\infty} F(\omega) e^{-i\omega x} \, d\omega$, then

$$\mathcal{F}_e\{2\pi F(-x); x \to \omega\} = \frac{1}{2\pi} \int_{-\infty}^{\infty} 2\pi F(-\xi) e^{i\omega\xi} \, d\xi = \int_{\infty}^{-\infty} F(x) e^{-i\omega x}(-dx) = \int_{-\infty}^{\infty} F(x) e^{-i\omega x} \, dx = f(\omega)$$

∎

The table 7-5 lists some important operational properties associated with the Fourier exponential transform while the table 7-6 is a short table of Fourier exponential transforms.

Closely related to the Fourier transform is the discrete Fourier transform (DFT), sometimes called the finite transform. This transform is used to analyze and process digital signals. One method of defining the discrete Fourier transform and its inverse is

$$F(k+1) = \sum_{k=0}^{N-1} f(n+1) e^{-i\left(\frac{2\pi k n}{N}\right)} \qquad (7.123)$$

with inverse transform

$$f(n+1) = \frac{1}{N} \sum_{k=0}^{N-1} F(k+1) e^{i\left(\frac{2\pi k n}{N}\right)} \qquad (7.124)$$

Here $f_1 = f(1), f_2 = f(2), \ldots, f_{N-1} = f(N-1)$ are a set of real or complex numbers which represent a discrete set of $N-1$ values. These numbers could represent a systematic sampling of a time function $f(t)$ over some interval of time. The DFT is a way to transform a sampled time function to a discrete frequency domain representation. In this way the DFT is said to represent a Fourier analysis of a finite set of discrete time function signal.

---
[6] There are other ways to define the Fourier transform pair.

## Table 7-5. Fourier Exponential Transform Properties

$f(x) = \int_{-\infty}^{\infty} F(\omega) e^{-i\omega x}\, d\omega = \mathcal{F}_e^{-1}\{F(\omega)\}$  $\qquad$ $F(\omega) = \frac{1}{2\pi}\int_{-\infty}^{\infty} f(x) e^{i\omega x}\, dx = \mathcal{F}_e\{f(x)\}$

| | $f(x) = \mathcal{F}_e^{-1}\{F(\omega)\}$ | $F(\omega) = \mathcal{F}_e\{f(x)\}$ | Comments |
|---|---|---|---|
| 1. | $2\pi F(-x)$ | $f(\omega)$ | Column interchange |
| 2. | $c_1 f(x) + c_2 g(x)$ | $c_1 F(\omega) + c_2 G(\omega)$ | Linearity property |
| 3. | $f'(x)$ | $-i\omega F(\omega)$ | Derivative property |
| 4. | $f''(x)$ | $(-i\omega)^2 F(\omega)$ | |
| 5. | $f^{(n)}(x)$ | $(-i\omega)^n F(\omega)$ | |
| 6. | $f(x-a)$ | $e^{i\omega a} F(\omega)$ | Shift property |
| 7. | $x f(x)$ | $-i\dfrac{dF}{d\omega} = -i F'(\omega)$ | Multiplication by $x$ property |
| 8. | $x^n f(x)$ | $(-i)^n \dfrac{d^n F(\omega)}{d\omega^n}$ | |
| 9. | $f * g = \dfrac{1}{2\pi}\int_{-\infty}^{\infty} f(\tau) g(x-\tau)\, d\tau$ | $F(\omega) G(\omega)$ | Convolution property |
| 10. | $\delta(x - x_0)$ | $\dfrac{1}{2\pi} e^{i\omega x_0}$ | Dirac delta function |
| 11. | $f(ax),\quad a > 0$ | $\dfrac{1}{a} F\left(\dfrac{\omega}{a}\right)$ | Scaling property |
| 12. | $f(ax) e^{ibx},\ a > 0$ | $\dfrac{1}{a} F\left(\dfrac{\omega + b}{a}\right)$ | Shift and scaling |
| 13. | $f(ax)\cos bx$ | $\dfrac{1}{2a}\left[F\left(\dfrac{\omega+b}{a}\right) + F\left(\dfrac{\omega-b}{a}\right)\right]$ | |
| 14. | $f(ax)\sin bx$ | $\dfrac{1}{2ia}\left[F\left(\dfrac{\omega+b}{a}\right) - F\left(\dfrac{\omega-b}{a}\right)\right]$ | |
| 15. | $f(x) e^{iax}$ | $F(\omega + a)$ | Shift property |

## Table 7-6. Fourier Exponential Transforms

| | $f(x) = \mathcal{F}_e^{-1}\{F(\omega)\} = \int_{-\infty}^{\infty} F(\omega) e^{-i\omega x}\, d\omega$ | $F(\omega) = \mathcal{F}_e\{f(x)\} = \frac{1}{2\pi} \int_{-\infty}^{\infty} f(x) e^{i\omega x}\, dx$ |
|---|---|---|
| 1. | $e^{-\alpha |x|}$ | $\dfrac{\alpha}{\pi(\alpha^2 + \omega^2)}$ |
| 2. | $\dfrac{2\alpha}{\alpha^2 + x^2}$ | $e^{-\alpha |\omega|}$ |
| 3. | $\begin{cases} 1, & |x| < L \\ 0, & |x| > L \end{cases}$ | $\dfrac{1}{\pi} \dfrac{\sin \omega L}{\omega}$ |
| 4. | $e^{-\alpha x} H(x)$ | $\dfrac{1}{2\pi(\alpha - i\omega)}$ |
| 5. | $e^{-\alpha x^2}$ | $\dfrac{1}{2\sqrt{\pi \alpha}} e^{-\omega^2/4\alpha}$ |
| 6. | $\operatorname{sgn}(x) = \begin{cases} 1, & x > 0 \\ 0, & x = 0 \\ -1, & x < 0 \end{cases}$ | $\dfrac{i}{\pi \omega}$ |
| 7. | $\dfrac{\sinh ax}{\sinh \pi x},\quad -\pi < a < \pi$ | $\dfrac{1}{2\pi} \dfrac{\sin a}{\cosh \omega + \cos a}$ |
| 8. | $\dfrac{x}{x^2 + \alpha^2}$ | $\dfrac{i}{2} e^{-\alpha |\omega|} \operatorname{sgn}(\omega)$ |
| 9. | $\sin \alpha x\, H(x)$ | $\dfrac{1}{2\pi} \dfrac{\alpha}{(-i\omega)^2 + \alpha^2}$ |
| 10. | $\cos \alpha x\, H(x)$ | $\dfrac{1}{2\pi} \dfrac{-i\omega}{(-i\omega)^2 + \alpha^2}$ |
| 11. | $\dfrac{2\alpha x}{(x^2 + \alpha^2)^2}$ | $\dfrac{i\omega}{2} e^{-\alpha |\omega|}$ |
| 12. | $\dfrac{1}{|x|}$ | $\dfrac{1}{\sqrt{2\pi}} \dfrac{1}{|\omega|}$ |
| 13. | $x e^{-\alpha |x|}\quad \alpha > 0$ | $\dfrac{2}{\pi} \dfrac{2\alpha i\omega}{(\alpha^2 + \omega^2)^2}\quad \omega > 0$ |
| 14. | $\dfrac{2 \sin \alpha x}{x}$ | $\begin{cases} 1 & |\omega| < \alpha \\ 0 & |\omega| > \alpha \end{cases}$ |
| 15. | $\operatorname{erf}\left(\dfrac{x}{2\sqrt{Kt}}\right)$ | $\dfrac{i}{\pi \omega} e^{-Kt\omega^2}$ |
| 16. | $\begin{cases} (a^2 - x^2)^{-1/2}, & |x| < a \\ 0, & |x| > a \end{cases}$ | $\dfrac{1}{2} J_0(a\omega)$ |

The Fourier sine transform is defined

$$\mathcal{F}_s\{f(x)\} = F_s(\omega) = \frac{2}{\pi}\int_0^\infty f(x)\sin\omega x\,dx \qquad (7.125)$$

with inverse transform

$$\mathcal{F}_s^{-1}\{F_s(\omega)\} = f(x) = \int_0^\infty F_s(\omega)\sin\omega x\,d\omega. \qquad (7.126)$$

**Table 7.7. Fourier Sine Transforms**

| | $f(x) = \mathcal{F}_s^{-1}\{F_s(\omega)\} = \int_0^\infty F_s(\omega)\sin\omega x\,d\omega$ | $F_s(\omega) = \mathcal{F}_s\{f(x)\} = \frac{2}{\pi}\int_0^\infty f(x)\sin\omega x\,dx$ |
|---|---|---|
| 1. | $f'(x)$ | $-\omega\mathcal{F}_c\{f(x)\}$ |
| 2. | $f''(x)$ | $-\omega^2 F_s(\omega) + \dfrac{2\omega}{\pi}f(0)$ |
| 3. | $f(ax),\quad a>0$ | $\dfrac{1}{a}F_s\left(\dfrac{\omega}{a}\right)$ |
| 4. | $f(ax)\cos bx\quad a>0,\,b>0$ | $\dfrac{1}{2a}\left[F_s\left(\dfrac{\omega+b}{a}\right) + F_s\left(\dfrac{\omega-b}{a}\right)\right]$ |
| 5. | $1$ | $\dfrac{2}{\pi}\cdot\dfrac{1}{\omega}$ |
| 6. | $\dfrac{1}{x}$ | $\operatorname{sgn}(\omega)$ |
| 7. | $e^{-\beta x}\quad \beta>0$ | $\dfrac{2}{\pi}\dfrac{\omega}{\beta^2+\omega^2}$ |
| 8. | $\dfrac{e^{-\beta x}}{x}\quad \beta>0$ | $\dfrac{2}{\pi}\arctan\left(\dfrac{\omega}{\beta}\right)$ |
| 9. | $\dfrac{1}{2}\operatorname{erfc}\left(\dfrac{x}{2\sqrt{Kt}}\right),\quad x>0$ | $\dfrac{1-e^{-Kt\omega^2}}{\pi\omega}$ |
| 10. | $\dfrac{1}{\pi}\int_0^\infty f(\xi)\left[g(x-\xi)-g(x+\xi)\right]d\xi$ | $F_s(\omega)G_c(\omega)$ |
| 11. | $\dfrac{1}{4\beta}\sqrt{\dfrac{\pi}{\beta}}xe^{-x^2/4\beta}$ | $\omega e^{-\beta\omega^2}$ |
| 12. | $\dfrac{x}{x^2+a^2}$ | $e^{-a\omega}$ |
| 13. | $\dfrac{1}{x(x^2+a^2)}$ | $\dfrac{1}{a^2}(1-e^{-a\omega})$ |
| 14. | $\dfrac{1}{e^{2\pi x}-1}$ | $\dfrac{1}{2\pi}\coth\dfrac{\omega}{2} - \dfrac{1}{\pi\omega}$ |

Define the Fourier cosine transform

$$\mathcal{F}_c\{f(x)\} = F_c(\omega) = \frac{2}{\pi}\int_0^\infty f(x)\cos\omega x\, dx \qquad (7.127)$$

with inverse transform

$$\mathcal{F}_c^{-1}\{F_c(\omega)\} = f(x) = \int_0^\infty F_c(\omega)\cos\omega x\, d\omega. \qquad (7.128)$$

The table 7.8 is a short table of Fourier cosine transforms.

| Table 7.8 Fourier Cosine Transforms | |
|---|---|
| $f(x) = \mathcal{F}_c^{-1}\{F_c(\omega)\} = \int_0^\infty F_c(\omega)\cos\omega x\, d\omega$ | $F_c(\omega) = \mathcal{F}_c\{f(x)\} = \frac{2}{\pi}\int_0^\infty f(x)\cos\omega x\, dx$ |
| 1. $f'(x)$ | $\omega\mathcal{F}_s\{f(x)\} - \frac{2}{\pi}f(0)$ |
| 2. $f''(x)$ | $-\omega^2 F_c(\omega) - \frac{2}{\pi}f'(0)$ |
| 3. $f(ax)$ | $\frac{1}{a}F_c\left(\frac{\omega}{a}\right)$ |
| 4. $f(ax)\cos bx \quad a>0,\ b>0$ | $\frac{1}{2a}\left[F_c\left(\frac{\omega+b}{a}\right) + F_c\left(\frac{\omega-b}{a}\right)\right]$ |
| 5. $e^{-\beta x} \quad \beta > 0$ | $\frac{2}{\pi}\frac{\beta}{\beta^2+\omega^2}$ |
| 6. $e^{-\beta^2 x^2}$ | $\frac{1}{\sqrt{\pi}}\frac{e^{-\omega^2/4\beta^2}}{\|\beta\|}$ |
| 7. $\frac{\beta}{x^2+\beta^2}$ | $e^{-\omega\beta}$ |
| 8. $\begin{cases} 1 & 0<x<L \\ 0 & L<x \end{cases}$ | $\frac{2}{\pi}\frac{\sin\omega L}{\omega}$ |
| 9. $\sqrt{\frac{Kt}{\pi}}e^{-x^2/4Kt} - \frac{x}{2}\mathrm{erfc}\left(\frac{x}{2\sqrt{Kt}}\right)$ | $\frac{1-e^{-\omega^2 Kt}}{\pi\omega^2}$ |
| 10. $\frac{1}{\pi}\int_0^\infty g(\xi)[f(x-\xi)+f(x+\xi)]\,d\xi$ | $G_c(\omega)F_c(\omega)$ |
| 11. $\frac{1}{2}\sqrt{\frac{\pi}{\beta}}e^{-x^2/4\beta}$ | $e^{-\beta\omega^2}$ |
| 12. $\begin{cases} 1, & 0<x<\beta \\ 0, & x>\beta \end{cases}$ | $\frac{2}{\pi}\frac{\sin\beta\omega}{\omega}$ |
| 13. $\frac{1}{\sqrt{x}}$ | $\sqrt{\frac{2}{\pi\omega}}$ |
| 14. $\frac{e^{\beta\sqrt{x}}}{\sqrt{x}}$ | $\sqrt{\frac{2}{\pi\omega}}\left[\cos(2\beta\sqrt{\omega}) - \sin(2\beta\sqrt{\omega})\right]$ |

# Chapter 8
# Probability and Statistics

### Introduction

The collecting of some type of data, organizing the data, determining how some characteristic of the data is to be presented as well conducting some type of analysis of the data, all comes under the category of probability and statistics. For example, to determine some characteristic associated with a very large group of objects, called the population, it is impractical to examine every member of the group in order to perform an analysis of the population. Instead a random selection of data associated with objects from the group is examined. This is called a random sample from the population. Populations can be finite or infinite and by selecting a sample from the population one expects that some characteristics of the population can be inferred from an analysis of the sample data.

Analysis of the sample data, without trying to infer conclusions about the population from which the sample data comes, is called descriptive or deductive statistics. An analysis of sample data which tries to predict some characteristic of the population is called inductive statistics or statistical inference.

### The Representation of Data

The data from a population can be either discrete or continuous. If $Y$ is a variable representing the characteristic being sampled and $Y$ can take on any value between two given values, then $Y$ is called a continuous variable. If $Y$ is not a continuous variable, then it is called a discrete variable.

Some examples of discrete data is presented in the table 8.1. These numbers can be plotted as vertical or horizontal bar graphs, either stacked or grouped or as a line graph. The figure 8-1 illustrates these type of graphs.

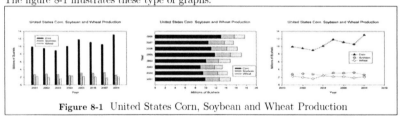

**Figure 8-1** United States Corn, Soybean and Wheat Production

| Table 8.1 |||
|---|---|---|---|
| Unite States Production of Corn, Soybeans and Wheat ||||
| (in millions of bushels) ||||
| Year | Corn | Soybean | Wheat |
| 2001 | 9.92 | 2.76 | 2.23 |
| 2002 | 9.50 | 2.89 | 1.95 |
| 2003 | 8.97 | 2.76 | 1.61 |
| 2004 | 10.09 | 2.45 | 2.35 |
| 2005 | 11.81 | 3.12 | 2.16 |
| 2006 | 11.11 | 3.06 | 2.11 |
| 2007 | 10.54 | 3.19 | 1.81 |
| 2008 | 13.07 | 2.59 | 2.07 |

**Tabular Representation of Data**

A statistical experiment usually consists of collecting data from a random selection of the population. For example, suppose the systolic blood pressure of two hundred individuals are taken from a random sample of the population. The systolic blood pressure is measured in units of mmHg and is the blood pressure as the heart begins to pump. The diastolic blood pressure being a measure of the blood pressure between heart beats. The data set collected consists of 200 numbers, representing the sample size. A representative set of such numbers is presented in the table 8.2.

```
127 115 132 117 138 138 152 121 142 120 104 116 139 165 150 132 142  94 124 145
157 137 118 163 138 159 140  87 162 132 156 148 159 136 164 103 125 136 136 146
102 111 142 116 145 156 167  95 148 143 120 130  95 171 115  87 139 119 148 132
169 121 138 128 129 143 143 128 108  77 120 128 157 109 173 125 159 100  97 144
119 129 131 124 161 144 154 119 125  97 123 129 113 119 109 112 156 168 135 136
135 145 156 125 140 130  86 101 139 184 144 118 150 149 142 118 134 124 154 142
186 130 127 168 122 139 156 146 107 168 117 100 134 113 104 115 149 148 133 128
121 148 133 144 127 127 168 102 117 123 156 129  89 138 136 100 153 110 112 150
104 148 124 114 121 126 153 128 114 137 131 104 135 124 146 115 152 127 113 143
139 147 134 142 133 124 149 156 142 109 147  96 142 163 120 118 180 125 157 118
```

**Table 8.2** Systolic Blood Pressure (mmHg) measurements taken from 200 Random Individuals

Examine the data in table 8.2 and order the data in a tally sheet to form a frequency table. Show that the smallest value is 74 and the largest value is 186. Divide the data into categories or class intervals of equal length by defining an upper limit and lower limit and midpoint for each class interval. This is called grouping the data. Examples of class intervals are given in the table 8.3 where 74 is the first midpoint with 16 midpoints to 186. If $74 + 16x = 186$, then $x = 7$ steps between midpoints or the class interval is of size 7. Go through the data and find the number of systolic blood pressures in each class interval. This is called determining the class frequency associated with the grouped data. Then calculate the relative frequency column, the cumulative frequency column and cumulative relative frequency column as illustrated in the table 8.3. The cumulative frequency associated with a value $x$ is just the sum of the frequencies less than or equal to $x$. The cumulative relative frequency is obtained by dividing the cumulative frequency by the sample size. Note that the cumulative frequency ends in the sample size and the cumulative relative frequency ends with 1.

| | | Table 8.3 | Frequency Table | | | |
|---|---|---|---|---|---|---|
| Class Interval | Class Midpoint | Tallies | Frequency | Relative Frequency | Cumulative Frequency | Cumulative Relative Frequency |
| 71-77 | 74 | / | 1 | $\frac{1}{200} = 0.005$ | 1 | 0.005 |
| 78-84 | 81 | | 0 | $\frac{0}{200} = 0.000$ | 1 | 0.005 |
| 85-91 | 88 | //// | 4 | $\frac{4}{200} = 0.020$ | 5 | 0.025 |
| 92-98 | 95 | ///// / | 6 | $\frac{6}{200} = 0.030$ | 11 | 0.055 |
| 98-105 | 102 | ///// ///// / | 11 | $\frac{11}{200} = 0.055$ | 22 | 0.110 |
| 106-112 | 109 | ///// //// | 9 | $\frac{9}{200} = 0.045$ | 31 | 0.155 |
| 113-119 | 116 | ///// ///// /// ///// ///// | 23 | $\frac{23}{200} = 0.115$ | 54 | 0.270 |
| 120-126 | 123 | ///// ///// /// ///// ///// | 23 | $\frac{23}{200} = 0.115$ | 77 | 0.385 |
| 128-133 | 130 | ///// ///// ///// ///// ///// / | 26 | $\frac{26}{200} = 0.130$ | 103 | 0.515 |
| 134-140 | 137 | ///// ///// ///// ///// ///// | 25 | $\frac{25}{200} = 0.125$ | 128 | 0.640 |
| 141-147 | 144 | ///// ///// //// ///// ///// | 24 | $\frac{24}{200} = 0.120$ | 152 | 0.760 |
| 148-154 | 151 | ///// ///// ///// /// | 18 | $\frac{18}{200} = 0.090$ | 170 | 0.850 |
| 155-161 | 158 | ///// //// ///// | 14 | $\frac{14}{200} = 0.070$ | 184 | 0.920 |
| 162-168 | 165 | ///// ///// | 10 | $\frac{10}{200} = 0.050$ | 194 | 0.970 |
| 168-175 | 172 | /// | 3 | $\frac{3}{200} = 0.015$ | 197 | 0.985 |
| 176-182 | 179 | / | 1 | $\frac{1}{200} = 0.005$ | 198 | 0.990 |
| 183-189 | 186 | // | 2 | $\frac{2}{200} = 0.010$ | 200 | 1.00 |

A graphical representation of the data in table 8.3 can be presented by defining a relative frequency function $f(x)$ and a cumulative relative frequency function $F(x)$ associated with the sample. These functions are defined

$$f(x) = \begin{cases} f_j & \text{when } x = X_j \\ 0 & \text{otherwise} \end{cases} \qquad F(x) = \sum_{x_j \le x} f(x_j) = \text{sum of all } f(x_j) \text{ for which } x_j \le x \qquad (8.1)$$

and are illustrated in the figure 8-2 .

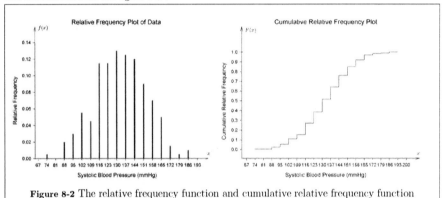

**Figure 8-2** The relative frequency function and cumulative relative frequency function

The results illustrated in the table 8.3 can be generalized. If $X_1, X_2, \ldots, X_k$ are $k$ different numerical values in a sample of size $N$ where $X_1$ occurs $\tilde{f}_1$ times and $X_2$ occurs $\tilde{f}_2$ times, ..., and $X_k$ occurs $\tilde{f}_k$ times, then $\tilde{f}_1, \tilde{f}_2, \ldots, \tilde{f}_k$ are called the frequencies associated with the data set and satisfy

$$\tilde{f}_1 + \tilde{f}_2 + \cdots + \tilde{f}_k = N = \text{sample size}$$

The relative frequencies associated with the data is defined by

$$f_1 = \frac{\tilde{f}_1}{N}, \quad f_2 = \frac{\tilde{f}_2}{N}, \quad \ldots, \quad f_k = \frac{\tilde{f}_k}{N}$$

which satisfy the summation property

$$\sum_{i=1}^{k} f_i = f_1 + f_2 + \cdots + f_k = 1$$

Define a frequency function associated with the sample using

$$f(x) = \begin{cases} f_j, & \text{when } x = X_j \\ 0, & \text{otherwise} \end{cases} \qquad \text{for } j = 1, \ldots, k$$

The frequency function determines how the numbers in the sample are distributed.

Also define a cumulative frequency function $F(x)$ for the sample, sometimes referred to as a sample distribution function. The cumulative frequency function is defined

$$F(x) = \sum_{t \leq x} f(t) = \text{sum of all relative frequencies less than or equal to } x$$

Whenever the data has too many numerical values then one usually defines class intervals and class midpoints with class frequencies as in table 8.3. This is called grouping of the data and the corresponding frequency function and cumulative frequency function are associated with the grouped data.

The relative frequency distribution $f(x)$ is also called a discrete probability distribution for the sample and the cumulative relative frequency function $F(x)$ or distribution function represents a probability. In particular,

$$\begin{aligned} F(x) &= P(X \leq x) = \text{Probability that population variable } X \text{ is less than or equal to } x \\ 1 - F(x) &= P(X > x) = \text{Probability that population variable } X \text{ is greater than } x \end{aligned} \quad (8.2)$$

### Arithmetic Mean or Sample Mean

Given a set of data points $X_1, X_2, \ldots, X_N$, define the arithmetic mean or sample mean of the data set by

$$\text{sample mean} = \overline{X} = \frac{X_1 + X_2 + \cdots + X_N}{N} = \frac{\sum_{j=1}^{N} X_j}{N} \quad (8.3)$$

If the frequency of the data points are known, say $X_1, X_2, \ldots, X_k$ occur with frequencies $\tilde{f}_1, \tilde{f}_2, \ldots, \tilde{f}_k$, then the arithmetic mean is calculated

$$\overline{X} = \frac{\tilde{f}_1 X_1 + \tilde{f}_2 X_2 + \cdots + \tilde{f}_k X_k}{\tilde{f}_1 + \tilde{f}_2 + \cdots + \tilde{f}_k} = \frac{\sum_{j=1}^{k} \tilde{f}_j X_j}{\sum_{j=1}^{k} \tilde{f}_j} = \frac{\sum_{j=1}^{k} \tilde{f}_j X_j}{N} \quad (8.4)$$

Note that the finite data collected is used to calculate an estimate of the true population mean $\mu$ associated with the total population.

### Median, Mode and Percentiles

After arranging the data from low to high, the median of the data set is the middle value or the arithmetic mean of the two middle values. This value divides the data set into two equal numbered parts. In a similar fashion find those points which divide the data set, arranged in order of magnitude, into four equal parts. These values are usually denoted $Q_1, Q_2, Q_3$ and are called the first, second and third quartiles. Note that $Q_2$ will be the same as the median. If the data set is divided into ten equal parts by numbers $D_1, D_2, D_3, D_4, D_5, D_6, D_7, D_8, D_9$, then these numbers are called deciles. If the data set is divided into one hundred equal parts by numbers $P_1, P_2, \ldots, P_{99}$, then these numbers are called percentiles. In general, if the data is divided up into quartiles, deciles, percentiles or some other equal subdivision, then the subdivisions created are called quantiles. The mode of the data set is that value which occurs with greatest frequency. Note that the mode may not exist or even if it does exist, it might not be a unique value. A unimodal data set is one which has a unique single mode.

## The Geometric and Harmonic Mean

The geometric mean $G$ associated with the data set $\{X_1, X_2, \ldots, X_N\}$ is the $N$th root of the product of the numbers in the set. The geometric mean is denoted

$$G = \sqrt[N]{X_1 X_2 \cdots X_N} \tag{8.5}$$

The harmonic mean $H$ associated with the above data set is obtain by first taking the arithmetic mean of the reciprocals and then taking the reciprocal of the result. The harmonic mean can be expressed using either of the relations

$$H = \frac{1}{\frac{1}{N}\sum_{i=1}^{N}\frac{1}{X_i}} \quad \text{or} \quad \frac{1}{H} = \frac{1}{N}\sum_{i=1}^{N}\frac{1}{X_i} \tag{8.6}$$

The arithmetic mean $\overline{X}$, geometric mean $G$ and harmonic mean $H$, satisfy the inequalities

$$H \leq G \leq \overline{X} \tag{8.8}$$

The equality sign being used when all the numbers in the data set are equal to one another.

## The Root Mean Square (RMS)

The root mean square (RMS), sometimes referred to as the quadratic mean, of the data set $\{X_1, X_2, \ldots, X_n\}$ is defined

$$RMS = \sqrt{\frac{\sum_{j=1}^{N} X_j^2}{N}} \tag{8.8}$$

## Mean Deviation and Sample Variance

The mean deviation (MD), sometimes called the average deviation, of a set of numbers $X_1, X_2, \ldots, X_N$ represents a measure of the data spread from the mean and is defined

$$MD = \frac{\sum_{j=1}^{N} | X_j - \overline{X} |}{N} = \frac{1}{N}\left[| X_1 - \overline{X} | + | X_2 - \overline{X} | + \cdots + | X_n - \overline{X} |\right] \tag{8.9}$$

where $\overline{X}$ is the arithmetic mean of the data. The mean deviation associated with numbers $X_1, X_2, \ldots, X_k$ occurring with frequencies $\tilde{f}_1, \tilde{f}_2, \ldots, \tilde{f}_k$ can be calculated

$$MD = \frac{\sum_{j=1}^{k} \tilde{f}_j | X_j - \overline{X} |}{N} = \frac{1}{N}\left[\tilde{f}_1 | X_1 - \overline{X} | + \tilde{f}_2 | X_2 - \overline{X} | + \cdots + \tilde{f}_k | X_k - \overline{X} |\right] \tag{8.10}$$

The sample variance of the data set $\{X_1, X_2, \ldots, X_n\}$ is denoted $s^2$ and is calculated using the relation

$$s^2 = \frac{1}{n-1}\sum_{j=1}^{n}(X_j - \overline{X})^2 = \frac{1}{n-1}\left[(x_1 - \overline{X})^2 + (x_2 - \overline{X})^2 + \cdots + (x_n - \overline{X})^2\right] \tag{8.11}$$

where $\overline{X}$ is the sample mean or arithmetic mean. The sample variance is a measure of how much dispersion or spread there is in the data. The positive square root of the sample variance is denoted $s$, which is called the standard deviation of the sample.

Note that some textbooks define the sample variance as

$$S^2 = \frac{1}{n}\sum_{j=1}^{n}(X_j - \overline{X})^2 \qquad (8.12)$$

where $n$ is used as a divisor instead of $n-1$. Why the different definitions for sample variance? The reason is that for small values for $n$, say $n < 30$, then equation (8.11) will produce a better estimate of the true standard deviation $\sigma$ associated with the total population from which the sample is taken. For sample sizes larger than $n = 30$ there will be very little difference in calculation of the standard deviation from either definition. To convert the standard deviation $S$, calculated using equation (8.12), to that determined by equation (8.11) one need only multiply $S$ by $\sqrt{\frac{n}{n-1}}$ to obtain $s$.

The sample variance given by equation (8.11) requires that one first calculate $\overline{X}$, then one must calculate all the terms $X_j - \overline{X}$. All this preliminary calculation introduces roundoff errors into the final result. The sample variance can be calculated using a short cut method of computing without having to do preliminary calculations. The short cut method is derived using the expansion

$$(X_j - \overline{X})^2 = X_j^2 - 2X_j\overline{X} + \overline{X}^2$$

and substituting it into equation (8.11). A summation of terms gives

$$\sum_{j=1}^{n}(X_j - \overline{X})^2 = \sum_{j=1}^{n}X_j^2 - 2\overline{X}\sum_{j=1}^{n}X_j + \overline{X}^2\sum_{j=1}^{n}(1)$$

The substitution $\overline{X} = \frac{1}{n}\sum_{j=1}^{n}X_j$ and using $\sum_{j=1}^{n}(1) = n$, gives the result

$$\sum_{j=1}^{n}(X_j - \overline{X})^2 = \sum_{j=1}^{n}X_j^2 - 2\frac{1}{n}\sum_{j=1}^{n}X_j\sum_{j=1}^{n}X_j + \left(\frac{1}{n}\sum_{j=1}^{n}X_j\right)^2 n$$

$$= \sum_{j=1}^{n}X_j^2 - \frac{2}{n}\left(\sum_{j=1}^{n}X_j\right)^2 + \frac{1}{n}\left(\sum_{j=1}^{n}X_n\right)^2$$

$$= \sum_{j=1}^{n}X_j^2 - \frac{1}{n}\left(\sum_{j=1}^{n}X_j\right)^2$$

This produces the shortcut formula for the sample variance

$$s^2 = \frac{1}{n-1}\left[\sum_{j=1}^{n}X_j^2 - \frac{1}{n}\left(\sum_{j=1}^{n}X_j\right)^2\right] \qquad (8.13)$$

If $X_1, \ldots, X_m$ are $m$ sample values occurring with frequencies $\tilde{f}_1, \ldots, \tilde{f}_m$, the equation (8.13) can be expressed in the form

$$s^2 = \frac{1}{n-1}\left[\sum_{j=1}^{m} X_j \tilde{f}_j - \frac{1}{n}\left(\sum_{j=1}^{m} X_j \tilde{f}_j\right)^2\right] \qquad (8.14)$$

In general, if the true population mean $\mu$ is known exactly, so that $\mu = \frac{1}{N}\sum_{j=1}^{N} X_j$, where $N$ is the population size, then the population standard deviation is given by

$$\sigma = \sqrt{\frac{\sum_{j=1}^{N}(X_j - \mu)^2}{N}} \qquad (8.15)$$

Use $N$ if the exact population mean is known and use $n-1$ if samples of size $n \ll N$ are selected from a population where the true mean $\mu$ is unknown.

## Probability

An experiment or observation produces samples from a population where the outcome recorded either belongs or does not belong to a prescribed collection of events being studied. A sample space $S$ is the set of all possible outcomes from an experiment. A sample space can be either finite or infinite. An example of a sample space with a finite collection of events is the roll of a single die. Here the sample space is $S = \{1, 2, 3, 4, 5, 6\}$ corresponding to the numbers on the six faces of the die. An example of an infinite sample space is that of an experiment where the outcome from an single event can be a real number within a specified range.

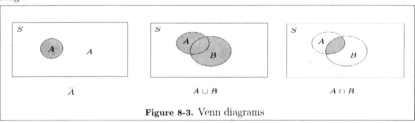

**Figure 8-3.** Venn diagrams

A Venn diagram consists of representing the sample space by a rectangle, then any event of $S$ can be represented by the interior of a circle $A$ within the rectangle. The set of all events not in $A$ is called the complement of $A$ and is denoted using the notation $\overline{A}$. The null set, empty set or impossible event is denoted by the symbol $\emptyset$. The union of two events $A$ and $B$ is denoted $A \cup B$ and represents all events or experiments of $S$ contained in $A$ or $B$ or both. The intersection of two events $A$ and $B$ is denoted $A \cap B$ and represents all events in $S$ contained in both $A$ and $B$. The concepts of a complement, union and intersection of sets is illustrated in the figure 8.3 If two sets $A$ and $B$ have no events in common, then this is denoted by $A \cap B = \emptyset$. In this case, the sets $A$ and $B$ are said to be mutually exclusive events or disjoint events. The notation $A \subset B$ is used to denote "all elements of $A$ are contained in the set $B$" This can also be expressed $B \supset A$, which is read " $B$ contains $A$". If the sample space contains $n$-sets $\{A_1, A_2, \ldots, A_n\}$, then the union of these sets is denoted

$$A_1 \cup A_2 \cup \cdots \cup A_n \quad \text{or} \quad \bigcup_{j=1}^{n} A_j$$

The intersection of these sets is denoted

$$A_1 \cap A_2 \cap \cdots \cap A_n \quad \text{or} \quad \bigcap_{j=1}^{n} A_j$$

If $A_j \cap A_k = \emptyset$, for all values of $j$ and $k$, with $k \neq j$, then the sets $\{A_1, A_2, \ldots, A_n\}$ are said to represent mutually exclusive events.

## Probability Fundamentals

Assuming that there are $h$ ways an event can happen and $f$ ways for the event to fail and these ways are all equally likely to happen, then the probability $p$ that an event will happen in a given trial is

$$p = \frac{h}{h+f} \tag{8.16}$$

and the probability $q$ that the event will fail is given by

$$q = \frac{f}{h+f} \tag{8.17}$$

These probabilities satisfy $p + q = 1$.

In general, given a finite sample space $S = \{e_1, e_2, \ldots, e_n\}$ containing $n$ simple events $e_1, \ldots, e_n$, assign to each element of $S$ a number $P(e_i)$, $i = 1, \ldots, n$ called the probability assigned to event $e_i$ of $S$. These probability numbers must satisfy the following conditions.

1. Each probability is a nonnegative number satisfying $0 \leq P(e_i) \leq 1$
2. The sum of the probabilities assigned to all simple events of the sample space must sum to unity or

$$\sum_{j=1}^{n} P(e_j) = P(e_1) + P(e_2) + \cdots + P(e_n) = 1$$

3. The probability assigned to the entire sample space is unity and one writes $P(S) = 1$.

## Probability of an Event

After assigning probabilities to each simple event of $S$, it is then possible to determine the probability of any event $E$ associated with events from $S$. Consider the following cases.

1. The event $E = \emptyset$ is the empty set.
2. The event $E$ is one of the simple events $e_i$ from $S$ ($i$ fixed, with $1 \leq i \leq n$)
3. The event $E$ is the union of two or more events from $S$.

For the case 1, define the probability of the empty set $\emptyset$ as zero and write $P(\emptyset) = 0$. In case 2, the probability of event $E$ is the same as the probability $P(e_i)$ so that, $P(E) = P(e_i)$. Consider now the case 3. If $E$ is an event associated with the sample space $S$ and $\overline{E}$ is its complement, then

$$P(E) = 1 - P(\overline{E}) \tag{8.18}$$

This is known as the complementation rule for probabilities.

If $E_1$ and $E_2$ are mutually exclusive events associated with $S$, then $E_1 \cap E_2 = \emptyset$ and

$$P(E_1 \cup E_2) = P(E_1) + P(E_2), \qquad E_1 \cap E_2 = \emptyset \tag{8.19}$$

In general, if $E = E_1 \cup E_2 \cup \cdots \cup E_m$ is an event and $E_1, E_2, \ldots, E_m$ are mutually exclusive events associated with $S$, then the intersection gives $E_1 \cap E_2 \cap \cdots \cap E_m = \emptyset$ and the probability of $E$ is

$$P(E) = P(E_1 \cup E_2 \cup \cdots \cup E_m) = P(E_1) + P(E_2) + \cdots + P(E_m) \tag{8.20}$$

This is known as the addition rule for mutually exclusive events.

If $E_1$ and $E_2$ are arbitrary events associated with a sample space $S$, and these events are not mutually exclusive, then

$$P(E_1 \cup E_2) = P(E_1) + P(E_2) - P(E_1 \cap E_2) \tag{8.21}$$

Here $P(E_1)$ is the sum of all the simple events defining $E_1$ and $P(E_2)$ is the sum of all the simple events defining $E_2$. If the events are not mutually exclusive, then the sum of the probabilities associated with the simple events common to both the events $E_1$ and $E_2$ are counted twice. The sum of these common probabilities of simple events which are counted twice is $P(E_1 \cap E_2)$ and so this value is subtracted from the sum $P(E_1) + P(E_2)$.

### Example 8-1.

Two coins are tossed. What is the probability that at least one tail occurs? Assume the coins are not trick coins so that there are four equally likely events that can occur. The sample space for the experiment is $S = \{HH, HT, TH, TT\}$. Because each event is equally likely, assign a probability of 1/4 to each event. For this example, the event $E$ to be investigated is

$$E = \{HT\} \cup \{TH\} \cup \{TT\}$$

That is, at least one tail occurs. Consequently,

$$P(E) = P(HT) + P(TH) + P(TT) = 1/4 + 1/4 + 1/4 = 3/4$$

■

## Example 8-2.

A pair of fair dice are rolled. The sample space associated with this experiment is a representation of all possible outcomes.

$$S = \{(1,1) \quad (2,1) \quad (3,1) \quad (4,1) \quad (5,1) \quad (6,1)$$
$$(1,2) \quad (2,2) \quad (3,2) \quad (4,2) \quad (5,2) \quad (6,2)$$
$$(1,3) \quad (2,3) \quad (3,3) \quad (4,3) \quad (5,3) \quad (6,3)$$
$$(1,4) \quad (2,4) \quad (3,4) \quad (4,4) \quad (5,4) \quad (6,4)$$
$$(1,5) \quad (2,5) \quad (3,5) \quad (4,5) \quad (5,5) \quad (6,5)$$
$$(1,6) \quad (2,6) \quad (3,6) \quad (4,6) \quad (5,6) \quad (6,6)\}$$

There are 36 equally likely possible outcomes. Assign a probability of 1/36 to each simple event.

If $E_1$ is the event that a 7 is rolled, then

$$P(E_1) = P((1,6)) + P((2,5)) + P((3,4)) + P((4,3)) + P((5,2)) + P((6,1)) = 6/36 = 1/6$$

If $E_2$ is the event that an 11 is rolled, then

$$P(E_2) = P((5,6)) + P((6,5)) = 2/36 = 1/18$$

If $E_3$ is the event doubles are rolled, then

$$P(E_3) = P((1,1)) + P((2,2)) + P((3,3)) + P((4,4)) + P((5,5)) + P((6,6)) = 6/36 = 1/6$$

If $E_4$ is the event that a 10 is rolled, then

$$P(E_4) = P((4,6)) + P((5,5)) + P((6,4)) = 3/36 = 1/12$$

If $E_5$ is the event a 10 is rolled or a double is rolled, the $E_5 = E_4 \cup E_3$. Note that the event $(5,5)$ is common to both events $E_4$ and $E_3$ with $E_4 \cap E_3 = (5,5)$ and $P(E_4 \cap E_3) = P((5,5)) = 1/36$. Hence,

$$P(E_5) = P(E_4 \cup E_3) = P(E_4) + P(E_3) - P(E_4 \cap E_3) = 3/36 + 6/36 - 1/36 = 8/36 = 2/9$$

If $E_6$ is the event that a 10 is rolled or a 7 is rolled, then $E_6 = E_4 \cup E_1$. Here $E_4 \cap E_1 = \emptyset$ and so these events are mutually exclusive. Consequently,

$$P(E_6) = P(E_4 \cup E_1) = P(E_4) + P(E_1) = 3/36 + 6/36 = 9/36 = 1/4$$

∎

Note that there are many situations where the sample space is not finite. In such cases the probabilities assigned to the events in $S$ are based upon employment of relative frequencies observed from taking a large number of trials from the population being studied.

## Conditional Probability

If two events $E_1$ and $E_2$ are related in some manner such that the probability of occurrence of event $E_1$ depends upon whether $E_2$ has or has not occurred, then this is called the conditional probability of $E_1$ given $E_2$ and it is denoted using the notation $P(E_1 \mid E_2)$. Here the vertical line is read as "given" and events to the right of the vertical line are treated as events which have occurred. The conditional probability of $E_1$ given $E_2$ is

$$P(E_1 \mid E_2) = \frac{P(E_1 \cap E_2)}{P(E_2)}, \qquad P(E_2) \neq 0 \tag{8.22}$$

The conditional probability of $E_2$ given $E_1$ is

$$P(E_2 \mid E_1) = \frac{P(E_1 \cap E_2)}{P(E_1)}, \qquad P(E_1) \neq 0 \tag{8.23}$$

The equations (8.22) and (8.23) imply that the probability of both events $E_1$ and $E_2$ occurring is given by

$$P(E_1 \cap E_2) = P(E_1)P(E_2 \mid E_1) = P(E_2)P(E_1 \mid E_2), \qquad P(E_1) \neq 0, \quad P(E_2) \neq 0 \tag{8.24}$$

If the events $E_1$ and $E_2$ are independent events, then

$$P(E_1 \cap E_2) = P(E_1)P(E_2) \tag{8.25}$$

and consequently

$$P(E_1 \mid E_2) = P(E_1), \qquad \text{and} \qquad P(E_2 \mid E_1) = P(E_2)$$

This condition occurs whenever the probability of $E_1$ does not depend upon the event $E_2$ and similarly, the probability of event $E_2$ does not depend upon $E_1$.

Two events $E_1$ and $E_2$ are said to be independent events if and only if the probability of occurrence of $E_1$ and $E_2$ is given by $P(E_1 \cap E_2) = P(E_1)P(E_2)$. That is, the probability of both $E_1$ and $E_2$ occurring is the product of the probabilities of occurrence of each event. Two events that are not independent are called dependent events.

In general, if $E_1, E_2, \ldots, E_m$ are all independent events, then

$$P(E_1 \cap E_2 \cap \cdots \cap E_m) = P(E_1)P(E_2)\cdots P(E_m) \tag{8.26}$$

This is sometime written in the form

$$P(\bigcap_{k=1}^{m} E_k) = \prod_{k=1}^{m} P(E_k) \tag{8.27}$$

and is known as the multiplication principle for independent events.

For example, in studying the occurrence or non occurrence of three events $E_1, E_2, E_3$ the probability is denoted

$$P(E_1 \cap E_2 \cap E_3) = P(E_1)P(E_2 \mid E_1)P(E_3 \mid E_1 \cap E_2) \tag{8.28}$$

and for independent events

$$P(E_1 \cap E_2 \cap E_3) = P(E_1)P(E_2)P(E_3) \tag{8.29}$$

## Example 8-3.

The dice table is completely surrounded with players so that you can see only a part of the table. A player rolls the dice and you see one die comes up a 6, but you can't see the other die. What is the probability the player has rolled a 7 or 11?

**Solution:** Here there are two events $E_1, E_2$ with

$$E_1 = \text{event one die is a 6}$$
$$E_2 = \text{event sum of dice is 7 or 11}$$

and we are to find $P(E_2 \mid E_1)$. To solve this problem write down the simple events as

$$E_1 = \{(6,1), (6,2), (6,3), (6,4), (6,5), (6,6)\}$$
$$E_2 = \{(1,6), (2,5), (3,4), (4,3), (5,2), (6,1), (5,6), (6,5)\}$$
$$\text{with} \quad E_1 \cap E_2 = \{(6,1), (6,5)\}$$

Recall the simple events all have equal probabilities of 1/36 and consequently

$$P(E_2 \mid E_1) = \frac{P(E_1 \cap E_2)}{P(E_1)} = \frac{2/36}{6/36} = 1/3$$

Observe that $P(E_2) = 8/36 = 2/9 \neq P(E_2 \mid E_1) = 1/3$. These events are not independent. That is, knowing one die is a 6 does effect the probability of the sum being 7 or 11.

∎

## Example 8-4.

Two cards are selected at random from an ordinary deck of 52 cards. Find the probability that both cards are spades. Find the probability that the first card is a spade and the second card is a heart. In performing this experiment assume that the first card selected is not replaced in the deck.

**Solution:** Examine the events

$$E_1 = \text{The event that the first card selected is a spade.}$$
$$E_2 = \text{The event that the second card selected is a spade.}$$
$$E_3 = \text{The event that the second card selected is a heart.}$$

Using elementary probability theory

$$P(E_1) = \frac{13}{52} = \frac{\text{Number of spades in deck}}{\text{Total number of cards in deck}}$$

Now if event $E_1$ has occurred, the deck now has only 51 cards with 12 spades. Consequently, the conditional probability is

$$P(E_2 \mid E_1) = \frac{12}{51} = \frac{\text{Number of spades in deck}}{\text{Total number of cards in deck}}$$
$$\text{and} \quad P(E_3 \mid E_1) = \frac{13}{51} = \frac{\text{Number of hearts in deck}}{\text{Total number of cards in deck}}$$

Calculate the probabilities

$$P(E_1 \cap E_2) = P(E_1)P(E_2 \mid E_1) = \frac{13}{52} \cdot \frac{12}{51} = \frac{3}{51}$$
$$\text{and} \quad P(E_1 \cap E_3) = P(E_1)P(E_3 \mid E_1) = \frac{13}{52} \cdot \frac{13}{51} = \frac{13}{204}$$

## Discrete and Continuous Probability Distributions

The estimated probability of an event is taken as the relative frequency of occurrence of the event. As the number of observations upon which the relative frequency is based increases, then the discrete probability is replace by a continuous function $f(x)$ called the probability function or probability density function of the distribution with the condition that the total area under the probability density function must equal unity. The figure 8-4 is a graphical representation illustrating this conversion.

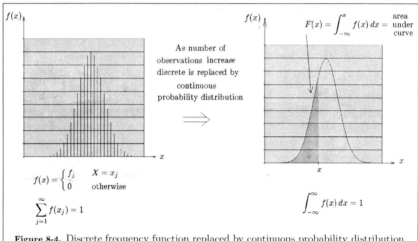

**Figure 8-4.** Discrete frequency function replaced by continuous probability distribution.

The function $F(x) = \sum_{x_j \leq x} f(x_j)$ is called the cumulative frequency function associated with the discrete sample. In the continuous case it is called the distribution function $F(x)$ and calculated by the integral $F(x) = \int_{-\infty}^{x} f(x)\,dx$ which represents the area from $-\infty$ to $x$ under the probability density function. These summation processes are illustrated in the figure 8-5.

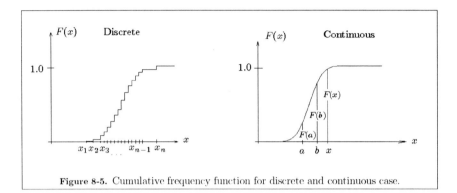

**Figure 8-5.** Cumulative frequency function for discrete and continuous case.

In both the discrete and continuous cases the cumulative frequency function represents the probability

$$F(x) = P(X \leq x) = \int_{-\infty}^{x} f(x)\,dx \quad \text{with} \quad 1 - F(x) = P(X > x) = \int_{x}^{\infty} f(x)\,dx \qquad (8.30)$$

with the property that if $a, b$ are points $x$ with $a < b$, then

$$P(a < x \leq b) = P(X \leq b) - P(X \leq a) = F(b) - F(a) \qquad (8.31)$$

which represents the probability that a random variable $X$ lies between $a$ and $b$. In the discrete case

$$P(a < X \leq b) = \sum_{a < x_j \leq b} f(x_j) = F(b) - F(a) \qquad (8.32)$$

and in the continuous case

$$P(a < X \leq b) = \int_{a}^{b} f(x)\,dx = F(b) - F(a) \qquad (8.33)$$

The continuous cumulative frequency function satisfies the properties

$$\frac{dF(x)}{dx} = f(x), \qquad F(-\infty) = 0, \qquad F(+\infty) = 1, \quad \text{and} \quad F(a) < F(b) \text{ if } a < b \qquad (8.34)$$

The table 8.4 illustrates the relationships of the mean and variance associated with the discrete and continuous probability densities.

If $X$ is a real random variable and $g(X)$ is any continuous function of $X$, then the numbers

$$E[g(X)] = \sum_{j=1}^{n} g(x_j) f(x_j) \qquad \text{discrete}$$

$$E[g(X)] = \int_{-\infty}^{\infty} g(x) f(x)\,dx \qquad \text{continuous} \qquad (8.35)$$

| Table 8.4 Mean and Variance for Discrete and Continuous Distributions | | |
|---|---|---|
| Discrete | | Continuous |
| $\mu = E[x] = \sum_{j=1}^{n} x_j f(x_j)$ | population mean $\mu$ | $\mu = E[x] = \int_{-\infty}^{\infty} x f(x)\, dx$ |
| $\sigma^2 = E[(x-\mu)^2] = \sum_{j=1}^{n}(x_j - \mu)^2 f(x_j)$ | population variance $\sigma^2$ | $\sigma^2 = E[(x-\mu)^2] = \int_{-\infty}^{\infty}(x-\mu)^2 f(x)\, dx$ |

associated with the probability density $f(x)$ are defined as the mathematical expectation of the function $g(X)$. In the special case $g(X) = X^k$ for $k = 1, 2, \ldots, n$ an integer, the equations (8.35) become

$$E[X^k] = \sum_j x_j^k f(x_j) \qquad \text{discrete}$$

$$E[X^k] = \int_{-\infty}^{\infty} x^k f(x)\, dx \qquad \text{continuous}$$

These equations are referred to as the $k$th moment of $X$. In the special case $g(X) = (X - \mu)^k$, the equations (8.35) become

$$E[(X-\mu)^k] = \sum_j (x_j - \mu)^k f(x_j) \qquad \text{discrete}$$

$$E[(X-\mu)^k] = \int_{-\infty}^{\infty} (x-\mu)^k f(x)\, dx \qquad \text{continuous}$$

and these quantiles are called the $k$th central moments of $X$. Note the special cases

$$E[1] = 1, \qquad \mu = E[X], \qquad \sigma^2 = E[(X-\mu)^2] \qquad (8.36)$$

The expectation of a sum of random variables $X_1, X_2, \ldots, X_n$ equals the sum of the expectations and consequently

$$E(X_1 + X_2 + \cdots X_n) = E(X_1) + E(X_2) + \cdots + E(X_n) \qquad (8.37)$$

The expectation of a product of independent random variables equals the product of the expectations which is expressed

$$E(X_1 X_2 \cdots X_n) = E(X_1) E(X_2) \cdots E(X_n) \qquad (8.38)$$

## Scaling

The probability density function $f(x)$ is said to be symmetric with respect a number $x = \mu$ if for all values of $x$ the density function satisfies the relation

$$f(\mu + x) = f(\mu - x) \tag{8.39}$$

A random variable $X$ having a mean $\mu$ and variance $\sigma^2$ can be scaled by introducing the new variable $Z = (X - \mu)/\sigma$. The variable $Z$ is referred to as the standardized variable corresponding to $X$.

Let $f(x)$ denote the probability density function associated with the random variable $X$ and define the function $f^*(z) = \sigma f(x) = \sigma f(\sigma z + \mu)$ as the probability function associated with the random variable $Z$. Using the scaling illustrated in the figure above, observe that $x = \sigma z + \mu$ with $dx = \sigma\, dz$ so that

$$f(x)\, dx = f(\sigma z + \mu)\sigma\, dz = f^*(z)\, dz$$

then the mean value on the Z-scale is given by

$$\mu^* = \int_{-\infty}^{\infty} z f^*(z)\, dz = \int_{-\infty}^{\infty} \left(\frac{x}{\sigma} - \frac{\mu}{\sigma}\right) f(x)\, dx$$
$$= \frac{1}{\sigma}\int_{-\infty}^{\infty} x f(x)\, dx - \frac{\mu}{\sigma}\int_{-\infty}^{\infty} f(x)\, dx$$
$$= \frac{1}{\sigma}\mu - \frac{\mu}{\sigma}(1) = 0$$

and the variance on the Z-scale is given by

$$\sigma^{*2} = \int_{-\infty}^{\infty} (z - \mu^*)^2 f^*(z)\, dz = \int_{-\infty}^{\infty} z^2 f^*(z)\, dz \quad \text{since } \mu^* = 0$$
$$\sigma^{*2} = \int_{-\infty}^{\infty} \left(\frac{x - \mu}{\sigma}\right)^2 f(x)\, dx = \frac{1}{\sigma^2}\int_{-\infty}^{\infty} (x - \mu)^2 f(x)\, dx = \frac{1}{\sigma^2}\sigma^2 = 1$$

This demonstrates that the introduction of a scaled variable Z centers the mean and introduces a variance of unity.

## The Normal Distribution

The normal probability distribution is continuous with two parameters $\mu$ and $\sigma > 0$, where $\sigma^2$ is the variance of the distribution. The normal probability distribution has the form

$$f(x) = \frac{1}{\sigma\sqrt{2\pi}} e^{-\frac{1}{2}(x-\mu)^2/\sigma^2} = \frac{1}{\sigma\sqrt{2\pi}} \exp\left[-\frac{1}{2}(x-\mu)^2/\sigma^2\right], \quad -\infty < x < \infty \tag{8.40}$$

and is illustrated in the figure 8-6. The parameter $\mu$ is known as the mean of the distribution and represents a location parameter for positioning the curve on the $x$-axis. The parameter $\sigma$

is called a scale parameter which is associated with the spread and height of the probability curve. The quantity $\sigma^2$ represents the variance of the distribution and $\sigma$ represents the standard deviation of the distribution. The total area under this curve is 1 with approximately 68.27% of the area between the lines $\mu \pm \sigma$, 95.45% of the total area is between the lines $\mu \pm 2\sigma$ and 99.73% of the total area is between the lines $\mu \pm 3\sigma$. In terms of probabilities

$$P(\mu-\sigma < X \leq \mu+\sigma) = .6827, \qquad P(\mu-2\sigma < X \leq \mu+2\sigma) = .9545, \qquad P(\mu-3\sigma < X \leq \mu+3\sigma) = .9973 \qquad (8.41)$$

The function

$$\phi(z) = \frac{1}{\sqrt{2\pi}} e^{-z^2/2} \qquad (8.42)$$

is called a normalized probability distribution with mean $\mu = 0$ and standard deviation of $\sigma = 1$.

The normal probability distribution satisfies the following properties.

$$\text{Mean} = \mu, \qquad \text{Variance} = \sigma^2, \qquad \text{Standard deviation} = \sigma$$

The distribution function $F(x)$ associated with the normal probability density is given by

$$F(x) = \int_{-\infty}^{x} f(x)\,dx = P(X \leq x) \qquad (8.43)$$

and represents the area under the probability curve form $-\infty$ to $x$. This area represents the probability $P(X \leq x)$. The total area under the normal density function is unity and so the area

$$1 - F(x) = \int_{x}^{\infty} f(x)\,dx = P(X > x) \qquad (8.44)$$

represents the probability $P(X > x)$. These areas are illustrated in the figure 8-7.

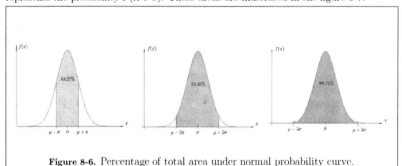

Figure 8-6. Percentage of total area under normal probability curve.

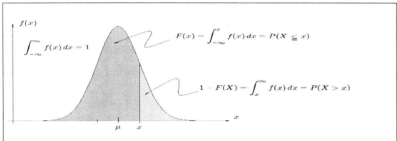

**Figure 8-7.** Area under normal probability curve representing probabilities.

## Standardization

The normal probability distribution has the form

$$N(x; \mu, \sigma^2) = f(x) = \frac{1}{\sigma\sqrt{2\pi}} e^{-\frac{1}{2}\left(\frac{x-\mu}{\sigma}\right)^2} \tag{8.45}$$

and the distribution function is given by

$$F(x) = \frac{1}{\sigma\sqrt{2\pi}} \int_{-\infty}^{x} e^{-\frac{1}{2}\left(\frac{\xi-\mu}{\sigma}\right)^2} d\xi \tag{8.46}$$

Introducing the standardized variable $z = \dfrac{x-\mu}{\sigma}$, with $dz = \dfrac{dx}{\sigma}$, the distribution function given by equation (8.46) is converted to the normalized form

$$\Phi(z) = \frac{1}{\sqrt{2\pi}} \int_{-\infty}^{z} e^{-z^2/2} dz, \tag{8.47}$$

where $z$ is called a random normal number.

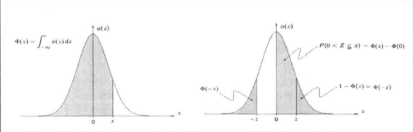

**Figure 8-8.** Standard normal probability curve and distribution function as area.

Note that $F(x) = \Phi\left(\frac{x-\mu}{\sigma}\right)$ so that in terms of probabilities

$$P(a < X \leq b) = F(b) - F(a) = \Phi\left(\frac{b-\mu}{\sigma}\right) - \Phi\left(\frac{a-\mu}{\sigma}\right) \tag{8.48}$$

In figure 8-8, the normal probability curve is symmetric about $z = 0$ and so the area under the curve between 0 and $z$ represents the probability $P(0 < Z \leq z) = \Phi(z) - \Phi(0)$ and quantities like $\Phi(-z)$, by symmetry, have the value $\Phi(-z) = 1 - \Phi(z)$. The standard normal curve has the properties that $\Phi(-\infty) = 0$, $\Phi(0) = 1/2$ and $\Phi(\infty) = 1$. The table 8.4 gives the area under the standard normal curve for values of $z \geq 0$, then it is possible to employ the symmetry of the standard normal curve to calculated specific areas associated with probabilities. For example, to find the area from $-\infty$ to $-1.65$ examine the table of values and find $\Phi(1.65) = .9505$ so that $\phi(-1.65) = 1 - .9505 = .0495$ or the area from $-1.65$ to $+\infty$ is .9505, then one can write the probability statement $P(Z > -1.65) = .9505$. As another example, to find the area under the standard normal curve between $z = -1.65$ and $z = 1$, first find the following values

Area from 0 to $1 = \Phi(1) - \Phi(0) = .8413 - 0.5000 = .3413$

Area from 0 to $1.65 = \Phi(1.65) - \Phi(0) = .9505 - .5000 = .4505$

Area from $-1.65$ to $0 = .4505$

Area from $-1.65$ to $1 = .4505 + .3413 = .7918 = P(-1.65 < Z \leq 1)$

The normal distribution function $\Phi(z) = \frac{1}{\sqrt{2\pi}} \int_{-\infty}^{z} e^{-\xi^2/2} d\xi$ and the error function which is defined

$$\text{erf}(z) = \frac{2}{\sqrt{\pi}} \int_{0}^{z} e^{-u^2} du \tag{8.49}$$

can be related by writing

$$\Phi(z) = \frac{1}{\sqrt{2\pi}} \int_{-\infty}^{0} e^{-\xi^2/2} d\xi + \frac{1}{\sqrt{2\pi}} \int_{0}^{z} e^{-\xi^2/2} d\xi = \frac{1}{2} + \frac{1}{\sqrt{2\pi}} \int_{0}^{z} e^{-\xi^2/2} d\xi$$

and then making the substitutions $u = \xi/\sqrt{2}$, $du = d\xi/\sqrt{2}$ to obtain

$$\Phi(z) = \frac{1}{2} + \frac{1}{\sqrt{2\pi}} \int_{0}^{z/\sqrt{2}} e^{-u^2} \sqrt{2}\, du = \frac{1}{2} + \frac{1}{2}\text{erf}\left(\frac{z}{\sqrt{2}}\right) \tag{8.50}$$

The normal probability functions given by equations (8.40) and (8.47) are known by other names such as Gaussian distribution, normal curve, bell shaped curve, etc. The normal distribution occurs in the study of various types of errors such as measurements in the quality and precision control of tools and equipment. The normal distribution arises in many different applied areas of the physical and social sciences because of the central limit theorem. The central limit theorem, sometimes called the law of large numbers, involves consequences of taking large samples from any kind of distribution and can be described as follows. Perform an experiment and select $n$ independent random variables $X$ from some population. If $x_1, x_2, \ldots, x_n$ represents the set of $n$ independent random variables selected, then the mean $m_1$ of this sample can be constructed. Perform the experiment again and calculate the mean $m_2$ of the second set of $n$ random independent variables. Continue doing this same experiment a large number of times and collect all the mean values from each experiment. This gives a set of average values $S = \{m_1, m_2, \ldots, m_N\}$ created from performing the experiment $N$ times. The central limit theorem says that the distribution of the set of average values $S$ approaches a normal probability distribution with mean $\mu_s$ and variance $\sigma_s^2$ given by

$$\mu_s = \text{Mean of the set of averages from a large number of samples} = \mu$$

$$\sigma_s^2 = \text{Variance of set of averages from large number of samples} = \frac{\sigma^2}{n}$$

where $\mu$ and $\sigma^2$ represent the true mean and true variance of the population being sampled. The central limit theorem always holds and does not depend upon the shape of the original distribution being sampled. The normal distribution is also related to least-square estimation. It is also used as the theoretical basis for the chi-square, student-t and F-distributions. The normal distribution can be used in many Monte Carlo methods.

## The Binomial Distribution

The binomial probability distribution is given by

$$b(x; n, p) = f(x) = \begin{cases} \binom{n}{x} p^x q^{n-x}, & x = 0, 1, 2, \ldots, n \\ 0, & \text{otherwise} \end{cases} \qquad q = 1 - p \qquad (8.51)$$

It is a discrete probability distribution with parameters $n$ and $p$ where $n$ represents the number of trials and $p$ represents the probability of success in a single trial with $q = 1 - p$ the probability of failure in a single trial. For large values of $n$ the binomial distribution approaches the normal distribution. In equation (8.51), the function $f(x)$ represents the probability of $x$ successes and $n - x$ failures in $n$-trials. The cumulative probabilities are given by

$$F(x) = B(x; n, p) = \sum_{k=0}^{x} b(k; n, p), \qquad \text{for } x = 0, 1, 2, \ldots, n \qquad (8.52)$$

As an exercise verify that

$$b(x; n, p) = b(n - x; n, 1 - p), \qquad B(x; n, p) = 1 - B(n - x - 1; n, 1 - p) \qquad (8.53)$$

The binomial probability law, sometimes called the Bernoulli distribution, occurs in those application areas where one of two possible outcomes can result in a single trial. For example, (yes,no), (success, failure), (left, right), (on, off), (defective, nondefective), etc. For example, if there are $d$ defective items in a bin of $N$ items and an item is selected at random from the bin, then the probability of obtaining a defective item in a single trial is $p = d/N$.

The binomial probability distribution involves sampling with replacement. Consequently, each time a sample of $n$ items is selected from the bin containing $N$ items, the probability of obtaining $x$ defective items is given by equation (8.51) with $p = d/N$ and $q = 1 - d/N$.

In the equation (8.51), the term

$$\binom{n}{x} = \frac{n!}{x!(n-x)!}, \qquad \binom{n}{0} = 1, \quad \text{and} \quad \binom{0}{0} = 1 \qquad (8.54)$$

represent the binomial coefficients in the binomial expansion

$$(p+q)^n = \binom{n}{0}p^n + \binom{n}{1}p^{n-1}q + \binom{n}{2}p^{n-2}q^2 + \binom{n}{3}p^{n-3}q^3 + \cdots + \binom{n}{x}p^x q^{n-x} + \cdots + \binom{n}{n}q^n = 1 \qquad (8.55)$$

In equations (8.51) and (8.55) the term $\binom{n}{x}$ represents the number of different way of selecting $x$-objects from a collection of $n$-objects and the term

$$p^x q^{n-x} = \underbrace{pp \cdots p}_{x \text{ times}} \underbrace{qq \cdots q}_{n-x \text{ times}} \qquad (8.56)$$

represents the probability of $x$ successes and $n - x$ failures in $n$-trials without regard to any ordering of the arrangements of how the successes or failures occur. Consequently, the equation (8.56) must be multiplied by the number of different arrangements of the successes and failures and this is what produces the binomial probability distribution.

The binomial distribution has the following properties

$$\text{mean} = \mu = np \quad \text{and} \quad \text{variance} = \sigma^2 = npq \qquad (8.57)$$

The figure 8-9 illustrates the binomial distribution for the parameter values $n = 10$ and $p = 0.2, 0.5$ and $0.9$.

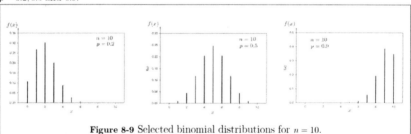

**Figure 8-9** Selected binomial distributions for $n = 10$.

## The Multinomial Distribution

The multinomial distribution occurs when many events can happen during a single trial. If only one event can result from $m$ mutually exclusive events $E_1, E_2, \ldots, E_m$ occurring in a single trial, where $p_1, p_2, \ldots, p_m$ are the probabilities assigned to the $m$-events, then the probability of getting $n_1$ $E_1's$, $n_2$ $E_2's, \ldots, n_m$ $E_m's$ is given by the multinomial probability function

$$f(n_1, n_2, \ldots, n_m) = \frac{n!}{n_1! n_2! \ldots n_m!} p_1^{n_1} p_2^{n_2} \cdots p_m^{n_m}$$

where $n_1 + n_2 + \cdots + n_m = n$ and $p_1 + p_2 + \cdots + p_m = 1$.

## The Poisson Distribution

The Poisson probability distribution has the form

$$f(x;\lambda) = \frac{\lambda^x e^{-\lambda}}{x!}, \qquad x = 0, 1, 2, 3, \ldots \tag{8.58}$$

with parameter $\lambda$ and having the properties,

1. $\sum_{x=0}^{\infty} \frac{\lambda^x e^{-\lambda}}{x!} = e^{-\lambda}\left(1 + \lambda + \frac{\lambda^2}{2!} + \frac{\lambda^3}{3!} + \cdots\right) = e^{-\lambda} e^{\lambda} = 1$
2. mean=$\mu = \lambda$
3. variance $\sigma^2 = \lambda$

The cumulative probability function is given by

$$F(x;\lambda) = \sum_{k=0}^{x} f(k;\lambda) \tag{8.59}$$

The Poisson distribution occurs in application areas which record isolated events over a period of time. For example, the number of cars entering an intersection in a ten minute interval, the number of telephone lines in use during different periods of the day, the number of customers waiting in line, the life expectancy of a light bulb, the number of transistors that fail in one year, etc.

The figure 8-10 illustrates the Poisson distribution for the parameter values $\lambda = 1/2, 1, 2$ and 3.

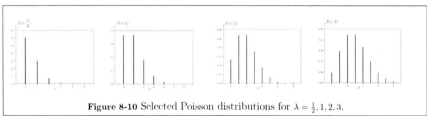

**Figure 8-10** Selected Poisson distributions for $\lambda = \frac{1}{2}, 1, 2, 3$.

## The Hypergeometric Distribution

The hypergeometric probability distribution has the form

$$f(x) = h(x; n, n_1, n_2) = \frac{\binom{n_1}{x}\binom{n_2}{n-x}}{\binom{n_1+n_2}{n}}, \qquad x = 0, 1, 2, 3, \ldots, n \tag{8.60}$$

where $x$ is an integer satisfying $0 \le x \le n$, $n_1$ represents the number of successes and $n_2$ represents the number of failures, where $n$ items are selected from $(n_1 + n_2)$ items without replacement. This is a probability distribution with three parameters, $n, n_1$ and $n_2$. The hypergeometric probability distribution is used in quality control, estimates of animal population size from capture-recapture data, the spread of an infectious disease when a fixed number of individuals are exposed to an illness.

Note that the binomial distribution is used in sampling with replacement while the hypergeometric distribution is applicable for problems where there is sampling without replacement. The hypergeometric distribution has mean

$$\mu = \frac{nn_1}{n_1 + n_2}$$

and variance given by

$$\sigma^2 = \frac{nn_1 n_2 (n_1 + n_2 - n)}{(n_1 + n_2)^2 (n_1 + n_2 - 1)}$$

The equation (8.60) represents the probability of $x$ successes and $n - x$ failures selected from $n_1 + n_2$ items where the sampling is without replacement. As an example, to find the probability of selecting two aces from a standard deck of 52 playing cards in 6 draws with no replacement of cards selected one would select the following parameters for the hypergeometric distribution. Here there are 6 draws so that $n = 6$. There are 4 aces in the deck so $n_1 = 4$ is the number of successes in the deck and $n_2 = 48$ is the number of failures in the deck, with $n_1 + n_2 = 52$ the total number of cards in the deck. The hypergeometric distribution gives the probability of $x = 2$ successes in $n = 6$ draws as

$$h(2; 6, 4, 48) = \frac{\binom{4}{2}\binom{48}{4}}{\binom{52}{6}} = \frac{621}{10829} = 0.0573$$

**The Exponential Distribution**

The exponential probability distribution has a parameter $\lambda > 0$ and is defined

$$f(x) = \begin{cases} \lambda e^{-\lambda x}, & \text{for } x > 0 \\ 0, & \text{otherwise} \end{cases} \quad (8.61)$$

The exponential distribution is used in studying time to failure of a piece of equipment, waiting time for next event to occur, like waiting time for an elevator, or time waiting in line to be served. This distribution has the mean

$$\mu = \lambda$$

and the variance is given by

$$\sigma^2 = \lambda^2$$

Note that the area under the probability curve $f(x)$, for $-\infty < x < \infty$ is equal to 1 or

$$\int_{-\infty}^{\infty} f(x)\, dx = \int_{0}^{\infty} \lambda e^{-\lambda x}\, dx = 1$$

## The Gamma Distribution

The gamma probability distribution is defined

$$f(x) = \begin{cases} \frac{1}{\theta^\alpha \Gamma(\alpha)} x^{\alpha-1} e^{-x/\theta}, & \text{for } x > 0 \\ 0, & \text{for } x \leq 0 \end{cases} \quad (8.62)$$

and has the two parameters $\alpha > 0$ and $\theta > 0$. This probability distribution arises in determining the waiting time for a given number of events to occur. For example, waiting for 10 calls to a switch board, or life testing until a failure occurs. It also occurs in weather prediction of precipitation processes. The gamma distribution has mean

$$\mu = \alpha \theta$$

and variance given by

$$\sigma^2 = \alpha \theta^2$$

The gamma distribution with parameters $\alpha = 1$ and $\theta = 1/\lambda$ produces the exponential distribution. The figure 8-11 illustrates the gamma distribution for selected values of the parameters $\alpha$ and $\theta$.

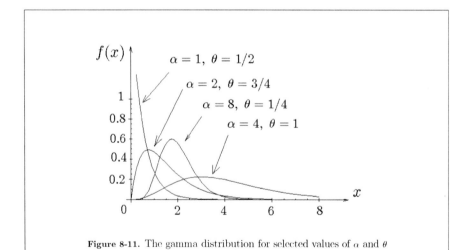

**Figure 8-11.** The gamma distribution for selected values of $\alpha$ and $\theta$

## Chi-Square $\chi^2$ Distribution

The chi-square probability distribution has the form

$$f(x) = \begin{cases} \dfrac{1}{2^{\nu/2}\Gamma(\nu/2)} x^{(\nu-2)/2} e^{-x/2}, & \text{for } x > 0 \\ 0, & \text{elsewhere} \end{cases} \qquad (8.63)$$

where $\nu = 1, 2, 3, \ldots$ is a parameter called the number of degrees of freedom. Note that the chi-square distribution is sometimes written as the $\chi^2$-distribution. It is a special case of the gamma distribution when the parameters of the gamma distribution take on the values $\alpha = \nu/2$ and $\theta = 2$. This distribution has the mean $\mu = \nu$ and variance $\sigma^2 = 2\nu$.

The chi-square distribution is used for testing of hypothesis, determining confidence intervals and testing differences in various statistics associated with independent samples.

## Student's t-Distribution

The student's[1] t-distribution with $n$ degrees of freedom is given by the probability density function

$$f(x) = \frac{1}{\sqrt{n\pi}} \frac{\Gamma\left(\dfrac{n+1}{2}\right)}{\Gamma\left(\dfrac{n}{2}\right)} \left(1 + \frac{x^2}{n}\right)^{-(n+1)/2}, \qquad -\infty < x < \infty \qquad (8.64)$$

where $n = 1, 2, 3, \ldots$ is a parameter. The student's t-distribution has the mean 0 for $n > 1$, otherwise the mean is undefined. Similarly, the variance is given by $\dfrac{n}{n-2}$ for $n > 2$, otherwise the variance is undefined.

The cumulative distribution function is given by

$$F(x) = \int_{-\infty}^{x} f(x)\,dx = \int_{-\infty}^{x} \frac{1}{\sqrt{n\pi}} \frac{\Gamma\left(\dfrac{n+1}{2}\right)}{\Gamma\left(\dfrac{n}{2}\right)} \left(1 + \frac{x^2}{n}\right)^{-(n+1)/2} dx \qquad (8.65)$$

The table 8.5 contains values of $t_{\alpha,n}$ which satisfy the equation

$$\int_{t_{\alpha,n}}^{\infty} f(x)\,dx = \alpha = 1 - F(t_{\alpha,n}) \qquad (8.66)$$

The normal distribution is related to the student's t-distribution as follows. If $\bar{x}$ and $s$ are the mean and standard deviation associated with a random sample of size $n$ from a normal distribution $N(x; \mu, \sigma^2)$, then the quantity $\dfrac{(\bar{x} - \mu)\sqrt{n}}{s}$ has a student-t-distribution with $n - 1$ degrees of freedom.

---

[1] Developed by W.S. Gosset who used the name "Student" as a pseudonym.

## The F-Distribution

The F-distribution has the probability density function

$$f(x) = f_{n,m}(x) = \begin{cases} \dfrac{\Gamma\left(\dfrac{m+n}{2}\right)}{\Gamma\left(\dfrac{m}{2}\right)\Gamma\left(\dfrac{n}{2}\right)} n^{n/2} m^{m/2} \dfrac{x^{n/2-1}}{(m+nx)^{(m+n)/2}}, & \text{for } x > 0 \\ 0, & \text{for } x < 0 \end{cases} \quad (8.67)$$

which is sometimes given in the form

$$f(x) = f_{n,m}(x) = \begin{cases} \dfrac{\Gamma\left(\dfrac{m+n}{2}\right)}{\Gamma\left(\dfrac{m}{2}\right)\Gamma\left(\dfrac{n}{2}\right)} (n/m)^{n/2} \dfrac{x^{n/2-1}}{(1+\dfrac{n}{m}x)^{(m+n)/2}}, & \text{for } x > 0 \\ 0, & \text{for } x < 0 \end{cases} \quad (8.68)$$

The F-distribution has the parameters $m = 1, 2, 3, \ldots$ and $n = 1, 2, 3, \ldots$.

If $X_1$ and $X_2$ are independent random variables associated with a chi-square distribution having respectively degrees of freedom $n$ and $m$, then the quantity $Y = \dfrac{X_1/n}{X_2/m}$ will have a F-distribution with $n$ and $m$ degrees of freedom.

The tables 8.6 (a)(b)(c)(d)(e) contain values of $F_{\alpha,n,m}$ such that $\int_{F_{(\alpha,n,m)}} f_{n,m}(x)\,dx = \alpha$ for $\alpha$ having the values $0.1, 0.05, .025, .01,$ and $.005$. Observe the symmetry of the F-distribution and note that in the use of the upper tail values from the tables it is customary to employ the relation

$$F(df_m, df_n, 1 - \alpha/2) = \dfrac{1}{F(df_n, df_m, \alpha/2)} \quad (8.69)$$

where $df_m$ and $df_n$ denote the degrees of freedom for $m$ and $n$.

The chi-square, student t and F distributions are used in testing of hypothesis, confidence intervals and testing differences or ratios of various statistics associated with independent samples. The degrees of freedom associated with these distributions can be thought of as a parameter representing an increase in reliability of the calculated statistic. That is, a statistic associated with one degree of freedom is less reliable than the same statistic calculated using a higher degree of freedom. In some cases the degrees of freedom are related to the number of data points used to calculate the statistic. In some cases the degrees of freedom is obtained by subtracting 1 from the sample size $n$.

## The Uniform Distribution

The uniform probability density function $f(x)$ and the associated distribution function $F(x)$ are given by

$$f(x) = \begin{cases} \dfrac{1}{b-a}, & a < x < b \\ 0, & \text{otherwise} \end{cases} \qquad F(x) = \int_{-\infty}^{x} f(x)\,dx = \int_{a}^{x} f(x)\,dx$$

It is sometimes referred to as the rectangular distribution on the interval $a < x < b$.

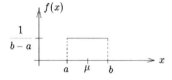 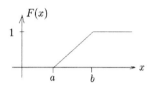

This distribution has the mean

$$\mu = \int_{-\infty}^{\infty} x f(x)\,dx = \int_{a}^{b} x \frac{1}{b-a}\,dx = \frac{1}{2}(a+b)$$

and variance

$$\sigma^2 = \int_{-\infty}^{\infty} (x-\mu)^2 f(x)\,dx = \frac{1}{12}(b-a)^2$$

The cumulative distribution function is given by

$$F(x) = \begin{cases} 0, & x < a \\ \frac{x-a}{b-a}, & a \le x \le b \\ 1, & x > b \end{cases}$$

The uniform probability density function is used in pseudo-random number generators with sampling is over the interval $0 \le x \le 1$.

## Confidence Intervals

Sampling theory is a study of the various relationships that exist between properties of a population and information obtained based upon samples from the population. For example, each sample collected from a population has associated with it a sample mean $\bar{x} = \mu_{\bar{x}}$ and sample variance $s^2 = \sigma_{\bar{x}}^2$. How do these quantities compare with the true population mean $\mu$ and true population variance $\sigma^2$? It would be nice to put limits on the calculated values $\mu_{\bar{x}}$ and write statements like

$$\mu_{\bar{x}} - \gamma_1 < \mu < \mu_{\bar{x}} + \gamma_2$$

and be able to say that there is a 90% probability that the true mean lies within the specified limits. Is there a way to alter the limits $\mu_{\bar{x}} - \gamma_1$ and $\mu_{\bar{x}} + \gamma_2$ to obtain a 95%,97%,or 99% probability? The values 90%, 95%, 97% or 99% are called confidence levels. To answer the above questions, employ the central limit theorem which says that if $n$ is the number of items in one sample (the sample size) and $n$ is large, the sample means $\mu_{\bar{x}}$ will be normally distributed and have a mean of $\mu_{\bar{x}} = \mu$ and standard deviation of $\sigma_{\bar{x}} = \sigma/\sqrt{n}$. The normal distribution can be scaled to standard form by making the change of variable $\bar{x} = \mu_{\bar{x}} + \sigma_{\bar{x}} Z$ as illustrated in the figure 8-12.

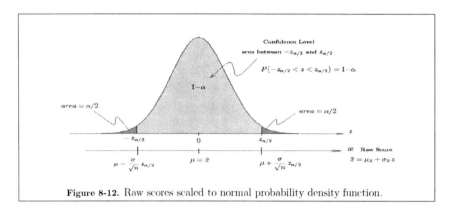

**Figure 8-12.** Raw scores scaled to normal probability density function.

To use the central limit theorem select a confidence level $\gamma = 1 - \alpha$ which represents the area between the limits $-z_{\alpha/2}$ and $z_{\alpha/2}$ associated with the normalized normal probability density function as illustrated in the figure 8-12. This determines $\alpha$ and $\alpha/2$ and the values $\pm z_{\alpha/2}$ can be obtained from the normalized probability table 8.4. Some examples are

| $1 - \alpha$ | .90 | .95 | .99 | .999 |
|---|---|---|---|---|
| $\alpha$ | .10 | .05 | .01 | .001 |
| $\alpha/2$ | .05 | .025 | .005 | .0005 |
| $z_{\alpha/2}$ | 1.645 | 1.960 | 2.576 | 3.291 |

If $\bar{x}$ is the mean of a sample $\{x_1, x_2, \ldots, x_n\}$ of size $n$, then confidence limits on the value $\bar{x}$ are determined as follows.

**Normal distribution with known variance $\sigma^2$**

If the variance of the population is known then use the central limit theorem to construct the following confidence interval for the mean $\mu$ of the population based upon a $1 - \alpha = \gamma$ level of confidence

$$CONF\{\bar{x} - \frac{\sigma}{\sqrt{n}}z_{\alpha/2} \leq \mu \leq \bar{x} + \frac{\sigma}{\sqrt{n}}z_{\alpha/2}\} \qquad (8.70)$$

**Normal distribution with unknown variance $\sigma^2$**

In the case where the population variance is unknown, then make use of the fact that $t = \dfrac{|\bar{x} - \mu|}{s/\sqrt{n}}$ follows a student t-distribution to construct a confidence interval. From the student t-distribution determine the value $t_{\alpha/2, n-1}$ based upon $n - 1$ degrees of freedom such that the right tailed area equals $\alpha/2$ as illustrated in the accompanying figure.

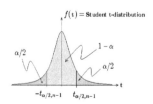

Some examples for a sample size of $n = 11$ and degrees of freedom $n - 1 = 10$ are given in the following table.

| $1 - \alpha$ | .90 | .95 | .99 | .999 |
|---|---|---|---|---|
| $\alpha$ | .10 | .05 | .01 | .001 |
| $\alpha/2$ | .05 | .025 | .005 | .0005 |
| $t_{\alpha/2,10}$ | 1.812 | 2.228 | 3.169 | 4.144 |

The confidence interval for the mean $\mu$ of the population uses the computed variance

$$s^2 = \frac{1}{n-1}\sum_{j=1}^{n}(x_j - \bar{x})^2 \tag{8.71}$$

to produce the $\gamma = 1 - \alpha$ confidence level

$$CONF\{\bar{x} - \frac{s}{\sqrt{n}}t_{\alpha/2,n-1} \leq \mu \leq \bar{x} + \frac{s}{\sqrt{n}}t_{\alpha/2,n-1}\} \tag{8.72}$$

where $n$ is the sample size.

**Confidence interval for the variance $\sigma^2$**

The confidence interval for the variance $\sigma^2$ of the population having a normal distribution is based upon the fact that the variable $Y = (n-1)s^2/\sigma^2$ follows a chi-square distribution with $n - 1$ degrees of freedom, where again $n$ represents the sample size. First select a level of confidence $\gamma = 1 - \alpha$ and then from a chi-square distribution table with $n - 1$ degrees of freedom determine the $\chi^2_{\alpha/2,n-1}$ and $\chi^2_{1-\alpha/2,n-1}$ values which represent the points corresponding to the tail areas of the chi-square probability density function as illustrated.

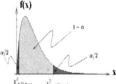

Secondly, one must calculate the variance squared $s^2$ using equation (8.71), then construct the confidence interval for the variance of a normal distribution given by

$$CONF\{(n-1)\frac{s^2}{\chi^2_{1-\alpha/2,n-1}} \leq \sigma^2 \leq (n-1)\frac{s^2}{\chi^2_{\alpha/2,n-1}}\} \tag{8.73}$$

## Least Squares Curve Fitting

A set of data points

$$(x_1, y_1), (x_2, y_2), (x_3, y_3), \ldots, (x_i, y_i), \ldots, (x_{n-1}, y_{n-1}), (x_n, y_n)$$

can be plotted on ordinary graph paper and then a line $y = \beta_0 + \beta_1 x$ can also be plotted to obtain a figure such as illustrated in the figure 8-13.

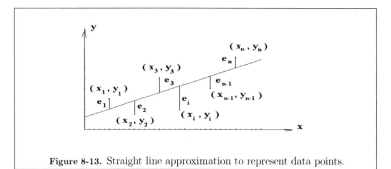

**Figure 8-13.** Straight line approximation to represent data points.

Assume that the data points are normally distributed about the straight line and that errors $e_1, e_2, \ldots, e_n$ occur in the $y$-variable, where the errors are defined as the differences between the $y$-value on the line and the $y$-value of the data point. What would be the "best" straight line to represent the given data points? There are many ways to define "best". By defining the error $e_i$ associated with the $i$th data point $(x_i, y_i)$ as

$$e_i = (y \text{ of line at } x_i) - (y \text{ data value at } x_i)$$
$$e_i = \beta_0 + \beta_1 x_i - y_i \tag{8.74}$$

then associated with the given set of data are the errors

$$\begin{aligned}
e_1 &= y(x_1) - y_1 = \beta_0 + \beta_1 x_1 - y_1 \\
e_2 &= y(x_2) - y_2 = \beta_0 + \beta_1 x_2 - y_2 \\
e_3 &= y(x_3) - y_3 = \beta_0 + \beta_1 x_3 - y_3 \\
&\vdots \\
e_n &= y(x_n) - y_n = \beta_0 + \beta_1 x_n - y_n.
\end{aligned} \tag{8.75}$$

One way to define the "best" straight line $y = \beta_0 + \beta_1 x$ is to select the constants $\beta_0$ and $\beta_1$ which minimize the sum of squares of the errors associated with the data set. That is, if

$$E = E(\beta_0, \beta_1) = \sum_{i=1}^{n} e_i^2 = \sum_{i=1}^{n} (\beta_0 + \beta_1 x_i - y_i)^2 \tag{8.76}$$

denotes the sum of squares of the errors, then $E$ has a minimum value when the conditions

$$\frac{\partial E}{\partial \beta_0} = 0 \quad \text{and} \quad \frac{\partial E}{\partial \beta_1} = 0 \tag{8.77}$$

are satisfied simultaneously. Hence, the constants $\beta_0$ and $\beta_1$ must be selected to satisfy the simultaneous equations

$$\begin{aligned}
\frac{\partial E}{\partial \beta_0} &= 2 \sum_{i=1}^{n} (\beta_0 + \beta_1 x_i - y_i)(1) = 0 \\
\frac{\partial E}{\partial \beta_1} &= 2 \sum_{i=1}^{n} (\beta_0 + \beta_1 x_i - y_i)(x_i) = 0.
\end{aligned} \tag{8.78}$$

The equations (8.78) simplify to the $2 \times 2$ linear system of equations

$$n\beta_0 + \left(\sum_{i=1}^n x_i\right)\beta_1 = \sum_{i=1}^n y_i$$
$$\left(\sum_{i=1}^n x_i\right)\beta_0 + \left(\sum_{i=1}^n x_i^2\right)\beta_1 = \sum_{i=1}^n x_i y_i$$
(8.79)

which can then be solved for the coefficients $\beta_0$ and $\beta_1$. This gives the "best" least squares straight line $y = y(x) = \beta_0 + \beta_1 x$.

Alternatively, set all of the equations (8.75) equal to zero, to obtain a system of equations having the matrix form

$$A\overline{\beta} = \overline{y}$$

$$\begin{bmatrix} 1 & x_1 \\ 1 & x_2 \\ 1 & x_3 \\ \vdots & \vdots \\ 1 & x_n \end{bmatrix} \begin{bmatrix} \beta_0 \\ \beta_1 \end{bmatrix} = \begin{bmatrix} y_1 \\ y_2 \\ y_3 \\ \vdots \\ y_n \end{bmatrix}.$$
(8.80)

By doing this the data set of errors, calculated from the difference in the data set $y$ values and the straight line $y$ values, is represented as an over determined system of equations for determining the constants $\beta_0$ and $\beta_1$. That is, there are more equations than there are unknowns and so the unknowns $\beta_0, \beta_1$ are selected to minimize the sum of squares error associated with the over determined system of equations. Observe that left multiplying both sides of equation (8.80) by the transpose matrix $A^T$ gives the new set of equations $A^T A\overline{\beta} = A^T \overline{y}$ or

$$\begin{bmatrix} 1 & 1 & 1 & \ldots & 1 \\ x_1 & x_2 & x_3 & \ldots & x_n \end{bmatrix} \begin{bmatrix} 1 & x_1 \\ 1 & x_2 \\ 1 & x_3 \\ \vdots & \vdots \\ 1 & x_n \end{bmatrix} \begin{bmatrix} \beta_0 \\ \beta_1 \end{bmatrix} = \begin{bmatrix} 1 & 1 & 1 & \ldots & 1 \\ x_1 & x_2 & x_3 & \ldots & x_n \end{bmatrix} \begin{bmatrix} y_1 \\ y_2 \\ y_3 \\ \vdots \\ y_n \end{bmatrix}$$

which simplifies to

$$\begin{bmatrix} n & \sum_{i=1}^n x_i \\ \sum_{i=1}^n x_i & \sum_{i=1}^n x_i^2 \end{bmatrix} \begin{bmatrix} \beta_0 \\ \beta_1 \end{bmatrix} = \begin{bmatrix} \sum_{i=1}^n y_i \\ \sum_{i=1}^n x_i y_i \end{bmatrix}$$
(8.81)

which is the matrix form of the equations (8.79). This presents an alternative way to solve for the coefficients $\beta_0$ and $\beta_1$

## Linear Regression

The least squares method applied to a straight line fit of data can be generalized to any linear combination of functions. Given a set of data points $(x_i, y_i)$, for $i = 1, 2, \ldots, n$, assume a curve fit function of the form

$$y = y(x) = \beta_0 f_0(x) + \beta_1 f_1(x) + \beta_2 f_2(x) + \cdots + \beta_k f_k(x) \tag{8.82}$$

where $\beta_0, \beta_1, \ldots, \beta_k$ are unknown coefficients and $f_0(x), f_1(x), f_2(x), \ldots, f_k(x)$ represent linearly independent functions, called the basis of the representation. Note that for the previous straight line fit the independent functions $f_0(x) = 1$ and $f_1(x) = x$ were used. In general, select any set of independent functions and select the $\beta$ coefficients such that the sum of squares error

$$E = E(\beta_0, \beta_1, \ldots, \beta_k) = \sum_{i=1}^{n} (y(x_i) - y_i)^2$$

$$E = E(\beta_0, \beta_1, \ldots, \beta_k) = \sum_{i=1}^{n} [\beta_0 f_0(x_i) + \beta_1 f_1(x_i) + \beta_2 f_2(x_i) + \cdots + \beta_k f_k(x_i) - y_i]^2 \tag{8.83}$$

is a minimum. The determination of the $\beta$-values requires a solution be found from the set of simultaneous least square equations

$$\frac{\partial E}{\partial \beta_0} = 0, \quad \frac{\partial E}{\partial \beta_1} = 0, \quad \cdots, \quad \frac{\partial E}{\partial \beta_k} = 0. \tag{8.84}$$

Another way to obtain the system of equations (8.84) is to first represent the data in the matrix form

$$A\overline{\beta} = \overline{y}$$

$$\begin{bmatrix} f_0(x_1) & f_1(x_1) & f_2(x_1) & \cdots & f_k(x_1) \\ f_0(x_2) & f_1(x_2) & f_2(x_2) & \cdots & f_k(x_2) \\ \vdots & \vdots & \vdots & \ddots & \vdots \\ f_0(x_n) & f_1(x_n) & f_2(x_n) & \cdots & f_k(x_n) \end{bmatrix} \begin{bmatrix} \beta_0 \\ \beta_1 \\ \beta_2 \\ \vdots \\ \beta_k \end{bmatrix} = \begin{bmatrix} y_1 \\ y_2 \\ \vdots \\ y_n \end{bmatrix} \tag{8.85}$$

Both sides of the equation (8.85) can be left multiplied by the transpose matrix $A^T$ and the resulting system can be solved for the unknown coefficients. In matrix notation write

$$A\overline{\beta} = \overline{y}$$
$$A^T A \overline{\beta} = A^T \overline{y} \tag{8.86}$$
$$\overline{\beta} = (A^T A)^{-1} A^T \overline{y}.$$

The solution of the system of equations (8.84) or (8.86) will produce the coefficients $\beta_i$, $i = 0, 1, \ldots, k$, which minimizes the sum of square error.

## Monte Carlo Methods

Monte Carlo methods is a term used to describe a wide variety of computer techniques which employ random number generators to simulate an event or events and then perform a statistical analysis of the results. Sometimes Monte Carlo methods are constructed to solve difficult problems where deterministic methods fail. If performed properly, Monte Carlo methods can give very accurate answers. The only drawback is that some Monte Carlo techniques take a very long time to run on the computer. An example of a simple Monte Carlo method is the calculation of the area $A$ of a circle using random numbers. Consider a circle with radius $1/2$ which is placed inside the unit square having vertices $(0,0), (1,0), (1,1)$ and $(0,1)$. The area of this circle is $\pi/4 = 0.7853981634...$.

Most computer languages have a uniform random number generator which generates pseudo-random numbers lying between 0 and 1. Construct a computer program which employs the uniform random number generator to generate two random numbers $(x_r, y_r)$, where $0 < x_r < 1$ and $0 < y_r < 1$, then imagine the circle inside the unit square as a circular dart board and the random number generated by the computer program $(x_r, y_r)$ is where the dart lands. Construct the computer program to perform a test as to whether the point $(x_r, y_r)$ is on or off the circular dart board. Perform this test N-times and record the number of hits which land on or inside the circle. To calculate the area use the ratio

$$\frac{\text{Number hits inside circle}}{\text{Total number of darts thrown}} = \frac{\text{Area of circle}}{\text{Area of square}} \implies \frac{H}{N} = \frac{A}{1} = A$$

where $H$ denotes the number of hits and $N$ represents the number of darts thrown.

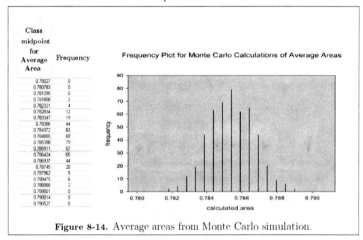

**Figure 8-14.** Average areas from Monte Carlo simulation.

Perform the above experiment $K$-times to calculate a set of approximate areas $\{A_1, A_2, \ldots, A_K\}$ having an average area $\overline{A} = \frac{1}{K}\sum_{i=1}^{K} A_i$. Put all of the above computer code in a loop and calculate $M$-averages $\{\overline{A}_1, \overline{A}_2, \ldots, \overline{A}_M\}$. The central limit theorem tells us that the set of averages must be normally distributed. By calculating the mean and standard deviation associated with all these averages it is possible to determine very accurate bounds on the area of the circle. Using the values $N = 1000$ throws, $K = 100$ areas, and $M = 500$ area averages, modern laptop computers can calculate the results in less than one minute. The data generated for the above values of N,K and M is presented in figure 8-14 as a bar chart having a mean 0.7853974 and standard deviation 0.001282.

## Linear Interpolation

Consider the set of values in a table as illustrated in the accompanying figure. Let $F_{11}$ denote the value in the table corresponding to the position $(x_1, y_1)$. Similarly, define the values $F_{12}, F_{21}$ and $F_{22}$ corresponding respectively to the points $(x_1, y_2)$, $(x_2, y_1)$ and $(x_2, y_2)$. Interpolation over this two dimensional array is the problem of determining the values $F_\alpha$, $F_\beta$ and $F_{\alpha,\beta}$ which are positioned on the boundaries and interior to the box connecting the known data values $F_{11}, F_{12}, F_{22}, F_{21}$. As an exercise verify that

$$F_\alpha = (1-\alpha)F_{11} + \alpha F_{12}, \qquad F_\beta = (1-\beta)F_{11} + \beta F_{21}$$
$$F_{\alpha,\beta} = (1-\alpha)(1-\beta)F_{11} + \alpha(1-\beta)F_{12} + \beta(1-\alpha)F_{21} + \alpha\beta F_{22}$$

Note how the $F_\alpha$ and $F_\beta$ values vary as the parameters $\alpha$ and $\beta$ vary from 0 to 1. This is a straight forward linear interpolation between the given values. The value $F_{\alpha,\beta}$ is obtained by first doing a linear interpolation in the $y$ directions at the columns $x_1$ and $x_2$, which is then followed by a linear interpolation in the $x$-direction.

An alternative method of interpolation is to use the Taylor series expansion in both the $x$ and $y$ directions to obtain the alternative interpolation formula

$$F_{\alpha,\beta} = (1 - \alpha - \beta)F_{11} + \beta F_{21} + \alpha F_{12}$$

|       | $x_1$ |       | $x_2$ |       |
|-------|-------|-------|-------|-------|
|       | 7.956 | 7.471 | 7.134 | 6.885 |
|       | 7.343 | 6.872 | 6.545 | 6.302 |
|       | 6.881 | 6.422 | 6.102 | 5.865 |
|       | 6.521 | 6.071 | 5.757 | 5.525 |
| $y_1$ | 6.233 | 5.791 | 5.482 | 5.253 |
| $y_2$ | 5.998 | 5.562 | 5.257 | 5.031 |
|       | 5.803 | 5.372 | 5.071 | 4.847 |
|       | 5.638 | 5.212 | 4.913 | 4.692 |

Sometimes it is necessary to modify the above interpolation formulas for application to entries in a three-dimensional array of numbers. The interpolation result is obtained by applying the one-dimensional interpolation formulas in each of the $x, y$ and $z$ directions.

## Table 8.4   Area Under Standard Normal Curve

$$Area = \Phi(z) = \frac{1}{\sqrt{2\pi}} \int_{-\infty}^{z} e^{-\xi^2/2}\, d\xi$$

| z | Area | z | Area | z | Area | z | Area | z | Area | z | Area | z | Area | z | Area |
|---|---|---|---|---|---|---|---|---|---|---|---|---|---|---|---|
| 0.00 | .5000 | .50 | .6915 | 1.00 | .8413 | 1.50 | .9332 | 2.00 | .9772 | 2.50 | .9938 | 3.00 | .9987 | | |
| .01 | .5040 | .51 | .6950 | 1.01 | .8438 | 1.51 | .9345 | 2.01 | .9778 | 2.51 | .9940 | 3.01 | .9987 | | |
| .02 | .5080 | .52 | .6985 | 1.02 | .8461 | 1.52 | .9357 | 2.02 | .9783 | 2.52 | .9941 | 3.02 | .9987 | | |
| .03 | .5120 | .53 | .7019 | 1.03 | .8485 | 1.53 | .9370 | 2.03 | .9788 | 2.53 | .9943 | 3.03 | .9988 | | |
| .04 | .5160 | .54 | .7054 | 1.04 | .8508 | 1.54 | .9382 | 2.04 | .9793 | 2.54 | .9945 | 3.04 | .9988 | | |
| .05 | .5199 | .55 | .7088 | 1.05 | .8531 | 1.55 | .9394 | 2.05 | .9798 | 2.55 | .9946 | 3.05 | .9989 | | |
| .06 | .5239 | .56 | .7123 | 1.06 | .8554 | 1.56 | .9406 | 2.06 | .9803 | 2.56 | .9948 | 3.06 | .9989 | | |
| .07 | .5279 | .57 | .7157 | 1.07 | .8577 | 1.57 | .9418 | 2.07 | .9808 | 2.57 | .9949 | 3.07 | .9989 | | |
| .08 | .5319 | .58 | .7190 | 1.08 | .8599 | 1.58 | .9429 | 2.08 | .9812 | 2.58 | .9951 | 3.08 | .9990 | | |
| .09 | .5359 | .59 | .7224 | 1.09 | .8621 | 1.59 | .9441 | 2.09 | .9817 | 2.59 | .9952 | 3.09 | .9990 | | |
| .10 | .5398 | .60 | .7257 | 1.10 | .8643 | 1.60 | .9452 | 2.10 | .9821 | 2.60 | .9953 | 3.10 | .9990 | | |
| .11 | .5438 | .61 | .7291 | 1.11 | .8665 | 1.61 | .9463 | 2.11 | .9826 | 2.61 | .9955 | 3.11 | .9991 | | |
| .12 | .5478 | .62 | .7324 | 1.12 | .8686 | 1.62 | .9474 | 2.12 | .9830 | 2.62 | .9956 | 3.12 | .9991 | | |
| .13 | .5517 | .63 | .7357 | 1.13 | .8708 | 1.63 | .9484 | 2.13 | .9834 | 2.63 | .9957 | 3.13 | .9991 | | |
| .14 | .5557 | .64 | .7389 | 1.14 | .8729 | 1.64 | .9495 | 2.14 | .9838 | 2.64 | .9959 | 3.14 | .9992 | | |
| .15 | .5596 | .65 | .7422 | 1.15 | .8749 | 1.65 | .9505 | 2.15 | .9842 | 2.65 | .9960 | 3.15 | .9992 | | |
| .16 | .5636 | .66 | .7454 | 1.16 | .8770 | 1.66 | .9515 | 2.16 | .9846 | 2.66 | .9961 | 3.16 | .9992 | | |
| .17 | .5675 | .67 | .7486 | 1.17 | .8790 | 1.67 | .9525 | 2.17 | .9850 | 2.67 | .9962 | 3.17 | .9992 | | |
| .18 | .5714 | .68 | .7517 | 1.18 | .8810 | 1.68 | .9535 | 2.18 | .9854 | 2.68 | .9963 | 3.18 | .9993 | | |
| .19 | .5753 | .69 | .7549 | 1.19 | .8830 | 1.69 | .9545 | 2.19 | .9857 | 2.69 | .9964 | 3.19 | .9993 | | |
| .20 | .5793 | .70 | .7580 | 1.20 | .8849 | 1.70 | .9554 | 2.20 | .9861 | 2.70 | .9965 | 3.20 | .9993 | | |
| .21 | .5832 | .71 | .7611 | 1.21 | .8869 | 1.71 | .9564 | 2.21 | .9864 | 2.71 | .9966 | 3.21 | .9993 | | |
| .22 | .5871 | .72 | .7642 | 1.22 | .8888 | 1.72 | .9573 | 2.22 | .9868 | 2.72 | .9967 | 3.22 | .9994 | | |
| .23 | .5910 | .73 | .7673 | 1.23 | .8907 | 1.73 | .9582 | 2.23 | .9871 | 2.73 | .9968 | 3.23 | .9994 | | |
| .24 | .5948 | .74 | .7704 | 1.24 | .8925 | 1.74 | .9591 | 2.24 | .9875 | 2.74 | .9969 | 3.24 | .9994 | | |
| .25 | .5987 | .75 | .7734 | 1.25 | .8944 | 1.75 | .9599 | 2.25 | .9878 | 2.75 | .9970 | 3.25 | .9994 | | |
| .26 | .6026 | .76 | .7764 | 1.26 | .8962 | 1.76 | .9608 | 2.26 | .9881 | 2.76 | .9971 | 3.26 | .9994 | | |
| .27 | .6064 | .77 | .7794 | 1.27 | .8980 | 1.77 | .9616 | 2.27 | .9884 | 2.77 | .9972 | 3.27 | .9995 | | |
| .28 | .6103 | .78 | .7823 | 1.28 | .8997 | 1.78 | .9625 | 2.28 | .9887 | 2.78 | .9973 | 3.28 | .9995 | | |
| .29 | .6141 | .79 | .7852 | 1.29 | .9015 | 1.79 | .9633 | 2.29 | .9890 | 2.79 | .9974 | 3.29 | .9995 | | |
| .30 | .6179 | .80 | .7881 | 1.30 | .9032 | 1.80 | .9641 | 2.30 | .9893 | 2.80 | .9974 | 3.30 | .9995 | | |
| .31 | .6217 | .81 | .7910 | 1.31 | .9049 | 1.81 | .9649 | 2.31 | .9896 | 2.81 | .9975 | 3.31 | .9995 | | |
| .32 | .6255 | .82 | .7939 | 1.32 | .9066 | 1.82 | .9656 | 2.32 | .9898 | 2.82 | .9976 | 3.32 | .9995 | | |
| .33 | .6293 | .83 | .7967 | 1.33 | .9082 | 1.83 | .9664 | 2.33 | .9901 | 2.83 | .9977 | 3.33 | .9996 | | |
| .34 | .6331 | .84 | .7995 | 1.34 | .9099 | 1.84 | .9671 | 2.34 | .9904 | 2.84 | .9977 | 3.34 | .9996 | | |
| .35 | .6368 | .85 | .8023 | 1.35 | .9115 | 1.85 | .9678 | 2.35 | .9906 | 2.85 | .9978 | 3.35 | .9996 | | |
| .36 | .6406 | .86 | .8051 | 1.36 | .9131 | 1.86 | .9686 | 2.36 | .9909 | 2.86 | .9979 | 3.36 | .9996 | | |
| .37 | .6443 | .87 | .8078 | 1.37 | .9147 | 1.87 | .9693 | 2.37 | .9911 | 2.87 | .9979 | 3.37 | .9996 | | |
| .38 | .6480 | .88 | .8106 | 1.38 | .9162 | 1.88 | .9699 | 2.38 | .9913 | 2.88 | .9980 | 3.38 | .9996 | | |
| .39 | .6517 | .89 | .8133 | 1.39 | .9177 | 1.89 | .9706 | 2.39 | .9916 | 2.89 | .9981 | 3.39 | .9997 | | |
| .40 | .6554 | .90 | .8159 | 1.40 | .9192 | 1.90 | .9713 | 2.40 | .9918 | 2.90 | .9981 | 3.40 | .9997 | | |
| .41 | .6591 | .91 | .8186 | 1.41 | .9207 | 1.91 | .9719 | 2.41 | .9920 | 2.91 | .9982 | 3.41 | .9997 | | |
| .42 | .6628 | .92 | .8212 | 1.42 | .9222 | 1.92 | .9726 | 2.42 | .9922 | 2.92 | .9982 | 3.42 | .9997 | | |
| .43 | .6664 | .93 | .8238 | 1.43 | .9236 | 1.93 | .9732 | 2.43 | .9925 | 2.93 | .9983 | 3.43 | .9997 | | |
| .44 | .6700 | .94 | .8264 | 1.44 | .9251 | 1.94 | .9738 | 2.44 | .9927 | 2.94 | .9984 | 3.44 | .9997 | | |
| .45 | .6736 | .95 | .8289 | 1.45 | .9265 | 1.95 | .9744 | 2.45 | .9929 | 2.95 | .9984 | 3.45 | .9997 | | |
| .46 | .6772 | .96 | .8315 | 1.46 | .9279 | 1.96 | .9750 | 2.46 | .9931 | 2.96 | .9985 | 3.46 | .9997 | | |
| .47 | .6808 | .97 | .8340 | 1.47 | .9292 | 1.97 | .9756 | 2.47 | .9932 | 2.97 | .9985 | 3.47 | .9997 | | |
| .48 | .6844 | .98 | .8365 | 1.48 | .9306 | 1.98 | .9761 | 2.48 | .9934 | 2.98 | .9986 | 3.48 | .9997 | | |
| .49 | .6879 | .99 | .8389 | 1.49 | .9319 | 1.99 | .9767 | 2.49 | .9936 | 2.99 | .9986 | 3.49 | .9998 | | |

Table 8.5(a). Critical Values for the Chi-square Distribution with $\nu$ Degrees of Freedom

$$\int_0^{\chi^2_{(1-\alpha)}} \frac{x^{(\nu/2)-1}}{2^{\nu/2}\Gamma(\nu/2)} e^{-x/2}\, dx = 1 - \alpha$$

| $\alpha$ | 0.995 | 0.990 | 0.975 | 0.950 | 0.900 |
|---|---|---|---|---|---|
| $1-\alpha$ | 0.005 | 0.010 | 0.025 | 0.050 | 0.100 |
| $\nu$ | $\chi^2_{0.005}$ | $\chi^2_{0.010}$ | $\chi^2_{0.025}$ | $\chi^2_{0.050}$ | $\chi^2_{0.100}$ |
| 1 | 0.0000 | 0.0002 | 0.0010 | 0.0039 | 0.0158 |
| 2 | 0.0100 | 0.0201 | 0.0506 | 0.1026 | 0.2107 |
| 3 | 0.0717 | 0.1148 | 0.2158 | 0.3518 | 0.5844 |
| 4 | 0.2070 | 0.2971 | 0.4844 | 0.7107 | 1.0636 |
| 5 | 0.4117 | 0.5543 | 0.8312 | 1.1455 | 1.6103 |
| 6 | 0.6757 | 0.8721 | 1.2373 | 1.6354 | 2.2041 |
| 7 | 0.9893 | 1.2390 | 1.6899 | 2.1673 | 2.8331 |
| 8 | 1.3444 | 1.6465 | 2.1797 | 2.7326 | 3.4895 |
| 9 | 1.7349 | 2.0879 | 2.7004 | 3.3251 | 4.1682 |
| 10 | 2.1559 | 2.5582 | 3.2470 | 3.9403 | 4.8652 |
| 11 | 2.6032 | 3.0535 | 3.8157 | 4.5748 | 5.5778 |
| 12 | 3.0738 | 3.5706 | 4.4038 | 5.2260 | 6.3038 |
| 13 | 3.5650 | 4.1069 | 5.0088 | 5.8919 | 7.0415 |
| 14 | 4.0747 | 4.6604 | 5.6287 | 6.5706 | 7.7895 |
| 15 | 4.6009 | 5.2293 | 6.2621 | 7.2609 | 8.5468 |
| 16 | 5.1422 | 5.8122 | 6.9077 | 7.9616 | 9.3122 |
| 17 | 5.6972 | 6.4078 | 7.5642 | 8.6718 | 10.0852 |
| 18 | 6.2648 | 7.0149 | 8.2307 | 9.3905 | 10.8649 |
| 19 | 6.8440 | 7.6327 | 8.9065 | 10.1170 | 11.6509 |
| 20 | 7.4338 | 8.2604 | 9.5908 | 10.8508 | 12.4426 |
| 21 | 8.0337 | 8.8972 | 10.2829 | 11.5913 | 13.2396 |
| 22 | 8.6427 | 9.5425 | 10.9823 | 12.3380 | 14.0415 |
| 23 | 9.2604 | 10.1957 | 11.6886 | 13.0905 | 14.8480 |
| 24 | 9.8862 | 10.8564 | 12.4012 | 13.8484 | 15.6587 |
| 25 | 10.5197 | 11.5240 | 13.1197 | 14.6114 | 16.4734 |
| 26 | 11.1602 | 12.1981 | 13.8439 | 15.3792 | 17.2919 |
| 27 | 11.8076 | 12.8785 | 14.5734 | 16.1514 | 18.1139 |
| 28 | 12.4613 | 13.5647 | 15.3079 | 16.9279 | 18.9392 |
| 29 | 13.1211 | 14.2565 | 16.0471 | 17.7084 | 19.7677 |
| 30 | 13.7867 | 14.9535 | 16.7908 | 18.4927 | 20.5992 |
| 40 | 20.7065 | 22.1643 | 24.4330 | 26.5093 | 29.0505 |
| 50 | 27.9907 | 29.7067 | 32.3574 | 34.7643 | 37.6886 |
| 60 | 35.5345 | 37.4849 | 40.4817 | 43.1880 | 46.4589 |
| 70 | 43.2752 | 45.4417 | 48.7576 | 51.7393 | 55.3289 |
| 80 | 51.1719 | 53.5401 | 57.1532 | 60.3915 | 64.2778 |
| 90 | 59.1963 | 61.7541 | 65.6466 | 69.1260 | 73.2911 |
| 100 | 67.3276 | 70.0649 | 74.2219 | 77.9295 | 82.3581 |

## Table 8.5(b). Critical Values for the Chi-square Distribution with $\nu$ Degrees of Freedom

$$\int_0^{\chi^2_{(1-\alpha)}} \frac{x^{(\nu/2)-1}}{2^{\nu/2}\Gamma(\nu/2)} e^{-x/2}\, dx = 1-\alpha$$

| $\alpha$ | 0.100 | 0.050 | 0.025 | 0.010 | 0.005 |
|---|---|---|---|---|---|
| $1-\alpha$ | 0.900 | 0.950 | 0.975 | 0.990 | 0.995 |
| $\nu$ | $\chi^2_{0.900}$ | $\chi^2_{0.950}$ | $\chi^2_{0.975}$ | $\chi^2_{0.990}$ | $\chi^2_{0.995}$ |
| 1 | 2.7055 | 3.8415 | 5.0239 | 6.6349 | 7.8794 |
| 2 | 4.6052 | 5.9915 | 7.3778 | 9.2103 | 10.5966 |
| 3 | 6.2514 | 7.8147 | 9.3484 | 11.3449 | 12.8382 |
| 4 | 7.7794 | 9.4877 | 11.1433 | 13.2767 | 14.8603 |
| 5 | 9.2364 | 11.0705 | 12.8325 | 15.0863 | 16.7496 |
| 6 | 10.6446 | 12.5916 | 14.4494 | 16.8119 | 18.5476 |
| 7 | 12.0170 | 14.0671 | 16.0128 | 18.4753 | 20.2777 |
| 8 | 13.3616 | 15.5073 | 17.5345 | 20.0902 | 21.9550 |
| 9 | 14.6837 | 16.9190 | 19.0228 | 21.6660 | 23.5894 |
| 10 | 15.9872 | 18.3070 | 20.4832 | 23.2093 | 25.1882 |
| 11 | 17.2750 | 19.6751 | 21.9200 | 24.7250 | 26.7568 |
| 12 | 18.5493 | 21.0261 | 23.3367 | 26.2170 | 28.2995 |
| 13 | 19.8119 | 22.3620 | 24.7356 | 27.6882 | 29.8195 |
| 14 | 21.0641 | 23.6848 | 26.1189 | 29.1412 | 31.3193 |
| 15 | 22.3071 | 24.9958 | 27.4884 | 30.5779 | 32.8013 |
| 16 | 23.5418 | 26.2962 | 28.8454 | 31.9999 | 34.2672 |
| 17 | 24.7690 | 27.5871 | 30.1910 | 33.4087 | 35.7185 |
| 18 | 25.9894 | 28.8693 | 31.5264 | 34.8053 | 37.1565 |
| 19 | 27.2036 | 30.1435 | 32.8523 | 36.1909 | 38.5823 |
| 20 | 28.4120 | 31.4104 | 34.1696 | 37.5662 | 39.9968 |
| 21 | 29.6151 | 32.6706 | 35.4789 | 38.9322 | 41.4011 |
| 22 | 30.8133 | 33.9244 | 36.7807 | 40.2894 | 42.7957 |
| 23 | 32.0069 | 35.1725 | 38.0756 | 41.6384 | 44.1813 |
| 24 | 33.1962 | 36.4150 | 39.3641 | 42.9798 | 45.5585 |
| 25 | 34.3816 | 37.6525 | 40.6465 | 44.3141 | 46.9279 |
| 26 | 35.5632 | 38.8851 | 41.9232 | 45.6417 | 48.2899 |
| 27 | 36.7412 | 40.1133 | 43.1945 | 46.9629 | 49.6449 |
| 28 | 37.9159 | 41.3371 | 44.4608 | 48.2782 | 50.9934 |
| 29 | 39.0875 | 42.5570 | 45.7223 | 49.5879 | 52.3356 |
| 30 | 40.2560 | 43.7730 | 46.9792 | 50.8922 | 53.6720 |
| 40 | 51.8051 | 55.7585 | 59.3417 | 63.6907 | 66.7660 |
| 50 | 63.1671 | 67.5048 | 71.4202 | 76.1539 | 79.4900 |
| 60 | 74.3970 | 79.0819 | 83.2977 | 88.3794 | 91.9517 |
| 70 | 85.5270 | 90.5312 | 95.0232 | 100.4252 | 104.2149 |
| 80 | 96.5782 | 101.8795 | 106.6286 | 112.3288 | 116.3211 |
| 90 | 107.5650 | 113.1453 | 118.1359 | 124.1163 | 128.2989 |
| 100 | 118.4980 | 124.3421 | 129.5612 | 135.8067 | 140.1695 |

### Table 8.6  Critical Values $t_{\alpha,n}$ for the Student's t Distribution with $n$ Degrees of Freedom

$$\int_{t_{\alpha,n}}^{\infty} \frac{1}{\sqrt{n\pi}} \frac{\Gamma\left(\frac{n+1}{2}\right)}{\Gamma\left(\frac{n}{2}\right)} \left(1+\frac{x^2}{n}\right)^{-(n+1)/2} dx = \alpha = 1 - F(t_{\alpha,n})$$

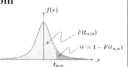

| n | $t_{0.1,n}$ | $t_{0.05,n}$ | $t_{0.025,n}$ | $t_{0.01,n}$ | $t_{0.005,n}$ | $t_{0.001}$ |
|---|---|---|---|---|---|---|
| 1 | 3.078 | 6.314 | 12.706 | 31.821 | 63.657 | 318.309 |
| 2 | 1.886 | 2.920 | 4.303 | 6.965 | 9.925 | 22.327 |
| 3 | 1.638 | 2.353 | 3.182 | 4.541 | 5.841 | 10.215 |
| 4 | 1.533 | 2.132 | 2.776 | 3.747 | 4.604 | 7.173 |
| 5 | 1.476 | 2.015 | 2.571 | 3.365 | 4.032 | 5.893 |
| 6 | 1.440 | 1.943 | 2.447 | 3.143 | 3.707 | 5.208 |
| 7 | 1.415 | 1.895 | 2.365 | 2.998 | 3.499 | 4.785 |
| 8 | 1.397 | 1.860 | 2.306 | 2.896 | 3.355 | 4.501 |
| 9 | 1.383 | 1.833 | 2.262 | 2.821 | 3.250 | 4.297 |
| 10 | 1.372 | 1.812 | 2.228 | 2.764 | 3.169 | 4.144 |
| 11 | 1.363 | 1.796 | 2.201 | 2.718 | 3.106 | 4.025 |
| 12 | 1.356 | 1.782 | 2.179 | 2.681 | 3.055 | 3.930 |
| 13 | 1.350 | 1.771 | 2.160 | 2.650 | 3.012 | 3.852 |
| 14 | 1.345 | 1.761 | 2.145 | 2.624 | 2.977 | 3.787 |
| 15 | 1.341 | 1.753 | 2.131 | 2.602 | 2.947 | 3.733 |
| 16 | 1.337 | 1.746 | 2.120 | 2.583 | 2.921 | 3.686 |
| 17 | 1.333 | 1.740 | 2.110 | 2.567 | 2.898 | 3.646 |
| 18 | 1.330 | 1.734 | 2.101 | 2.552 | 2.878 | 3.610 |
| 19 | 1.328 | 1.729 | 2.093 | 2.539 | 2.861 | 3.579 |
| 20 | 1.325 | 1.725 | 2.086 | 2.528 | 2.845 | 3.552 |
| 21 | 1.323 | 1.721 | 2.080 | 2.518 | 2.831 | 3.527 |
| 22 | 1.321 | 1.717 | 2.074 | 2.508 | 2.819 | 3.505 |
| 23 | 1.319 | 1.714 | 2.069 | 2.500 | 2.807 | 3.485 |
| 24 | 1.318 | 1.711 | 2.064 | 2.492 | 2.797 | 3.467 |
| 25 | 1.316 | 1.708 | 2.060 | 2.485 | 2.787 | 3.450 |
| 26 | 1.315 | 1.706 | 2.056 | 2.479 | 2.779 | 3.435 |
| 27 | 1.314 | 1.703 | 2.052 | 2.473 | 2.771 | 3.421 |
| 28 | 1.313 | 1.701 | 2.048 | 2.467 | 2.763 | 3.408 |
| 29 | 1.311 | 1.699 | 2.045 | 2.462 | 2.756 | 3.396 |
| 30 | 1.310 | 1.697 | 2.042 | 2.457 | 2.750 | 3.385 |
| 40 | 1.303 | 1.684 | 2.021 | 2.423 | 2.704 | 3.307 |
| 50 | 1.299 | 1.676 | 2.009 | 2.403 | 2.678 | 3.261 |
| 60 | 1.296 | 1.671 | 2.000 | 2.390 | 2.660 | 3.232 |
| 70 | 1.294 | 1.667 | 1.994 | 2.381 | 2.648 | 3.211 |
| 80 | 1.292 | 1.664 | 1.990 | 2.374 | 2.639 | 3.195 |
| 90 | 1.291 | 1.662 | 1.987 | 2.368 | 2.632 | 3.183 |
| 100 | 1.290 | 1.660 | 1.984 | 2.364 | 2.626 | 3.174 |
| ∞ | 1.282 | 1.645 | 1.960 | 2.326 | 2.576 | 3.098 |

## Table 8.7(a) Critical Values of the F-Distribution for $\alpha = 0.1$

$$\int_{F_{(\alpha,n,m)}}^{\infty} \frac{\Gamma\left(\frac{m+n}{2}\right)}{\Gamma\left(\frac{m}{2}\right)\Gamma\left(\frac{n}{2}\right)} n^{n/2} m^{m/2} \frac{x^{n/2-1}}{(m+nx)^{(m+n)/2}} dx = \alpha$$

$n$ is degrees of freedom for numerator and $m$ is degrees of freedom for denominator

| $m$\$n$ | 1 | 2 | 3 | 4 | 5 | 6 | 7 | 8 | 9 | 10 | 15 | 20 | 25 | 30 | 40 | 60 | 120 |
|---|---|---|---|---|---|---|---|---|---|---|---|---|---|---|---|---|---|
| 1 | 39.863 | 49.500 | 53.593 | 55.833 | 57.240 | 58.204 | 58.906 | 59.439 | 59.858 | 60.195 | 61.220 | 61.740 | 62.055 | 62.265 | 62.529 | 62.794 | 63.061 |
| 2 | 8.526 | 9.000 | 9.162 | 9.243 | 9.293 | 9.326 | 9.349 | 9.367 | 9.381 | 9.392 | 9.425 | 9.441 | 9.451 | 9.458 | 9.466 | 9.475 | 9.483 |
| 3 | 5.538 | 5.462 | 5.391 | 5.343 | 5.309 | 5.285 | 5.266 | 5.252 | 5.240 | 5.230 | 5.200 | 5.184 | 5.175 | 5.168 | 5.160 | 5.151 | 5.143 |
| 4 | 4.545 | 4.325 | 4.191 | 4.107 | 4.051 | 4.010 | 3.979 | 3.955 | 3.936 | 3.920 | 3.870 | 3.844 | 3.828 | 3.817 | 3.804 | 3.790 | 3.775 |
| 5 | 4.060 | 3.780 | 3.619 | 3.520 | 3.453 | 3.405 | 3.368 | 3.339 | 3.316 | 3.297 | 3.238 | 3.207 | 3.187 | 3.174 | 3.157 | 3.140 | 3.123 |
| 6 | 3.776 | 3.463 | 3.289 | 3.181 | 3.108 | 3.055 | 3.014 | 2.983 | 2.958 | 2.937 | 2.871 | 2.836 | 2.815 | 2.800 | 2.781 | 2.762 | 2.742 |
| 7 | 3.589 | 3.257 | 3.074 | 2.961 | 2.883 | 2.827 | 2.785 | 2.752 | 2.725 | 2.703 | 2.632 | 2.595 | 2.571 | 2.555 | 2.535 | 2.514 | 2.493 |
| 8 | 3.458 | 3.113 | 2.924 | 2.806 | 2.726 | 2.668 | 2.624 | 2.589 | 2.561 | 2.538 | 2.464 | 2.425 | 2.400 | 2.383 | 2.361 | 2.339 | 2.316 |
| 9 | 3.360 | 3.006 | 2.813 | 2.693 | 2.611 | 2.551 | 2.505 | 2.469 | 2.440 | 2.416 | 2.340 | 2.298 | 2.272 | 2.255 | 2.232 | 2.208 | 2.184 |
| 10 | 3.285 | 2.924 | 2.728 | 2.605 | 2.522 | 2.461 | 2.414 | 2.377 | 2.347 | 2.323 | 2.244 | 2.201 | 2.174 | 2.155 | 2.132 | 2.107 | 2.082 |
| 11 | 3.225 | 2.860 | 2.660 | 2.536 | 2.451 | 2.389 | 2.342 | 2.304 | 2.274 | 2.248 | 2.167 | 2.123 | 2.095 | 2.076 | 2.052 | 2.026 | 2.000 |
| 12 | 3.177 | 2.807 | 2.606 | 2.480 | 2.394 | 2.331 | 2.283 | 2.245 | 2.214 | 2.188 | 2.105 | 2.060 | 2.031 | 2.011 | 1.986 | 1.960 | 1.932 |
| 13 | 3.136 | 2.763 | 2.560 | 2.434 | 2.347 | 2.283 | 2.234 | 2.195 | 2.164 | 2.138 | 2.053 | 2.007 | 1.978 | 1.958 | 1.931 | 1.904 | 1.876 |
| 14 | 3.102 | 2.726 | 2.522 | 2.395 | 2.307 | 2.243 | 2.193 | 2.154 | 2.122 | 2.095 | 2.010 | 1.962 | 1.933 | 1.912 | 1.885 | 1.857 | 1.828 |
| 15 | 3.073 | 2.695 | 2.490 | 2.361 | 2.273 | 2.208 | 2.158 | 2.119 | 2.086 | 2.059 | 1.972 | 1.924 | 1.894 | 1.873 | 1.845 | 1.817 | 1.787 |
| 16 | 3.048 | 2.668 | 2.462 | 2.333 | 2.244 | 2.178 | 2.128 | 2.088 | 2.055 | 2.028 | 1.940 | 1.891 | 1.860 | 1.839 | 1.811 | 1.782 | 1.751 |
| 17 | 3.026 | 2.645 | 2.437 | 2.308 | 2.218 | 2.152 | 2.102 | 2.061 | 2.028 | 2.001 | 1.912 | 1.862 | 1.831 | 1.809 | 1.781 | 1.751 | 1.719 |
| 18 | 3.007 | 2.624 | 2.416 | 2.286 | 2.196 | 2.130 | 2.079 | 2.038 | 2.005 | 1.977 | 1.887 | 1.837 | 1.805 | 1.783 | 1.754 | 1.723 | 1.691 |
| 19 | 2.990 | 2.606 | 2.397 | 2.266 | 2.176 | 2.109 | 2.058 | 2.017 | 1.984 | 1.956 | 1.865 | 1.814 | 1.782 | 1.759 | 1.730 | 1.699 | 1.666 |
| 20 | 2.975 | 2.589 | 2.380 | 2.249 | 2.158 | 2.091 | 2.040 | 1.999 | 1.965 | 1.937 | 1.845 | 1.794 | 1.761 | 1.738 | 1.708 | 1.677 | 1.643 |
| 21 | 2.961 | 2.575 | 2.365 | 2.233 | 2.142 | 2.075 | 2.023 | 1.982 | 1.948 | 1.920 | 1.827 | 1.776 | 1.742 | 1.719 | 1.689 | 1.657 | 1.623 |
| 22 | 2.949 | 2.561 | 2.351 | 2.219 | 2.128 | 2.060 | 2.008 | 1.967 | 1.933 | 1.904 | 1.811 | 1.759 | 1.726 | 1.702 | 1.671 | 1.639 | 1.604 |
| 23 | 2.937 | 2.549 | 2.339 | 2.207 | 2.115 | 2.047 | 1.995 | 1.953 | 1.919 | 1.890 | 1.796 | 1.744 | 1.710 | 1.686 | 1.655 | 1.622 | 1.587 |
| 24 | 2.927 | 2.538 | 2.327 | 2.195 | 2.103 | 2.035 | 1.983 | 1.941 | 1.906 | 1.877 | 1.783 | 1.730 | 1.696 | 1.672 | 1.641 | 1.607 | 1.571 |
| 25 | 2.918 | 2.528 | 2.317 | 2.184 | 2.092 | 2.024 | 1.971 | 1.929 | 1.895 | 1.866 | 1.771 | 1.718 | 1.683 | 1.659 | 1.627 | 1.593 | 1.557 |
| 30 | 2.881 | 2.489 | 2.276 | 2.142 | 2.049 | 1.980 | 1.927 | 1.884 | 1.849 | 1.819 | 1.722 | 1.667 | 1.632 | 1.606 | 1.573 | 1.538 | 1.499 |
| 40 | 2.835 | 2.440 | 2.226 | 2.091 | 1.997 | 1.927 | 1.873 | 1.829 | 1.793 | 1.763 | 1.662 | 1.605 | 1.568 | 1.541 | 1.506 | 1.467 | 1.425 |
| 50 | 2.809 | 2.412 | 2.197 | 2.061 | 1.966 | 1.895 | 1.840 | 1.796 | 1.760 | 1.729 | 1.627 | 1.568 | 1.529 | 1.502 | 1.465 | 1.424 | 1.379 |
| 60 | 2.791 | 2.393 | 2.177 | 2.041 | 1.946 | 1.875 | 1.819 | 1.775 | 1.738 | 1.707 | 1.603 | 1.543 | 1.504 | 1.476 | 1.437 | 1.395 | 1.348 |
| 100 | 2.756 | 2.356 | 2.139 | 2.002 | 1.906 | 1.834 | 1.778 | 1.732 | 1.695 | 1.663 | 1.557 | 1.494 | 1.453 | 1.423 | 1.382 | 1.336 | 1.282 |
| 120 | 2.748 | 2.347 | 2.130 | 1.992 | 1.896 | 1.824 | 1.767 | 1.722 | 1.684 | 1.652 | 1.545 | 1.482 | 1.440 | 1.409 | 1.368 | 1.320 | 1.265 |

**Table 8.7(b) Critical Values of the F-Distribution for $\alpha = 0.05$**

$$\int_{F_{(\alpha,n,m)}}^{\infty} \frac{\Gamma\left(\frac{m+n}{2}\right)}{\Gamma\left(\frac{m}{2}\right)\Gamma\left(\frac{n}{2}\right)} n^{n/2} m^{m/2} \frac{x^{n/2-1} dx}{(m+nx)^{(m+n)/2}} = \alpha$$

$n$ is degrees of freedom for numerator and $m$ is degrees of freedom for denominator

| $m$ \ $n$ | 1 | 2 | 3 | 4 | 5 | 6 | 7 | 8 | 9 | 10 | 15 | 20 | 25 | 30 | 40 | 60 | 120 |
|---|---|---|---|---|---|---|---|---|---|---|---|---|---|---|---|---|---|
| 1 | 161.446 | 199.500 | 215.707 | 224.583 | 230.162 | 233.986 | 236.768 | 238.883 | 240.543 | 241.882 | 245.950 | 248.013 | 249.260 | 250.095 | 251.143 | 252.196 | 253.253 |
| 2 | 18.513 | 19.000 | 19.164 | 19.247 | 19.296 | 19.330 | 19.353 | 19.371 | 19.385 | 19.396 | 19.429 | 19.446 | 19.456 | 19.462 | 19.471 | 19.479 | 19.487 |
| 3 | 10.128 | 9.552 | 9.277 | 9.117 | 9.013 | 8.941 | 8.887 | 8.845 | 8.812 | 8.786 | 8.703 | 8.660 | 8.634 | 8.617 | 8.594 | 8.572 | 8.549 |
| 4 | 7.709 | 6.944 | 6.591 | 6.388 | 6.256 | 6.163 | 6.094 | 6.041 | 5.999 | 5.964 | 5.858 | 5.803 | 5.769 | 5.746 | 5.717 | 5.688 | 5.658 |
| 5 | 6.608 | 5.786 | 5.409 | 5.192 | 5.050 | 4.950 | 4.876 | 4.818 | 4.772 | 4.735 | 4.619 | 4.558 | 4.521 | 4.496 | 4.464 | 4.431 | 4.398 |
| 6 | 5.987 | 5.143 | 4.757 | 4.534 | 4.387 | 4.284 | 4.207 | 4.147 | 4.099 | 4.060 | 3.938 | 3.874 | 3.835 | 3.808 | 3.774 | 3.740 | 3.705 |
| 7 | 5.591 | 4.737 | 4.347 | 4.120 | 3.972 | 3.866 | 3.787 | 3.726 | 3.677 | 3.637 | 3.511 | 3.445 | 3.404 | 3.376 | 3.340 | 3.304 | 3.267 |
| 8 | 5.318 | 4.459 | 4.066 | 3.838 | 3.687 | 3.581 | 3.500 | 3.438 | 3.388 | 3.347 | 3.218 | 3.150 | 3.108 | 3.079 | 3.043 | 3.005 | 2.967 |
| 9 | 5.117 | 4.256 | 3.863 | 3.633 | 3.482 | 3.374 | 3.293 | 3.230 | 3.179 | 3.137 | 3.006 | 2.936 | 2.893 | 2.864 | 2.826 | 2.787 | 2.748 |
| 10 | 4.965 | 4.103 | 3.708 | 3.478 | 3.326 | 3.217 | 3.135 | 3.072 | 3.020 | 2.978 | 2.845 | 2.774 | 2.730 | 2.700 | 2.661 | 2.621 | 2.580 |
| 11 | 4.844 | 3.982 | 3.587 | 3.357 | 3.204 | 3.095 | 3.012 | 2.948 | 2.896 | 2.854 | 2.719 | 2.646 | 2.601 | 2.570 | 2.531 | 2.490 | 2.448 |
| 12 | 4.747 | 3.885 | 3.490 | 3.259 | 3.106 | 2.996 | 2.913 | 2.849 | 2.796 | 2.753 | 2.617 | 2.544 | 2.498 | 2.466 | 2.426 | 2.384 | 2.341 |
| 13 | 4.667 | 3.806 | 3.411 | 3.179 | 3.025 | 2.915 | 2.832 | 2.767 | 2.714 | 2.671 | 2.533 | 2.459 | 2.412 | 2.380 | 2.339 | 2.297 | 2.252 |
| 14 | 4.600 | 3.739 | 3.344 | 3.112 | 2.958 | 2.848 | 2.764 | 2.699 | 2.646 | 2.602 | 2.463 | 2.388 | 2.341 | 2.308 | 2.266 | 2.223 | 2.178 |
| 15 | 4.543 | 3.682 | 3.287 | 3.056 | 2.901 | 2.790 | 2.707 | 2.641 | 2.588 | 2.544 | 2.403 | 2.328 | 2.280 | 2.247 | 2.204 | 2.160 | 2.114 |
| 16 | 4.494 | 3.634 | 3.239 | 3.007 | 2.852 | 2.741 | 2.657 | 2.591 | 2.538 | 2.494 | 2.352 | 2.276 | 2.227 | 2.194 | 2.151 | 2.106 | 2.059 |
| 17 | 4.451 | 3.592 | 3.197 | 2.965 | 2.810 | 2.699 | 2.614 | 2.548 | 2.494 | 2.450 | 2.308 | 2.230 | 2.181 | 2.148 | 2.104 | 2.058 | 2.011 |
| 18 | 4.414 | 3.555 | 3.160 | 2.928 | 2.773 | 2.661 | 2.577 | 2.510 | 2.456 | 2.412 | 2.269 | 2.191 | 2.141 | 2.107 | 2.063 | 2.017 | 1.968 |
| 19 | 4.381 | 3.522 | 3.127 | 2.895 | 2.740 | 2.628 | 2.544 | 2.477 | 2.423 | 2.378 | 2.234 | 2.155 | 2.106 | 2.071 | 2.026 | 1.980 | 1.930 |
| 20 | 4.351 | 3.493 | 3.098 | 2.866 | 2.711 | 2.599 | 2.514 | 2.447 | 2.393 | 2.348 | 2.203 | 2.124 | 2.074 | 2.039 | 1.994 | 1.946 | 1.896 |
| 21 | 4.325 | 3.467 | 3.072 | 2.840 | 2.685 | 2.573 | 2.488 | 2.420 | 2.366 | 2.321 | 2.176 | 2.096 | 2.045 | 2.010 | 1.965 | 1.916 | 1.866 |
| 22 | 4.301 | 3.443 | 3.049 | 2.817 | 2.661 | 2.549 | 2.464 | 2.397 | 2.342 | 2.297 | 2.151 | 2.071 | 2.020 | 1.984 | 1.938 | 1.889 | 1.838 |
| 23 | 4.279 | 3.422 | 3.028 | 2.796 | 2.640 | 2.528 | 2.442 | 2.375 | 2.320 | 2.275 | 2.128 | 2.048 | 1.996 | 1.961 | 1.914 | 1.865 | 1.813 |
| 24 | 4.260 | 3.403 | 3.009 | 2.776 | 2.621 | 2.508 | 2.423 | 2.355 | 2.300 | 2.255 | 2.108 | 2.027 | 1.975 | 1.939 | 1.892 | 1.842 | 1.790 |
| 25 | 4.242 | 3.385 | 2.991 | 2.759 | 2.603 | 2.490 | 2.405 | 2.337 | 2.282 | 2.236 | 2.089 | 2.007 | 1.955 | 1.919 | 1.872 | 1.822 | 1.768 |
| 30 | 4.171 | 3.316 | 2.922 | 2.690 | 2.534 | 2.421 | 2.334 | 2.266 | 2.211 | 2.165 | 2.015 | 1.932 | 1.878 | 1.841 | 1.792 | 1.740 | 1.683 |
| 40 | 4.085 | 3.232 | 2.839 | 2.606 | 2.449 | 2.336 | 2.249 | 2.180 | 2.124 | 2.077 | 1.924 | 1.839 | 1.783 | 1.744 | 1.693 | 1.637 | 1.577 |
| 50 | 4.034 | 3.183 | 2.790 | 2.557 | 2.400 | 2.286 | 2.199 | 2.130 | 2.073 | 2.026 | 1.871 | 1.784 | 1.727 | 1.687 | 1.634 | 1.576 | 1.511 |
| 60 | 4.001 | 3.150 | 2.758 | 2.525 | 2.368 | 2.254 | 2.167 | 2.097 | 2.040 | 1.993 | 1.836 | 1.748 | 1.690 | 1.649 | 1.594 | 1.534 | 1.467 |
| 100 | 3.936 | 3.087 | 2.696 | 2.463 | 2.305 | 2.191 | 2.103 | 2.032 | 1.975 | 1.927 | 1.768 | 1.676 | 1.616 | 1.573 | 1.515 | 1.450 | 1.376 |
| 120 | 3.920 | 3.072 | 2.680 | 2.447 | 2.290 | 2.175 | 2.087 | 2.016 | 1.959 | 1.910 | 1.750 | 1.659 | 1.598 | 1.554 | 1.495 | 1.429 | 1.352 |

425

Table 8.7(c) Critical Values of the F-Distribution for $\alpha = 0.025$

$$\int_{F_{(\alpha,n,m)}}^{\infty} \frac{\Gamma\left(\frac{m+n}{2}\right)}{\Gamma\left(\frac{m}{2}\right)\Gamma\left(\frac{n}{2}\right)} n^{n/2} m^{m/2} \frac{x^{n/2-1} dx}{(m+nx)^{(m+n)/2}} = \alpha$$

$n$ is degrees of freedom for numerator and $m$ is degrees of freedom for denominator

| $n$ \ $m$ | 1 | 2 | 3 | 4 | 5 | 6 | 7 | 8 | 9 | 10 | 15 | 20 | 25 | 30 | 40 | 60 | 120 |
|---|---|---|---|---|---|---|---|---|---|---|---|---|---|---|---|---|---|
| 1 | 647.789 | 799.500 | 864.163 | 899.583 | 921.848 | 937.111 | 948.217 | 956.656 | 963.285 | 968.627 | 984.867 | 993.103 | 998.081 | 1001.414 | 1005.598 | 1009.800 | 1014.020 |
| 2 | 38.506 | 39.000 | 39.165 | 39.248 | 39.298 | 39.331 | 39.355 | 39.373 | 39.387 | 39.398 | 39.431 | 39.448 | 39.458 | 39.465 | 39.473 | 39.481 | 39.490 |
| 3 | 17.443 | 16.044 | 15.439 | 15.101 | 14.885 | 14.735 | 14.624 | 14.540 | 14.473 | 14.419 | 14.253 | 14.167 | 14.115 | 14.081 | 14.037 | 13.992 | 13.947 |
| 4 | 12.218 | 10.649 | 9.979 | 9.605 | 9.364 | 9.197 | 9.074 | 8.980 | 8.905 | 8.844 | 8.657 | 8.560 | 8.501 | 8.461 | 8.411 | 8.360 | 8.309 |
| 5 | 10.007 | 8.434 | 7.764 | 7.388 | 7.146 | 6.978 | 6.853 | 6.757 | 6.681 | 6.619 | 6.428 | 6.329 | 6.268 | 6.227 | 6.175 | 6.123 | 6.069 |
| 6 | 8.813 | 7.260 | 6.599 | 6.227 | 5.988 | 5.820 | 5.695 | 5.600 | 5.523 | 5.461 | 5.269 | 5.168 | 5.107 | 5.065 | 5.012 | 4.959 | 4.904 |
| 7 | 8.073 | 6.542 | 5.890 | 5.523 | 5.285 | 5.119 | 4.995 | 4.899 | 4.823 | 4.761 | 4.568 | 4.467 | 4.405 | 4.362 | 4.309 | 4.254 | 4.199 |
| 8 | 7.571 | 6.059 | 5.416 | 5.053 | 4.817 | 4.652 | 4.529 | 4.433 | 4.357 | 4.295 | 4.101 | 3.999 | 3.937 | 3.894 | 3.840 | 3.784 | 3.728 |
| 9 | 7.209 | 5.715 | 5.078 | 4.718 | 4.484 | 4.320 | 4.197 | 4.102 | 4.026 | 3.964 | 3.769 | 3.667 | 3.604 | 3.560 | 3.505 | 3.449 | 3.392 |
| 10 | 6.937 | 5.456 | 4.826 | 4.468 | 4.236 | 4.072 | 3.950 | 3.855 | 3.779 | 3.717 | 3.522 | 3.419 | 3.355 | 3.311 | 3.255 | 3.198 | 3.140 |
| 11 | 6.724 | 5.256 | 4.630 | 4.275 | 4.044 | 3.881 | 3.759 | 3.664 | 3.588 | 3.526 | 3.330 | 3.226 | 3.162 | 3.118 | 3.061 | 3.004 | 2.944 |
| 12 | 6.554 | 5.096 | 4.474 | 4.121 | 3.891 | 3.728 | 3.607 | 3.512 | 3.436 | 3.374 | 3.177 | 3.073 | 3.008 | 2.963 | 2.906 | 2.848 | 2.787 |
| 13 | 6.414 | 4.965 | 4.347 | 3.996 | 3.767 | 3.604 | 3.483 | 3.388 | 3.312 | 3.250 | 3.053 | 2.948 | 2.882 | 2.837 | 2.780 | 2.720 | 2.659 |
| 14 | 6.298 | 4.857 | 4.242 | 3.892 | 3.663 | 3.501 | 3.380 | 3.285 | 3.209 | 3.147 | 2.949 | 2.844 | 2.778 | 2.732 | 2.674 | 2.614 | 2.552 |
| 15 | 6.200 | 4.765 | 4.153 | 3.804 | 3.576 | 3.415 | 3.293 | 3.199 | 3.123 | 3.060 | 2.862 | 2.756 | 2.689 | 2.644 | 2.585 | 2.524 | 2.461 |
| 16 | 6.115 | 4.687 | 4.077 | 3.729 | 3.502 | 3.341 | 3.219 | 3.125 | 3.049 | 2.986 | 2.788 | 2.681 | 2.614 | 2.568 | 2.509 | 2.447 | 2.383 |
| 17 | 6.042 | 4.619 | 4.011 | 3.665 | 3.438 | 3.277 | 3.156 | 3.061 | 2.985 | 2.922 | 2.723 | 2.616 | 2.548 | 2.502 | 2.442 | 2.380 | 2.315 |
| 18 | 5.978 | 4.560 | 3.954 | 3.608 | 3.382 | 3.221 | 3.100 | 3.005 | 2.929 | 2.866 | 2.667 | 2.559 | 2.491 | 2.445 | 2.384 | 2.321 | 2.256 |
| 19 | 5.922 | 4.508 | 3.903 | 3.559 | 3.333 | 3.172 | 3.051 | 2.956 | 2.880 | 2.817 | 2.617 | 2.509 | 2.441 | 2.394 | 2.333 | 2.270 | 2.203 |
| 20 | 5.871 | 4.461 | 3.859 | 3.515 | 3.289 | 3.128 | 3.007 | 2.913 | 2.837 | 2.774 | 2.573 | 2.464 | 2.396 | 2.349 | 2.287 | 2.223 | 2.156 |
| 21 | 5.827 | 4.420 | 3.819 | 3.475 | 3.250 | 3.090 | 2.969 | 2.874 | 2.798 | 2.735 | 2.534 | 2.425 | 2.356 | 2.308 | 2.246 | 2.182 | 2.114 |
| 22 | 5.786 | 4.383 | 3.783 | 3.440 | 3.215 | 3.055 | 2.934 | 2.839 | 2.763 | 2.700 | 2.498 | 2.389 | 2.320 | 2.272 | 2.210 | 2.145 | 2.076 |
| 23 | 5.750 | 4.349 | 3.750 | 3.408 | 3.183 | 3.023 | 2.902 | 2.808 | 2.731 | 2.668 | 2.466 | 2.357 | 2.287 | 2.239 | 2.176 | 2.111 | 2.041 |
| 24 | 5.717 | 4.319 | 3.721 | 3.379 | 3.155 | 2.995 | 2.874 | 2.779 | 2.703 | 2.640 | 2.437 | 2.327 | 2.257 | 2.209 | 2.146 | 2.080 | 2.010 |
| 25 | 5.686 | 4.291 | 3.694 | 3.353 | 3.129 | 2.969 | 2.848 | 2.753 | 2.677 | 2.613 | 2.411 | 2.300 | 2.230 | 2.182 | 2.118 | 2.052 | 1.981 |
| 30 | 5.568 | 4.182 | 3.589 | 3.250 | 3.026 | 2.867 | 2.746 | 2.651 | 2.575 | 2.511 | 2.307 | 2.195 | 2.124 | 2.074 | 2.009 | 1.940 | 1.866 |
| 40 | 5.424 | 4.051 | 3.463 | 3.126 | 2.904 | 2.744 | 2.624 | 2.529 | 2.452 | 2.388 | 2.182 | 2.068 | 1.994 | 1.943 | 1.875 | 1.803 | 1.724 |
| 50 | 5.340 | 3.975 | 3.390 | 3.054 | 2.833 | 2.674 | 2.553 | 2.458 | 2.381 | 2.317 | 2.109 | 1.993 | 1.919 | 1.866 | 1.796 | 1.721 | 1.639 |
| 60 | 5.286 | 3.925 | 3.343 | 3.008 | 2.786 | 2.627 | 2.507 | 2.412 | 2.334 | 2.270 | 2.061 | 1.944 | 1.869 | 1.815 | 1.744 | 1.667 | 1.581 |
| 100 | 5.179 | 3.828 | 3.250 | 2.917 | 2.696 | 2.537 | 2.417 | 2.321 | 2.244 | 2.179 | 1.968 | 1.849 | 1.770 | 1.715 | 1.640 | 1.558 | 1.463 |
| 120 | 5.152 | 3.805 | 3.227 | 2.894 | 2.674 | 2.515 | 2.395 | 2.299 | 2.222 | 2.157 | 1.945 | 1.825 | 1.746 | 1.690 | 1.614 | 1.530 | 1.433 |

## Table 8.7(d) Critical Values of the $F$-Distribution for $\alpha = 0.01$

$$\int_{F_{(\alpha,m,n)}}^{\infty} \frac{\Gamma\left(\frac{m+n}{2}\right)}{\Gamma\left(\frac{m}{2}\right)\Gamma\left(\frac{n}{2}\right)} n^{n/2} m^{m/2} \frac{x^{n/2-1}}{(m+nx)^{(m+n)/2}} dx = \alpha$$

$m$ is degrees of freedom for numerator and $n$ is degrees of freedom for denominator

| $n$\\$m$ | 1 | 2 | 3 | 4 | 5 | 6 | 7 | 8 | 9 | 10 | 15 | 20 | 25 | 30 | 40 | 60 | 120 |
|---|---|---|---|---|---|---|---|---|---|---|---|---|---|---|---|---|---|
| 1 | 4052.181 | 4999.500 | 5403.352 | 5624.583 | 5763.650 | 5858.986 | 5928.356 | 5981.070 | 6022.473 | 6055.847 | 6157.285 | 6208.730 | 6239.825 | 6260.649 | 6286.782 | 6313.030 | 6339.391 |
| 2 | 98.503 | 99.000 | 99.166 | 99.249 | 99.299 | 99.333 | 99.356 | 99.374 | 99.388 | 99.399 | 99.433 | 99.449 | 99.459 | 99.466 | 99.474 | 99.482 | 99.491 |
| 3 | 34.116 | 30.817 | 29.457 | 28.710 | 28.237 | 27.911 | 27.672 | 27.489 | 27.345 | 27.229 | 26.872 | 26.690 | 26.579 | 26.505 | 26.411 | 26.316 | 26.221 |
| 4 | 21.198 | 18.000 | 16.694 | 15.977 | 15.522 | 15.207 | 14.976 | 14.799 | 14.659 | 14.546 | 14.198 | 14.020 | 13.911 | 13.838 | 13.745 | 13.652 | 13.558 |
| 5 | 16.258 | 13.274 | 12.060 | 11.392 | 10.967 | 10.672 | 10.456 | 10.289 | 10.158 | 10.051 | 9.722 | 9.553 | 9.449 | 9.379 | 9.291 | 9.202 | 9.112 |
| 6 | 13.745 | 10.925 | 9.780 | 9.148 | 8.746 | 8.466 | 8.260 | 8.102 | 7.976 | 7.874 | 7.559 | 7.396 | 7.296 | 7.229 | 7.143 | 7.057 | 6.969 |
| 7 | 12.246 | 9.547 | 8.451 | 7.847 | 7.460 | 7.191 | 6.993 | 6.840 | 6.719 | 6.620 | 6.314 | 6.155 | 6.058 | 5.992 | 5.908 | 5.824 | 5.737 |
| 8 | 11.259 | 8.649 | 7.591 | 7.006 | 6.632 | 6.371 | 6.178 | 6.029 | 5.911 | 5.814 | 5.515 | 5.359 | 5.263 | 5.198 | 5.116 | 5.032 | 4.946 |
| 9 | 10.561 | 8.022 | 6.992 | 6.422 | 6.057 | 5.802 | 5.613 | 5.467 | 5.351 | 5.257 | 4.962 | 4.808 | 4.713 | 4.649 | 4.567 | 4.483 | 4.398 |
| 10 | 10.044 | 7.559 | 6.552 | 5.994 | 5.636 | 5.386 | 5.200 | 5.057 | 4.942 | 4.849 | 4.558 | 4.405 | 4.311 | 4.247 | 4.165 | 4.082 | 3.996 |
| 11 | 9.646 | 7.206 | 6.217 | 5.668 | 5.316 | 5.069 | 4.886 | 4.744 | 4.632 | 4.539 | 4.251 | 4.099 | 4.005 | 3.941 | 3.860 | 3.776 | 3.690 |
| 12 | 9.330 | 6.927 | 5.953 | 5.412 | 5.064 | 4.821 | 4.640 | 4.499 | 4.388 | 4.296 | 4.010 | 3.858 | 3.765 | 3.701 | 3.619 | 3.535 | 3.449 |
| 13 | 9.074 | 6.701 | 5.739 | 5.205 | 4.862 | 4.620 | 4.441 | 4.302 | 4.191 | 4.100 | 3.815 | 3.665 | 3.571 | 3.507 | 3.425 | 3.341 | 3.255 |
| 14 | 8.862 | 6.515 | 5.564 | 5.035 | 4.695 | 4.456 | 4.278 | 4.140 | 4.030 | 3.939 | 3.656 | 3.505 | 3.412 | 3.348 | 3.266 | 3.181 | 3.094 |
| 15 | 8.683 | 6.359 | 5.417 | 4.893 | 4.556 | 4.318 | 4.142 | 4.004 | 3.895 | 3.805 | 3.522 | 3.372 | 3.278 | 3.214 | 3.132 | 3.047 | 2.959 |
| 16 | 8.531 | 6.226 | 5.292 | 4.773 | 4.437 | 4.202 | 4.026 | 3.890 | 3.780 | 3.691 | 3.409 | 3.259 | 3.165 | 3.101 | 3.018 | 2.933 | 2.845 |
| 17 | 8.400 | 6.112 | 5.185 | 4.669 | 4.336 | 4.102 | 3.927 | 3.791 | 3.682 | 3.593 | 3.312 | 3.162 | 3.068 | 3.003 | 2.920 | 2.835 | 2.746 |
| 18 | 8.285 | 6.013 | 5.092 | 4.579 | 4.248 | 4.015 | 3.841 | 3.705 | 3.597 | 3.508 | 3.227 | 3.077 | 2.983 | 2.919 | 2.835 | 2.749 | 2.660 |
| 19 | 8.185 | 5.926 | 5.010 | 4.500 | 4.171 | 3.939 | 3.765 | 3.631 | 3.523 | 3.434 | 3.153 | 3.003 | 2.909 | 2.844 | 2.761 | 2.674 | 2.584 |
| 20 | 8.096 | 5.849 | 4.938 | 4.431 | 4.103 | 3.871 | 3.699 | 3.564 | 3.457 | 3.368 | 3.088 | 2.938 | 2.843 | 2.778 | 2.695 | 2.608 | 2.517 |
| 21 | 8.017 | 5.780 | 4.874 | 4.369 | 4.042 | 3.812 | 3.640 | 3.506 | 3.398 | 3.310 | 3.030 | 2.880 | 2.785 | 2.720 | 2.636 | 2.548 | 2.457 |
| 22 | 7.945 | 5.719 | 4.817 | 4.313 | 3.988 | 3.758 | 3.587 | 3.453 | 3.346 | 3.258 | 2.978 | 2.827 | 2.733 | 2.667 | 2.583 | 2.495 | 2.403 |
| 23 | 7.881 | 5.664 | 4.765 | 4.264 | 3.939 | 3.710 | 3.539 | 3.406 | 3.299 | 3.211 | 2.931 | 2.781 | 2.686 | 2.620 | 2.535 | 2.447 | 2.354 |
| 24 | 7.823 | 5.614 | 4.718 | 4.218 | 3.895 | 3.667 | 3.496 | 3.363 | 3.256 | 3.168 | 2.889 | 2.738 | 2.643 | 2.577 | 2.492 | 2.403 | 2.310 |
| 25 | 7.770 | 5.568 | 4.675 | 4.177 | 3.855 | 3.627 | 3.457 | 3.324 | 3.217 | 3.129 | 2.850 | 2.699 | 2.604 | 2.538 | 2.453 | 2.364 | 2.270 |
| 30 | 7.562 | 5.390 | 4.510 | 4.018 | 3.699 | 3.473 | 3.304 | 3.173 | 3.067 | 2.979 | 2.700 | 2.549 | 2.453 | 2.386 | 2.299 | 2.208 | 2.111 |
| 40 | 7.314 | 5.179 | 4.313 | 3.828 | 3.514 | 3.291 | 3.124 | 2.993 | 2.888 | 2.801 | 2.522 | 2.369 | 2.271 | 2.203 | 2.114 | 2.019 | 1.917 |
| 50 | 7.171 | 5.057 | 4.199 | 3.720 | 3.408 | 3.186 | 3.020 | 2.890 | 2.785 | 2.698 | 2.419 | 2.265 | 2.167 | 2.098 | 2.007 | 1.909 | 1.803 |
| 60 | 7.077 | 4.977 | 4.126 | 3.649 | 3.339 | 3.119 | 2.953 | 2.823 | 2.718 | 2.632 | 2.352 | 2.198 | 2.098 | 2.028 | 1.936 | 1.836 | 1.726 |
| 100 | 6.895 | 4.824 | 3.984 | 3.513 | 3.206 | 2.988 | 2.823 | 2.694 | 2.590 | 2.503 | 2.223 | 2.067 | 1.965 | 1.893 | 1.797 | 1.692 | 1.572 |
| 120 | 6.851 | 4.787 | 3.949 | 3.480 | 3.174 | 2.956 | 2.792 | 2.663 | 2.559 | 2.472 | 2.192 | 2.035 | 1.932 | 1.860 | 1.763 | 1.656 | 1.533 |

427

Table 8.7(e) Critical Values of the F-Distribution for α = 0.005

$$\int_{F_{(\alpha,n,m)}}^{\infty} \frac{\Gamma\left(\frac{m+n}{2}\right)}{\Gamma\left(\frac{m}{2}\right)\Gamma\left(\frac{n}{2}\right)} n^{n/2} m^{m/2} \frac{x^{n/2-1}\,dx}{(m+nx)^{(m+n)/2}} = \alpha$$

n is degrees of freedom for numerator and m is degrees of freedom for denominator

| m\n | 1 | 2 | 3 | 4 | 5 | 6 | 7 | 8 | 9 | 10 | 15 | 20 | 25 | 30 | 40 | 60 | 120 |
|---|---|---|---|---|---|---|---|---|---|---|---|---|---|---|---|---|---|
| 1 | 16210.723 | 19999.500 | 21614.741 | 22499.583 | 23055.798 | 23437.111 | 23714.566 | 23925.406 | 24091.004 | 24224.487 | 24630.205 | 24835.971 | 24960.540 | 25043.625 | 25148.153 | 25253.137 | 25358.573 |
| 2 | 198.501 | 199.000 | 199.166 | 199.250 | 199.300 | 199.333 | 199.357 | 199.375 | 199.388 | 199.400 | 199.433 | 199.450 | 199.460 | 199.466 | 199.475 | 199.483 | 199.491 |
| 3 | 55.552 | 49.799 | 47.467 | 46.195 | 45.392 | 44.838 | 44.434 | 44.126 | 43.882 | 43.686 | 43.085 | 42.778 | 42.591 | 42.466 | 42.308 | 42.149 | 41.989 |
| 4 | 31.333 | 26.284 | 24.259 | 23.155 | 22.456 | 21.975 | 21.622 | 21.352 | 21.139 | 20.967 | 20.438 | 20.167 | 20.002 | 19.892 | 19.752 | 19.611 | 19.468 |
| 5 | 22.785 | 18.314 | 16.530 | 15.556 | 14.940 | 14.513 | 14.200 | 13.961 | 13.772 | 13.618 | 13.146 | 12.903 | 12.755 | 12.656 | 12.530 | 12.402 | 12.274 |
| 6 | 18.635 | 14.544 | 12.917 | 12.028 | 11.464 | 11.073 | 10.786 | 10.566 | 10.391 | 10.250 | 9.814 | 9.589 | 9.451 | 9.358 | 9.241 | 9.122 | 9.001 |
| 7 | 16.236 | 12.404 | 10.882 | 10.050 | 9.522 | 9.155 | 8.885 | 8.678 | 8.514 | 8.380 | 7.968 | 7.754 | 7.623 | 7.534 | 7.422 | 7.309 | 7.193 |
| 8 | 14.688 | 11.042 | 9.596 | 8.805 | 8.302 | 7.952 | 7.694 | 7.496 | 7.339 | 7.211 | 6.814 | 6.608 | 6.482 | 6.396 | 6.288 | 6.177 | 6.065 |
| 9 | 13.614 | 10.107 | 8.717 | 7.956 | 7.471 | 7.134 | 6.885 | 6.693 | 6.541 | 6.417 | 6.032 | 5.832 | 5.708 | 5.625 | 5.519 | 5.410 | 5.300 |
| 10 | 12.826 | 9.427 | 8.081 | 7.343 | 6.872 | 6.545 | 6.302 | 6.116 | 5.968 | 5.847 | 5.471 | 5.274 | 5.153 | 5.071 | 4.966 | 4.859 | 4.750 |
| 11 | 12.226 | 8.912 | 7.600 | 6.881 | 6.422 | 6.102 | 5.865 | 5.682 | 5.537 | 5.418 | 5.049 | 4.855 | 4.736 | 4.654 | 4.551 | 4.445 | 4.337 |
| 12 | 11.754 | 8.510 | 7.226 | 6.521 | 6.071 | 5.757 | 5.525 | 5.345 | 5.202 | 5.085 | 4.721 | 4.530 | 4.412 | 4.331 | 4.228 | 4.123 | 4.015 |
| 13 | 11.374 | 8.186 | 6.926 | 6.233 | 5.791 | 5.482 | 5.253 | 5.076 | 4.935 | 4.820 | 4.460 | 4.270 | 4.153 | 4.073 | 3.970 | 3.866 | 3.758 |
| 14 | 11.060 | 7.922 | 6.680 | 5.998 | 5.562 | 5.257 | 5.031 | 4.857 | 4.717 | 4.603 | 4.247 | 4.059 | 3.942 | 3.862 | 3.760 | 3.655 | 3.547 |
| 15 | 10.798 | 7.701 | 6.476 | 5.803 | 5.372 | 5.071 | 4.847 | 4.674 | 4.536 | 4.424 | 4.070 | 3.883 | 3.766 | 3.687 | 3.585 | 3.480 | 3.372 |
| 16 | 10.575 | 7.514 | 6.303 | 5.638 | 5.212 | 4.913 | 4.692 | 4.521 | 4.384 | 4.272 | 3.920 | 3.734 | 3.618 | 3.539 | 3.437 | 3.332 | 3.224 |
| 17 | 10.384 | 7.354 | 6.156 | 5.497 | 5.075 | 4.779 | 4.559 | 4.389 | 4.254 | 4.142 | 3.793 | 3.607 | 3.492 | 3.412 | 3.311 | 3.206 | 3.097 |
| 18 | 10.218 | 7.215 | 6.028 | 5.375 | 4.956 | 4.663 | 4.445 | 4.276 | 4.141 | 4.030 | 3.683 | 3.498 | 3.382 | 3.303 | 3.201 | 3.096 | 2.987 |
| 19 | 10.073 | 7.093 | 5.916 | 5.268 | 4.853 | 4.561 | 4.345 | 4.177 | 4.043 | 3.933 | 3.587 | 3.402 | 3.287 | 3.208 | 3.106 | 3.000 | 2.891 |
| 20 | 9.944 | 6.986 | 5.818 | 5.174 | 4.762 | 4.472 | 4.257 | 4.090 | 3.956 | 3.847 | 3.502 | 3.318 | 3.203 | 3.123 | 3.022 | 2.916 | 2.806 |
| 21 | 9.830 | 6.891 | 5.730 | 5.091 | 4.681 | 4.393 | 4.179 | 4.013 | 3.880 | 3.771 | 3.427 | 3.243 | 3.128 | 3.049 | 2.947 | 2.841 | 2.730 |
| 22 | 9.727 | 6.806 | 5.652 | 5.017 | 4.609 | 4.322 | 4.109 | 3.944 | 3.812 | 3.703 | 3.360 | 3.176 | 3.061 | 2.982 | 2.880 | 2.774 | 2.663 |
| 23 | 9.635 | 6.730 | 5.582 | 4.950 | 4.544 | 4.259 | 4.047 | 3.882 | 3.750 | 3.642 | 3.300 | 3.116 | 3.001 | 2.922 | 2.820 | 2.713 | 2.602 |
| 24 | 9.551 | 6.661 | 5.519 | 4.890 | 4.486 | 4.202 | 3.991 | 3.826 | 3.695 | 3.587 | 3.246 | 3.062 | 2.947 | 2.868 | 2.765 | 2.658 | 2.546 |
| 25 | 9.475 | 6.598 | 5.462 | 4.835 | 4.433 | 4.150 | 3.939 | 3.776 | 3.645 | 3.537 | 3.196 | 3.013 | 2.898 | 2.819 | 2.716 | 2.609 | 2.496 |
| 30 | 9.180 | 6.355 | 5.239 | 4.623 | 4.228 | 3.949 | 3.742 | 3.580 | 3.450 | 3.344 | 3.006 | 2.823 | 2.708 | 2.628 | 2.524 | 2.415 | 2.300 |
| 40 | 8.828 | 6.066 | 4.976 | 4.374 | 3.986 | 3.713 | 3.509 | 3.350 | 3.222 | 3.117 | 2.781 | 2.598 | 2.482 | 2.401 | 2.296 | 2.184 | 2.064 |
| 50 | 8.626 | 5.902 | 4.826 | 4.232 | 3.849 | 3.579 | 3.376 | 3.219 | 3.092 | 2.988 | 2.653 | 2.470 | 2.353 | 2.272 | 2.164 | 2.050 | 1.925 |
| 60 | 8.495 | 5.795 | 4.729 | 4.140 | 3.760 | 3.492 | 3.291 | 3.134 | 3.008 | 2.904 | 2.570 | 2.387 | 2.270 | 2.187 | 2.079 | 1.962 | 1.834 |
| 100 | 8.241 | 5.589 | 4.542 | 3.963 | 3.589 | 3.325 | 3.127 | 2.972 | 2.847 | 2.744 | 2.411 | 2.227 | 2.108 | 2.024 | 1.912 | 1.790 | 1.652 |
| 120 | 8.179 | 5.539 | 4.497 | 3.921 | 3.548 | 3.285 | 3.087 | 2.933 | 2.808 | 2.705 | 2.373 | 2.188 | 2.069 | 1.984 | 1.871 | 1.747 | 1.606 |

# Chapter 9
# Selected Applied Mathematics Topics

The following is a collection of miscellaneous topics from mathematics, physics and chemistry. Some topics represent fundamental concepts and are presented as review subject matter. Other topics represent applications of undergraduate mathematics applied to problems from the sciences and engineering. Students are encouraged to seek out additional references and applications from their libraries and to pursue more advance information concerning topics of interest. This chapter is a very small sampling from the wide variety of applications involving the use of mathematics.

## Motion of a Particle (Dynamics)

Dynamics is the study of the motion of particles and rigid bodies. In studying the motion of a single particle, Newton's laws of motion are employed.

**Newton's First Law:** If the resultant external forces acting on a body is zero, then the body moves with a constant velocity.

**Newton's Second Law:** The time rate of change of linear momentum is equal to the sum of the external forces acting on a body. This can be expressed $\vec{F} = \frac{d}{dt}(m\vec{V})$ where $m\vec{V}$ is the linear momentum of the particle.

**Newton's Third Law:** When two bodies interact, the force of the first body on the second is equal to the force of the second body on the first but oppositely directed.

Newton's second law of motion is used to model many dynamics problems and leads to a variety of differential equations. Introducing the symbols $s$ for distance, $v$ for velocity, $a$ for acceleration and $t$ for time, the velocity can be represented as the time rate of change of distance and $v = \frac{ds}{dt} = \dot{s}$. The acceleration can be represented as the time rate of change of velocity and can be represented in any of the forms

$$a = \dot{v} = \frac{dv}{dt} = \frac{d^2s}{dt^2} = \ddot{s}. \tag{9.1}$$

Each of these forms assume that the velocity and the distance can be expressed as functions of time in a form $v = v(t)$ and $s = s(t)$. If the velocity $v$ is a function of the distance $s$, then $v = v(s)$ and consequently the acceleration can be represented by using the chain rule from calculus

$$a = \frac{dv}{dt} = \frac{dv}{ds}\frac{ds}{dt} = v\frac{dv}{ds} \tag{9.2}$$

Newton's second law of motion states that the time rate of change of linear momentum is equal to the sum of the forces acting on the body. The motion is along a straight line in the direction of the resultant force $\vec{F}$ and the linear momentum is given by $m\vec{v}$, where $m$ is the mass and $\vec{v}$ is the velocity. The mathematical model for straight line motion of an object with a constant mass can be represented in one of the forms

$$F = ma = mv\frac{dv}{ds} = m\frac{dv}{dt} = m\frac{d^2s}{dt^2}.$$

Here $[F]$ is pounds (lb), $[a]$ is feet per second squared (ft/sec²), $[m]$ is slugs (lb/ft/sec²) or any other equivalent set of consistent units may be used such as given in figure 9-1.

| System | Force | Mass | Acceleration | Approximate acceleration of gravity |
|---|---|---|---|---|
| MKS | 1 Newton = 1 Kg · meter/sec² | Kg | meter/sec² | 9.8 meter/sec² |
| FPS | 1 lb = 1 slug · ft/sec² | slug | ft/sec² | 32 ft/sec² |
| CGS | 1 dyne = gram · cm/sec² | gram | cm/sec² | 980 cm/sec² |

1 Kg-force = 1000 grams = 9.807 Newtons = 2.505 lbs = $9.807(10)^5$ dynes

1 meter = 100 centimeters = 39.37 inches = 3.281 feet

**Figure 9-1.** Systems of units for Newton's second law.

### Example 9-1.

A cannon ball of mass $m$ is fired from a cannon with an initial velocity $v_0$ inclined at an angle $\theta$ with the horizontal as illustrated. Neglect air resistance and find the equations of motion, maximum height, and range of the cannon ball.

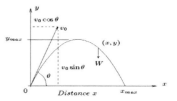

**Solution:** Let $y = y(t)$ denote the vertical height at any time $t$ and let $x = x(t)$ denote the horizontal distance at any time $t$. Consider the cannon ball at a position $(x,y)$ and examine the forces acting on it. In the $y$-direction the force due to the weight of the cannon ball is $W = mg$, $(g = 32.2 \, \text{ft/sec}^2)$. The equation of motion in the $y$-direction is represented as

$$m\frac{d^2y}{dt^2} = -W = -mg. \tag{9.3}$$

Forces in the $x$-direction like air resistance are neglected. Newton's second law can then be expressed

$$m\frac{d^2x}{dt^2} = 0. \tag{9.4}$$

Equations (9.3) and (9.4) are solved subject to the initial conditions:

$$x(0) = 0, \qquad y(0) = 0$$
$$\dot{x}(0) = v_0 \cos\theta, \qquad \dot{y}(0) = v_0 \sin\theta,$$

where $v_0$ is the initial speed and $\theta$ is the angle of inclination of the cannon. Solving the differential equations (9.3) and (9.4) by successive integrations gives

$$\dot{y} = -gt + c_1, \qquad \dot{x} = c_3 \qquad (3.15)$$

$$y = -g\frac{t^2}{2} + c_1 t + c_2, \qquad x = c_3 t + c_4. \qquad (3.16)$$

where $c_1, c_2, c_3, c_4$ are constants of integration. The solution satisfying the initial conditions can be expressed as

$$y = y(t) = -\frac{g}{2}t^2 + (v_0 \sin\theta)\, t$$
$$x = x(t) = (v_0 \cos\theta)\, t. \qquad (9.5)$$

These are parametric equations describing the position of the cannon ball. They can be written in the vector form as

$$\vec{r} = \vec{r}(t) = x(t)\,\hat{e}_1 + y(t)\,\hat{e}_2 = (v_0 \cos\theta)\, t\, \hat{e}_1 + \left[-\frac{g}{2}t^2 + (v_0 \sin\theta)\, t\right]\hat{e}_2. \qquad (9.6)$$

to describe the position of the cannon ball as a function of time. The maximum height occurs where the derivative $\frac{dy}{dt}$ is zero, and the maximum range occurs when the height $y$ returns to zero at some time $t > 0$. The velocity is zero when $t$ has the value $t_1 = v_0 \sin\theta/g$, and at this time,

$$y_{max} = y(t_1) = \frac{v_0^2 \sin^2\theta}{2g}, \qquad x = x(t_1) = \frac{v_0^2 \sin 2\theta}{2g} \qquad (9.7)$$

The maximum range occurs when $y = 0$ at time $t_2 = 2\frac{v_0 \sin\theta}{g}$, and at this time,

$$x_{max} = x(t_2) = \frac{v_0^2 \sin 2\theta}{g}.$$

Eliminating $t$ from the parametric equations (9.5), demonstrates that the trajectory of the cannon ball is a parabola.

■

**Example 9-2.** A weight is catapulted vertically into the air with a large initial velocity denoted by $v_0$. Neglect air resistance and describe the motion of the weight.

**Solution:** Let $t$ denote time and let $y = y(t)$ represent the displacement above the Earths surface. Neglecting air resistance, the only force acting on the object is its own weight, which acts downward, so that

$$m\frac{d^2y}{dt^2} = -w = -mg \quad \text{or} \quad \frac{d^2y}{dt^2} = -g. \qquad (9.8)$$

Integrating this equation twice produces

$$\frac{dy}{dt} = -gt + c_1, \qquad y = -\frac{g}{2}t^2 + c_1 t + c_2$$

where $c_1$ and $c_2$ are constants of integration. Applying the initial conditions $y(0) = 0$, and $\dot{y}(0) = v_0$, gives the equation describing the position of the weight as a function of time for $y \geq 0$,

$$y = y(t) = -\frac{g}{2}t^2 + v_0 t. \tag{9.9}$$

This models the dynamics of a particle near the surface of the earth.

∎

**Example 9-3.** Solve the previous example using Newton's law of gravitation, where the acceleration of a particle is inversely proportional to the square of its distance from the center of the Earth.

**Solution:** This example differs from the previous one in the assumption made concerning the physical law which governs the phenomenon. In general, the model in example 9-2 is valid for objects near the Earths surface while the model in this example is more accurate for objects far from the Earth. Both models neglect the effect of air resistance on the motion of the object.

Let $r = r(t)$ denote the distance from the center of the Earth with $R$ denoting the radius of the Earth. The results of this example can then by related to the previous example by letting $y = y(t) = r(t) - R$ denote the distance above the Earth. Newton's law of gravitation is written

$$\frac{d^2 r}{dt^2} = \frac{k}{r^2}, \tag{9.10}$$

where $k$ is a proportionality constant. Using the condition that when $r = R$, $\frac{d^2 r}{dt^2} = -g$, the acceleration of gravity, equation (9.10) gives

$$-g = \frac{k}{R^2} \quad \text{or} \quad k = -gR^2, \tag{9.11}$$

and the differential equation of motion becomes

$$\frac{d^2 r}{dt^2} = -\frac{gR^2}{r^2} \quad \text{or} \quad v\frac{dv}{dr} = -\frac{gR^2}{r^2}, \tag{9.12}$$

where $\frac{dr}{dt} = v$, $\frac{d^2 r}{dt^2} = \frac{dv}{dr}\frac{dr}{dt} = v\frac{dv}{dr}$. Separate the variables in equation (9.12) and integrate both sides of the equation, using the initial conditions, $t = 0$, $r = R$, and $\frac{dr}{dt} = v_0$, to obtain

$$v^2 = \left(\frac{dr}{dt}\right)^2 = \frac{2gR^2}{r} + v_0^2 - 2gR. \tag{9.13}$$

Assume $v_0^2 < 2gR$ in equation (9.13), and write the equation in the form

$$\sqrt{r}\,dr = \sqrt{2gR^2 + (v_0^2 - 2gR)r}\,dt. \tag{9.14}$$

Separate the variables in equation (9.14) and then integrate both sides of the equation using the initial condition $t = 0, r = R$ to obtain the solution

$$\sqrt{R(A^2 - R)} - \sqrt{r(A^2 - r)} + A^2 \left[\arcsin(\frac{\sqrt{r}}{A}) - \arcsin(\frac{\sqrt{R}}{A})\right] = \sqrt{(2gR - v_0^2)}\, t, \qquad (9.15)$$

where $A^2 = \frac{2gR^2}{2gR - v_0^2}$ is a constant. This solution is much more complicated than the solution to example 9-2. In order for equation (9.13) to give real answers, it is required that the square of the velocity remain positive. Note that for large values of the variable $r$ the term involving $1/r$ in equation (9.13) can be neglected, then it is necessary that $v_0^2 \geq 2gR$, if $v^2$ is to remain positive. The value $v_e = \sqrt{2gR}$ is called the escape velocity and represents the velocity necessary to escape the gravitational pull of the Earth. Our assumption $v_0^2 < 2gR$ produced the solution equation (9.15) which is only valid for a certain interval of time.

Make the substitution $r = y + R$, with $y \geq 0$, in equation (9.15) to compare the solution to that obtained in the example 9-2. The two solutions can be calculated and plotted using numerical methods and a digital computer. You will find that the two solutions are close to one another. This comparison study is left as an exercise. ∎

## Kepler's Laws

Johannes Kepler,[1] an astronomer and mathematician, discovered three laws concerning the motion of the planets. He discovered these laws from an analysis of experimental data without the aid of calculus. Newton, using calculus, verified these laws with the model for the inverse square law of attraction. Let $m$ denote the mass of a planet and $M$ denote the mass of the Sun and use the Newton law of gravitation, $\vec{F} = -\frac{GmM}{r^2}\hat{e}_r = m\frac{d^2\vec{r}}{dt^2} = m\frac{d\vec{V}}{dt}$. Take the cross product of this equation with $\vec{r}$ and show one obtains

$$\frac{d}{dt}\left(\vec{r} \times \frac{d\vec{r}}{dt}\right) = -\frac{GM}{r^2}\vec{r} \times \hat{e}_r = \vec{0} \qquad (9.16)$$

since $\vec{r}$ and $\hat{e}_r$ have the same direction. An integration of this equation produces the result

$$\vec{r} \times \frac{d\vec{r}}{dt} = \vec{h} = \text{Constant} \qquad (9.17)$$

The quantity $\vec{H} = \vec{r} \times m\vec{V}$ is defined as the angular momentum so that the quantity $\vec{h} = \vec{r} \times \frac{d\vec{r}}{dt}$ appearing in equation (9.17) is called the angular momentum per unit mass. Equation (9.17) states that the angular momentum per unit mass is a constant for the two-body system under consideration. Since $\vec{h}$ is a constant vector, it can be verified that

$$\begin{aligned}\frac{d}{dt}\left(\vec{V} \times \vec{h}\right) &= \frac{d\vec{V}}{dt} \times \vec{h} = -\frac{GM}{r^2}\hat{e}_r \times \left(\vec{r} \times \frac{d\vec{r}}{dt}\right) = -\frac{GM}{r^2}\hat{e}_r \times \left[r\hat{e}_r \times \left(r\frac{d\hat{e}_r}{dt} + \frac{dr}{dt}\hat{e}_r\right)\right] \\ &= -GM\,\hat{e}_r \times \left(\hat{e}_r \times \frac{d\hat{e}_r}{dt}\right) = GM\frac{d\hat{e}_r}{dt}.\end{aligned} \qquad (9.18)$$

---

[1] Johannes Kepler (1571–1630), German astronomer and mathematician.

An integration of equation (9.18) gives

$$\vec{V} \times \vec{h} = GM\hat{e}_r + \vec{C}, \qquad (9.19)$$

where $\vec{C}$ is a constant vector of integration. Using the triple scalar product formula it is readily verified that

$$\vec{r} \cdot (\vec{V} \times \vec{h}) = \vec{h} \cdot \left(\vec{r} \times \frac{d\vec{r}}{dt}\right) = h^2 = \vec{r} \cdot (\vec{V} \times \vec{h}) = GM\vec{r} \cdot \hat{e}_r + \vec{r} \cdot \vec{C}$$

or

$$h^2 = GMr + Cr\cos\theta, \qquad (9.21)$$

where $\theta$ is the angle between the vectors $\vec{C}$ and $\vec{r}$. Solve for $r$ from equation (9.21) and find

$$r = \frac{p}{1 + \epsilon \cos\theta}, \qquad (9.21)$$

where $p = h^2/GM$ and $\epsilon = C/GM$. This result is known as Kepler's first law and implies that all the planets of the solar system describe elliptical paths with the sun at one focus.

Kepler's second law states that the position vector $\vec{r}$ sweeps out equal areas in equal time intervals. Consider the area swept out by the position vector of a planet during a time interval $\Delta t$. This element of area, in polar coordinates, is written as

$$dA = \frac{1}{2}r^2 d\theta$$

and therefore the rate of change of this area with respect to time is

$$\frac{dA}{dt} = \frac{1}{2}r^2 \frac{d\theta}{dt}.$$

If the position vector is given by $\vec{r} = r\cos\theta\,\hat{e}_1 + r\sin\theta\,\hat{e}_2$, then the angular momentum per unit mass has components which can be calculated from the determinant

$$\vec{h} = \vec{r} \times \frac{d\vec{r}}{dt} = \begin{vmatrix} \hat{e}_1 & \hat{e}_2 & \hat{e}_3 \\ r\cos\theta & r\sin\theta & 0 \\ -r\sin\theta\,\dot\theta + \dot r\cos\theta & r\cos\theta\,\dot\theta + \dot r\sin\theta & 0 \end{vmatrix}$$

Expanding the above determinant and simplifying gives

$$\vec{h} = r^2 \frac{d\theta}{dt}\hat{e}_3 = h\,\hat{e}_3 \qquad (9.22)$$

If $h$ is a constant, then

$$\frac{dA}{dt} = \frac{1}{2}r^2 \frac{d\theta}{dt} = \frac{1}{2}h = \text{Constant}. \qquad (9.23)$$

This result is known as Kepler's second law. Analysis of the second law informs us that the position vector sweeps out equal areas during equal time intervals.

The time it takes for mass $m$ to complete one orbit about mass $M$ is called the period of the motion. Denote this period by the Greek letter $\tau$. Note that equation (9.23) implies that when $r^2$ is small $\frac{d\theta}{dt}$ becomes large and, conversely, when $\frac{d\theta}{dt}$ is small $r^2$ becomes large. The resulting motion is for planets to move faster when they are closer to the Sun and slower when they are farther away. Integration of equation (9.23) from $t = 0$ to $t = \tau$, produces

$$A = \frac{h}{2}\tau, \tag{9.24}$$

where $A$ is the area of the ellipse and $\tau$ is the period of one orbit. From calculus, the area of an ellipse is given by the formula $A = \pi ab$, where $a$ is the semi-major axis and $b = a\sqrt{1-\epsilon^2}$ is the semi-minor axis. Equation (9.24) can therefore be expressed in the form

$$A = \pi a^2 \sqrt{1-\epsilon^2} = \frac{h}{2}\tau$$

from which the period of the orbit is $\tau = \frac{2\pi a^2}{h}\sqrt{1-\epsilon^2}$. Use the substitutions $p = \frac{h^2}{GM}$ and $1-\epsilon^2 = \frac{p}{a}$ to show the period of the orbit can be expressed

$$\tau = \frac{2\pi a^{3/2}}{\sqrt{GM}} \quad \text{or} \quad \tau^2 = \frac{4\pi^2 a^3}{GM}. \tag{9.25}$$

This result is known as Kepler's third law and depicts the fact that the square of the period of one revolution is proportional to the cube of the semi-major axis of the elliptical orbit.

## Moment of a Force

To calculate the moment of a force $\vec{F}$, of magnitude $F$, with respect to a line $L_1$ construct a line $L_2$ containing the force vector. This line is called the line of action of the force. The line $L_1$ about which the force acts is called the moment axis.

If $\ell$ is the perpendicular distance between the two lines $L_1$ and $L_2$, then the moment of the force about the moment axis has magnitude $M$ of force times perpendicular distance or $M = F\ell$. This moment is a vector quantity as it produces either a clockwise (-) or counterclockwise (+) rotation about the moment axis. The above description can be visualize by considering a plane containing the force $\vec{F}$ which is perpendicular to the line $L_1$ as illustrated in the accompanying figure.

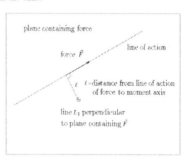

The moment of $\vec{F}$ about the moment axis is usually represented as a vector along the moment axis with arrowhead of the moment vector pointing in the direction of a right-hand screw.

The moment of a force with respect to a line or axis is a measure of the forces tendency to produce a rotation about the line. Let the force be resolved into components parallel and perpendicular to the coordinate axes having the form

$$\vec{F} = F_1\,\hat{e}_1 + F_2\,\hat{e}_2 + F_3\,\hat{e}_3.$$

The component of the force which is parallel to an axis has no tendency to produce a rotation about that axis. For a chosen axis, the moment about that axis is the product of the force component times the perpendicular distance of the force from the axis. By using the right-hand screw rule, assign a negative sign to the moment if it acts clockwise and a positive sign to the moment if it acts counterclockwise. The moment of a force about a point 0 is a vector quantity given by $\vec{M}_0 = \vec{r} \times \vec{F}$, where $\vec{r}$ is a position vector from point 0 to some point along the line of action of $\vec{F}$. The moment of a force $F$, acting at a point $\vec{r} = x\,\hat{e}_1 + y\,\hat{e}_2 + z\,\hat{e}_3$, about the $x$-, $y$- and $z$-axes is calculated in the figure 9-2(a).

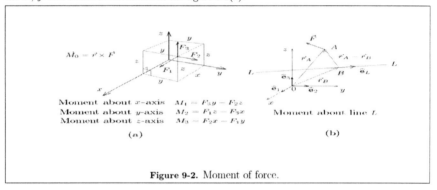

**Figure 9-2.** Moment of force.

(a) For the moment about the $x$-axis:

$F_1$ component parallel to $x$-axis does not produce a moment

(Force)($\perp$ distance) $= +F_3 y$ (Counterclockwise rotation)

(Force)($\perp$ distance) $= -F_2 z$ (Clockwise rotation)

The total moment about the $x$-axis is therefore $\quad M_1 = F_3 y - F_2 z$.

(b) For the moment about the $y$-axis:

(Force)($\perp$ distance) $= +F_1 z$ (Counterclockwise rotation)

$F_2$ component parallel to the $y$-axis does not produce a moment

(Force)($\perp$ distance) $= -F_3 x$ (Clockwise rotation)

The total moment about the $y$-axis is therefore $\quad M_2 = F_1 z - F_3 x$.

(c) For the moment about the $z$-axis:

(Force)($\perp$ distance) $= -F_1 y$ (Clockwise rotation)

(Force)($\perp$ distance) $= +F_2 x$ (Counterclockwise rotation)

$F_3$ component parallel to the $z$-axis does not produce a moment

The total moment about the $z$-axis is given by $\quad M_3 = F_2 x - F_1 y$.

The total moment about the origin is a vector quantity and is the vector sum of the above moments. The moment about the origin is given by

$$\vec{M}_0 = M_1\,\hat{e}_1 + M_2\,\hat{e}_2 + M_3\,\hat{e}_3 = (F_3y - F_2z)\,\hat{e}_1 + (F_1z - F_3x)\,\hat{e}_2 + (F_2x - F_1y)\,\hat{e}_3 \qquad (9.26)$$

This moment about the origin can be written as the cross product

$$\vec{M}_0 = \vec{r} \times \vec{F} = \begin{vmatrix} \hat{e}_1 & \hat{e}_2 & \hat{e}_3 \\ x & y & z \\ F_1 & F_2 & F_3 \end{vmatrix} \qquad (9.27)$$

To find the moment of a force $\vec{F}$ about a line $L$, choose any point on the line and find the moment about this point. The moment about the line $L$ is the projection of the vector moment $\vec{M}$ on this line. Changing the above notation, let $\vec{r}_A$ denote the position vector of the origin of force $F$ and denote by $\vec{r}_B$ the position vector of a point $B$ on the line $L$. This situation is illustrated in figure 9-2(b)

The moment about the point $B$ in figure 9-2(b) is given by

$$\vec{M}_B = (\vec{r}_A - \vec{r}_B) \times \vec{F}$$

(i.e., the position vector of the force with respect to the point $B$ crossed with the force). The moment about the line $L$ is then the projection of this moment onto the line $L$ and is given by

$$\vec{M}_B \cdot \hat{e}_L = \text{Projection of moment vector on line } L,$$

where $\hat{e}_L$ is a unit vector in the direction of the line $L$. The direction of the unit vector on the line $L$ is arbitrary. However, once $\hat{e}_L$ has been chosen one must be careful to analyze the dot product $\vec{M}_B \cdot \hat{e}_L$ as its algebraic sign determines the rotation sense produced by the moment (i.e., clockwise or counterclockwise).

## Center of Mass for System of Particles

In one-dimension, if masses $m_1$ and $m_2$ are located at positions $x_1$ and $x_2$ on the $x$-axis, then the center of mass of the system is that point $\bar{x}$ where the sum of the moments about the point $\bar{x}$ is zero or

$$+m_1(\bar{x} - x_1) - m_2(x_2 - \bar{x}) = 0$$

Solving for $\bar{x}$ gives the center of mass

$$\bar{x} = \frac{m_1 x_2 + m_2 x_2}{m_1 + m_2} \qquad (9.28)$$

If there were $n$-masses $m_1, m_2, \ldots, m_n$ at positions $x_1, x_2, \ldots, x_n$, then a summation of the moments about the center of mass $\bar{x}$ produces the result that

$$\bar{x} = \frac{m_1 x_1 + m_2 x_2 + \cdots + m_n x_n}{m_1 + m_2 + \cdots + m_n} = \frac{\sum_{j=1}^n m_j x_j}{\sum_{j=1}^n m_j} \qquad (9.29)$$

In two-dimensions, consider a system of masses $m_1, m_2, \ldots, m_n$ located at positions $(x_1, y_1), (x_2, y_2), \ldots, (x_n, y_n)$ and then sum moments about the center of mass $(\bar{x}, \bar{y})$ of the system to obtain the relations

$$\bar{x} = \frac{\sum_{j=1}^n m_j x_j}{\sum_{j=1}^n m_j}, \qquad \bar{y} = \frac{\sum_{j=1}^n m_j y_j}{\sum_{j=1}^n m_j} \qquad (9.30)$$

This is just a repeat of the one-dimensional problem in both the $x$ and $y$ directions.

Similarly, in three-dimensions, consider a system of particles $m_1, m_2, \ldots, m_n$ having position vectors $\vec{r}_j = x_j \hat{e}_1 + y_j \hat{e}_2 + z_j \hat{e}_3$ for $j = 1, \ldots, n$. If $\vec{r}_c$ is the vector denoting the position of the center of mass, then the center of mass is given by

$$\vec{r}_c = \frac{\sum_{j=1}^n m_j \vec{r}_j}{\sum_{j=1}^n m_j} \qquad (9.31)$$

Note that this is just a repeat of the one-dimensional problem placed in a vector form. The equations (9.30) are a special case of equation (9.31) when all the $z$-values are zero. The equation (9.31) can also be generalized using vectors in an $n$-dimensional space.

### Center of Mass for Plane Areas

The center of mass of a plane area or solid is also found by taking moments. An $n$th moment of an element of area, element of volume or element of mass involves the perpendicular distance $\ell$ to an axis, where $\ell$ is raised to the power $n$ and then multiplied by the element of area, volume or mass. A first moment of area is $dM = \ell\, dA$, a second moment of area is $dI_{\ell\ell} = \ell^2\, dA$, a third moment of area is $dI = \ell^3\, dA$, etc.

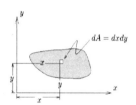

For a plane area, the summation of first moments of an element of area $dA$ about the $x$ and $y$-axes are given by the integrals

$$M_x = \int y\, dA, \qquad \text{and} \qquad M_y = \int x\, dA$$

The centroid is then the point $(\bar{x}, \bar{y})$ such that moment equations

$$M_x = A\bar{y}, \qquad M_y = A\bar{x} \qquad (9.32)$$

are satisfied. This implies that the centroid is a point where all the area $A$ can be concentrated to produce the same moments obtained by integration. Solving the equations (9.32) for $\bar{x}$ and $\bar{y}$ gives

$$\bar{x} = \frac{M_y}{A} = \frac{\int x\, dA}{\int dA}, \qquad \text{and} \qquad \bar{y} = \frac{M_x}{A} = \frac{\int y\, dA}{\int dA} \qquad (9.33)$$

The following are some common shapes where the centroid is well known.

### The triangle

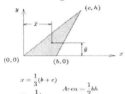

$$\bar{x} = \frac{1}{3}(b+c)$$
$$\bar{y} = \frac{1}{3}h \qquad Area = \frac{1}{2}bh$$

### The quadrant of a circle

$$\bar{x} = \frac{4R}{3\pi}$$
$$\bar{y} = \frac{4R}{3\pi} \qquad Area = \frac{1}{4}\pi R^2$$

### The quadrant of an ellipse

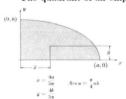

$$\bar{x} = \frac{4a}{3\pi}$$
$$\bar{y} = \frac{4b}{3\pi} \qquad Area = \frac{\pi}{4}ab$$

### The quadrant of a parabola

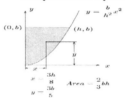

$$y = \frac{b}{h^2}x^2$$

$$\bar{x} = \frac{3h}{8}$$
$$\bar{y} = \frac{3b}{5} \qquad Area = \frac{2}{3}bh$$

### Right circular cone

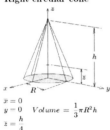

$$\bar{x} = 0$$
$$\bar{y} = 0 \qquad Volume = \frac{1}{3}\pi R^2 h$$
$$\bar{z} = \frac{h}{4}$$

### Hemisphere

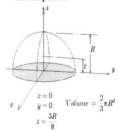

$$\bar{x} = 0$$
$$\bar{y} = 0 \qquad Volume = \frac{2}{3}\pi R^3$$
$$\bar{z} = \frac{3R}{8}$$

### Parallel-axis theorem

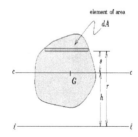

The parallel-axis theorem for areas assumes that the second moment of area $I_{cc}$ with respect to an axis $c$ through the centroid of the area is known and it is desired to find the second moment of area with respect to an axis $\ell$ parallel to $c$. The second moment of area about the axis $\ell$ is given by

$$dI_{\ell\ell} = r^2\, dA = (s+h)^2\, dA, \qquad r \text{ is } \perp \text{ distance of } dA \text{ from axis}$$
$$dI_{\ell\ell} = s^2\, dA + 2sh\, dA + h^2\, dA$$
$$I_{\ell\ell} = \int s^2\, dA + 2\int sh\, dA + \int h^2\, dA, \qquad I_{cc} = \int s^2\, dA, \text{ h is constant}$$
$$I_{\ell\ell} = I_{cc} + h^2 A$$

If axis $\ell$ is a distance $h$ from the axis $c$, then $I_{\ell\ell} = I_{cc} + Ah^2$ where $A$ is the area. Note that the integral $\int sh\, dA$ is zero because the axis $c$ passes through the centroid.

## Centroids and Volumes

Let $\rho(x,y,z)$ denote a density function and $d\tau = dx\,dy\,dz$ element of volume within a solid of mass $m$, then $dm = \rho\, d\tau$ is element of mass and the total mass of the solid is obtained a triple integral over the volume occupied by the solid. This written

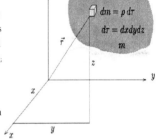

$$\text{total mass} = m = \iiint_V \rho(x,y,z)\, d\tau$$

The distance of the element of mass $dm$ from the coordin planes $yz$, $xz$, and $xy$ are respectively, $x, y, z$.

The quantities $M_{yz}, M_{xz}, M_{xy}$ defined by the triple integrals

$$M_{yz} = \iiint_V x\rho(x,y,z)\, d\tau, \quad M_{xz} = \iiint_V y\rho(x,y,z)\, d\tau, \quad M_{xy} = \iiint_V z\rho(x,y,z)\, d\tau$$

are called moments about the three coordinate planes and the center of mass of the solid is located at the point $(\bar{x}, \bar{y}, \bar{z})$ determined from the relations

$$\bar{x} = \frac{M_{yz}}{m}, \qquad \bar{y} = \frac{M_{xz}}{m}, \qquad \bar{z} = \frac{M_{xy}}{m}, \qquad m = \iiint_V \rho(x,y,z)\, d\tau \qquad (9.34)$$

## Second Moments or Moments of Inertia

The moment of inertia $I$, associated with a body of mass $m$, about an axis is defined as the scalar quantity $I = m\ell^2$ where $\ell$ is the perpendicular distance of the mass $m$ from the axis or rotation.

The second moments of inertia are related to the distribution of mass of a rigid body relative to an orthogonal reference frame and are sometimes called mass inertia components of the rigid body with mass $m$ about the reference axes $xyz$. In three dimensions these quantities are defined

$$I_{xx} = \iiint_V (y^2 + z^2)\rho\, d\tau \qquad I_{xy} = \iiint_V xy\rho\, d\tau$$
$$I_{yy} = \iiint_V (x^2 + z^2)\rho\, d\tau \qquad I_{xz} = \iiint_V xz\rho\, d\tau \qquad (9.35)$$
$$I_{zz} = \iiint_V (x^2 + y^2)\rho\, d\tau \qquad I_{yz} = \iiint_V yz\rho\, d\tau$$

where $\rho$ is the density of the volume element $d\tau$. The quantities $I_{xx}, I_{yy}, I_{zz}$ are called the mass moments of inertia about the $x,y$ and $z$ axes respectively. The terms $I_{xy} = I_{yx}$, $I_{xz} = I_{zx}$ and $I_{yz} = I_{zy}$ are called the products of inertia. Note that the mass moments of inertia are triple integrals of the element of mass $dm$ multiplied by a distance squared from an axes and so are some times referred to as second moments of inertia. The inertia matrix

$$I = \begin{pmatrix} I_{xx} & -I_{xy} & -I_{xz} \\ -I_{yz} & I_{yy} & -I_{yz} \\ -I_{zx} & -I_{zy} & I_{zz} \end{pmatrix}$$

is a quantity that often arises in the study of elasticity and the dynamics of rigid body motion. By a rotation of the $(x, y, z)$ axes to a set $(\bar{x}, \bar{y}, \bar{z})$ of three mutually perpendicular axes, which causes the inertia matrix to become diagonalize and take on the form

$$\bar{I} = \begin{pmatrix} I_{\bar{x}\bar{x}} & 0 & 0 \\ 0 & I_{\bar{y}\bar{y}} & 0 \\ 0 & 0 & I_{\bar{z}\bar{z}} \end{pmatrix}$$

then the axes $(\bar{x}, \bar{y}, \bar{z})$ are called principal axes.

The parallel-axis theorem for masses is similar to the parallel-axis theorem for areas. Just replace area by mass. The moment of inertia or second moment with respect to any axis equals the moment of inertia with respect to a parallel axis through the mass center plus the product of the mass times the square of the distance between the two axes.

**Angular Velocity**

A rigid body is one where any two distinct points remain a constant distance apart for all time. A rigid body in motion can be studied by considering both translational and rotational motion of the points within the body. Assume there is no translational motion but only rotational motion of the rigid body. A simple rotation of every point in the rigid body, about a line through the body, can be described by (a) an axis of rotation $L$ and (b) an angular velocity vector $\vec{\omega}$. If the axis of rotation remains fixed in space, then all points in the rigid body must move in circular arcs about the line $L$. Consider a point $P$ revolving about $L$ in a circular path of radius $a$ as illustrated in figure 9-3.

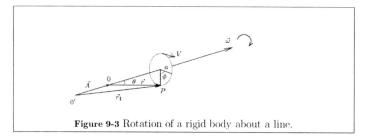

Figure 9-3 Rotation of a rigid body about a line.

The average angular speed of the point $P$ is given by $\frac{\Delta \phi}{\Delta t}$, where $\Delta \phi$ is the angle swept out by $P$ in a time interval $\Delta t$. The instantaneous angular speed is a scalar quantity $\omega$ determined by

$$\omega = \frac{d\phi}{dt} = \lim_{\Delta t \to 0} \frac{\Delta \phi}{\Delta t}.$$

There is a direction associated with the angular motion of $P$ about the line $L$ and thus an angular velocity vector $\vec{\omega}$ is introduced and defined so that

1. $\vec{\omega}$ has a magnitude or length equal to the angular speed $\omega$,
2. $\vec{\omega}$ is perpendicular to the plane of the circular path,
3. The direction of $\vec{\omega}$ is in the direction of advance of a right-hand screw when turned in the direction of rotation.

Choose any point $O$ on the line $L$ and construct the position vector $\vec{r}$ from $O$ to an arbitrary point $P$ inside the rigid body. The arc length $s$ swept out as $P$ moves through the angle $\phi$ is given by $s = a\phi$. The magnitude of the linear speed $V$, of the point $P$, is given by

$$V = \frac{ds}{dt} = a\frac{d\phi}{dt} = a\omega = |\vec{V}|.$$

From the geometry in figure 9-3, one finds $a = |\vec{r}|\sin\theta$, and hence the magnitude of the velocity can be represented as

$$\frac{ds}{dt} = |\vec{V}| = |\vec{\omega}||\vec{r}|\sin\theta.$$

The velocity vector is always normal to the plane containing the position vector and the angular velocity vector. Therefore the velocity vector can be expressed as

$$\frac{d\vec{r}}{dt} = \vec{V} = \vec{\omega} \times \vec{r} = |\vec{\omega}||\vec{r}|\sin\theta \, \hat{e}_n,$$

where $\hat{e}_n$ is a unit vector perpendicular to the plane containing the vectors $\vec{\omega}$ and $\vec{r}$. The expression for the velocity of a rotating vector is independent of the orientation of the Cartesian $x$-,$y$-,$z$-axes as long as the origin of the coordinate system lies on the axis of rotation. To prove this result let $O'$ denote the origin of some new $x', y', z'$ Cartesian reference frame with its origin on the axis of rotation. If $\vec{r}_1$ is the position vector from this origin to the same point $P$ considered earlier, then

$$\frac{d\vec{r}_1}{dt} = \vec{V}_1 = \vec{\omega} \times \vec{r}_1.$$

It therefore remains to show that $\vec{V}_1 = \vec{V}$. From the geometry of figure 9-3, observe that the vectors $\vec{r}_1$ and $\vec{r}$ are related by the vector equation

$$\vec{r}_1 = \vec{A} + \vec{r},$$

where $\vec{A}$ is a vector from the origin of one system to the origin of the other system and lying along the axis of rotation and in the same direction as $\vec{\omega}$. Consequently, there follows the result

$$\frac{d\vec{r}_1}{dt} = \vec{V}_1 = \vec{\omega} \times \vec{r}_1 = \vec{\omega} \times (\vec{A} + \vec{r}) = \vec{\omega} \times \vec{A} + \vec{\omega} \times \vec{r} = \vec{\omega} \times \vec{r} = \vec{V} = \frac{d\vec{r}}{dt}.$$

Note that $\vec{\omega} \times \vec{A}$ is zero since $\vec{\omega}$ and $\vec{A}$ have the same direction.

In general,

$$\frac{d\vec{B}}{dt} = \vec{\omega} \times \vec{B}, \qquad (9.36)$$

where $\vec{B}$ is any vector connecting two fixed points within a rigid body which is rotating about a line $L$ with an angular velocity $\vec{\omega}$.

**Angular Momentum**

The angular momentum $\vec{H}$ of a particle of mass $m$ rotating about an axis is defined as the moment of the linear momentum or

$$\vec{H} = \vec{r} \times (m\vec{V}) \quad \text{with units } \mathrm{Kg\,m^2/s} \text{ or } \mathrm{N\,m\,s}$$

For angular motion the velocity is given by $\vec{V} = \dfrac{d\vec{r}}{dt} = \vec{\omega} \times \vec{r}$ so that $\vec{H} = \vec{r} \times (m\vec{\omega} \times \vec{r})$. Employing the vector identity $\vec{A} \times (\vec{B} \times \vec{C}) = \vec{B}(\vec{A} \cdot \vec{C}) - \vec{C}(\vec{A} \cdot \vec{B})$ verify that

$$\vec{H} = mr^2\vec{\omega} = I\vec{\omega}, \qquad \text{since} \quad \vec{r} \cdot \vec{\omega} = 0,$$

where $I = mr^2$ is the moment of inertia about the rotation axis.

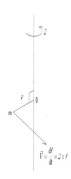

The law of conservation of energy states that if there are no external resultant torques acting on the system, then the total angular momentum remains constant.

**Moments and Newton's Second Law**

Recall that a moment causes a rotational motion. Let us investigate what happens when Newton's second law is applied to rotational motion.

The angular momentum of a particle is defined as the moment of the linear momentum. Let $\vec{H}$ denote the angular momentum, $m\vec{V}$, the linear momentum, and $\vec{r}$, the position vector of the particle, and write

$$\vec{H} = \vec{r} \times (m\vec{V}) = \vec{r} \times \left(m\frac{d\vec{r}}{dt}\right). \qquad (4.5)$$

Differentiating this relation produces

$$\frac{d\vec{H}}{dt} = \vec{r} \times \left(m\frac{d^2\vec{r}}{dt^2}\right) + \frac{d\vec{r}}{dt} \times \left(m\frac{d\vec{r}}{dt}\right).$$

Observe that the second cross product term is zero because the vectors are parallel. Also note that by using Newton's second law, involving a constant mass, there results

$$\vec{F} = m\vec{a} = m\frac{d\vec{V}}{dt} = m\frac{d^2\vec{r}}{dt^2}.$$

Comparing these last two equations, one finds the time rate of change of angular momentum is expressible in terms of the force $\vec{F}$ acting upon the particle. This relationship is found to be

$$\frac{d\vec{H}}{dt} = \vec{r} \times \vec{F} = \vec{M}.$$

The symbols in our last equation tell us about a fundamental principal in Newtonian dynamics that the time rate of change of angular momentum equals the moment of the force acting on the particle.

## Impulse-Momentum Laws

The quantity $\int_{t_1}^{t_2} \vec{F} dt$ is called the linear impulse and since

$$\vec{F} = \frac{d}{dt}(m\vec{V}), \quad \text{then} \quad \int_{t_1}^{t_2} \vec{F} dt = m\vec{V} \Big|_{t_1}^{t_2} = m\vec{V}_2 - m\vec{V}_1$$

which states that the impulse equals the change in linear momentum. The quantity $\int_{t_1}^{t_2} \vec{M} dt$ is called the angular impulse and since

$$\vec{M} = \frac{d\vec{H}}{dt}, \quad \text{then} \quad \int_{t_1}^{t_2} \vec{M} dt = \vec{H} \Big|_{t_1}^{t_2} = \vec{H}_2 - \vec{H}_1$$

The above equation states that the angular impulse equals the change in angular momentum.

## Euler Angles

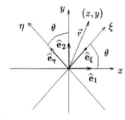

Consider a rotation of axes in two dimensions as depicted in the accompanying figure. Let $x, y$ denote the fixed axes and let $\xi, \eta$ denote the axes which have been rotated through an angle $\theta$ with respect to the fixed $x, y$ axes. A position vector $\vec{r}$ of a general point $(x, y)$ can be referenced with respect to either set of axes. If $(x, y)$ is referenced with respect to the $x, y$ axes, then write

$$\vec{r} = x\,\hat{e}_1 + y\,\hat{e}_2 \tag{9.37}$$

using the basis vectors $\hat{e}_1$ and $\hat{e}_2$.

If the same position vector is referenced with respect to the $\xi, \eta$ axes write

$$\vec{r} = \xi\,\hat{e}_\xi + \eta\,\hat{e}_\eta \tag{9.38}$$

using the basis vectors $\hat{e}_\xi$ and $\hat{e}_\eta$ which rotate with the $\xi, \eta$ axes.

The transformation equations between the two sets of axes can be determined by considering projections of $\vec{r}$ onto the respective axes. This produces the projections

$$\vec{r} \cdot \hat{e}_1 = x = \xi\,\hat{e}_\xi \cdot \hat{e}_1 + \eta\,\hat{e}_\eta \cdot \hat{e}_1$$
$$\vec{r} \cdot \hat{e}_2 = y = \xi\,\hat{e}_\xi \cdot \hat{e}_2 + \eta\,\hat{e}_\eta \cdot \hat{e}_2$$

where

$$\hat{e}_1 \cdot \hat{e}_\xi = \cos\theta \quad \hat{e}_1 \cdot \hat{e}_\eta = \cos(\frac{\pi}{2} - \theta) = -\sin\theta$$
$$\hat{e}_2 \cdot \hat{e}_\eta = \cos\theta \quad \hat{e}_2 \cdot \hat{e}_\xi = \cos(\frac{\pi}{2} - \theta) = \sin\theta$$

In these equations the dot product of the unit vectors gives the cosine of the angles between the vectors so that

$$x = \xi \cos\theta - \eta \sin\theta$$
$$y = \xi \sin\theta + \eta \cos\theta$$

These equations can be represented in the matrix form as

$$\begin{pmatrix} x \\ y \end{pmatrix} = \begin{pmatrix} \cos\theta & -\sin\theta \\ \sin\theta & \cos\theta \end{pmatrix} \begin{pmatrix} \xi \\ \eta \end{pmatrix} \quad (9.39)$$

Observe that the coefficient matrix is an orthogonal matrix, and its inverse is its transpose. The inverse transformation can be represented

$$\begin{pmatrix} \xi \\ \eta \end{pmatrix} = \begin{pmatrix} \cos\theta & \sin\theta \\ -\sin\theta & \cos\theta \end{pmatrix} \begin{pmatrix} x \\ y \end{pmatrix}. \quad (9.40)$$

The concepts associated with rotation in two dimensions can be extend to three-dimensional rotation by considering sequences of two dimensional rotations. Let $x_1, y_1, z_1$ denote a set of rectangular axes which initially coincide with another set of rectangular axes $x, y, z$ which are fixed in space. The $x, y, z$ axes is called an inertial reference frame. The following rotation of axes define the Euler angles $\Omega, i, \omega$.

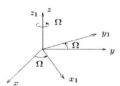

Rotation of $(x, y)$ axes through angle $\Omega$

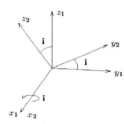

Rotation of $(y_1, z_1)$ axes through angle $i$

1. Rotate the $x_1, y_1$ axes in the $x, y$ plane through an angle $\Omega$ as illustrated in accompanying figure. This rotation is represented by the matrix equation

$$\begin{pmatrix} x_1 \\ y_1 \\ z_1 \end{pmatrix} = \begin{pmatrix} \cos\Omega & \sin\Omega & 0 \\ -\sin\Omega & \cos\Omega & 0 \\ 0 & 0 & 1 \end{pmatrix} \begin{pmatrix} x \\ y \\ z \end{pmatrix}. \quad (9.41)$$

2. Now let $x_2, y_2, z_2$ denote a reference frame which initially coincides with the $x_1, y_1, z_1$ axes. Rotate the $y_2, z_2$ axes in the plane $y_1, z_1$ through an angle $i$ as illustrated, This transformation is represented by the matrix equation

$$\begin{pmatrix} x_2 \\ y_2 \\ z_2 \end{pmatrix} = \begin{pmatrix} 1 & 0 & 0 \\ 0 & \cos i & \sin i \\ 0 & -\sin i & \cos i \end{pmatrix} \begin{pmatrix} x_1 \\ y_1 \\ z_1 \end{pmatrix}. \quad (9.42)$$

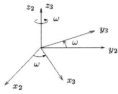

Rotation of $(x_2, y_2)$ axes through angle $\omega$

3. Finally, let $x_3, y_3, z_3$ denote a set of axes which initially coinciding with the $x_2, y_2, z_2$ axes. Rotate the $x_3, y_3$ axes, in the plane of $x_2, y_2$, through an angle $\omega$ as illustrated. The resulting transformation is represented

$$\begin{pmatrix} x_3 \\ y_3 \\ z_3 \end{pmatrix} = \begin{pmatrix} \cos\omega & \sin\omega & 0 \\ -\sin\omega & \cos\omega & 0 \\ 0 & 0 & 1 \end{pmatrix} \begin{pmatrix} x_2 \\ y_2 \\ z_2 \end{pmatrix}. \qquad (9.43)$$

Observe that the matrices that appear in the transformations (9.41), (9.42) and (9.43) are orthogonal matrices. By combining the successive transformations given above, there results the transformation illustrated in figure 9-4.

The sequence of transformations is represented by the matrix equation

$$\begin{pmatrix} x_3 \\ y_3 \\ z_3 \end{pmatrix} = \begin{pmatrix} \cos\omega & \sin\omega & 0 \\ -\sin\omega & \cos\omega & 0 \\ 0 & 0 & 1 \end{pmatrix} \begin{pmatrix} 1 & 0 & 0 \\ 0 & \cos i & \sin i \\ 0 & -\sin i & \cos i \end{pmatrix} \begin{pmatrix} \cos\Omega & \sin\Omega & 0 \\ -\sin\Omega & \cos\Omega & 0 \\ 0 & 0 & 1 \end{pmatrix} \begin{pmatrix} x \\ y \\ z \end{pmatrix}. \qquad (9.44)$$

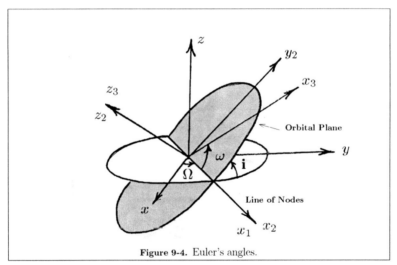

Figure 9-4. Euler's angles.

## Physical Interpretation for Euler Angles

In celestial mechanics, the stars in the heavens are assumed to reside on the inner surface of a great sphere of infinite radius called the celestial sphere. The Earth is considered as being placed at the exact center of this great sphere. The projection of the plane of the Earths equator intersects the celestial sphere and defines the equator of the celestial sphere.

The basic motion of the Earth is that it rotates on its axes and revolves about the Sun. The Earth makes one complete orbit of the Sun in approximately $365\frac{1}{4}$ days. The orbit is an ellipse with the Sun at one focus. The figure 9-5 depicts several possible positions of the Earth in its orbit about the Sun. The Sun, as seen from the Earth, "appears" to have different positions on the celestial sphere (with respect to the stars which are considered as points on the celestial sphere). The apparent motion of the Sun, on the celestial sphere, traces out a great circle path on the celestial sphere called the ecliptic. The plane of the ecliptic is inclined to the equatorial plane at an angle of approximately $23\frac{1}{2}$ degrees. Some nominal positions of the Sun as it moves along the ecliptic are illustrated in figure 9-5. In figure 9-6, the Sun arrives at position $A$ on approximately March 21st of each year. The point $A$ is called the vernal equinox (equal days and equal nights). The vernal equinox is denoted by the symbol $\Upsilon$ which is the symbol for Aries, the ram, because at one time, the vernal equinox passed through the constellation of Aries.

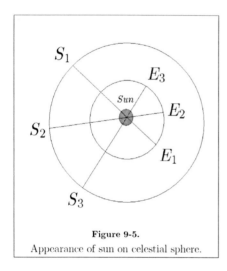

Figure 9-5.
Appearance of sun on celestial sphere.

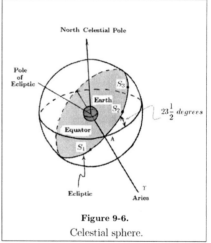

Figure 9-6.
Celestial sphere.

The orbit of a space vehicle or satellite is determined by specifying the geometric parameters of its elliptic orbit and by specifying the orientation of its orbital plane with respect to the celestial sphere.

In celestial mechanics and space mechanics, one particular coordinate system which is used is called the right ascension declination system. This coordinate system has three orthogonal unit vectors $\hat{e}_1$, $\hat{e}_2$, $\hat{e}_3$. The unit vector $\hat{e}_1$ lies in the equatorial plane and points toward the vernal equinox. The vector $\hat{e}_2$ also lies in the equatorial plane and is rotated 90 degrees counterclockwise with respect to $\hat{e}_1$. The vector $\hat{e}_3$ is in the direction of the north celestial pole and forms a right-hand system of coordinates. The angles which specify the orientation of an orbit are the Euler angles $\Omega$, $i$, $\omega$. The angle $\Omega$ is called the longitude of the ascending node and is measured in the equatorial plane. It is the angle between the line of nodes (the line of intersection of the orbit plane and the equatorial plane) and the principal direction $\hat{e}_1$. The angle $\omega$ is called the argument of perigee and is measured in the orbit plane between the line of nodes and the line which passes through the point of perigee and focus of the orbit. The angle $i$ is called the angle of inclination of the orbital plane and is the angle between the orbital plane and equatorial plane. These angles are illustrated in figure 9-4.

## Space Curves, Curvature, and Torsion

A curve in three-dimensional space can be described by a position vector

$$\vec{r} = \vec{r}(t) = x(t)\,\hat{e}_1 + y(t)\,\hat{e}_2 + z(t)\,\hat{e}_3, \tag{4.22}$$

where $t$ is a parameter. Such a curve has an element of arc length given by

$$ds^2 = d\vec{r} \cdot d\vec{r} = dx^2 + dy^2 + dz^2, \tag{4.23}$$

and the rate of change of arc length with respect to the parameter $t$ can be written as

$$\frac{ds}{dt} = \sqrt{\left(\frac{dx}{dt}\right)^2 + \left(\frac{dy}{dt}\right)^2 + \left(\frac{dz}{dt}\right)^2}. \tag{4.24}$$

When the vector equation (4.22) is used to represent the position of a particle in motion, the derivative of the position vector with respect to time $t$ represents the velocity of the particle and

$$\vec{V} = \frac{d\vec{r}}{dt} = \frac{dx}{dt}\,\hat{e}_1 + \frac{dy}{dt}\,\hat{e}_2 + \frac{dz}{dt}\,\hat{e}_3.$$

The speed of the particle is related to the rate of change of arc length with respect to time and is obtained by taking the square root of the dot product of the velocity with itself, since

$$\vec{V} \cdot \vec{V} = \frac{d\vec{r}}{dt} \cdot \frac{d\vec{r}}{dt} = \left(\frac{ds}{dt}\right)^2 = V^2,$$

where $V$ is the magnitude of the velocity.

The unit tangent vector to the curve is given by $\hat{e}_t = \dfrac{d\vec{r}}{ds}$, where $s$ is the arc length. This derivative can be evaluated using chain rule differentiation to obtain

$$\frac{d\vec{r}}{dt} = \frac{d\vec{r}}{ds}\frac{ds}{dt} = \hat{e}_t \sqrt{\left(\frac{dx}{dt}\right)^2 + \left(\frac{dy}{dt}\right)^2 + \left(\frac{dz}{dt}\right)^2} = \hat{e}_t V.$$

Here the unit tangent vector can be determined from the formula

$$\hat{e}_t = \frac{d\vec{r}}{ds} = \frac{1}{\sqrt{\left(\frac{dx}{dt}\right)^2 + \left(\frac{dy}{dt}\right)^2 + \left(\frac{dz}{dt}\right)^2}} \left(\frac{dx}{dt}\hat{e}_1 + \frac{dy}{dt}\hat{e}_2 + \frac{dz}{dt}\hat{e}_3\right).$$

Differentiate the unit tangent vector relation $\hat{e}_t \cdot \hat{e}_t = 1$ with respect to arc length $s$ to obtain $2\hat{e}_t \cdot \frac{d\hat{e}_t}{ds} = 0$, which implies the vector $\frac{d\hat{e}_t}{ds}$ is perpendicular to the unit tangent vector. The vector $\hat{e}_n = \rho \frac{d\hat{e}_t}{ds}$ is defined to be the unit normal vector to the curve, where $\rho$ is a scalar chosen such that $|\rho \frac{d\hat{e}_t}{ds}| = 1$. The scalar $\rho$ is called the radius of curvature and its reciprocal $\kappa = 1/\rho$ is called the curvature.

Note that there are an infinite number of normal vectors to a point on a three dimensional curve in space. The vector $\hat{e}_n = \rho \frac{d\hat{e}_t}{ds}$ is called the principal unit normal to the curve. The vector normal to the plane containing $\hat{e}_t$ and $\hat{e}_n$ and which forms a right-handed system is called the binormal vector to the curve. The unit binormal vector $\hat{e}_b$ is determined from the cross product relation $\hat{e}_b = \hat{e}_t \times \hat{e}_n$. The plane containing the tangent vector and principal normal vector is called the osculating plane. The plane containing the normal vector and binormal vector is called the normal plane and the plane determined by the binormal and tangent vectors is called the rectifying plane. These planes are illustrated in figure 9-7.

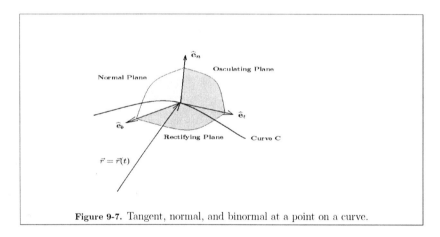

**Figure 9-7.** Tangent, normal, and binormal at a point on a curve.

Differential relationships exist between the unit normal, unit tangent, and unit binormal vectors to a space curve given by

$$\frac{d\hat{e}_t}{ds} = \kappa\,\hat{e}_n, \qquad \frac{d\hat{e}_b}{ds} = -\tau\,\hat{e}_n, \qquad \frac{d\hat{e}_n}{ds} = \tau\,\hat{e}_b - \kappa\,\hat{e}_t \qquad (9.45)$$

where $\kappa$ and $\tau$ are scalars called the curvature and torsion. The differential relations given by equations (9.45) are known as the Frenet-Serret formulas.

**Velocity and Acceleration**

A physical example illustrating the use of the unit tangent and normal vectors is found in determining the normal and tangential components of the velocity and acceleration vectors as a particle moves along a curve. If $\vec{r}$ denotes the position vector of the particle, $\vec{V}$ its velocity, and $\vec{a}$ its acceleration, then

$$\vec{V} = \frac{d\vec{r}}{dt}, \qquad \vec{a} = \frac{d\vec{V}}{dt} = \frac{d^2\vec{r}}{dt^2}, \qquad V^2 = \vec{V} \cdot \vec{V} = \frac{d\vec{r}}{dt} \cdot \frac{d\vec{r}}{dt} = \left(\frac{ds}{dt}\right)^2, \qquad (9.46)$$

where $s$ is the arc length along the curve. Using chain rule differentiation one finds

$$\vec{V} = \frac{d\vec{r}}{dt} = \frac{d\vec{r}}{ds}\frac{ds}{dt} = \hat{e}_t V.$$

This equation demonstrates that the velocity vector $\vec{V}$ is directed along the tangent vector to the curve at any time $t$ and has the magnitude given by $V = \frac{ds}{dt}$ which represents the speed of the particle.

The derivative of the velocity vector with respect to time $t$ is the acceleration and

$$\vec{a} = \frac{d\vec{V}}{dt} = \hat{e}_t \frac{dV}{dt} + \frac{d\hat{e}_t}{dt} V.$$

The time rate of change of the unit tangent vector can be expressed as

$$\frac{d\hat{e}_t}{dt} = \frac{d\hat{e}_t}{ds}\frac{ds}{dt} = \frac{V}{\rho}\hat{e}_n.$$

Substituting this result into the acceleration vector gives

$$\vec{a} = \frac{dV}{dt}\hat{e}_t + \frac{V^2}{\rho}\hat{e}_n$$

which shows the acceleration vector lies in the osculating plane. The tangential component of the acceleration is $\frac{dV}{dt}$ and the normal component of the acceleration is $V^2/\rho$.

**Example 9-4.**

Consider a particle moving on a unit circle with constant angular speed $\omega$. Assume the position vector of the particle at any time $t$ is

$$\vec{r} = \vec{r}(t) = \cos\omega t\ \hat{e}_1 + \sin\omega t\ \hat{e}_2.$$

Calculate the velocity and acceleration of the particle and show

$$\vec{V} = \frac{d\vec{r}}{dt} = -\omega \sin\omega t\ \hat{e}_1 + \omega \cos\omega t\ \hat{e}_2$$

$$\vec{a} = \frac{d\vec{V}}{dt} = \frac{d^2\vec{r}}{dt^2} = -\omega^2 \cos\omega t\ \hat{e}_1 - \omega^2 \sin\omega t\ \hat{e}_2 = -\omega^2 \vec{r}.$$

The above equations demonstrate that the velocity vector is normal to the position vector and $\vec{V} \cdot \vec{r} = 0$ at any time $t$. Also from these equations, note that the acceleration vector is directed along the position vector towards the origin. The name centripetal acceleration is applied to this type of motion.

For circular motion, the velocity can also be represented by the cross product relation

$$\vec{V} = \frac{d\vec{r}}{dt} = \vec{\omega} \times \vec{r}.$$

where $\vec{\omega} = \omega \hat{e}_3$. For $\vec{\omega} = \omega \hat{e}_3$ constant, in both magnitude and direction, the acceleration can be obtained by differentiation the velocity vector and employing the triple product formula to obtain

$$\vec{a} = \frac{d\vec{V}}{dt} = \vec{\omega} \times \frac{d\vec{r}}{dt} + \frac{d\vec{\omega}}{dt} \times \vec{r} = \vec{\omega} \times \vec{V} = \vec{\omega} \times (\vec{\omega} \times \vec{r}) = -\omega^2 \vec{r}$$

■

## Relative Motion

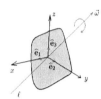

Consider a rigid body which is rotating about a line $\ell$ with an angular velocity $\vec{\omega}$ as illustrated in the accompanying figure. If $\vec{r}$ denotes the position vector of any point in the rigid body, with reference to a fixed point 0 on the axis of rotation, then the velocity of that point is given by

$$\vec{V} = \frac{d\vec{r}}{dt} = \vec{\omega} \times \vec{r}. \qquad (9.47)$$

Let $x, y, z$ denote a set of Cartesian axes with origin 0 on the line $\ell$. Let these axes be fixed in the rigid body and consider them as rotating about the line $\ell$. The unit vectors $\hat{e}_1$, $\hat{e}_2$, $\hat{e}_3$ attached to these axes are also rotating about the line $\ell$ and the rate of change of these vectors with time with respect to the fixed set of axes are given by the relations

$$\frac{d\hat{e}_1}{dt} = \vec{\omega} \times \hat{e}_1, \qquad \frac{d\hat{e}_2}{dt} = \vec{\omega} \times \hat{e}_2, \qquad \frac{d\hat{e}_3}{dt} = \vec{\omega} \times \hat{e}_3 \qquad (9.48)$$

These relations are special cases of equation (9.36) and result when $\vec{B}$ is replaced by the unit vectors $\hat{e}_1$, $\hat{e}_2$ and $\hat{e}_3$ respectively.

Let $\bar{x}, \bar{y}, \bar{z}$ be a fixed or inertial set of axes, and let $x, y, z$ denote a set of axes fixed in the rigid body with origin on the axis of rotation and rotating with respect to the fixed set of axes. A position vector

$$\vec{r} = x\,\hat{e}_1 + y\,\hat{e}_2 + z\,\hat{e}_3,$$

in the moving set of $x, y, z$ axes, has the time rate of change

$$\frac{d\vec{r}}{dt} = \frac{dx}{dt}\hat{e}_1 + \frac{dy}{dt}\hat{e}_2 + \frac{dz}{dt}\hat{e}_3 + x\frac{d\hat{e}_1}{dt} + y\frac{d\hat{e}_2}{dt} + z\frac{d\hat{e}_3}{dt}$$
$$\frac{d\vec{r}}{dt} = \frac{dx}{dt}\hat{e}_1 + \frac{dy}{dt}\hat{e}_2 + \frac{dz}{dt}\hat{e}_3 + x(\vec{\omega} \times \hat{e}_1) + y(\vec{\omega} \times \hat{e}_2) + z(\vec{\omega} \times \hat{e}_3) \qquad (9.49)$$
$$\frac{d\vec{r}}{dt} = \frac{\delta \vec{r}}{\delta t} + \vec{\omega} \times \vec{r}.$$

Here the time rate of change of the position vector $\vec{r}$ relative to a fixed set of axes is composed of the two terms $\frac{\delta \vec{r}}{\delta t}$ and $\vec{\omega} \times \vec{r}$. The term $\frac{\delta \vec{r}}{\delta t}$ is just like an ordinary derivative, where $\hat{e}_1, \hat{e}_2$ and $\hat{e}_3$ are treated as constant vectors. Such a derivative is sometimes referred to as a frame derivative. The second term $\vec{\omega} \times \vec{r}$ represents the time derivative of the position vector $\vec{r}$ due to the angular velocity $\vec{\omega}$ of the body. Note: If the position vector $\vec{r}$ points to a fixed point $(x, y, z)$ within the rotating body, then $(\dot{x} = \dot{y} = \dot{z} = 0)$ and equation (9.49) reduces to equation (9.47).

Let $\alpha$ and $\beta$ denote two points relative to the fixed set of axes $\bar{x}, \bar{y}, \bar{z}$ and let $\vec{r}_\alpha$ and $\vec{r}_\beta$ denote the position vectors of these points. The vector equation

$$\vec{r}_\beta = \vec{r}_{\beta/\alpha} + \vec{r}_\alpha \qquad (9.50)$$

is called the relative position equation, where $\vec{r}_{\beta/\alpha} = \vec{r}_\beta - \vec{r}_\alpha$ is called the position vector of $\beta$ relative to $\alpha$. Differentiation of the relative position equation produces a relative velocity equation and

$$\frac{d\vec{r}_\beta}{dt} = \frac{d\vec{r}_{\beta/\alpha}}{dt} + \frac{d\vec{r}_\alpha}{dt} \qquad \text{or} \qquad (9.51)$$
$$\vec{V}_\beta = \vec{V}_{\beta/\alpha} + \vec{V}_\alpha.$$

Similarly a differentiation of the relative velocity equation produces a relative acceleration equation and

$$\vec{a}_\beta = \vec{a}_{\beta/\alpha} + \vec{a}_\alpha. \qquad (9.52)$$

If the origins of the fixed $\bar{x}, \bar{y}, \bar{z}$ axes and the rotating $x, y, z$ axes do not coincide and the origin $O$ experiences a translational velocity $\vec{V}_O$ relative to the fixed origin $\bar{O}$, then the relative position equation of a point $P(x, y, z)$ as seen by an observer situated at the fixed origin can be expressed as

$$\vec{r}_P = \vec{r}_{P/O} + \vec{r}_O,$$

where it is understood that $\vec{r}_P$ and $\vec{r}_O$ are referenced with respect to the fixed axes. The situation is illustrated in the figure 9-8.

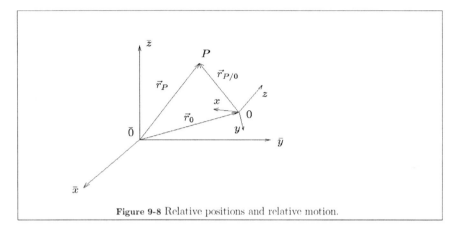

**Figure 9-8** Relative positions and relative motion.

Differentiating the relative position expression results in the relative velocity equation

$$\frac{d\vec{r}_P}{dt} = \frac{d\vec{r}_{P/O}}{dt} + \frac{d\vec{r}_O}{dt}.$$

With the use of equations (9.49) this simplifies to

$$\vec{V}_P = \frac{\delta \vec{r}_{P/O}}{\delta t} + \vec{\omega} \times \vec{r}_{P/O} + \vec{V}_O \quad \text{or}$$

$$\vec{V}_P = \vec{V}_{P/O} + \vec{\omega} \times \vec{r}_{P/O} + \vec{V}_O,$$
(9.53)

where

$\vec{V}_{P/O}$ is the velocity of $P$ relative to the moving origin $O$,
$\vec{V}_O$ is the translational velocity of $O$, and
$\vec{\omega} \times \vec{r}_{P/O}$ is the velocity of $P$ due to the angular motion $\vec{\omega}$.

The relative acceleration equation results when the relative velocity equation is differentiated giving

$$\frac{d\vec{V}_P}{dt} = \frac{d}{dt}\left(\vec{V}_{P/O}\right) + \vec{\omega} \times \frac{d\vec{r}_{P/O}}{dt} + \frac{d\vec{\omega}}{dt} \times \vec{r}_{P/O} + \frac{d\vec{V}_O}{dt}.$$
(9.54)

In this expression

$$\vec{V}_{P/O} = \frac{\delta \vec{r}_{P/O}}{\delta t} = \frac{dx}{dt}\hat{e}_1 + \frac{dy}{dt}\hat{e}_2 + \frac{dz}{dt}\hat{e}_3$$

and consequently

$$\frac{d}{dt}\left(\vec{V}_{P/O}\right) = \frac{d^2x}{dt^2}\hat{e}_1 + \frac{d^2y}{dt^2}\hat{e}_2 + \frac{d^2z}{dt^2}\hat{e}_3 + \frac{dx}{dt}\frac{d\hat{e}_1}{dt} + \frac{dy}{dt}\frac{d\hat{e}_2}{dt} + \frac{dz}{dt}\frac{d\hat{e}_3}{dt}.$$

Therefore, equation (9.54) simplifies to the form

$$\vec{a}_P = \vec{a}_{P/O} + \vec{\omega} \times \vec{V}_{P/O} + \vec{\omega} \times \left(\vec{V}_{P/O} + \vec{\omega} \times \vec{r}_{P/O}\right) + \frac{d\vec{\omega}}{dt} \times \vec{r}_{P/O} + \vec{a}_O$$

or

$$\vec{a}_P = \vec{a}_{P/O} + 2\vec{\omega} \times \vec{V}_{P/O} + \vec{\omega} \times \left(\vec{\omega} \times \vec{r}_{P/O}\right) + \frac{d\vec{\omega}}{dt} \times \vec{r}_{P/O} + \vec{a}_O. \quad (9.55)$$

In this relative acceleration equation

$\vec{a}_O$ is the translational acceleration of $O$ as observed from $\overline{O}$,

$\frac{d\vec{\omega}}{dt} \times \vec{r}_{P/O}$ is the acceleration due to the rotational motion of the $x, y, z$ axes,

$\vec{\omega} \times (\vec{\omega} \times \vec{r}_{P/O})$ is called the centripetal acceleration,

$2\vec{\omega} \times \vec{V}_{P/O}$ is called the Coriolis acceleration, and

$\vec{a}_{P/O}$ is the acceleration of $P$ as observed from the $x, y, z$ reference frame origin $O$.

## Mechanical Vibrations

Figure 9-9 illustrates a point $A$, on the circumference of a unit circle, which is rotating with a constant angular velocity of $\omega = \frac{d\theta}{dt}$ about its center. As point $A$ rotates around the circle, plot the amplitude $y$ versus the angle $\theta$ to obtain the circular function $y = \sin\theta$.

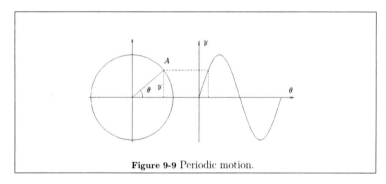

**Figure 9-9** Periodic motion.

Here $\dot{\theta} = \omega$ is the angular velocity where $[\omega]^{\ddagger}$ = radians/sec. An integration of this equation gives, $\theta = \omega t + \phi$, where $\phi$ is a constant called the phase angle and represents the initial value of $\theta$ at time $t = 0$. The motion of point $A$ is a periodic motion with $A$ returning to its initial position after $\theta$ has completed one revolution or $\theta$ has changed by $2\pi$ radians.

‡ The notation [ ] is read "The dimensions of" what is inside the brackets.

In terms of $\theta$, the period $P$ is $2\pi$ radians, whereas in terms of time $t$ the period is $2\pi/\omega$, since $\theta$ changes from $\phi$ to $\phi + 2\pi$ as $t$ changes from 0 to $2\pi/\omega$.

The frequency of the motion is defined as

$$f = \frac{1}{P} = \frac{\omega}{2\pi} \quad \text{cycles/second}$$

and represents the number of complete vibrations in 1 second. One vibration per second is called a hertz[2] denoted by the symbol $Hz$. For example, if the frequency were 10 kilohertz = 10,000 cycles/second $or$ (10 kHz), then in one second the point $A$ would complete 10,000 revolutions.

In the study of sound, frequency is referred to as pitch and the greater the frequency, the higher the pitch. The human ear can discern frequencies between 24 and 24,000 Hz. Musical instruments have frequencies between 40 and 4,000 Hz.

## Phenomenon of Beats

For harmonic vibrations of the form

$$y = A\sin(\omega t + \phi) \quad \text{or} \quad y = A\cos(\omega t + \phi)$$

the quantity $A$ is called the amplitude of the vibration, $\omega$ is called the angular frequency of the motion, $P = 2\pi/\omega$ is the period of the vibration, and $f = \omega/2\pi = 1/P$ is called the frequency of the vibration. Given two simple harmonic motions of the form

$$y_1 = A\cos(\omega_1 t + \phi_1) \quad \text{and} \quad y_2 = A\cos(\omega_2 t + \phi_2)$$

where $\omega_1$ is different from $\omega_2$, then the superposition of these motions gives

$$y = y_1 + y_2 = A\cos(\omega_1 t + \phi_1) + A\cos(\omega_2 t + \phi_2). \tag{9.56}$$

By using the trigonometric identities

$$\cos(\alpha + \beta) = \cos\alpha\cos\beta - \sin\alpha\sin\beta$$
$$\cos(\alpha - \beta) = \cos\alpha\cos\beta + \sin\alpha\sin\beta$$

with

$$\alpha - \beta = \omega_1 t + \phi_1 \quad \text{and} \quad \alpha + \beta = \omega_2 t + \phi_2,$$

equation (9.56) can be written in the form

$$y = y_1 + y_2 = 2A\cos\left[\frac{1}{2}(\omega_1 + \omega_2)t + \frac{1}{2}(\phi_1 + \phi_2)\right]\cos\left[\frac{1}{2}(\omega_2 - \omega_1)t + \frac{1}{2}(\phi_2 - \phi_1)\right]. \tag{9.58}$$

---

[2] Heinrich R. Hertz (1857–1894) German physicist.

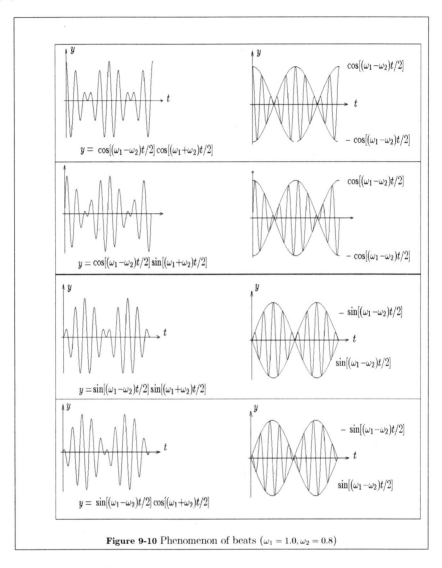

**Figure 9-10** Phenomenon of beats ($\omega_1 = 1.0, \omega_2 = 0.8$)

Observe that for $\omega_1$ near $\omega_2$ the quantity $|\omega_2 - \omega_1|$ is small, which implies the period of the second cosine term in equation (9.58) is large (or its frequency is small) and equation (9.58) can be represented as

$$y = B(t) \cos\left[\frac{1}{2}(\omega_2 - \omega_1)t + \frac{1}{2}(\phi_2 - \phi_1)\right]$$

where $\quad B(t) = 2A \cos\left[\frac{1}{2}(\omega_1 + \omega_2)t + \frac{1}{2}(\phi_1 + \phi_2)\right]$

is the time varying amplitude (envelope) of the motion. When $\omega_1$ is close to $\omega_2$, this superposition of two vibrations of slightly different pitch is referred to as the phenomenon of beats, and the resulting vibration is characterized by a rapidly changing vibration with a slowly changing envelope. This phenomenon of beats is illustrated in figure 9-10 for various combinations of sine and cosine products.

## Vibrations of a Spring Mass System

Consider a vertical spring which is suspended from a support as illustrated in figure 9-11(a). If a weight $W$ is attached to the spring, the spring stretches some distance $s_0$, and the weight remains at rest in an equilibrium position as illustrated in the figure 9-11(b). If the weight is displaced from this equilibrium position and then released, it undergoes a vibratory motion with respect to a set of reference axes constructed at the equilibrium position as illustrated in figure 9-11(c).

In order to model the above problem, the following assumptions are made:
(a) No motion exists in the horizontal direction.
(b) A downward displacement is considered as positive.
(c) The spring is a linear spring and obeys Hooke's[3] law which states that the restoring force of the spring is proportional to the spring displacement.

Using Hooke's law the spring force holding the weight in equilibrium can be calculated. In figure 9-11(b), there is no motion because the weight $W$ acting down is offset by the spring force acting upward. Let $f_s$ denote the spring force illustrated in figure 9-11(d). Using Hooke's law, the spring force $f_s$ is proportional to the displacement $s_0$ and so is written $f_s = Ks_0$, where $K$ is the proportionality constant called the spring constant. Summation of the forces in equilibrium is represented in figure 9-11(b) which illustrates the spring force equal to the weight acting down or $f_s = Ks_0 = W$. This determines the spring constant $K$ as

$$K = \frac{W}{s_0}. \tag{9.58}$$

In figure 9-11(c), the spring force acting on the weight is given by $f_s = K(s_0 + y)$ where $y$ is the displacement from the equilibrium position. By Newton's second law, the motion of the motion of the weight is modeled by summing the forces in the $y$ direction to give

$$m\frac{d^2y}{dt^2} = W - f_s = W - Ks_0 - Ky = -Ky, \qquad m = \frac{W}{g} \tag{9.59}$$

---
[3] Robert Hooke (1635–1703) English physicist.

or
$$\frac{d^2y}{dt^2} + \omega^2 y = 0. \tag{9.60}$$

Here $\omega^2 = K/m$ or $\omega = \sqrt{K/m}$. The quantity $\omega$ is called the natural frequency of the undamped system.

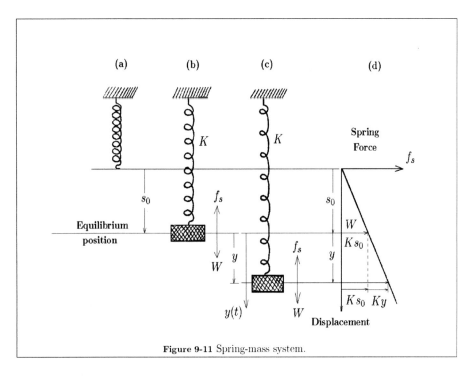

**Figure 9-11** Spring-mass system.

Observe the sign of the spring force in equation (9.59). If $y > 0$, the restoring force is in the negative direction. If $y < 0$, the restoring force is in the positive direction. The directions of the forces are important because forces are vector quantities and must have both a magnitude and a direction. The direction of the forces is one check that the problem is correctly modeled.

If additional forces are added to the spring mass system, such as damping forces and external forces, then equation (9.59) must be modified to include these additional forces. In figure 9-12, assume a damper and an external force are attached to the spring as illustrated.

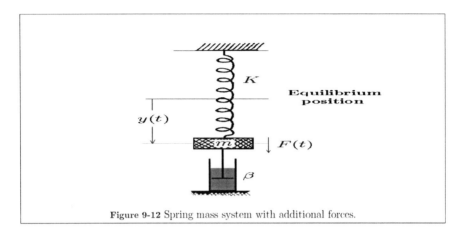

**Figure 9-12** Spring mass system with additional forces.

If there is a damping force $F_D$ which opposes the motion of the mass and the magnitude of the damping force is proportional to the velocity[4], then this can be represented by

$$F_D = -\beta \frac{dy}{dt} \tag{9.61}$$

where $\beta$ is the proportionality constant called the damping coefficient. The sign of the damping force tells us that if $y$ is increasing and $\frac{dy}{dt} > 0$, the damping force is in the negative direction, whereas, if $y$ is decreasing and $\frac{dy}{dt} < 0$, the damping force acts in the positive direction. In figure 9-12, the quantity $F(t)$ denotes an external force. Newton's second law and the summation of forces produces the following mathematical model describing the motion of the spring mass system illustrated in the figure 9-12.

$$m\frac{d^2y}{dt^2} = -Ky - \beta\frac{dy}{dt} + F(t) \tag{9.62}$$

or

$$m\ddot{y} + \beta\dot{y} + Ky = F(t). \tag{9.65}$$

Each term in equation (9.65) represents a force term. The quantity $m\ddot{y}$ is called the inertial force, $\beta\dot{y}$ is the damping force, $Ky$ is the spring force and $F(t)$ is an external force. Here $m$, $\beta$ and $K$ are all positive constants with dimensions $[m] = \frac{[W]}{[g]} = \frac{\text{lbs}}{\text{ft/sec}^2}$, $[\beta] = \frac{\text{lbs}}{\text{ft/sec}}$, $[K] = \frac{\text{lbs}}{\text{ft}}$ with $y$ and $t$ having the dimensions $[y] = $ ft and $[t] = $ seconds.

---

[4] Various assumptions can be made to model other types of damping

## Mechanical Resonance

In equation (9.65), let $F(t) = F_0 \cos \lambda t$, with $\lambda$ is a constant, and then construct the general solution to equation (9.65) for this special case. Recall that to solve

$$L(y) = m\ddot{y} + \beta \dot{y} + Ky = F_0 \cos \lambda t \tag{9.66}$$

it is customary to first solve the homogeneous equation

$$L(y) = m\ddot{y} + \beta \dot{y} + Ky = 0. \tag{9.65}$$

This is an ordinary differential equation with constant coefficients and so assume an exponential solution $y = \exp(\gamma t)$ and obtain the characteristic equation

$$m\gamma^2 + \beta\gamma + K = 0$$

with characteristic roots

$$\gamma = \frac{-\beta \pm \sqrt{\beta^2 - 4mK}}{2m} = -\frac{\beta}{2m} \pm \sqrt{\left(\frac{\beta}{2m}\right)^2 - \frac{K}{m}}. \tag{9.66}$$

In equation (9.66), the discriminant determines the type of motion that results. The following cases are considered.

**CASE I (Homogeneous Equation and Overdamping)**

If $(\beta/2m)^2 - K/m > 0$, equation (9.66) has two distinct roots $\gamma_1$ and $\gamma_2$ where both $\gamma_1$ and $\gamma_2$ are negative, then the corresponding complementary solution of equation (9.65) has transient terms and is of the form

$$y_c = c_1 e^{\gamma_1 t} + c_2 e^{\gamma_2 t}, \quad \gamma_1 < 0, \; \gamma_2 < 0 \tag{9.67}$$

where $c_1, c_2$ are arbitrary constants. This type of solution illustrates that if $\beta$ is too large, then no oscillatory motion can exist. In such a situation, the system is said to be overdamped.

**CASE II (Homogeneous equation and underdamping)**

For the condition $(\beta/2m)^2 - K/m < 0$, let $\omega_0^2 = K/m - (\beta/2m)^2$ and obtain from the characteristic equation (9.66) the two complex characteristic roots

$$\gamma_1 = -\beta/2m + i\omega_0 \quad \text{and} \quad \gamma_2 = -\beta/2m - i\omega_0.$$

These roots produce a complementary solution of the form

$$y_c = e^{-\beta t/2m}(c_1 \sin \omega_0 t + c_2 \cos \omega_0 t)$$

or

$$y_c = \sqrt{c_1^2 + c_2^2}\, e^{-\beta t/2m} \cos(\omega_0 t - \phi) \tag{9.69}$$

with $c_1, c_2$ arbitrary constants. Here $\omega_0$ represents the damped natural frequency of the system. If $\beta$ is small, $\omega_0$ is approximately the natural frequency of the undamped system given by $\omega = \sqrt{K/m}$. The complementary solution equation (9.69) denotes a damped oscillatory solution which can be visualized by plotting the curves

$$y_1 = \sqrt{c_1^2 + c_2^2}\, e^{-\beta t/2m} \quad \text{and} \quad y_2 = -y_1$$

as envelopes of the oscillation $\cos(\omega_0 t - \phi)$ as is illustrated in figure 9-13.

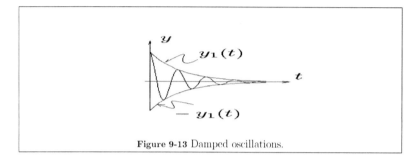

**Figure 9-13** Damped oscillations.

**CASE III (Homogeneous equation and critical damping)** If $(\beta/2m)^2 - K/m = 0$, equation (9.66) has the repeated roots $\gamma_1 = \gamma_2 = -\beta/2m$ which produces the complementary solution

$$y_c = (c_1 + c_2 t) e^{-\beta t/2m}. \tag{9.69}$$

Since this case separates the overdamping from the underdamping, it is called the critically damped case, and the value of $\beta$ which produces this critical case is $\beta_c = 2m\omega$ where $\omega = \sqrt{K/m}$ is the natural frequency of the undamped system.

Associated with the complementary solution from one of the cases I, II, or III, is the particular solution of the nonhomogeneous equation (9.65). The particular solution can be determined by the method of undetermined coefficients by substituting

$$y_p = A \cos \lambda t + B \sin \lambda t \tag{9.70}$$

with $A$ and $B$ unknown constants, into the differential equation (9.66) to obtain the equations

$$\begin{aligned}(K - m\lambda^2)A + \beta\lambda B &= F_0 \\ -\beta\lambda A + (K - m\lambda^2)B &= 0,\end{aligned} \tag{9.71}$$

which determine the constants $A$ and $B$. Solving equations (9.71) gives

$$A = \frac{(K - m\lambda^2)F_0}{\Delta} \quad \text{and} \quad B = \frac{\beta\lambda F_0}{\Delta} \tag{9.72}$$

where
$$\Delta = (K - m\lambda^2)^2 + \beta^2\lambda^2.$$

The particular solution can then be expressed as

$$y_p = \frac{F_0}{\sqrt{\Delta}} \cos(\lambda t - \phi) \tag{9.73}$$

where $\phi$ is a phase angle defined by $\tan\phi = \beta\lambda/(K - m\lambda^2)$ for $m\lambda^2 \neq K$. The general solution to equation (9.66) can then be written $y = y_c + y_p$. In the general solution, the complementary solutions are transient solutions, and the particular solution represents the steady state oscillations. The amplitude of the steady state oscillations is given by

$$\text{Amp} = \frac{F_0}{\sqrt{\Delta}} = \frac{F_0}{\sqrt{(K - m\lambda^2)^2 + \beta^2\lambda^2}} = \frac{F_0}{m\sqrt{(\omega^2 - \lambda^2)^2 + 4\lambda^2\omega^2(\beta/\beta_c)^2}} \tag{9.74}$$

where $\omega$ is the natural frequency of the undamped system and $\beta_c = 2m\omega$ is the critical value of the damping. For $\beta = 0$ (no damping), the denominator in equation (9.74) becomes $m|\omega^2 - \lambda^2|$ and approaches zero as $\lambda$ tends toward $\omega$. Thus, with no damping, as the angular frequency $\lambda$ of the forcing term approaches the natural frequency $\omega$ of the system, the denominator in equation (9.66) approaches zero, which in turn causes the amplitude of the oscillations to increase without bound. This is known as the phenomenon of resonance. For $\beta \neq 0$, there can still be a resonance-type behavior whereby the amplitude of the oscillations become large for some specific value of the forcing frequency $\lambda$.

Define the resonance frequency as the value of $\lambda$ which produces the maximum amplitude of the oscillation, if an oscillation exists. This amplitude, given by equation (9.74), has a maximum value when the denominator is a minimum. Let

$$H = (\omega^2 - \lambda^2)^2 + 4\lambda^2\omega^2(\beta/\beta_c)^2$$

denote this denominator. The quantity $H$ has a minimum value with respect to $\lambda$ when the derivative of $H$ with respect to $\lambda$ is zero. Calculating this derivative gives

$$\frac{dH}{d\lambda} = 2(\omega^2 - \lambda^2)(-2\lambda) + 8\lambda\omega^2(\beta/\beta_c)^2 = 0$$

when

$$\lambda^2 = \omega^2\left[1 - 2(\beta/\beta_c)^2\right]. \tag{9.75}$$

The phenomenon of resonance can be seen graphically by plotting the amplitude, as given by equation (9.75), versus $\lambda$ for various values of the ratio $\beta/\beta_c$ as in figure 9-14.

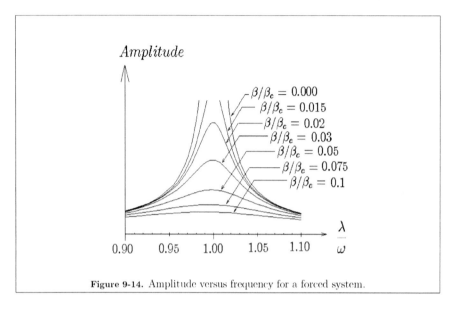

**Figure 9-14.** Amplitude versus frequency for a forced system.

In practical problems, it is important to be able to design vibratory structures to avoid resonance. For example, printing presses vibrating at the correct frequency can act as forcing functions to cause large vibrations and even collapse of the supporting floor. High winds can act as forcing functions to cause resonance oscillations of structures. Flutter of aircraft wings, if not controlled properly, can result in dynamic instability of an aircraft.

Resonance can also be a desired phenomenon such as in tuning an electrical circuit for a maximum response of a voltage of a specified frequency. In the study of electrical circuits where frequency is a variable, it is desirable to have frequency response characteristics for the circuit in a graphical form similar to figure 9-14.

Train yourself to look for curves which have the shape of the curves in figure 9-14. Chances are some kind of resonance phenomenon is taking place. There are lots of curves having the general shape of the curves illustrated in figure 9-14. These type of curves usually result from the vibration models used to study a wide variety of subjects and are the design basis of a large number of measuring devices. The following is a brief list of subject areas where it is possible to find additional applications of the basic equations of vibratory phenomena and resonance. Examine topics listed under mechanical vibrations, earthquake modeling, atomic vibrations, atomic cross sections, vibrations of atoms in crystals, scattering of atoms, particles, and waves from crystal surfaces, sound waves, string instruments, tidal motions, lasers, electron spin resonance, nuclear magnetic resonance, and behavior of viscoelastic materials.

## Torsional Vibrations

Torsional vibrations are similar in form to the spring mass system and the differential equation of the motion can be obtained from the relations

$$\sum \text{Torques} = I\alpha = I\frac{d^2\theta}{dt^2} \tag{9.77}$$

where $\theta$ denotes the angular displacement, $I$ is the moment of inertia of the body, and $\alpha = \ddot{\theta}$ is the angular acceleration. Consider a disk attached to a fixed rod as in figure 9-15.

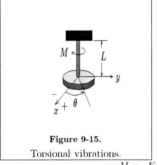

Figure 9-15.
Torsional vibrations.

If the disk is rotated through an angle $\theta$, there is a restoring moment $M$ produced by the rod. Hooke's law states that the restoring moment is proportional to the angular displacement and

$$M = -K_T\theta \tag{9.77}$$

where $K_T$ is called the spring constant of the shaft and is the proportionality constant associated with the angular displacement $\theta$. From the relation in equation (9.77) there results

$$M = -K_T\theta = I\frac{d^2\theta}{dt^2} \quad \text{or} \quad I\frac{d^2\theta}{dt^2} + K_T\theta = 0 \tag{9.78}$$

as the equation of motion describing the angular displacement $\theta$.

By adding a linear damper and external force to equation (9.78), a more general equation results

$$I\frac{d^2\theta}{dt^2} + \beta\frac{d\theta}{dt} + K_T\theta = F(t) \tag{9.79}$$

From strength of materials, the constant $K_T$ is given by the relation $K_T = GJ/L$, where $G$ is called the shearing modulus of the rod material, $J$ is the polar moment of inertia of the rod cross section, and $L$ is the length of the shaft.

### The simple pendulum

For the pendulum illustrated the forces about 0 are the weight of the mass and the radial force along the string. The radial force passes through the origin and so does not produce a moment about the origin. The moment of inertia of the mass $m$ about 0 is given by $I = m\ell^2$ and the torque about 0 is given by $T = -(mg)(\ell\sin\theta)$ and consequently the equation of motion can be expressed

$$T = -mg\ell\sin\theta = m\ell^2\frac{d^2\theta}{dt^2}$$

## Solid Angles

A cone is described as a family of intersecting lines. A right circular cone is an example which is easily recognized, however, this is only one special kind of a cone. A general cone is described by a line having one point fixed in space which is free to rotate. Figure 9-16 illustrates two cones which differ from a right circular cone.

Let $dS$ denote an element of surface area on a surface $S$ and connect all the points forming the boundary of $dS$ to a common point 0, thus forming a cone. Then construct a sphere of radius $r = |\vec{r}|$ centered at 0, where $\vec{r}$ is a position vector from point 0 to a point on the element of surface area $dS$. This cone intersects the sphere in an element of surface area $d\Omega$. The solid angle subtended by the cone constructed from point 0 to the element of surface area $dS$ is defined

$$d\omega = \frac{d\Omega}{r^2} = \frac{\vec{r} \cdot \hat{e}_n \, dS}{r^3} = \frac{\vec{r} \cdot d\vec{S}}{r^3}.$$

Physically this represents the surface area on that portion of a unit sphere which is intersected by the cone. The situation is illustrated in figure 9-17.

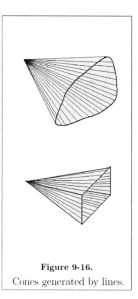

**Figure 9-16.** Cones generated by lines.

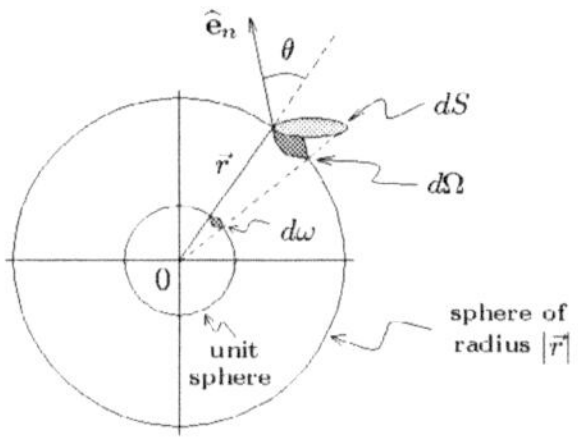

**Figure 9-17.** Solid angle as surface area on unit sphere.

Solid angles are measured in units called steradians. One steradian is the angle subtended when $d\Omega = r^2$. The total solid angle about point 0 is equal to the area of the unit sphere or $4\pi$ steradians. Construct a unit outward normal $\hat{e}_n$ to the surface where the position vector and element of surface area $dS$ are positioned. Let $\theta$ denote the angle between the outward normal and the extended position vector $\vec{r}$ as illustrated in figure 9-17.

From the dot product relation

$$\hat{e}_n \cdot \vec{r} = r \cos\theta \quad \text{there results} \quad \cos\theta = \frac{\hat{e}_n \cdot \vec{r}}{r}$$

so that the projection of the element $dS$ on the sphere of radius $r$ produces the result

$$d\Omega = \cos\theta \, dS = \frac{\hat{e}_n \cdot \vec{r}}{r} \, dS. \tag{9.80}$$

The algebraic sign of $\cos\theta$ depends upon the relation of the outward normal to the sphere of radius $r$.

Consider the surface integral

$$\iint_S d\omega = \iint_S \frac{d\Omega}{r^2} = \iint_S \frac{\hat{e}_n \cdot \vec{r}}{r^3} dS = \iint_S \frac{\vec{r}}{r^3} d\vec{S},$$

where the surface $S$ encloses a bounded, closed, simply connected region. Surface integrals of this type represent the total sum of the solid angles subtended by the element $dS$, summed over the surface $S$. For the solid angle summed about a point $0'$ outside the surface, the resulting sum of the solid angles is zero. This is because for each positive sum $+d\omega$ there is a corresponding negative sum $-d\omega$, and these add to zero in pairs. If the solid angle is summed about a point 0 inside the surface, the resulting sum is not zero. Here the sum of the areas $d\omega$ on the unit sphere, subtended by the elements $dS$, do not add together in pairs to produce zero but instead give the total surface area of the unit sphere which is $4\pi$ steradians. From these results a theorem which can be written

$$\iint_S d\omega = \iint_S \frac{\vec{r}}{r^3} d\vec{S} = \begin{cases} 0 & \text{if origin is outside surface} \\ 4\pi & \text{if origin is inside surface} \end{cases} \qquad (9.81)$$

This theorem is utilized in the study of inverse square law potentials and is known as Gauss' theorem.

**Laplace's Equation**

For $\vec{F} = \vec{F}(x, y, z)$, a vector field which is both irrotational and solenoidal, the curl and divergence of $\vec{F}$ are zero

$$\text{curl } \vec{F} = \vec{0} \quad \text{and} \quad \text{div } \vec{F} = 0 \qquad (9.82)$$

Under these conditions $\vec{F}$ is derivable from a scalar function $\Phi$ called the scalar potential, where $\vec{F} = \text{grad } \Phi = \nabla \Phi$. Hence, $\Phi$ must be a solution of Laplace's equation $\nabla^2 \Phi = 0$. That is,

$$\text{div } \vec{F} = \nabla \cdot (\nabla \Phi) = \nabla^2 \Phi = 0.$$

This partial differential equation has many physical applications associated with it and is considered one of the more important equations arising in science, engineering and physics. Recall that the Laplacian can be expressed in different forms depending upon the coordinate system in which it is represented. In a rectangular coordinate system, the Laplacian is expressed as

$$\nabla^2 U = \frac{\partial^2 U}{\partial x^2} + \frac{\partial^2 U}{\partial y^2} + \frac{\partial^2 U}{\partial z^2}. \qquad (9.83)$$

In cylindrical coordinates $(r, \theta, z)$, the Laplacian takes the form

$$\nabla^2 U = \frac{\partial^2 U}{\partial r^2} + \frac{1}{r}\frac{\partial U}{\partial r} + \frac{1}{r^2}\frac{\partial^2 U}{\partial \theta^2} + \frac{\partial^2 U}{\partial z^2}, \qquad (9.84)(2.37)$$

and in spherical coordinates $(r, \theta, \phi)$, it has the form

$$\nabla^2 U = \frac{\partial^2 U}{\partial r^2} + \frac{2}{r}\frac{\partial U}{\partial r} + \frac{1}{r^2}\frac{\partial^2 U}{\partial \theta^2} + \frac{\cot\theta}{r^2}\frac{\partial U}{\partial \theta} + \frac{1}{r^2 \sin^2\theta}\frac{\partial^2 U}{\partial \phi^2}. \tag{9.85}$$

## The Periodic Table of Elements

The periodic table of the elements is given on the following page. A tremendous amount of study of the physical and chemical properties of the elements has brought our knowledge of atomic physics and chemistry to the state it is in today. The periodic table of the elements is our present aid to understanding the chemical elements and how these elements react chemically. The periodic table of the elements has undergone many changes since it was first introduced by the Russian chemist Dimitri Mendeleev.

The elements in the periodic table are listed by atomic number. Recall that the atomic number represents the number of protons in the nucleus of the atom. Most elements have naturally occurring isotopes. Isotopes are atoms having the same atomic number but differing in the number of neutrons in the nucleus. Below the atomic number in the periodic table of the elements is the chemical symbol for the element and below this is the average atomic weight associated with the element. The atomic weight is defined as the relative weight rather than the actual absolute weight since atoms and molecules are too small to weigh. The gram atomic weight of an element is denoted by the weight in grams that contains the same number of atoms as twelve grams of carbon-12. The molar mass of an element is defined as the mass in grams that equals the average atomic weight. The term gram-molecular weight is abbreviated mole and the molecular weight of a compound is the sum of the atomic weights. For example, water ($H_2O$) has 2 Hydrogen plus 1 Oxygen giving it weight $2(1.008) + 16.000 = 18.016$ as the molecular weight of water.

Having $x$-grams of an element $j$, with atomic weight $A_j$, then $x/A_j$ is the number of moles of the element $j$. The number $N_A$ given by

$$N_A = \frac{\text{molar mass of an element}}{\text{weight in grams of one atom}} = 6.0221415 \times 10^{23}$$

is called Avogadro's[5] number and represents the number of atoms in one mole of an element. The weight in grams of a single atom is the molar mass divided by Avogadro's number or $A_j/N_A$. For example, the weight of a molecule of water is given by $A_j/N_A = 18.016/6.023(10^{23})\,\text{g}$.

---

[5] Amadeo Avogadro (1776-1856) Italian chemist.

# Periodic Table of the Elements

| Period | 1A | 2A | 3B | 4B | 5B | 6B | 7B | 8B | 8B | 8B | 1B | 2B | 3A | 4A | 5A | 6A | 7A | 8A |
|---|---|---|---|---|---|---|---|---|---|---|---|---|---|---|---|---|---|---|
| | 1 | 2 | 3 | 4 | 5 | 6 | 7 | 8 | 9 | 10 | 11 | 12 | 13 | 14 | 15 | 16 | 17 | 18 |
| 1 | 1 H 1.008 | | | | | | | | | | | | | | | | | 2 He 4.003 |
| 2 | 3 Li 6.941 | 4 Be 9.012 | | | | | | | | | | | 5 B 10.81 | 6 C 12.011 | 7 N 14.01 | 8 O 16.00 | 9 F 19.00 | 10 Ne 20.18 |
| 3 | 11 Na 22.99 | 12 Mg 24.31 | | | | | | | | | | | 13 Al 26.98 | 14 Si 28.09 | 15 P 30.97 | 16 S 32.07 | 17 Cl 35.45 | 18 Ar 39.95 |
| 4 | 19 K 39.10 | 20 Ca 40.08 | 21 Sc 44.96 | 22 Ti 47.88 | 23 V 50.94 | 24 Cr 52.00 | 25 Mn 54.94 | 26 Fe 55.85 | 27 Co 58.93 | 28 Ni 58.69 | 29 Cu 63.55 | 30 Zn 65.39 | 31 Ga 69.72 | 32 Ge 72.59 | 33 As 74.92 | 34 Se 78.95 | 35 Br 79.90 | 36 Kr 83.80 |
| 5 | 37 Rb 85.47 | 38 Sr 87.62 | 39 Y 88.91 | 40 Zr 91.22 | 41 Nb 92.91 | 42 Mo 95.94 | 43 Tc (98) | 44 Ru 101.1 | 45 Rh 102.9 | 46 Pd 106.4 | 47 Ag 107.9 | 48 Cd 112.4 | 49 In 114.8 | 50 Sn 118.7 | 51 Sb 121.8 | 52 Te 127.6 | 53 I 126.9 | 54 Xe 131.3 |
| 6 | 55 Cs 132.9 | 56 Ba 137.3 | 57 La *138.9 | 72 Hf 178.5 | 73 Ta 180.9 | 74 W 183.9 | 75 Re 186.2 | 76 Os 190.2 | 77 Ir 192.2 | 78 Pt 195.1 | 79 Au 197.0 | 80 Hg 200.5 | 81 Tl 204.4 | 82 Pb 207.2 | 83 Bi 209.0 | 84 Po (210) | 85 At (210) | 86 Rn (222) |
| 7 | 87 Fr (223) | 88 Ra (226) | 89 Ac **(227) | 104 Rf (257) | 105 Db (260) | 106 Sg (263) | 107 Bh (262) | 108 Hs (265) | 109 Ds (271) | 110 Ds (271) | 111 Uuu (272) | 112 Uub (277) | 113 Uut | 114 Uuq (289) | 115 Uup | 116 Uuh | 117 Uus | 118 Uuo |

*Lanthanoids

| 58 Ce 140.1 | 59 Pr 140.9 | 60 Nd 144.2 | 61 Pm (147) | 62 Sm 150.4 | 63 Eu 152.0 | 64 Gd 157.3 | 65 Tb 158.9 | 66 Dy 162.5 | 67 Ho 163.9 | 68 Er 167.3 | 69 Tm 168.9 | 70 Yb 173.0 | 71 Lu 175.0 |
|---|---|---|---|---|---|---|---|---|---|---|---|---|---|

**Actinoids

| 90 Th 232.0 | 91 Pa (231) | 92 U (238) | 93 Np (237) | 94 Pu (242) | 95 Am (243) | 96 Cm (247) | 97 Bk (247) | 98 Cf (249) | 99 Es (254) | 100 Fm (253) | 101 Md (256) | 102 No (254) | 103 Lr (257) |
|---|---|---|---|---|---|---|---|---|---|---|---|---|---|

Table 9.1 The Chemical Elements by Atomic Number

| Element | Abbreviated Symbol | Atomic Number | Atomic Weight |
|---|---|---|---|
| Hydrogen | H | 1 | 1.0080 |
| Helium | He | 2 | 4.003 |
| Lithium | Li | 3 | 6.940 |
| Beryllium | Be | 4 | 9.013 |
| Boron | B | 5 | 10.82 |
| Carbon | C | 6 | 12.010 |
| Nitrogen | N | 7 | 14.008 |
| Oxygen | O | 8 | 16.000 |
| Fluorine | F | 9 | 19.00 |
| Neon | Ne | 10 | 20.183 |
| Sodium | Na | 11 | 22.997 |
| Magnesium | Mg | 12 | 24.32 |
| Aluminum | Al | 13 | 26.97 |
| Silicon | Si | 14 | 28.06 |
| Phosphorus | P | 15 | 30.98 |
| Sulfur | S | 16 | 32.066 |
| Chlorine | Cl | 17 | 35.457 |
| Argon | A | 18 | 39.944 |
| Potassium | K | 19 | 39.096 |
| Calcium | Ca | 20 | 40.08 |
| Scandium | Sc | 21 | 45.10 |
| Titanium | Ti | 22 | 47.90 |
| Vanadium | V | 23 | 50.95 |
| Chromium | Cr | 24 | 52.01 |
| Manganese | Mn | 25 | 54.93 |
| Iron | Fe | 26 | 55.85 |
| Cobalt | Co | 27 | 58.94 |
| Nickel | Ni | 28 | 58.69 |
| Copper | Cu | 29 | 63.54 |
| Zinc | Zn | 30 | 65.38 |
| Gallium | Ga | 31 | 69.72 |
| Germanium | Ge | 32 | 72.60 |
| Arsenic | As | 33 | 74.91 |
| Selenium | Se | 34 | 78.96 |
| Bromine | Br | 35 | 79.916 |
| Krypton | Kr | 36 | 83.70 |
| Rubidium | Rb | 37 | 85.48 |
| Strontium | Sr | 38 | 87.63 |
| Yttrium | Y | 39 | 88.92 |
| Zirconium | Zr | 40 | 91.22 |
| Niobium[2] | Nb | 41 | 92.91 |
| Molybdenum | Mo | 42 | 95.95 |

[2] Also called Columbium (Cb)

| Table 9.1 The Chemical Elements by Atomic Number ||||
|---|---|---|---|
| Element | Abbreviated Symbol | Atomic Number | Atomic Weight |
| Technetium | Tc | 43 | 99[†] |
| Ruthenium | Ru | 44 | 101.7 |
| Rhodium | Rh | 45 | 102.91 |
| Palladium | Pd | 46 | 106.7 |
| Silver | Ag | 47 | 107.880 |
| Cadmium | Cd | 48 | 112.41 |
| Indium | In | 49 | 114.76 |
| Tin | Sn | 50 | 118.70 |
| Antimony | Sb | 51 | 121.76 |
| Tellurium | Te | 52 | 127.61 |
| Iodine | I | 53 | 126.92 |
| Xenon | Xe | 54 | 131.3 |
| Cesium | Cs | 55 | 132.91 |
| Barium | Ba | 56 | 137.36 |
| Lanthanum | La | 57 | 138.92 |
| Cerium | Ce | 58 | 140.13 |
| Praseodymium | Pr | 59 | 140.92 |
| Neodymium | Nd | 60 | 144.27 |
| Promethium | Pm | 61 | 147[†] |
| Samarium | Sm | 62 | 150.43 |
| Europium | Eu | 63 | 152.0 |
| Gadolinium | Gd | 64 | 156.9 |
| Terbium | Tb | 65 | 159.2 |
| Dysprosium | Dy | 66 | 162.46 |
| Holmium | Ho | 67 | 164.94 |
| Erbium | Er | 68 | 167.2 |
| Thulium | Tm | 69 | 169.4 |
| Ytterbium | Yb | 70 | 173.04 |
| Lutetium | Lu | 71 | 174.99 |
| Hafnium | Hf | 72 | 178.6 |
| Tantalum | Ta | 73 | 180.88 |
| Tungsten[3] | W | 74 | 183.92 |
| Rhenium | Re | 75 | 186.31 |
| Osmium | Os | 76 | 190.2 |
| Iridium | Ir | 77 | 193.1 |
| Platinum | Pt | 78 | 195.23 |
| Gold | Au | 79 | 197.2 |
| Mercury | Hg | 80 | 200.61 |

[†] Atomic weight uncertain.
[3] Also called Wolfram (W)

| Table 9.1 The Chemical Elements by Atomic Number ||||
|---|---|---|---|
| Element | Abbreviated Symbol | Atomic Number | Atomic Weight |
| Thallium | Tl | 81 | 204.3833 |
| Lead | Pb | 82 | 207.2 |
| Bismuth | Bi | 83 | 208.9804 |
| Polonium | Po | 84 | 209 † |
| Astatine | At | 85 | 210 † |
| Radon | Rn | 86 | 222 † |
| Francium | Fr | 87 | 223 † |
| Radium | Ra | 88 | 226 † |
| Actinium | Ac | 89 | 227 † |
| Thorium | Th | 90 | 232.0381 |
| Protactinium | Pa | 91 | 231.0359 |
| Uranium | U | 92 | 238.0289 |
| Neptunium | Np | 93 | 237 † |
| Plutonium | Pu | 94 | 244 † |
| Americium | Am | 95 | 243 † |
| Curium | Cm | 96 | 247 † |
| Berkelium | Bk | 97 | 247 † |
| Californium | Cf | 98 | 251 † |
| Einsteinium | Es | 99 | 252 † |
| Fermium | Fm | 100 | 257 † |
| Mendelevium | Md | 101 | 258 † |
| Nobelium | No | 102 | 259 † |
| Lawrencium | Lr | 103 | 262 † |
| Rutherfordium | Rf | 104 | 261 † |
| Dubnium | Db | 105 | 262 † |
| Seaborgium | Sg | 106 | 266 † |
| Bohrium | Bh | 107 | 264 † |
| Hassium | Hs | 108 | 277 † |
| Meitnerium | Mt | 109 | 268 † |
| Darmstadtium | Ds | 110 | 261.9 † |
| Roentgenium | Rg | 111 | 280 † |
| Ununbium | Uub | 112 | 285 † |
| Ununtrium | Uut | 113 | 284 † |
| Ununquadium | Uuq | 114 | 289 † |
| Ununpentium | Uup | 115 | 288 † |
| Ununhexium | Uuh | 116 | 292 † |
| Ununseptium | Uus | 117 | † |
| Ununoctium | Uuo | 118 | 294 † |

† Atomic weights uncertain.

The elements in the periodic table are listed in order of increasing atomic number. The periodic table has "groups" 1, 2, 3, ..., 18 listed across the top of the table. Elements with similar properties are listed in vertical columns under the group number. The rows in the periodic table are listed 1, 2, 3, ..., 7 and are called periods. Each horizontal row or period represents the filling of a quantum shell by electrons. Elements in the same group have a similar electron configuration in their outermost shell, the so called valence shell. The electrons in the outer shell having the largest energy occur in elements which are more chemically reactive.

The electron shells are labeled K,L,M,N,O,P and Q or levels 1,2,3,4,5,6,7 from the innermost shell to the outer shells. Each shell has subshells labeled s,p,d,f. The s subshell can have a maximum of 2 electrons. The p subshell can have a maximum of 6 electrons. The d subshell can have a maximum of 10 electrons and the f subshell can have a maximum of 14 electrons. The K shell contains only an s subshell and so can have up to two electrons. The L shell can contain s and p subshells and so can have up to $2+6=8$ electrons. The M shell can contain s,p and d subshells and can have up to $2+6+10=18$ electrons. The N and O shells can have s,p,d and f subshells and so can have a maximum of $2+6+10+14=32$ electrons. The P shell can have s,p and d subshells and the Q shell can have s and p subshells. A summary of the above information is listed in the table 9.2.

| Table 9.2 Electron Shells and Subshells | | | | | | | |
|---|---|---|---|---|---|---|---|
| Level | 1 | 2 | 3 | 4 | 5 | 6 | 7 |
| Old Notation | K | L | M | N | O | P | Q |
| Subshells | 1s | 2s,2p | 3s,3p,3d | 4s,4p,4d,4f | 5s,5p,5d,5f | 6s,6p,6d | 7s,7p |
| Maximum Electrons | 2 | 8 | 18 | 32 | 32 | 18 | 8 |

The region of space occupied by an electron is called an orbital. Think of orbitals as a probability region having variable density which can be viewed as an electron cloud which surrounds the nucleus of an atom.

In the figure illustrated the darker regions have a higher probability of finding an electron than the lighter regions. Each orbital has its own name and electron cloud shape. The notation for depicting electronic configurations can be illustrated as follows. The electronic configurations can be built one electron at a time using the periodic table. The subshells are filled by counting across periods, starting with Hydrogen up to the element of interest. Thus

$$\begin{array}{ll} H & 1s^1 \\ He & 1s^2 \\ Li & 1s^22s^1 \\ Be & 1s^22s^2 \\ B & 1s^22s^22p^1 \end{array}$$

Here the notation $1s^22s^22p^63s^23p^5$ denotes 2 electrons in subshell 1s, 2 electrons in subshell 2s, 6 electrons is subshell 2p, 2 electrons in subshell 3s and 5 electrons in subshell 3p. This is the ground state configuration for the element Cl.

For example, take the group 1 in the periodic table of elements which contains the elements H, Li, Na , K, Rb, Cs, Fr and list the electron configuration associated with each element.

$$\begin{array}{ll} H & 1s^1 \\ Li & 1s^22s^1 \\ Na & 1s^22s^22p^63s^1 \\ K & 1s^22s^22p^63s^23p^64s^1 \\ Rb & 1s^22s^22p^63s^23p^63d^{10}4s^24p^65s^1 \\ Cs & 1s^22s^22p^63s^23p^63d^{10}4s^24p^64d^{10}5s^25p^66s^1 \\ Fr & 1s^22s^22p^63s^23p^63d^{10}4s^24p^64d^{10}4f^{14}5s^25p^65d^{10}6s^26p^67s^1 \end{array}$$

Note that there is one unpaired electron in the outer shell of each element. These elements are said to have a valence of +1. The notation used for how electrons fill up the shells and subshells starts with the level number followed by a superscript which denotes the number of electrons in the subshell. The proper order of the subshells, listed in order of increasing energy, is

$$1s, 2s, 2p, 3s, 3p, 4s, 3d, 4p, 5s, 4d, 5p, 6s, 4f, 5d, 6p, 7s, 5f, 6d$$

which is taken from the periodic table of elements listed in the figure 9-18.

In contrast the elements in column 17 of the periodic table of elements are called the Halogens and have the electronic structure

$$\begin{array}{ll} F & 1s^22s^22p^5 \\ Cl & 1s^22s^22p^63s^23p^5 \\ Br & 1s^22s^22p^63s^23p^63d^{10}4s^24p^5 \\ I & 1s^22s^22p^63s^23p^63d^{10}4s^24p^64d^{10}5s^25p^5 \\ At & 1s^22s^22p^63s^23p^63d^{10}4s^24p^64d^{10}4f^{14}5s^25p^65d^{10}6s^26p^5 \end{array}$$

Note that these elements are shy 1 electron from completing the outer shell and hence are said to have a valence of −1.

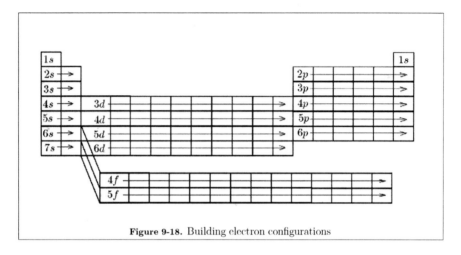

**Figure 9-18.** Building electron configurations

The above rules for constructing electronic configurations have exceptions when approaching the bottom of the periodic table.

## Modeling of Chemical Kinetics

In physical chemistry there are many applications of differential equations. Chemical kinetics is concerned with rates of chemical changes. Consider the reaction equation between carbon monoxide and nitrogen dioxide

$$CO + NO_2 \xrightarrow{k} CO_2 + NO.$$

In this equation, the compounds CO and $NO_2$ are called reactants and $CO_2$ and NO are called the product elements and $k$ is called the rate constant. The rate or speed of reaction tells how the concentration of a reactant (or product) changes with time. Concentrations are expressed in gram-moles per liter and the dimension of a reaction rate is (mole − liter$^{-1}$ time$^{-1}$).

In general, chemical reactions can be expressed by equations having the form

$$i_1 A_1 + i_2 A_2 + i_3 A_3 + \ldots \xrightarrow{k} j_1 B_1 + j_2 B_2 + j_3 B_3 + \ldots,$$

where $A_1, A_2, A_3, \ldots$ denote molecules of the reactants and $B_1, B_2, B_3, \ldots$ denote molecules of the products formed and $k$ is the rate constant. The quantities $i_1, i_2, i_3, \ldots, j_1, j_2, j_3, \ldots$ are positive integers denoting the proportion of elements that combine during the reaction (stoichiometry). The reaction rate is the change in concentration of a substance per unit of time and is modeled after the law of mass action at constant temperature, which can be expressed

$$-\frac{d[A_1]}{dt} = \text{Reaction rate} = K[A_1]^\alpha [A_2]^\beta [A_3]^\gamma \cdots,$$

where $K$ is a proportionality constant called the rate constant of the reaction, and the notation $[A_k]$ denotes the concentration of the $k$th reactant in units of moles per liter. The powers $\alpha, \beta, \gamma, \ldots$ to which the concentrations of the reactants are raised determine the order of the reaction. Experiments show the powers, $\alpha, \beta, \gamma, \ldots$ are usually one of the numbers $0, \frac{1}{2}, 1, \frac{3}{2}$ or $2$. Reactions are classified according to the sum of these exponents. The reaction is of the first order if $\alpha + \beta + \gamma + \ldots = 1$; it is of the second order if the sum of the exponents is 2; it is of the third order if the sum of the exponents is 3. Sometimes the sum of the exponents is fractional. Note that the order of a chemical reaction is arrived at through experimentation. It is not possible to determine the order of a reaction by looking at the reaction equation.

The rate law easiest to solve is derived from a first-order reaction equation having the form

$$A \xrightarrow{k_1} P,$$

where $A$ denotes the reactant which decays to the product element $P$ and $k_1$ is a rate constant. An example of a first order reaction of this type is

$$C_2H_6 \xrightarrow{k_1} 2CH_3$$

If $[A]$ denotes the concentration of the reactant $A$, then the differential equation describing the reaction is written

$$-\frac{d[A]}{dt} = k_1[A],$$

which describes the rate law of a first-order reaction. This equation depicts the concentration of $A$ decreasing at a rate proportional to the concentration of $A$. Such an equation is easily solved by separation of variables. If at time $t = 0$ the initial concentration is $[A]_0$, then at any time $t$ the concentration is given by

$$[A] = [A]_0 e^{-k_1 t}.$$

Consider a second-order reaction of the form

$$A + B \xrightarrow{k_2} P,$$

where $A$ and $B$ are the reacting elements, $P$ represents the product element and $k_2$ is the rate constant. Two examples of a second order reaction having this form are

(a) $H^+ + OH^- \xrightarrow{k_2} H_2O$

(b) $H_2 + I_2 \xrightarrow{k_2} 2HI$.

Let $[A]$ and $[B]$ denote the concentrations of the reactants $A$ and $B$. The rate law for the above reaction can be expressed as

$$-\frac{d[A]}{dt} = k_2[A][B].$$

Note that the negative sign shows that the concentration of $A$ decreases at a rate proportional to the product of the concentration of the reacting elements. Let $[A]_0$ and $[B]_0$ denote the initial concentrations of the reactants $A$ and $B$ in moles/liter. If $x = x(t)$ denotes the number of moles/liter which have reacted after time $t$, then the concentration of $A$ decreases to $([A]_0 - x)$ and the concentration of $B$ decreases to $([B]_0 - x)$. The change in concentration is due to the fact that one molecule of $A$ reacts with one molecule of $B$ and if a molecule of $A$ is lost, then a molecule of $B$ also must be lost. This allows the above differential equation for the reaction rate to be expressed in the form

$$-\frac{d[A]}{dt} = -\frac{d([A]_0 - x)}{dt} = k_2([A]_0 - x)([B]_0 - x).$$

This differential equation can be solved by the method of separation of variables.

Experimentally it has been found that the reaction

$$CO + NO_2 \xrightarrow{k} CO_2 + NO$$

is a second-order reaction and the reaction rate is given by the product $K[CO][NO_2]$. There are several choices for the reaction rate which can be taken to be the change in concentration of $CO_2$ per unit of time or as the change in concentration of $CO$ per unit of time. The reaction rate is a positive quantity, therefore by selecting the reaction rate in terms of the change in concentration of CO per unit of time, there results

$$\text{Reaction rate} = -\frac{\Delta[CO]}{\Delta t}.$$

Alternatively, it is possible to represent the reaction rate in terms of the change in concentration of $CO_2$ per unit of time, then one would write

$$\text{Reaction rate} = \frac{\Delta[CO_2]}{\Delta t}.$$

In the above reaction, note that $[CO]$ is decreasing while $[CO_2]$ is increasing and one mole of $CO_2$ is formed for each mole of $CO$ reacting, hence the minus sign in the above reaction rate equation. One could also represent the reaction rate in terms of a negative change in concentration of $NO_2$ per unit time or in terms of a positive change in the concentration of NO per unit of time.

Assume initially there are $a$ moles/liter of CO and $b$ moles/liter of $NO_2$ which are combined. If $x = x(t)$ is used to denote the number of moles/liter which have reacted after time $t$, then the concentration of CO at time $t$ can be represented as $[CO] = (a-x)$ moles/liter and the concentration of $NO_2$ at time $t$ can be represented as $[NO_2] = (b-x)$ moles/liter. The law of mass action enables us to model this reaction with the differential equation

$$-\frac{d}{dt}(a-x) = -\frac{d}{dt}(b-x) = \frac{dx}{dt} = k(a-x)(b-x). \tag{9.86}$$

Separating the variables and using the technique of partial fractions gives

$$\frac{dx}{(a-x)(b-x)} = k\,dt = \frac{1}{(b-a)}\frac{dx}{(a-x)} - \frac{1}{(b-a)}\frac{dx}{(b-x)}, \quad b \neq a, \tag{9.87}$$

which is an equation easily integrated and results in the solution

$$\frac{b-x}{a-x} = A\exp[(b-a)kt], \tag{9.89}$$

where $A$ is a constant of integration. Solving equation (9.89) for the variable $x$ gives

$$x = x(t) = \frac{b - aA\exp[(b-a)kt]}{1 - A\exp[(b-a)kt]}, \quad b \neq a. \tag{9.89}$$

Applying the initial condition, requires that $x = 0$ at $t = 0$.

Solve for the constant of integration to show $A = b/a$. The limit given by $\lim_{t\to\infty} x(t)$ is of some interest as this limiting value produces the limiting reagent of the reaction. Observe

$$\lim_{t\to\infty} x(t) = \begin{cases} a, & a < b \\ b, & b < a \end{cases} = \min(a,b).$$

### Thermodynamics

Experiments on a fixed mass of gas has established the following gas laws which relate the pressure $p$, volume $V$ and absolute temperature $T$.

**Boyle's Law** If $T$ is held constant, then $pV = constant$.
**Charles's Law** If $p$ is held constant, then $\frac{V}{T} = constant$.
**Gay-Lussac Law** If $V$ is held constant, then $\frac{p}{T} = constant$.

These laws are summarized using the gas equation

$$\frac{p_1 V_1}{T_1} = \frac{p_2 V_2}{T_2}$$

where pressure $p$ can be measured in units $[N/m^2]$, volume $V$ can be measured in units $[m^3]$ and absolute temperature $T$ is measured in units $[K]$.

The ideal gas absolute temperature is defined using Boyles law which states $pV \propto T$ which produces the equation of state for an ideal gas, which can be expressed in the form

$$pV = nRT$$

$$[\frac{N}{m^2}][m^3] = [mol]\left[\frac{J}{mol\,K}\right][K]$$

where $n$ is the number of moles of gas and $R = 8.314472$ $[\frac{J}{mol\,K}]$ is the ideal gas constant or universal molar gas constant. Note that real gases may or may not obey the ideal gas law. For gases which are imperfect, there are many other proposed equations of state. Some of these proposed equations are valid over selected ranges and conditions and can be found under such names as *Van der Waals equation, Berthelot equation, Dieterici equation, Beattie-Bridgeman equation, Virial equation.*

The zeroth law of thermodynamics states that if two bodies are in thermal equilibrium with a third body, then the two bodies must be in thermal equilibrium with each other. The zeroth law is used to develop the concept of temperature. Here thermodynamic equilibrium infers that the system is (i) in chemical equilibrium and (ii) there are no pressure or temperature gradients which would cause the system to change with time.

The first law of thermodynamics is an energy conservation principle which can be expressed $dQ = dU + dW$ where $dQ$ is the heat supplied to a gas, $dU$ is the change in internal energy of the gas and $dW$ is the external work done.

The second law of thermodynamics examines processes that can happen in an isolated system and states that the only processes which can occur are those for which the entropy either increases or remains constant. Here entropy $S$ is related to the ability or inability of a systems energy to do work. The change in entropy is defined as $dS = dQ/T$ where $dQ$ is the heat absorbed in an isothermal and reversible process and $T$ denotes the absolute temperature.

Recall that the ability of gases to change when subjected to pressure and temperature variations can be described by the equation of state of an ideal gas

$$PV = nRT, \qquad (9.90)$$

where $P$ is the pressure [N/m²], V is the volume [m³], $n$ is the amount of gas [moles], $R$ is the universal gas constant [J/mol · K], and $T$ is the temperature [K]. For an ideal gas, the gas constant $R$ can also be expressed in terms of the specific heat at constant pressure $C_p$, [J/mol · K] and the specific heat at constant volume $C_v$, [J/mol · K] by Mayer's equation $R = C_p - C_v$. Equation (9.90) is illustrated in the pressure-volume diagram of figure 9-19.

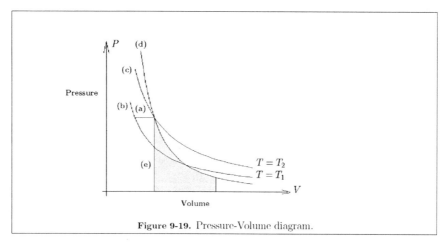

**Figure 9-19.** Pressure-Volume diagram.

The curves where $T$ is a constant are called isothermal curves and are the hyperbola (b) and (c) illustrated in figure 9-19 for the values $T_1$ and $T_2$. When a gas undergoes changes of state it can do so by an isobaric process (P is a constant) illustrated by line (a) in figure 9-19, an isovolumetric process (V is a constant) illustrated by the line (e) in figure 9-19, an isothermal process (T is a constant) illustrated by the hyperbolas with $T = T_1$ and $T = T_2$ in figure 9-19, and an adiabatic process (no heat is transferred) represented by the curve (d) in figure 9-19.

The first law of thermodynamics states that when a gas undergoes a change, the equation $dU = dQ + dW$ must be satisfied, where $dU$ is the change in internal energy, $dQ$ is the change in heat supplied to the gas, and $dW$ is the work done. An adiabatic process is one in which $dQ = 0$. For an adiabatic process, the first law of thermodynamics requires $dU = dW$. The work done $dW$ is related to the volume change by the relation $dW = -PdV$, and the change in internal energy is related to the temperature change by the relation $dU = \mu C_p dT$. For an adiabatic process

$$\frac{dP}{P} + \gamma \frac{dV}{V} = 0$$

which implies the adiabatic curve (d) in figure 9-19 can be described by any of the equations

$$TV^{\gamma-1} = \text{Constant}, \quad TP^{\frac{1-\gamma}{\gamma}} = \text{Constant}, \quad or \quad PV^{\gamma} = \text{Constant},$$

where $\gamma = C_p/C_v$ is the ratio of the specific heat at constant pressure to the specific heat at constant volume. Also note that during an adiabatic process $dQ = 0$ so that the work done by the system undergoing a change in volume is given by the integral of $dW$ which is represented by the shaded area in the figure 9-19.

## Electrical Circuits

The basic elements needed to study electrical circuits are as follows:

(a) Resistance $R$ is denoted by the symbol ⏚⟋⟍⟋⟍⟋⏚
    The dimension of resistance is ohms[6] and written $[R]$ = ohms.

(b) Inductance $L$ is denoted by the symbol ⏚〇〇〇〇〇〇⏚
    The dimension of inductance is henries[7] and written $[L]$ = henries.

(c) Capacitance $C$ denoted by the symbol ⊣⊢
    The dimension of capacitance is farads[8] and written $[C]$ = farads.

(d) Electromotive force (emf) $E$ or $V$ denoted by the symbols ⊙ or ⊣∣⊢
    The dimensions of electromotive force is volts[9] and written $[E] = [V]$ = volts.

(d) Current $I$ is a function of time, denoted $I = I(t)$, with dimensions of amperes[10] and written $[I]$ = amperes.

(f) Charge $Q$ on the capacitance is a function of time and written $Q = Q(t)$, with dimensions $[Q]$ = coulombs.

The basic laws associated with electrical circuits are as follows: The current is the time rate of change of charge. This can be represented with the above notation as

$$I = \frac{dQ}{dt} \quad \text{with} \quad [I] = \text{amperes}, \quad [\frac{dQ}{dt}] = \text{coulombs/second} \tag{9.91}$$

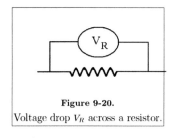

Figure 9-20.
Voltage drop $V_R$ across a resistor.

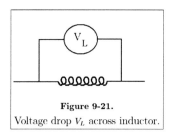

Figure 9-21.
Voltage drop $V_L$ across inductor.

The voltage drop $V_R$ across a resistance, see figure 9-20, is proportional to the current through the resistance. The proportionality constant is called the resistance $R$. In symbols this can be represented as $V_R = RI$ where

$$[V_R] = \text{volts} = [R][I] = (\text{ohm})(\text{ampere}) \tag{9.92}$$

---

[6] George Simon Ohm (1787–1854), German physicist.
[7] Joseph Henry (1797–1878), American physicist.
[8] Michael Faraday (1791–1867) English physicist.
[9] Alessandro Volta (1745–1827) Italian scientist.
[10] André Marie Ampére (1775–1836) French physicist.

The voltage drop $V_L$ across an inductance, see figure 9-21, is proportional to the time rate of change of current through the inductance. The proportionality constant is called the inductance $L$. In symbols this can be represented as

$$V_L = L\frac{dI}{dt} \quad \text{where} \quad [V_L] = \text{volts} = [L]\left[\frac{dI}{dt}\right] = (\text{henry})(\text{ampere/second}) \tag{9.93}$$

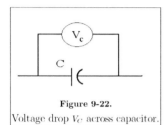

Figure 9-22.
Voltage drop $V_C$ across capacitor.

The voltage drop $V_C$ across a capacitance, see figure 9-22, is proportional to the charge $Q$ of the capacitance. The proportionality constant is denoted $1/C$. In symbols this is represented as

$$V_C = \frac{Q}{C} \tag{9.94}$$

where $[V_C] = \text{volts} = \left[\frac{1}{C}\right][Q] = \text{coulombs/farad}$

The Kirchoff laws for an electric circuit are.

**Kirchhoff's[11] first law :**
The sum of the voltage drops around a closed circuit must equal zero.
**Kirchhoff's second law:**
The amount of current into a junction must equal the current leaving the junction.

The place in an electrical circuit where two or more circuit elements are joined together is called a junction. A closed circuit or loop occurs whenever a path constructed through connected elements within a circuit closes upon itself. Voltage drops are selected as positive, whereas voltage rises are selected as negative.

**Example 9-5.** For the RC-circuit illustrated in figure 9-23, set up the differential equation describing the rate of change of the charge $Q$ on the capacitor. Make the assumption that $Q(0) = 0$.

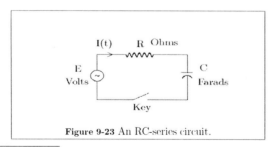

Figure 9-23 An RC-series circuit.

---

[11] Gustav Robert Kirchhoff (1924-1887) German physicist.

**Solution:** For a path around the circuit illustrated in figure 9-23, the Kirchhoff's voltage law would be written

$$V_R + V_C - E = 0.$$

Let $I = I(t) = \frac{dQ}{dt}$ denote the current in the circuit at any time $t$. By Kirchhoff's first law:

$$\begin{pmatrix} \text{Voltage drop} \\ \text{across } R \end{pmatrix} + \begin{pmatrix} \text{Voltage drop} \\ \text{across } C \end{pmatrix} = \begin{pmatrix} \text{Applied} \\ \text{emf} \end{pmatrix}$$

$$V_R + V_C = E$$
$$RI + \frac{Q}{C} = E.$$

This gives the differential equation

$$L(Q) = R\frac{dQ}{dt} + \frac{1}{C}Q = E \tag{9.95}$$

where $R, C$ and $E$ are constants. The solution of the homogeneous equation $L(Q) = 0$ is $Q_c = c_1 \exp(-t/RC)$ and a particular solution is $Q_p = CE$. The general solution of equation is represented by the sum $Q = Q_c + Q_p$ and the solution satisfying $Q(0) = 0$ is given by

$$Q = Q(t) = EC(1 - e^{-t/RC}) \tag{9.96}$$

The relation (9.96) is employed to determine the current $I$ and voltages $V_C$ and $V_R$ as

$$I = I(t) = \frac{dQ}{dt} = \frac{E}{R}e^{-t/RC}$$
$$V_C = \frac{Q}{C} = E(1 - e^{-t/RC}) \tag{9.97}$$
$$V_R = RI = Ee^{-t/RC}$$

In equations (9.96) and ((9.97) the term $\exp(-t/RC)$ is called a transient term and the constant $\tau = RC$ is called the time constant for the circuit. In general, terms of the form $\exp(-t/\alpha)$ are transient terms, and such terms are short lived and quickly or slowly decay, depending upon the time constant $\tau = \alpha$. The following table illustrates values of $\exp(-t/\alpha)$ for $t$ equal to various values of the time constant.

| Time $t$ | $\exp(-t/\alpha)$ |
|---|---|
| $\alpha$ | 0.3679 |
| $2\alpha$ | 0.1353 |
| $3\alpha$ | 0.0498 |
| $4\alpha$ | 0.0183 |
| $5\alpha$ | 0.0067 |

The values in the above table gives us valuable information concerning equations such as (9.96) and (9.97). The table shows that decaying exponential terms are essentially zero after five time constants. This is because the values of the exponential terms are less than 1 percent of their initial values.

Solutions to circuit problems are usually divided into two parts, called transient terms and steady state terms. Transient terms eventually decay and disappear and do not contribute to the solution after about 5 time constants. The steady state terms are the part of the solution which remains after the transient terms become negligible. ∎

**Example 9-6.** For the parallel circuit illustrated in figure 9-24, apply Kirchhoff's first law to each of the three closed circuits.

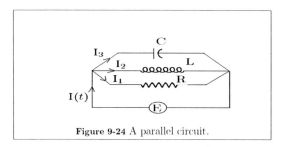

**Figure 9-24** A parallel circuit.

Note that each closed circuit has the same voltage drop. This produces the following equations.
$$E = RI_1$$
$$E = L\frac{dI_2}{dt} \tag{9.98}$$
$$E = \frac{1}{C}\int I_3\,dt.$$

Kirchhoff's second law applied to the given circuit tells us

$$I = I_1 + I_2 + I_3. \tag{9.99}$$

If the impressed current $I$ is given, the above four equations can be reduced to one ordinary differential equation from which the impressed voltage $E$ can be found. Write equation (9.99) in the form
$$I = \frac{E}{R} + \frac{1}{L}\int E\,dt + C\frac{dE}{dt}.$$
called a differential integral equation. By differentiation of this equation there results an ordinary linear second-order differential equation:
$$\frac{dI}{dt} = \frac{1}{R}\frac{dE}{dt} + \frac{E}{L} + C\frac{d^2E}{dt^2},$$
where $E$ is the dependent variable to be determined.

Conversely, if $E$ is given and $I$ is unknown, then equations (9.98) give us $I_1$, $I_2$, and $I_3$, and equation (9.99) can be used to determine the current $I$.

∎

### Four-terminal networks

Consider the electrical networks illustrated in figure 9-25. No matter how complicated the circuit inside the boxes, there are only two input and two output terminals. Such devices are called four terminal networks and are represented by a box like those illustrated in figure

9-25, where the quantities $Z$, $Z_1$, and $Z_2$ are called impedances. Impedances $Z$ are used in alternating current (a.c.) circuits and are analogous to the resistance $R$ use in direct current (d.c.) circuits.

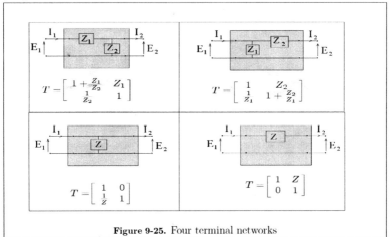

**Figure 9-25.** Four terminal networks

In figure 9-25 the quantities $I_1$, $E_1$ and $I_2$, $E_2$ are the input and output current and voltages. The vectors $S_1 = \text{col}[E_1, I_1]$ and $S_2 = \text{col}[E_2, I_2]$ are called the input state vector and output state vector of the network. The networks are assumed to be linear so that the general relation between the input and output states can be expressed as

$$S_1 = TS_2, \quad \text{where} \quad T = \begin{bmatrix} T_{11} & T_{12} \\ T_{21} & T_{22} \end{bmatrix}$$

is called the transmission matrix. The element $T_{11}, T_{12}, T_{21}, T_{22}$ are in general complex numbers which satisfy the property $\det T = T_{11}T_{22} - T_{21}T_{12} = 1$. Solving for $S_2$ in terms of $S_1$ gives

$$S_2 = T^{-1}S_1 = PS_1$$

where the matrix $P$ is called the transfer matrix of the network and is given by

$$P = \begin{bmatrix} T_{22} & -T_{12} \\ -T_{21} & T_{11} \end{bmatrix}$$

If the input output current and voltages are linearly related, then it is easy to solve for the currents in terms of the voltages or the voltages in terms of the currents to obtain

$$\begin{bmatrix} I_1 \\ I_2 \end{bmatrix} = \begin{bmatrix} \frac{T_{22}}{T_{12}} & -\frac{1}{T_{12}} \\ \frac{1}{T_{12}} & -\frac{T_{11}}{T_{12}} \end{bmatrix} \begin{bmatrix} E_1 \\ E_2 \end{bmatrix} \qquad \begin{bmatrix} E_1 \\ E_2 \end{bmatrix} = \begin{bmatrix} \frac{T_{11}}{T_{21}} & -\frac{1}{T_{21}} \\ \frac{1}{T_{21}} & -\frac{T_{22}}{T_{21}} \end{bmatrix} \begin{bmatrix} I_1 \\ I_2 \end{bmatrix}$$

Applying Kirchoff's laws to the short-circuit condition $E_2 = 0$ and the open-circuit condition $I_2 = 0$ allows for the determination of the transmission matrices given in the figure 9-25.

## Prisms

Snell's law of refraction states that when a ray of light hits an interface between two media, where the speed of light $c_1$ in media 1 differs from the speed of light $c_2$ in media 2, then refraction occurs and the ray entering the second material is deviated in direction such that

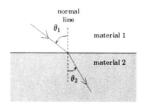

$$n = \frac{c_1}{c_2} = \frac{\sin\theta_1}{\sin\theta_2}$$

where $\theta_1$ and $\theta_2$ are the entering and leaving angles of the ray measured with respect to the normal line to the media interface. The quantity $n$ is called the relative refractive index.

Consider a ray of light impinging upon a prism as illustrated in the figure 9-26. Let the refracting angle of the prism be denoted by $\alpha$ and let $n$ be the refractive index of the prism. Extend the leaving ray in a straight line to intersect the extension of the entering ray. The angle of intersection $\delta$ between these extended lines is called the angle of deviation. To solve for the angle of deviation as a function of the entering angle $\theta_1$ and the refracting angle $\alpha$ of the prism make use of the fact that the exterior angle of a triangle is equal to the sum of the opposite interior angles.

An examination of figure 9-26 shows that

$$\delta = (\theta_1 - \theta_2) + (\theta_4 - \theta_3) \quad \text{and} \quad \alpha = \theta_2 + \theta_3 \quad \Longrightarrow \quad \delta = \theta_1 + \theta_4 - \alpha \tag{9.100}$$

Note that by Snell's law

$$\begin{aligned}\sin\theta_4 &= n\sin\theta_3 \\ \sin\theta_1 &= n\sin\theta_2\end{aligned} \quad \Longrightarrow \quad \begin{aligned}\theta_4 &= \sin^{-1}(n\sin\theta_3) \\ \sin\theta_2 &= \frac{1}{n}\sin\theta_1\end{aligned} \quad \Longrightarrow \quad \delta = \theta_1 + \sin^{-1}(n\sin\theta_3) - \alpha \tag{9.101}$$

From equation (9.100) one finds $\theta_3 = \alpha - \theta_2$ and consequently

$$\sin\theta_3 = \sin(\alpha - \theta_2) = \sin\alpha\cos\theta_2 - \cos\alpha\sin\theta_2 \tag{9.102}$$

Also from equation (9.101) note that

$$\cos\theta_2 = \sqrt{1 - \sin^2\theta_2} = \sqrt{1 - \frac{\sin^2\theta_1}{n^2}} \tag{9.103}$$

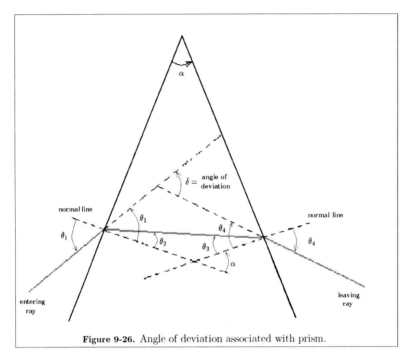

Figure 9-26. Angle of deviation associated with prism.

Using the results from equations (9.101), (9.102) and (9.103) verify that

$$n \sin \theta_3 = \sin \alpha \sqrt{n^2 - \sin^2 \theta_1} - \cos \alpha \sin \theta_1 \tag{9.104}$$

Substitute this last result into the right-hand side of equation (9.101) to show the angle of deviation can be expressed

$$\delta = \theta_1 + \sin^{-1}\left[\sin \alpha \sqrt{n^2 - \sin^2 \theta_1} - \cos \alpha \sin \theta_1\right] - \alpha \tag{9.105}$$

As a calculus problem, show that $\delta$ has a minimum value when $\theta_1 = \theta_4$. Also note that in general $n$ is a function of wavelength with red light having the least deflection and violet light having the maximum deflection. This is what produces the colors of the rainbow when white light enters the prism.

# References

- Abramowiz, M., Stegun, I.A., **Handbook of Mathematical Functions with Formulas, Graphs, and Mathematical Tables**, Dover Publications, Inc., New York, 1970.
- Atkins, P.W., **Physical Chemistry**, Second Edition, W.H. Freeman and Company, New York, 1982.
- Bell, E.T., **Men of Mathematics**, Simon and Schuster, New York, 1937.
- Bowditch, N., **American Practical Navigator**, H.O. Pub. No. 9, United States Printing Office, Washington, 1962.
- Brown, T.L., LeMay, Jr., H.E., Bursten, B.E., **Chemistry, The Central Science**, Fifth Edition, Prentice Hall, Englewood Cliffs, New Jersey, 1991.
- Cajori, F. **A History of Mathematics**, second edition, The MacMillian Company, New York, 1893.
- Coolidge, J.L., **A History of Geometrical Methods**, Dover Publications, New York, 1963.
- Erdélyi, A., Editor, **Higher Transcendental Functions**, Volumes I,II,III, Robert E. Krieger Publishing Company, Malabar, Florida, 1981.
- Eves, H., **An Introduction to the History of Mathematics**, Revised Edition, Holt, Rinehart and Winston, New York, 1964.
- Eves, H., **Great Moments in Mathematics (Before 1650)**, Mathematical Association of America, 1980.
- Eves, H., **Great Moments in Mathematics (After 1650)**, Mathematical Association of America, 1981.
- Gradshteyn, I.S., Ryzhik, I.M., **Table of Integrals, Series, and Products**, Academic Press, New York, 1980.
- Harris, J.W., Stocker, H., **Handbook of Mathematics and Computational Science**, Springer-Verlag, New York, 1998.
- Heath, Sir Thomas, **An Introduction to the History of Mathematics**, Volumes I,II, Oxford University Press, Glasgow, 1921.
- Hodgman, C.D, Weast, R.C., Selby, S.M., Editors, **Handbook of Chemistry and Physics**, 39th Edition, Chemical Rubber Publishing Company, Cleveland, Ohio, 1957.
- Johnson, D.E., Johnson, J.R., **Mathematical Methods in Engineering and Physics**, The Ronald Press Company, New York, 1965.

- Knuth, D.E., **The TeXbook**, Addison-Wesley Publishing Company, Reading, Massachusetts, 1984.
- Korn, G.A., Korn, T.M., **Mathematical Handbook for Scientists and Engineers**, McGraw Hill Book Company, New York, 1968.
- Maclean, Andrew J.P., **Parametric Equations for Surfaces**, Centre for Autonomous Systems, an ARC centre of Excellence, The University of Sydney, March 2005, p.30.
- Magnus, W., Oberhettinger, F., **Formulas and Theorems for the Functions of Mathematical Physics**, Chelsea Publishing Company, New York, 1954.
- Maor, E., **The Pythagorean Theorem**, Princeton University Press, Princeton and Oxford, 2007.
- Masterton, W.L., Slowinski, E.J., **Chemical Principles**, Second Edition, W.B. Saunders Company, Philadelphia, 1969.
- Netz, R., Noel, W., **The Archimedes Codex**, Da Capo Press, Philadelphia, PA, 2007.
- Newman, J.R., **The World of Mathematics**, Simon and Schuster, New York, 1956.
- Rektorys, K., **Survey of Applicable Mathematics**, Iliffe books Ltd, London, 1969.
- Salomon, D., **The Advanced TeXbook**, Springer-Verlag, New York, 1995.
- Sawyer, S.A., Krantz, S.G., **A TeX Primer for Scientists**, CRC Press, Boca Raton, Florida, 1995.
- Smith, D.E., **History of Modern Mathematics**, Mathematical Monographs, No.1, Fourth Edition, John Wiley and Sons, New York, 1906.
- Spiegel, M.R., **Mathematical Handbook of Formulas and Tables**, Schaum's Outline Series, McGraw Hill Book Company, New York, 1968.
- Spiegel, M.R., **Theory and Problems of Statistics**, Schaum Publishing Company, New York, 1961.
- Stewart, J., **Calculus Early Transcendentals**, Brooks/Cole Publishing Company, Boston, 1999.
- **Trachtenberg Speed System of Basic Mathematics**, translated by Ann Cutler and Rudolph Mchane, Doubleday & Company, Inc, Garden City, New York, 1967.
- Whittaker, E.T., Watson, G.N., **A Course of Modern Analysis**, Fourth Edition, Cambridge University Press, Cambridge, 1962.

# APPENDIX A
# Units of Measurement

The following units, abbreviations and prefixes are from the Système International d'Unitès (designated SI in all Languages.)

**Prefixes.**

| | Abbreviations | |
|---|---|---|
| Prefix | Multiplication factor | Symbol |
| tera | $10^{12}$ | T |
| giga | $10^{9}$ | G |
| mega | $10^{6}$ | M |
| kilo | $10^{3}$ | K |
| hecto | $10^{2}$ | h |
| deka | 10 | da |
| deci | $10^{-1}$ | d |
| centi | $10^{-2}$ | c |
| milli | $10^{-3}$ | m |
| micro | $10^{-6}$ | $\mu$ |
| nano | $10^{-9}$ | n |
| pico | $10^{-12}$ | p |

**Basic Units.**

| Basic units of measurement | | |
|---|---|---|
| Unit | Name | Symbol |
| Length | meter | m |
| Mass | kilogram | kg |
| Time | second | s |
| Electric current | ampere | A |
| Temperature | degree Kelvin | °K |
| Luminous intensity | candela | cd |

| Supplementary units | | |
|---|---|---|
| Unit | Name | Symbol |
| Plane angle | radian | rad |
| Solid angle | steradian | sr |

| DERIVED UNITS | | |
|---|---|---|
| Name | Units | Symbol |
| Area | square meter | $m^2$ |
| Volume | cubic meter | $m^3$ |
| Frequency | hertz | Hz ($s^{-1}$) |
| Density | kilogram per cubic meter | $kg/m^3$ |
| Velocity | meter per second | $m/s$ |
| Angular velocity | radian per second | $rad/s$ |
| Acceleration | meter per second squared | $m/s^2$ |
| Angular acceleration | radian per second squared | $rad/s^2$ |
| Force | newton | N ($kg \cdot m/s^2$) |
| Pressure | newton per square meter | $N/m^2$ |
| Kinematic viscosity | square meter per second | $m^2/s$ |
| Dynamic viscosity | newton second per square meter | $N \cdot s/m^2$ |
| Work, energy, quantity of heat | joule | J ($N \cdot m$) |
| Power | watt | W (J/s) |
| Electric charge | coulomb | C ($A \cdot s$) |
| Voltage, Potential difference | volt | V (W/A) |
| Electromotive force | volt | V (W/A) |
| Electric force field | volt per meter | V/m |
| Electric resistance | ohm | $\Omega$ (V/A) |
| Electric capacitance | farad | F ($A \cdot s/V$) |
| Magnetic flux | weber | Wb ($V \cdot s$) |
| Inductance | henry | H ($V \cdot s/A$) |
| Magnetic flux density | tesla | T ($Wb/m^2$) |
| Magnetic field strength | ampere per meter | A/m |
| Magnetomotive force | ampere | A |

**Physical Constants:**

- $4 \arctan 1 = \pi = 3.14159\,26535\,89793\,23846\,2643\,\ldots$
- $\lim_{n \to \infty} \left(1 + \frac{1}{n}\right)^n = e = 2.71828\,18284\,59045\,23536\,0287\,\ldots$
- Euler's constant $\quad \gamma = 0.57721\,56649\,01532\,86060\,6512\,\ldots$
- $\gamma = \lim_{n \to \infty} \left(1 + \frac{1}{2} + \frac{1}{3} + \cdots + \frac{1}{n} - \log n\right) \quad$ Euler's constant
- Speed of light in vacuum $= 2.997925(10)^8 \ m \ s^{-1}$
- Electron charge $= 1.60210(10)^{-19} \ C$
- Avogadro's constant $= 6.0221415(10)^{23} \ mol^{-1}$
- Plank's constant $= 6.6256(10)^{-34} \ J\,s$
- Universal gas constant $= 8.3143 \ J\,K^{-1}\,mol^{-1} = 8314.3 \ J\,Kg^{-1}\,K^{-1}$
- Boltzmann constant $= 1.38054(10)^{-23} \ J\,K^{-1}$
- Stefan–Boltzmann constant $= 5.6697(10)^{-8} \ W\,m^{-2}\,K^{-4}$
- Gravitational constant $= 6.67(10)^{-11} \ N\,m^2 kg^{-2}$

**Appendix A**

# Appendix B
## Table of Integrals
### Indefinite Integrals

#### General Integration Properties

1. If $\dfrac{dF(x)}{dx} = f(x)$, then $\int f(x)\,dx = F(x) + C$

2. If $\int f(x)\,dx = F(x) + C$, then the substitution $x = g(u)$ gives $\int f(g(u))\,g'(u)\,du = F(g(u)) + C$

   For example, if $\int \dfrac{dx}{x^2 + \beta^2} = \dfrac{1}{\beta}\tan^{-1}\dfrac{x}{\beta} + C$, then $\int \dfrac{du}{(u+\alpha)^2 + \beta^2} = \dfrac{1}{\beta}\tan^{-1}\dfrac{u+\alpha}{\beta} + C$

3. Integration by parts. If $v_1(x) = \int v(x)\,dx$, then $\int u(x)v(x)\,dx = u(x)v_1(x) - \int u'(x)v_1(x)\,dx$

4. Repeated integration by parts or generalized integration by parts.
   If $v_1(x) = \int v(x)\,dx$, $v_2(x) = \int v_1(x)\,dx, \ldots, v_n(x) = \int v_{n-1}(x)\,dx$, then
   $$\int u(x)v(x)\,dx = uv_1 - u'v_2 + u''v_3 - u'''v_4 + \cdots + (-1)^{n-1}u^{n-1}v_n + (-1)^n \int u^{(n)}(x)v_n(x)\,dx$$

5. If $f^{-1}(x)$ is the inverse function of $f(x)$ and if $\int f(x)\,dx$ is known, then
   $$\int f^{-1}(x)\,dx = zf(z) - \int f(z)\,dz, \qquad \text{where} \quad z = f^{-1}(x)$$

6. Fundamental theorem of calculus.
   If the indefinite integral of $f(x)$ is known, say $\int f(x)\,dx = F(x) + C$, then the definite integral
   $$\int_a^b dA = \int_a^b f(x)\,dx = F(x)\big|_a^b = F(b) - F(a)$$
   represents the area bounded by the x-axis, the curve $y = f(x)$ and the lines $x = a$ and $x = b$.

   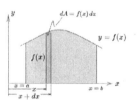

7. Inequalities.
   (i) If $f(x) \leq g(x)$ for all $x \in (a,b)$, then $\int_a^b f(x)\,dx \leq \int_a^b g(x)\,dx$
   (ii) If $|f(x)| \leq M$ for all $x \in (a,b)$ and $\int_a^b f(x)\,dx$ exists, then
   $$\left|\int_a^b f(x)\,dx\right| \leq \int_a^b f(x)\,dx \leq M(b-a)$$

8. $\int \dfrac{u'(x)\,dx}{u(x)} = \ln|u(x)| + C$

9. $\int (\alpha u(x) + \beta)^n\, u'(x)\,dx = \dfrac{(\alpha u(x)+\beta)^{n+1}}{\alpha(n+1)} + C$

10. $\int \dfrac{u'(x)v(x) - v'(x)u(x)}{v^2(x)}\,dx = \dfrac{u(x)}{v(x)} + C$

11. $\int \dfrac{u'(x)v(x) - u(x)v'(x)}{u(x)v(x)}\,dx = \ln\left|\dfrac{u(x)}{v(x)}\right| + C$

12. $\int \dfrac{u'(x)v(x) - u(x)v'(x)}{u^2(x) + v^2(x)}\,dx = \tan^{-1}\dfrac{u(x)}{v(x)} + C$

13. $\int \dfrac{u'(x)v(x) - u(x)v'(x)}{u^2(x) - v^2(x)}\,dx = \dfrac{1}{2}\ln\left|\dfrac{u(x)-v(x)}{u(x)+v(x)}\right| + C$

14. $\int \dfrac{u'(x)\,dx}{\sqrt{u^2(x)+\alpha}} = \ln|u(x) + \sqrt{u^2(x)+\alpha}| + C$

15. $\int \dfrac{u(x)\,dx}{(u(x)+\alpha)(u(x)+\beta)} = \begin{cases} \dfrac{\alpha}{\alpha-\beta}\int \dfrac{dx}{u(x)+\alpha} - \dfrac{\beta}{\alpha-\beta}\int \dfrac{dx}{u(x)+\beta}, & \alpha \neq \beta \\ \int \dfrac{dx}{u(x)+\alpha} - \alpha \int \dfrac{dx}{(u(x)+\alpha)^2}, & \beta = \alpha \end{cases}$

16. $\int \dfrac{u'(x)\,dx}{\alpha u^2(x) + \beta u(x)} = \dfrac{1}{\beta}\ln\left|\dfrac{u(x)}{\alpha u(x)+\beta}\right| + C$

17. $\int \dfrac{u'(x)\,dx}{u(x)\sqrt{u^2(x)-\alpha^2}} = \dfrac{1}{\alpha}\sec^{-1}\dfrac{u(x)}{\alpha} + C$

18. $\int \dfrac{u'(x)\,dx}{\alpha^2 + \beta^2 u^2(x)} = \dfrac{1}{\alpha\beta}\tan^{-1}\dfrac{\beta u(x)}{\alpha} + C$

19. $\int \dfrac{u'(x)\,dx}{\alpha^2 u^2(x) - \beta^2} = \dfrac{1}{2\alpha\beta}\ln\left|\dfrac{\alpha u(x)-\beta}{\alpha u(x)+\beta}\right| + C$

20. $\int f(\sin x)\,dx = 2\int f\left(\dfrac{2u}{1+u^2}\right)\dfrac{du}{1+u^2}, \quad u = \tan\dfrac{x}{2}$

21. $\int f(\sin x)\,dx = \int f(u)\dfrac{du}{\sqrt{1-u^2}}, \quad u = \sin x$

22. $\int f(\cos x)\,dx = 2\int f\left(\dfrac{1-u^2}{1+u^2}\right)\dfrac{du}{1+u^2}, \quad u = \tan\dfrac{x}{2}$

23. $\int f(\cos x)\,dx = -\int f(u)\dfrac{du}{\sqrt{1-u^2}}, \quad u = \cos x$

24. $\int f(\sin x, \cos x)\,dx = \int f(u, \sqrt{1-u^2})\dfrac{du}{\sqrt{1-u^2}}, \quad u = \sin x$

25. $\int f(\sin x, \cos x)\,dx = 2\int f\left(\dfrac{2u}{1+u^2}, \dfrac{1-u^2}{1+u^2}\right)\dfrac{du}{1+u^2}, \quad u = \tan\dfrac{x}{2}$

26. $\int f(x, \sqrt{\alpha+\beta x})\,dx = \dfrac{2}{\beta}\int f\left(\dfrac{u^2-\alpha}{\beta}, u\right)u\,du, \quad u^2 = \alpha + \beta x$

27. $\int f(x, \sqrt{\alpha^2 - x^2})\,dx = \alpha \int f(\alpha\sin u, \alpha\cos u)\cos u\,du, \quad x = \alpha\sin u$

### General Integrals

28. $\int c\,u(x)\,dx = c\int u(x)\,dx$

29. $\int [u(x) + v(x)]\,dx = \int u(x)\,dx + \int v(x)\,dx$

30. $\int u(x)\,u'(x)\,dx = \frac{1}{2}\mid u(x)\mid^2 + C$

31. $\int [u(x) - v(x)]\,dx = \int u(x)\,dx - \int v(x)\,dx]$

32. $\int u^n(x)\,u'(x)\,dx = \frac{[u(x)]^{n+1}}{n+1} + C$

33. $\int u(x)\,v'(x)\,dx = u(x)\,v(x) - \int u'(x)\,v(x)\,dx$

34. $\int F'[u(x)]\,u'(x)\,dx = F[u(x)] + C$

35. $\int \frac{u'(x)}{u(x)}\,dx = \ln\mid u(x)\mid + C$

36. $\int \frac{u'}{2\sqrt{u}}\,dx = \sqrt{u} + C$

37. $\int 1\,dx = x + C$

38. $\int x^n\,dx = \frac{x^{n+1}}{n+1} + C$

39. $\int \frac{1}{x}\,dx = \ln\mid x\mid + C$

40. $\int e^{au}\,u'\,dx = \frac{1}{a}e^{au} + C$

41. $\int a^u\,u'\,dx = \frac{1}{\ln a}a^u + C$

42. $\int \sin u\,u'\,dx = \cos u + C$

43. $\int \cos u\,u'\,dx = -\sin u + C$

44. $\int \tan u\,u'\,dx = \ln\mid\sec u\mid + C$

45. $\int \cot u\,u'\,dx = \ln\mid\sin u\mid + C$

46. $\int \sec u\,u'\,dx = \ln\mid\sec u + \tan u\mid + C$

47. $\int \csc u\,u'\,dx = \ln\mid\csc u - \cot u\mid + C$

48. $\int \sinh u\,u'\,dx = \cosh u + C$

49. $\int \cosh u\,u'\,dx = \sinh u + C$

50. $\int \tanh u\,u'\,dx = \ln\cosh u + C$

51. $\int \coth u\,u'\,dx = \ln\sinh u + C$

52. $\int \text{sech}\,u\,u'\,dx = \sin^{-1}(\tanh u) + C$

53. $\int \text{csch}\,u\,u'\,dx = \ln\tanh\frac{u}{2} + C$

54. $\int \sin^2 u\,u'\,dx = \frac{1}{2}u - \frac{1}{4}\sin 2u + C$

55. $\int \cos^2 u\,u'\,dx = \frac{u}{2} + \frac{1}{4}\sin 2u + C$

56. $\int \tan^2 u\,u'\,dx = \tan u - u + C$

57. $\int \cot^2 u\,u'\,dx = -\cot u - u + C$

58. $\int \sec^2 u\,u'\,dx = \tan u + C$

59. $\int \csc^2 u\,u'\,dx = -\cot u + C$

60. $\int \sinh^2 u\,u'\,dx = \frac{1}{4}\sinh 2u - \frac{1}{2}u + C$

61. $\int \cosh^2 u\,u'\,dx = \frac{1}{4}\sinh 2u + \frac{1}{2}u + C$

62. $\int \tanh^2 u\,u'\,dx = u - \tanh u + C$

63. $\int \coth^2 u\,u'\,dx = u - \coth u + C$

64. $\int \text{sech}^2 u\,u'\,dx = \tanh u + C$

65. $\int \text{csch}^2 u\,u'\,dx = -\coth u + C$

66. $\int \sec u\tan u\,u'\,dx = \sec u + C$

67. $\int \csc u\cot u\,u'\,dx = -\csc u + C$

68. $\int \text{sech}\,u\tanh u\,u'\,dx = -\text{sech}\,u + C$

69. $\int \text{csch}\,u\coth u\,u'\,dx = -\text{csch}\,u + C$

### Integrals containing $X = a + bx$, $a \neq 0$ and $b \neq 0$

70. $\int X^n\,dx = \frac{X^{n+1}}{b(n+1)} + C,\quad n \neq -1$

71. $\int xX^n\,dx = \frac{X^{n+2}}{b^2(n+2)} - \frac{aX^{n+1}}{b^2(n+1)} + C,\quad n \neq -1, n \neq -2$

72. $\int X(x+c)^n\,dx = \frac{b}{n+2}(x+c)^{n+2} + \frac{a-bc}{n+1}(x+c)^{n+1} + C$

73. $\int x^2 X^n \, dx = \frac{1}{b^3}\left[\frac{X^{n+3}}{n+3} - \frac{2aX^{n+2}}{n+2} + \frac{a^2 X^{n+1}}{n+1}\right] + C$

74. $\int x^{n-1} X^m \, dx = \frac{1}{n+m} x^n X^m + \frac{am}{m+n}\int x^{n-1} X^{m-1} \, dx$

75. $\int \frac{X^m}{x^{n+1}} \, dx = -\frac{1}{na}\frac{X^{m+1}}{x^n} + \frac{m-n+1}{n}\frac{b}{a}\int \frac{X^m}{x^n} \, dx$

76. $\int \frac{dx}{X} = \frac{1}{b}\ln X + C$

77. $\int \frac{x \, dx}{X} = \frac{1}{b^2}(X - a\ln|X|) + C$

78. $\int \frac{x^2 \, dx}{X} = \frac{1}{2b^3}\left(X^2 - 4aX + 2a^2\ln|X|\right) + C$

79. $\int \frac{dx}{xX} = \frac{1}{a}\ln\left|\frac{x}{X}\right| + C$

80. $\int \frac{dx}{x^3 X} = -\frac{a+2bx}{a^2 x X} + \frac{2b}{a^3}\ln\left|\frac{X}{x}\right| + C$

81. $\int \frac{dx}{X^2} = -\frac{1}{bX} + C$

82. $\int \frac{x \, dx}{X^2} = \frac{1}{b^2}\left[\ln|X| + \frac{a}{X}\right] + C$

83. $\int \frac{x^2 \, dx}{X^2} = \frac{1}{b^3}\left[X - 2a\ln|X| - \frac{a^2}{X}\right] + C$

84. $\int \frac{dx}{xX^2} = \frac{1}{aX} - \frac{1}{a^2}\ln\left|\frac{X}{x}\right| + C$

85. $\int \frac{dx}{x^2 X^2} = -\frac{a+2bx}{a^2 x X} + \frac{2b}{a^3}\ln\left|\frac{X}{x}\right| + C$

86. $\int \frac{dx}{X^3} = -\frac{1}{2bX^2} + C$

87. $\int \frac{x \, dx}{X^3} = \frac{1}{b^2}\left[\frac{-1}{X} + \frac{a}{2X^2}\right] + C$

88. $\int \frac{x^2 \, dx}{X^3} = \frac{1}{b^3}\left[\ln|X| + \frac{2a}{X} - \frac{a^2}{2X^2}\right] + C$

89. $\int \frac{dx}{xX^3} = \frac{1}{2aX^2} + \frac{1}{aX} - \ln\left|\frac{X}{x}\right| + C$

90. $\int \dfrac{dx}{x^2 X^3} = \dfrac{-b}{2a^2 X} - \dfrac{2b}{a^3 X} - \dfrac{1}{a^3 x} + \dfrac{3b}{a^4} \ln \left| \dfrac{X}{x} \right|$

91. $\int \dfrac{x\,dx}{X^n} = \dfrac{1}{b^2} \left[ \dfrac{-1}{(n-2)X^{n-2}} + \dfrac{a}{(n-1)X^{n-1}} \right] + C, \quad n \neq 1, 2$

92. $\int \dfrac{x^2\,dx}{X^n} = \dfrac{1}{b^3} \left[ \dfrac{-1}{(n-3)X^{n-3}} + \dfrac{2a}{(n-2)X^{n-2}} - \dfrac{a^2}{(n-1)X^{n-1}} \right] + C, \quad n \neq 1, 2, 3$

93. $\int \sqrt{X}\,dx = \dfrac{2}{3b} X^{3/2} + C$

94. $\int x\sqrt{X}\,dx = \dfrac{2}{15b^2}(3bx - 2a)X^{3/2} + C$

95. $\int x^2\sqrt{X}\,dx = \dfrac{2}{105b^3}(8a^2 - 12abx + 15b^2 x^2)X^{3/2} + C$

96. $\int \dfrac{\sqrt{X}}{x}\,dx = 2\sqrt{X} + a\int \dfrac{dx}{x\sqrt{X}}$

97. $\int \dfrac{\sqrt{X}}{x^2}\,dx = -\dfrac{\sqrt{X}}{x} + \dfrac{b}{2}\int \dfrac{dx}{x\sqrt{X}}$

98. $\int \dfrac{dx}{\sqrt{X}} = \dfrac{2}{b}\sqrt{X} + C$

99. $\int \dfrac{x\,dx}{\sqrt{X}} = \dfrac{2}{3b^2}(bx - 2a)\sqrt{X} + C$

100. $\int \dfrac{x^2\,dx}{\sqrt{X}} = \dfrac{2}{15b^3}(8a^2 - 4abx + 3b^2 x^2)\sqrt{X} + C$

101. $\int \dfrac{dx}{x\sqrt{X}} = \begin{cases} \dfrac{1}{\sqrt{a}} \ln \left| \dfrac{\sqrt{X} - \sqrt{a}}{\sqrt{X} + \sqrt{a}} \right| + C_1, & a > 0 \\ \dfrac{2}{\sqrt{-a}} \tan^{-1} \sqrt{\dfrac{X}{-a}} + C_2, & a < 0 \end{cases}$

102. $\int \dfrac{dx}{x^2 \sqrt{X}} = -\dfrac{\sqrt{X}}{ax} - \dfrac{b}{2a} \int \dfrac{dx}{x\sqrt{X}}$

103. $\int x^n \sqrt{X}\,dx = \dfrac{2}{(2n+3)b} x^n X^{3/2} - \dfrac{2na}{(2n+3)b} \int x^{n-1} \sqrt{X}\,dx$

104. $\int \dfrac{\sqrt{X}}{x^n}\,dx = \dfrac{-1}{(n-1)a} \dfrac{X^{3/2}}{x^{n-1}} - \dfrac{(2n-5)b}{2(n-1)a} \int \dfrac{\sqrt{X}}{x^{n-1}}\,dx$

105. $\int x^{m-1} X^n\,dx = \dfrac{x^m X^n}{m+n} + \dfrac{an}{m+n} \int x^{m-1} X^{n-1}\,dx + C$

106. $\int \dfrac{X^n}{x^{m+1}}\,dx = -\dfrac{X^{n+1}}{ma\,x^m} + \dfrac{n-m+1}{m} \dfrac{b}{a} \int \dfrac{X^n}{x^m}\,dx$

**107.** $\int \dfrac{X^n}{x}\, dx = \dfrac{X^n}{n} + a \int \dfrac{X^{n-1}}{x}\, dx$

Integrals containing $X = a + bx$ and $Y = \alpha + \beta x$, $(b \neq 0,\ \beta \neq 0,\ \Delta = a\beta - \alpha b \neq 0)$

**108.** $\int \dfrac{dx}{XY} = \dfrac{1}{\Delta} \ln\left|\dfrac{Y}{X}\right| + C$

**109.** $\int \dfrac{x\,dx}{XY} = \dfrac{1}{\Delta}\left[\dfrac{a}{b}\ln|X| - \dfrac{\alpha}{\beta}\ln|Y|\right] + C$

**110.** $\int \dfrac{x^2\,dx}{XY} = \dfrac{x}{b\beta} - \dfrac{a^2}{b^2\Delta}\ln|X| + \dfrac{\alpha^2}{\beta^2\Delta}\ln|Y| + C$

**111.** $\int \dfrac{dx}{X^2 Y} = \dfrac{1}{\Delta}\left(\dfrac{1}{X} + \dfrac{\beta}{\Delta}\ln\left|\dfrac{Y}{X}\right|\right) + C$

**112.** $\int \dfrac{x\,dx}{X^2 Y} = -\dfrac{a}{b\Delta X} - \dfrac{\alpha}{\Delta^2}\ln\left|\dfrac{Y}{X}\right| + C$

**113.** $\int \dfrac{x^2\,dx}{X^2 Y} = \dfrac{a^2}{b^2 \Delta X} + \dfrac{1}{\Delta^2}\left[\dfrac{\alpha^2}{\beta}\ln|Y| + \dfrac{a(a\beta - 2\alpha b)}{b^2}\ln|X|\right] + C$

**114.** $\int \dfrac{X}{Y}\, dx = \dfrac{b}{\beta} x + \dfrac{\Delta}{\beta^2} \ln\left|\dfrac{Y}{X}\right| + C$

**115.** $\int \sqrt{XY}\, dx = \dfrac{\Delta + 2bY}{4b\beta}\sqrt{XY} - \dfrac{\Delta^2}{8b\beta}\int \dfrac{dx}{\sqrt{XY}}$

**116.** $\int \dfrac{dx}{X^n Y^m} = \dfrac{-1}{(m-1)\Delta X^{n-1} Y^{m-1}} + \dfrac{(m+n-2)b}{(m-1)\Delta}\int \dfrac{dx}{X^n Y^{m-1}},\quad m \neq 1$

**117.** $\int \dfrac{dx}{Y\sqrt{X}} = \begin{cases} \dfrac{2}{\sqrt{-\Delta\beta}} \tan^{-1} \dfrac{\beta\sqrt{X}}{\sqrt{-\Delta\beta}} + C_1, & \Delta\beta < 0 \\[6pt] \dfrac{1}{\sqrt{\Delta\beta}} \ln\left|\dfrac{\beta\sqrt{X} - \sqrt{\Delta\beta}}{\beta\sqrt{X} + \sqrt{\Delta\beta}}\right| + C_2, & \Delta\beta > 0 \end{cases}$

**118.** $\int \dfrac{dx}{\sqrt{XY}} = \begin{cases} \dfrac{2}{\sqrt{-b\beta}} \tan^{-1} \sqrt{\dfrac{-\beta X}{bY}} + C_1, & b\beta < 0,\ bY > 0 \\[6pt] \dfrac{2}{\sqrt{b\beta}} \tanh^{-1} \sqrt{\dfrac{\beta X}{bY}} + C_2, & b\beta > 0,\ bY > 0 \end{cases}$

**119.** $\int \dfrac{x\,dx}{\sqrt{XY}} = \dfrac{1}{b\beta}\sqrt{XY} - \dfrac{(b\alpha + a\beta)}{2b\beta}\int \dfrac{dx}{\sqrt{XY}}$

**120.** $\int \dfrac{\sqrt{Y}}{\sqrt{X}}\, dx = \dfrac{1}{b}\sqrt{XY} - \dfrac{\Delta}{2b}\int \dfrac{dx}{\sqrt{XY}}$

**121.** $\int \dfrac{\sqrt{X}}{Y}\, dx = \dfrac{2}{\beta}\sqrt{X} + \dfrac{\Delta}{\beta}\int \dfrac{dx}{Y\sqrt{X}}$

| Integrals containing terms of the form $a + bx^n$ |

122. $\displaystyle\int \frac{dx}{a+bx^2} = \begin{cases} \dfrac{1}{\sqrt{ab}}\tan^{-1}\left(\sqrt{\dfrac{b}{a}}\,x\right) + C, & ab > 0 \\[1ex] \dfrac{1}{2\sqrt{-ab}}\ln\left|\dfrac{a+\sqrt{-ab}\,x}{a-\sqrt{-ab}\,x}\right| + C, & ab < 0 \end{cases}$

123. $\displaystyle\int \frac{x\,dx}{a+bx^2} = \frac{1}{2b}\ln\left|x^2 + \frac{a}{b}\right| + C$

124. $\displaystyle\int \frac{x^2\,dx}{a+bx^2} = \frac{x}{b} - \frac{a}{b}\int \frac{dx}{a+bx^2}$

125. $\displaystyle\int \frac{dx}{(a+bx^2)^2} = \frac{x}{2a(a+bx^2)} + \frac{1}{2a}\int \frac{dx}{a+bx^2}$

126. $\displaystyle\int \frac{dx}{x(a+bx^2)} = \frac{1}{2a}\ln\left|\frac{x^2}{a+bx^2}\right| + C$

127. $\displaystyle\int \frac{dx}{x^2(a+bx^2)} = -\frac{1}{ax} - \frac{b}{a}\int \frac{dx}{a+bx^2}$

128. $\displaystyle\int \frac{dx}{(a+bx^2)^{n+1}} = \frac{1}{2na}\frac{x}{(a+bx^2)^n} + \frac{2n-1}{2na}\int \frac{dx}{(a+bx^2)^n}$

129. $\displaystyle\int \frac{dx}{\alpha^3+\beta^3 x^3} = \frac{1}{6\alpha^2\beta}\left[2\sqrt{3}\tan^{-1}\left(\frac{2\beta x - \alpha}{\sqrt{3}\alpha}\right) + \ln\left|\frac{(\alpha+\beta x)^2}{\alpha^2 - \alpha\beta x + \beta^2 x^2}\right|\right] + C$

130. $\displaystyle\int \frac{x\,dx}{\alpha^3+\beta^3 x^3} = \frac{1}{6\alpha\beta^2}\left[2\sqrt{3}\tan^{-1}\left(\frac{2\beta x - \alpha}{\sqrt{3}\alpha}\right) - \ln\left|\frac{(\alpha+\beta x)^2}{\alpha^2 - \alpha\beta x + \beta^2 x^2}\right|\right] + C$

If $X = a + bx^n$, then

131. $\displaystyle\int x^{m-1}X^p\,dx = \frac{x^m X^p}{m+pn} + \frac{apn}{m+pn}\int x^{m-1}X^{p-1}\,dx$

132. $\displaystyle\int x^{m-1}X^p\,dx = -\frac{x^m X^{p+1}}{an(p+1)} + \frac{m+pn+n}{an(p+1)}\int x^{m-1}X^{p+1}\,dx$

133. $\displaystyle\int x^{m-1}X^p\,dx = \frac{x^{m-n}X^{p+1}}{b(m+pn)} - \frac{(m-n)a}{b(m+pn)}\int x^{m-n-1}X^p\,dx$

134. $\displaystyle\int x^{m-1}X^p\,dx = \frac{x^m X^{p+1}}{am} - \frac{(m+pn+n)b}{am}\int x^{m+n-1}X^p\,dx$

135. $\displaystyle\int x^{m-1}X^p\,dx = \frac{x^{m-n}X^{p+1}}{bn(p+1)} - \frac{m-n}{bn(p+1)}\int x^{m-n-1}X^{p+1}\,dx$

136. $\displaystyle\int x^{m-1}X^p\,dx = \frac{x^m X^p}{m} - \frac{bpn}{m}\int x^{m+n-1}X^{p-1}\,dx$

**Appendix B**

**Integrals containing $X = 2ax - x^2$, $a \neq 0$**

**137.** $\displaystyle \int \sqrt{X}\, dx = \frac{(x-a)}{2}\sqrt{X} + \frac{a^2}{2}\sin^{-1}\left(\frac{x-a}{|a|}\right) + C$

**138.** $\displaystyle \int \frac{dx}{\sqrt{X}} = \sin^{-1}\left(\frac{x-a}{|a|}\right) + C$

**139.** $\displaystyle \int x\sqrt{X}\, dx = \sin^{-1}\left(\frac{x-a}{|a|}\right) + C$

**140.** $\displaystyle \int \frac{x\, dx}{\sqrt{X}} = -\sqrt{X} + a\sin^{-1}\left(\frac{x-a}{|a|}\right) + C$

**141.** $\displaystyle \int \frac{dx}{X^{3/2}} = \frac{x-a}{a^2\sqrt{X}} + C$

**142.** $\displaystyle \int \frac{x\, dx}{X^{3/2}} = \frac{x}{a\sqrt{X}} + C$

**143.** $\displaystyle \int \frac{dx}{X} = \frac{1}{2a}\ln\left|\frac{x}{x-2a}\right| + C$

**144.** $\displaystyle \int \frac{x\, dx}{X} = -\ln|x - 2a| + C$

**145.** $\displaystyle \int \frac{dx}{X^2} = -\frac{1}{4ax} - \frac{1}{4a^2(x-2a)} + \frac{1}{4a^2}\ln\left|\frac{x}{x-2a}\right| + C$

**146.** $\displaystyle \int \frac{x\, dx}{X^2} = -\frac{1}{2a(x-2a)} + \frac{1}{4a^2}\ln\left|\frac{x}{x-2a}\right| + C$

**147.** $\displaystyle \int x^n\sqrt{X}\, dx = -\frac{1}{n+2}x^{n-1}X^{3/2} + \frac{(2n+1)a}{n+2}\int x^{n-1}\sqrt{X}\, dx, \quad n \neq -2$

**148.** $\displaystyle \int \frac{\sqrt{X}\, dx}{x^n} = \frac{1}{(3-2n)a}\frac{X^{3/2}}{x^n} + \frac{n-3}{(2n-3)a}\int \frac{\sqrt{X}}{x^{n-1}}\, dx, \quad n \neq 3/2$

**Integrals containing $X = ax^2 + bx + c$ with $\Delta = 4ac - b^2$, $\Delta \neq 0$, $a \neq 0$**

**149.** $\displaystyle \int \frac{dx}{X} = \begin{cases} \dfrac{1}{\sqrt{-\Delta}}\ln\left(\dfrac{2ax+b-\sqrt{-\Delta}}{2ax+b+\sqrt{-\Delta}}\right) + C_1, & \Delta < 0 \\[6pt] \dfrac{2}{\sqrt{\Delta}}\tan^{-1}\dfrac{2ax+b}{\sqrt{\Delta}} + C_2, & \Delta > 0 \\[6pt] -\dfrac{1}{a(x+b/2a)} + C_3, & \Delta = 0 \end{cases}$

**150.** $\displaystyle \int \frac{x\, dx}{X} = \frac{1}{2c}\ln|X| - \frac{b}{2a}\int \frac{1}{X}\, dx$

**151.** $\displaystyle \int \frac{x^2\, dx}{X} = \frac{x}{a} - \frac{b}{2a^2}\ln|X| + \frac{2ac-\Delta}{2a^2}\int \frac{dx}{X}$

**499**

152. $\int \dfrac{dx}{xX} = \dfrac{1}{2c}\ln\left|\dfrac{x^2}{X}\right| - \dfrac{b}{2c}\int \dfrac{dx}{X}$

153. $\int \dfrac{dx}{x^2 X} = \dfrac{b}{2c^2}\ln\left|\dfrac{X}{x^2}\right| - \dfrac{1}{cx} + \dfrac{2ac-\Delta}{2c^2}\int \dfrac{dx}{X}$

154. $\int \dfrac{dx}{X^2} = \dfrac{bx+2c}{\Delta X} - \dfrac{b}{\Delta}\int \dfrac{dx}{X}$

155. $\int \dfrac{x\,dx}{X^2} = -\dfrac{bx+2c}{\Delta X} - \dfrac{b}{\Delta}\int \dfrac{dx}{X}$

156. $\int \dfrac{x^2\,dx}{X^2} = \dfrac{(2ac-\Delta)x+bc}{a\Delta X} + \dfrac{2c}{\Delta}\int \dfrac{dx}{X}$

157. $\int \dfrac{dx}{xX^2} = \dfrac{1}{2cX} - \dfrac{b}{2c}\int \dfrac{dx}{X^2} + \dfrac{1}{c}\int \dfrac{dx}{xX}$

158. $\int \dfrac{dx}{x^2 X^2} = -\dfrac{1}{cxX} - \dfrac{3a}{c}\int \dfrac{dx}{X^2} - \dfrac{2b}{c}\int \dfrac{dx}{xX^2}$

159. $\int \dfrac{dx}{\sqrt{X}} = \begin{cases} \dfrac{1}{\sqrt{a}}\ln|2\sqrt{aX}+2ax+b|+C_1, & a>0 \\ \dfrac{1}{\sqrt{a}}\sinh^{-1}\left(\dfrac{2ax+b}{\sqrt{\Delta}}\right)+C_2, & a>,\ \Delta>0 \\ -\dfrac{1}{\sqrt{-a}}\sin^{-1}\left(\dfrac{2ax+b}{\sqrt{-\Delta}}\right)+C_3, & a<0,\ \Delta<0 \end{cases}$

160. $\int \dfrac{x\,dx}{\sqrt{X}} = \dfrac{1}{a}\sqrt{X} - \dfrac{b}{2a}\int \dfrac{dx}{\sqrt{X}}$

161. $\int \dfrac{x^2\,dx}{\sqrt{X}} = \left(\dfrac{x}{2a} - \dfrac{3b}{4a^2}\right)\sqrt{X} + \dfrac{2b^2-\Delta}{8a^2}\int \dfrac{dx}{\sqrt{X}}$

162. $\int \dfrac{dx}{x\sqrt{X}} = \begin{cases} -\dfrac{1}{\sqrt{c}}\ln\left|\dfrac{2\sqrt{cX}}{x}+\dfrac{2c}{x}+b\right|+C_1, & c>0 \\ -\dfrac{1}{\sqrt{c}}\sinh^{-1}\left(\dfrac{bx+2c}{x\sqrt{\Delta}}\right)+C_2, & c>0,\ \Delta>0 \\ \dfrac{1}{\sqrt{-c}}\sin^{-1}\left(\dfrac{bx+2c}{x\sqrt{-\Delta}}\right)+C_3, & c<0,\ \Delta<0 \end{cases}$

163. $\int \dfrac{dx}{x^2\sqrt{X}} = -\dfrac{\sqrt{X}}{cx} - \dfrac{b}{2c}\int \dfrac{dx}{x\sqrt{X}}$

164. $\int \sqrt{X}\,dx = \dfrac{1}{4a}(2ax+b)\sqrt{X} + \dfrac{\Delta}{8a}\int \dfrac{dx}{\sqrt{X}}$

165. $\int x\sqrt{X}\,dx = \dfrac{1}{3a}X^{3/2} - \dfrac{b(2ax+b)}{8a^2}\sqrt{X} - \dfrac{b\Delta}{16a^2}\int \dfrac{dx}{\sqrt{X}}$

166. $\int x^2\sqrt{X}\,dx = \dfrac{6ax-5b}{24a^2}X^{3/2} + \dfrac{4b^2-\Delta}{16a^2}\int \sqrt{X}\,dx$

**Appendix B**

**167.** $\int \dfrac{\sqrt{X}}{x} dx = \sqrt{X} + \dfrac{b}{2}\int \dfrac{dx}{\sqrt{X}} + c\int \dfrac{dx}{x\sqrt{X}}$

**168.** $\int \dfrac{\sqrt{X}}{x^2} dx = -\dfrac{\sqrt{X}}{x} + a\int \dfrac{dx}{\sqrt{X}} + \dfrac{b}{2}\int \dfrac{dx}{x\sqrt{X}}$

**169.** $\int \dfrac{dx}{X^{3/2}} = \dfrac{2(2ax+b)}{\Delta\sqrt{X}} + C$

**170.** $\int \dfrac{x\,dx}{X^{3/2}} = \dfrac{-2(bx+2c)}{\Delta\sqrt{X}} + C$

**171.** $\int \dfrac{x^2\,dx}{X^{3/2}} = \dfrac{(b^2-\Delta)x + 2bc}{a\Delta\sqrt{X}} + \dfrac{1}{a}\int \dfrac{dx}{\sqrt{X}}$

**172.** $\int \dfrac{dx}{xX^{3/2}} = \dfrac{1}{x\sqrt{X}} + \dfrac{1}{c}\int \dfrac{dx}{x\sqrt{X}} - \dfrac{b}{2c}\int \dfrac{dx}{X^{3/2}}$

**173.** $\int \dfrac{dx}{x^2 X^{3/2}} = -\dfrac{ax^2 + 2bx + c}{c^2 x\sqrt{X}} + \dfrac{b^2 - 2ac}{2c^2}\int \dfrac{dx}{X^{3/2}} - \dfrac{3b}{2c^2}\int \dfrac{dx}{x\sqrt{X}}$

**174.** $\int \dfrac{dx}{X\sqrt{X}} = \dfrac{2(2ax+b)}{\Delta\sqrt{X}} + C$

**175.** $\int \dfrac{dx}{X^2\sqrt{X}} = \dfrac{2(2ax+b)}{3\Delta\sqrt{X}}\left(\dfrac{1}{X} + \dfrac{8a}{\Delta}\right) + C$

**176.** $\int X\sqrt{X}\,dx = \dfrac{(2ax+b)}{8a}\sqrt{X}\left(X + \dfrac{3\Delta}{8a}\right) + \dfrac{3\Delta^2}{128a^2}\int \dfrac{dx}{\sqrt{X}}$

**177.** $\int X^2\sqrt{X}\,dx = \dfrac{(2ax+b)}{8a}\sqrt{X}\left(X^2 + \dfrac{5\Delta}{16a}X + \dfrac{15\Delta^2}{128a^2}\right) + \dfrac{5\Delta^3}{1024a^3}\int \dfrac{dx}{\sqrt{X}}$

**178.** $\int \dfrac{x\,dx}{X\sqrt{X}} = -\dfrac{2(bx+2c)}{\Delta\sqrt{X}} + C$

**179.** $\int \dfrac{x^2\,dx}{X\sqrt{X}} = \dfrac{(b^2-\Delta)x + 2bc}{a\Delta\sqrt{X}} + \dfrac{1}{a}\int \dfrac{dx}{\sqrt{X}}$

**180.** $\int xX\sqrt{X}\,dx = \dfrac{X^2\sqrt{X}}{5a} - \dfrac{b}{2a}\int X\sqrt{X}\,dx$

**181.** $\int f(x, \sqrt{ax^2 + bx + c})\,dx$ Try substitutions (i) $\sqrt{ax^2 + bx + c} = \sqrt{a}(x+z)$

(ii) $\sqrt{ax^2 + bx + c} = xz + \sqrt{c}$ and if $ax^2 + bx + c = a(x-x_1)(x-x_2)$, then (iii) let $(x - x_2) = z^2(x - x_1)$

**Integrals containing** $X = x^2 + a^2$

**182.** $\int \dfrac{dx}{X} = \dfrac{1}{a}\tan^{-1}\dfrac{x}{a} + C$ or $\dfrac{1}{a}\cos^{-1}\dfrac{a}{\sqrt{x^2+a^2}} + C$ or $\dfrac{1}{a}\sec^{-1}\dfrac{\sqrt{x^2+a^2}}{a} + C$

**183.** $\int \dfrac{x\,dx}{X} = \dfrac{1}{2}\ln X + C$

**184.** $\int \dfrac{x^2\,dx}{X} = x - a\tan^{-1}\dfrac{x}{a} + C$

**185.** $\int \dfrac{x^3\,dx}{X} = \dfrac{x^2}{2} - \dfrac{a^2}{2}\ln|x^2 + a^2| + C$

**186.** $\int \dfrac{dx}{xX} = \dfrac{1}{2a^2}\ln\left|\dfrac{x^2}{X}\right| + C$

**187.** $\int \dfrac{dx}{x^2 X} = -\dfrac{1}{a^2 x} - \dfrac{1}{a^3}\tan^{-1}\dfrac{x}{a} + C$

**188.** $\int \dfrac{dx}{x^3 X} = -\dfrac{1}{2a^2 x^2} - \dfrac{1}{2a^4}\ln\left|\dfrac{x^2}{X}\right| + C$

**189.** $\int \dfrac{dx}{X^2} = \dfrac{x}{2a^2 X} + \dfrac{1}{2a^3}\tan^{-1}\dfrac{x}{a} + C$

**190.** $\int \dfrac{x\,dx}{X^2} = -\dfrac{1}{2X} + C$

**191.** $\int \dfrac{x^2\,dx}{X^2} = -\dfrac{x}{2X} + \dfrac{1}{2a}\tan^{-1}\dfrac{x}{a} + C$

**192.** $\int \dfrac{x^3\,dx}{X^2} = \dfrac{a^2}{2X} + \dfrac{1}{2}\ln|X| + C$

**193.** $\int \dfrac{dx}{xX^2} = \dfrac{1}{2a^2 X} + \dfrac{1}{2a^4}\ln\left|\dfrac{x}{X}\right| + C$

**194.** $\int \dfrac{dx}{x^2 X^2} = -\dfrac{1}{a^4 x} - \dfrac{x}{2a^4 X} - \dfrac{3}{2a^5}\tan^{-1}\dfrac{x}{a} + C$

**195.** $\int \dfrac{dx}{x^3 X^2} = -\dfrac{1}{2a^4 x^2} - \dfrac{1}{2a^4 X} - \dfrac{1}{a^6}\ln\left|\dfrac{x^2}{X}\right| + C$

**196.** $\int \dfrac{dx}{X^3} = \dfrac{x}{4a^2 X^2} + \dfrac{3x}{8a^4 X} + \dfrac{3}{8a^5}\tan^{-1}\dfrac{x}{a} + C$

**197.** $\int \dfrac{dx}{X^n} = \dfrac{x}{2(n-1)a^2 X^{n-1}} + \dfrac{2n-3}{(2(n-1)a^2}\int \dfrac{dx}{X^{n-1}},\qquad n>1$

**198.** $\int \dfrac{x\,dx}{X^n} = -\dfrac{1}{2(n-1)X^{n-1}} + C$

**199.** $\int \dfrac{dx}{xX^n} = \dfrac{1}{2(n-1)a^2 X^{n-1}} + \dfrac{1}{a^2}\int \dfrac{dx}{xX^{n-1}}$

---

| Integrals containing the square root of $X = x^2 + a^2$ |
|---|

**200.** $\int \sqrt{X}\,dx = \dfrac{1}{2}xX + \dfrac{a^2}{2}\ln|x + \sqrt{X}| + C$

**Appendix B**

**201.** $\int x\sqrt{X}\,dx = \frac{1}{3}X^{3/2} + C$

**202.** $\int x^2\sqrt{X}\,dx = \frac{1}{4}xX^{3/2} - \frac{1}{8}a^2x\sqrt{X} - \frac{a^2}{8}\ln|x+\sqrt{X}| + C$

**203.** $\int x^3\sqrt{X}\,dx = \frac{1}{5}X^{5/2} - \frac{a^2}{3}X^{3/2} + C$

**204.** $\int \frac{\sqrt{X}}{x}\,dx = \sqrt{X} - a\ln\left|\frac{a+\sqrt{X}}{x}\right| + C$

**205.** $\int \frac{\sqrt{X}}{x^2}\,dx = -\frac{\sqrt{X}}{x} + \ln|x+\sqrt{X}| + C$

**206.** $\int \frac{\sqrt{X}}{x^3}\,dx = -\frac{\sqrt{X}}{2x^2} - \frac{1}{2a}\ln\left|\frac{a+\sqrt{X}}{x}\right| + C$

**207.** $\int \frac{dx}{\sqrt{X}} = \ln|x+\sqrt{X}| + C \quad \text{or} \quad \sinh^{-1}\frac{x}{a} + C$

**208.** $\int \frac{x\,dx}{\sqrt{X}} = \sqrt{X} + C$

**209.** $\int \frac{x^2\,dx}{\sqrt{X}} = \frac{x}{2}\sqrt{X} - \frac{a^2}{2}\ln|x+\sqrt{X}| + C$

**210.** $\int \frac{x^3\,dx}{\sqrt{X}} = \frac{1}{3}X^{3/2} - a^2\sqrt{X} + C$

**211.** $\int \frac{dx}{x\sqrt{X}} = -\frac{1}{a}\ln\left|\frac{a+\sqrt{X}}{x}\right| + C$

**212.** $\int \frac{dx}{x^2\sqrt{X}} = -\frac{\sqrt{X}}{a^2 x} + C$

**213.** $\int \frac{dx}{x^3\sqrt{X}} = -\frac{\sqrt{X}}{2a^2x^2} + \frac{1}{2a^3}\ln\left|\frac{a+\sqrt{X}}{x}\right| + C$

**214.** $\int X^{3/2}\,dx = \frac{1}{4}X^{3/2} + \frac{3}{8}a^2x\sqrt{X} + \frac{3}{8}a^4\ln|x+\sqrt{X}| + C$

**215.** $\int xX^{3/2}\,dx = \frac{1}{5}X^{5/2} + C$

**216.** $\int x^2X^{3/2}\,dx = \frac{1}{6}X^{5/2} - \frac{1}{24}a^2xX^{3/2} - \frac{1}{16}a^4x\sqrt{X} - \frac{1}{16}a^6\ln|x+\sqrt{X}| + C$

**217.** $\int x^3X^{3/2}\,dx = \frac{1}{7}X^{7/2} - \frac{1}{5}a^2X^{5/2} + C$

Appendix B

218. $\int \dfrac{X^{3/2}}{x}\,dx = \dfrac{1}{3}X^{3/2} + a^2\sqrt{X} - a^3 \ln\left|\dfrac{a+\sqrt{X}}{x}\right| + C$

219. $\int \dfrac{X^{3/2}}{x^2}\,dx = -\dfrac{X^{3/2}}{x} + \dfrac{3}{2}x\sqrt{X} + \dfrac{3}{2}a^2 \ln|x+\sqrt{X}| + C$

220. $\int \dfrac{X^{3/2}}{x^3}\,dx = -\dfrac{X^{3/2}}{2x^2} + \dfrac{3}{2}\sqrt{X} - \dfrac{3}{2}a\ln\left|\dfrac{a+\sqrt{X}}{x}\right| + C$

221. $\int \dfrac{dx}{X^{3/2}} = \dfrac{x}{a^2\sqrt{X}} + C$

222. $\int \dfrac{x\,dx}{X^{3/2}} = \dfrac{-1}{\sqrt{X}} + C$

223. $\int \dfrac{x^2\,dx}{X^{3/2}} = \dfrac{-x}{\sqrt{X}} + \ln|x+\sqrt{X}| + C$

224. $\int \dfrac{x^3\,dx}{X^{3/2}} = \sqrt{X} + \dfrac{a^2}{\sqrt{X}} + C$

225. $\int \dfrac{dx}{xX^{3/2}} = \dfrac{1}{a^2\sqrt{X}} - \dfrac{1}{a^3}\ln\left|\dfrac{a+\sqrt{X}}{x}\right| + C$

226. $\int \dfrac{dx}{x^2 X^{3/2}} = -\dfrac{\sqrt{X}}{a^4 x} - \dfrac{x}{a^4\sqrt{X}} + C$

227. $\int \dfrac{dx}{x^3 X^{3/2}} = \dfrac{-1}{2a^2 x^2\sqrt{X}} - \dfrac{3}{2a^4\sqrt{X}} + \dfrac{3}{2a^5}\ln\left|\dfrac{a+\sqrt{X}}{x}\right| + C$

228. $\int f(x,\sqrt{X})\,dx = a\int f(a\tan u, a\sec u)\sec^2 u\,du, \quad x = a\tan u$

---

**Integrals containing $X = x^2 - a^2$ with $x^2 > a^2$**

229. $\int \dfrac{dx}{X} = \dfrac{1}{2a}\ln\left(\dfrac{x-a}{x+a}\right) + C \quad \text{or} \quad -\dfrac{1}{a}\coth^{-1}\dfrac{x}{a} + C \quad \text{or} \quad -\dfrac{1}{a}\tanh^{-1}\dfrac{a}{x} + C$

230. $\int \dfrac{x\,dx}{X} = \dfrac{1}{2}\ln X + C$

231. $\int \dfrac{x^2\,dx}{X} = x + \dfrac{a}{2}\ln\left|\dfrac{x-a}{x+a}\right| + C$

232. $\int \dfrac{x^3\,dx}{X} = \dfrac{x^2}{2} + \dfrac{a^2}{2}\ln|X| + C$

233. $\int \dfrac{dx}{xX} = \dfrac{1}{2a^2}\ln\left|\dfrac{X}{x^2}\right| + C$

234. $\int \dfrac{dx}{x^2 X} = \dfrac{1}{a^2 x} + \dfrac{1}{2a^3}\ln\left|\dfrac{x-a}{x+a}\right| + C$

Appendix B

**235.** $\int \dfrac{dx}{x^3 X} = \dfrac{1}{2a^2 x} - \dfrac{1}{2a^4} \ln\left|\dfrac{x^2}{X}\right| + C$

**236.** $\int \dfrac{dx}{X^2} = \dfrac{-x}{2a^2 X} - \dfrac{1}{4a^3} \ln\left|\dfrac{x-a}{x+a}\right| + C$

**237.** $\int \dfrac{x\,dx}{X^2} = \dfrac{-1}{2X} + C$

**238.** $\int \dfrac{x^2\,dx}{X^2} = \dfrac{-x}{2X} + \dfrac{1}{4a} \ln\left|\dfrac{x-a}{x+a}\right| + C$

**239.** $\int \dfrac{x^3\,dx}{X^2} = \dfrac{-a^2}{2X} + \dfrac{1}{2} \ln|X| + C$

**240.** $\int \dfrac{dx}{xX^2} = \dfrac{-1}{2a^2 X} + \dfrac{1}{2a^4} \ln\left|\dfrac{x^2}{X}\right| + C$

**241.** $\int \dfrac{dx}{x^2 X^2} = -\dfrac{1}{a^4 x} - \dfrac{x}{2a^4 X} - \dfrac{3}{4a^5} \ln\left|\dfrac{x-a}{x+a}\right| + C$

**242.** $\int \dfrac{dx}{x^3 X^2} = -\dfrac{1}{2a^4 x^2} - \dfrac{1}{2a^4 X} + \dfrac{1}{a^6} \ln\left|\dfrac{x^2}{X}\right| + C$

**243.** $\int \dfrac{dx}{X^n} = \dfrac{-x}{2(n-1)a^2 X^{n-1}} - \dfrac{2n-3}{2(n-1)a^2} \int \dfrac{dx}{X^{n-1}}, \quad n > 1$

**244.** $\int \dfrac{x\,dx}{X^n} = \dfrac{-1}{2(n-1)X^{n-1}} + C$

**245.** $\int \dfrac{dx}{xX^n} = \dfrac{-1}{2(n-1)a^2 X^{n-1}} - \dfrac{1}{a^2} \int \dfrac{dx}{xX^{n-1}}$

---

**Integrals containing the square root of $X = x^2 - a^2$ with $x^2 > a^2$**

**246.** $\int \sqrt{X}\,dx = \dfrac{1}{2}x\sqrt{X} - \dfrac{a^2}{2} \ln|x + \sqrt{X}| + C$

**247.** $\int x\sqrt{X}\,dx = \dfrac{1}{3}X^{3/2} + C$

**248.** $\int x^2 \sqrt{X}\,dx = \dfrac{1}{4}xX^{3/2} + \dfrac{1}{8}a^2 x\sqrt{X} - \dfrac{a^4}{8} \ln|x + \sqrt{X}| + C$

**249.** $\int x^3 \sqrt{X}\,dx = \dfrac{1}{5}X^{5/2} + \dfrac{1}{3}a^2 X^{3/2} + C$

**250.** $\int \dfrac{X}{x}\,dx = \sqrt{X} - a\sec^{-1}\left|\dfrac{x}{a}\right| + C$

**251.** $\int \frac{X}{x^2} dx = -\frac{\sqrt{X}}{x} + \ln|x + \sqrt{X}| + C$

**252.** $\int \frac{X}{x^3} dx = -\frac{\sqrt{X}}{2x^2} + \frac{1}{2a} \sec^{-1}|\frac{x}{a}| + C$

**253.** $\int \frac{dx}{\sqrt{X}} = \ln|x + \sqrt{X}| + C$

**254.** $\int \frac{x\, dx}{\sqrt{X}} = \sqrt{X} + C$

**255.** $\int \frac{x^2\, dx}{\sqrt{X}} = \frac{1}{2} x\sqrt{X} + \frac{a^2}{2} \ln|x + \sqrt{X}| + C$

**256.** $\int \frac{x^3\, dx}{\sqrt{X}} = \frac{1}{3} X^{3/2} + a^2 \sqrt{X} + C$

**257.** $\int \frac{dx}{x\sqrt{X}} = \frac{1}{a} \sec^{-1}|\frac{x}{a}| + C$

**258.** $\int \frac{dx}{x^2\sqrt{X}} = \frac{\sqrt{X}}{a^2 x} + C$

**259.** $\int \frac{dx}{x^3\sqrt{X}} = \frac{\sqrt{X}}{2a^2 x^2} + \frac{1}{2a^3} \sec^{-1}|\frac{x}{a}| + C$

**260.** $\int X^{3/2}\, dx = \frac{x}{4} X^{3/2} - \frac{3}{8} a^2 x\sqrt{X} + \frac{3}{8} a^4 \ln|x + \sqrt{X}| + C$

**261.** $\int x X^{3/2}\, dx = \frac{1}{5} X^{5/2} + C$

**262.** $\int x^2 X^{3/2}\, dx = \frac{1}{6} x X^{5/2} + \frac{1}{24} a^2 x X^{3/2} - \frac{1}{16} a^4 x\sqrt{X} + \frac{a^6}{16} \ln|x + \sqrt{X}| + C$

**263.** $\int x^3 X^{3/2}\, dx = \frac{1}{7} X^{7/2} + \frac{1}{5} a^2 X^{5/2} + C$

**264.** $\int \frac{X^{3/2}}{x}\, dx = \frac{1}{3} X^{3/2} - a^2 \sqrt{X} + a^3 \sec^{-1}|\frac{x}{a}| + C$

**265.** $\int \frac{X^{3/2}}{x^2}\, dx = -\frac{X^{3/2}}{x} + \frac{3}{2} x\sqrt{X} - \frac{3}{2} a^2 \ln|x + \sqrt{X}| + C$

**266.** $\int \frac{X^{3/2}}{x^3}\, dx = -\frac{X^{3/2}}{2x^2} + \frac{3}{2}\sqrt{X} - \frac{3}{2} a \sec^{-1}|\frac{x}{a}| + C$

**267.** $\int \frac{dx}{X^{3/2}} = -\frac{x}{a^2\sqrt{X}} + C$

Appendix B

**268.** $\int \dfrac{x\,dx}{X^{3/2}} = \dfrac{-1}{\sqrt{X}} + C$

**269.** $\int \dfrac{x^2\,dx}{X^{3/2}} = -\dfrac{x}{\sqrt{X}} - \dfrac{a^2}{\sqrt{X}} + C$

**270.** $\int \dfrac{x^3\,dx}{X^{3/2}} = \sqrt{X} + \ln|x + \sqrt{X}| + C$

**271.** $\int \dfrac{dx}{xX^{3/2}} = \dfrac{-1}{a^2\sqrt{X}} - \dfrac{1}{a^3}\sec^{-1}\left|\dfrac{x}{a}\right| + C$

**272.** $\int \dfrac{dx}{x^2 X^{3/2}} = -\dfrac{\sqrt{X}}{a^4 x} - \dfrac{x}{a^4\sqrt{X}} + C$

**273.** $\int \dfrac{dx}{x^3 X^{3/2}} = \dfrac{1}{2a^2 x^2 \sqrt{X}} - \dfrac{3}{2a^4 \sqrt{X}} - \dfrac{3}{2a^5}\sec^{-1}\left|\dfrac{x}{a}\right| + C$

---

**Integrals containing $X = a^2 - x^2$ with $x^2 < a^2$**

**274.** $\int \dfrac{dx}{X} = \dfrac{1}{2a}\ln\left(\dfrac{a+x}{a-x}\right) + C \quad \text{or} \quad \dfrac{1}{a}\tanh^{-1}\dfrac{x}{a} + C$

**275.** $\int \dfrac{x\,dx}{X} = -\dfrac{1}{2}\ln X + C$

**276.** $\int \dfrac{x^2\,dx}{X} = -x + \dfrac{a}{2}\ln\left|\dfrac{a+x}{a-x}\right| + C$

**277.** $\int \dfrac{x^3\,dx}{X} = -\dfrac{1}{2}x^2 - \dfrac{a^2}{2}\ln|X| + C$

**278.** $\int \dfrac{d}{xX} = \dfrac{1}{2a^2}\ln\left|\dfrac{x^2}{X}\right| + C$

**279.** $\int \dfrac{dx}{x^2 X} = -\dfrac{1}{a^2 x} + \dfrac{1}{2a^3}\ln\left|\dfrac{a+x}{a-x}\right| + C$

**280.** $\int \dfrac{dx}{x^3 X} = -\dfrac{1}{2a^2 x^2} + \dfrac{1}{2a^4}\ln\left|\dfrac{x^2}{X}\right| + C$

**281.** $\int \dfrac{dx}{X^2} = \dfrac{x}{2a^2 X} + \dfrac{1}{4a^3}\ln\left|\dfrac{a+x}{a-x}\right| + C$

**282.** $\int \dfrac{x\,dx}{X^2} = \dfrac{1}{2X} + C$

**283.** $\int \dfrac{x^2\,dx}{X^2} = \dfrac{x}{2X} - \dfrac{1}{4a}\ln\left|\dfrac{a+x}{a-x}\right| + C$

Appendix B

**284.** $\int \dfrac{x^3\, dx}{X^2} = \dfrac{a^2}{2X} + \dfrac{1}{2}\ln|X| + C$

**285.** $\int \dfrac{dx}{xX^2} = \dfrac{1}{2a^2 X} + \dfrac{1}{2a^4}\ln\left|\dfrac{x^2}{X}\right| + C$

**286.** $\int \dfrac{dx}{x^2 X^2} = -\dfrac{1}{a^4 x} + \dfrac{x}{2a^4 X} + \dfrac{3}{4a^5}\ln\left|\dfrac{a+x}{a-x}\right| + C$

**287.** $\int \dfrac{dx}{x^3 X^2} = -\dfrac{1}{2a^4 x^2} + \dfrac{1}{2a^4 X} + \dfrac{1}{a^6}\ln\left|\dfrac{x^2}{X}\right| + C$

**288.** $\int \dfrac{dx}{X^n} = \dfrac{x}{2(n-1)a^2 X^{n-1}} + \dfrac{2n-3}{2(n-1)a^2}\int \dfrac{dx}{X^{n-1}}$

**289.** $\int \dfrac{x\, dx}{X^n} = \dfrac{1}{2(n-1)X^{n-1}} + C$

**290.** $\int \dfrac{dx}{xX^n} = \dfrac{1}{2(n-1)a^2 X^{n-1}} + \dfrac{1}{a^2}\int \dfrac{dx}{xX^{n-1}}$

---

**Integrals containing the square root of $X = a^2 - x^2$ with $x^2 < a^2$**

**291.** $\int \sqrt{X}\, dx = \dfrac{1}{2}x\sqrt{X} + \dfrac{a^2}{2}\sin^{-1}\dfrac{x}{a} + C$

**292.** $\int x\sqrt{X}\, dx = -\dfrac{1}{3}X^{3/2} + C$

**293.** $\int x^2\sqrt{X}\, dx = -\dfrac{1}{4}xX^{3/2} + \dfrac{1}{8}a^2 x\sqrt{X} + \dfrac{1}{8}a^4 \sin^{-1}\dfrac{x}{a} + C$

**294.** $\int x^3\sqrt{X}\, dx = \dfrac{1}{5}X^{5/2} - \dfrac{1}{3}a^2 X^{3/2} + C$

**295.** $\int \dfrac{\sqrt{X}}{x}\, dx = \sqrt{X} - a\ln\left|\dfrac{a+\sqrt{X}}{x}\right| + C$

**296.** $\int \dfrac{\sqrt{X}}{x^2}\, dx = -\dfrac{\sqrt{X}}{x} - \sin^{-1}\dfrac{x}{a} + C$

**297.** $\int \dfrac{\sqrt{X}}{x^3}\, dx = -\dfrac{\sqrt{X}}{2x^2} + \dfrac{1}{2a}\ln\left|\dfrac{a+\sqrt{X}}{x}\right| + C$

**298.** $\int \dfrac{dx}{\sqrt{X}} = \sin^{-1}\dfrac{x}{a} + C$

**299.** $\int \dfrac{x\, dx}{\sqrt{X}} = -\sqrt{X} + C$

**Appendix B**

300. $\int \dfrac{x^2\,dx}{\sqrt{X}} = -\dfrac{1}{2}x\sqrt{X} + \dfrac{a^2}{2}\sin^{-1}\dfrac{x}{a} + C$

301. $\int \dfrac{x^3\,dx}{\sqrt{X}} = \dfrac{1}{3}X^{3/2} - a^2\sqrt{X} + C$

302. $\int \dfrac{dx}{x\sqrt{X}} = -\dfrac{1}{a}\ln\left|\dfrac{a+\sqrt{X}}{x}\right| + C$

303. $\int \dfrac{dx}{x^2\sqrt{X}} = -\dfrac{\sqrt{X}}{a^2 x} + C$

304. $\int \dfrac{dx}{x^3\sqrt{X}} = -\dfrac{\sqrt{X}}{2a^2 x^2} - \dfrac{1}{2a^3}\ln\left|\dfrac{a+\sqrt{X}}{x}\right| + C$

305. $\int X^{3/2}\,dx = \dfrac{1}{4}xX^{3/2} + \dfrac{3}{8}a^2 x\sqrt{X} + \dfrac{3}{8}a^4 \sin^{-1}\dfrac{x}{a} + C$

306. $\int xX^{3/2}\,dx = -\dfrac{1}{5}X^{5/2} + C$

307. $\int x^2 X^{3/2}\,dx = -\dfrac{1}{6}xX^{5/2} + \dfrac{1}{24}a^2 x X^{3/2} + \dfrac{1}{16}a^4 x\sqrt{X} + \dfrac{a^6}{16}\sin^{-1}\dfrac{x}{a} + C$

308. $\int x^3 X^{3/2}\,dx = \dfrac{1}{7}X^{7/2} - \dfrac{1}{5}a^2 X^{5/2} + C$

309. $\int \dfrac{X^{3/2}}{x}\,dx = \dfrac{1}{3}X^{3/2}a^2\sqrt{X} - a^3 \ln\left|\dfrac{a+\sqrt{X}}{x}\right| + C$

310. $\int \dfrac{X^{3/2}}{x^2}\,dx = -\dfrac{X^{3/2}}{x} - \dfrac{3}{2}x\sqrt{X} - \dfrac{3}{2}a^2 \sin^{-1}\dfrac{x}{a} + C$

311. $\int \dfrac{X^{3/2}}{x^3}\,dx = -\dfrac{X^{3/2}}{2x^2} - \dfrac{3}{2}\sqrt{X} + \dfrac{3}{2}a\ln\left|\dfrac{a+\sqrt{X}}{x}\right| + C$

312. $\int \dfrac{dx}{X^{3/2}} = \dfrac{x}{a^2\sqrt{X}} + C$

313. $\int \dfrac{x\,dx}{X^{3/2}} = \dfrac{1}{\sqrt{X}} + C$

314. $\int \dfrac{x^2\,dx}{X^{3/2}} = \dfrac{x}{\sqrt{X}} - \sin^{-1}\dfrac{x}{a} + C$

315. $\int \dfrac{x^3\,dx}{X^{3/2}} = \sqrt{X} + \dfrac{a^2}{\sqrt{X}} + C$

316. $\int \dfrac{dx}{xX^{3/2}} = \dfrac{1}{a^2\sqrt{X}} - \dfrac{1}{a^3}\ln\left|\dfrac{a+\sqrt{X}}{x}\right| + C$

317. $\int \dfrac{dx}{x^2 X^{3/2}} = -\dfrac{\sqrt{X}}{a^4 x} + \dfrac{x}{a^4 \sqrt{X}} + C$

318. $\int \dfrac{dx}{x^3 X^{3/2}} = -\dfrac{1}{2a^2 x^2 \sqrt{X}} + \dfrac{3}{2a^4 \sqrt{X}} - \dfrac{3}{2a^5} \ln\left|\dfrac{a + \sqrt{X}}{x}\right| + C$

**Integrals Containing** $X = x^3 + a^3$

319. $\int \dfrac{dx}{X} = \dfrac{1}{6a^2} \ln\left|\dfrac{(x+a)^3}{X}\right| + \dfrac{1}{\sqrt{3}a^2} \tan^{-1}\left(\dfrac{2x - a}{\sqrt{3}a}\right) + C$

320. $\int \dfrac{x\,dx}{X} = \dfrac{1}{6a} \ln\left|\dfrac{X}{(x+a)^3}\right| + \dfrac{1}{\sqrt{3}a} \tan^{-1}\left(\dfrac{2x - a}{\sqrt{3}a}\right) + C$

321. $\int \dfrac{x^2\,dx}{X} = \dfrac{1}{3} \ln|X| + C$

322. $\int \dfrac{dx}{xX} = \dfrac{1}{3a^3} \ln\left|\dfrac{x^3}{X}\right| + C$

323. $\int \dfrac{dx}{x^2 X} = -\dfrac{1}{a^3 x} - \dfrac{1}{6a^4} \ln\left|\dfrac{X}{(x+a)^3}\right| - \dfrac{1}{\sqrt{3}a^4} \tan^{-1}\left(\dfrac{2x - a}{\sqrt{3}a}\right) + C$

324. $\int \dfrac{dx}{X^2} = \dfrac{x}{3a^3 X} + \dfrac{1}{9a^5} \ln\left|\dfrac{(x+a)^3}{X}\right| + \dfrac{2}{3\sqrt{3}a^5} \tan^{-1}\left(\dfrac{2x - a}{\sqrt{3}a}\right) + C$

325. $\int \dfrac{x\,dx}{X^2} = \dfrac{x^2}{3a^3 X} + \dfrac{1}{18a^4} \ln\left|\dfrac{X}{(x+a)^3}\right| + \dfrac{1}{3\sqrt{3}a^4} \tan^{-1}\left(\dfrac{2x - a}{\sqrt{3}a}\right) + C$

326. $\int \dfrac{x^2\,dx}{X^2} = -\dfrac{1}{3X} + C$

327. $\int \dfrac{dx}{xX^2} = \dfrac{1}{3a^3 X} + \dfrac{1}{3a^6} \ln\left|\dfrac{x^3}{X}\right| + C$

328. $\int \dfrac{dx}{x^2 X^2} = -\dfrac{1}{a^6 x} - \dfrac{x^2}{3a^6 X} - \dfrac{4}{3a^6} \int \dfrac{x\,dx}{X}$

329. $\int \dfrac{dx}{X^3} = \dfrac{1}{54a^3}\left[\dfrac{9a^5 x}{X^2} + \dfrac{15a^2 x}{X} + 10\sqrt{3}\tan^{-1}\left(\dfrac{2x-a}{\sqrt{3}a}\right) + 10\ln|x+a| - 5\ln|x^2 - ax + a^2|\right] + C$

**Integrals containing** $X = x^4 + a^4$

330. $\int \dfrac{dx}{X} = \dfrac{1}{4\sqrt{2}a^3} \ln\left|\dfrac{X}{(x^2 - \sqrt{2}ax + a^2)^2}\right| - \dfrac{1}{2\sqrt{2}a^3} \tan^{-1}\left(\dfrac{\sqrt{2}ax}{x^2 - a^2}\right) + C$

331. $\int \dfrac{x\,dx}{X} = \dfrac{1}{2a^2} \tan^{-1}\left(\dfrac{x^2}{a^2}\right) + C$

332. $\int \dfrac{x^2\,dx}{X} = \dfrac{1}{4\sqrt{2}a} \ln\left|\dfrac{X}{(x^2 + \sqrt{2}ax + a^2)^2}\right| - \dfrac{1}{2\sqrt{2}a} \tan^{-1}\left(\dfrac{\sqrt{2}ax}{x^2 - a^2}\right) + C$

333. $\int \dfrac{x^3\,dx}{X} = \dfrac{1}{4} \ln|X| + C$

334. $\int \dfrac{dx}{xX} = \dfrac{1}{4a^4} \ln\left|\dfrac{x^4}{X}\right| + C$

335. $\int \dfrac{dx}{x^2 X} = -\dfrac{1}{a^4 x} - \dfrac{1}{\sqrt{2}4a^5} \ln\left|\dfrac{(x^2 - \sqrt{2}ax + a^2)^2}{X}\right| + \dfrac{1}{2\sqrt{2}a^5} \tan^{-1}\left(\dfrac{\sqrt{2}ax}{x^2 - a^2}\right) + C$

336. $\int \dfrac{dx}{x^3 X} = -\dfrac{1}{2a^4 x^2} - \dfrac{1}{2a^6} \tan^{-1}\left(\dfrac{x^2}{a^2}\right) + C$

### Integrals containing $X = x^4 - a^4$

**337.** $\displaystyle\int \frac{dx}{X} = \frac{1}{4a^3} \ln\left|\frac{x-a}{x+a}\right| - \frac{1}{2a^3} \tan^{-1}\left(\frac{x}{a}\right) + C$

**338.** $\displaystyle\int \frac{x\,dx}{X} = \frac{1}{4a^2} \ln\left|\frac{x^2-a^2}{x^2+a^2}\right| + C$

**339.** $\displaystyle\int \frac{x^2\,dx}{X} = \frac{1}{4a} \ln\left|\frac{x-a}{x+a}\right| + \frac{1}{2a} \tan^{-1}\left(\frac{x}{a}\right) + C$

**340.** $\displaystyle\int \frac{x^3\,dx}{X} = \frac{1}{4} \ln|X| + C$

**341.** $\displaystyle\int \frac{dx}{xX} = \frac{1}{4a^4} \ln\left|\frac{X}{x^4}\right| + C$

**342.** $\displaystyle\int \frac{dx}{x^2 X} = \frac{1}{a^4 x} + \frac{1}{4a^5} \ln\left|\frac{x-a}{x+a}\right| + \frac{1}{2a^5} \tan^{-1}\left(\frac{x}{a}\right) + C$

**343.** $\displaystyle\int \frac{dx}{x^3 X} = \frac{1}{2a^4 x^2} + \frac{1}{4a^6} \ln\left|\frac{x^2-a^2}{x^2+a^2}\right| + C$

### Miscelaneous algebraic integrals

**344.** $\displaystyle\int \frac{dx}{b^2 + (x+a)^2} = \frac{1}{b} \tan^{-1}\frac{x+a}{b} + C$

**345.** $\displaystyle\int \frac{dx}{b^2 - (x+a)^2} = \frac{1}{b} \tanh^{-1}\frac{x+a}{b} + C$

**346.** $\displaystyle\int \frac{dx}{(x+a)^2 - b^2} = -\frac{1}{b} \coth^{-1}\frac{x+a}{b} + C$

**347.** $\displaystyle\int \frac{dx}{\sqrt{x(a-x)}} = 2\sin^{-1}\sqrt{\frac{x}{a}} + C$

**348.** $\displaystyle\int \frac{dx}{\sqrt{x(a+x)}} = 2\sinh^{-1}\sqrt{\frac{x}{a}} + C$

**349.** $\displaystyle\int \frac{dx}{\sqrt{x(x-a)}} = 2\cosh^{-1}\sqrt{\frac{x}{a}} + C$

**350.** $\displaystyle\int \frac{dx}{(b+x)(a-x)} = 2\tan^{-1}\sqrt{\frac{b+x}{a-x}} + C, \quad a > x$

**351.** $\displaystyle\int \frac{dx}{(x-b)(a-x)} = 2\tan^{-1}\sqrt{\frac{x-b}{a-x}} + C, \quad a > x > b$

**352.** $\displaystyle\int \frac{dx}{(x+b)(x+a)} = \begin{cases} 2\tanh^{-1}\sqrt{\frac{x+b}{x+a}} + C_1, & a > b \\ 2\tanh^{-1}\sqrt{\frac{x+a}{x+b}} + C_2, & a < b \end{cases}$

353. $\int \dfrac{dx}{x\sqrt{x^{2n}-a^{2n}}} = -\dfrac{1}{na^n}\sin^{-1}\left(\dfrac{a^n}{x^n}\right) + C$

354. $\int \sqrt{\dfrac{x+a}{x-a}}\,dx = \sqrt{x^2-a^2} + a\cosh^{-1}\dfrac{x}{a} + C$

355. $\int \sqrt{\dfrac{a+x}{a-x}}\,dx = a\sin^{-1}\dfrac{x}{a} - \sqrt{a^2-x^2} + C$

356. $\int x\sqrt{\dfrac{a-x}{a+x}}\,dx = \dfrac{a^2}{2}\cos^{-1}\left(\dfrac{x}{a}\right) + \dfrac{(x-2a)}{2}\sqrt{a^2-x^2} + C, \quad a > x$

357. $\int x\sqrt{\dfrac{a+x}{a-x}}\,dx = \dfrac{a^2}{2}\sin^{-1}\dfrac{x}{a} - \dfrac{x+2a}{2}\sqrt{a^2-x^2} + C$

358. $\int (x+a)\sqrt{\dfrac{x+b}{x-b}}\,dx = (x+a+b)\sqrt{x^2-b^2} + \dfrac{b}{2}(2a+b)\cosh^{-1}\dfrac{x}{b} + C$

359. $\int \dfrac{dx}{\sqrt{2ax+x^2}} = \ln|x+a+\sqrt{2ax+x^2}| + C$

360. $\int \sqrt{ax^2+c}\,dx = \begin{cases} \dfrac{1}{2}x\sqrt{ax^2+c} + \dfrac{c}{2\sqrt{a}}\ln|\sqrt{a}x+\sqrt{ax^2+c}| + c, & a > 0 \\ \dfrac{1}{2}x\sqrt{ax^2+c} + \dfrac{c}{2\sqrt{-a}}\sin^{-1}\left(\sqrt{\dfrac{-a}{c}}\,x\right) + C, & a < 0 \end{cases}$

361. $\int \sqrt{\dfrac{1+ax}{1-ax}}\,dx = \dfrac{1}{a}\sin^{-1}x - \dfrac{1}{a}\sqrt{1-x^2} + C$

362. $\int \dfrac{dx}{(ax+b)^2 + (cx+d)^2} = \dfrac{1}{ad-bc}\tan^{-1}\left[\dfrac{(a^2+c^2)x+(ab+cd)}{ad-bc}\right] + C, \quad ad-bc \neq 0$

363. $\int \dfrac{dx}{(ax+b)^2 - (cx+d)^2} = \dfrac{1}{2(bc-ad)}\ln\left|\dfrac{(a+c)x+(b+d)}{(a-c)x+(b-d)}\right| + C, \quad ad-bc \neq 0$

364. $\int \dfrac{x\,dx}{(ax^2+b)^2 + (cx^2+d)^2} = \dfrac{1}{2(ad-bc)}\tan^{-1}\left[\dfrac{(a^2+c^2)x^2+(ab+cd)}{ad-bc}\right] + C, \quad ad-bc \neq 0$

365. $\int \dfrac{dx}{(x^2+a^2)(x^2+b^2)} = \dfrac{1}{b^2-a^2}\left(\dfrac{1}{a}\tan^{-1}\dfrac{x}{a} - \dfrac{1}{b}\tan^{-1}\dfrac{x}{b}\right) + C$

366. $\int \dfrac{(x^2+a^2)(x^2+b^2)}{(x^2+c^2)(x^2+d^2)}\,dx = x + \dfrac{1}{d^2-c^2}\left[\dfrac{(a^2-c^2)(b^2-c^2)}{c}\tan^{-1}\dfrac{x}{c} - \dfrac{(a^2-d^2)(b^2-d^2)}{d}\tan^{-1}\dfrac{x}{d}\right] + C$

367. $\int \dfrac{ax^2+b}{(cx^2+d)(ex^2+f)}\,dx = \dfrac{1}{\sqrt{cd}}\left(\dfrac{ad-bc}{ed-fc}\right)\tan^{-1}\left(\sqrt{\dfrac{c}{d}}\,x\right) + \dfrac{1}{\sqrt{ef}}\left(\dfrac{af-be}{fc-ed}\right)\tan^{-1}\left(\sqrt{\dfrac{e}{f}}\,x\right) + C$

368. $\int \dfrac{x\,dx}{(ax^2+bx+c)^2 + (ax^2-bx+c)^2} = \dfrac{1}{4b\sqrt{b^2+4ac}}\ln\left|\dfrac{2a^2x^2+2ac+b^2-b\sqrt{b^2+4ac}}{2a^2x^2+2ac+b^2+b\sqrt{b^2+4ac}}\right| + C, \quad b^2+4ac > 0$

**Appendix B**

369. $\int \dfrac{dx}{(x^2+a^2)(x^2+b^2)} = \dfrac{1}{b^2-a^2}\left(\dfrac{1}{a}\tan^{-1}\dfrac{x}{a} - \dfrac{1}{b}\tan^{-1}\dfrac{x}{b}\right) + C$

370. $\int \dfrac{(x^2+\alpha^2)(x^2+\beta^2)}{(x^2+\gamma^2)(x^2+\delta^2)}\,dx = x + \dfrac{1}{\delta^2-\gamma^2}\left[\dfrac{(\alpha^2-\gamma^2)(\beta^2-\gamma^2)}{\gamma}\tan^{-1}\dfrac{x}{\gamma} - \dfrac{(\alpha^2-\delta^2)(\beta^2-\delta^2)}{\delta}\tan^{-1}\dfrac{x}{\delta}\right] + C$

371. $\int \dfrac{\alpha x^2+\beta}{(\gamma x^2+\delta)(\epsilon x^2+\zeta)}\,dx = \dfrac{1}{\sqrt{\gamma\delta}}\dfrac{\alpha\delta-\beta\gamma}{\epsilon\delta-\zeta\gamma}\tan^{-1}\left(\sqrt{\dfrac{\gamma}{\delta}}x\right) + \dfrac{1}{\sqrt{\epsilon\zeta}}\dfrac{\alpha\zeta-\beta\epsilon}{\zeta\gamma-\epsilon\delta}\tan^{-1}\left(\sqrt{\dfrac{\epsilon}{\zeta}}x\right) + C$

372. $\int \dfrac{dx}{\sqrt{(x+a)(x+b)}} = \cosh^{-1}\left(\dfrac{2x+a+b}{a-b}\right) + C,\quad a\neq b$

373. $\int \dfrac{dx}{\sqrt{(x-b)(u-x)}} = 2\tan^{-1}\sqrt{\dfrac{x-b}{u-x}} + C$

374. $\int \dfrac{dx}{(\alpha x+\beta)^2+(\gamma x+\delta)^2} = \dfrac{1}{\alpha\delta-\beta\gamma}\tan^{-1}\left[\dfrac{(\alpha^2+\gamma^2)x+(\alpha\beta+\gamma\delta)}{\alpha\delta-\beta\gamma}\right] + C$

375. $\int \dfrac{x\,dx}{(a^2+b^2-x^2)\sqrt{(a^2-x^2)(x^2-b^2)}} = \dfrac{1}{2ab}\sin^{-1}\left[\dfrac{(a^2+b^2)x^2-(a^4+b^4)}{(a^2-b^2)(a^2+b^2-x^2)}\right] + C$

376. $\int \dfrac{(x+b)\,dx}{(x^2+a^2)\sqrt{x^2+c^2}} = \dfrac{1}{\sqrt{a^2-c^2}}\sin^{-1}\sqrt{\dfrac{x^2+c^2}{x^2+a^2}} + \dfrac{b}{a\sqrt{a^2-c^2}}\cosh^{-1}\left[\dfrac{a}{c}\sqrt{\dfrac{x^2+c^2}{x^2+a^2}}\right] + C$

377. $\int \dfrac{px+q}{ax^2+bx+c}\,dx = \dfrac{p}{2a}\ln|ax^2+bx+c| + \left(q-\dfrac{pb}{2a}\right)\int\dfrac{dx}{ax^2+bx+c}$

378. $\int \dfrac{(\sqrt{a}-\sqrt{x})^2}{(a^2+ax+x^2)\sqrt{x}}\,dx = \dfrac{2\sqrt{3}}{\sqrt{a}}\tan^{-1}\dfrac{2\sqrt{x}+\sqrt{a}}{\sqrt{3a}} - \dfrac{2}{\sqrt{3a}}\tan^{-1}\dfrac{2\sqrt{x}-\sqrt{a}}{\sqrt{3a}} + C$

379. $\int (a+x)\sqrt{a^2+x^2}\,dx = \dfrac{1}{6}(2x^2+3ax+2a^2)\sqrt{a^2+x^2} + \dfrac{1}{2}a^2\sinh^{-1}\dfrac{x}{a} + C$

380. $\int \dfrac{x^2+a^2}{x^4+a^2x^2+a^4}\,dx = \dfrac{1}{a\sqrt{3}}\tan^{-1}\dfrac{ax\sqrt{3}}{a^2-x^2} + C$

381. $\int \dfrac{x^2-a^2}{x^4+a^2x^2+a^4}\,dx = \dfrac{1}{2a^3}\ln\dfrac{x^2-ax+a^2}{x^2+ax+a^2} + C$

**Integrals containing $\sin ax$**

382. $\int \sin ax\,dx = -\dfrac{1}{a}\cos ax + C$

383. $\int x\sin ax\,dx = \dfrac{1}{a^2}\sin ax - \dfrac{x}{a}\cos ax + C$

384. $\int x^2\sin ax\,dx = \dfrac{2}{a^2}x\sin ax + \left(\dfrac{2}{a^3}-\dfrac{x^2}{a}\right)\cos ax + C$

**385.** $\int x^3 \sin ax\, dx = \left(\dfrac{3x^2}{a^2} - \dfrac{6}{a^4}\right)\sin ax + \left(\dfrac{6x}{a} - \dfrac{x^3}{a}\right)\cos ax + C$

**386.** $\int x^n \sin ax\, dx = -\dfrac{1}{a}x^n \cos ax + \dfrac{n}{a^2}x^{n-1}\sin ax - \dfrac{n(n-1)}{a^2}\int x^{n-2}\sin ax\, dx$

**387.** $\int \dfrac{\sin ax}{x}\, dx = ax - \dfrac{a^3 x^3}{3\cdot 3!} + \dfrac{a^5 x^5}{5\cdot 5!} - \dfrac{a^7 x^7}{7\cdot 7!} + \cdots + \dfrac{(-1)^n x^{2n+1} x^{2n+1}}{(2n+1)\cdot (2n+1)!} + \cdots$

**388.** $\int \dfrac{\sin ax}{x^2}\, dx = -\dfrac{1}{a}\sin ax + a\int \dfrac{\cos ax}{x}\, dx$

**389.** $\int \dfrac{\sin ax}{x^3}\, dx = -\dfrac{a}{2x}\cos ax - \dfrac{1}{2x^2}\sin ax - \dfrac{a^2}{2}\int \dfrac{\sin ax}{x}\, dx$

**390.** $\int \dfrac{\sin ax}{x^n}\, dx = -\dfrac{\sin ax}{(n-1)x^{n-1}} + \dfrac{a}{n-1}\int \dfrac{\cos ax}{x^{n-1}}\, dx$

**391.** $\int \dfrac{dx}{\sin ax} = \dfrac{1}{a}\ln|\csc ax - \cot ax| + C$

**392.** $\int \dfrac{x\, dx}{\sin ax} = \dfrac{1}{a^2}\left[ax + \dfrac{a^3 x^3}{18} + \dfrac{7a^5 x^5}{1800} + \cdots + \dfrac{2(2^{2n-1}-1)\mathfrak{B}_n a^{2n+1} x^{2n+1}}{(2n+1)!} + \cdots\right] + C$

where $\mathfrak{B}_n$ is the $n$th Bernoulli number $\mathfrak{B}_1 = 1/6, \mathfrak{B}_2 = 1/30, \ldots$

**393.** $\int \dfrac{dx}{x\sin ax} = -\dfrac{1}{ax} + \dfrac{ax}{6} + \dfrac{7a^3 x^3}{1080} + \cdots + \dfrac{2(2^{2n-1}-1)\mathfrak{B}_n a^{2n+1} x^{2n+1}}{(2n-1)(2n)!} + \cdots + C$

**394.** $\int \sin^2 ax\, dx = \dfrac{x}{2} - \dfrac{\sin 2ax}{4a} + C$

**395.** $\int x\sin^2 ax\, dx = \dfrac{x^2}{4} - \dfrac{x\sin 2ax}{4a} - \dfrac{\cos 2ax}{8a^2} + C$

**396.** $\int x^2 \sin^2 ax\, dx = \dfrac{1}{6a} - \dfrac{1}{4a^2}\cos 2ax + \dfrac{1}{24a^3}(3 - 6a^2 x^2)\sin 2ax + C$

**397.** $\int \sin^3 ax\, dx = -\dfrac{\cos ax}{a} + \dfrac{\cos^2 ax}{3a} + C$

**398.** $\int x\sin^3 ax\, dx = \dfrac{1}{12a}x\cos 3ax - \dfrac{1}{36a^2}\sin 3ax - \dfrac{3}{4a}x\cos ax + \dfrac{3}{4a^2}\sin ax + C$

**399.** $\int \sin^4 ax\, dx = \dfrac{3}{8}x - \dfrac{\sin 2ax}{4a} + \dfrac{\sin 4ax}{32a} + C$

**400.** $\int \dfrac{dx}{\sin^2 ax} = -\dfrac{1}{a}\cot ax + C$

**401.** $\displaystyle\int \frac{x\,dx}{\sin^2 ax} = -\frac{x}{a}\cot ax + \frac{1}{a^2}\ln|\sin ax| + C$

**402.** $\displaystyle\int \frac{dx}{\sin^3 ax} = -\frac{\cos ax}{2a\sin^2 ax} + \frac{1}{2a}\ln\left|\tan\frac{ax}{2}\right| + C$

**403.** $\displaystyle\int \frac{dx}{\sin^n ax} = \frac{-\cos ax}{(n-1)a\sin^{n-1} ax} + \frac{n-2}{n-1}\int \frac{dx}{\sin^{n-2} ax}$

**404.** $\displaystyle\int \frac{dx}{1-\sin ax} = \frac{1}{a}\tan\left(\frac{\pi}{4} - \frac{ax}{2}\right) + C$

**405.** $\displaystyle\int \frac{dx}{a - \sin ax} = \frac{2}{a\sqrt{a^2-1}}\tan^{-1}\left[\frac{a\tan(ax/2) - 1}{\sqrt{a^2-1}}\right] + C, \quad a > 1$

**406.** $\displaystyle\int \frac{x\,dx}{1-\sin ax} = \frac{x}{a}\tan\left(\frac{\pi}{4} - \frac{ax}{2}\right) + \frac{2}{a^2}\ln\left|\sin\left(\frac{\pi}{4} - \frac{ax}{2}\right)\right| + C$

**407.** $\displaystyle\int \frac{dx}{1+\sin ax} = -\frac{1}{a}\tan\left(\frac{\pi}{4} - \frac{ax}{2}\right) + C$

**408.** $\displaystyle\int \frac{dx}{a+\sin ax} = \frac{2}{a\sqrt{a^2-1}}\tan^{-1}\left[\frac{1 + a\tan(ax/2)}{\sqrt{a^2-1}}\right] + C, \quad a > 1$

**409.** $\displaystyle\int \frac{x\,dx}{1+\sin ax} = \frac{x}{a}\tan\left(\frac{\pi}{4} - \frac{ax}{2}\right) + \frac{2}{a^2}\ln\left|\sin\left(\frac{\pi}{4} - \frac{ax}{2}\right)\right| + C$

**410.** $\displaystyle\int \frac{dx}{1+\sin^2 x} = \frac{1}{\sqrt{2}}\tan^{-1}(\sqrt{2}\tan x) + C$

**411.** $\displaystyle\int \frac{dx}{1-\sin^2 x} = \tan x + C$

**412.** $\displaystyle\int \frac{dx}{(1-\sin ax)^2} = \frac{1}{2a}\tan\left(\frac{\pi}{4} - \frac{ax}{2}\right) + \frac{1}{6a}\tan^3\left(\frac{\pi}{4} - \frac{ax}{2}\right) + C$

**413.** $\displaystyle\int \frac{dx}{(1+\sin ax)^2} = -\frac{1}{2a}\tan\left(\frac{\pi}{4} - \frac{ax}{2}\right) - \frac{1}{6a}\tan^3\left(\frac{\pi}{4} - \frac{ax}{2}\right) + C$

**414.** $\displaystyle\int \frac{dx}{\alpha + \beta\sin ax} = \begin{cases} \dfrac{2}{a\sqrt{\alpha^2-\beta^2}}\tan^{-1}\left(\alpha\tan\dfrac{ax}{2} + \beta\right) + C, & \alpha^2 > \beta^2 \\[2mm] \dfrac{1}{a\sqrt{\beta^2-\alpha^2}}\ln\left|\dfrac{\alpha\tan\frac{ax}{2} + \beta - \sqrt{\beta^2-\alpha^2}}{\alpha\tan\frac{ax}{2} + \beta + \sqrt{\beta^2-\alpha^2}}\right| + C, & \alpha^2 < \beta^2 \\[2mm] \dfrac{1}{a\alpha}\tan\left(\dfrac{ax}{2} \pm \dfrac{\pi}{4}\right) + C, & \beta = \pm\alpha \end{cases}$

**415.** $\displaystyle\int \frac{dx}{\alpha^2 + \beta^2\sin^2 ax} = \frac{1}{a\alpha\sqrt{\beta^2+\alpha^2}}\tan^{-1}\left(\frac{\sqrt{\beta^2+\alpha^2}}{\alpha}\tan ax\right) + C$

416. $\displaystyle\int \frac{dx}{\alpha^2 - \beta^2 \sin^2 ax} = \begin{cases} \dfrac{1}{a\alpha\sqrt{\alpha^2 - \beta^2}} \tan^{-1}\left(\dfrac{\sqrt{\alpha^2 - \beta^2}}{\alpha} \tan ax\right) + C, & \alpha^2 > \beta^2 \\[2pt] \dfrac{1}{2a\alpha\sqrt{\beta^2 - \alpha^2}} \ln\left|\dfrac{\sqrt{\beta^2 - \alpha^2}\tan ax + \alpha}{\sqrt{\beta^2 - \alpha^2}\tan ax - \alpha}\right| + C, & \alpha^2 < \beta^2 \end{cases}$

417. $\displaystyle\int \sin^n ax\, dx = -\frac{1}{an}\sin^{n-1} ax \cos ax + \frac{n-1}{n}\int \sin^{n-2} ax\, dx$

418. $\displaystyle\int \frac{dx}{\sin^n ax} = \frac{-\cos ax}{(n-1)a\sin^{n-1} ax} + \frac{n-2}{n-1}\int \frac{dx}{\sin^{n-2} ax}$

419. $\displaystyle\int x^n \sin ax\, dx = -\frac{1}{a}x^n \cos ax + \frac{n}{a}\int x^{n-1}\cos ax\, dx$

420. $\displaystyle\int \frac{\alpha + \beta \sin ax}{1 \pm \sin ax}\, dx = \beta x + \frac{\alpha \mp \beta}{a}\tan\left(\frac{\pi}{4} \mp \frac{ax}{2}\right) + C$

421. $\displaystyle\int \frac{\alpha + \beta \sin ax}{a + b \sin ax}\, dx = \frac{\beta}{b}x + \frac{\alpha b - a\beta}{b}\int \frac{dx}{a + b \sin ax}$

422. $\displaystyle\int \frac{dx}{\alpha + \frac{\beta}{\sin ax}} = \frac{x}{\alpha} - \frac{\beta}{\alpha}\int \frac{dx}{\beta + \alpha \sin ax}$

**Integrals containing $\cos ax$**

423. $\displaystyle\int \cos ax\, dx = \frac{1}{a}\sin ax + C$

424. $\displaystyle\int x \cos ax\, dx = \frac{1}{a^2}\cos ax + \frac{x}{a}\sin ax + C$

425. $\displaystyle\int x^2 \cos ax\, dx = \frac{2x}{a^2}\cos ax + \left(\frac{x^2}{a} - \frac{2}{a^3}\right)\sin ax + C$

426. $\displaystyle\int x^n \cos ax\, dx = \frac{1}{a}x^n \sin ax + \frac{n}{a^2}x^{n-1}\cos ax - \frac{n(n-1)}{a^2}\int x^{n-2}\cos ax\, dx$

427. $\displaystyle\int \frac{\cos ax}{x}\, dx = \ln|x| - \frac{a^2 x^2}{2 \cdot 2!} + \frac{a^4 x^4}{4 \cdot 4!} - \frac{a^6 x^6}{6 \cdot 6!} + \cdots + \frac{(-1)^n a^{2n} x^{2n}}{(2n)\cdot(2n)!} + \cdots + C$

428. $\displaystyle\int \frac{\cos ax\, dx}{x^n} = -\frac{\cos ax}{(n-1)x^{n-1}} - \frac{a}{n-1}\int \frac{\sin ax}{x^{n-1}}\, dx$

429. $\displaystyle\int \frac{dx}{\cos ax} = \frac{1}{a}\ln|\sec ax + \tan ax| + C$

430. $\displaystyle\int \frac{x\, dx}{\cos ax} = \frac{1}{a^2}\left[\frac{a^2 x^2}{2} + \frac{a^4 x^4}{4 \cdot 2!} + \frac{5a^6 x^6}{6 \cdot 4!} + \cdots + \frac{\mathfrak{E}_n a^{2n+2} x^{2n+2}}{(2n+2)\cdot(2n)!} + \cdots\right] + C$

431. $\displaystyle\int \frac{dx}{x \cos ax} = \ln|x| + \frac{a^2 x^2}{4} + \frac{5a^4 x^4}{96} + \cdots + \frac{\mathfrak{E}_n a^{2n} x^{2n}}{2n(2n)!} + \cdots + C$

where $\mathfrak{E}_n$ is the $n$th Euler number $\mathfrak{E}_1 = 1, \mathfrak{E}_2 = 5, \mathfrak{E}_3 = 61, \ldots$.

**Appendix B**

432. $\displaystyle\int \frac{dx}{1+\cos ax} = \frac{1}{a}\tan\frac{ax}{2} + C$

433. $\displaystyle\int \frac{dx}{1-\cos ax} = -\frac{1}{a}\cot\frac{ax}{2} + C$

434. $\displaystyle\int \sqrt{1-\cos ax}\,dx = -2\sqrt{2}\cos\frac{ax}{2} + C$

435. $\displaystyle\int \sqrt{1+\cos ax}\,dx = 2\sqrt{2}\sin\frac{ax}{2} + C$

436. $\displaystyle\int \cos^2 ax\,dx = \frac{x}{2} + \frac{\sin 2ax}{4a} + C$

437. $\displaystyle\int x\cos^2 ax\,dx = \frac{x^2}{4} + \frac{1}{4a}x\sin 2ax + \frac{1}{8a^2}\cos 2ax + C$

438. $\displaystyle\int \cos^3 ax\,dx = \frac{\sin ax}{a} - \frac{\sin^3 ax}{3a} + C$

439. $\displaystyle\int \cos^4 ax\,dx = \frac{3}{8}x + \frac{1}{4a}\sin 2ax + \frac{1}{32a}\sin 4ax + C$

440. $\displaystyle\int \frac{dx}{\cos^2 ax} = \frac{1}{a}\tan ax + C$

441. $\displaystyle\int \frac{x\,dx}{\cos^2 ax} = \frac{x}{a}\tan ax + \frac{1}{a^2}\ln|\cos ax| + C$

442. $\displaystyle\int \frac{dx}{\cos^3 ax} = \frac{1}{2a}\frac{\sin ax}{\cos^2 ax} + \frac{1}{2a}\ln\left|\tan\left(\frac{\pi}{4}+\frac{ax}{2}\right)\right| + C$

443. $\displaystyle\int \frac{dx}{1-\cos ax} = -\frac{1}{a}\cot\frac{ax}{2} + C$

444. $\displaystyle\int \frac{x\,dx}{1-\cos ax} = -\frac{x}{a}\cot\frac{ax}{2} + \frac{2}{a^2}\ln\left|\sin\frac{ax}{2}\right| + C$

445. $\displaystyle\int \frac{dx}{1+\cos ax} = \frac{1}{a}\tan\frac{ax}{2} + C$

446. $\displaystyle\int \frac{x\,dx}{1+\cos ax} = \frac{x}{a}\tan\frac{ax}{2} + \frac{2}{a^2}\ln\left|\cos\frac{ax}{2}\right| + C$

447. $\displaystyle\int \frac{dx}{1+\cos^2 ax} = -\frac{1}{\sqrt{2}a}\tan^{-1}(\sqrt{2}\cot ax) + C$

448. $\displaystyle\int \frac{dx}{1-\cos^2 ax} = -\frac{1}{a}\cot ax + C$

**Appendix B**

**449.** $\int \dfrac{dx}{(1-\cos ax)^2} = -\dfrac{1}{2a}\cot\dfrac{ax}{2} - \dfrac{1}{6a}\cot^3\dfrac{ax}{2} + C$

**450.** $\int \dfrac{dx}{(1+\cos ax)^2} = \dfrac{1}{2a}\tan\dfrac{ax}{2} + \dfrac{1}{6a}\tan^2\dfrac{ax}{2} + C$

**451.** $\int \dfrac{dx}{\alpha + \beta\cos ax} = \begin{cases} \dfrac{2}{a\sqrt{\alpha^2-\beta^2}} \tan^{-1}\left(\sqrt{\dfrac{\alpha-\beta}{\alpha+\beta}}\tan\dfrac{ax}{2}\right) + C, & \alpha^2 > \beta^2 \\ \dfrac{1}{a\sqrt{\beta^2-\alpha^2}} \ln\left|\dfrac{\sqrt{\beta+\alpha}+\sqrt{\beta-\alpha}\tan\frac{ax}{2}}{\sqrt{\beta+\alpha}-\sqrt{\beta-\alpha}\tan\frac{ax}{2}}\right| + C, & \alpha^2 < \beta^2 \end{cases}$

**452.** $\int \dfrac{dx}{\alpha + \frac{\beta}{\cos ax}} = \dfrac{x}{\alpha} - \dfrac{\beta}{\alpha}\int \dfrac{dx}{\beta + \alpha\cos ax}$

**453.** $\int \dfrac{dx}{(\alpha + \beta\cos ax)^2} = \dfrac{\alpha\sin ax}{a(\beta^2-\alpha^2)(\alpha+\beta\cos ax)} - \dfrac{\alpha}{\beta^2-\alpha^2}\int \dfrac{dx}{\alpha+\beta\cos ax}, \quad \alpha \neq \beta$

**454.** $\int \dfrac{dx}{\alpha^2 + \beta^2\cos^2 ax} = \dfrac{1}{a\alpha\sqrt{\alpha^2+\beta^2}}\tan^{-1}\left(\dfrac{\alpha\tan ax}{\sqrt{\alpha^2+\beta^2}}\right) + C$

**455.** $\int \dfrac{dx}{\alpha^2 - \beta^2\cos^2 ax} = \begin{cases} \dfrac{1}{a\alpha\sqrt{\alpha^2-\beta^2}}\tan^{-1}\left(\dfrac{\alpha\tan ax}{\sqrt{\alpha^2-\beta^2}}\right) + C, & \alpha^2 > \beta^2 \\ \dfrac{1}{2a\alpha\sqrt{\beta^2-\alpha^2}}\ln\left|\dfrac{\alpha\tan ax - \sqrt{\beta^2-\alpha^2}}{\alpha\tan ax + \sqrt{\beta^2-\alpha^2}}\right| + C, & \alpha^2 < \beta^2 \end{cases}$

**456.** $\int \dfrac{dx}{\cos^n ax} = \dfrac{\sec^{(n-2)} ax \tan ax}{(n-1)a} + \dfrac{n-2}{n-1}\int \sec^{n-2} ax\, dx + C$

---

**Integrals containing both sine and cosine functions**

**457.** $\int \sin ax \cos ax\, dx = \dfrac{1}{2a}\sin^2 ax + C$

**458.** $\int \dfrac{dx}{\sin ax \cos ax} = -\dfrac{1}{a}\ln|\cot ax| + C$

**459.** $\int \sin ax \cos bx\, dx = -\dfrac{\cos(a-b)x}{2(a-b)} - \dfrac{\cos(a+b)x}{2(a+b)} + C, \quad a \neq b$

**460.** $\int \sin ax \sin bx\, dx = \dfrac{\sin(a-b)x}{2(a-b)} - \dfrac{\sin(a+b)x}{2(a+b)} + C$

**461.** $\int \cos ax \cos bx\, dx = \dfrac{\sin(a-b)x}{2(a-b)} + \dfrac{\sin(a+b)x}{2(a+b)} + C$

**462.** $\int \sin^n ax \cos ax\, dx = \dfrac{\sin^{n+1} ax}{(n+1)a} + C$

**463.** $\int \cos^n ax \sin ax\, dx = -\dfrac{\cos^{n+1} ax}{(n+1)a} + C$

Appendix B

**464.** $\displaystyle\int \frac{\sin ax \, dx}{\cos ax} = \frac{1}{a} \ln|\sec ax| + C$

**465.** $\displaystyle\int \frac{\cos ax \, dx}{\sin ax} = \frac{1}{a} \ln|\sin ax| + C$

**466.** $\displaystyle\int \frac{x \sin ax \, dx}{\cos ax} = \frac{1}{a^2}\left[\frac{a^3 x^3}{3} + \frac{a^5 x^5}{5} + \frac{2a^7 x^7}{105} + \cdots + \frac{2^{2n}(2^{2n}-1)\mathfrak{B}_n a^{2n+1} x^{2n+1}}{(2n+1)!}\right] + C$

**467.** $\displaystyle\int \frac{x \cos ax \, dx}{\sin ax} = \frac{1}{a^2}\left[ax - \frac{a^3 x^3}{9} - \frac{a^5 x^5}{225} - \cdots - \frac{2^{2n}\mathfrak{B}_n a^{2n+1} x^{2n+1}}{(2n+1)!} - \cdots\right] + C$

**468.** $\displaystyle\int \frac{\cos ax \, dx}{x \sin ax} = -\frac{1}{ax} - \frac{ax}{2} - \frac{a^3 x^3}{135} - \cdots - \frac{2^{2n}\mathfrak{B}_n a^{2n-1} x^{2n-1}}{(2n-1)(2n)!} - \cdots + C$

**469.** $\displaystyle\int \frac{\sin ax}{x \cos ax} dx = ax + \frac{a^3 x^3}{9} + \frac{2a^5 x^5}{75} + \cdots + \frac{2^{2n}(2^{2n}-1)\mathfrak{B}_n a^{2n-1} x^{2n-1}}{(2n-1)(2n)!} + \cdots + C$

**470.** $\displaystyle\int \frac{\sin^2 ax}{\cos^2 ax} dx = \frac{1}{a}\tan ax - x + C$

**471.** $\displaystyle\int \frac{\cos^2 ax}{\sin^2 ax} dx = -\frac{1}{a}\cot ax - x + C$

**472.** $\displaystyle\int \frac{x \sin^2 ax}{\cos^2 ax} dx = \frac{1}{a} x \tan ax + \frac{1}{a^2}\ln|\cos ax| - \frac{1}{2} x^2 + C$

**473.** $\displaystyle\int \frac{x \cos^2 ax}{\sin^2 ax} dx = -\frac{1}{a} x \cot ax + \frac{1}{a^2}\ln|\sin ax| - \frac{1}{2} x^2 + C$

**474.** $\displaystyle\int \frac{\cos ax}{\sin ax} dx = \frac{1}{a}\ln|\sin ax| + C$

**475.** $\displaystyle\int \frac{\sin^3 ax}{\cos^3 ax} dx = \frac{1}{2a}\tan^2 ax + \frac{1}{a}\ln|\cos ax| + C$

**476.** $\displaystyle\int \frac{\cos^3 ax}{\sin^3 ax} dx = -\frac{1}{2a}\cot^2 ax - \frac{1}{a}\ln|\sin ax| + C$

**477.** $\displaystyle\int \sin(ax+b)\sin(ax+\beta)\,dx = \frac{x}{2}\cos(b-\beta) - \frac{1}{4a}\sin(2ax+b+\beta) + C$

**478.** $\displaystyle\int \sin(ax+b)\cos(ax+\beta)\,dx = \frac{x}{2}\sin(b-\beta) - \frac{1}{4a}\cos(2ax+b+\beta) + C$

**479.** $\displaystyle\int \cos(ax+b)\cos(ax+\beta)\,dx = \frac{x}{2}\cos(b-\beta) + \frac{1}{4a}\sin(2ax+b+\beta) + C$

**480.** $\displaystyle\int \sin^2 ax \cos^2 bx \, dx = \begin{cases} \dfrac{x}{4} - \dfrac{\sin 2ax}{8a} + \dfrac{\sin 2bx}{8b} - \dfrac{\sin 2(a-b)x}{16(a-b)} - \dfrac{\sin 2(a+b)x}{16(a+b)} + C, & b \neq a \\ \dfrac{x}{8} - \dfrac{\sin 4ax}{32a} + C, & b = a \end{cases}$

**481.** $\int \dfrac{dx}{\sin ax \cos ax} = \dfrac{1}{a} \ln |\tan ax| + C$

**482.** $\int \dfrac{dx}{\sin^2 ax \cos ax} = \dfrac{1}{a} \ln \left| \tan \left( \dfrac{\pi}{4} + \dfrac{ax}{2} \right) \right| - \dfrac{1}{a \sin ax} + C$

**483.** $\int \dfrac{dx}{\sin ax \cos^2 ax} = \dfrac{1}{a} \ln \left| \tan \dfrac{ax}{2} \right| + \dfrac{1}{a \cos ax} + C$

**484.** $\int \dfrac{dx}{\sin^2 ax \cos^2 ax} = -\dfrac{2 \cos 2ax}{a} + C$

**485.** $\int \dfrac{\sin^2 ax}{\cos ax} dx = -\dfrac{\sin ax}{a} + \dfrac{1}{a} \ln \left| \tan \left( \dfrac{ax}{2} + \dfrac{\pi}{4} \right) \right| + C$

**486.** $\int \dfrac{\cos^2 ax}{\sin ax} dx = \dfrac{\cos ax}{a} + \dfrac{1}{a} \ln \left| \tan \dfrac{ax}{2} \right| + C$

**487.** $\int \dfrac{dx}{\cos ax \,(1 + \sin ax)} = \dfrac{1}{2a(1 + \sin ax)} \left[ -1 + (1 + \sin ax) \ln \left| \dfrac{\cos \frac{ax}{2} + \sin \frac{ax}{2}}{\cos \frac{ax}{2} - \sin \frac{ax}{2}} \right| \right] + C$

**488.** $\int \dfrac{dx}{\sin ax \,(1 + \cos ax)} = \dfrac{1}{4a} \sec^2 \dfrac{ax}{2} + \dfrac{1}{2a} \ln \left| \tan \dfrac{ax}{2} \right| + C$

**489.** $\int \dfrac{dx}{\sin ax \,(\alpha + \beta \sin ax)} = \dfrac{1}{a\alpha} \ln \left| \tan \dfrac{ax}{2} \right| - \dfrac{\beta}{\alpha} \int \dfrac{dx}{\alpha + \beta \sin ax}$

**490.** $\int \dfrac{dx}{\cos ax \,(\alpha + \beta \sin ax)} = \dfrac{1}{\alpha^2 - \beta^2} \left[ \dfrac{\alpha}{a} \ln \left| \tan \left( \dfrac{\pi}{4} + \dfrac{ax}{2} \right) \right| - \dfrac{\beta}{\alpha} \ln \left| \dfrac{\alpha + \beta \sin ax}{\cos ax} \right| \right] + C, \qquad \beta \neq \alpha$

**491.** $\int \dfrac{dx}{\sin ax \,(\alpha + \beta \cos ax)} = \dfrac{1}{\alpha^2 - \beta^2} \left[ \dfrac{\alpha}{a} \ln \left| \tan \dfrac{ax}{2} \right| + \dfrac{\beta}{a} \ln \left| \dfrac{\alpha + \beta \cos ax}{\sin ax} \right| \right] + C, \qquad \beta \neq \alpha$

**492.** $\int \dfrac{dx}{\cos ax \,(\alpha + \beta \cos ax)} = \dfrac{1}{a\alpha} \ln \left| \tan \left( \dfrac{\pi}{4} + \dfrac{ax}{2} \right) \right| - \dfrac{\beta}{\alpha} \int \dfrac{dx}{\alpha + \beta \cos ax}$

**493.** $\int \dfrac{dx}{\alpha + \beta \cos ax + \gamma \sin ax} = \begin{cases} \dfrac{2}{a\sqrt{-R}} \tan^{-1} \left( \dfrac{\gamma + (\alpha - \beta) \tan \frac{ax}{2}}{\sqrt{-R}} \right) + C, & \alpha^2 > \beta^2 + \gamma^2 \\ & R = \beta^2 + \gamma^2 - \alpha^2 \\ \dfrac{1}{a\sqrt{R}} \ln \left| \dfrac{\gamma - \sqrt{R} + (\alpha - \beta) \tan \frac{ax}{2}}{\gamma + \sqrt{R} + (\alpha - \beta) \tan \frac{ax}{2}} \right| + C, & \alpha^2 < \beta^2 + \gamma^2 \\ \dfrac{1}{a\beta} \ln \left| \beta + \gamma \tan \dfrac{ax}{2} \right| + C, & \alpha = \beta \\ \dfrac{1}{a\beta} \ln \left| \dfrac{\cos \frac{ax}{2} + \sin \frac{ax}{2}}{(\beta + \gamma) \cos \frac{ax}{2} + (\gamma - \beta) \sin \frac{ax}{2}} \right| + C, & \alpha = \gamma \\ \dfrac{1}{a\gamma} \ln \left| 1 + \tan \dfrac{ax}{2} \right| + C, & \alpha = \beta = \gamma \end{cases}$

**494.** $\int \dfrac{dx}{\sin ax \pm \cos ax} = \dfrac{1}{\sqrt{2}a} \ln \left| \tan \left( \dfrac{ax}{2} \pm \dfrac{\pi}{8} \right) \right| + C$

**495.** $\displaystyle\int \frac{\sin ax\, dx}{\sin ax \pm \cos ax} = \frac{x}{2} \mp \ln|\sin ax \pm \cos ax| + C$

**496.** $\displaystyle\int \frac{\cos ax\, dx}{\sin ax \pm \cos ax} = \pm\frac{x}{2} + \frac{1}{2a}\ln|\sin axx \pm \cos ax| + C$

**497.** $\displaystyle\int \frac{\sin ax\, dx}{\alpha + \beta \sin ax} = \frac{1}{a\beta}\ln|\alpha + \beta \sin ax| + C$

**498.** $\displaystyle\int \frac{\cos ax\, dx}{\alpha + \beta \sin ax} = \frac{1}{a\beta}\ln|\alpha + \beta \sin ax| + C$

**499.** $\displaystyle\int \frac{\sin ax \cos ax\, dx}{\alpha^2 \cos^2 ax + \beta^2 ax} = \frac{1}{2a(\beta^2 - \alpha^2)}\ln|\alpha^2 \cos^2 ax + \beta^2 \sin^2 ax| + C, \quad \beta \neq \alpha$

**500.** $\displaystyle\int \frac{dx}{\alpha^2 \sin^2 ax + \beta^2 \cos^2 ax} = \frac{1}{a\alpha\beta}\tan^{-1}\left(\frac{\alpha}{\beta}\tan ax\right) + C$

**501.** $\displaystyle\int \frac{dx}{\alpha^2 \sin^2 ax - \beta^2 \cos^2 ax} = \frac{1}{2a\alpha\beta}\ln\left|\frac{\alpha \tan ax - \beta}{\alpha \tan ax + \beta}\right| + C$

**502.** $\displaystyle\int \frac{\sin^n ax}{\cos^{(n+2)} ax}\, dx = \frac{\tan^{n+1} ax}{(n+1)a} + C$

**503.** $\displaystyle\int \frac{\cos^n ax}{\sin^{(n+2)} ax}\, dx = -\frac{\cot^{(n+1)} ax}{(n+1)a} + C$

**504.** $\displaystyle\int \frac{dx}{\alpha + \beta\frac{\sin ax}{\cos ax}} = \frac{\alpha x}{\alpha^2 + \beta^2} + \frac{\beta}{a(\alpha^2 + \beta^2)}\ln|\beta \sin ax + \alpha \cos ax| + C$

**505.** $\displaystyle\int \frac{dx}{\alpha + \beta\frac{\cos ax}{\sin ax}} = \frac{\alpha x}{\alpha^2 + \beta^2} - \frac{\beta}{a(\alpha^2 + \beta^2)}\ln|\alpha \sin ax + \beta \cos ax| + C$

**506.** $\displaystyle\int \frac{\cos^n ax}{\sin^n ax}\, dx = -\frac{\cot^{(n-1)} ax}{(n-1)a} - \int \cot^{(n-2)} ax\, dx$

**507.** $\displaystyle\int \frac{\sin^n ax}{\cos^n ax}\, dx = \frac{\tan^{n-1} ax}{(n-1)a} - \int \frac{\sin^{n-2} ax}{\cos^{n-2} ax}\, dx$

**508.** $\displaystyle\int \frac{\sin ax}{\cos^{(n+1)} ax}\, dx = \frac{1}{na}\sec^n ax + C$

**509.** $\displaystyle\int \frac{\alpha \sin x + \beta \cos x}{\gamma \sin x + \delta \cos x}\, dx = \frac{[(\alpha\gamma + \beta\delta)x + (\beta\gamma - \alpha\gamma)\ln|\gamma \sin x + \delta \cos x|]}{\gamma^2 + \delta^2} + C$

**510.** $\displaystyle\int \frac{\alpha + \beta \sin x}{a + b \cos x}\, dx = \begin{cases} \dfrac{2\alpha}{\sqrt{a^2 - b^2}}\tan^{-1}\sqrt{\dfrac{a-b}{a+b}}\tan\dfrac{x}{2} - \dfrac{\beta}{b}\ln|a + b\cos x| + C, & a > b \\[1em] \dfrac{2\alpha}{\sqrt{b^2 - a^2}}\tanh^{-1}\sqrt{\dfrac{b-a}{b+a}}\tan\dfrac{x}{2} - \dfrac{\beta}{b}\ln|a + b\cos x| + C, & a < b \end{cases}$

Appendix B

511. $\displaystyle\int \frac{dx}{a^2 - b^2 \cos^2 x} = \begin{cases} \frac{1}{a\sqrt{a^2-b^2}} \tan^{-1}\left(\frac{a}{\sqrt{a^2-b^2}} \tan x\right) + C, & a > b \\ \frac{-1}{a\sqrt{b^2-a^2}} \tanh^{-1}\left(\frac{a}{\sqrt{b^2-a^2}} \tan x\right) + C, & b > a \end{cases}$

512. $\displaystyle\int \frac{dx}{(a \cos x + b \sin x)^2} = \frac{1}{a^2 + b^2} \tan\left(x - \tan^{-1} \frac{b}{a}\right) + C$

513. $\displaystyle\int \frac{\sin x\, dx}{\sqrt{a\cos^2 x + 2b\cos x + c}} = \begin{cases} \frac{-1}{\sqrt{-a}} \sin^{-1}\left(\frac{\sqrt{-a(a\cos^2 x + 2b\cos x + c)}}{\sqrt{b^2 - ac}}\right) + C, & b^2 > ac,\ a < 0 \\ \frac{-1}{\sqrt{a}} \sinh^{-1}\left(\frac{\sqrt{a(a\cos^2 x + 2b\cos x + c)}}{\sqrt{b^2 - ac}}\right) + C, & b^2 > ac,\ a > 0 \\ \frac{-1}{\sqrt{a}} \cosh^{-1}\left(\frac{\sqrt{a(a\cos^2 x + 2b\cos x + c)}}{\sqrt{ac - b^2}}\right) + C, & b^2 < ac,\ a > 0 \end{cases}$

514. $\displaystyle\int \frac{\cos x\, dx}{\sqrt{a\sin^2 x + 2b\sin x + c}} = \begin{cases} \frac{1}{\sqrt{-a}} \sin^{-1}\left(\frac{\sqrt{-a(a\sin^2 x + 2b\sin x + c)}}{\sqrt{b^2 - ac}}\right) + C, & b^2 > ac,\ a < 0 \\ \frac{1}{\sqrt{a}} \sinh^{-1}\left(\frac{\sqrt{a(a\sin^2 x + 2b\sin x + c)}}{\sqrt{b^2 - ac}}\right) + C, & b^2 > ac,\ a > 0 \\ \frac{1}{\sqrt{a}} \cosh^{-1}\left(\frac{\sqrt{a(a\sin^2 x + 2b\sin x + c)}}{\sqrt{ac - b^2}}\right) + C, & b^2 < ac,\ a > 0 \end{cases}$

**Integrals containing $\tan ax$, $\cot ax$, $\sec ax$, $\csc ax$**

Write integrals in terms of $\sin ax$ and $\cos ax$ and see previous listings.

**Integrals containing inverse trigonometric functions**

515. $\displaystyle\int \sin^{-1} \frac{x}{a}\, dx = x \sin^{-1} \frac{x}{a} + \sqrt{a^2 - x^2} + C$

516. $\displaystyle\int \cos^{-1} \frac{x}{a}\, dx = x \cos^{-1} \frac{x}{a} - \sqrt{a^2 - x^2} + C$

517. $\displaystyle\int \tan^{-1} \frac{x}{a}\, dx = x \tan^{-1} \frac{x}{a} - \frac{a}{2} \ln|x^2 + a^2| + C$

518. $\displaystyle\int \cot^{-1} \frac{x}{a}\, dx = x \cot^{-1} \frac{x}{a} + \frac{a}{2} \ln|x^2 + a^2| + C$

519. $\displaystyle\int \sec^{-1} \frac{x}{a}\, dx = \begin{cases} x \sec^{-1} \frac{x}{a} - a \ln|x + \sqrt{x^2 - a^2}| + C, & 0 < \sec^{-1} \frac{x}{a} < \pi/2 \\ x \sec^{-1} \frac{x}{a} + a \ln|x + \sqrt{x^2 - a^2}| + C, & \pi/2 < \sec^{-1} \frac{x}{a} < \pi \end{cases}$

520. $\displaystyle\int \csc^{-1} \frac{x}{a}\, dx = \begin{cases} x \csc^{-1} \frac{x}{a} + a \ln|x + \sqrt{x^2 - a^2}| + C, & 0 < \csc^{-1} \frac{x}{a} < \pi/2 \\ x \csc^{-1} \frac{x}{a} - a \ln|x + \sqrt{x^2 - a^2}| + C, & -\pi/2 < \csc^{-1} \frac{x}{a} < 0 \end{cases}$

521. $\displaystyle\int x \sin^{-1} \frac{x}{a}\, dx = \left(\frac{x^2}{2} - \frac{a^2}{4}\right) \sin^{-1} \frac{x}{a} + \frac{1}{4} x \sqrt{a^2 - x^2} + C$

522. $\displaystyle\int x \cos^{-1} \frac{x}{a}\, dx = \left(\frac{x^2}{2} - \frac{a^2}{4}\right) \cos^{-1} \frac{x}{a} - \frac{1}{4} x \sqrt{a^2 - x^2} + C$

**523.** $\int x \tan^{-1} \dfrac{x}{a}\, dx = \dfrac{1}{2}(x^2 + a^2)\tan^{-1}\dfrac{x}{a} - \dfrac{a}{2}\ln|x^2 + a^2| + C$

**524.** $\int x \cot^{-1} \dfrac{x}{a}\, dx = \dfrac{1}{2}(x^2 + a^2)\cot^{-1}\dfrac{x}{a} + \dfrac{a}{2}x + C$

**525.** $\int x \sec^{-1} \dfrac{x}{a}\, dx = \begin{cases} \dfrac{1}{2}x^2 \sec^{-1}\dfrac{x}{a} - \dfrac{a}{2}\sqrt{x^2 - a^2} + C, & 0 < \sec^{-1}\dfrac{x}{a} < \pi/2 \\ \dfrac{1}{2}x^2 \sec^{-1}\dfrac{x}{a} + \dfrac{a}{2}\sqrt{x^2 - a^2} + C, & \pi/2 < \sec^{-1}\dfrac{x}{a} < \pi \end{cases}$

**526.** $\int x \csc^{-1} \dfrac{x}{a}\, dx = \begin{cases} \dfrac{1}{2}x^2 \csc^{-1}\dfrac{x}{a} + \dfrac{a}{2}\sqrt{x^2 - a^2} + C, & 0 < \csc^{-1}\dfrac{x}{a} < \pi/2 \\ \dfrac{1}{2}x^2 \csc^{-1}\dfrac{x}{a} - \dfrac{a}{2}\sqrt{x^2 - a^2} + C, & -\pi/2 < \csc^{-1}\dfrac{x}{a} < 0 \end{cases}$

**527.** $\int x^2 \sin^{-1} \dfrac{x}{a}\, dx = \dfrac{1}{3}x^3 \sin^{-1}\dfrac{x}{a} + \dfrac{1}{9}(x^2 + 2a^2)\sqrt{a^2 - x^2} + C$

**528.** $\int x^2 \cos^{-1} \dfrac{x}{a}\, dx = \dfrac{1}{3}x^3 \cos^{-1}\dfrac{x}{a} - \dfrac{1}{9}(x^2 + 2a^2)\sqrt{a^2 - x^2} + C$

**529.** $\int x^2 \tan^{-1} \dfrac{x}{a}\, dx = \dfrac{1}{3}\tan^{-1}\dfrac{x}{a} - \dfrac{a}{6}x^2 + \dfrac{a^3}{6}\ln|x^2 + a^2| + C$

**530.** $\int x^2 \cot^{-1} \dfrac{x}{a}\, dx = \dfrac{1}{3}\cot^{-1}\dfrac{x}{a} + \dfrac{a}{6}x^2 - \dfrac{a^3}{6}\ln|a^2 + x^2| + C$

**531.** $\int x^2 \sec^{-1} \dfrac{x}{a}\, dx = \begin{cases} \dfrac{1}{3}x^3 \sec^{-1}\dfrac{x}{a} - \dfrac{a}{6}x\sqrt{x^2 - a^2} - \dfrac{a^3}{6}\ln|x + \sqrt{x^2 - a^2}| + C, & 0 < \sec^{-1}\dfrac{x}{a} < \pi/2 \\ \dfrac{1}{3}x^3 \sec^{-1}\dfrac{x}{a} + \dfrac{a}{6}x\sqrt{x^2 - a^2} + \dfrac{a^3}{6}\ln|x + \sqrt{x^2 - a^2}| + c, & \pi/2 < \sec^{-1}\dfrac{x}{a} < \pi \end{cases}$

**532.** $\int x^2 \csc^{-1} \dfrac{x}{a}\, dx = \begin{cases} \dfrac{1}{3}x^3 \csc^{-1}\dfrac{x}{a} + \dfrac{a}{6}x\sqrt{x^2 - a^2} + \dfrac{a^3}{6}\ln|x + \sqrt{x^2 - a^2}| + C, & 0 < \csc^{-1}\dfrac{x}{a} < \pi/2 \\ \dfrac{1}{3}x^3 \csc^{-1}\dfrac{x}{a} - \dfrac{a}{6}x\sqrt{x^2 - a^2} - \dfrac{a^3}{6}\ln|x + \sqrt{x^2 - a^2}| + C, & -\pi/2 < \csc^{-1}\dfrac{x}{a} < 0 \end{cases}$

**533.** $\int \dfrac{1}{x} \sin^{-1} \dfrac{x}{a}\, dx = \dfrac{x}{a} + \dfrac{1}{2 \cdot 3 \cdot 3}\left(\dfrac{x}{a}\right)^3 + \dfrac{1 \cdot 3}{2 \cdot 4 \cdot 5 \cdot 5}\left(\dfrac{x}{a}\right)^5 + \dfrac{1 \cdot 3 \cdot 5}{2 \cdot 4 \cdot 6 \cdot 7 \cdot 7}\left(\dfrac{x}{a}\right)^7 + \cdots + C$

**534.** $\int \dfrac{1}{x} \cos^{-1} \dfrac{x}{a}\, dx = \dfrac{\pi}{2}\ln|x| + - \int \dfrac{1}{x}\sin^{-1}\dfrac{x}{a}\, dx$

**535.** $\int \dfrac{1}{x} \tan^{-1} \dfrac{x}{a}\, dx = \dfrac{x}{a} - \dfrac{1}{3^2}\left(\dfrac{x}{a}\right)^3 + \dfrac{1}{5^2}\left(\dfrac{x}{a}\right)^5 - \dfrac{1}{7^2}\left(\dfrac{x}{a}\right)^7 + \cdots + C$

**536.** $\int \dfrac{1}{x} \cot^{-1} \dfrac{x}{a}\, dx = \dfrac{\pi}{2}\ln|x| - \int \dfrac{1}{x}\tan^{-1}\dfrac{x}{a}\, dx$

**537.** $\int \dfrac{1}{x} \sec^{-1} \dfrac{x}{a}\, dx = \dfrac{\pi}{2}\ln|x| + \dfrac{a}{x} + \dfrac{1}{2 \cdot 3 \cdot 3}\left(\dfrac{x}{a}\right)^3 + \dfrac{1 \cdot 3}{2 \cdot 4 \cdot 5 \cdot 5}\left(\dfrac{x}{a}\right)^5 + \dfrac{1 \cdot 3 \cdot 5}{2 \cdot 4 \cdot 6 \cdot 7 \cdot 7}\left(\dfrac{x}{a}\right)^7 + \cdots + C$

Appendix B

**538.** $\displaystyle\int \frac{1}{x}\csc^{-1}\frac{x}{a}\,dx = -\left(\frac{a}{x} + \frac{1}{2\cdot 3\cdot 3}\left(\frac{x}{a}\right)^3 + \frac{1\cdot 3}{2\cdot 4\cdot 5\cdot 5}\left(\frac{x}{a}\right)^5 + \frac{1\cdot 3\cdot 5}{2\cdot 4\cdot 6\cdot 7\cdot 7}\left(\frac{x}{a}\right)^7 + \cdots\right) + C$

**539.** $\displaystyle\int \frac{1}{x^2}\sin^{-1}\frac{x}{a}\,dx = -\frac{1}{x}\sin^{-1}\frac{x}{a} - \frac{1}{a}\ln\left|\frac{a+\sqrt{a^2-x^2}}{a}\right| + C$

**540.** $\displaystyle\int \frac{1}{x^2}\cos^{-1}\frac{x}{a}\,dx = -\frac{1}{x}\cos^{-1}\frac{x}{a} + \frac{1}{a}\ln\left|\frac{a+\sqrt{a^2-x^2}}{a}\right| + C$

**541.** $\displaystyle\int \frac{1}{x^2}\tan^{-1}\frac{x}{a}\,dx = -\frac{1}{x}\tan^{-1}\frac{x}{a} - \frac{1}{2a}\ln\left|\frac{x^2+a^2}{a^2}\right| + C$

**542.** $\displaystyle\int \frac{1}{x^2}\cot^{-1}\frac{x}{a}\,dx = -\frac{1}{x}\cot^{-1}\frac{x}{a} + \frac{1}{2a}\int \frac{1}{x}\tan^{-1}\frac{x}{a}\,dx$

**543.** $\displaystyle\int \frac{1}{x^2}\sec^{-1}\frac{x}{a}\,dx = \begin{cases} -\frac{1}{x}\sec^{-1}\frac{x}{a} + \frac{1}{ax}\sqrt{x^2-a^2} + C, & 0 < \sec^{-1}\frac{x}{a} < \pi/2 \\ -\frac{1}{x}\sec^{-1}\frac{x}{a} - \frac{1}{ax}\sqrt{x^2-a^2} + C, & \pi/2 < \sec^{-1}\frac{x}{a} < \pi \end{cases}$

**544.** $\displaystyle\int \frac{1}{x^2}\csc^{-1}\frac{x}{a}\,dx = \begin{cases} -\frac{1}{x}\csc^{-1}\frac{x}{a} - \frac{1}{ax}\sqrt{x^2-a^2} + C, & 0 < \csc^{-1}\frac{x}{a} < \pi/2 \\ -\frac{1}{x}\csc^{-1}\frac{x}{a} + \frac{1}{ax}\sqrt{x^2-a^2} + C, & -\pi/2 < \csc^{-1}\frac{x}{a} < 0 \end{cases}$

**545.** $\displaystyle\int \sin^{-1}\sqrt{\frac{x}{a+x}}\,dx = (a+x)\tan^{-1}\sqrt{\frac{x}{a}} - \sqrt{ax} + C$

**546.** $\displaystyle\int \cos^{-1}\sqrt{\frac{x}{a+x}}\,dx = (2a+x)\tan^{-1}\sqrt{\frac{x}{2a}} - \sqrt{2ax} + C$

**Integrals containing the exponential function**

**547.** $\displaystyle\int e^{ax}\,dx = \frac{1}{a}e^{ax} + C$

**548.** $\displaystyle\int xe^{ax}\,dx = \left(\frac{x}{a} - \frac{1}{a^2}\right)e^{ax} + C$

**549.** $\displaystyle\int x^2 e^{ax}\,dx = \left(\frac{x^2}{a} - \frac{2x}{a^2} + \frac{2}{a^3}\right)e^{ax} + C$

**550.** $\displaystyle\int x^n e^{ax}\,dx = \frac{1}{a}x^n e^{ax} - \frac{n}{a}\int x^{n-1} e^{ax}\,dx$

**551.** $\displaystyle\int \frac{1}{x}e^{ax}\,dx = \ln|x| + \frac{ax}{1\cdot 1!} + \frac{(ax)^2}{2\cdot 2!} + \frac{(ax)^3}{3\cdot 3!} + \cdots + C$

**552.** $\displaystyle\int \frac{1}{x^n}e^{ax}\,dx = -\frac{1}{(n-1)x^{n-1}}e^{ax} + \frac{a}{n-1}\int \frac{1}{x^{n-1}}e^{ax}\,dx$

**553.** $\displaystyle\int \frac{e^{ax}}{\alpha + \beta e^{ax}}\,dx = \frac{1}{a\beta}\ln|\alpha + \beta e^{ax}| + C$

554. $\int e^{ax} \sin bx \, dx = \left( \dfrac{a \sin bx - b \cos bx}{a^2 + b^2} \right) e^{ax} + C$

555. $\int e^{ax} \cos bx \, dx = \left( \dfrac{a \cos bx + b \sin bx}{a^2 + b^2} \right) e^{ax} + C$

556. $\int e^{ax} \sin^n bx \, dx = \left( \dfrac{a \sin bx - nb \cos bx}{a^2 + n^2 b^2} \right) e^{ax} \sin^{n-1} bx + \dfrac{n(n-1)b^2}{a^2 + n^2 b^2} \int e^{ax} \sin^{n-2} bx \, dx$

557. $\int e^{ax} \cos^n bx \, dx = \left( \dfrac{a \cos bx + nb \sin bx}{a^2 + n^2 b^2} \right) e^{ax} \cos^{n-1} bx + \dfrac{n(n-1)b^2}{a^2 + n^2 b^2} \int e^{ax} \cos^{n-2} bx \, dx$

Another way to express the above integrals is to define

$$C_n = \int e^{ax} \cos^n bx \, dx \text{ and } S_n = \int e^{ax} \sin^n bx \, dx,$$ then one can write the reduction formulas

$$C_n = \dfrac{a \cos bx + nb \sin bx}{a^2 + n^2 b^2} e^{ax} \cos^{n-1} bx + \dfrac{n(n-1)b^2}{a^2 + n^2 b^2} C_{n-2}$$

$$S_n = \dfrac{a \sin bx - nb \cos bx}{a^2 + n^2 b^2} e^{ax} \sin^{n-1} bx + \dfrac{n(n-1)b^2}{a^2 + n^2 b^2} S_{n-2}$$

558. $\int x e^{ax} \sin bx \, dx = \left( \dfrac{[2ab - b(a^2+b^2)x]\cos bx + [a(a^2+b^2)x - a^2 + b^2]\sin bx}{(a^2+b^2)^2} \right) e^{ax} + C$

559. $\int x e^{ax} \cos bx \, dx = \left( \dfrac{[a(a^2+b^2)x - a^2 + b^2]\cos bx + [b(a^2+b^2)x - 2ab]\sin bx}{(a^2+b^2)^2} \right) e^{ax} + C$

560. $\int e^{ax} \ln x \, dx = \dfrac{1}{a} e^{ax} \ln x - \dfrac{1}{a} \int \dfrac{1}{x} e^{ax} \, dx$

561. $\int e^{ax} \sinh bx \, dx = \left[ \dfrac{a \sinh bx - b \cosh bx}{(a-b)(a+b)} \right] e^{ax} + C, \quad a \neq b$

562. $\int e^{ax} \sinh ax \, dx = \dfrac{1}{4a} e^{2ax} - \dfrac{x}{2} + C$

563. $\int e^{ax} \cosh bx \, dx = \left[ \dfrac{a \cosh bx - b \sinh bx}{(a-b)(a+b)} \right] e^{ax} + C, \quad a \neq b$

564. $\int e^{ax} \cosh ax \, dx = \dfrac{1}{4a} e^{2ax} + \dfrac{x}{2} + C$

565. $\int \dfrac{dx}{\alpha + \beta e^{ax}} = \dfrac{x}{\alpha} - \dfrac{1}{a\alpha} \ln|\alpha + \beta e^{ax}| + C$

566. $\int \dfrac{dx}{(\alpha + \beta e^{ax})^2} = \dfrac{x}{\alpha^2} + \dfrac{1}{a\alpha(\alpha + \beta e^{ax})} - \dfrac{1}{a\alpha^2} \ln|\alpha + \beta e^{ax}| + C$

567. $\int \dfrac{dx}{\alpha e^{ax} + \beta e^{-ax}} = \begin{cases} \dfrac{1}{a\sqrt{\alpha\beta}} \tan^{-1}\left(\sqrt{\dfrac{\alpha}{\beta}} e^{ax}\right) + C, & \alpha\beta > 0 \\ \dfrac{1}{2a\sqrt{-\alpha\beta}} \ln\left|\dfrac{e^{ax} - \sqrt{-\beta/\alpha}}{e^{ax} + \sqrt{-\beta/\alpha}}\right| + C, & \alpha\beta < 0 \end{cases}$

568. $\int e^{ax} \sin^2 bx\, dx = \left( \dfrac{a^2 + 4b^2 - a^2 \cos(2bx) - 2ab\sin(2bx)}{2a(a^2 + 4b^2)} \right) e^{ax} + C$

569. $\int e^{ax} \cos^2 bx\, dx = \left( \dfrac{a^2 + 4b^2 + a^2 \cos(2bx) + 2ab\sin(2bx)}{2a(a^2 + 4b^2)} \right) e^{ax} + C$

**Integrals containing the logarithmic function**

570. $\int \ln x\, dx = x \ln|x| + C$

571. $\int x \ln x\, dx = \dfrac{1}{2} x^2 \ln|x| - \dfrac{1}{4} x^2 + C$

572. $\int x^n \ln x\, dx = -\dfrac{1}{(n+1)^2} x^{n+1} + \dfrac{1}{n+1} x^{n+1} \ln|x| + C, \quad n \neq -1$

573. $\int \dfrac{1}{x} \ln x\, dx = \dfrac{1}{2} (\ln|x|)^2 + C$

574. $\int \dfrac{dx}{x \ln x} = \ln|\ln|x|| + C$

575. $\int \dfrac{1}{x^2} \ln x\, dx = -\dfrac{1}{x} - \dfrac{1}{x} \ln|x| + C$

576. $\int (\ln|x|)^2\, dx = x(\ln|x|)^2 - 2x\ln|x| + 2x + C$

577. $\int \dfrac{1}{x}(\ln|x|)^n\, dx = \dfrac{1}{n+1}(\ln|x|)^{n+1} + C, \quad n \neq -1$

578. $\int (\ln|x|)^n\, dx = x(\ln|x|)^n - n \int (\ln|x|)^{n-1}\, dx$

579. $\int \ln|x^2 + a^2|\, dx = x \ln|x^2 + a^2| - 2x + 2a \tan^{-1} \dfrac{x}{a} + C$

580. $\int \ln|x^2 - a^2|\, dx = x \ln|x^2 - a^2| - 2x + a \ln\left|\dfrac{x+a}{x-a}\right| + C$

581. $\int (ax+b)\ln(\beta x + \gamma)\, dx = \dfrac{\beta^2(ax+b)^2 - (b\beta - a\gamma)^2}{2a\beta^2} \ln(\beta x + \gamma) - \dfrac{a}{4\beta^2}(\beta x + \gamma)^2 - \dfrac{1}{\beta}(b\beta - a\gamma)x + C$

582. $\int (\ln ax)^2\, dx = x(\ln ax)^2 - 2x \ln ax + 2x + C$

**Integrals containing the hyperbolic function $\sinh ax$**

583. $\int \sinh ax\, dx = \dfrac{1}{a} \cosh ax + C$

Appendix B

**584.** $\int x \sinh ax \, dx = \frac{1}{a} x \cosh ax - \frac{1}{a^2} \sinh ax + C$

**585.** $\int x^2 \sinh ax \, dx = \left( \frac{x^2}{a} + \frac{2}{a^3} \right) \cosh ax - \frac{2x}{a^2} \sinh ax + C$

**586.** $\int x^n \sinh ax \, dx = \frac{1}{a} x^n \cosh ax - \frac{n}{a} \int x^{n-1} \cosh ax \, dx$

**587.** $\int \frac{1}{x} \sinh ax \, dx = ax + \frac{(ax)^3}{3 \cdot 3!} + \frac{(ax)^5}{4 \cdot 5!} + \cdots + C$

**588.** $\int \frac{1}{x^2} \sinh ax \, dx = -\frac{1}{x} \sinh ax + a \int \frac{1}{x} \cosh ax \, dx$

**589.** $\int \frac{1}{x^n} \sinh ax \, dx = -\frac{\sinh ax}{(n-1)x^{n-1}} + \frac{a}{n-1} \int \frac{1}{x^{n-1}} \cosh ax \, dx$

**590.** $\int \frac{dx}{\sinh ax} = \frac{1}{a} \ln | \tanh \frac{ax}{2} | + C$

**591.** $\int \frac{x \, dx}{\sinh ax} = \frac{1}{a^2} \left[ ax - \frac{(ax)^3}{18} + frac7(ax)^51800 + \cdots + (-1)^n \frac{2(2^{2n}-1)\mathfrak{B}_n \, a^{2n+1} x^{2n+1}}{(2n+1)!} + \cdots \right] + C$

**592.** $\int \sinh^2 ax \, dx = \frac{1}{2a} x \sinh 2ax - \frac{1}{2} x + C$

**593.** $\int \sinh^n ax \, dx = \frac{1}{na} \sinh^{n-1} ax \cosh ax - \frac{n-1}{n} \int \sinh^{n-2} ax \, dx$

**594.** $\int x \sinh^2 ax \, dx = \frac{1}{4a} x \sinh 2ax - \frac{1}{8a^2} \cosh 2ax - \frac{1}{4} x^2 + C$

**595.** $\int \frac{dx}{\sinh^2 ax} = -\frac{1}{a} \coth ax + C$

**596.** $\int \frac{dx}{\sinh^3 ax} = -\frac{1}{2a} \operatorname{csch} ax \coth ax - \frac{1}{2a} \ln | \tanh \frac{ax}{2} | + C$

**597.** $\int \frac{x \, dx}{\sinh^2 ax} = -\frac{1}{a} x \coth ax + \frac{1}{a^2} \ln | \sinh ax | + C$

**598.** $\int \sinh ax \sinh bx \, dx = \frac{1}{2(a+b)} \sinh(a+b)x - \frac{1}{2(a-b)} \sinh(a-b)x + C$

**599.** $\int \sinh ax \sin bx \, dx = \frac{1}{a^2+b^2} [a \cosh ax \sin bx - b \sinh ax \cos bx] + C$

**600.** $\int \sinh ax \cos bx \, dx = \frac{1}{a^2+b^2} [a \cosh ax \cos bx + b \sinh ax \sin bx] + C$

601. $\int \dfrac{dx}{\alpha + \beta \sinh ax} = \dfrac{1}{\sqrt{\alpha^2 + \beta^2}} \ln \left| \dfrac{\beta e^{ax} + \alpha - \sqrt{\alpha^2 + \beta^2}}{\beta e^{ax} + \alpha + \sqrt{\alpha^2 + \beta^2}} \right| + C$

602. $\int \dfrac{dx}{(\alpha + \beta \sinh ax)^2} = \dfrac{-\beta}{a(\alpha^2 + \beta^2)} \dfrac{\cosh ax}{\alpha + \beta \sinh ax} + \dfrac{\alpha}{\alpha^2 + \beta^2} \int \dfrac{dx}{\alpha + \beta \sinh ax}$

603. $\int \dfrac{dx}{\alpha^2 + \beta^2 \sinh^2 ax} = \begin{cases} \dfrac{1}{a\alpha \sqrt{\beta^2 - \alpha^2}} \tan^{-1}\left(\dfrac{\sqrt{\beta^2 - \alpha^2} \tanh ax}{\alpha}\right) + C, & \beta^2 > \alpha^2 \\[6pt] \dfrac{1}{2a\alpha \sqrt{\alpha^2 - \beta^2}} \ln \left| \dfrac{\alpha + \sqrt{\alpha^2 - \beta^2} \tanh ax}{\alpha - \sqrt{\alpha^2 - \beta^2} \tanh ax} \right| + C, & \beta^2 < \alpha^2 \end{cases}$

604. $\int \dfrac{dx}{\alpha^2 - \beta^2 \sinh^2 ax} = \dfrac{1}{2a\alpha \sqrt{\alpha^2 + \beta^2}} \ln \left| \dfrac{\alpha + \sqrt{\alpha^2 + \beta^2} \tanh ax}{\alpha - \sqrt{\alpha^2 + \beta^2} \tanh ax} \right| + C$

**Integrals containing the hyperbolic function $\cosh ax$**

605. $\int \cosh ax \, dx = \dfrac{1}{a} \sinh ax + C$

606. $\int x \cosh ax \, dx = \dfrac{1}{a} x \sinh ax - \dfrac{1}{a^2} \cosh ax + C$

607. $\int x^2 \cosh ax \, dx = -\dfrac{2}{a^2} x \cosh ax + \left(\dfrac{x^2}{a} + \dfrac{2}{a^3}\right) \sinh ax + C$

608. $\int x^n \cosh ax \, dx = \dfrac{1}{a} x^n \sinh ax - \dfrac{n}{a} \int x^{n-1} \sinh ax \, dx$

609. $\int \dfrac{1}{x} \cosh ax \, dx = \ln|x| + \dfrac{(ax)^2}{2 \cdot 2!} + \dfrac{(ax)^4}{4 \cdot 4!} + \dfrac{(ax)^6}{6 \cdot 6!} + \cdots + C$

610. $\int \dfrac{1}{x^2} \cosh ax \, dx = -\dfrac{1}{x} \cosh ax + a \int \dfrac{1}{x} \sinh ax \, dx$

611. $\int \dfrac{1}{x^n} \cosh ax \, dx = -\dfrac{1}{n-1} \dfrac{\cosh ax}{x^{n-1}} + \dfrac{a}{n-1} \int \dfrac{\sinh ax}{x^{n-1}} \, dx, \qquad n > 1$

612. $\int \dfrac{dx}{\cosh ax} = \dfrac{2}{a} \tan^{-1} e^{ax} + C$

613. $\int \dfrac{x \, dx}{\cosh ax} = \dfrac{1}{a^2} \left[ \dfrac{a^2 x^2}{2} - \dfrac{a^4 x^4}{8} + \dfrac{5 a^6 x^6}{144} + \cdots + (-1)^n \dfrac{\mathcal{E}_n a^{2n+2} x^{2n+2}}{(2n+2) \cdot (2n)!} + \cdots \right] + C$

614. $\int \cosh^2 ax \, dx = \dfrac{1}{2} x + \dfrac{1}{2} \sinh ax \cosh ax + C$

615. $\int \cosh^n ax \, dx = \dfrac{1}{na} \cosh^{n-1} ax \sinh ax + \dfrac{n-1}{n} \int \cosh^{n-2} ax \, dx$

616. $\int x \cosh^2 ax \, dx = \dfrac{1}{4} x^2 + \dfrac{1}{4a} x \sinh 2ax - \dfrac{1}{8a^2} \cosh 2ax + C$

**Appendix B**

**617.** $\displaystyle\int \frac{dx}{\cosh^2 ax} = \frac{1}{a}\tanh ax + C$

**618.** $\displaystyle\int \frac{x\,dx}{\cosh^2 ax} = \frac{1}{a}x\tanh ax - \frac{1}{a^2}\ln|\cosh ax| + C$

**619.** $\displaystyle\int \frac{dx}{\cosh^n ax} = \frac{1}{(n-1)a}\frac{x\sinh ax}{\cosh^{n-1} ax} + \frac{n-2}{n-1}\int \frac{dx}{\cosh^{n-2} ax}, \quad n > 1$

**620.** $\displaystyle\int \cosh ax \cosh bx\,dx = \frac{1}{2(a-b)}\sinh(a-b)x + \frac{1}{2(a+b)}\sinh(a+b)x + C$

**621.** $\displaystyle\int \cosh ax \sin bx\,dx = \frac{1}{a^2+b^2}[a\sinh ax \sin bx - b\cosh ax \cos bx] + C$

**622.** $\displaystyle\int \cosh ax \cos bx\,dx = \frac{1}{a^2+b^2}[a\sinh ax \cos bx + b\cosh ax \sin bx] + C$

**623.** $\displaystyle\int \frac{dx}{\alpha + \beta \cosh ax} = \begin{cases} \dfrac{2}{\sqrt{\beta^2-\alpha^2}}\tan^{-1}\dfrac{\beta e^{ax}+\alpha}{\sqrt{\beta^2-\alpha^2}} + C, & \beta^2 > \alpha^2 \\[1em] \dfrac{1}{a\sqrt{\alpha^2-\beta^2}}\ln\left|\dfrac{\beta e^{ax}+\alpha-\sqrt{\alpha^2-\beta^2}}{\beta e^{ax}+\alpha+\sqrt{\alpha^2-\beta^2}}\right| + C, & \beta^2 < \alpha^2 \end{cases}$

**624.** $\displaystyle\int \frac{dx}{1+\cosh ax} = \frac{1}{a}\tanh ax + C$

**625.** $\displaystyle\int \frac{x\,dx}{1+\cosh ax} = \frac{x}{a}\tanh\frac{ax}{2} - \frac{2}{a^2}\ln\left|\cosh\frac{ax}{2}\right| + C$

**626.** $\displaystyle\int \frac{dx}{-1+\cosh ax} = -\frac{1}{a}\coth\frac{ax}{2} + C$

**627.** $\displaystyle\int \frac{dx}{(\alpha+\beta\cosh ax)^2} = \frac{\beta\sinh ax}{a(\beta^2-\alpha^2)(\alpha+\beta\cosh ax)} - \frac{\alpha}{\beta^2-\alpha^2}\int \frac{dx}{\alpha+\beta\cosh ax}$

**628.** $\displaystyle\int \frac{dx}{\alpha^2-\beta^2\cosh^2 ax} = \begin{cases} \dfrac{1}{2a\alpha\sqrt{\alpha^2-\beta^2}}\ln\left|\dfrac{\alpha\tanh ax+\sqrt{\alpha^2-\beta^2}}{\alpha\tanh ax-\sqrt{\alpha^2-\beta^2}}\right| + C, & \alpha^2 > \beta^2 \\[1em] \dfrac{-1}{a\alpha\sqrt{\beta^2-\alpha^2}}\tan^{-1}\dfrac{\alpha\tanh ax}{\sqrt{\beta^2-\alpha^2}} + C, & \alpha^2 < \beta^2 \end{cases}$

**629.** $\displaystyle\int \frac{dx}{\alpha^2+\beta^2\cosh^2 ax} = \frac{1}{a\alpha\sqrt{\alpha^2+\beta^2}}\tanh^{-1}\left(\frac{\alpha\tanh ax}{\sqrt{\alpha^2+\beta^2}}\right) + C$

| Integrals containing the hyperbolic functions $\sinh ax$ and $\cosh ax$ |
|---|

**630.** $\displaystyle\int \sinh ax \cosh ax\,dx = \frac{1}{2a}\sinh^2 ax + C$

**631.** $\displaystyle\int \sinh ax \cosh bx\,dx = \frac{1}{2(a+b)}\cosh(a+b)x + \frac{1}{2(a-b)}\cosh(a-b)x + C$

632. $\int \sinh^2 ax \cosh^2 ax\, dx = \frac{1}{32a}\sinh 4ax - \frac{1}{8}x + C$

633. $\int \sinh^n ax \cosh ax\, dx = \frac{1}{(n+1)a}\sinh^{n+1} ax + C, \quad n \neq -1$

634. $\int \cosh^n ax \sinh ax\, dx = \frac{1}{(n+1)a}\cosh^{n+1} ax + C, \quad n \neq -1$

635. $\int \frac{\sinh ax}{\cosh ax}\, dx = \frac{1}{a}\ln|\cosh ax| + C$

636. $\int \frac{\cosh ax}{\sinh ax}\, dx = \frac{1}{a}\ln|\sinh ax| + C$

637. $\int \frac{dx}{\sinh ax \cosh ax} = \frac{1}{a}\ln|\tanh ax| + C$

638. $\int \frac{x \sinh ax}{\cosh ax}\, dx = \frac{1}{a^2}\left[\frac{a^3 x^3}{3} - \frac{a^5 x^5}{15} + \cdots + (-1)^n \frac{2^{2n}(2^{2n}-1)\mathfrak{B}_n a^{2n+1} x^{2n+1}}{(2n+1)!} + \cdots\right] + C$

639. $\int \frac{x \cosh ax}{\sinh ax}\, dx = \frac{1}{a^2}\left[ax + \frac{a^3 x^3}{9} - \frac{a^5 x^5}{225} + \cdots + (-1)^{n-1}\frac{2^{2n}\mathfrak{B}_n a^{2n+1} x^{2n+1}}{(2n+1)!} + \cdots\right] + C$

640. $\int \frac{\sinh^2 ax}{\cosh^2 ax}\, dx = x - \frac{1}{a}\tanh ax + C$

641. $\int \frac{\cosh^2 ax}{\sinh^2 ax}\, dx = x - \frac{1}{a}\coth ax + C$

642. $\int \frac{x \sinh^2 ax}{\cosh^2 ax}\, dx = \frac{1}{2}x^2 - \frac{1}{a}x \tanh ax + \frac{1}{a^2}\ln|\cosh ax| + C$

643. $\int \frac{x \cosh^2 ax}{\sinh^2 ax}\, dx = \frac{1}{2}x^2 - \frac{1}{a}x \coth ax + \frac{1}{a^2}\ln|\sinh ax| + C$

644. $\int \frac{\sinh ax}{x \cosh ax}\, dx = ax - \frac{a^3 x^3}{9} + \cdots + (-1)^{n-1}\frac{2^{2n}(2^{2n}-1)\mathfrak{B}_n a^{2n-1} x^{2n-1}}{(2n-1)(2n)!} + \cdots + C$

645. $\int \frac{\cosh ax}{x \sinh ax}\, dx = -\frac{1}{ax} + \frac{ax}{3} - \frac{a^3 x^3}{135} + \cdots + (-1)^n \frac{2^{2n}\mathfrak{B}_n a^{2n-1} x^{2n-1}}{(2n-1)(2n)!} + \cdots + C$

646. $\int \frac{\sinh^3 ax}{\cosh^3 ax}\, dx = \frac{1}{a}\ln|\cosh ax| - \frac{1}{2a}\tanh^2 ax + C$

647. $\int \frac{\cosh^3 ax}{\sinh^3 ax}\, dx = \frac{1}{a}\ln|\sinh ax| - \frac{1}{2a}\coth^2 ax + C$

648. $\int \frac{dx}{\sinh ax \cosh^2 ax} = \frac{1}{a}\operatorname{sech} ax + \frac{1}{a}\ln\tanh\frac{ax}{2}| + C$

**649.** $\displaystyle\int \frac{dx}{\sinh^2 ax \cosh ax} = -\frac{1}{a}\tan^{-1}(\sinh ax) - \frac{1}{a}\operatorname{csch} ax + C$

**650.** $\displaystyle\int \frac{dx}{\sinh^2 ax \cosh^2 ax} = -\frac{2}{a}\coth ax + C$

**651.** $\displaystyle\int \frac{\sinh^2 ax}{\cosh ax}\,dx = \frac{1}{a}\sinh ax - \frac{1}{a}\tan^{-1}(\sinh ax) + C$

**652.** $\displaystyle\int \frac{\cosh^2 ax}{\sinh ax}\,dx = \frac{1}{a}\cosh ax + \frac{1}{a}\ln\left|\tanh\frac{ax}{2}\right| + C$

**653.** $\displaystyle\int \frac{dx}{\cosh ax\,(1+\sinh ax)} = \frac{1}{2a}\ln\left|\frac{1+\sinh ax}{\cosh ax}\right| + \frac{1}{a}\tan^{-1}e^{ax} + C$

**654.** $\displaystyle\int \frac{dx}{\sinh ax\,(\cosh ax + 1)} = \frac{1}{2a}\ln\left|\tanh\frac{ax}{2}\right| + \frac{1}{2a(\cosh ax + 1)} + C$

**655.** $\displaystyle\int \frac{dx}{\sinh ax\,(\cosh ax - 1)} = -\frac{1}{2a}\ln\left|\tanh\frac{ax}{2}\right| - \frac{1}{2a(\cosh ax - 1)} + C$

**656.** $\displaystyle\int \frac{dx}{\alpha + \beta\,\frac{\sinh ax}{\cosh ax}} = \frac{\alpha x}{\alpha^2 - \beta^2} - \frac{\beta}{a(\alpha^2-\beta^2)}\ln|\beta\sinh ax + \alpha\cosh ax| + C$

**657.** $\displaystyle\int \frac{dx}{\alpha + \beta\,\frac{\cosh ax}{\sinh ax}} = \frac{\alpha x}{\alpha^2 - \beta^2} + \frac{\beta}{a(\alpha^2-\beta^2)}\ln|\alpha\sinh ax + \beta\cosh ax| + C$

**658.** $\displaystyle\int \frac{dx}{b\cosh ax + c\sinh ax} = \begin{cases} \dfrac{1}{a\sqrt{b^2-c^2}}\sec^{-1}\left[\dfrac{b\cosh ax + c\sinh ax}{\sqrt{b^2-c^2}}\right] + C, & b^2 > c^2 \\[2mm] \dfrac{-1}{a\sqrt{c^2-b^2}}\operatorname{csch}^{-1}\left[\dfrac{b\cosh ax + c\sinh ax}{\sqrt{c^2-b^2}}\right] + C, & b^2 < c^2 \end{cases}$

**Integrals containing the hyperbolic functions $\tanh ax$, $\coth ax$, $\operatorname{sech} ax$, $\operatorname{csch} ax$**

Express integrals in terms of $\sinh ax$ and $\cosh ax$ and see previous listings.

**Integrals containing inverse hyperbolic functions**

**659.** $\displaystyle\int \sinh^{-1}\frac{x}{a}\,dx = x\sinh^{-1}\frac{x}{a} - \sqrt{x^2+a^2} + C$

**660.** $\displaystyle\int \cosh^{-1}\frac{x}{a}\,dx = \begin{cases} x\cosh^{-1}(x/a) - \sqrt{x^2-a^2}, & \cosh^{-1}(x/a) > 0 \\ x\cosh^{-1}(x/a) + \sqrt{x^2-a^2}, & \cosh^{-1}(x/a) < 0 \end{cases}$

**661.** $\displaystyle\int \tanh^{-1}\frac{x}{a}\,dx = x\tanh^{-1}\frac{x}{a} + \frac{a}{2}\ln|a^2 - x^2| + C$

**662.** $\displaystyle\int \coth^{-1}\frac{x}{a}\,dx = x\coth^{-1}\frac{x}{a} + \frac{a}{2}\ln|x^2 - a^2| + C$

**663.** $\displaystyle\int \operatorname{sech}^{-1}\frac{x}{a}\,dx = \begin{cases} x\operatorname{sech}^{-1}\dfrac{x}{a} + a\sin^{-1}\dfrac{x}{a} + C, & \operatorname{sech}^{-1}(x/a) > 0 \\ x\operatorname{sech}^{-1}\dfrac{x}{a} - a\sin^{-1}\dfrac{x}{a} + C, & \operatorname{sech}^{-1}(x/a) < 0 \end{cases}$

**664.** $\int \operatorname{csch}^{-1}\frac{x}{a}\,dx = x\operatorname{csch}^{-1}\frac{x}{a} \pm a\sinh^{-1}\frac{x}{a}, \quad + \text{ for } x > 0 \text{ and } - \text{ for } x < 0$

**665.** $\int x\sinh^{-1}\frac{x}{a}\,dx = \left(\frac{x^2}{2} + \frac{a^2}{4}\right)\sinh^{-1}\frac{x}{a} - \frac{1}{4}xx\sqrt{x^2 + a^2} + C$

**666.** $\int x\cosh^{-1}\frac{x}{a}\,dx = \begin{cases} \frac{1}{4}(2x^2 - a^2)\cosh^{-1}\frac{x}{a} - \frac{1}{4}x\sqrt{x^2 - a^2} + C, & \cosh^{-1}(x/a) > 0 \\ \frac{1}{4}(2x^2 - a^2)\cosh^{-1}\frac{x}{a} + \frac{1}{4}x\sqrt{x^2 - a^2} + C, & \cosh^{-1}(x/a) < 0 \end{cases}$

**667.** $\int x\tanh^{-1}\frac{x}{a}\,dx = \frac{ax}{2} + \frac{1}{2}(x^2 - a^2)\tanh^{-1}\frac{x}{a} + C$

**668.** $\int x\coth^{-1}\frac{x}{a}\,dx = \frac{ax}{2} + \frac{1}{2}(x^2 - a^2)\coth^{-1}\frac{x}{a} + C$

**669.** $\int x\operatorname{sech}^{-1}\frac{x}{a}\,dx = \begin{cases} \frac{1}{2}x^2\operatorname{sech}^{-1}\frac{x}{a} - \frac{1}{2}a\sqrt{a^2 - x^2}, & \operatorname{sech}^{-1}(x/a) > 0 \\ \frac{1}{2}x\operatorname{sech}^{-1}\frac{x}{a} + \frac{1}{2}a\sqrt{a^2 - x^2} + C, & \operatorname{sech}^{-1}(x/a) < 0 \end{cases}$

**670.** $\int x\operatorname{csch}^{-1}\frac{x}{a}\,dx = \frac{1}{2}x^2\operatorname{csch}^{-1}\frac{x}{a} \pm \frac{a}{2}\sqrt{x^2 + a^2} + C, \quad + \text{ for } x > 0 \text{ and } - \text{ for } x < 0$

**671.** $\int x^2\sinh^{-1}\frac{x}{a}\,dx = \frac{1}{3}x^3\sinh^{-1}\frac{x}{a} + \frac{1}{9}(2a^2 - x^2)\sqrt{x^2 + a^2} + C$

**672.** $\int x^2\cosh^{-1}\frac{x}{a}\,dx = \begin{cases} \frac{1}{3}x^3\cosh^{-1}\frac{x}{a} - \frac{1}{9}(x^2 + 2a^2)\sqrt{x^2 - a^2} + C, & \cosh^{-1}(x/a) > 0 \\ \frac{1}{3}x^3\cosh^{-1}\frac{x}{a} + \frac{1}{9}(x^2 + 2a^2)\sqrt{x^2 - a^2} + C, & \cosh^{-1}(x/a) < 0 \end{cases}$

**673.** $\int x^2\tanh^{-1}\frac{x}{a}\,dx = \frac{a}{6}x^2 + \frac{1}{3}x^3\tanh^{-1}\frac{x}{a} + \frac{1}{6}a^3\ln|a^2 - x^2| + C$

**674.** $\int x^2\coth^{-1}\frac{x}{a}\,dx = \frac{a}{6}x^2 + \frac{1}{3}x^3\coth^{-1}\frac{x}{a} + \frac{1}{6}a^3\ln|x^2 - a^2| + C$

**675.** $\int x^2\operatorname{sech}^{-1}\frac{x}{a}\,dx = \frac{1}{3}x^3\operatorname{sech}^{-1}\frac{x}{a} - \frac{1}{3}\int\frac{x^3\,dx}{\sqrt{x^2 + a^2}}$

**676.** $\int x^2\operatorname{csch}^{-1}\frac{x}{a}\,dx = \frac{1}{3}x^3\operatorname{csch}^{-1}\frac{x}{a} \pm \frac{a}{3}\int\frac{x^2\,dx}{\sqrt{x^2 + a^2}}$

**677.** $\int x^n\sinh^{-1}\frac{x}{a}\,dx = \frac{1}{n+1}x^{n+1}\sinh^{-1}\frac{x}{a} - \frac{1}{n+1}\int\frac{x^{n+1}\,dx}{\sqrt{x^2 - a^2}}$

**678.** $\int x^n\cosh^{-1}\frac{x}{a}\,dx = \begin{cases} \frac{1}{n+1}x^{n+1}\cosh^{-1}\frac{x}{a} - \frac{1}{n+1}\int\frac{x^{n+1}}{\sqrt{x^2 - a^2}}, & \cosh^{-1}(x/a) > 0 \\ \frac{1}{n+1}x^{n+1}\cosh^{-1}\frac{x}{a} + \frac{1}{n+1}\int\frac{x^{n+1}\,dx}{\sqrt{x^2 - a^2}}, & \cosh^{-1}(x/a) < 0 \end{cases}$

**679.** $\int x^n\tanh^{-1}\frac{x}{a}\,dx = \frac{1}{n+1}x^{n+1}\tanh^{-1}\frac{x}{a} - \frac{a}{n+1}\int\frac{x^{n+1}\,dx}{a^2 - x^2}$

**680.** $\int x^n \coth^{-1} \frac{x}{a} dx = \frac{1}{n+1} x^{n+1} \coth^{-1} \frac{x}{a} - \frac{a}{n+1} \int \frac{x^{n+1} dx}{a^2 - x^2}$

**681.** $\int x^n \operatorname{sech}^{-1} \frac{x}{a} dx = \begin{cases} \frac{1}{n+1} x^{n+1} \operatorname{sech}^{-1} \frac{x}{a} + \frac{a}{n+1} \int \frac{x^n dx}{\sqrt{a^2 - x^2}}, & \operatorname{sech}^{-1}(x/a) > 0 \\ \frac{1}{n+1} x^{n+1} \operatorname{sech}^{-1} \frac{x}{a} - \frac{a}{n+1} \int \frac{x^n dx}{\sqrt{a^2 - x^2}}, & \operatorname{sech}^{-1}(x/a) < 0 \end{cases}$

**682.** $\int x^n \operatorname{csch}^{-1} \frac{x}{a} dx = \frac{1}{n+1} x^{n+1} \operatorname{csch}^{-1} \frac{x}{a} \pm \frac{a}{n+1} \int \frac{x^n dx}{\sqrt{x^2 + a^2}}, \quad +\text{ for } x > 0, -\text{ for } x < 0$

**683.** $\int \frac{1}{x} \sinh^{-1} \frac{x}{a} dx = \begin{cases} \frac{x}{a} - \frac{(x/a)^3}{2 \cdot 3 \cdot 3} + \frac{1 \cdot 3(x/a)^5}{2 \cdot 4 \cdot 4 \cdot 5} - \frac{1 \cdot 3 \cdot 5(x/a)^7}{2 \cdot 4 \cdot 6 \cdot 7 \cdot 7} + \cdots + C, & |x| > a \\ \frac{1}{2} \left(\ln\left|\frac{2x}{a}\right|\right)^2 - \frac{(a/x)^2}{2 \cdot 2 \cdot 2} + \frac{1 \cdot 3(a/x)^4}{2 \cdot 4 \cdot 4 \cdot 4} - \frac{1 \cdot 3 \cdot 5(a/x)^6}{2 \cdot 4 \cdot 6 \cdot 6 \cdot 6} + \cdots + C, & x > a \\ -\frac{1}{2} \left(\ln\left|\frac{-2x}{a}\right|\right)^2 + \frac{(a/x)^2}{2 \cdot 2 \cdot 2} - \frac{1 \cdot 3(a/x)^4}{2 \cdot 4 \cdot 4 \cdot 4} + \frac{1 \cdot 3 \cdot 5(a/x)^6}{2 \cdot 4 \cdot 6 \cdot 6 \cdot 6} + \cdots + C, & x < -a \end{cases}$

**684.** $\int \frac{1}{x} \cosh^{-1} \frac{x}{a} dx = \pm \left[ \frac{1}{2} \left(\ln\left|\frac{2x}{a}\right|\right)^2 + \frac{(a/x)^2}{2 \cdot 2 \cdot 2} + \frac{1 \cdot 3(a/x)^4}{2 \cdot 4 \cdot 4 \cdot 4} + \frac{1 \cdot 3 \cdot 5(a/x)^6}{2 \cdot 4 \cdot 6 \cdot 6 \cdot 6} + \cdots \right] + C$
$+ \text{ for } \cosh^{-1}(x/a) > 0, - \text{ for } \cosh^{-1}(x/a) < 0$

**685.** $\int \frac{1}{x} \tanh^{-1} \frac{x}{a} dx = \frac{x}{a} + \frac{(x/a)^3}{3^2} + \frac{(x/a)^5}{5^2} + \cdots + C$

**686.** $\int \frac{1}{x} \coth^{-1} \frac{x}{a} dx = \frac{ax}{2} + \frac{1}{2}(x^2 - a^2) \coth^{-1} \frac{x}{a} + C$

**687.** $\int \frac{1}{x} \operatorname{sech}^{-1} \frac{x}{a} dx = \begin{cases} -\frac{1}{2} \ln\left|\frac{a}{x}\right| \ln\left|\frac{4a}{x}\right| - \frac{(x/a)^2}{2 \cdot 2 \cdot 2} - \frac{1 \cdot 3(x/a)^4}{2 \cdot 4 \cdot 4 \cdot 4} - \cdots + C, & \operatorname{sech}^{-1}(x/a) > 0 \\ \frac{1}{2} \ln\left|\frac{a}{x}\right| \ln\left|\frac{4a}{x}\right| + \frac{(x/a)^2}{2 \cdot 2 \cdot 2} + \frac{1 \cdot 3(x/a)^4}{2 \cdot 4 \cdot 4 \cdot 4} + \cdots, & \operatorname{sech}^{-1}(x/a) < 0 \end{cases}$

**688.** $\int \frac{1}{x} \operatorname{csch}^{-1} \frac{x}{a} dx = \begin{cases} \frac{1}{2} \ln\left|\frac{x}{a}\right| \ln\left|\frac{4a}{x}\right| + \frac{(x/a)^2}{2 \cdot 2 \cdot 2} - \frac{1 \cdot 3(x/a)^4}{2 \cdot 4 \cdot 4 \cdot 4} + \cdots + C, & 0 < x < a \\ \frac{1}{2} \ln\left|\frac{-x}{a}\right| \ln\left|\frac{-x}{4a}\right| - \frac{(x/a)^2}{2 \cdot 2 \cdot 2} + \frac{1 \cdot 3(x/a)^4}{2 \cdot 4 \cdot 4 \cdot 4} - \cdots, & -a < x < 0 \\ -\frac{a}{x} + \frac{(a/x)^3}{2 \cdot 3 \cdot 3} - \frac{1 \cdot 3(a/x)^5}{2 \cdot 4 \cdot 5 \cdot 5} + \cdots + C, & |x| > a \end{cases}$

---

**Integrals evaluated by reduction formula**

**689.** If $S_n = \int \sin^n x \, dx$, then $S_n = -\frac{1}{n} \sin^{n-1} x \cos x + \frac{n-1}{n} S_{n-2}$

**690.** If $C_n = \int \cos^n x \, dx$, then $C_n = \frac{1}{n} \sin x \cos^{n-1} x + \frac{n-1}{n} C_{n-2}$

**691.** If $I_n = \int \frac{\sin^n ax}{\cos ax} dx$, then $I_n = \frac{-1}{(n-1)a} \sin^{n-1} ax + I_{n-2}$

**692.** If $I_n = \int \frac{\cos^n ax}{\sin ax} dx$, then $I_n = \frac{1}{(n-1)a} \cos^{n-1} ax + I_{n-2}$

**693.** If $S_m = \int x^m \sin nx \, dx$ and $C_m = \int x^m \cos nx \, dx$, then

$$S_m = \frac{-1}{n} x^m \cos nx + \frac{m}{n} C_{m-1} \quad \text{and} \quad C_m = \frac{1}{n} x^m \sin nx - \frac{m}{n} S_{m-1}$$

**694.** If $I_1 = \int \tan x \, dx$, and $I_n = \int \tan^n x \, dx$, then $I_n = \frac{1}{n-1} \tan^{n-1} x - I_{n-2}, \quad n = 2, 3, 4, \ldots$

**695.** If $I_n = \int \frac{\sin^n ax}{\cos ax} \, dx$, then $I_n = -\frac{\sin^{n-1} ax}{(n-1)a} + I_{n-2}$

**696.** If $I_n = \int \frac{\cos^n ax}{\sin ax} \, dx$, then $I_n = \frac{\cos^{n-1} ax}{(n-1)a} + I_{n-2}$

**697.** If $I_{n,m} = \int \sin^n x \cos^m x \, dx$, then

$$I_{n,m} = \frac{-1}{n+m} \sin^{n-1} x \cos^{m+1} x + \frac{n-1}{n+m} I_{n-2,m}$$
$$I_{n,m} = \frac{1}{n+1} \sin^{n+1} x \cos^{m+1} x + \frac{n+m+2}{n+1} I_{n+2,m}$$
$$I_{n,m} = \frac{1}{n+m} \sin^{n+1} x \cos^{m-1} x + \frac{m-1}{n+m} I_{n,m+2}$$
$$I_{n,m} = \frac{-1}{m+1} \sin^{n+1} x \cos^{m+1} x + \frac{n+m+2}{m+1} I_{n,m+2}$$
$$I_{n,m} = \frac{-1}{m+1} \sin^{n-1} x \cos^{m+1} x + \frac{n-1}{m+1} I_{n-2,m+2}$$
$$I_{n,m} = \frac{1}{n+1} \sin^{n+1} x \cos^{m-1} x + \frac{m-1}{n+1} I_{n+2,m-2}$$

**698.** If $S_n = \int e^{ax} \sin^n bx \, dx$ and $C_n = \int e^{ax} \cos^n bx \, dx$, then

$$C_n = e^{ax} \cos^{n-1} bx \left[ \frac{a \cos bx + nb \sin bx}{a^2 + n^2 b^2} \right] + \frac{n(n-1)b^2}{a^2 + n^2 b^2} C_{n-2}$$
$$S_n = e^{ax} \sin^{n-1} ax \left[ \frac{a \sin bx - nb \cos nx}{a^2 + n^2 b^2} \right] + \frac{n(n-1)b^2}{a^2 + n^2 b^2} S_{n-2}$$

**699.** If $I_n = \int x^m (\ln x)^n \, dx$, then $I_n = \frac{1}{m+1} x^{m+1} (\ln x)^n - \frac{n}{m+1} I_{n-1}$

---

| Integrals involving Bessel functions |
|---|

**700.** $\int J_1(x) \, dx = -J_0(x) + C$

**701.** $\int x J_1(x) \, dx = -x J_0(x) + \int J_0(x) \, dx$

**702.** $\int x^n J_1(x) \, dx = -x^n J_0(x) + n \int x^{n-1} J_0(x) \, dx$

**Appendix B**

**703.** $\int \frac{J_1(x)}{x}\,dx = -J_1(x) + \int J_0(x)\,dx$

**704.** $\int x^\nu J_{\nu-1}(x)\,dx = x^\nu J_\nu(x) + C$

**705.** $\int x^{-\nu} J_{\nu+1}(x)\,dx = x^{-\nu} J_\nu(x) + C$

**706.** $\int \frac{J_1(x)}{x^n}\,dx = \frac{-1}{n}\frac{J_1(x)}{x^{n-1}} + \frac{1}{n}\int \frac{J_0(x)}{x^{n-1}}\,dx$

**707.** $\int xJ_0(x)\,dx = xJ_1(x) + C$

**708.** $\int x^2 J_0(x)\,dx = x^2 J_1(x) + xJ_0(x) - \int J_0(x)\,dx$

**709.** $\int x^n J_0(x)\,dx = x^n J_1(x) + (n-1)x^{n-1}J_0(x) - (n-1)^2 \int x^{n-2} J_0(x)\,dx$

**710.** $\int \frac{J_0(x)}{x^n}\,dx = \frac{J_1(x)}{(n-1)^2 x^{n-2}} - \frac{J_0(x)}{(n-1)x^{n-1}} - \frac{1}{(n-1)^2}\int \frac{J_0(x)}{x^{n-2}}\,dx$

**711.** $\int J_{n+1}(x)\,dx = \int J_{n-1}(x)\,dx - 2J_n(x)$

**712.** $\int xJ_n(\alpha x)J_n(\beta x)\,dx = \frac{x}{\beta^2 - \alpha^2}[\alpha J_n'(\alpha x)J_n(\beta x) - \beta J_n'(\beta x)J_n(\alpha x)] + C$

**713.** If $I_{m,n} = \int x^m J_n(x)\,dx, \quad m \geq -n$, then

$$I_{m,n} = -x^m J_{n-1}(x) + (m+n-1)I_{m-1,n-1}$$

**714.** If $I_{n,0} = \int x^n J_0(x)\,dx$, then $I_{n,0} = x^n J_1(x) + (n-1)x^{n-1}J_0(x) - (n-1)^2 I_{n-2,0}$ Note that $I_{1,0} = \int xJ_0(x)\,dx = xJ_1(x) + C$ and $I_{0,1} = \int J_1(x)\,dx = -J_0(x) + C$ Note also that the integral $I_{0,0} = \int J_0(x)\,dx$ cannot be given in closed form.

**Appendix B**

## Definite integrals

### General integration properties

1. If $\frac{dF(x)}{dx} = f(x)$, then $\int_a^b f(x)\,dx = F(x)|_a^b = F(b) - F(a)$

2. 
$$\int_0^\infty f(x)\,dx = \lim_{b \to \infty} \int_0^b f(x)\,dx, \qquad \int_{-\infty}^a f(x)\,dx = \lim_{b \to -\infty} \int_b^a f(x)\,dx$$

3. If $f(x)$ has a singular point at $x = b$, then $\int_a^b f(x)\,dx = \lim_{\epsilon \to 0} \int_a^{b-\epsilon} f(x)\,dx$

4. If $f(x)$ has a singular point at $x = a$, then $\int_a^b f(x)\,dx = \lim_{\epsilon \to 0} \int_{a+\epsilon}^b f(x)\,dx$

5. If $f(x)$ has a singular point at $x = c$, $a < c < b$, then $\int_a^b f(x)\,dx = \int_a^{c-\epsilon} f(x)\,dx + \int_{c+\epsilon}^b f(x)\,dx$

6. 
$$\int_a^b cf(x)\,dx = c \int_a^b f(x)\,dx, \quad c \text{ constant}$$
$$\int_a^a f(x)\,dx = 0,$$
$$\int_0^b f(x)\,dx = \int_0^b f(b-x)\,dx$$
$$\int_a^b f(x)\,dx = -\int_b^a f(x)\,dx,$$
$$\int_a^b f(x)\,dx = \int_a^c f(x)\,dx + \int_c^b f(x)\,dx$$

7. Mean value theorems

$$\int_a^b f(x)\,dx = f(c)(b-a), \quad a \leq c \leq b$$

$$\int_a^b f(x)g(x)\,dx = f(c)\int_a^b g(x)\,dx, \quad g(x) \geq 0,\ a \leq c \leq b$$

$$\int_a^b f(x)g(x)\,dx = f(a)\int_a^\xi g(x)\,dx \qquad \int_a^b f(x)g(x)\,dx = f(b)\int_\eta^b g(x)\,dx$$
$$a < \xi < b \qquad\qquad\qquad\qquad a < \eta < b$$

The last mean value theorem requires that $f(x)$ be monotone increasing and nonnegative throughout the interval $(a, b)$.

8. Numerical integration

Divide the interval $(a,b)$ into $n$ equal parts by defining a step size $h = \frac{b-a}{n}$.

Two numerical integration schemes are

(a) Trapezoidal rule with global error $-\frac{(b-a)}{12} h^2 f''(\xi)$ for $a < \xi < b$.

$$\int_a^b f(x)\,dx = \frac{h}{2}[f(x_0) + 2f(x_1) + 2f(x_2) + \cdots 2f(x_{n-1}) + f(x_n)]$$

(b) Simpson's 1/3 rule with global error $-\frac{(b-a)}{90} h^4 f^{(iv)}(\xi)$ for $a < \xi < b$.

$$\int_a^b f(x)\,dx = \frac{2h}{3}[f(x_0) + 4f(x_1) + 2f(x_2) + 4f(x_3) + 2f(x_4) + \cdots + 2f(x_{n-2}) + 4f(x_{n-1}) + f(x_n)]$$

**Appendix B**

**9.** If $f(x)$ is periodic with period $L$, then $f(x+L) = f(x)$ for all $x$ and $\int_0^{nL} f(x)\,dx = n\int_0^L f(x)\,dx$, for integer values of $n$.

**10.** $\underbrace{\int_0^x dx \int_0^x dx \cdots \int_0^x dx}_{n \text{ integration signs}} f(x) = \frac{1}{(n-1)!} \int_0^x (x-u)^{n-1} f(u)\,du$

### Integrals containing algebraic terms

**11.** $\displaystyle\int_0^1 x^{m-1}(1-x)^{n-1}\,dx = B(m,n) = \frac{\Gamma(m)\Gamma(n)}{\Gamma(m+n)}, \quad m>0, n>0$

**12.** $\displaystyle\int_0^1 \frac{dx}{\sqrt{1-x^4}} = \frac{1}{4\sqrt{2\pi}}\left[\Gamma\!\left(\tfrac{1}{4}\right)\right]^2$

**13.** $\displaystyle\int_0^1 \frac{dx}{(1-x^{2n})^{n/2}} = \frac{\pi}{2n \sin\frac{\pi}{2n}}$

**14.** $\displaystyle\int_0^1 \frac{1}{\beta-\alpha x}\frac{dx}{\sqrt{x(1-x)}} = \frac{\pi}{\sqrt{\beta(\beta-\alpha)}}$

**15.** $\displaystyle\int_0^1 \frac{x^p - x^{-p}}{x^q - x^{-q}}\frac{dx}{x} = \frac{\pi}{2q}\tan\frac{p\pi}{2q}, \quad |p|<q$

**16.** $\displaystyle\int_0^1 \frac{x^p + x^{-p}}{x^q + x^{-q}}\frac{dx}{x} = \frac{\pi}{2q}\sec\frac{p\pi}{2q}, \quad |p|<q$

**17.** $\displaystyle\int_0^1 \frac{x^{p-1} - x^{1-p}}{1-x^2}\,dx = \frac{\pi}{2}\cot\frac{p\pi}{2}, \quad 0<p<2$

**18.** $\displaystyle\int_0^a \frac{dx}{\sqrt{a^2-x^2}} = \frac{\pi}{2}$

**19.** $\displaystyle\int_0^a \sqrt{a^2-x^2}\,dx = \frac{\pi}{4}a^2$

**20.** $\displaystyle\int_0^\infty \frac{dx}{x^2+a^2} = \frac{\pi}{2a}$

**21.** $\displaystyle\int_0^\infty \frac{x^{\alpha-1}}{1+x}\,dx = \frac{\pi}{\sin\alpha\pi}, \quad 0<\alpha<1$

**22.** $\displaystyle\int_0^1 \frac{x^{\alpha-1} + x^{-\alpha}}{1+x}\,dx = \frac{\pi}{\sin\alpha\pi}, \quad 0<\alpha<1$

**23.** $\displaystyle\int_0^\infty \frac{x^m\,dx}{1+x^2} = \frac{\pi}{2}\sec\frac{m\pi}{2}$

**24.** $\displaystyle\int_0^\infty \frac{x^{\alpha-1}}{1-x^2}\,dx = \frac{\pi}{2}\cot\frac{\alpha\pi}{2}$

Appendix B

25. $\displaystyle\int_0^\infty \frac{dx}{1-x^n} = \frac{\pi}{n}\cot\frac{\pi}{n}$

26. $\displaystyle\int_0^\infty \frac{dx}{(a^2x^2+c^2)(x^2+b^2)} = \frac{\pi}{2bc}\frac{1}{c+ab}$

27. $\displaystyle\int_0^\infty \frac{dx}{(a^2+x^2)(b^2+x^2)} = \frac{\pi}{2}\frac{1}{ab(a+b)}$

28. $\displaystyle\int_0^\infty \frac{dx}{(a^2-x^2)(x^2+p^2)} = \frac{\pi}{2p}\frac{1}{a^2+p^2}$

29. $\displaystyle\int_0^\infty \frac{x^2\,dx}{(a^2-x^2)(x^2+p^2)} = \frac{\pi}{2}\frac{p}{a^2+p^2}$

30. $\displaystyle\int_0^\infty \frac{x^2\,dx}{(x^2+a^2)(x^2+b^2)(x^2+c^2)} = \frac{\pi}{2(a+b)(b+c)(c+a)}$

31. $\displaystyle\int_0^\infty \frac{\sqrt{x}\,dx}{1+x^2} = \frac{\pi}{\sqrt{2}}$

32. $\displaystyle\int_0^\infty \frac{x\,dx}{(1+x)(1+x^2)} = \frac{\pi}{4}$

| Integrals containing trigonometric terms |

33. $\displaystyle\int_0^1 \frac{\sin^{-1} x}{x}\,dx = \frac{\pi}{2}\ln 2$

34. $\displaystyle\int_0^{\pi/2} \frac{\tan^{-1}(\frac{b}{a}\tan\theta)\,d\theta}{\tan\theta} = \frac{\pi}{2}\ln\left|1+\frac{b}{a}\right|$

35. $\displaystyle\int_0^{\pi/2} \sin^2 x\,dx = \frac{\pi}{4}$

36. $\displaystyle\int_0^{\pi/2} \cos^2 x\,dx = \frac{\pi}{4}$

37. $\displaystyle\int_0^{\pi/2} \frac{dx}{a+b\cos x} = \frac{\cos^{-1}(b/a)}{\sqrt{a^2-b^2}}$

38. $\displaystyle\int_0^{\pi/2} \sin^{2m-1} x \cos^{2n-1} x\,dx = B(m,n) = \frac{\Gamma(m)\Gamma(n)}{\Gamma(m+n)}, \quad m>0, n>0$

39. $\displaystyle\int_0^{\pi/2} \sin^p x \cos^q x\,dx = \frac{\Gamma(\frac{p+1}{2})\Gamma(\frac{q+1}{2})}{2\Gamma(\frac{p+q}{2}+1)}$

40. $\displaystyle\int_0^{\pi/2} \frac{dx}{1+\tan^m x} = \frac{\pi}{4}$

41. $\displaystyle\int_0^\pi \cos p\theta \cos q\theta\,d\theta = \begin{cases} 0, & p\neq q \\ \dfrac{\pi}{2}, & p=q \end{cases}$

Appendix B

**42.** $\int_0^\pi \sin p\theta \sin q\theta \, d\theta = \begin{cases} 0, & p \neq q \\ \dfrac{\pi}{2}, & p = q \end{cases}$

**43.** $\int_0^\pi \sin p\theta \cos q\theta \, d\theta = \begin{cases} 0, & p+q \text{ even} \\ \dfrac{2p}{p^2-q^2}, & p+q \text{ odd} \end{cases}$

**44.** $\int_0^\pi \dfrac{x \, dx}{a^2 - \cos^2 x} = \dfrac{\pi^2}{2a\sqrt{a^2-1}}$

**45.** $\int_0^\pi \dfrac{dx}{a + b\cos x} = \dfrac{\pi}{\sqrt{a^2 - b^2}}$

**46.** $\int_0^\pi \dfrac{\sin \theta \, d\theta}{1 - 2a\cos\theta + a^2} = \dfrac{2}{a} \tanh^{-1} a$

**47.** $\int_0^\pi \dfrac{\sin 2\theta \, d\theta}{1 - 2a\cos\theta + a^2} = \dfrac{2}{a^2}(1+a^2)\tanh^{-1} a - \dfrac{2}{a}$

**48.** $\int_0^\pi \dfrac{x \sin x \, dx}{1 - 2a\cos x + a^2} = \begin{cases} \dfrac{\pi}{a} \ln(1+a), & |a| < 1 \\ \pi \ln\left(1 + \dfrac{1}{a}\right), & |a| > 1 \end{cases}$

**49.** $\int_0^\pi \dfrac{\cos p\theta \, d\theta}{1 - 2a\cos\theta + a^2} = \begin{cases} \dfrac{\pi a^p}{1 - a^2}, & a^2 < 1 \\ \dfrac{\pi a^{-p}}{a^2 - 1}, & a^2 > 1 \end{cases}$

**50.** $\int_0^\pi \dfrac{\cos p\theta \, d\theta}{(1 - 2a\cos\theta + a^2)^2} = \begin{cases} \dfrac{\pi a^p}{(1-a^2)^3}[(p+1) - (p-1)a^2], & a^2 < 1 \\ \dfrac{\pi a^{-p}}{(a^2-1)^3}[(1-p) + (1+p)a^2], & a^2 > 1 \end{cases}$

**51.** $\int_0^\pi \dfrac{\cos p\theta \, d\theta}{(1 - 2a\cos\theta + a^2)^3} = \begin{cases} \dfrac{\pi a^p}{2(1-a^2)^5}\left[(p+2)(p+1) + 2(p+2)(p-2)a^2 + (p-2)(p-1)a^4\right], & a^2 < 1 \\ \dfrac{\pi a^{-p}}{2(a^2-1)^5}\left[(1-p)(2-p) + 2(2-p)(2+p)a^2 + (2+p)(1+p)a^4\right], & a^2 > 1 \end{cases}$

**52.** $\int_0^{2\pi} \dfrac{dx}{(a + b\sin x)^2} = \dfrac{2\pi a}{(a^2 - b^2)^{3/2}}$

**53.** $\int_0^{2\pi} \dfrac{dx}{a + b\sin x} = \dfrac{2\pi}{\sqrt{a^2 - b^2}}$

**54.** $\int_0^{2\pi} \dfrac{dx}{a + b\cos x} = \dfrac{2\pi}{\sqrt{a^2 - b^2}}$

**55.** $\int_0^{2\pi} \dfrac{dx}{(a + b\sin x)^2} = \dfrac{2\pi a}{(a^2 - b^2)^{3/2}}$

**56.** $\int_0^{2\pi} \dfrac{dx}{(a + b\cos x)^2} = \dfrac{2\pi a}{(a^2 - b^2)^{3/2}}$

**57.** $\int_{-L}^{L} \sin\dfrac{m\pi x}{L} \sin\dfrac{n\pi x}{L} \, dx = \begin{cases} 0, & m \neq n \\ \dfrac{L}{2}, & m = n \end{cases} \quad m, n \text{ integers}$

**58.** $\int_{-L}^{L} \cos \frac{m\pi x}{L} \sin \frac{n\pi x}{L} \, dx = 0 \quad$ for all integer $m, n$ values

**59.** $\int_{-L}^{L} \cos \frac{m\pi x}{L} \cos \frac{n\pi x}{L} \, dx = \begin{cases} 0, & m \neq n \\ \frac{L}{2}, & m = n \neq 0 \\ L, & m = n = 0 \end{cases}$

**60.** $\int_{0}^{\infty} \frac{x^m \, dx}{1 + 2x \cos \beta + x^2} = \frac{\pi}{\sin m\pi} \frac{\sin m\beta}{\sin \beta}$

**61.** $\int_{0}^{\infty} \frac{\sin \alpha x}{x} \, dx = \begin{cases} \pi/2, & \alpha > 0 \\ 0, & \alpha = 0 \\ -\pi/2, & \alpha < 0 \end{cases}$

**62.** $\int_{0}^{\infty} \frac{\sin \alpha x \sin \beta x}{x} \, dx = \begin{cases} 0, & \alpha > \beta > 0 \\ \pi/2, & 0 < \alpha < \beta \\ \pi/4, & \alpha = \beta > 0 \end{cases}$

**63.** $\int_{0}^{\infty} \frac{\sin \alpha x \sin \beta x}{x^2} \, dx = \begin{cases} \frac{\pi \alpha}{2}, & 0 < \alpha \leq \beta \\ \frac{\pi \beta}{2}, & \alpha \geq \beta > 0 \end{cases}$

**64.** $\int_{0}^{\infty} \frac{\sin^2 \alpha x}{x^2} \, dx = \frac{\pi \alpha}{2}$

**65.** $\int_{0}^{\infty} \frac{1 - \cos \alpha x}{x^2} \, dx = \frac{\pi \alpha}{2}$

**66.** $\int_{0}^{\infty} \frac{\cos \alpha x}{x^2 + a^2} \, dx = \frac{\pi}{2a} e^{-\alpha a}$

**67.** $\int_{0}^{\infty} \frac{x \sin \alpha x}{x(x^2 + a^2)} \, dx = \frac{\pi}{2} e^{-\alpha a}$

**68.** $\int_{0}^{\infty} \frac{\sin x}{x^p} \, dx = \frac{\pi}{2 \Gamma(p) \sin(p\pi/2)}$

**69.** $\int_{0}^{\infty} \frac{\cos x}{x^p} \, dx = \frac{\pi}{2 \Gamma(p) \cos(p\pi/2)}$

**70.** $\int_{0}^{\infty} \frac{\tan x}{x} \, dx = \frac{\pi}{2}$

**71.** $\int_{0}^{\infty} \frac{\sin \alpha x}{x(x^2 + a^2)} \, dx = \frac{\pi}{2a^2}(1 - e^{-\alpha a})$

**72.** $\int_{0}^{\infty} \frac{\sin^2 x}{x^2} \, dx = \frac{\pi}{2}$

**73.** $\int_{0}^{\infty} \frac{\sin^3 x}{x^3} \, dx = \frac{3\pi}{8}$

**74.** $\int_{0}^{\infty} \frac{\sin^4 x}{x^4} \, dx = \frac{\pi}{3}$

75. $\int_0^\infty \sin ax^2 \cos 2bx\, dx = \frac{1}{2}\sqrt{\frac{\pi}{2a}}\left(\cos\frac{b^2}{a} - \sin\frac{b^2}{a}\right)$

76. $\int_0^\infty \cos ax^2 \cos 2bx\, dx = \frac{1}{2}\sqrt{\frac{\pi}{2a}}\left(\cos\frac{b^2}{a} + \sin\frac{b^2}{a}\right)$

77. $\int_0^\infty \frac{dx}{x^4 + 2a^2 x^2 \cos 2\beta + a^4} = \frac{\pi}{4a^3 \cos\beta}$

78. $\int_0^\infty \cos\left(x^2 + \frac{a^2}{x^2}\right) dx = \frac{\sqrt{\pi}}{2}\cos\left(\frac{\pi}{4} + 2a\right)$

79. $\int_0^\infty \sin\left(x^2 + \frac{a^2}{x^2}\right) dx = \frac{\sqrt{\pi}}{2}\sin\left(\frac{\pi}{4} + 2a\right)$

80. $\int_0^\infty \frac{\tan bx\, dx}{x(p^2 + x^2)} = \frac{\pi}{2p^2}\tanh bp$

81. $\int_0^\infty \frac{x \tan bx\, dx}{p^2 + x^2} = \frac{\pi}{2} - \frac{\pi}{2}\tanh bp$

82. $\int_0^\infty \frac{x \cot bx\, dx}{p^2 + x^2} = \frac{\pi}{2}\coth bp$

83. $\int_0^\infty \frac{\sin ax}{\sin bx}\frac{dx}{(p^2 + x^2)} = \frac{\pi}{2p}\frac{\sinh ap}{\sinh bp}, \quad a < b$

84. $\int_0^\infty \frac{\cos ax}{\cos bx}\frac{dx}{(p^2 + x^2)} = \frac{\pi}{2p}\frac{\cosh ap}{\cosh bp}, \quad a < b$

85. $\int_0^\infty \frac{\sin ax}{\cos bx}\frac{dx}{(p^2 + x^2)} = \frac{\pi}{2p^2}\frac{\sinh ap}{\cosh bp}, \quad a < b$

86. $\int_0^\infty \frac{\sin ax}{\cos bx}\frac{x\, dx}{(x^2 + p^2)} = -\frac{\pi}{2}\frac{\sinh ap}{\cosh bp}, \quad a < b$

87. $\int_0^\infty \frac{\cos ax}{\sin bx}\frac{x\, dx}{(p^2 + x^2)} = \frac{\pi}{2}\frac{\cosh ap}{\sinh bp}, \quad a < b$

**Integrals containing exponential and logarithmic terms**

88. $\int_0^1 \frac{\ln\frac{1}{x}}{1+x}\, dx = \frac{\pi^2}{12}$

89. $\int_0^1 \frac{\ln\frac{1}{x}}{(1-x)}\, dx = \frac{\pi^2}{6}$

90. $\int_0^1 \frac{(\ln\frac{1}{x})^3}{1-x}\, dx = \frac{\pi^4}{15}$

91. $\int_0^1 \frac{\ln(1+x)}{x}\, dx = \frac{\pi^2}{12}$

92. $\int_0^1 \frac{\ln(1-x)}{x}\,dx = -\frac{\pi^2}{6}$

93. $\int_0^1 (ax^2+bx+c)\frac{\ln\frac{1}{x}}{1-x}\,dx = (a+b+c)\frac{\pi^2}{6} - (a+b) - \frac{a}{4}$

94. $\int_0^1 \frac{\ln\frac{1}{x}}{\sqrt{1-x^2}}\,dx = \frac{\pi}{2}\ln 2$

95. $\int_0^1 \frac{1-x^{p-1}}{(1-x)(1-x^p)}(\ln\frac{1}{x})^{2n-1}\,dx = \frac{1}{4n}(1-\frac{1}{p^{2n}})(2\pi)^{2n}\mathfrak{B}_{2n-1}$

96. $\int_0^1 \frac{x^m - x^n}{\ln x}\,dx = \ln\left|\frac{1+m}{1+n}\right|$

97. $\int_0^1 x^p(\ln x)^n\,dx = \begin{cases} (-1)^n \dfrac{n!}{(p+1)^{n+1}}, & \text{n an integer} \\ (-1)^n \dfrac{\Gamma(n+1)}{(p+1)^{n+1}}, & \text{n noninteger} \end{cases}$

98. $\int_0^{\pi/4} \ln(1+\tan x)\,dx = \frac{\pi}{8}\ln 2$

99. $\int_0^{\pi/2} \ln\sin\theta\,d\theta = \frac{\pi}{2}\ln(\frac{1}{2})$

100. $\int_0^\pi \ln(a+b\cos x)\,dx = \pi\ln\left|\frac{a+\sqrt{a^2+b^2}}{2}\right|$

101. $\int_0^{2\pi} \ln(a+b\cos x)\,dx = 2\pi\ln|a+\sqrt{a^2-b^2}|$

102. $\int_0^{2\pi} \ln(a+b\sin x)\,dx = 2i\ln|a+\sqrt{a^2-b^2}|$

103. $\int_0^\infty e^{-ax}\,dx = \frac{1}{a}$

104. $\int_0^\infty x^n e^{-ax}\,dx = \frac{\Gamma(n+1)}{a^{n+1}}$

105. $\int_0^\infty e^{-a^2 x^2}\,dx = \frac{1}{2a}\sqrt{\pi} = \frac{1}{2a}\Gamma(\frac{1}{2})$

106. $\int_0^\infty x^n e^{-a^2 x^2}\,dx = \frac{\Gamma(\frac{m+1}{2})}{2a^{m+1}}$

107. $\int_0^\infty e^{-ax}\cos bx\,dx = \frac{a}{a^2+b^2}$

108. $\int_0^\infty e^{-ax}\sin bx\,dx = \frac{b}{a^2+b^2}$

**Appendix B**

109. $\int_0^\infty e^{-ax} \dfrac{\sin bx}{x} \, dx = \tan^{-1} \dfrac{b}{a}$

110. $\int_0^\infty \dfrac{e^{-ax} - e^{-bx}}{x} \, dx = \ln \dfrac{b}{a}$

111. $\int_0^\infty e^{-a^2 x^2} \cos bx \, dx = \dfrac{\sqrt{\pi}}{2a} e^{-b^2/4a^2}$

112. $\int_0^\infty e^{-(ax^2 + b/x^2)} \, dx = \dfrac{1}{2}\sqrt{\dfrac{\pi}{a}} e^{-2\sqrt{ab}}$

113. $\int_0^\infty x^{2n} e^{-\beta x^2} \, dx = \dfrac{(2n-1)(2n-3)\cdots 5 \cdot 3 \cdot 1}{2^{n+1} \beta^n} \sqrt{\dfrac{\pi}{\beta}}$

114. $\int_0^\infty e^{-k\left(\frac{x^2}{a^2} + \frac{b^2}{x^2}\right)} \, dx = \dfrac{\sqrt{\pi}}{2} \dfrac{a}{\sqrt{k}} e^{-2kb/a}$

115. $\int_0^\infty \dfrac{\sin rx \, dx}{x(x^4 + 2a^2 x^2 \cos 2\beta + a^4)} = \dfrac{\pi}{2a^4}\left[1 - \dfrac{\sin(ar \sin \beta + 2\beta)}{\sin 2\beta} e^{-\beta r \cos \beta}\right]$

116. $\int_0^\infty \dfrac{\cos rx \, dx}{x^4 + 2a^2 x^2 \cos 2\beta + a^4} = \dfrac{\pi}{2a^3} \dfrac{\sin(\beta + ar \sin \beta)}{\sin 2\beta} e^{-ar \cos \beta}$

117. $\int_0^\infty \dfrac{\sin rx \, dx}{x(x^6 + a^6)} = \dfrac{\pi}{6a^6}\left[3 - e^{-ar} - 2e^{-ar/2} \cos \dfrac{ar\sqrt{3}}{2}\right]$

118. $\int_0^\infty \dfrac{\cos rx \, dx}{x^6 + a^6} = \dfrac{\pi}{6a^5}\left[e^{-ar} - 2e^{-ar/2} \cos\left(\dfrac{ar\sqrt{3}}{2} + \dfrac{2\pi}{3}\right)\right]$

119. $\int_0^\infty \dfrac{\sin \pi x \, dx}{x(1 - x^2)} = \pi$

120. $\int_0^\infty \dfrac{e^{-qx} - e^{-px}}{x} \cos bx \, dx = \dfrac{1}{2} \ln \left| \dfrac{p^2 + b^2}{q^2 + b^2} \right|$

121. $\int_0^\infty \dfrac{e^{-qx} - e^{-px}}{x} \sin bx \, dx = \tan^{-1} \dfrac{p}{b} - \tan^{-1} \dfrac{q}{b}$

122. $\int_0^\infty e^{-ax} \dfrac{\sin px - \sin qx}{x} \, dx = \tan^{-1} \dfrac{p}{a} - \tan^{-1} \dfrac{q}{b}$

123. $\int_0^\infty e^{-ax} \dfrac{\cos px - \cos qx}{x} \, dx = \dfrac{1}{2} \ln \left| \dfrac{a^2 + a^2}{a^2 + p^2} \right|$

124. $\int_0^\infty x e^{-x^2} \sin ax \, dx = \dfrac{a\sqrt{\pi}}{4} e^{-a^2/4}$

125. $\int_0^\infty x^2 e^{-x^2} \cos ax \, dx = \dfrac{\sqrt{\pi}}{4}\left(1 - \dfrac{a^2}{2}\right) e^{-a^2/4}$

126. $\int_0^\infty x^3 e^{-x^2} \sin ax \, dx = \dfrac{\sqrt{\pi}}{8}\left(3a - \dfrac{a^3}{2}\right) e^{-a^2/4}$

**543**

127. $\int_0^\infty x^4 e^{-x^2} \cos ax\, dx = \dfrac{\sqrt{\pi}}{8}\left(3 - 3a^2 + \dfrac{a^4}{4}\right) e^{-a^2/4}$

128. $\int_0^\infty \left(\dfrac{\ln x}{x-1}\right)^3 dx = \pi^2$

129. $\int_{-\infty}^\infty \dfrac{x \sin rx\, dx}{(x-b)^2 + a^2} = \dfrac{\pi}{a}(a \cos br + b \sin br)\, e^{-ar}$

130. $\int_{-\infty}^\infty \dfrac{\sin rx\, dx}{x[(x-b)^2 + a^2]} = \dfrac{\pi}{a(a^2+b^2)}\left[a - (\cos br - b \sin br)\, e^{-ar}\right]$

131. $\int_{-\infty}^\infty \dfrac{\cos rx\, dx}{(x-b)^2 + a^2} = \dfrac{\pi}{a} e^{-ar} \cos br$

132. $\int_{-\infty}^\infty \dfrac{\sin rx\, dx}{(x-b)^2 + a^2} = \dfrac{\pi}{a} e^{-ar} \sin br$

133. $\int_{-\infty}^\infty e^{-x^2} \cos 2nx\, dx = \sqrt{\pi}\, e^{-n^2}$

134. $\int_0^\infty \dfrac{x^{p-1} \ln x}{1+x}\, dx = \dfrac{-\pi^2}{\sin p\pi} \cot p\pi, \quad 0 < p < 1$

135. $\int_0^\infty e^{-x} \ln x\, dx = -\gamma$

136. $\int_0^\infty e^{-x^2} \ln x\, dx = -\dfrac{\sqrt{\pi}}{4}(\gamma + 2\ln 2)$

137. $\int_0^\infty \ln\left(\dfrac{e^x+1}{e^x-1}\right) dx = \dfrac{\pi^2}{4}$

138. $\int_0^\infty \dfrac{x\, dx}{e^x - 1} = \dfrac{\pi^2}{6}$

139. $\int_0^\infty \dfrac{x\, dx}{e^x + 1} = \dfrac{\pi^2}{12}$

| Integrals containing hyperbolic terms |

140. $\int_0^1 \dfrac{\sinh(m \ln x)}{\sinh(\ln x)}\, dx = \dfrac{\pi}{2} \tan\dfrac{m\pi}{2}, \quad |m| < 1$

141. $\int_0^\infty \dfrac{\sin ax}{\sinh bx}\, dx = \dfrac{\pi}{2b} \tanh\left(\dfrac{\pi a}{2b}\right)$

142. $\int_0^\infty \dfrac{\cos ax}{\cosh bx}\, dx = \dfrac{\pi}{2b} \text{sech}\left(\dfrac{\pi a}{2b}\right)$

143. $\int_0^\infty \dfrac{x\, dx}{\sinh ax} = \dfrac{\pi^2}{4a^2}$

Appendix B

**144.** $\int_0^\infty \dfrac{\sinh px}{\sinh qx}\, dx = \dfrac{\pi}{2q}\tan(\dfrac{\pi p}{2q}),\quad |p|<q$

**145.** $\int_0^\infty \dfrac{\cosh ax - \cosh bx}{\sinh \pi x}\, dx = \ln\left|\dfrac{\cos\frac{b}{2}}{\cos\frac{a}{2}}\right|,\quad -\pi<b<a<\pi$

**146.** $\int_0^\infty \dfrac{\sinh px}{\sinh qx}\cos mx\, dx = \dfrac{\pi}{2q}\dfrac{\sin\frac{\pi p}{q}}{\cos\frac{\pi p}{q}+\cosh\frac{\pi m}{q}},\quad q>0, p^2<q^2$

**147.** $\int_0^\infty \dfrac{\sinh px}{\cosh qx}\sin mx\, dx = \dfrac{\pi}{q}\dfrac{\sin\frac{p\pi}{2q}\sinh\frac{m\pi}{2q}}{\cos\frac{p\pi}{q}+\cosh\frac{m\pi}{q}}$

**148.** $\int_0^\infty \dfrac{\cosh px}{\cosh qx}\cos mx\, dx = \dfrac{\pi}{q}\dfrac{\cos\frac{p\pi}{2q}\cosh\frac{m\pi}{2q}}{\cos\frac{p\pi}{q}+\cosh\frac{m\pi}{q}}$

**Miscellaneous Integrals**

**149.** $\int_0^x \xi^{\lambda-1}[1-\xi^\mu]^\nu\, d\xi = \dfrac{x^\lambda}{\lambda} F(-\nu,\dfrac{\lambda}{\mu};\dfrac{\lambda}{\mu}+1;x^\mu)$   See hypergeometric function

**150.** $\int_0^\pi \cos(n\phi - x\sin\phi)\, d\phi = \pi J_n(x)$

**151.** $\int_{-a}^a (a+x)^{m-1}(a-x)^{n-1}\, dx = (2a)^{m+n-1}\dfrac{\Gamma(m)\Gamma(n)}{\Gamma(m+n)}$

**152.** If $f'(x)$ is continuous and $\int_1^\infty \dfrac{f(x)-f(\infty)}{x}\, dx$ converges, then
$$\int_0^\infty \dfrac{f(ax)-f(bx)}{x}\, dx = [f(0)-f(\infty)]\ln\dfrac{b}{a}$$

**153.** If $f(x)=f(-x)$ so that $f(x)$ is an even function, then
$$\int_0^\infty f\left(x-\dfrac{1}{x}\right) dx = \int_0^\infty f(x)\, dx$$

**154.** Elliptic integral of the first kind
$$\int_0^\theta \dfrac{d\theta}{\sqrt{1-k^2\sin^2\theta}} = F(\theta,k),\quad 0<k<1$$

**155.** Elliptic integral of the second kind
$$\int_0^\theta \sqrt{1-k^2\sin^2\theta}\, d\theta = E(\theta,k)$$

**156.** Elliptic integral of the third kind
$$\int_0^\theta \dfrac{d\theta}{(1+n\sin^2\theta)\sqrt{1-k^2\sin^2\theta}} = \Pi(\theta,k,n)$$

Appendix B

# APPENDIX C
# Miscellaneous Topics

## Conversion Factors

Conversion factors are readily available from various internet web sites and from some software mathematics packages. To find a site on the internet just go to any convenient search engine and type in CONVERSION FACTORS to obtain a long listing of web sites giving access to thousands of conversion factors.

## How to Study

The secret to success in studying mathematics (or any other topic) is developing an understanding of definitions and fundamental concepts. By developing a complete familiarity with the fundamental concepts and definitions associated with a subject, a strong foundation for further study is constructed. Learn to recognize and apply the basic principles and fundamentals to both familiar and unfamiliar situations. This is achieved with constant practice and continued use of fundamentals over a period of time. Acquire textbooks on specific topics that go into detail and provide a thorough explanation of a subject. This will aid in developing a better understanding of fundamental principles and their applications.

Knowledge is like a jigsaw puzzle with an infinite number of pieces. The more pieces that are acquired, the larger the picture is, that can be constructed. Making discoveries is like finding new pieces of the puzzle, and then, having the ability to put the new pieces together to create something previously unknown.

## TeX

TeX is an extremely flexible computer typesetting program developed by Donald E. Knuth at Stanford University. It is being used extensively to prepare text for the publication of mathematical formulas, tables and just plain text for the publication of reports and books. It is the lithographers tool of the twenty-first century that replaces old printing techniques by use of a computer. If you intend to enter the field of mathematics, science or engineering, then now is the time to acquire some introductory books on TeX and how to use it. After learning the fundamentals, then one can acquire more advanced TeX books.

LaTeX is a subset of TeX, which has fewer and supposedly simpler commands to learn. You don't have to know TeX to use LaTeX. Many scientific journals require that articles be submitted using the LaTeX format.

# Index

## A

Abel's formula 318
abscissa 74, 101
absolute value 16, 41, 221
absolutely convergent 204
acceleration 226, 429, 450
acute angle 67
adding vectors 210
addition 27, 29
addition rule 394
additional operators 260
additive constant 171
adjoint operators 313
algebra 5, 39
alternating series 205
amortization 44
amplitude 75, 375
amplitude versus frequency 463
analytic function 317
analytic geometry 145
angular frequency 81
angular momentum 433
angular velocity 254, 441
antiderivative operator 170
arbitrary constants 275
arbitrary events 394
arbitrary path of integration 249
arc length 227
arc length formula 184
arc length parameter 223
Archimedes spiral 141
area formulas 185
area in polar coordinates 194
area inside simple closed curve 250
area of circle 253
area of the elemental parallelogram 235
area of the parallelogram 220
arithmetic 27, 29
arithmetic mean 389
arithmetic progression 53
arrays 203
associated Laguerre polynomials 350, 370
Associated Legendre functions 348, 367
associative law 28, 39, 105
asymptotic lines 107, 137
atomic number 467
Avogadro's number 467
axis intercepts 106

## B

Babylonian mathematics 2
base 2 numbers 25
base vectors 213
basic laws of arithmetic 27
basis solutions 286
beats 456
belongs to 16
Bernoulli distribution 405
Bernoulli number 164, 346
Bernoulli polynomials 346
Bernoulli's equation 279
Bessel differential equation 333, 361
Bessel differential equations 338
Bessel function of the first kind 333
Bessel function properties 335
Bessel functions 333, 365
Bessel functions of the second kind 334
Bessel functions orthogonality 362
Beta function 333
bilinear concomitant 315
binary system 26
binomial formula 52
binomial probability distribution 405
binormal vector 232, 449
Bliss's theorem 184
Bonnet's theorem 182
boundary conditions 358
boundary operator 296
boundary value problems (BVP) 295
bounded 273
Boyle's law 477
business mathematics 44

## C

Cairo papyrus 3
calculators 35
calculus 153
Cartesian axes 213
Cartesian coordinate system 101
Cartesian graph paper 108
Cauchy equations 311
Cauchy existence theorem 273
Cauchy's generalized mean value theorem 161
celestial mechanics 446
center of mass 437
central limit theorem 413

central moments 400
centroid 439
chain rule 280
change of variable 172, 199, 239, 251
characteristic 42
characteristic equation 287
Charles's law 477
Chebyshev polynomials first kind 351, 370
Chebyshev polynomials second kind 351, 371
chemical kinetics 474
chemical symbol 467
Chinese proof of Pythagorean theorem 65
chi-square probability distribution 410
circle 120
circular functions 71
circulation 252
circulation density 254
class frequencies 389
class intervals 387
class midpoints 389
clay tablets 1
clockwise sense 232
closed interval 18
closed surface 225
cofunction formulas 77
cofunctions 70
colinear vectors 211
combinations 49
common denominator 40
commutative group 211
commutative law 28, 39, 105
comparison tests 204
complementary angles 67
complementary error function 374
complementary solution 298, 461
complementation rule 394
complete elliptic integral of the first kind 375
complete elliptic integral of the second kind 376
complete elliptic integral of the third kind 376
complex form Fourier series 360
complex numbers 90, 272
complex roots 288
component form 216
component of a vector 103
components of the differential 235
composite function 165, 169, 280
compound period 44
concave downward 163
concave upward 163
concavity 163
conditional probability 396
conditionally convergent 204
confidence intervals 410

congruences 38
conic sections in polar coordinates 138
conjugate axis 137
conjugate hyperbola 137
conjunct 315
consistent units 430
constant coefficients 282
constellation of Aries 447
continued fraction 17
continuity 154, 273
continuous function 154
continuous probability distributions 398
continuous variable 385
contour plots 222
conversion period 44
coordinate curves 234
cosine function 67, 346
cosine integral 374
counting board 2
counterclockwise sense 232
counting systems 2
coversine 68
Cramer's rule 61
critical damping 461
cross product 218
cross product formula 219
cube roots 152
cubic equation 46
cumulative distribution function 410
cumulative frequency 387
cumulative frequency function 398
cumulative relative frequency 387
curl 252, 268
current loops 229
curtate cycloid 143
curvature 448
curve sketching 105, 123
curve sketching polar coordinates 122
curves in space 241
curvilinear coordinate 261, 267
cycloids 143
cylinder 148
cylindrical coordinates 196, 240, 265, 267

D

D'Alembertian operator 23
damping force 459
data spread 390
decimal system 26
definite integral 171, 177, 226
del operator 260
density of the field lines 241

dependent variable 275
dependent variable absent 280
derivative of determinant 203
derivatives 155
determinant 64
determinants and conic sections 138
determinants and parabola 136
diastolic blood pressure 386
difference equation 45
difference formula 79, 92
differential equations 225
differential equations with variable coefficients 311
differential operator 354
differentials 157
differentiation 172
differentiation of arrays 203
differentiation operator 171
dimensionally consistant 15
Dirac delta function 331
direction cosines 103, 216
direction of integration 232
directional derivatives 223
directrix of the parabola 135
Dirichlet boundary conditions 297
discontinuous at that point 154
discrete Fourier transform (DFT) 380
discrete probability distribution 389, 398
discrete variable 385
disjoint events 393
distance between two points 66
distinct products 49
distinct real roots 287
distributive law 28, 105, 211
divergence 240, 246, 268
division 27, 32
dot product of vectors 214
double angle formula 78
double integrals 192
double-division method 33
dynamics 429

E

e-permutation symbol 219
early mathematicians 8
ecliptic 447
Egyptian mathematics 3
eigenfunction 358
eigenfunctions 353, 358
electric flux 242
electrical circuits 480
electron shells 472
electrostatic intensity 242

element of surface area 235
elemental parallelogram 235
elementary probability theory 397
elements in the periodic table 472
elimination method 61
elliptic cone 149
elliptic cylinder 148
elliptic integral of first kind 375
elliptic integral of second kind 376
elliptic orbit 447
elliptic paraboloid 150
empty set $\emptyset$ 16, 393
entire sample space 393
epicycloids 144
equally likely 51
equating coefficients 217
equation of circle 120
equation of plane 218
equation of state 478
equilateral triangle 97
equi-level curves 222
equi-level surfaces 222
error function 374
escape velocity 433
Euclidean three space 215
Euler angles 444
Euler equations 311
Euler identities 94
Euler numbers 347
Euler polynomials 347
Euler's identity 274
evaluation of integrals 198
even and odd numbers 37
even function 73
exact differential equation 248, 277, 313
existence of solution 272
expectation 400
explicit form 234, 270
exponential function 43, 169, 176, 287, 346
exponential integral 374
exponential probability distribution 408
exponential terms 299
exponents 41, 344

F

factored form 46
factorization 47
family of curves 270
family of solutions 272
F-distribution 411
field 39, 48
field characteristic 222

# 549

field lines 225, 241
finite sample space 393
first derivative test 163
first order differential equation 272
fixed points 136
flow of fluid 255
flux 241
flux integral 242
foci of the hyperbola 136
focus of the parabola 135
for all 17
four terminal networks 483
Fourier associated Legendre series 367
Fourier coefficients 355
Fourier cosine transform table 384
Fourier exponential transform 380
Fourier series 360
Fourier sine transform table 383
Fourier trigonometric series 359
Fourier-associated Laguerre series 370
Fourier-Bessel series 363
Fourier-Chebyshev series 370
Fourier-Hermite series 369
Fourier-Laguerre series 369
Fourier-Legendre series 366
fractional part 26
fractional-part function 331
fractions 40
Fredholm integral equation 373
free electron 473
free vectors 103, 209, 222
frequency function 388
frequency of the motion 455
Fresnel integrals 372
Frobenius series 324
Frobenius solution 333
Frobenius type solution 340
frustum of pyramid 3
function of position 225
function to a power 169
functional form 270
functional relation 139
functions 139
fundamental principal 49
fundamental system 284
fundamental theorem of arithmetic 37

## G

Gamma function 332
gamma probability distribution 409
Gauss differential equation 340
Gauss divergence theorem 244

Gauss' test 205
gay-Lussac law 477
general position vector 214
general series 204
general solution 270, 319
generalized first mean value theorem 182
generalized Fourier series 355
generalized hypergeometric function 343
generalized mean value theorem 182
generating function 344
geometric mean 390
geometric progression 53
geometry 97
geometry 152
Golenischev papyrus 3
gradient 20, 222, 268
gradient function 222
graphical method 61
graphical representation 388
graphics 152
graphs 139
graphs of trigonometric functions 74
greatest rate of change 223
Greek alphabet 39
Green's first identity 259
Green's identity 316, 354
Green's second identity 259
Green's theorem in the plane 247
group 48, 210
grouping the data 387

## H

half angle formulas for triangle 100
half-angle formula 78
harmonic mean 390
harmonic series 205
haversine 68
Heaviside step function 331
Hermite polynomials 349
Hermite polynomials 369
hertz 455
hexadecimal number 25
hexagonal graph paper 111
higher derivatives 156
history 1
homogeneous differential equation 270
homogeneous function 276
Hooke's law 457
Hurwitz zeta function 377
hyperbola 136
hyperbolic functions 84, 94, 175
hyperbolic identities 85, 91

hyperbolic paraboloid 150
hyperbolic spiral 141
hyperboloid of one sheet 150
hyperboloid of two sheets 149
hypergeometric function 340
hypergeometric probability distribution 407
hypocycloid 144

I

if and only if 17, 27
imaginary cuts 259
imaginary part 90
imaginary roots 288
imaginary unit 90, 274
implicit differentiation 159, 174
implicit form 224, 234, 270
implicit function theorem 159
implies 18, 41
improper integrals 183
impulse function 331
inclusion 16
incomplete beta function 333
incomplete elliptic integral of the first kind 375
incomplete elliptic integral of the second kind 376
incomplete elliptic integral of the third kind 376
increases without bound 153
indefinite integral 170, 225
independent events 396
independent random variables 401
independent variable 275
independent variable absent 280
indeterminate forms 161
indicial roots 329
indirect proof 27, 38
inductive statistics 385
inequalities 41
inequalities involving integrals 204
infinite series 322
initial point of the vector 103
initial value problem (IVP) 271
inner product 354
instantaneous flux per unit volume 246
integer function 331
integer part 26
integral equations 373
integral form 276
integral operator 170
integral substitutions 199
integral theorems 240, 257
integrating factors 278
integration 172
integration by parts 182
integration of arrays 203

integration of vectors 225
integration operator 171
intensity of a vector field 241
intercept formula 119
interest 44
interest rate 44
intermediate value theorem 155
interpolation 419
intersection 393
intersection of the coordinate surfaces 263
invariance 76
inverse element 210
inverse function 81, 169
inverse hyperbolic functions 88, 175
inverse trigonometric functions 82, 93, 173, 332
involute of a circle 141
inward normal 225
irrational numbers 37
irregular singular point 317, 325
irrotational vector field 254
isothermal curves 479
isotopes 467

J

Jacobi elliptic functions 375
Jacobian determinant 240, 251
jump discontinuity 153

K

Kepler's laws 433
kinetic energy 231
Kirchoff laws 481
Klein bottle 151
Kronecker delta 20

L

Lagrange's identity 315
Laguerre polynomials 350, 369
Laplace equation 466
Laplace transform 377
Laplace transform properties 378
Laplace transform table 379
Laplacian 268
Laplacian operator 18, 260
latus rectum 137
Laurent series 316
law of cosines 100
law of sines 99
law of sines for triangle 221

law of tangents 100
laws of algebra 39
laws of logarithms 43
least square curve fitting 414
left distributive law 40
left-hand limit 153
left-handed coordinate system 267
Legendre equation 366
Legendre functions of the first kind 348
Leibnitz formula 181
Leibnitz rule 158, 275
lemniscate 145
level curves 224, 277
level surfaces 224
L'Hospital's rule 161
limit 153
limiting process 227
line 217
line integral 228, 231
linear combination 211, 290
linear dependence 211, 283
linear differential equation 270
linear equations 45, 61
linear first order differential equations 278
linear impulse 227
linear independence 211, 283
linear interpolation 419
linear momentum 227, 429
linear operator 269
linear regression 417
linearly independent solutions 329
local maximum 162
local minimum 162
local region 272
logarithmic function 43
logarithmic function 176
log-log graph paper 115
lower incomplete gamma function 333

## M

Maclaurin series 164, 324
magnetic dipole moments 229
magnitude 209, 235
mantissa 42
mapping 225
mass moments of inertia 440
mathematical abbreviations 15
mathematical events 4
mathematical expectation 400
mathematical induction 27
mathematical notation 15
maximum values 162

mean 402
mean deviation 390
mean value theorem 180
mean value theorem for derivatives 160, 180
mean value theorem for integrals 179
mechanical resonance 460
mechanical vibrations 454
median 212, 389
method of interpolation 419
method of reduction of order 318
method of undetermined coefficients 299
method of variation of parameters 305
metric components 262
minimum values 162
mirror image 83
miscellaneous topics 429
mixed boundary value problem 297
mixed partial derivative 277
Möbius strip 151
mode 389
modification rule 290
modified Bessel differential equation 340
modified Bessel functions of the first kind 339
modified Bessel functions of the second kind 339
mole 467
molecular weight 467
moment of a force 435
moment vector 435
moments 443
moments of inertia 440
Monte Carlo methods 418
Moscow papyrus 3
motion along a curve 230
motion of a particle 429
motion of weight 431
multinomial distribution 406
multinomial expansion 52
multinomial probability function 406
multiple angle formulas 91
multiple-valued functions 96
multiplication 27, 31
multiplicative inverse 39
multiply connected region 259
multi-valued functions 82
mutually exclusive events 393

## N

$n$-parameter family 271
natural logarithm 43, 170
necessary and sufficient conditions 27
necessary condition 249
negative numbers 28

negative sense 233
Neumann boundary conditions 297
Neumann function 334
Newton's law of gravitation 432
Newton's laws 429
Newton's second law 227, 230, 443
noncolinear vectors 211
nonhomgeneous equations 319
nonhomogeneous differential equation 270
nonhomogeneous linear differential equations 298
nonlinear differential equation 270
nonlinear operator 269
norm squared 356, 362
normal distribution 401
normal form for line 119
normal probability curve 404
normal probability distribution 401
normal to surface 234
normal vector 224, 449
normalization 266
normalized fundamental set 320
normalized vector 213
notation 15
number conversion 26
number of field lines 241
number representation 25
number theory 36

O

oblate spheroid 149
obtuse 67
odd function 73
one parameter family 270
one-point Green's function 310, 320
open interval 18
operations with complex numbers 90
operator $\nabla$
operator concept 269
operator form 222
operators 260
order of arrangement 49
order of integration 194
order properties 41
ordinary differential equations 269
ordinary point 317
ordinate 101
orientated surfaces 233
origin 209
Orthogonal functions 351, 367
orthogonal system 267
orthogonal trigonometric functions 356
Orthogonality 354

orthonormal functions 371
outer product 218
outward normal 225
ovals of Cassini 144
over determined system of equations 416
overdamping 460

P

papyri 3
parabola 135
parallel circuit 483
parallel-axis theorem 439
parallelogram 220
parallelogram law 103
parallelogram law for vector addition 210
parametric equation for line 120
parametric equations 169, 217
parametric equations for ellipse 135
parametric form 234
parametric solutions 280
partial derivatives 158
partial differential equation 269
partial fractions 200
particular solution 271, 298
partitioning 178
percentiles 389
perimeter of circle 120
periodic boundary conditions 352, 358
periodic function 75
periodic payments 45
periodic Sturm-Liouville system 352
periodic table of elements 467
permutations 49
permutation symbol 219
phase angle 81, 289
phenomenon of beats 456
physical interpretation of curl 254
physical interpretation of divergence 246
piecewise continuous 353
piecewise continuous functions 359
piecewise smooth functions 367
pitch 455
plane 218
plane curve 139, 233
plane curves parametric form 141
planimeter 251
P-notation 344
point characteristic of the vector field 242
point slope formula 155
Poisson distribution 407
Poisson probability distribution 407
polar coordinates 122

polar equation of circle 124
polar equation of line 124
polar graph paper 109
pole of order M 316
Polish notation 35
polygon 122
polynomial equation 46
polynomial set 345
polynomial terms 299
population 385
position vector 227
positive sense 233
power series 326
pressure-volume diagram 478
prime numbers 37
principal part 317
principal value properties 84
principal values 93
prisms 485
probability 51
probability 385
probability curve 402
probability density function 398
probability density function 401
probability function 398
probability fundamentals 393
probability graph paper 118
product formula 79, 92
product rule 169
projection 215, 229
prolate cycloid 143
properties of cross product 219
properties of definite integral 179
properties of differentials 157
properties of Fourier exponential transform 380
properties of hyperbolic functions 86
properties of Laplace transforms 378
properties of limits 154
properties of series 321
properties of the Bessel functions. 336
properties of vectors 209
proportionality constant 225
Pythagoras of Samos 5
Pythagorean identities 71
Pythagorean theorem 65, 70

Q

quadratic equation 45
quality control 407
quantum shell 472
quartic equation 47

R

Raabe's test 205
radicals 54
radius of convergence 325, 342
random 397
random variable 401
rate law 475
rate of change of a scalar field 222
ratio test 205
rational numbers 37
reactant 474
real part 90
real random variable 400
reciprocal functions 75
rectangular coordinates 101
recurrence relation 327, 345
recursion formula hypergeometric functions 342
reductio ad absurdum 27
reduction formula 201
reflexive law 39
region of integration 193, 258
regular points 272
regular singular point 317, 325, 328, 333, 344
regular Sturm-Liouville problem 355
regular system 353
related integral theorem 257
relative frequency 52, 395
relative maximum 162
relative minimum 162
relative motion 451
relative weights 467
repeated roots 289
representation of data 385
resonance 460
reverse Polish notation 35
Rhind papyrus 3
Riemann differential equation 343
Riemann zeta function 377
right angled 97
right ascension declination system 448
right distributive law 40
right-hand limit 153
right-hand rule 218, 255
right-handed coordinate system 267
rigid bodies 429, 441
Rolle's theorem 161
root mean square 390
rose curves 141
rotation of a curve 190
rotation of area about axis 186
rotation of axes 101
rotational field 254

rotational motion 443
rule of nine 29
rule of signs 28
rules for differentiation 167
rules for integration 171

S

sample mean 389
sample space 392
sample variance 390
sample variance 391
sampling theory 412
scalar 209
scalar field 222
scalar multiplication 209
scalar product of vectors 214
scalar quantity 242
scalar relation 244
scale parameter 402
scalene triangle 97
scaling 106, 401
secant line 155
second derivative 156
second moments 440
second order reaction 475
second-order equations 280
self adjoint form 319
self-adjoint operator 319
semi-log graph paper 112
separation of variables 275, 477
series circuit 482
series representation 321
series solutions 316
set of solutions 288
sexagesimal number system 2
shifting of axes 279
short cut method 391
sign variation 71
similar triangles 97
simple closed curve 232
simple harmonic motion 80
simply connected region 259
sine function 67, 346
sine integral 374
single valued 81
singular integral equation 373
singular point 317, 341
singular point of the surface 235
singular points 272
singular solution 272
singular Sturm-Liouville problem 355
singular system 354

sink 246
smooth surface 234
solenoidal 246
solid angles 465
solution family 270, 277
solution of differential equations 248, 270
source 246
space curves 448
special functions 332
special graph paper 107
special products 47
special right triangles 69
specialty areas 13
spectrum of the problem 353
sphere 147
spherical coordinates 197, 240, 267
Spherical harmonics 371
spirals 141
spring-mass system 457
square root 34, 42
standard deviation 402
standardization 403
standardized variable 401
statistical inference 385
statistics 385
steradians 465
Stoke's theorem 255
straight line 119
straight line approximation 415
straight line using determinants 120
student's t-distribution 410
Sturm-Liouville systems 352
Sturm-Liouville theorem 366
subshells 472
subtraction 27, 30
successive transformations 446
sufficient and necessary conditions 27
sufficient conditions 273
Sumerians 2
summation 237
summation formula 92
superposition principle 283
supplementary angles 67
surface area 189, 242
surface area from parametric form 238
surface coordinates 235
surface integral 233
surfaces 145
symmetric form of line 218
symmetric law 39
symmetric matrix 262
symmetry 93, 105
systolic blood pressure 386

# 555

**T**

Table of Fourier sine transforms, 383
tabular representation of data 386
tally stick 4
tangent function 67
tangent line 156
tangent vector 449
tangential component 231
Taylor series 164
Taylor's theorem 323
tensor calculus 262
terminal point of vector 103
terminus 209
test for symmetry 106
testing of hypothesis 410
theory of proportions 53
thermodynamics 477
time line 4
torsion 448
torsional vibrations 464
torus 151
total flux 241
Trachtenberg system 29
transformation of coordinates 102
transformations 93
transitive law 39
translation of axes 101
transverse axis 137
triangles 97, 100, 121
triangular graph paper 110
trigonometric functions 70, 74, 358
trigonometric identities 76, 91
trigonometry 66
triple integrals 195
triple scalar product 220, 240, 434
trisection point 213
true population mean 389
tube of field lines 246

**U**

underdamping 460
undetermined coefficients 299
Uniform convergence 206
uniform probability density function 411
union 393
unique solution 287
uniqueness of solution 272
unit normal vector 224
unit outward normal 232
unit step function 331

unit tangent vector 215, 224, 232
unit vectors 103, 213
upper incomplete gamma function 333
use of determinants 121

**V**

variable coefficients 311
variables separable 276
variance of the distribution 402
variation of parameters 305
vector 209
vector addition 103
vector element of surface area 233
vector equation of line 120
vector field 222, 225, 237
vector field plots 225
vector identities 220
vector properties 105
vectors 102
velocity 226, 429, 450
Venn diagrams 392
vernal equinox 447
versine 68
vertices 212
vibrations 454
voltage drop 481
Volterra integral equation 373
volume integral 233, 239
volume of the parallelepiped 221
volume rate of flow 242
volumes 186

**W**

wavelength 486
Weber-Schlafli ratio 334
Weierstrass M-test 206
work done 229
Wronskian determinant 285

**Z**

zero 4
zeros of Bessel function 334, 362

Index

Printed in the United Kingdom by
Lightning Source UK Ltd., Milton Keynes
137321UK00002B/60/P